Lecture Notes in Networks and Systems 800

The series "Lecture Notes in Networks and Systems" publishes the latest developments in Networks and Systems—quickly, informally and with high quality. Original research reported in proceedings and post-proceedings represents the core of LNNS.

Volumes published in LNNS embrace all aspects and subfields of, as well as new challenges in, Networks and Systems.

The series contains proceedings and edited volumes in systems and networks, spanning the areas of Cyber-Physical Systems, Autonomous Systems, Sensor Networks, Control Systems, Energy Systems, Automotive Systems, Biological Systems, Vehicular Networking and Connected Vehicles, Aerospace Systems, Automation, Manufacturing, Smart Grids, Nonlinear Systems, Power Systems, Robotics, Social Systems, Economic Systems and other. Of particular value to both the contributors and the readership are the short publication timeframe and the world-wide distribution and exposure which enable both a wide and rapid dissemination of research output.

The series covers the theory, applications, and perspectives on the state of the art and future developments relevant to systems and networks, decision making, control, complex processes and related areas, as embedded in the fields of interdisciplinary and applied sciences, engineering, computer science, physics, economics, social, and life sciences, as well as the paradigms and methodologies behind them.

Indexed by SCOPUS, INSPEC, WTI Frankfurt eG, zbMATH, SCImago.

All books published in the series are submitted for consideration in Web of Science.

For proposals from Asia please contact Aninda Bose (aninda.bose@springer.com).

Alvaro Rocha · Hojjat Adeli ·
Gintautas Dzemyda · Fernando Moreira ·
Valentina Colla
Editors

Information Systems and Technologies

WorldCIST 2023, Volume 2

Editors
Alvaro Rocha
ISEG
Universidade de Lisboa
Lisbon, Cávado, Portugal

Hojjat Adeli
College of Engineering
The Ohio State University
Columbus, OH, USA

Gintautas Dzemyda
Institute of Data Science and Digital Technologies
Vilnius University
Vilnius, Lithuania

Fernando Moreira
DCT
Universidade Portucalense
Porto, Portugal

Valentina Colla
TeCIP Institute
Scuola Superiore Sant'Anna
Pisa, Italy

ISSN 2367-3370 ISSN 2367-3389 (electronic)
Lecture Notes in Networks and Systems
ISBN 978-3-031-45644-2 ISBN 978-3-031-45645-9 (eBook)
https://doi.org/10.1007/978-3-031-45645-9

This Springer imprint is published by the registered company Springer Nature Switzerland AG
The registered company address is: Gewerbestrasse 11, 6330 Cham, Switzerland

Paper in this product is recyclable.

Preface

This book contains a selection of papers accepted for presentation and discussion at the 2023 World Conference on Information Systems and Technologies (WorldCIST'23). This conference had the scientific support of the Sant'Anna School of Advanced Studies, Pisa, University of Calabria, Information and Technology Management Association (ITMA), IEEE Systems, Man, and Cybernetics Society (IEEE SMC), Iberian Association for Information Systems and Technologies (AISTI), and Global Institute for IT Management (GIIM). It took place in Pisa city, Italy, 4–6 April 2023.

The World Conference on Information Systems and Technologies (WorldCIST) is a global forum for researchers and practitioners to present and discuss recent results and innovations, current trends, professional experiences and challenges of modern Information Systems and Technologies research, technological development, and applications. One of its main aims is to strengthen the drive toward a holistic symbiosis between academy, society, and industry. WorldCIST'23 was built on the successes of: WorldCIST'13 held at Olhão, Algarve, Portugal; WorldCIST'14 held at Funchal, Madeira, Portugal; WorldCIST'15 held at São Miguel, Azores, Portugal; WorldCIST'16 held at Recife, Pernambuco, Brazil; WorldCIST'17 held at Porto Santo, Madeira, Portugal; WorldCIST'18 held at Naples, Italy; WorldCIST'19 held at La Toja, Spain; WorldCIST'20 held at Budva, Montenegro; WorldCIST'21 held at Terceira Island, Portugal; and WorldCIST'22, which took place online at Budva, Montenegro.

The Program Committee of WorldCIST'23 was composed of a multidisciplinary group of 339 experts and those who are intimately concerned with Information Systems and Technologies. They have had the responsibility for evaluating, in a 'blind review' process, and the papers received for each of the main themes proposed for the Conference were: A) Information and Knowledge Management; B) Organizational Models and Information Systems; C) Software and Systems Modeling; D) Software Systems, Architectures, Applications, and Tools; E) Multimedia Systems and Applications; F) Computer Networks, Mobility, and Pervasive Systems; G) Intelligent and Decision Support Systems; H) Big Data Analytics and Applications; I) Human-Computer Interaction; J) Ethics, Computers & Security; K) Health Informatics; L) Information Technologies in Education; M) Information Technologies in Radiocommunications; and N) Technologies for Biomedical Applications.

The conference also included workshop sessions taking place in parallel with the conference ones. Workshop sessions covered themes such as: Novel Computational Paradigms, Methods, and Approaches in Bioinformatics; Artificial Intelligence for Technology Transfer; Blockchain and Distributed Ledger Technology (DLT) in Business; Enabling Software Engineering Practices Via Latest Development's Trends; Information Systems and Technologies for the Steel Sector; Information Systems and Technologies for Digital Cultural Heritage and Tourism; Recent Advances in Deep Learning Methods and Evolutionary Computing for Health Care; Data Mining and Machine Learning in Smart Cities; Digital Marketing and Communication, Technologies, and Applications;

Digital Transformation and Artificial Intelligence; and Open Learning and Inclusive Education Through Information and Communication Technology.

WorldCIST'23 and its workshops received about 400 contributions from 53 countries around the world. The papers accepted for oral presentation and discussion at the conference are published by Springer (this book) in four volumes and will be submitted for indexing by WoS, Scopus, EI-Compendex, DBLP, and/or Google Scholar, among others. Extended versions of selected best papers will be published in special or regular issues of leading and relevant journals, mainly JCR/SCI/SSCI and Scopus/EI-Compendex indexed journals.

We acknowledge all of those that contributed to the staging of WorldCIST'23 (authors, committees, workshop organizers, and sponsors). We deeply appreciate their involvement and support that was crucial for the success of WorldCIST'23.

April 2023

Alvaro Rocha
Hojjat Adeli
Gintautas Dzemyda
Fernando Moreira
Valentina Colla

Organization

Honorary Chair

Hojjat Adeli	The Ohio State University, USA

General Chair

Álvaro Rocha	ISEG, University of Lisbon, Portugal

Co-chairs

Gintautas Dzemyda	Vilnius University, Lithuania
Sandra Costanzo	University of Calabria, Italy

Workshops Chair

Fernando Moreira	Portucalense University, Portugal

Local Organizing Committee

Valentina Cola (Chair)	Scuola Superiore Sant'Anna—TeCIP Institute, Italy
Marco Vannucci	Scuola Superiore Sant'Anna—TeCIP Institute, Italy
Vincenzo Iannino	Scuola Superiore Sant'Anna—TeCIP Institute, Italy
Stefano Dettori	Scuola Superiore Sant'Anna—TeCIP Institute, Italy

Advisory Committee

Ana Maria Correia (Chair)	University of Sheffield, UK
Brandon Randolph-Seng	Texas A&M University, USA

Chris Kimble	KEDGE Business School & MRM, UM2, Montpellier, France
Damian Niwiński	University of Warsaw, Poland
Florin Gheorghe Filip	Romanian Academy, Romania
Janusz Kacprzyk	Polish Academy of Sciences, Poland
João Tavares	University of Porto, Portugal
Jon Hall	The Open University, UK
John MacIntyre	University of Sunderland, UK
Karl Stroetmann	Empirica Communication & Technology Research, Germany
Majed Al-Mashari	King Saud University, Saudi Arabia
Miguel-Angel Sicilia	University of Alcalá, Spain
Mirjana Ivanovic	University of Novi Sad, Serbia
Paulo Novais	University of Minho, Portugal
Wim Van Grembergen	University of Antwerp, Belgium
Mirjana Ivanovic	University of Novi Sad, Serbia
Reza Langari	Texas A&M University, USA
Wim Van Grembergen	University of Antwerp, Belgium

Program Committee

Abderrahmane Ez-Zahout	Mohammed V University, Morocco
Adriana Gradim	University of Aveiro, Portugal
Adriana Peña Pérez Negrón	Universidad de Guadalajara, Mexico
Adriani Besimi	South East European University, Macedonia
Agostinho Sousa Pinto	Polythecnic of Porto, Portugal
Ahmed El Oualkadi	Abdelmalek Essaadi University, Morocco
Akex Rabasa	University Miguel Hernandez, Spain
Alba Córdoba-Cabús	University of Malaga, Spain
Alberto Freitas	FMUP, University of Porto, Portugal
Aleksandra Labus	University of Belgrade, Serbia
Alessio De Santo	HE-ARC, Switzerland
Alexandru Vulpe	University Politehnica of Bucharest, Romania
Ali Idri	ENSIAS, University Mohamed V, Morocco
Alicia García-Holgado	University of Salamanca, Spain
Amélia Badica	Universti of Craiova, Romania
Amélia Cristina Ferreira Silva	Polytechnic of Porto, Portugal
Amit Shelef	Sapir Academic College, Israel
Alanio de Lima	UFC, Brazil
Almir Souza Silva Neto	IFMA, Brazil
Álvaro López-Martín	University of Malaga, Spain

Ana Carla Amaro	Universidade de Aveiro, Portugal
Ana Isabel Martins	University of Aveiro, Portugal
Anabela Tereso	University of Minho, Portugal
Anabela Gomes	University of Coimbra, Portugal
Anacleto Correia	CINAV, Portugal
Andrew Brosnan	University College Cork, Ireland
Andjela Draganic	University of Montenegro, Montenegro
Aneta Polewko-Klim	University of Białystok, Institute of Informatics, Poland
Aneta Poniszewska-Maranda	Lodz University of Technology, Poland
Angeles Quezada	Instituto Tecnologico de Tijuana, Mexico
Anis Tissaoui	University of Jendouba, Tunisia
Ankur Singh Bist	KIET, India
Ann Svensson	University West, Sweden
Anna Gawrońska	Poznański Instytut Technologiczny, Poland
Antoni Oliver	University of the Balearic Islands, Spain
Antonio Jiménez-Martín	Universidad Politécnica de Madrid, Spain
Aroon Abbu	Bell and Howell, USA
Arslan Enikeev	Kazan Federal University, Russia
Beatriz Berrios Aguayo	University of Jaen, Spain
Benedita Malheiro	Polytechnic of Porto, ISEP, Portugal
Bertil Marques	Polytechnic of Porto, ISEP, Portugal
Boris Shishkov	ULSIT/IMI-BAS/IICREST, Bulgaria
Borja Bordel	Universidad Politécnica de Madrid, Spain
Branko Perisic	Faculty of Technical Sciences, Serbia
Carla Pinto	Polytechnic of Porto, ISEP, Portugal
Carlos Balsa	Polythecnic of Bragança, Portugal
Carlos Rompante Cunha	Polytechnic of Bragança, Portugal
Catarina Reis	Polytechnic of Leiria, Portugal
Célio Gonçalo Marques	Polythenic of Tomar, Portugal
Cengiz Acarturk	Middle East Technical University, Turkey
Cesar Collazos	Universidad del Cauca, Colombia
Christine Gruber	K1-MET, Austria
Christophe Guyeux	Universite de Bourgogne Franche Comté, France
Christophe Soares	University Fernando Pessoa, Portugal
Christos Bouras	University of Patras, Greece
Christos Chrysoulas	London South Bank University, UK
Christos Chrysoulas	Edinburgh Napier University, UK
Ciro Martins	University of Aveiro, Portugal
Claudio Sapateiro	Polytechnic of Setúbal, Portugal
Cosmin Striletchi	Technical University of Cluj-Napoca, Romania
Costin Badica	University of Craiova, Romania

Cristian García Bauza	PLADEMA-UNICEN-CONICET, Argentina
Cristina Caridade	Polytechnic of Coimbra, Portugal
David Cortés-Polo	University of Extremadura, Spain
David Kelly	University College London, UK
Daria Bylieva	Peter the Great St.Petersburg Polytechnic University, Russia
Dayana Spagnuelo	Vrije Universiteit Amsterdam, Netherlands
Dhouha Jaziri	University of Sousse, Tunisia
Dmitry Frolov	HSE University, Russia
Dulce Mourato	ISTEC - Higher Advanced Technologies Institute Lisbon, Portugal
Edita Butrime	Lithuanian University of Health Sciences, Lithuania
Edna Dias Canedo	University of Brasilia, Brazil
Egils Ginters	Riga Technical University, Latvia
Ekaterina Isaeva	Perm State University, Russia
Eliana Leite	University of Minho, Portugal
Enrique Pelaez	ESPOL University, Ecuador
Eriks Sneiders	Stockholm University, Sweden
Esperança Amengual	Universitat de les Illes Balears, Spain
Esteban Castellanos	ESPE, Ecuador
Fatima Azzahra Amazal	Ibn Zohr University, Morocco
Fernando Bobillo	University of Zaragoza, Spain
Fernando Molina-Granja	National University of Chimborazo, Ecuador
Fernando Moreira	Portucalense University, Portugal
Fernando Ribeiro	Polytechnic Castelo Branco, Portugal
Filipe Caldeira	Polythecnic of Viseu, Portugal
Filipe Portela	University of Minho, Portugal
Filippo Neri	University of Naples, Italy
Firat Bestepe	Republic of Turkey Ministry of Development, Turkey
Francesco Bianconi	Università degli Studi di Perugia, Italy
Francisco García-Peñalvo	University of Salamanca, Spain
Francisco Valverde	Universidad Central del Ecuador, Ecuador
Frederico Branco	University of Trás-os-Montes e Alto Douro, Portugal
Galim Vakhitov	Kazan Federal University, Russia
Gayo Diallo	University of Bordeaux, France
Gema Bello-Orgaz	Universidad Politecnica de Madrid, Spain
George Suciu	BEIA Consult International, Romania
Ghani Albaali	Princess Sumaya University for Technology, Jordan

Gian Piero Zarri	University Paris-Sorbonne, France
Giovanni Buonanno	University of Calabria, Italy
Gonçalo Paiva Dias	University of Aveiro, Portugal
Goreti Marreiros	ISEP/GECAD, Portugal
Graciela Lara López	University of Guadalajara, Mexico
Habiba Drias	University of Science and Technology Houari Boumediene, Algeria
Hafed Zarzour	University of Souk Ahras, Algeria
Haji Gul	City University of Science and Information Technology, Pakistan
Hakima Benali Mellah	Cerist, Algeria
Hamid Alasadi	Basra University, Iraq
Hatem Ben Sta	University of Tunis at El Manar, Tunisia
Hector Fernando Gomez Alvarado	Universidad Tecnica de Ambato, Ecuador
Hector Menendez	King's College London, UK
Hélder Gomes	University of Aveiro, Portugal
Helia Guerra	University of the Azores, Portugal
Henrique da Mota Silveira	University of Campinas (UNICAMP), Brazil
Henrique S. Mamede	University Aberta, Portugal
Henrique Vicente	University of Évora, Portugal
Hicham Gueddah	University Mohammed V in Rabat, Morocco
Hing Kai Chan	University of Nottingham Ningbo China, China
Igor Aguilar Alonso	Universidad Nacional Tecnológica de Lima Sur, Peru
Inês Domingues	University of Coimbra, Portugal
Isabel Lopes	Polytechnic of Bragança, Portugal
Isabel Pedrosa	Coimbra Business School - ISCAC, Portugal
Isaías Martins	University of Leon, Spain
Issam Moghrabi	Gulf University for Science and Technology, Kuwait
Ivan Armuelles Voinov	University of Panama, Panama
Ivan Dunđer	University of Zabreb, Croatia
Ivone Amorim	University of Porto, Portugal
Jaime Diaz	University of La Frontera, Chile
Jan Egger	IKIM, Germany
Jan Kubicek	Technical University of Ostrava, Czech Republic
Jeimi Cano	Universidad de los Andes, Colombia
Jesús Gallardo Casero	University of Zaragoza, Spain
Jezreel Mejia	CIMAT, Unidad Zacatecas, Mexico
Jikai Li	The College of New Jersey, USA
Jinzhi Lu	KTH-Royal Institute of Technology, Sweden
Joao Carlos Silva	IPCA, Portugal

João Manuel R. S. Tavares	University of Porto, FEUP, Portugal
João Paulo Pereira	Polytechnic of Bragança, Portugal
João Reis	University of Aveiro, Portugal
João Reis	University of Lisbon, Portugal
João Rodrigues	University of the Algarve, Portugal
João Vidal Carvalho	Polythecnic of Coimbra, Portugal
Joaquin Nicolas Ros	University of Murcia, Spain
John W. Castro	University de Atacama, Chile
Jorge Barbosa	Polythecnic of Coimbra, Portugal
Jorge Buele	Technical University of Ambato, Ecuador
Jorge Gomes	University of Lisbon, Portugal
Jorge Oliveira e Sá	University of Minho, Portugal
José Braga de Vasconcelos	Universidade Lusófona, Portugal
Jose M Parente de Oliveira	Aeronautics Institute of Technology, Brazil
José Machado	University of Minho, Portugal
José Paulo Lousado	Polythecnic of Viseu, Portugal
Jose Quiroga	University of Oviedo, Spain
Jose Silvestre Silva	Academia Militar, Portugal
Jose Torres	Universidty Fernando Pessoa, Portugal
Juan M. Santos	University of Vigo, Spain
Juan Manuel Carrillo de Gea	University of Murcia, Spain
Juan Pablo Damato	UNCPBA-CONICET, Argentina
Kalinka Kaloyanova	Sofia University, Bulgaria
Kamran Shaukat	The University of Newcastle, Australia
Karima Moumane	ENSIAS, Morocco
Katerina Zdravkova	University Ss. Cyril and Methodius, North Macedonia
Khawla Tadist	Marocco
Khalid Benali	LORIA—University of Lorraine, France
Khalid Nafil	Mohammed V University in Rabat, Morocco
Korhan Gunel	Adnan Menderes University, Turkey
Krzysztof Wolk	Polish-Japanese Academy of Information Technology, Poland
Kuan Yew Wong	Universiti Teknologi Malaysia (UTM), Malaysia
Kwanghoon Kim	Kyonggi University, South Korea
Laila Cheikhi	Mohammed V University in Rabat, Morocco
Laura Varela-Candamio	Universidade da Coruña, Spain
Laurentiu Boicescu	E.T.T.I. U.P.B., Romania
Lbtissam Abnane	ENSIAS, Morocco
Lia-Anca Hangan	Technical University of Cluj-Napoca, Romania
Ligia Martinez	CECAR, Colombia
Lila Rao-Graham	University of the West Indies, Jamaica

Łukasz Tomczyk	Pedagogical University of Cracow, Poland
Luis Alvarez Sabucedo	University of Vigo, Spain
Luís Filipe Barbosa	University of Trás-os-Montes e Alto Douro
Luis Mendes Gomes	University of the Azores, Portugal
Luis Pinto Ferreira	Polytechnic of Porto, Portugal
Luis Roseiro	Polytechnic of Coimbra, Portugal
Luis Silva Rodrigues	Polythencic of Porto, Portugal
Mahdieh Zakizadeh	MOP, Iran
Maksim Goman	JKU, Austria
Manal el Bajta	ENSIAS, Morocco
Manuel Antonio Fernández-Villacañas Marín	Technical University of Madrid, Spain
Manuel Ignacio Ayala Chauvin	University Indoamerica, Ecuador
Manuel Silva	Polytechnic of Porto and INESC TEC, Portugal
Manuel Tupia	Pontifical Catholic University of Peru, Peru
Manuel Au-Yong-Oliveira	University of Aveiro, Portugal
Marcelo Mendonça Teixeira	Universidade de Pernambuco, Brazil
Marciele Bernardes	University of Minho, Brazil
Marco Ronchetti	Universita' di Trento, Italy
Mareca María PIlar	Universidad Politécnica de Madrid, Spain
Marek Kvet	Zilinska Univerzita v Ziline, Slovakia
Maria João Ferreira	Universidade Portucalense, Portugal
Maria José Sousa	University of Coimbra, Portugal
María Teresa García-Álvarez	University of A Coruna, Spain
Maria Sokhn	University of Applied Sciences of Western Switzerland, Switzerland
Marijana Despotovic-Zrakic	Faculty Organizational Science, Serbia
Marilio Cardoso	Polythecnic of Porto, Portugal
Mário Antunes	Polythecnic of Leiria & CRACS INESC TEC, Portugal
Marisa Maximiano	Polytechnic Institute of Leiria, Portugal
Marisol Garcia-Valls	Polytechnic University of Valencia, Spain
Maristela Holanda	University of Brasilia, Brazil
Marius Vochin	E.T.T.I. U.P.B., Romania
Martin Henkel	Stockholm University, Sweden
Martín López Nores	University of Vigo, Spain
Martin Zelm	INTEROP-VLab, Belgium
Mazyar Zand	MOP, Iran
Mawloud Mosbah	University 20 Août 1955 of Skikda, Algeria
Michal Adamczak	Poznan School of Logistics, Poland
Michal Kvet	University of Zilina, Slovakia
Miguel Garcia	University of Oviedo, Spain

Miguel Melo	INESC TEC, Portugal
Mihai Lungu	University of Craiova, Romania
Mircea Georgescu	Al. I. Cuza University of Iasi, Romania
Mirna Muñoz	Centro de Investigación en Matemáticas A.C., Mexico
Mohamed Hosni	ENSIAS, Morocco
Monica Leba	University of Petrosani, Romania
Nadesda Abbas	UBO, Chile
Narjes Benameur	Laboratory of Biophysics and Medical Technologies of Tunis, Tunisia
Natalia Grafeeva	Saint Petersburg University, Russia
Natalia Miloslavskaya	National Research Nuclear University MEPhI, Russia
Naveed Ahmed	University of Sharjah, United Arab Emirates
Neeraj Gupta	KIET group of institutions Ghaziabad, India
Nelson Rocha	University of Aveiro, Portugal
Nikola S. Nikolov	University of Limerick, Ireland
Nicolas de Araujo Moreira	Federal University of Ceara, Brazil
Nikolai Prokopyev	Kazan Federal University, Russia
Niranjan S. K.	JSS Science and Technology University, India
Noemi Emanuela Cazzaniga	Politecnico di Milano, Italy
Noureddine Kerzazi	Polytechnique Montréal, Canada
Nuno Melão	Polytechnic of Viseu, Portugal
Nuno Octávio Fernandes	Polytechnic of Castelo Branco, Portugal
Nuno Pombo	University of Beira Interior, Portugal
Olga Kurasova	Vilnius University, Lithuania
Olimpiu Stoicuta	University of Petrosani, Romania
Patricia Zachman	Universidad Nacional del Chaco Austral, Argentina
Paula Serdeira Azevedo	University of Algarve, Portugal
Paula Dias	Polytechnic of Guarda, Portugal
Paulo Alejandro Quezada Sarmiento	University of the Basque Country, Spain
Paulo Maio	Polytechnic of Porto, ISEP, Portugal
Paulvanna Nayaki Marimuthu	Kuwait University, Kuwait
Paweł Karczmarek	The John Paul II Catholic University of Lublin, Poland
Pedro Rangel Henriques	University of Minho, Portugal
Pedro Sobral	University Fernando Pessoa, Portugal
Pedro Sousa	University of Minho, Portugal
Philipp Jordan	University of Hawaii at Manoa, USA
Piotr Kulczycki	Systems Research Institute, Polish Academy of Sciences, Poland

Prabhat Mahanti	University of New Brunswick, Canada
Rabia Azzi	Bordeaux University, France
Radu-Emil Precup	Politehnica University of Timisoara, Romania
Rafael Caldeirinha	Polytechnic of Leiria, Portugal
Raghuraman Rangarajan	Sequoia AT, Portugal
Raiani Ali	Hamad Bin Khalifa University, Qatar
Ramadan Elaiess	University of Benghazi, Libya
Ramayah T.	Universiti Sains Malaysia, Malaysia
Ramazy Mahmoudi	University of Monastir, Tunisia
Ramiro Gonçalves	University of Trás-os-Montes e Alto Douro & INESC TEC, Portugal
Ramon Alcarria	Universidad Politécnica de Madrid, Spain
Ramon Fabregat Gesa	University of Girona, Spain
Ramy Rahimi	Chungnam National University, South Korea
Reiko Hishiyama	Waseda University, Japan
Renata Maria Maracho	Federal University of Minas Gerais, Brazil
Renato Toasa	Israel Technological University, Ecuador
Reyes Juárez Ramírez	Universidad Autonoma de Baja California, Mexico
Rocío González-Sánchez	Rey Juan Carlos University, Spain
Rodrigo Franklin Frogeri	University Center of Minas Gerais South, Brazil
Ruben Pereira	ISCTE, Portugal
Rui Alexandre Castanho	WSB University, Poland
Rui S. Moreira	UFP & INESC TEC & LIACC, Portugal
Rustam Burnashev	Kazan Federal University, Russia
Saeed Salah	Al-Quds University, Palestine
Said Achchab	Mohammed V University in Rabat, Morocco
Sajid Anwar	Institute of Management Sciences Peshawar, Pakistan
Sami Habib	Kuwait University, Kuwait
Samuel Sepulveda	University of La Frontera, Chile
Snadra Costanzo	University of Calabria, Italy
Sandra Patricia Cano Mazuera	University of San Buenaventura Cali, Colombia
Sassi Sassi	FSJEGJ, Tunisia
Seppo Sirkemaa	University of Turku, Finland
Shahnawaz Talpur	Mehran University of Engineering & Technology Jamshoro, Pakistan
Silviu Vert	Politehnica University of Timisoara, Romania
Simona Mirela Riurean	University of Petrosani, Romania
Slawomir Zolkiewski	Silesian University of Technology, Poland
Solange Rito Lima	University of Minho, Portugal
Sonia Morgado	ISCPSI, Portugal

Contents

Information and Knowledge Management

Information Technologies in Education

Information and Knowledge Management

Knowledge Discovery in Wikidata with Machine Learning in Graph

Stalin Figueroa(✉)

University of Alcala, 28801 Alcalá de Henares, Spain
stalin.figueroa@edu.uah.es

Abstract. Wikipedia contains a large amount of information distributed on many web pages and in several languages. As for a particular topic, the same thing happens, there is a lot of information on the same topic, and it is distributed on many web pages.

When you need to know the number of elements of a nominal categorical variable or to count the elements of a category that are distributed in many wikidata web pages, then there is a difficulty, and you need to apply machine learning techniques.

With Graph Machine learning, you can harness the power of representational learning and apply it to data extracted from wikidata, using graphs to find underlying structural similarities effectively and efficiently. In addition, the graphs provide us with an additional dimension to analyze the data.

In this scientific article, the methodology to reach that goal is studied, and a particular topic has been taken to apply this methodology, and obtain the answers.

The theme is "Search for similar Italian painters with their similar paintings in terms of material used and genre of art that are in museums around the world."

Keywords: knowledge · machine learning · graphs

1 Introduction

Wikipedia contains a large amount of information distributed on many web pages and in several languages. As for a particular topic, the same thing happens, there is a lot of information on the same topic, and it is distributed on many web pages.

As you can see, the amount of information is immensely large and spread over many web pages. The problem occurs when it is necessary to obtain counted, categorized, and qualified information from the analyzed data set.

When it is necessary to know the number of elements of a nominal categorical variable, to count the elements of a category that are distributed in many wikidata web pages, or to find similar elements and subelements, then there is a difficulty, and techniques need to be applied. Machine learning.

In this scientific article, the methodology to reach that goal is studied, and a particular topic has been taken to apply this methodology, and obtain the answers.

A. Rocha et al. (Eds.): WorldCIST 2023, LNNS 800, pp. 3–12, 2024.
https://doi.org/10.1007/978-3-031-45645-9_1

The theme is "Find similar Italian painters with their similar paintings in terms of material used and genre of art that is in the world's museums".

This research work is based on a methodology that starts by using the wikidata query service tool to download the data; executing instructions of the SPARQL language, then the data is prepared, the exploratory analysis of the data is done, and the graphs are generated, to extract the knowledge of the graphs with machine learning in graph techniques, and reach the objective of the answers.

2 Research Questions

Please note that the first paragraph of a section or subsection is not indented. The first paragraphs that follows a table, figure, equation etc. does not have an indent, either.

Subsequent paragraphs, however, are indented.

The research questions addressed by this study are:

RQ1. How do search in wikidata knowledge bases, similar elements, and sub-elements with characteristics that associate them?

RQ2. When is it necessary to apply machine learning techniques to graphs, to graphs of Knowledge of wikidata?

Regarding RQ1, Wikidata knowledge bases contain many topics and subtopics, which can be categorized, and categories usually have subcategories and many items, which are spread over many wikidata web pages, and can be counted, and similar elements and subelements can also be discovered, applying graph machine learning techniques, through which nodes can be embedded with several dimensions and adjacent nodes embedded to obtain similar elements and subelements based on characteristics that associate them.

To respond to RQ2, it will be considered to apply machine learning in graph techniques, for Wikidata knowledge bases and graphs, in cases of wanting to obtain condensed data from the result of processing several web pages, such as categories, subcategories, elements, and subelements. Similar in their features.

Graphs greatly help advanced machine learning models to understand the existing relationships between data more quickly so that knowledge is increasingly more and of higher quality.

Contribution
The contribution of this article is the discovery of knowledge in the wikidata, knowledge bases through graphs, machine learning techniques, representation learning with Node2Vec, Edge2Vec, Graph2Vec models, and machine learning in graphs.

3 Methodology

In the first place, the discovery of the knowledge of the responses to the wikidata topic "Search for similar Italian painters with their similar paintings in terms of material used and genre of art that are in the world's museums". It is obtained by following these seven procedures (Fig. 1):

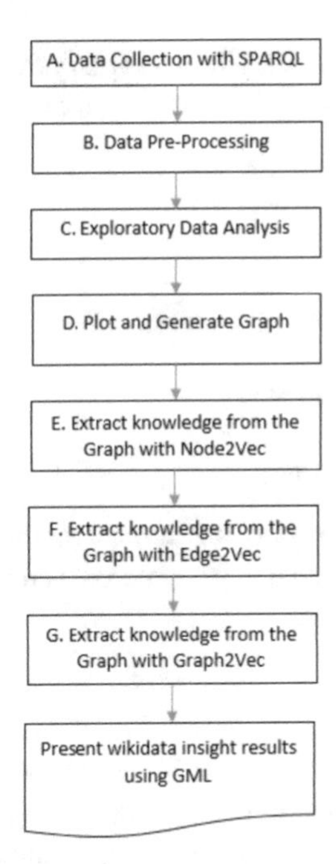

Fig. 1. Framework for knowledge discovery in Wikidata with machine learning in graph.

4 Data Collection with SPARQL

In this stage, a portion of the wikidata knowledge base is downloaded or a query of data related to a specific topic is carried out, through the Wikidata Query Service tool, and the exploratory analysis and pre-processing of the downloaded data.

With the sparql query that follows, the query of the subject is made: Pictures of Paintings by Italian Painters and Painters in the Museums of the World.

```
SELECT ?pintura ?pinturaLabel ?creador ?creadorLabel ?imagen ?g_nero
?g_neroLabel ?museoLabel ?museoLocation ?museoLocationLabel ?coordenadas
?sexo_o_g_nero ?sexo_o_g_neroLabel ?pa_s ?pa_sLabel ?material_empleado ?mate-
rial_empleadoLabel ?movimiento ?movimientoLabel
   WHERE {?pintura (wdt:P31/(wdt:P279*)) wd:Q3305213;
     wdt:P170 ?creador.
    ?creador wdt:P27 wd:Q38.
    ?pintura wdt:P276 ?museo.
    SERVICE wikibase:label { bd:serviceParam wikibase:language
    "[AUTO_LANGUAGE],en,es". }
    OPTIONAL { ?museo wdt:P625 ?coordenadas. }
    OPTIONAL { ?museo wdt:P131 ?museoLocation. }
    OPTIONAL { ?pintura wdt:P18 ?imagen. }
    OPTIONAL { ?pintura wdt:P136 ?g_nero. }
    OPTIONAL { ?creador wdt:P21 ?sexo_o_g_nero. }
    OPTIONAL { ?museo wdt:P17 ?pa_s. }
    OPTIONAL { ?pintura wdt:P186 ?material_empleado. }
    OPTIONAL { ?pintura wdt:P135 ?movimiento. }}
```

Running this sparql query on the Wikidata Query Service returns 3856 results.

With this information extracted from wikidata, queries can also be made, with the following classifications:

- Painting Frames by Painter
- Painting frames by art genre.
- Pictures painting material
- Period of the artistic movement by painting.
- Painters by genre of art.
- Museums with paintings on display
- Museums by country.
- Gender or sex of the painter artist.

Once the data extraction and query are done, with the Wikidata Query Service tool, we download the data in CSV format to pre-process it and leave the data set standardized and normalized, for later analysis.

4.1 Pre-processing of the Data

The data frame of the downloaded file has 3856 rows and 19 columns.

Each row contains the information of a painting frame, for example, the name of the painting frame, name of the painter, genre of art, the museum in which it is exhibited, etc. In conclusion, this dataframe contains information on 3,856 paintings by Italian painters, which are exhibited in museums around the world (Fig. 2).

```
df.info()

<class 'pandas.core.frame.DataFrame'>
RangeIndex: 3856 entries, 0 to 3855
Data columns (total 19 columns):
 #   Column                    Non-Null Count  Dtype
---  ------                    --------------  -----
 0   pintura                   3856 non-null   object
 1   pinturaLabel              3856 non-null   object
 2   creador                   3856 non-null   object
 3   creadorLabel              3856 non-null   object
 4   imagen                    477 non-null    object
 5   g_nero                    934 non-null    object
 6   g_neroLabel               934 non-null    object
 7   museoLabel                3856 non-null   object
 8   museoLocation             3818 non-null   object
 9   museoLocationLabel        3818 non-null   object
 10  coordenadas               3820 non-null   object
 11  sexo_o_g_nero             3849 non-null   object
 12  sexo_o_g_neroLabel        3849 non-null   object
 13  pa_s                      3849 non-null   object
 14  pa_sLabel                 3849 non-null   object
 15  movimiento                154 non-null    object
 16  movimientoLabel           154 non-null    object
 17  material_empleado         2555 non-null   object
 18  material_empleadoLabel    2555 non-null   object
dtypes: object(19)
memory usage: 572.5+ KB
```

Fig. 2. General dataframe information

In this output, you can see the number of columns 19, the total number of rows 3856, and the number of rows without nulls. Also the data types.

Columns that do not have 3856 non-null means that the difference; are null values, which are represented by NaN "The Wikidata repository consists mainly of items, each with a tag, a description, and any number of aliases. Elements are uniquely identified by a prefix Q followed by a number" [1]. As an example, you can see the first record of the dataframe. The painting element, with title or label Crucifix with Stories of the Passion of Christ identified by the prefix (Q 3698266), and the creator or painter element, with the name of the painter, or Maestro Della Croce label, identified by the prefix (Q 3842987) (Fig. 3).

	pintura	pinturaLabel	creador	creadorLabel
0	http://www.wikidata.org/entity/Q3698266	Crucifix with Stories of the Passion of Christ	http://www.wikidata.org/entity/Q3842987	Maestro della Croce 434

Fig. 3. First record of the dataframe

The declarations describe the detailed characteristics of the element and consist of a property and a value. Properties are identified by a prefix P followed by a number.

"The following statements, ownership, and value of the Master Painter Della Croce, taken from the wikidata page. https://www.wikidata.org/wiki/Q3842987 (Table 1).

Table 1. Identification of the element and property in wikidata.

	Item Element Painter	Property	Worth
ID	Q3842987	P27	Q38
Label	Croce Master	country of citizenship	Italy

In conclusion, wikidata elements are uniquely identified by a prefix Q, followed by a number. The characteristics of the elements are known as properties and are identified by a prefix P followed by a number, and the value of the property is identified by a prefix Q followed by a number.

Properties can also link to external databases. A property that links an item to an external database, such as an authority control database used by libraries and archives, is called an identifier. Site links connect an item to corresponding content on client wikis, such as Wikipedia, Wikibooks, or Wikiquote.

All of this information can be displayed in any language, even if the data originated in a different language. By accessing these values, the client's wikis will display the most up-to-date data." [1].

4.2 Exploratory Data Analysis

At the end of the exploratory data analysis, it is possible to see the scope of the discovery of the information, which could be obtained, then with the main entities, the topology of the graph is designed, from which information is discovered with automatic graph learning techniques will be obtained (Fig. 4).

```
df.groupby(['pintor'])['genero_arte'].value_counts()

pintor                              genero_arte
Achille Funi                        alegoría                   8
                                    pintura mitológica         4
                                    pintura de historia        3
Adone Comboni                       pintura del paisaje        2
Adriano Spilimbergo                 paisaje urbano             4
Afro Basaldella                     pintura del paisaje        6
Alberto Burri                       arte abstracto             2
Alberto Caligiani                   bodegón                    2
Alberto Manfredi                    autorretrato               1
Alberto Martini                     autorretrato               6
Alberto Salietti                    pintura del paisaje        1
Aldo Carpi                          pintura del paisaje        2
Aldo Conti                          pintura del paisaje        2
Aldo Raimondi                       paisaje urbano             4
Alessandro Abate                    autorretrato               2
Alessandro Algardi                  arte abstracto             2
Alfonso Corradi                     pintura del paisaje        2
Alfonso Fratteggiani Bianchi        arte abstracto            16
```

Fig. 4. Number of paintings by Painter and Genre of Art

4.3 Study of the Topology of the Graph

In part B, it was shown that the Wikidata repository consists mainly of items, each with a tag, a description, and any number of aliases. Elements are uniquely identified by an ID, which is made up of a prefix Q, followed by a number. The characteristics of these elements are known as properties and are identified by a prefix P followed by a number, and the value of the property is identified by a prefix Q followed by a number (Fig. 5).

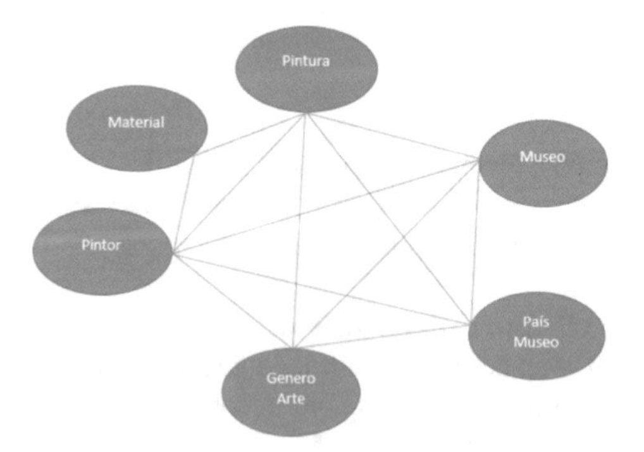

Fig. 5. Topology of the Graph.

The nodes shown in the graph topology are the main fields that will be taken from the data frame downloaded from wikidata to make the graph. These fields are the element labels, and as you know, the label is tied to the element ID, which is made up of a prefix Q followed by a number. So when it comes to working with the labels there is no problem, because it is like working with the same ID of the element. Fields containing element IDs must not be removed from the data frame.

By visualizing the topology of the undirected graph of the image, we can get a great idea of what kind of data we are working with.

Understanding the topology of the graph, you can see and say the following:

- The painter uses painting materials to paint the painting frame.
- The paintings are classified by genre of art
- Painters are also classified by genre of art, due to their paintings.
- The genre of art is a style of art with properties that differentiate it from others
- The paintings are exhibited in the museums of the different countries
- Some museums exhibit paintings by genre of art and materials used.

In this chart, each of these columns or fields will need to be converted to create a From, To format. There are a few different ways to do this and the context of the data may dictate the best route to handle it.

In this section, the graph is generated, and the knowledge regarding the subject of Italian Painters is discovered, through automatic graph learning techniques, with programming language python 3.7.6, libraries: networkx 2.6.3, karateclub 1.2 .3 and node2vec 0.4.3.

The nodes shown in the graph topology are the main fields that will be taken from the dataframe to make the graph.

Then the records containing NaN fields will be deleted for a better performance of the graph. Then, with the data frame pre-processed, each of the columns will be converted into a From - To format, in order to draw the graph, and ob-tain the required knowledge through machine learning techniques on graphs.

4.4 Extract Knowledge From the Graph with Representation Learning for Nodes Using Node2Vec

"Node2Vec is a variant of the Deep Walk model and has a noticeable difference in how random walks are generated. Deep Walk has the limitation of preserving local neighborhood information of a node, so Node2Vec employs a flexible and powerful combination of Breadth-First Search (BFS) and Depth-First Search (DFS) to perform graph exploration. The combination of these two algorithms is regularized by the following probabilities:

p: the probability that a random walk returns to the previous node

q: the probability that a random walk can pass through a part of the graph never explored before.

This model introduces bias walks that can trade-off between the local microscopic view (BFS) and the macroscopic global view (DFS) of the network.

The Node2Vec algorithm works as follows:

- Calculate the probabilities of the random walks.
- Simulate random walks starting from each node.
- Optimizes the model goal using stochastic gradient descent." [2]

With the Node2Vec model, the nodes will be embedded in the graph, and painters similar to the painter Achille Funi will be sought, in terms of art genre.

The data_graph parameter contains the graph data link, the dimensions = 2 parameter indicates the two dimensions with which the nodes will be embedded in the graph.

When executing this finished program, the result shown in Fig. 6 is obtained.

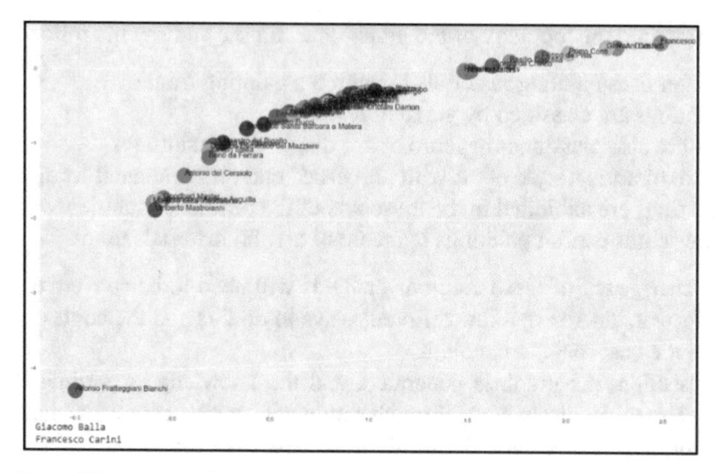

Fig. 6. Embedding nodes with 2 dimensions to search for painters similar to Achille Funi

Giacomo Balla and Francesco Carini, are painters similar to Achille Funi, due to their genre of art and materials used in the paintings.

4.5 Extract Knowledge From the Graph with Representation Learning for Nodes Using Edge2Vec

"Edge2Vec is a relatively simple extension of Node2Vec, in this model, the node embedding of two adjacent nodes is used to perform some basic mathematical operations to extract the embedding from the edge that connects them.

Depending on the nature of your data, and how they relate to each other, different embeddings will be valuable. An effective evaluation method may be to test each of them and see which one has the most sensitive edge separation." [2].

To visualize the first two dimensions of border inlays between the painters Achille Funi and his similar ones, with their similar paintings, it will be done using a python program. Where the node embeddings of two adjacent nodes will be traced and the paintings of painters similar to Achille Funi will be searched.

The results show that the similar paintings of Achille Funi with his similar painters are:

- Goddess Roma by Achille Funi
- Minerva by Achille Funi
- Glory by Achille Funi
- Roman Soldier by Achille Funi
- Flowers by Francesco Carini
- Changeable Weather by Francesco Carini
- Lampada sd arch by Giacomo Balla
- Abstract speed + sound by Giacomo Balla

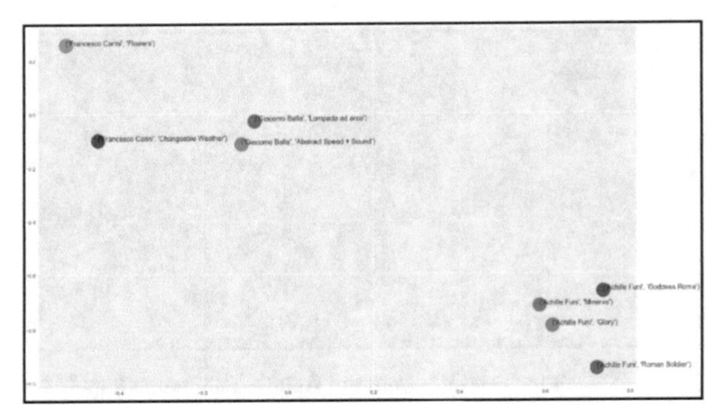

Fig. 7. Node embeddings of two adjacent nodes to search for paintings by painters similar to Achille Funi

4.6 Extract Knowledge from the Graph with Representation Learning for Nodes Using Graph2Vec

"This is the highest generalization of node and edge representation learning. In Natural Language Processing, this model is called Doc2Vec because it is the embedding of multiple documents instead of words or sentences. When dealing with graphs, they are divided into sub-graphs, and then the model is trained to learn a representation of each of these sub-graphs as an embedding.

By understanding the Doc2Vec model, the Graph2Vec algorithm can be described. The main idea behind this model is to see a complete graph as a document and each of its generated sub-graphs, as an ego graph of each node, as words that make up the document. In other words, a graph is made up of sub-graphs, just as a document is made up of sentences. According to this description, the algorithm can be summarized in the following steps:

1. A set of rooted subgraphs is generated around each node.
2. The skip-gram Doc2Vec model is trained using the generated sub-graphs.
3. The information contained in the hidden layers of the trained Doc2Vec model is used to extract the optimized embedding of each node, through stochastic gradient descent." [2].

For this case, you can create row-only subcharts that contain Achille Funi and the artists that the model found similar to him. All the information in it can be represented in a much more condensed and efficient way using embeddings.

Note that we need to convert the node labels to integers for the model to fit correctly, and it will be done with 2 dimensions, using a python program (Fig. 8).

With the Graph2Vec representation for graphs, all the information is represented in a much more condensed and efficient way using embeddings. As you can see in the previous image, the 2 painters similar to Achille Funi are presented: Giacomo Balla and Francesco Carini.

Fig. 8. Search for painters similar to Achille Funi with Graph2Vec

5 Results

The results obtained meet the objectives of the research topic "Search for similar Italian painters with their similar paintings in terms of material used and genre of art that are in museums around the world".

With Node2Vec, a variant of the Deep Walk model, node embedding with 2 dimensions is made, in the generated knowledge graph, to find painters similar to a required painter, in this case, the painter Achille Funi, and find the required results.

Giacomo Balla and Francesco Carini, are painters similar to Achille Funi, due to their genre of art and the material used in the paintings.

With Edge2Vec, which is a relatively simple extension of Node2Vec, the node embeddings of two adjacent nodes are plotted, in the generated knowledge graph, and you will find the paintings of painters similar to Achille Funi, as can be seen in Fig. 7.

References

1. Wikipedia: Introduction to Wikidata (2019). https://www.wikidata.org/wiki/Wikidata:Introduction. Accessed 01 Apr 2019
2. Stamile, C., Marzullo, A., Deusebio, E.: Graph Machine Learning. Packt Publishing (2021)
3. Hamilton, W.L., Ying, R., Leskovec, J.: Representation learning on graphs: Methods and applications. arXiv preprint arXiv:1709.05584 (2017)
4. Hamilton, W.L.: Graph representation learning. Synthesis Lectures on Artificial. Intelligence and Machine Learning, 14(3), 1–159 (2020)
5. Mora-Cantallops, M., Sánchez-Alonso, S., García-Barriocanal, E.: A systematic literature review on Wikidata. Data Technol. Appl. **53**(3), 250–268 (2019). https://doi.org/10.1108/DTA-12-2018-0110
6. Needham, M., s, A.E.: Graph algorithms: practical examples in Apache Spark and Neo4j. O'Reilly Media (2019)

Optimizing Topic Modelling for Comments on Social Networks: Reactions to Science Communication on COVID

Bernardo Cerqueira de Lima[1], Renata Maria Abrantes Baracho[1], and Thomas Mandl[2](✉)

[1] Federal University of Minas Gerais, UFMG, Belo Horizonte, Brazil
bernardolima95@gmail.com, renatabaracho@arq.ufmg.br

[2] University of Hildesheim, Universitätsplatz 1, Hildesheim, Germany
mandl@uni-hildesheim.de

Abstract. Channels on social media which disseminate scientific information ot the public have been of great importance during the COVID crisis. The authors of videos and texts as well as researchers who study the impact of scientific information online are interested in the reactions on the design of such information resources. This study aims to obtain insights into the perception of scientific information for the public. This is related to the relationship between the information behaviour of individuals and science communication (e.g. through videos or texts) during the COVID-19 crisis. To analyze the reactions to scientific information for the public, we selected Twitter users who are doctors, researchers, science communicators or represent research institutes, processing their replies for 20 months from the beginning of the pandemic, and performing topic modeling on the textual data. The goal was to find topics that relate to the theme of reactions to scientific communication. We present a method that filters such reactions of consumers of science communication. Various topic modelling experiments show that topics can support the search for relevant online reactions, defined by sentences such as: "This was very informative! Thanks".

Keywords: Topic modelling · Corona Crisis · Science Communication · Social Media

1 Introduction

Like many other crises, the Corona pandemic is affecting the information behaviour of citizens. As Web popularity rankings have shown, the pandemic makes citizens turn to different sources of information since they develop specific needs in periods of crisis communication [10]. As a consequence, they are confronted with challenges regarding the reliability and trustworthiness of information [4]. Channels on social media create diverse content for science communication to accommodate these needs [14]. Due to the importance of the science

A. Rocha et al. (Eds.): WorldCIST 2023, LNNS 800, pp. 13–22, 2024.
https://doi.org/10.1007/978-3-031-45645-9_2

understanding during a crisis, it is of particular importance to understand patterns of information behaviour (e.g. [18]). Media creators need to understand the quality criteria which users apply in order to select their resources and which determine their preferences. Information resources in general and multimodal science communication in particular differ in the way they portray scientific information [14]. However, research does not look in detail into how scientific information is disseminated successfully and which approaches seem to work well.

In this work, we intend to analyse the online discourse on the Corona crisis from the perspective of science communication. We are interested in extracting the subset of comments which users post in science communication channel as reactions to the content. Most comments which follow on scientific channels are not related to the content or design of the format but rather follow up on the general Corona discourse about political views of the crisis.

The aim of this study could be described as follows: "Find all tweets which are reactions to science communication formats which comment on posts containing science communication for the public. Examples for relevant tweets are comments on the quality of explanations, expressing gratitude or discussing the quality of references." In order to find such documents, aggressive filtering is necessary. A first approach on searching with keywords was not successful and led to low diversity. Thus, topic modelling was explored as a method for finding the relevant subset from a crawled collection. After success with a small dataset, we explored this method further and present the methodology and results for a collection of tweets from Brazilian Twitter accounts in this paper.

The contribution of this work is in the development of a methodology including optimized topic modeling for finding science communication related tweets and for presenting an application of this methodology for the Brazilian science communication market. With this result, it is possible to explore these tweets and to observe which science communication strategies are received positively. Such a dataset can contribute to the improvement of dissemination of information in future crises (Fig. 1).

2 Related Work

The analysis of popular science communication channels for the German market (such as the NDR podcast with the virologist Christian Drosten or Mai Thi Nguyen-Kim's MaiLab on Youtube) or the Brazilian market (e.g. governmental bodies such as the Butantã Institute or communicators such as the microbiologist Átila Iamarino) shows that they use similar communicative strategies but with different emphasis. Factors relevant for the analysis are, among others, the communication of scientific insecurity, the degree of factuality, the degree of complexity, the potential use of emotionalisation, or the (self-)presentation of experts [12]. As communication on social media is essentially multimodal, visual information, location and body language are also important. To gain insights on how epidemiological information is communicated to non-expert audiences and

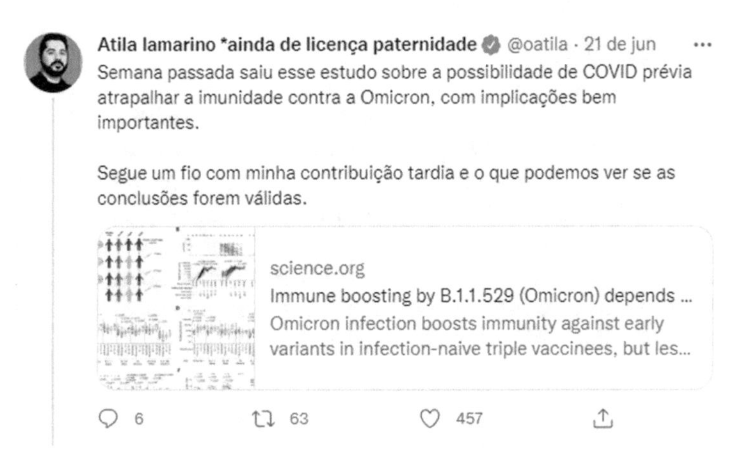

Fig. 1. Screenshot of a tweet from a science communicator from Brazil. Text: "Last week this new study came out about the possibility that COVID might mess up immunity against Omicron, with quite important implications."

how they are perceived, we intend to extract relevant comments by the online audience which can be interesting to study public communication of science [13].

However, so far the analysis of the content and design of Corona-related scientific information for the public has been carried out mainly qualitatively and with little consideration of the reactions of the online audiences (e.g. [8]).

The analysis of complex information needs during crisis using social media has been studied in the track Information Retrieval from Microblogs during Disasters (IRMiDis) [5]. Tweets about the Nepal earthquake were collected and provided. Systems were requested to extract e.g. information that is helpful for rescue workers or which reports on specific needs. Information propagation on Social Media during the Corona crisis has typically been focused on general trends, political attitudes or the dissemination of misinformation. Datasets available for social media communication on Corona are available but they tend to collect information very broadly [22]. It was shown, that potentially incorrect information is disseminated in a substantial amount [25]. An analysis of a large amount of claims has shown that misinformation spreads faster than true information already before the Corona crisis [24]. These results show that the quality of science communication should be studied. Several studies have already applied topic modelling [7, 16, 26] or classification [21] but did not specifically discuss reactions to science communication to the public.

In the context of analyzing short text data such as tweets, utilizing topic modeling techniques can facilitate the identification and categorization of various themes within a corpus. Empirical studies have demonstrated that the Latent Dirichlet Allocation (LDA) algorithm exhibits a strong performance in terms of the lexical diversity of the generated topics and the accuracy of topic categorization, compared to other methods [1].

3 Methodology

This study was conducted in several steps: First, we identified relevant science communication channels on Twitter; then, data from these sources from content made in the COVID-19 pandemic were collected and went into several textual data processing techniques, resulting into a final corpus prepared for natural language processing. Lastly, a topic model was created to help organise the massive text data into certain themes, with the goal of looking for topics that grouped terms related to reactions to scientific communication. The topic model was evaluated by the NPMI metric (Fig. 2).

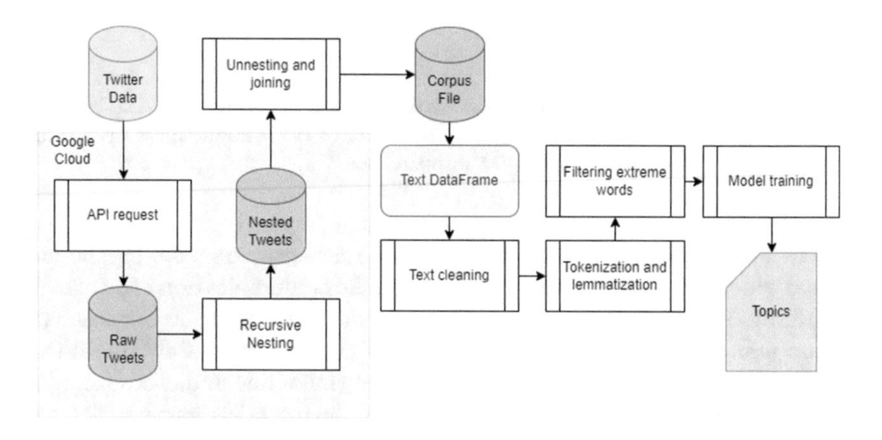

Fig. 2. Processing pipeline.

3.1 Tweet Collection

For the Brazilian market, 46 sources were manually chosen by their relevance to COVID discussion, most of them being doctors and institutes involved in COVID-19 research and discussion. The other channels are comprised by the most popular science communicators and news aggregators during the COVID 19 pandemic. Requests were sent to the Twitter API[1] to retrieve the tweets and retweets from these sources, dated from March 1st, 2020 to March 1st, 2022, as well as all of the replies contained inside their respective threads. Around 1.3 million tweets were collected and were organised in nested JSON files that reflected the complex structure on the website. Only tweets that were flagged by Twitter with the Portuguese language tag were collected and used in this study.

Because of Twitter's API and computing power limitation's in face of the massive amount of data collected, this step was conducted on a Google Cloud cluster with three virtual machines. This architectural decision was taken because, for the collection step and the nesting step, it was necessary to keep

[1] https://developer.twitter.com/en/docs/api-reference-index (21-07-2021).

continuous usage of computing resource for more than a week. Abstracting the codebase to the Cloud was necessary to maintain fault-tolerance during the crawling and collection process.

3.2 Text Processing

The first steps taken were to amass every tweet into a single pool and remove any tweets made by the original source, leaving only replies. The text content of every tweet in this dataset was then stripped of URLs, special characters, emojis and mentions, leaving only the actual text content of each tweet. The next steps were taken with the objective of reducing noise in the text data: the first was the removal of stopwords. We used the union of four stopword lists: a handmade list crafted for the study, the Spacy Portuguese News [11] stopwords list, the NLP Python library Gensim [20] and the Wordcloud [19] stopwords.

After the stopwords were removed, the text was tokenized using Spacy's rule-based function. After this step, another filter was applied in order to guarantee that only words larger than 3 characters were allowed as valid tokens. These tokens were then reduced to their root forms with lemmatization, performed by the Spacy library, which switches between rule and lookup-based methods to reduce words to their lemmas. In order to reduce overfitting and noise, words that appeared in less than two documents and that appeared in more than 99% of the documents were filtered.

3.3 Topic Modelling

Topic modelling is a computational method for content analysis for applications in which the content of a large collection of documents needs to be analyzed. Since topic modelling works unsupervised, it requires no assumptions about content words and can be applied for exploring a collection without bias.

Latent Dirichlet Allocation (LDA). LDA is one of the statistical topic modelling methods which tries to uncover topics inherent in a collection of documents [15]. The ultimate goal of this study was to develop a topic model, a probabilistic model that discovers recurring patterns in tweets with a specific focus on searching for a topic that would indicate the theme of reactions to scientific communication. There are several existing algorithms for topic modeling, and in this study the one chosen was LDA, which, given a fixed number of topics, estimates how much of each document is comprised of each topic, based on the probability distribution of each word belonging to a certain topic.

$$P(w|\alpha,\beta) = \prod_{d=1}^{M} \int p(\theta_d|\alpha)(\prod_{d=1}^{N_d} \sum_{Z_{dn}} p(z_{dn}|\theta_d)p(w_{dn}|z_{dn},\beta))d\theta_d \qquad (1)$$

The Eq. 1 represents, along with the Fig. 3, the three levels in LDA. The variables α and β represent document-topic density and topic-word density, respectively, and are parameters defined for the whole corpus. Θ is the topic distribution of the document, z is a set of N topics, and w is a set of N words.

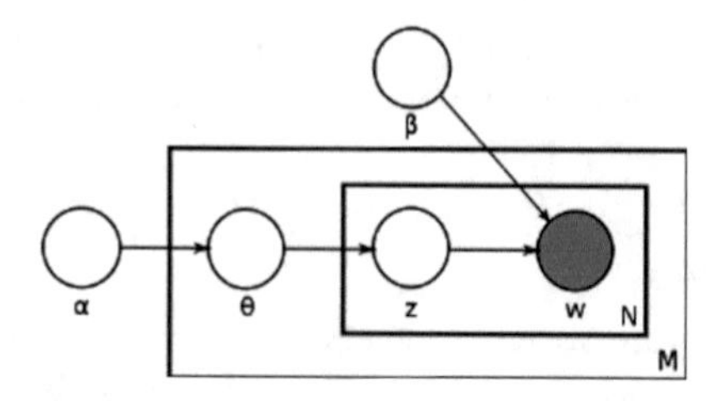

Fig. 3. Plate notation for the LDA algorithm. Image from [6].

Normalized Pointwise Mutual Information. Models that perform topic modeling include several hyperparameters which need to be set heuristically. One of them is the number of topics. It is important to choose a number of topics in which the results are still interpretable, but granular enough to avoid very general topics. Because this study aims at identifying a very specific and narrow topic, it is necessary to experiment with a higher numbers of topics.

One way to explore the optimal number of topics is the normalized pointwise mutual information metric, which evaluates the topic model by representing the top-n words of each topic as a vector in a semantic space. NPMI calculates their probability of co-ocurrence with each other, weighting these vectors by the NPMI of each term.

$$NPMI(w_i, w_j) = \frac{PMI(wi, wj)}{-log(p(wi, wj))} \tag{2}$$

The pointwise mutual information (PMI) is computed by the following formula:

$$PMI(w_i, w_j) = \frac{p(wi, wj)}{p(wi)p(wj)} \tag{3}$$

The values *p(wi)* and *p(wj)* represent the probability of each words ocurrence in a topic and *p(wi,wj)* is the probability of their co-ocurrence. The NPMI metric was selected because it tends to perform well when compared to human classification of topics [2].

4 Results

We decided on an ideal number of topics by evaluating the NPMI metric and by intellectual curation. Figure 4 shows the values for a set of topic number values.

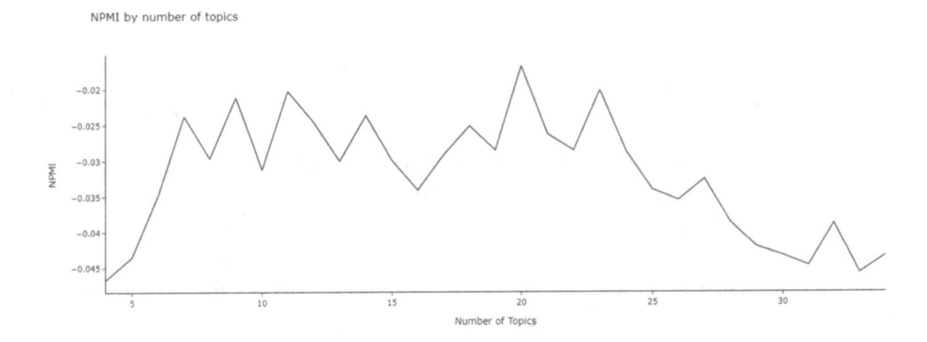

Fig. 4. NPMI coherence search by number of topics.

In our experiments, 20 topics led to a peak for the NPMI value, with the metric decreasing steadily after that number. Fewer topics also had a good NPMI result but were not able to properly specify different niches in the collection as it would have been necessary. Since the topic we were looking for was not a common theme in the tweets, a higher number of topics would allow for more specific topics and would make it easier to filter out tweets that were not directly related to the theme of reaction to scientific communication.

After execution of the LDA algorithm, the topics were formed as they can be seen on Fig. 5, plotted by their intertopic distance between each other. It can be observed that several of the topics are concentrated around the same area in the reduced two-dimensional space. This is one indicator for their similarity. A smaller number of topics could cover a similar area in the intertopic map. There would be less overlap between topics. However, would not be adequate for our search for a specific theme.

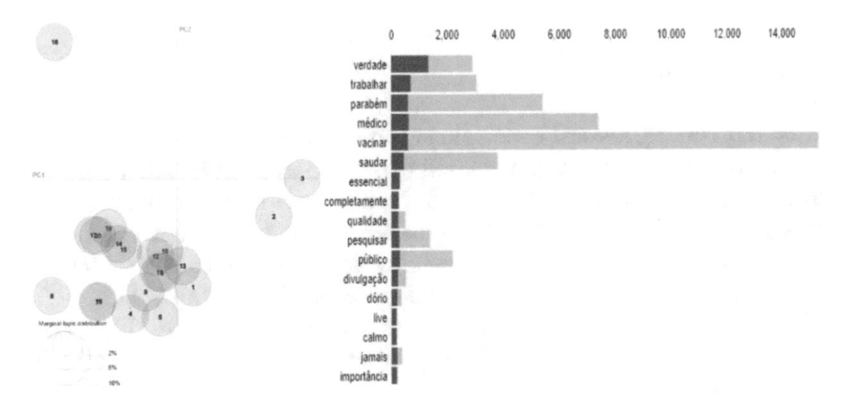

Fig. 5. Intertopic distance between 20 topics and the most relevant words found for topic 7. Visualization produced with the tool LDAvis [23].

With this selection of topics, we managed to find some topics that were related to the theme of reaction to scientific communication, but in none of the topics this seemed like the dominant theme. This is somewhat expected, as the nature of discussion in Twitter comments is generally not related to how the content was presented, but to the content at hand. In this particular experiment, we found two interesting topics, which can be seen in the Table 1. Since the content we are looking for is relatively rare, to find two topics out of the twenty topcis in which this theme can be found. Such a reduction helps with the goal of filtering a massive dataset of tweets in search for tweets directly related to science communication. Although these two topics are not only comprised of such content, they narrow down search.

Ultimately, the selected topics and the associated documents are a valid result. It can be further analysed and explored qualitatively to study the comments on science communication for the public.

Table 1. Two topics containing sentences relevant to scientific communication. Translation to English by the authors.

Topic No.	Most relevant words	Example sentences
7	truth, essential, working, completely, live, calm, importance, development, version, quality	"This 7th paragraph is not there for applying exceptions for Brazilians." "very good news" "This young guy did not talk about being cured, but only that he witnessed that NO ONE went to the hospital."
13	vacinnate, coronavirus, anxious, priority, reading, context, absolutely, layman, immune, entity	"Congratulations for this material. Excellent !" "Excellent news for those who believe in the Brazilian scientists." "And I have not seen that! I do not believe it!!"

5 Conclusion

Based on our results, several conclusions can be drawn.

Our analysis shows that topic modelling is an appropriate method for filtering massive information to a point at which humans can perform manual analysis. For this purpose, it was possible to reduce a set of 1.1 million tweets to four thousand, which can be processed manually with regards to their relevance to the theme. This shows that the methodology applied is overall valid.

For science communication channels, this result shows that they can filter relevant feedback form social media with a adequate effort and include it into their communication strategies.

Our methodology opens many alternatives which can be further explored. The selection of channels could have been more broad to cover also news outlets. Using other models than LDA, such as Neural Network based models, could lead to an improvement in NPMI, perplexity and human interpretability [9].

Transformer architectures and further NLP methods tested on social media data [17] might also be worth exploring. Other techniques such as different pooling and sampling methods could also change the result.

Future steps will include a sentiment analysis of topics and channels over the timeline of the pandemic: how did people feel about certain types of content as the pandemic went on? These questions can be answered with the given dataset and may ultimately lead to suggestions for improving the communication of scientists, institutes and governmental bodies to the public.

With these topics and tweets identified, further research is enabled. After manually tagging this filtered dataset according to relevance to the content of reactions to scientific communication, there might be sufficient data to train a text classifier model to detect such content in other datasets. This could contribute to provide better feedback to scientific institutions interested in keeping the general population well informed. Furthermore, it could be interesting to connect the amount of reactions to the overall trends of the pandemic [3].

Acknowledgements. This work was enabled and financed by the Volkswagen Foundation in Germany with the grant A133902 (Project Information Behavior and Media Discourse during the Corona Crisis: An interdisciplinary Analysis - InDisCo). Further financial support was provided by the Coordination for the Improvement of Higher Education Personnel (CAPES) from Brazil.

References

1. Albalawi, R., Yeap, T., Benyoucef, M.: Using topic modeling methods for short-text data: a comparative analysis. Front. Artif. Intell. **3** (07 2020). https://doi.org/10.3389/frai.2020.00042
2. Aletras, N., Stevenson, M.: Evaluating topic coherence using distributional semantics. In: Intl. Conf. on Computational Semantics (IWCS), pp. 13–22. ACl, Potsdam, Germany (2013). https://aclanthology.org/W13-0102
3. Ali, S., Khusro, S., Anwar, S., Ullah, A.: Ontocovid: ontology for semantic modeling of covid19 statistical data. In: Intl. Conf. on Information Technology and Applications, pp. 183–194. Springer (2022)
4. Barnwal, D., Ghelani, S., Krishna, R., Basu, M., Ghosh, S.: Identifying fact-checkable microblogs during disasters: a classification-ranking approach. In: Intl. Conf. on Distributed Computing and Networking, ICDCN, Bangalore, January 4–7, pp. 389–392. ACM (2019). doi:https://doi.org/10.1145/3288599.3295587
5. Basu, M., Ghosh, S., Ghosh, K.: Overview of the FIRE 2018 track: Information retrieval from microblogs during disasters (IRMiDis). In: Forum for Information Retrieval Evaluation, Gandhinagar, India, Dec. 6-9 (2018), http://ceur-ws.org/Vol-2266/T1-1.pdf
6. Blei, D.M., Ng, A.Y., Jordan, M.I.: Latent Dirichlet allocation. J. Mach. Learn. Res. **3**(null), 993–1022 (2003)
7. Boon-Itt, S., Skunkan, Y., et al.: Public perception of the COVID-19 pandemic on twitter: sentiment analysis and topic modeling study. JMIR Public Health Surveill. **6**(4), e21978 (2020)
8. Bucher, H.J., Boy, B., Christ, K.: Audiovisuelle Wissenschaftskommunikation auf YouTube: Eine Rezeptionsstudie zur Vermittlungsleistung von Wissenschaftsvideos. Springer (2021). https://doi.org/10.1007/978-3-658-35618-7

9. Ding, R., Nallapati, R., Xiang, B.: Coherence-aware neural topic modeling. In: Proceedings Conference on Empirical Methods in Natural Language Processing, pp. 830–836. ACL, Brussels (2018). https://doi.org/10.18653/v1/D18-1096
10. Dreisiebner, S., März, S., Mandl, T.: Information behavior during the Covid-19 crisis in German-speaking countries. J. Documentation **78**(7), 160–175 (2022). https://doi.org/10.1108/JD-12-2020-0217
11. Honnibal, M., Montani, I.: spaCy 2: Natural language understanding with Bloom embeddings, convolutional neural networks and incremental parsing (2017). github
12. Jaki, S.: The (Self-)Presentation of Researchers in TV Documentary Formats - A Multimodal Perspective. Zeitschrift für Semiotik **40**(3–4), 63–82 (2020)
13. Jaki, S.: This is simplified to the point of banality: Social-Media-Kommentare zu Gestaltungsweisen von TV-Dokus. J. für Medienlinguistik **4**(1), 54-87 (2021). https://doi.org/10.21248/jfml.2021.36. https://jfml.org/article/view/36
14. Jaki, S., Sabban, A.: Wissensformate in den Medien: Analysen aus Medienlinguistik und Medienwissenschaft, vol. 25. Frank & Timme GmbH (2016)
15. Kalepalli, Y., Tasneem, S., Teja, P.D.P., Manne, S.: Effective comparison of LDA with LSA for topic modelling. In: Intl. Conf. on Intelligent Computing and Control Systems (ICICCS), pp. 1245–1250. IEEE (2020)
16. de Melo, T., Figueiredo, C.M., et al.: Comparing news articles and tweets about COVID-19 in Brazil: sentiment analysis and topic modeling approach. JMIR Public Health Surveill. **7**(2), e24585 (2021)
17. Modha, S., Majumder, P., Mandl, T.: An empirical evaluation of text representation schemes to filter the social media stream. J. Exp. Theoretical Artif. Intell. **34**(3), 499–525 (2022). https://doi.org/10.1080/0952813x.2021.1907792
18. Montesi, M.: Human information behavior during the COVID-19 health crisis. a literature review. Library Inf. Sci. Res. **43**(4), 101122 (2021). 10.1016/j.lisr.2021.101122
19. Oesper, L., Merico, D., Isserlin, R., Bader, G.D.: Wordcloud: a cytoscape plugin to create a visual semantic summary of networks. Source Code Biol. Med. **6**(1), 7 (2011)
20. Rehurek, R., Sojka, P.: Gensim–python framework for vector space modelling. NLP Centre, Faculty of Informatics, Masaryk University, Brno 3(2) (2011)
21. Samuel, J., Ali, G.G.M.N., Rahman, M.M., Esawi, E., Samuel, Y.: COVID-19 public sentiment insights and machine learning for tweets classification. Inf. **11**(6), 314 (2020). https://doi.org/10.3390/info11060314
22. Shuja, J., Alanazi, E., Alasmary, W., Alashaikh, A.: COVID-19 open source data sets: a comprehensive survey. Appl. Intell. **51**(3), 1296–1325 (2021)
23. Sievert, C., Shirley, K.: Ldavis: a method for visualizing and interpreting topics (2014). https://doi.org/10.13140/2.1.1394.3043
24. Vosoughi, S., Roy, D., Aral, S.: The spread of true and false news online. Science **359**(6380), 1146–1151 (2018). https://doi.org/10.1126/science.aap9559
25. Yang, K., Torres-Lugo, C., Menczer, F.: Prevalence of low-credibility information on Twitter during the COVID-19 outbreak. CoRR abs/2004.14484 (2020). https://arxiv.org/abs/2004.14484
26. Yin, H., Song, X., Yang, S., Li, J.: Sentiment analysis and topic modeling for COVID-19 vaccine discussions. World Wide Web **25**(3), 1067–1083 (2022)

Pragmatic Statistics: Likert-Type Questionnaires Processing

Egils Ginters[1,2](✉)

[1] Riga Technical University, Riga 1658, Latvia
egils.ginters@rtu.lv
[2] Sociotechnical Systems OU, Sakala Street 7-2, 10141 Tallinn, Estonia
soctech@soctech.net

Abstract. Today, almost all systems are sociotechnical. The validation of the technical component is based on quantitative measurements and comparisons with various standards. On the other hand, the assessment of social factors is not so simple and is based on measurements of the audience's attitude. Usually, Likert-type surveys are used for such purposes. In order to obtain adequate information about the attitude of the audience, first of all, the survey must be properly prepared and a suitable audience must be selected, but the obtained results must be statistically processed. There are dozens of different statistical methods, but it is important to determine the most useful set of methods for different situations. The article has been developed in the form of a short manual/handbook, including calculation examples and recommendations. The article will be useful for both students and researchers, as well as project managers who need to measure the attitude of the target audience.

Keywords: Likert-Type Questionnaire · Survey Processing · Data Understanding · Reliability · Dichotomous Data · Shapiro-Wilk Test · Cronbach's Alpha · Attitude Measurements · Descriptive Statistics · Statistical Processing · Kolmogorov-Smirnov Test · Fleiss Kappa · Statistical Methods · Meaningfulness · Interrelated Influence · Pearson Correlation · Adjusted Wald Technique

1 Introduction

"If it looks like a duck, swims like a duck, and quacks like a duck, then it probably is a duck."

/The duck test/

Any work done must have a certain result, otherwise it is not work. There is no problem if the work is done for oneself, in this case everyone can draw clear and unambiguous conclusions. If the work is done for others, feedback and evaluation of the quality of the work is required. This process is sometimes called validation. During validation, the result achieved is compared with the set requirements, and the compliance of the result

A. Rocha et al. (Eds.): WorldCIST 2023, LNNS 800, pp. 23–51, 2024.
https://doi.org/10.1007/978-3-031-45645-9_3

with certain criteria is assessed. Sometimes the result is compared with similar results achieved by others. Technical systems are characterized by tangible deliverables that can be verified using quantitative measurements. However, the result of the work, which will be visible only after a longer time, can only be predicted and not measured.

In the same way, it is difficult to quantitatively measure the expected impact of some recommendations, or the consequences of received training. It will be similar if there are doubts about the correctness of the action or the work done, but there are no specific criteria at all, or they are very vague. In sociotechnical systems, instead of comparing with a typical standard, measurements of the audience's attitude are used. When performing these measurements, one should not forget that they are measurements of people's attitudes, which are valid only at the time of measurement, because the characteristic attributes of social systems are evolution and cognition. This means that attitude measurements, like the simulation of processes and behaviors, will never provide an exact result, but will only show a possible trend.

One of the most popular and widely used forms of attitude measurement are Likert questionnaires [1], which use a multi-point scale. There are a number of conditions for a truly representative survey to be carried out. First, a meaningful survey must be prepared, in which the questions must be interrelated and meet the purpose of the survey. Secondly, the correct goal audience must be selected, thirdly, the correct planning of the duration of the survey must be carried out, and finally, the correct statistical processing of the obtained survey data must be implemented.

There are many different statistical methods that are more or less suitable for processing Likert-type survey data. The choice is far from easy, but subjective discussions on this topic have been ongoing for almost a century. However, finding simple and focused recommendations in the form of a short guide is very problematic. The author aims to propose some specific statistical parameters and methods that can be used to analyze Likert-type survey data. The information is provided in the form of practical recommendations, adding calculation examples to the selected methods.

The methods and steps described in the article will be useful to any person whose work responsibilities include the task of evaluating the results of the activities performed. This short guide will be useful for students, researchers, project managers and other stakeholders.

2 Tips for Implementing Likert-Type Questionnaires

The Likert scale serves as a psychometric measurement of the audience's attitude and is named after American social scientist Rensis Likert, who devised the approach in 1932 [1]. Likert approach is one of the ways of quantification and give quantitative value, for example, 1, 2, 3 etc. to qualitative data in conformity with the selected scale. These data are ordinal and can be considered to be of interval type. This is an important assumption, however, it is difficult to imagine objectively defined and completely equal intervals in the measurement of people's attitudes, because their boundaries will always be subjective.

A Likert-style questionnaire consists of a set of Q_K questions, where the subjective answers of N respondents to each k-question are measured according to a Likert point scale.

According to the Likert scale can be measured agreement (*"Strongly Agree / Agree / Undecided / Disagree / Strongly Disagree"*), value (*"High / Moderate / Low / None"*), relevance (*"Excellent / Somewhat / Poor"*), frequency (*"Always / Very Frequently / Occasionally / Rarely / Very Rarely / Never"*), importance (*"Very Important / Moderately / Not Important"*), quality (*"Very Poor / Poor / Fair / Good / Very Good / Excellent / Exceptional"*) and other attitudes. Dichotomous psychometric measurements ("*Agree / Disagree" or "Yes / No"*) can also be used, but they cannot be recommended.

The Likert-type answer scale can be from 2 to 7 points, that is, each question can have 2 to 7 possible answers. However, it is desirable to create odd values set that includes a midpoint of values, or a neutral attitude. This would contribute to the compliance of survey results with reality, as well as improve the quality of statistical data processing. Opinions differ on the length of the response scale. A longer scale could provide a statistically better result. However, it is possible that it would make it difficult for the respondent to choose an answer and worsen the correspondence of the survey results to reality. Today, the 5-point Likert scale is considered one of the most popular.

To ensure the statistical reliability of the results, it is desirable to receive answers from *at least 50 respondents*.

The reliability of the results and the correspondence to reality is determined by the consistency of the questionnaire. Therefore, *a useful questionnaire* must comply with certain recommendations:

- The questionnaire must be narrowly focused and must provide an assessment of one specific research problem or topic;
- All questions must be thematically unified. One questionnaire cannot include question Q_1: : Is there life on Mars? and Q_2: Is the ice cream tasty??
- Controversial questions may not be included in the survey;
- Formulate the questions in such a way that respondents can easily answer all questions according to the chosen Likert scale;
- Respect the homogeneity of the value scale, for example, do not include frequency, quality and importance values in one common scale;
- If you want the statistics to help you know the more or less true attitudes of the respondents, never include dichotomous *"Yes / No"* questions in the survey;
- The question should be brief and grammatically simple;
- The question must not be ambiguous, it must be clearly understood and answered;
- Easier questions should be included at the beginning of the survey, and more complex questions should be placed at the end of the survey;
- The wording of the questions and the chosen Likert scale must be such that no questions remain unanswered;
- The survey should include as many questions as can be thoughtfully answered within 30 min (for interested respondents) or 3 min (for random respondents).

Before conducting a real survey, it is advisable to conduct a *warm-up survey* to familiarize the audience with the Likert scale. It is desirable to include some screening or classification questions in this pre-test survey, which would allow determining the respondent's competence, as well as the current physical readiness to answer the basic survey questions.

Each survey is conducted to find out the attitude of the audience. It should not be forgotten that human beings are not technical objects, that is, each answer is given with a certain amount of doubt, as well as based on the answers already given earlier. In order to display the results of the survey, statistical processing of the data is required.

3 Determination of Suitable Number of Respondents

It is important to find out the minimum number of submitted questionnaires or surveyed respondents in order to draw reasonable conclusions. Besides, in this case, other characteristics of the audience are not respected: gender, language, cultural and historical traditions, religion, etc. It is considered that the survey organizer will adhere to specific attributes if it will be relevant to the study. It is assumed that the audience is homogeneous.

It is often said that there are just lies and there are statistics. One of the reasons is the sensitivity of statistical data processing methods to the number of respondents. If the number of respondents is small, then statistical processing can give wrong results. A smaller number of measurements can be used in technical systems, but in attitudinal measurements, which are typical of sociotechnical systems, it is desirable that the number of respondents is not less than 50.

However, the basic question is about the required number of respondents N, which would allow to apply the results of the survey to a wider audience N_A than the one surveyed.

It is useful to use ***Slovin's formula*** [2], which determines the number of respondents N, if it is necessary to find out the opinion of N_A (see Table 1):

$$N = \frac{N_A}{\left(1 + N_A * e^2\right)} \tag{1}$$

where:

N – number of questionnaires submitted;
N_A– potential audience of users/stakeholders;
e – 0.05 for confidence 95%.

Table 1. Audience calculated by Slovin's formula.

N	0	1	5	10	19	44	80	133
N_A	0	1	5	10	20	50	100	200
N	171	286	370	385	398	400	400	
N_A	300	1000	5000	10000	100000	500000	1000000	

The origin of Slovin's formula has been the subject of debate for many years, as there is no convincing evidence of its correctness or authorship. Using this formula can make a lot of enemies, but even statistics gurus like Cochran respected this formula, so we can consider Slovin's formula useful enough and it can be used subject to some restrictions. In order to avoid massive attacks, we can call it tuned *Cochran's formula* [3] *with special restrictions.*

Slovin's formula shows that in order to find out the opinion of 5000 or 1 million respondents, a similar number of respondents should be polled, that is no more than 400 respondents. However, let us remind that the result is valid for the *homogeneous audience and confidence level 95%*. If the confidence level is different, then it would be more correct to use Cochran's formula.

And yet, we dare to say, if the sufficient condition of the data sample normality test is not reached at 400 measurements, then it will be a problem of the researcher's work quality.

4 Data Understanding

A sociotechnical system is open, quite stochastic, capable of evolution, cognition, and is also influenced by many unanticipated and hidden factors. This system is the result of the interaction of social/natural and technical structures. A system consists of a set of objects connected by various links. However, the nature of the system is determined not only by the structure of links, but also by heterogeneous objects. As a result, such a sociotechnical system can be both discrete and continuous, both deterministic and stochastic. Data flow is the result of system operation, so different types of data occur within one system.

The data type determines the possibilities of further statistical processing of the results:

- ***Nominal*** data – strictly defined and *equal categories* that cannot be arranged in any order, for example *"Women / Men"* or *"Fish / Amphibians / Birds / Reptiles / Mammals"* or *"Planets / Stars"* etc.;
- ***Ordinal*** data – variables *can be arranged* according to a certain characteristic in one of the selected sequences (ascending or descending). For example, medal can be *"Bronze / Silver / Gold"*, military rank – *"Lieutenant / Captain / Colonel / General"*. A more detail statistical evaluation is possible with ordinal data;
- ***Discrete*** data – countable variables whose value can be determined at strictly *defined moments of time*. This means that the value of these variables between these moments of time is unknown. For example, the number of boxes loaded on a truck or the number of students in an auditorium is an integer. However, to simplify data processing, it is often useful to view discrete data as continuous, but then round off its values. In attitude measurements, which are rather uncertain, such a decision is not objectionable;
- ***Parametric*** data – the possible *distribution* of variables *is clearly predictable*, but the possible distribution of ***nonparametric*** data is uncertain. It is safer to assume that the distribution of measurement results is not clear and to use nonparametric data tests;
- ***Continuous*** data – the variable value can be ascertained at *any moment of time* that belongs to the measurement time interval. Data variables may not be integers. If the measurement scale is divided into certain intervals (e.g., *"Bad / Satisfactory / Good"*) according to some characteristic, then this data is called ***interval*** data. Even in the

absence of formally defined strict interval criteria, data boundaries for Likert-type measurements are determined intuitively;

- ***Dichotomous*** data in attitude measurements is a specific case of ordinal distribution. A dichotomous variable indicates one of only two possible states or membership in one group or the other, e.g., "*Yes / No*". Sometimes dichotomous data is equated with ***binary*** or ***binomial*** data, but in attitude measurements this is fundamentally wrong. Binary data can usually have only two mutually exclusive states "*0*" or "*1*". In electronics, for example, it is determined by transistor switching – open / closed. Measurements are not taken during transistor switching, as this is considered transistor noise, so it can only be "*0*" or "*1*". On the other hand, "*Yes / No*" in attitude measurements will not be such unambiguous and nominal, and mutually exclusive concepts. Here the "*Yes / No*" answers can be treated more subjectively, it is more on the "*Yes*" or "*No*" side.

A Likert questionnaire [1] can be described as a set of Q_K questions, where the answers have been submitted by N respondents. The answer set for each question Q_k is a one-dimensional array with length N. The scale of the answer values is determined by the questionnaire preparer. It is possible that the 5-point Likert scale is one of the most widely used in creating questionnaires [4]. The questionnaire should be devoted to one specific problem of the study. It should not contain controversial questions. The survey questions must be interrelated and the questionnaire must be meaningful. Conditionally Likert-type data can be analysed as ordinal and interval type data.

The following sequence can be recommended for statistical processing of response data in Likert-type questionnaire:

- **Descriptive statistics**

Basic statistical characteristics or *descriptive statistics* [5] can be determined for each Q_K question answer set: ***mean, standard deviation, standard error, variance, median, mode, frequency, range, minimum, maximum***. If the data is normally distributed and N is large enough, then the ***skewness*** and ***kurtosis*** of the response set can be estimated, although these characterizations will reveal only minor nuances of the distribution;

- **Estimating the response distribution**

First of all, it is necessary to evaluate the appropriateness of the distribution of answers by one of the ***normality tests*** [6], for example, ***Kolmogorov-Smirnov*** or ***Shapiro-Wilk*** tests. If the survey is properly designed and implemented, the response data should be normally distributed, or at least close to the normal distribution [7]. Unless they are dichotomous data, which are handled slightly differently;

- **Confidence intervals assessment**

According to the confidence level, an interval around the mean is determined, in which the received answer is reliable. If the data is normally distributed, ***z-values*** [8] corresponding to the confidence level are used. In the case of dichotomous data, for example, the ***Adjusted Wald technique*** [9] is applied;

- **Pairwise inter-relation checking**

Pairwise inter-relation of questions is tested. If the data is normally distributed, then ***Pearson correlation*** [10] is used. To assess the interdependence of the questions the ***covariance*** [11] can be calculated. In the case of dichotomous data Pearson's correlation is replaced by ***Cohen's Kappa*** [12] or ***Fleiss Kappa*** [13] calculation to determine the interdependence of pairs of question answers;

- **Survey reliability or consistency assessment**

The inter-relations between the answers to the questions is calculated, which characterizes the questionnaire meaningfulness. In the case of normally distributed responses, for example, ***Cronbach's alpha*** [14] can be used. In case of dichotomous data, ***Fleiss Kappa*** [13] estimation is used.

What if the test confirms that the distribution of answers is far from a normal distribution? There are two options:

- The first option is to accept failure, modify the questionnaire, increase the number of respondents and repeat the survey. Of course, it is also possible to carry out correlation tests and calculate Cronbach's alpha, which will only confirm initial suspicions about errors in the preparation and implementation of the survey;
- The second option is to try to confuse and surprise the audience with such statistical parameters as median, mode, frequency, range, maximum, minimum, as well as prepare an attractive visualization of data. Although it should be noted that the use of these statistical parameters is not well suitable for processing Likert-type data.

Including ***dichotomous questions*** [15] in a Likert survey is the most direct route to a false result. First of all, the attitude of the interviewed person will be determined incorrectly, that is, it will not be possible to determine how convincing the expressed "*Yes*" or "*No*" is. Second, there are no convincing statistical techniques for handling dichotomous responses. The above-mentioned processing methods can be used, but nothing good will come of it anyway. Do not use dichotomous questions in attitude measurements!

5 Clarifying the Distribution of Results

Most statistical methods are based on the normal distribution of data [7], so the first and foremost task is to understand the distribution of the resulting data. The analysis of dichotomous data is a special case, so it will be discussed separately.

If the same parameter is qualitatively measured several times, then sooner or later the results should fit to a normal distribution. If such conformity can be achieved, the subsequent statistical processing procedures of the data will be greatly simplified. Researchers will be able to use means and standard deviations and other well-tested statistical methods comfortably and without much doubt.

There are many different tests for normality of data distributions [6], but differences in methods could produce somewhat nuanced results. And yet, in the case of modelling or simulation of sociotechnical systems, these are only irrelevant nuances, because in any case the modelling result will show a trend and not give an exact result. It is determined by the nature of sociotechnical systems – a large number of stochastic and unanticipated factors, as well as the possibilities of cognition, self-organization, etc.

There are many similar tests for determining probability distributions. For example, Chi-Square goodness-of-fit test [16], Kolmogorov-Smirnov test [17], Lilliefors test [18], Shapiro-Wilk test [19], Anderson-Darling test [20], Cramer von Mises test [21], D'Agostino skewness test [22], Anscombe-Glynn kurtosis test [23], Jarque-Bera test [24] etc. Several of these tests are just a modified original Kolmogorov-Smirnov test.

Perhaps the most popular are the ***Kolmogorov-Smirnov*** (if the number of respondents is *greater than 50*) and ***Shapiro-Wilk*** (if the number of respondents is *less than or equal to 50*) tests, but any of the above will be useful enough for evaluating distributions of Likert questionnaire results.

Likewise, it makes no sense to engage in endless theoretical discussions about the type of Likert survey data - ordinal, interval, continuous or discrete data. It should be remembered that these data are, firstly, quantitative, and secondly, the attitude of the respondents is measured with these data, which is a rather fuzzy concept.

5.1 Kolmogorov-Smirnov Test

Respecting the above-mentioned considerations, the ***Kolmogorov-Smirnov test*** [6] can be recommended. This test is widely used in engineering sciences.

The Kolmogorov-Smirnov test (K-S test) provides a comparison of the distribution of answer values $F_{Q_k}(n), n \in [1, N]$ of questions Q_k and $k \in [1, K]$ with one of the selected theoretical probability distributions $F_o(n)$. K-S test is a non-parametric test and therefore suitable for testing data distributions whose nature is not sufficiently predictable. In the case of meaningful Likert questionnaire, this theoretical probability distribution must be the normal distribution of the data.

There are two conditions for the K-S test - ***necessary*** and ***sufficient***:

- ***Necessary Condition***. If the $Asymp.Sig.(2 - tailed)\ \alpha > 0.05$ (confidence 95%) calculated in the test, then *the hypothesis* that $F_{Q_k}(n)$ corresponds to the theoretical probability distribution $F_o(n)$ *cannot be rejected*;
- ***Sufficient Condition***. The K-S test statistically calculates the *largest vertical distance* Z_{Q_k} between $F_o(n)$ and $F_{Q_k}(n)$:

$$Z_{Q_k} = \sup_{n} \left| F_o(n) - F_{Q_k}(n) \right| \quad (2)$$

where *sup* ("supremum") - the largest.

If the calculated Z_{Q_k} is greater than the critical vertical distance Z_{crit} (see Table 2) [25] from the theoretical probability distribution at the corresponding number of respondents N, then $F_{Q_k}(n)$ *does not sufficiently reliably correspond* to the theoretical probability distribution $F_o(n)$.

Table 2. Critical values of one-sample Kolmogorov-Smirnov test statistic [25].

	Alpha				
n	0.20	0.10	0.05	0.02	0.01
1	0.900	0.950	0.975	0.990	0.995
2	0.684	0.776	0.842	0.900	0.929
3	0.565	0.636	0.708	0.785	0.829
4	0.493	0.656	0.624	0.689	0.734
5	0.447	0.509	0.563	0.627	0.669
6	0.410	0.468	0.519	0.577	0.617
7	0.381	0.436	0.483	0.538	0.576
8	0.358	0.410	0.454	0.507	0.542
9	0.339	0.387	0.430	0.480	0.513
10	0.323	0.369	0.409	0.457	0.489
11	0.308	0.352	0.391	0.437	0.468
12	0.296	0.338	0.375	0.419	0.449
13	0.285	0.325	0.361	0.404	0.432
14	0.275	0.314	0.349	0.390	0.418
15	0.266	0.304	0.338	0.377	0.404
16	0.258	0.295	0.327	0.366	0.392
17	0.250	0.286	0.318	0.355	0.381
18	0.244	0.279	0.309	0.346	0.371
19	0.237	0.271	0.301	0.337	0.361
20	0.232	0.265	0.294	0.329	0.352
21	0.226	0.259	0.287	0.321	0.344
22	0.221	0.253	0.281	0.314	0.337
23	0.216	0.247	0.275	0.307	0.330
24	0.212	0.242	0.269	0.301	0.323
25	0.208	0.238	0.264	0.295	0.317
26	0.204	0.233	0.259	0.290	0.311
27	0.200	0.229	0.254	0.284	0.305
28	0.197	0.225	0.250	0.279	0.300
29	0.193	0.221	0.246	0.275	0.295
30	0.190	0.218	0.242	0.270	0.290
31	0.187	0.214	0.238	0.266	0.285
32	0.184	0.211	0.234	0.262	0.281
33	0.182	0.208	0.231	0.258	0.277
34	0.179	0.205	0.227	0.254	0.273
35	0.177	0.202	0.224	0.251	0.269
36	0.174	0.199	0.221	0.247	0.265
37	0.172	0.196	0.218	0.244	0.262
38	0.170	0.194	0.215	0.241	0.258
39	0.168	0.191	0.213	0.238	0.255
40	0.165	0.189	0.210	0.235	0.252
n > 40 approx.	$\frac{1.07}{\sqrt{n}}$	$\frac{1.22}{\sqrt{n}}$	$\frac{1.36}{\sqrt{n}}$	$\frac{1.52}{\sqrt{n}}$	$\frac{1.63}{\sqrt{n}}$

What if only the necessary condition is met, but not the sufficient condition? Do not panic, but remember that the aim of modelling in this case is to find out the tendency, not to get an exact result. A sociotechnical system has many unanticipated and stochastic influence factors, so the confirming result of the sufficient condition will be a rare situation. Likert-type measurements are attitudinal measurements, not technical measurements in the laboratory. It can be sometimes assumed that a successful result of the first condition is sufficient. However, in order to improve accuracy and be convinced, the number of respondents N can be increased, and it is also desirable to think about more precise wording of Q_k questions. It is definitely desirable to pay attention to the calculated *standard deviation*. It should *not exceed the permissible limits* of the normal distribution, that is, about 30–34% of calculated mean.

Example. The Likert answers on question Q_1 is sample set (3,2,2,3,3,4,5,2,2), but N is 9. If α is 0.05, then Z_{crit} is 0.430 (see Table 2). The test results see in Table 3.

$Asymp.Sig.(2 - tailed)$ is 0.653 and higher than 0.05. So, the necessary condition is met. It is possible that the results of the answers to Q_1 follow a normal distribution. However, Z_{Q_1} is 0.73 that is higher than $Z_{crit} = 0.43$ (see Table 2). This means that the sufficient condition is not met and it is not possible to claim with 95% probability that the responses to question Q_1 are normally distributed. On the other hand, the standard deviation 1.05 is around 36% of mean, and it is close to normal distribution border. Of course, it is impossible to achieve a convincing normal distribution under natural conditions with such a small number of measurements. Therefore, it is desirable to involve no less than 50 respondents in Likert-type surveys.

Table 3. K-S Q_1 test results *(PSPP software: Non-Parametric Statistics: 1 Sample K-S)* [26].

		Q1
N		9
Normal Parameters	Mean	2.89
	Std. Deviation	1.05
Most Extreme Differences	Absolute	.24
	Positive	.24
	Negative	-.20
Kolmogorov-Smirnov Z		.73
Asymp. Sig. (2-tailed)		.653

5.2 Shapiro-Wilk Test

The ***Shapiro-Wilk test*** [27] is recommended for testing the distributions of small data sets ($N \leq 50$). It was presented by Samuel Sanford Shapiro and Martin Wilk in 1965. Currently, the prevailing opinion is that this test is more accurate than the Kolmogorov-Smirnov test for small data samples. This statement might be true for borderline cases. This statement will be tested later using the survey Q_K test example discussed above.

The Shapiro-Wilk test is quite complicated [28]. There is a lot of room for error when performing manual calculations. During the implementation of the test, different tables of test parameters must be used, data samples must be sorted, and test result interpolation must be done. The test is very sensitive to rounding of values. Many free access statistical software packages do not include the Shapiro-Wilk test. It is advisable to be suspicious of various on-line web calculators, as there are often rounding errors here, which can lead to wrong test result. However, the use of this test may be useful in some cases.

The basis of the Shapiro-Wilk test is the calculation of the *value of test statistics* W_{calc} and the determination of the *corresponding probability interval* $p^W_{min} \leq p^W \leq p^W_{max}$, respecting 95% confidence.

The test can be divided into separate steps:

- First, the set of answer values of the question Q_k should be *arranged in ascending order*:

$$Q_k^1 \leq Q_k^2 \leq \ldots Q_k^n \leq \ldots Q_k^N \tag{3}$$

  where $n \in [1, N]$.
- The mean of the Q_k answer values must then be calculated:

$$\overline{Q_k} = \frac{\sum_{n=1}^{N} Q_k^n}{N} \tag{4}$$

- The parameter SS must be calculated:

$$SS = \sum_{n=1}^{N} \left(Q_k^n - \overline{Q_k}\right)^2$$

- Further the parameter m must be calculated as follows. If N is *even* then $m = N/2$, otherwise$m = (N - 1)/2$;
- To calculate the test parameter b, it is necessary to select the values of a_i corresponding to the number of respondents N from the Shapiro-Wilk test N/a_i table (see Table 4), where $i \in [1, m]$. The parameter b can be calculated as follows:

$$b = \sum_{i=1}^{m} a_i (Q_k^{N+1-i} - Q_k^i)$$

- The Shapiro-Wilk test value W_{calc} can now be calculated as:

$$W_{calc} = b^2/SS \tag{7}$$

- Next, in the Shapiro-Wilk test N/W_{calc} and $(p_{min}^W; p_{max}^W)$ table (see Table 5), using W_{calc} and N, the *corresponding probability interval* p_{min}^W and p_{max}^W for W_{min} and W_{max} should be found. In this table $W_{min} \leq W_{calc} \leq W_{max}$.

The Shapiro-Wilk test can be considered successfully completed if W_{calc} corresponds to $p_{min}^W \geq 0.05$, because then the sample data is considered to be normally distributed.

In the following, the corresponding p^W value of W_{calc} can be refined by linear interpolation:

$$p^W = p_{max}^W - \left(p_{max}^W - p_{min}^W\right) * (W_{max} - W_{calc})/(W_{max} - W_{min}) \tag{8}$$

If $p^W \geq 0.05$ then the answer values of question Q_k are considered to be normally distributed.

Example. The Likert answers on question Q_1 is sample set (3,2,2,3,3,4,5,2,2), but N is 9. $\overline{Q_1}$ is 2.8889. Question Q_1 values ordered sample set is (2,2,2,2,3,3,3,4,5).

$SS = \sum_{n=1}^{9} \left(Q_1^n - \overline{Q_1}\right)^2$ = (2–2.8889) ^2 + (2–2.8889) ^2 + (2–2.8889) ^2 + (2–2.8889) ^2 + (3–2.8889) ^2 + (3–2.8889) ^2 + (3–2.8889) ^2 + (4–2.8889) ^2 + (5–2.8889) ^2 = 8.8889. $m = (9 - 1)/2$ is 4. From Table 4 a_i values are selected for $N = 9$ as follows: a_1 is 0.5888, a_2 is 0.3244, a_3 is 0.1976, but a_4 is 0.0947. Then $b = \sum_{i=1}^{4} a_i (Q_1^{N+1-i} - Q_1^i) = 0.5888*(5–2) + 0.3244*(4–2) + 0.1976*(3–2) + 0.0947*(3–2) = 2.7075$. Test statistic W_{calc} is calculated as $W_{calc} = b^2/SS = 2.7075^2/8.8889 = 0.8247$. From Table 5 for $W_{calc} = 0.8247$ probabilities $p_{min}^W = 0.02$ and $p_{max}^W = 0.05$ are identified for $W_{min} = 0.791$ and $W_{max} = 0.829$. Since corresponding $p_{max}^W = 0.05$ the question Q_1 answers distribution normality is doubtful. However, interpolation to specify the p^W value can be done as follows:

$p^W = p_{max}^W - \left(p_{max}^W - p_{min}^W\right) * (W_{max} - W_{calc})/(W_{max} - W_{min})$ = 0.05 – (0.05–0.01) *(0.829–0.8247) / (0.829–0.791) = 0.05 – 0.04*0.0043 / 0.038 = 0.045. Since $p^W = 0.045 < 0.05$ then question Q_1 answers *are not normally distributed.*

This is a classic case of a borderline situation where the Shapiro-Wilk test works well. The first necessary condition of the Kolmogorov-Smirnov test convinced well, because $Asymp.Sig.(2 - tailed)$ was 0.653 and significantly higher than 0.05, so the hypothesis that the answers to the question Q_1 are normally distributed could not be rejected. However, the second and sufficient condition was not fulfilled because Z_{Q_1} was 0.73 that is higher than $Z_{crit} = 0.43$. The result of the Shapiro-Wilk test confirms that, although quite close, unfortunately Q_1 answers are not normally distributed. The results can be verified using web calculator [29].

6 Processing of the Results of the Likert-Type 5-point Questionnaire

Likert-type questionnaire Q_K includes $K = 5$ questions Q_k, where $k \in [1, K]$, but the answer scale includes five possible answers that describe the attitude of the respondent (1- "Strongly agree", 2- "Disagree", 3- "Neutral / Undecided", 4- "Agree", 5- "Strongly agree"). Only 9 valid answers were received in the test survey, $N = 9$ (see Table 6).

P.S. The number of respondents in the survey test example is insufficient for real attitude measurements and is used only to demonstrate statistical processing methods.

6.1 Question Q_k Descriptive Statistics

For each question Q_k can be calculated ***mean*** $\overline{Q_k}$:

$$\overline{Q_k} = \frac{\sum_{n=1}^{N} Q_k^n}{N} \tag{9}$$

where:

$\overline{Q_k}$ – mean of Likert answers on question Q_k;
K – number of questions;
N – number of questionnaires submitted;
Q_k^n – answer given on k-question by Likert scale.

Example. Mean for answers on question Q_1 is following: $\overline{Q_1} = (3 + 2 + 2 + 3 + 3 + 4 + 5 + 2 + 2)/9 = 26/9 = 2.89$.

For each Q_k ***standard deviation*** S_k [5] can be calculated:

$$S_k = \sqrt{\frac{\sum_{n=1}^{N} \left(Q_k^n - \overline{Q_k}\right)^2}{N - 1}} \tag{10}$$

but ***variance*** for each Q_k is ${S_k}^2$.

Example. Variance for answers on question Q_1 is following: ${S_1}^2 = [(3–2.89)^2 + (2–2.89)^2 + (2–2.89)^2 + (3–2.89)^2 + (3–2.89)^2 + (4–2.89)^2 + (5–2.89)^2 + (2–2.89)^2 + (2–2.89)^2]/8 = [0.0121 + 0.7921 + 0.7921 + 0.0121 + 0.0121 + 1.2321 + 4.4521 + 0.7921 + 0.7921]/8 = 8.8889/8 = 1.11$ Standard deviation for answers on question Q_1 is $\sqrt{{S_1}^2}$ and specifically 1.05.

Table 4. Shapiro-Wilk N/a_i values ($n = N$) [30].

i \ n	2	3	4	5	6	7	8	9	10
1	0·7071	0·7071	0·6872	0·6646	0·6431	0·6233	0·6052	0·5888	0·5739
2	—	·0000	·1677	·2413	·2806	·3031	·3164	·3244	·3291
3	—	—	—	·0000	·0875	·1401	·1743	·1976	·2141
4	—	—	—	—	—	·0000	·0561	·0947	·1224
5	—	—	—	—	—	—	—	·0000	·0399

i \ n	11	12	13	14	15	16	17	18	19	20
1	0·5601	0·5475	0·5359	0·5251	0·5150	0·5056	0·4968	0·4886	0·4808	0·4734
2	·3315	·3325	·3325	·3318	·3306	·3290	·3273	·3253	·3232	·3211
3	·2260	·2347	·2412	·2460	·2495	·2521	·2540	·2553	·2561	·2565
4	·1429	·1586	·1707	·1802	·1878	·1939	·1988	·2027	·2059	·2085
5	·0695	·0922	·1099	·1240	·1353	·1447	·1524	·1587	·1641	·1686
6	0·0000	0·0303	0·0539	0·0727	0·0880	0·1005	0·1109	0·1197	0·1271	0·1334
7	—	—	·0000	·0240	·0433	·0593	·0725	·0837	·0932	·1013
8	—	—	—	—	·0000	·0196	·0359	·0496	·0612	·0711
9	—	—	—	—	—	—	·0000	·0163	·0303	·0422
10	—	—	—	—	—	—	—	—	·0000	·0140

i \ n	21	22	23	24	25	26	27	28	29	30
1	0·4643	0·4590	0·4542	0·4493	0·4450	0·4407	0·4366	0·4328	0·4291	0·4254
2	·3185	·3156	·3126	·3098	·3069	·3043	·3018	·2992	·2968	·2944
3	·2578	·2571	·2563	·2554	·2543	·2533	·2522	·2510	·2499	·2487
4	·2119	·2131	·2139	·2145	·2148	·2151	·2152	·2151	·2150	·2148
5	·1736	·1764	·1787	·1807	·1822	·1836	·1848	·1857	·1864	·1870
6	0·1399	0·1443	0·1480	0·1512	0·1539	0·1563	0·1584	0·1601	0·1616	0·1630
7	·1092	·1150	·1201	·1245	·1283	·1316	·1346	·1372	·1395	·1415
8	·0804	·0878	·0941	·0997	·1046	·1089	·1128	·1162	·1192	·1219
9	·0530	·0618	·0696	·0764	·0823	·0876	·0923	·0965	·1002	·1036
10	·0263	·0368	·0459	·0539	·0610	·0672	·0728	·0778	·0822	·0862
11	0·0000	0·0122	0·0228	0·0321	0·0403	0·0476	0·0540	0·0598	0·0650	0·0697
12	—	—	·0000	·0107	·0200	·0284	·0358	·0424	·0483	·0537
13	—	—	—	—	·0000	·0094	·0178	·0253	·0320	·0381
14	—	—	—	—	—	—	·0000	·0084	·0159	·0227
15	—	—	—	—	—	—	—	—	·0000	·0076

i \ n	31	32	33	34	35	36	37	38	39	40
1	0·4220	0·4188	0·4156	0·4127	0·4096	0·4068	0·4040	0·4015	0·3989	0·3964
2	·2921	·2898	·2876	·2854	·2834	·2813	·2794	·2774	·2755	·2737
3	·2475	·2463	·2451	·2439	·2427	·2415	·2403	·2391	·2380	·2368
4	·2145	·2141	·2137	·2132	·2127	·2121	·2116	·2110	·2104	·2098
5	·1874	·1878	·1880	·1882	·1883	·1883	·1883	·1881	·1880	·1878
6	0·1641	0·1651	0·1660	0·1667	0·1673	0·1678	0·1683	0·1686	0·1689	0·1691
7	·1433	·1449	·1463	·1475	·1487	·1496	·1505	·1513	·1520	·1526
8	·1243	·1265	·1284	·1301	·1317	·1331	·1344	·1356	·1366	·1376
9	·1066	·1093	·1118	·1140	·1160	·1179	·1196	·1211	·1225	·1237
10	·0899	·0931	·0961	·0988	·1013	·1036	·1056	·1075	·1092	·1108
11	0·0739	0·0777	0·0812	0·0844	0·0873	0·0900	0·0924	0·0947	0·0967	0·0986
12	·0585	·0629	·0669	·0706	·0739	·0770	·0798	·0824	·0848	·0870
13	·0435	·0485	·0530	·0572	·0610	·0645	·0677	·0706	·0733	·0759
14	·0289	·0344	·0395	·0441	·0484	·0523	·0559	·0592	·0622	·0651
15	·0144	·0206	·0262	·0314	·0361	·0404	·0444	·0481	·0515	·0546
16	0·0000	0·0068	0·0131	0·0187	0·0239	0·0287	0·0331	0·0372	0·0409	0·0444
17	—	—	·0000	·0062	·0119	·0172	·0220	·0264	·0305	·0343
18	—	—	—	—	·0000	·0057	·0110	·0158	·0203	·0244
19	—	—	—	—	—	—	·0000	·0053	·0101	·0146
20	—	—	—	—	—	—	—	—	·0000	·0049

(*continued*)

Table 4. *(continued)*

$i \backslash n$	41	42	43	44	45	46	47	48	49	50
1	0·3940	0·3917	0·3894	0·3872	0·3850	0·3830	0·3808	0·3789	0·3770	0·3751
2	·2719	·2701	·2684	·2667	·2651	·2635	·2620	·2604	·2589	·2574
3	·2357	·2345	·2334	·2323	·2313	·2302	·2291	·2281	·2271	·2260
4	·2091	·2085	·2078	·2072	·2065	·2058	·2052	·2045	·2038	·2032
5	·1876	·1874	·1871	·1868	·1865	·1862	·1859	·1855	·1851	·1847
6	0·1693	0·1694	0·1695	0·1695	0·1695	0·1695	0·1695	0·1693	0·1692	0·1691
7	·1531	·1535	·1539	·1542	·1545	·1548	·1550	·1551	·1553	·1554
8	·1384	·1392	·1398	·1405	·1410	·1415	·1420	·1423	·1427	·1430
9	·1249	·1259	·1269	·1278	·1286	·1293	·1300	·1306	·1312	·1317
10	·1123	·1136	·1149	·1160	·1170	·1180	·1189	·1197	·1205	·1212
11	0·1004	0·1020	0·1035	0·1049	0·1062	0·1073	0·1085	0·1095	0·1105	0·1113
12	·0891	·0909	·0927	·0943	·0959	·0972	·0986	·0998	·1010	·1020
13	·0782	·0804	·0824	·0842	·0860	·0876	·0892	·0906	·0919	·0932
14	·0677	·0701	·0724	·0745	·0765	·0783	·0801	·0817	·0832	·0846
15	·0575	·0602	·0628	·0651	·0673	·0694	·0713	·0731	·0748	·0764
16	0·0476	0·0506	0·0534	0·0560	0·0584	0·0607	0·0628	0·0648	0·0667	0·0685
17	·0379	·0411	·0442	·0471	·0497	·0522	·0546	·0568	·0588	·0608
18	·0283	·0318	·0352	·0383	·0412	·0439	·0465	·0489	·0511	·0532
19	·0188	·0227	·0263	·0296	·0328	·0357	·0385	·0411	·0436	·0459
20	·0094	·0136	·0175	·0211	·0245	·0277	·0307	·0335	·0361	·0386
21	0·0000	0·0045	0·0087	0·0126	0·0163	0·0197	0·0229	0·0259	0·0288	0·0314
22		—	·0000	·0042	·0081	·0118	·0153	·0185	·0215	·0244
23	—	—	—	—	·0000	·0039	·0076	·0111	·0143	·0174
24	—	—	—	—	—	—	·0000	·0037	·0071	·0104
25	—	—	—	—	—	—	—	—	·0000	·0035

In this case, the standard deviation is about 36% of the mean. This suggests that the distribution of the data is not clear and may not follow a normal distribution. Usually, the standard deviation should not exceed 25%-34%. This deviation, expressed as a percentage, is called the ***coefficient of variation***.

Variance for answers on question Q_1 is ${S_1}^2 = 1.11$. Variance describes the dispersion of data around the mean. If the goal is to find out the trend of the process and not to achieve an exact result, then in practice there is no fundamental difference between standard deviation and variance. If the standard deviation characterizes the distance of the majority of measurements from the mean, then the variance describes more the dispersion of measurements around the mean. If there is no strong confidence that the distribution of the data is normal, then it is probably more appropriate to use the variance instead for standard deviation.

It is possible to calculate the ***standard error*** SE_k:

$$SE_k = \frac{S_k}{\sqrt{N}} \tag{11}$$

Example. Standard error for Q_1 is $SE_1 = 1.05/\sqrt{9} = 0.35$.

Standard error is used to calculate the ***margin of error***, which is $z * SE_k$, where z corresponds to appropriate confidence level if it is voluntary assumed that answer values are normally distributed (see Table 7).

Example. Margin of error for Q_1 is 1.96 x 0.35 = 0.69 with confidence level 95%.

Table 5. Shapiro-Wilk N/W_{calc} and $(p_{min}^{W}; p_{max}^{W})$ values $(n = N)$ [30].

	Level								
n	**0·01**	**0·02**	**0·05**	**0·10**	**0·50**	**0·90**	**0·95**	**0·98**	**0·99**
3	0·753	0·756	0·767	0·789	0·959	0·998	0·999	1·000	1·000
4	·687	·707	·748	·792	·935	·987	·992	·996	·997
5	·686	·715	·762	·806	·927	·979	·986	·991	·993
6	0·713	0·743	0·788	0·826	0·927	0·974	0·981	0·986	0·989
7	·730	·760	·803	·838	·928	·972	·979	·985	·988
8	·749	·778	·818	·851	·932	·972	·978	·984	·987
9	·764	·791	·829	·859	·935	·972	·978	·984	·986
10	·781	·806	·842	·869	·938	·972	·978	·983	·986
11	0·792	0·817	0·850	0·876	0·940	0·973	0·979	0·984	0·986
12	·805	·828	·859	·883	·943	·973	·979	·984	·986
13	·814	·837	·866	·889	·945	·974	·979	·984	·986
14	·825	·846	·874	·895	·947	·975	·980	·984	·986
15	·835	·855	·881	·901	·950	·975	·980	·984	·987
16	0·844	0·863	0·887	0·906	0·952	0·976	0·981	0·985	0·987
17	·851	·869	·892	·910	·954	·977	·981	·985	·987
18	·858	·874	·897	·914	·956	·978	·982	·986	·988
19	·863	·879	·901	·917	·957	·978	·982	·986	·988
20	·868	·884	·905	·920	·959	·979	·983	·986	·988
21	0·873	0·888	0·908	0·923	0·960	0·980	0·983	0·987	0·989
22	·878	·892	·911	·926	·961	·980	·984	·987	·989
23	·881	·895	·914	·928	·962	·981	·984	·987	·989
24	·884	·898	·916	·930	·963	·981	·984	·987	·989
25	·888	·901	·918	·931	·964	·981	·985	·988	·989
26	0·891	0·904	0·920	0·933	0·965	0·982	0·985	0·988	0·989
27	·894	·906	·923	·935	·965	·982	·985	·988	·990
28	·896	·908	·924	·936	·966	·982	·985	·988	·990
29	·898	·910	·926	·937	·966	·982	·985	·988	·990
30	·900	·912	·927	·939	·967	·983	·985	·988	·900
31	0·902	0·914	0·929	0·940	0·967	0·983	0·986	0·988	0·990
32	·904	·915	·930	·941	·968	·983	·986	·988	·990
33	·906	·917	·931	·942	·968	·983	·986	·989	·990
34	·908	·919	·933	·943	·969	·983	·986	·989	·990
35	·910	·920	·934	·944	·969	·984	·986	·989	·990
36	0·912	0·922	0·935	0·945	0·970	0·984	0·986	0·989	0·990
37	·914	·924	·936	·946	·970	·984	·987	·989	·990
38	·916	·925	·938	·947	·971	·984	·987	·989	·990
39	·917	·927	·939	·948	·971	·984	·987	·989	·991
40	·919	·928	·940	·949	·972	·985	·987	·989	·991
41	0·920	0·929	0·941	0·950	0·972	0·985	0·987	0·989	0·991
42	·922	·930	·942	·951	·972	·985	·987	·989	·991
43	·923	·932	·943	·951	·973	·985	·987	·990	·991
44	·924	·933	·944	·952	·973	·985	·987	·990	·991
45	·926	·934	·945	·953	·973	·985	·988	·990	·991
46	0·927	0·935	0·945	0·953	0·974	0·985	0·988	0·990	0·991
47	·928	·936	·946	·954	·974	·985	·988	·990	·991
48	·929	·937	·947	·954	·974	·985	·988	·990	·991
49	·929	·937	·947	·955	·974	·985	·988	·990	·991
50	·930	·938	·947	·955	·974	·985	·988	·990	·991

Confidence interval for the answers on Q_k question can be calculated as: $[\overline{Q_k} - MarginoferrorQ_k; \overline{Q_k}; \overline{Q_k} + MarginoferrorQ_k]$ (12).

Table 6. Questionnaire Q_K results *(PSPP software)* [26].

File Edit View Data Transform Analyse

3

Case	Q1	Q2	Q3	Q4	Q5	Score
1	3	2	4	1	4	14
2	2	2	3	1	2	10
3	2	3	1	1	3	10
4	3	1	2	1	1	8
5	3	1	3	3	3	13
6	4	5	4	3	5	21
7	5	2	5	2	1	15
8	2	1	1	5	4	13
9	2	3	1	5	3	14

Table 7. Confidence level and critical values (z-value) [8].

Confidence level (%)	z-value
80	1.28
85	1.44
90	1.64
95	1.96
98	2.33
99	2.58

Example. Confidence interval for the answers on Q_1 question can be determined as [2.2; 3.58] with confidence level 95%.

The ***minimum*** and ***maximum*** value of responses are 1 and 5 on a 5-point Likert scale.

Example. Minimum for answers value on question Q_1 is 2. Maximum for answers value on question Q_1 is 5.

When determining the minimum and maximum value, the ***range*** also can be calculated. Range is the difference between the maximum and minimum values in the given sample.

Example. Range for answers on question Q_1 is 5–2 = 3.

If the correspondence of the results to one of the statistical probability distributions is not approved first, then it is useful to calculate the ***median***, which is the middle element of a sample set arranged in ascending or descending order. If the number of respondents or submitted questionnaires is an *even* number, then the average value of adjacent elements in the middle of the sample is calculated.

Example. The Likert answers on question Q_1 is sample set (3,2,2,3,3,4,5,2,2), but ordered sample set is (2,2,2,2,3,3,3,4,5). The median of Q_1 sample set is 3.

It is also possible to calculate the ***mode***, which is the value of the most common Likert answer value in the given sample.

Example. The Likert answers on question Q_1 is sample set (3,2,2,3,3,4,5,2,2). Response value "2" has been used 4 times, "3" - 3 times, "4" - 1 time, and "5" - 1 time. The mode of Q_1 sample set is 2.

Descriptive statistics for Q_k can be calculated using different non-commercial software tools for data statistical processing, for example, PSPP software [26] (see Table 8).

Table 8. Descriptive statistics for question Q_1 *(PSPP software: Descriptive Statistics: Frequencies).*

Q1

Value Label	*Value*	*Frequency*	*Percent*	*Valid Percent*	*Cum Percent*
	2	4	44.44	44.44	44.44
	3	3	33.33	33.33	77.78
	4	1	11.11	11.11	88.89
	5	1	11.11	11.11	100.00
	Total	9	100.0	100.0	

Q1

N	*Valid*	9
	Missing	0
Mean		2.89
S.E. Mean		.35
Mode		2.00
Std Dev		1.05
Variance		1.11
Kurtosis		.61
Skewness		1.09
Range		3.00
Minimum		2.00
Maximum		5.00
Percentiles	50 (Median)	3

6.2 Relationship and Paired Influence of Questions

Traditionally, each questionnaire is thematically unified and presents some current problems/research evaluation. This means that survey Q_K questions Q_k must be related to each other. The overall goal is to find out the meaningfulness of the questionnaire or to be more precise - *consistency*.

One of the techniques that allows determining a linear relationship between two sets of data, i.e., the received answers according to the Likert scale and corresponding to any two interview questions Q_k un Q_{k+t}, is ***Pearson correlation coefficient*** r_k^{k+t}, where $k + t \leq K$ [10]:

$$r_k^{k+t} = \frac{\sum_{n=1}^{N}(Q_k^n - \overline{Q_k}) * (Q_{k+t}^n - \overline{Q_{k+t}})}{\sqrt{\sum_{n=1}^{N}\left(Q_k^n - \overline{Q_k}\right)^2 * \sum_{n=1}^{N}\left(Q_{k+t}^n - \overline{Q_{k+t}}\right)^2}} \quad (13)$$

where:

$\overline{Q_k}$ – mean of Likert answers on question Q_k;
$\overline{Q_{k+t}}$ – mean of Likert answers on question Q_{k+t};
K – number of questions;
N – number of questionnaires submitted;
Q_k^n – answer value given on k - question by Likert scale;
Q_{k+t}^n – answer value given on $k + t$ - question by Likert values scale;
$k + t \leq K$.

The range of Pearson correlation coefficient values is [-1,1], where the extreme states indicate complete linear interdependence of data sets, which is an impossibly ideal situation. On the other hand, 0 indicates that there is no linear dependence between the samples (see Table 9).

Table 9. Pearson correlation interpretation [31].

Scale of correlation coefficient $\left\|r_k^{k+t}\right\|$	Correlation value
]0, 0.19]	Very Low
[0.2, 0.39]	Low
[0.4, 0.59]	Moderate
[0.6, 0.79]	High
[0.8, 1.0]	Very High

A positive correlation coefficient means that as the value of the variable increases, the value of the dependent variable increases linearly. If the coefficient is negative, then the value of the dependent variable decreases.

Example. The paired influence of questions Q_1 and Q_2 included in the test questionnaire Q_K (see Table 6) is calculated as follows: $\overline{Q_1} = 2.89$, $\overline{Q_2} = 2.22$, $N = 9$, then $\sum_{n=1}^{9}\left(Q_1^n - \overline{Q_1}\right)^2 = 8.89$ and $\sum_{n=1}^{9}\left(Q_2^n - \overline{Q_2}\right)^2 = 13.56$. Then $\sqrt{\sum_{n=1}^{9}\left(Q_1^n - \overline{Q_1}\right)^2 * \sum_{n=1}^{9}\left(Q_2^n - \overline{Q_2}\right)^2} = 10.98$. Numerator $\sum_{n=1}^{9}(Q_1^n - \overline{Q_1}) * (Q_2^n - \overline{Q_2}) = 2.22$. Finally, 2.22 / 10.98 = 0.20.

Pearson correlation can be calculated for whole questionnaire Q_K using PSPP software (see Table 10).

Table 10. Pearson correlation Q_K calculated *(PSPP software: Bivariate Correlation).*

Correlations

		Q1	Q2	Q3	Q4	Q5
Q1	Pearson Correlation	1.00	.20	.84	-.18	-.18
	Sig. (2-tailed)		.601	.004	.640	.636
	N	9	9	9	9	9
Q2	Pearson Correlation	.20	1.00	.17	.06	.51
	Sig. (2-tailed)	.601		.661	.870	.162
	N	9	9	9	9	9
Q3	Pearson Correlation	.84	.17	1.00	-.38	-.08
	Sig. (2-tailed)	.004	.661		.308	.835
	N	9	9	9	9	9
Q4	Pearson Correlation	-.18	.06	-.38	1.00	.41
	Sig. (2-tailed)	.640	.870	.308		.274
	N	9	9	9	9	9
Q5	Pearson Correlation	-.18	.51	-.08	.41	1.00
	Sig. (2-tailed)	.636	.162	.835	.274	
	N	9	9	9	9	9

Pearson correlation r_1^2 is equal to r_2^1. Coefficient r_1^2 is 0.20, which suggests a small but positive Q_1 and Q_2 linear interdependence of answers. On the other hand, there is a significant and positive interdependence between the answers on Q_1 and Q_3, because r_1^3 is 0.84. However, this outcome is unlikely, because $Sig.(2 - tailed)$ is only 0.004.

On the other hand, the interdependence of Q_1 and Q_5 is negative (the higher the rating of Q_1, the lower the rating will be received on the Q_5 question). However, this dependence is not significantly expressed, since r_1^5 is only -0.18.

The weak interdependence of any two samples in test questionnaire Q_K may raise suspicions about the overall consistency of the conducted survey. Therefore, it is necessary to check the *overall reliability* of the entire questionnaire, not only the mutual influence of pairs of questions.

6.3 Internal Consistency of the Questionnaire

Cronbach's alpha (α) [14] is used to check the *internal consistency or reliability* of the questionnaire Q_K. The test checks the interrelationship of the answers to the questions and confirms the meaningfulness of the questionnaire. Cronbach's alpha is in the interval [0,1], where the rating [0, 0.5[indicates an unsuccessful interview, and [0.9, 1] indicates excellent internal consistency (see Table 11).

Both variance and covariance parameters can be used to calculate Cronbach's alpha. Major part of free access tools for data statistical processing does not include covariance calculations, so it is more useful to use Cronbach's alpha formula based on variance assessment.

Table 11. Cronbach's alpha interpretation [32].

Cronbach's alpha (α)	Internal consistency
$\alpha \geq 0.9$	Excellent
$0.9 > \alpha \geq 0.8$	Good
$0.8 > \alpha \geq 0.7$	Acceptable
$0.7 > \alpha \geq 0.6$	Questionable
$0.6 > \alpha \geq 0.5$	Poor
$0.5 > \alpha$	Unacceptable

Cronbach's alpha α can be calculated as follows:

$$\alpha = \left(\frac{K}{K-1}\right) * \left(1 - \frac{\sum_{k=1}^{K} S_k^2}{S_{score}^2}\right) \tag{14}$$

where:

K – number of questions;

S_k^2– variance of all answers $n \in [1, N]$ on each question Q_k;

S_{score}^2– variance of the total score array $n \in [1, N]$ of the answers submitted by each respondent.

Example. Descriptive statistics for test questionnaire Q_K are shown in Table 12.

Table 12. Descriptive statistics of Q_K*(PSPP software: Descriptive).*

Valid cases = 9; cases with missing value(s) = 0.

Variable	*N*	*Mean*	*Std Dev*	*Variance*	*Range*	*Minimum*	*Maximum*
Q1	9	2.89	1.05	1.11	3.00	2.00	5.00
Q2	9	2.22	1.30	1.69	4.00	1.00	5.00
Q3	9	2.67	1.50	2.25	4.00	1.00	5.00
Q4	9	2.44	1.67	2.78	4.00	1.00	5.00
Q5	9	2.89	1.36	1.86	4.00	1.00	5.00

Valid cases = 9; cases with missing value(s) = 0.

Variable	*N*	*Mean*	*Std Dev*	*Variance*	*Minimum*	*Maximum*
Score	9	13.11	3.76	14.11	8.00	21.00

Cronbach's alpha for the questionnaire Q_k mentioned above can be calculated as follows: $\sum_{k=1}^{K} S_k^2 = 1.11 + 1.69 + 2.25 + 2.78 + 1.86 = 9.69$; $S_{score}^2 = 14.11$; $K = 5$. Then Cronbach's alpha is α= 5 / (5 - 1) * (1 - 9.69 / 14.11) = 1.25 * (1 - 0.6867) = 0.39.

The same result can be calculated using PSPP software (see Table 13).

The calculated results confirm the earlier suspicions that the questionnaire Q_K was unprofessionally prepared. Respondents either give false answers or do not understand

Table 13. Cronbach's alpha calculation of Q_K *(PSPP software: Reliability).*

Reliability Statistics

Cronbach's Alpha	*K of Items*
.39	5

the questions asked. It is possible that the questionnaire includes controversial questions. Perhaps, although it is less likely, the number of respondents should be increased, or the survey should be conducted with a different audience. In any case, $\alpha < 0.5$, which means that the survey results cannot be used for validation purposes.

7 Dichotomous Answers Processing

The inclusion of dichotomous questions [15] in Likert-type surveys is fundamentally wrong. Of course, nominal answers can be included in the warming-up survey to classify the respondents, but the dichotomous scale should not be used to determine the attitude, because in attitude measurements *"Yes / No"* becomes ordinal and/or interval data. An intelligent person is neither a trigger nor a cellular automaton with only binary states. A person's *"Yes"* or *"No"* has many shades of grey. The *"Yes / No"* scale is not suitable for clarifying such attitudes, and there are no convincing statistical methods for processing such sensitive data.

Another important problem is that most statistical processing methods rely on the normal distribution of data, or refer to various indicators and constants inherited from processing normally distributed data. Usually, new methods are designed by modifying and adapting already well-known processing methods, so there will always be a bell-type probability distribution somewhere backstage.

When a 2-point Likert scale is used, the rate values may be debatable. In dichotomous distributions, it would be useful to use *"0"* and *"1"* rates, which improves the transparency of statistical processing. However, the distribution of these 2-point Likert scale answers cannot be analysed as binomial, as they are characteristics of the persons' attitudes.

7.1 Descriptive Statistics for Dichotomous Distributions

For each question Q_k can be calculated descriptive statistics: mean (*"Agree / Disagree"* proportion), standard deviation, standard error, frequencies, median, mode, range and other.

Example. Questions number K is 5 (from Q_1 to Q_5), Likert 2-point scale is 0 – *"Disagree (No)"* and 1 – *"Agree (Yes)"*, but respondents (questionnaires submitted) number N is 9 (see Table 14).

Example. For each question Q_k descriptive statistics can be calculated (see Table 15). Frequency *"Agree (Yes)"* is 7, but frequency *"Disagree (No)"* is 2 and mean (proportion) $\overline{Q_1}$ is 0.78 or 78%.

Table 14. Dichotomous questionnaire Q_K results *(PSPP software)*.

File Edit View Data Transform Analyse

1 : Q1 | 1

Case	Q1	Q2	Q3	Q4	Q5	Va
1	1	0	0	0	1	
2	1	1	0	0	1	
3	1	0	1	0	1	
4	1	0	0	0	1	
5	0	0	0	0	1	
6	0	1	1	0	1	
7	1	0	0	0	1	
8	1	0	0	1	1	
9	1	0	0	0	1	

Table 15. Descriptive statistics of Q_1 *(PSPP software: Descriptive Statistics: Frequencies)*.

Q1

Value Label	*Value*	*Frequency*	*Percent*	*Valid Percent*	*Cum Percent*
	0	2	22.22	22.22	22.22
	1	7	77.78	77.78	100.00
	Total	9	100.0	100.0	

Q1

N	*Valid*	9
	Missing	0
Mean		.78
S.E. Mean		.15
Mode		1.00
Std Dev		.44
Variance		.19
Kurtosis		.73
Skewness		-1.62
Range		1.00
Minimum		.00
Maximum		1.00
Percentiles	50 (Median)	1

The descriptive data calculated above can be used without any doubts. However, further actions with dichotomous data are already associated with a slight touch of shamanism, as there will inevitably be echoes of normal distribution processing ideas somewhere. However, using these methods is better than using nothing.

7.2 Adjusted Wald Technique

For dichotomous data, it is useful to determine a *confidence interval*, which can be calculated using, for example, the ***Adjusted Wald*** technique [33].

Confidence interval of answers on Q_k question in dichotomous distribution can be calculated by Adjusted Wald method:

$[\overline{Q_k^{mod}} - marSE_k^{mod}; \overline{Q_k^{mod}}; \overline{Q_k^{mod}} + marSE_k^{mod}]$ (15).

where:

$\overline{Q_k^{mod}}$ – mean (proportion) of Likert answers on question Q_k adjusted:

$$\overline{Q_k^{mod}} = \frac{FrequencyYes + z}{N + z^2} \tag{16}$$

where:

N – number of questionnaires submitted;

z – calculated for appropriate confidence level (see Table 7);

$FrequencyYes$ - number of *"Agree (Yes)"* on Q_k question.

However, $marSE_k^{mod}$– margin of standard error adjusted for mean (proportion) of Likert answers on question Q_k:

$$marSE_k^{mod} = z * \sqrt{\frac{\overline{Q_k^{mod}} * (1 - \overline{Q_k^{mod}})}{N + z^2}} \tag{17}$$

Example. Confidence interval with 95% confidence of answers on Q_1 question in dichotomous distribution can be calculated as follows: $\overline{Q_k^{mod}} = 7 + 1.96 / (9 + 1.96^2) = 8.96 / 12.84 = 0.7$, but $marSE_k^{mod} = 1.96 * \sqrt{\frac{0.7*(1-0.7)}{9+1.96^2}} = 1.96 * \sqrt{\frac{0.21}{12.84}} = 1.96 * 0.13 = 0.26$. Confidence interval of answers on Q_1 question is $0.44 \leq 0.7 \leq 0.96$. Since the *"Disagree"* answer does not belong to the confidence interval, it can be considered that the respondents answered *"Agree"* to question Q_1. The specific assumption is valid at the confidence level 95%.

Although the Adjusted Wald technique is recommended for the analysis of dichotomous distributions, it should be noted that the method still has links with the paradigm of normal data distribution. This is determined by the use of the constant z (see Table 7) which is applied to determine the confidence level.

7.3 Evaluating the Meaningfulness of a Dichotomous Questionnaire

An important problem is determining survey ***reliability***. Since the results of dichotomous surveys bear no resemblance to the normal distribution, other methods should be used instead of Cronbach's alpha. Various quantitative statistical methods are offered for evaluating dichotomous questionnaire's reliability, for example *Kappa calculation* [34], *ANOVA Binary ICC* [35], *Kuder-Richardson 20* [36], however, the suitability of all the above-mentioned methods in each specific case, as well as the significant differences between the methods, are debatable. There are no convincing recommendations for choosing valid reliability assessment method. In this case, one of the methods is offered

- ***Fleiss Kappa*** calculation [37], which the author considers useful enough for the analysis of dichotomous distributions of 2-point Likert-type survey results.

The Fleiss Kappa is a measure of how reliably three or more respondents measure the same question. In the case of ordinal data, ***Kendall's W*** coefficient [38] of concordance also can be used. If there are only two respondents, ***Cohen's Kappa*** [12] can be applied as a specific case.

Since there is neither a normal distribution nor a linear dependence here, the concept of inter-related assessments is used instead of the concept of correlation. The Fleiss Kappa measurement confirms the homogeneity of the audience, that is, the absence of incompetent persons among the respondents. At the same time, Fleiss Kappa is an indicator that shows the unambiguity of the meaning and wording of the questions. The Fleiss Kappa indicator confirms whether the survey data can be used to confirm an opinion, that is, whether the survey is meaningful (see Table 16). If the achieved result is less than 0.41, then the survey can be considered unsuccessful. However, the Fleiss Kappa method [13], like any other statistic, is very sensitive to the number of respondents. It is desirable that the number of respondents should be at least 50 and more. But even better, as argued earlier, Likert-type surveys should not include dichotomous questions at all.

Table 16. Fleiss Kappa F_K interpretation [39].

F_K-value	Strength of Agreement
1	Ideal
0.81–0.99	Almost perfect
0.61–0.8	Substantial
0.41–0.6	Moderate
0.21–0.4	Fair
0–0.2	Slight
< 0	No agreement

The formula for Fleiss Kappa F_K calculation is analogous to Cohen's Kappa:

$$F_K = (p_0 - p_e)/(1 - p_e) \tag{18}$$

where:

p_0 - relative observed agreement among respondents;

p_e – probability of expected agreement if random judgement;

F_K – interrelated reliability.

The questionnaire Q_K consists of K questions, where $k \in [1, K]$. Answers to K questions have been submitted by N respondents, where $n \in [1, N]$. The questionnaire includes dichotomous questions *"Agree (Yes) / Disagree (No)"*.

$\sum_{n=1}^{N} YQ_k^n$ – the total number of *"Agree"* given by N respondents on question k;

$\sum_{n=1}^{N} NQ_k^n$ – the total number of *"Disagree"* given by N respondents of question k;

SumY – the total number of *"Agree"* given by all respondents *N* on all questions *K*:

$$SumY = \sum_{k=1}^{K} \sum_{n=1}^{N} YQ_k^n$$

SumN - total number of *"Disagree"* given by all the respondents *N* on all the questions *K*:

$$SumN = \sum_{k=1}^{K} \sum_{n=1}^{N} NQ_k^n$$

ANS – total number of answers given by all the respondents:
$ANS = SumY + SumN$ or $ANS = K * N$ *(for checking)* (21).
ProbabilityY – probability of *"Agree"*:

$$ProbabilityY = SumY / ANS$$

ProbabilityN – probability of *"Disagree"*:
$ProbabilityN = SumN / ANS$.
Then

$$p_e = ProbabilityY^2 + ProbabilityN^2 \tag{22}$$

$$p_0 = \frac{1}{K * N(N-1)} * \left[\left(\sum_{k=1}^{K} \left(\sum_{n=1}^{N} YQ_k^n \right)^2 + \left(\sum_{n=1}^{N} NQ_k^n \right)^2 \right) - K * N \right]$$

Example. The given questionnaire Q_K (see Table 14) consists of *K* questions, where $k \in [1, 5]$. Answers to *K* questions have been submitted by *N* respondents, where $n \in [1, 9]$. The questionnaire includes answers on dichotomous questions 1 - *"Agree (Yes)"* and 0 - *"Disagree (No)"*. Fleiss Kappa calculation is shown in Table 17.

According to the evaluation results (F_K is *"Moderate"*), it can be concluded that the conducted survey only partially reflects the opinion of the audience. Whether the reason is the quality of the questionnaire, or whether the audience selection methods are misleading, it is within the competence of the survey implementers. That is why it is recommended to conduct a warm-up test before the survey, which would allow selecting a homogeneous audience.

Table 17. Fleiss Kappa F_K calculation for dichotomous questionnaire Q_K.

N	Q_1	Q_2	Q_3	Q_4	Q_5		
1	1	0	0	0	1		
2	1	1	0	0	1		
3	1	0	1	0	1		
4	1	0	0	0	1		
5	0	0	0	0	1		
6	0	1	1	0	1		
7	1	0	0	0	1		
8	1	0	0	1	1		
9	1	0	0	0	1		
	7	2	2	1	9	21	$\sum_{k=1}^{5}\sum_{n=1}^{9} YQ_k^n$
Y_k	$\sum_{n=1}^{9} YQ_1^n$	$\sum_{n=1}^{9} YQ_2^n$	$\sum_{n=1}^{9} YQ_3^n$	$\sum_{n=1}^{9} YQ_4^n$	$\sum_{n=1}^{9} YQ_5^n$		
	2	7	7	8	0	24	$\sum_{k=1}^{5}\sum_{n=1}^{9} NQ_k^n$
N_k	$\sum_{n=1}^{9} NQ_1^n$	$\sum_{n=1}^{9} NQ_2^n$	$\sum_{n=1}^{9} NQ_3^n$	$\sum_{n=1}^{9} NQ_4^n$	$\sum_{n=1}^{9} NQ_5^n$		
						45	ANS
					$ProbabilityY$	0.47	
					$ProbabilityN$	0.53	
					p_e	0.50	
	53	53	53	65	81	305	$\sum_{k=1}^{5} Y_k^2 + N_k^2$
	$Y_1^2 + N_1^2$	$Y_2^2 + N_2^2$	$Y_3^2 + N_3^2$	$Y_4^2 + N_4^2$	$Y_5^2 + N_5^2$		
					p_0	0.73	
					F_K	0.46	

8 Conclusion

Evaluating sociotechnical systems is not a simple task, as not only specific measurable technical and quantitative parameters must be respected, but also the attitude of users, which determines the acceptance and sustainability development of the given system.

Since the social system is open and affected by various stochastic factors, despite the perfectly implemented technical specification, the final result of the designer's work is often unsuccessful. The reason for the fiasco is the ignorance of the users' attitude. In order to find out this attitude, qualitative measurements of the audience's opinions are carried out using Likert-type surveys.

In order for the survey to provide an adequate representation of the situation, it is not only necessary to properly prepare the questionnaire, but also to statistically process the survey results. The set of statistical methods is very broad and nuanced.

Likert-type surveys will soon be 100 years old. They are widely used in all kinds of attitude measurements, but short and simple instructions for users on how to process the results are very difficult to find. Why? It's like there's nothing complicated here. If we repeat the same measurement several times, then sooner or later the results must correspond to the normal distribution, on which the majority of statistical methods are based. Use whatever method you want, but sooner or later you'll notice a suspicious bell type image somewhere in the background. However, there will still be problems, because attitude measurements are not technical manipulations where the object is compared to some standard. Every person's opinion has at least *"fifty shades of gray"*. Things can get really bad if a surveyor tries to use a dichotomous Likert scale. Then an endless discussion can follow about the nature of the data, whether it is nominal or interval data and what to do with this data next.

It is easy to get lost in this forest, therefore the author's goal was to provide a concentrated and short set of recommendations that could ensure a sufficiently successful processing of the results of a Likert-type survey.

The use of statistical methods in the processing of Likert-type surveys has a dual role. On the one hand, with the help of statistics, the reliability of the attitude measurements can be ascertained. And that is the basic purpose of the statistics application. However, at the same time, the results of statistical processing show the quality of the survey implementation. The results reveal the interrelationship of the questions and the overall meaningfulness of the survey tables. The main problem is that it is difficult to get an answer to the question of who is to blame if the survey results are not reliable enough. The reason may be both the quality of the survey content and the heterogeneity and inappropriateness of the audience, but maybe the duration of the survey or the conditions were inappropriate. Based on previous experience, the surveyor will have to make a decision either to repeat the survey with another audience, or to redo the questionnaire.

The author sincerely hopes that the recommendations on the choice of data processing methods and the implementation of Likert-type surveys will be useful to a wide range of users, both students and researchers, as well as any other surveyors who need to provide more or less reliable attitude measurements.

References

1. Likert, R.: A technique for the measurement of attitudes. Arch. Psychol. **140**, 5–55 (1932)
2. Tejada, J.J., Punzalan, J.R.B.: On the misuse of Slovin's formula. Phil. Stat. **61**(1), 129–136 (2012)
3. Cochran, W.G.: Sampling Techniques, 3rd edn. Wiley (1977)
4. Wanjohi, A. M., Syokau, P.: How to Conduct Likert Scale Analysis. https://www.kenpro.org/how-to-conduct-likert-scale-analysis/. Accessed 15 Oct 2022
5. Curtin University, Introduction to Statistics. Descriptive Statistics. https://uniskills.library.curtin.edu.au/numeracy/statistics/descriptive/. Accessed 12 Oct 2022

6. Razali, N.M., Wah, Y.B.: Power comparisons of shapiro-wilk, kolmogorov-smirnov, lilliefors and anderson-darling tests. J. Stat. Modeling Anal. **2**(1), 21–33 (2011)
7. Musselwhite, D.J., Wesolowski, B.C.: Normal distribution. In: The SAGE Encyclopedia of Educational Research, Measurement, and Evaluation, pp. 1155–1157 (2018)
8. Hazra, A.: Using the confidence interval confidently. J. Thorac. Dis. **9**(10), 4125–4130 (2017)
9. Sauro, J.: Confidence Interval Calculator for a Completion Rate. https://measuringu.com/calculators/wald/#:~:text=It%20uses%20the%20Wald%20Formula,of%201.96%20or%20approximately%202. Accessed 20 Oct 2022
10. Schober, P., Boer, C., Schwarte, L.A.: Correlation coefficients: appropriate use and interpretation. Anesth. Analg. **126**(5), 1763–1768 (2018)
11. Brede, M.: FEEG6017 Lecture: Relationship Between Two Variables: Correlation, Covariance and R-squared. https://www.southampton.ac.uk/~mb1a10/stats/FEEG6017_6.pdf. Accessed 15 Oct 2022
12. Cohen, J.: Weighted kappa: nominal scale agreement with provision for scaled disagreement or partial credit. Psychol. Bull. **70**, 213–220 (1968)
13. Fleiss, J.L.: Measuring nominal scale agreement among many raters. Psychol. Bull. **76**(5), 378–382 (1971)
14. Gliem, J.A., Gliem, R.R.: Calculating, interpreting, and reporting cronbach's alpha reliability coefficient for Likert-type scales. In: Proceedings of 2003 Midwest Research to Practice Conference in Adult, Continuing, and Community Education, pp. 82–88 (2003). https://scholarworks.iupui.edu/bitstream/handle/1805/344/gliem+&+gliem.pdf?sequence=1. Accessed 17 Oct 2022
15. Greenwald, H.J., O'Connell, S.M.: Comparison of dichotomous and likert formats. Psychol. Rep. **27**(2), 481–482 (1970)
16. Delucchi, K.L.: On the use and misuse of chi-square. In: Gideon, K., Lewis, C. (eds.) A Handbook for Data Analysis in the Behavioral Sciences: Statistical Issues, pp. 295–319 (1993)
17. Conover, W. J.: Practical Nonparametric Statistical. 3rd edn. pp. 428–433. John Wiley & Sons, New York (1999)
18. Lilliefors, H.W.: On the Kolmogorov-Smirnov test for normality with mean and variance unknown. J. Am. Stat. Assoc. **62**(318), 399–402 (1967)
19. King, A. P., Eckersley, R. J.: Inferential statistics IV: choosing a hypothesis test. In: Statistics for Biomedical Engineers and Scientists, pp. 147–171 (2019)
20. Arshad, M., Rasool, M.T., Ahmad, M.I.: Anderson darling and modified Anderson darling tests for generalized pareto distribution. J. Appl. Sci. **3**, 85–88 (2003)
21. Choulakian, V., Lockhart, R. A., Stephens, M. A.: Cramér-Von Mises Statistics for Discrete Distributions. In: International Encyclopaedia of Statistical Science (2011)
22. D'Agostino, R.B., Belanger, A.A.: Suggestion for using powerful and informative tests of normality. Am. Stat. **44**(4), 316–321 (1990)
23. Anscombe, F.J., Glynn, W.J.: Distribution of the kurtosis statistic b_2 for normal samples. Biometrika **70**(1), 227–234 (1983)
24. Bera, A. K., Jarque, C.M.: An efficient large-sample test for normality of observations and regression residuals. In: Working Papers in Economics and Econometrics **49** (1981)
25. Critical Values, Critical Values of One-Sample Kolmogorov-Smirnov Test Statistic D. http://oak.ucc.nau.edu/rh83/Statistics/ks1/. Accessed 17 Oct 2022
26. GNU PSPP. https://www.gnu.org/software/pspp/. Accessed 17 Oct 2022
27. Zaiontz, C.: Real Statistics: Shapiro-Wilk Original Test. https://www.real-statistics.com/tests-normality-and-symmetry/statistical-tests-normality-symmetry/shapiro-wilk-test/, last accessed 2022/10/12
28. Prabhaker, M., Chandra, P.M., Uttam, S., Anshul, G., Chinmoy, S., Amit, K.: Descriptive statistics and normality tests for statistical data. Ann. Card. Anaesth. **22**(1), 67–72 (2019)

29. Statistics Kingdom. Shapiro-Wilk Test Calculator. https://www.statskingdom.com/shapiro-wilk-test-calculator.html. Accessed 15 Oct 2022
30. Shapiro, S.S., Wilk, M.B.: An Analysis of Variance for Normality (Complete Samples). Biometrika **52**(3/4) (1965). http://webspace.ship.edu/pgmarr/Geo441/Readings/Shapiro%20and%20Wilk%201965%20-%20An%20Analysis%20of%20Variance%20Test%20for%20Normality.pdf. Accessed 20 Oct 2022
31. Chegg, Pearson Correlation Coefficient. https://www.chegg.com/homework-help/definitions/pearson-correlation-coefficient-pcc-31. Accessed 19 Oct 2022
32. Statistics How To. Cronbach's Alpha: Definition, Interpretation, SPSS. https://www.statisticshowto.com/probability-and-statistics/statistics-definitions/cronbachs-alpha-spss/. Accessed 7 Oct 2022
33. Agresti, A., Coull, B.: Approximate is better than 'Exact' for interval estimation of binomial proportions. Am. Stat. **52**(2), 119–126 (1998)
34. Sim, J., Wright, C.C.: The kappa statistic in reliability studies: use, interpretation, and sample size requirements. Phys. Ther. **85**(3), 257–268 (2005)
35. Liljequist, D., Elfving, B., Skavberg, R. K.: Intraclass correlation - a discussion and demonstration of basic features. PLoS ONE **14**(7) (2019)
36. Davey, J.W., Gugiu, P.C., Coryn, C.L.S.: Quantitative methods for estimating the reliability of qualitative data. J. Multi-Discipl. Eval. **6**(13), 140–162 (2010)
37. DATAtab. Fleiss Kappa. https://datatab.net/tutorial/fleiss-kappa. Accessed 17 Oct 2022
38. Kendall, M.G., Smith, B.B.: The problem of mm rankings. Ann. Math. Statist. **10**(3), 275–287 (1939)
39. Landis, J.R., Koch, G.G.: The measurement of observer agreement for categorical data. Biometrics **33**(1), 159–174 (1977)

The Influence of Knowing the Source of Information on Readers' Perceptions. An Exploratory Study on Twitter

Eleana Jerez-Villota[1,2(✉)], Francisco Jurado[2], and Jaime Moreno-Llorena[2]

[1] Departamento de Ciencias de la Computación, Universidad de las Fuerzas Armadas ESPE, Sangolquí, Ecuador
eijerez@espe.edu.ec

[2] Departamento de Ingeniería Informática, Universidad Autónoma de Madrid, Madrid, Spain
eleana.jerez@estudiante.uam.es, {Francisco.Jurado, Jaime.Moreno}@uam.es

Abstract. Disinformation, fake news or misinformation are well-known problems in social networks. Nevertheless, the readers play an active role-part because they consume and propagate such information. In this paper, we wonder how the readers perceive the information by identifying the source of information. The work presented in this paper details the exploratory study we conducted to check whether knowing the source of information influence people's decision when identifying types of information, using a classification of information based on argumentation. Thus, we analysed the information that is spread on Twitter, a digital social network that works as a micro-blog where users can post short open messages in real-time, which we know as tweets, and researchers can access information about them. The results obtained through the application of statistical tests and their contrast with graphic methods show that the decision of people that know the source does not influence when identifying as "fact or data", "opinion or judgment" and "counterargument". However, the opposite occurs for the type of information labelled as "argument". In addition, the results reveal that opinions spread more than facts on Twitter.

Keywords: Types of information · Influence of source · Argumentation

1 Introduction

Digital 2022 July Global Statshot Report [1], indicates that 63.1% of the world's population uses the Internet. According to a technical bulletin published by the National Institute of Statistics and Census of Ecuador (INEC) [2], 70.7% of people use the Internet in the country. A service that, for example, is normally used to disseminate information through tools such as Online Social Networks (OSNs). People spread all kinds of information related to their own lives, to important events occurring in the world, their country, city, or near environment. They often do so unconsciously, without knowing the source of

A. Rocha et al. (Eds.): WorldCIST 2023, LNNS 800, pp. 52–62, 2024.
https://doi.org/10.1007/978-3-031-45645-9_4

information or verifying the authenticity thereof. This issue has become a topic of interest for researchers in the field of Online Social Network Analysis (OSNA). For instance, in [3] researchers analyze the veracity of the information that people disseminate in OSNs, evaluating the trust, reputation and influence of the users who spread the information, which are important measures in network-based communication mechanisms. In another work, Ureña et al. [4] identify information as "authoritative information", which has been confirmed as true and published by authoritative sources such as organizations and media.

In OSNs, pieces of information called "rumours" are spread, and their veracity is often not determined at the time of publication [5]. According to Zhang et al. [6], rumours are a type of unconfirmed information that is often generated based on people's subjective willingness, rather than their objectivity. Individuals who spread rumours and then receive true or authoritative information may update their beliefs and go on to spread true information, even if they have been spreading the rumour for a long time. Furthermore, the spread of rumours with malicious intent has a significant influence on people's lives because can induce collective panic [7].

Another common term that often confuses readers is "fake news", which implies an intentional distortion of the truth. This term is usually used by politicians when referring to information that does not support their positions, but that is not necessarily false information [8]. Bad advice can also be considered "fake news", especially in the context of health, as its dissemination can exacerbate infectious disease outbreaks [9]. In an exploratory study about the propagation of fake news on Twitter, results showed that fake news spread faster than news with a part of truth [10]. Another study was focused on classifying the news according to their type: intentionality, the main topic addressed, the networks in which they circulated, and the country of origin among other variables. The results of the study showed that the most common intention of people when publishing fake news was that of an ideological nature, associated with topics such as false announcements from government, organizations, or public figures [11]. The World Health Organization (WHO) has defined the term "infodemic" to describe the dangers of false information during the management of disease outbreaks [12]. According to Pham et al. [13], all this information is referred to as "misinformation".

On the other hand, all false information that is intentionally spread in OSNs to cause chaos and panic in a population can be classified as "disinformation". Disinformation can cause tensions; tensions can lead to confrontations; confrontations can cause conflicts and conflicts can cause violence; and finally, public unrest leads to riots and disturbances [7]. According to the definitions presented in [14]: "true information" is content whose veracity can be verified; "disinformation" is defined as "deliberate falsehood", while "misinformation" is defined as "accidental falsehood".

Despite the fluid academic literature on types of information, classifying information that is intentionally propagated by people in OSNs can't be considered an easy task, because when people publish a single piece of information, that piece may initially contain several ideas, which may change during the propagation process. Each of them may then be classified as a type of information, which does not necessarily represent the initial piece of information. In other words, it is not possible to define such a generalized classification of information. We think that a more specific and less generalized classification, which is also easy to identifiable by people, will allow us to carry out a better analysis of the information propagated in OSNs. Thus, we considered interest to conduct an exploratory study on the influence that knowing or not the source of information spread around a topic of interest and published on Twitter, has on the decisions of people when identifying types of information, expecting promising results.

This paper is structured as follows: in Sect. 2, we present a classification of information that is intentionally propagated in OSNs based on argumentation; in Sect. 3, we describe the exploratory study; in Sect. 4, we present the data cleaning and analysis of results; finally, this paper ends with the conclusions, limitations and future works in Sect. 5.

2 Proposal of Classification of Information Based on Argumentation

Important events occurring either globally or locally are topics of interest to people, who often flood OSNs to share their own "opinions". The opinions that are propagated constitute points of view that are sometimes confronted or refuted by other users in the network. This is how the propagation of the difference of opinions takes place. Argumentation arises in response to, or in anticipation of, a difference in opinion, that may contain one or several pieces of evidence or proof called premises, which in term support an idea and which the individuals who postulate them represent "facts" [15].

The argumentation is composed of propositions presented to defend a point of view. The simple propositions are presented as sentences (subject + verb + predicate). When a negative position is expressed concerning a proposition, the point of view is negative, and thus, the propositions compose the "counterargumentation" [16].

Argumentation can refer to points of view and differences of opinion related to all kinds of topics and manifest itself in all kinds of communicative settings such as OSNs. Identifying types of information that are propagated in OSNs based on argumentation may prove to be a less complicated task for readers than identifying true or false information. To identify the latter, readers, as mentioned in [4, 9], should have certain characteristics such as a high level of education or cultural background. Readers may also need more details of the information that is published to identify it as true or false, but, currently, most OSNs have limitations on the size of the information that users can share in their profiles. As mentioned above in the argumentation, sentences can be taken as the simplest representation of a proposition, these being relatively short if we compare them with long paragraphs made up of several sentences.

Based on concepts about argumentation introduced in this section, we think that it would be easier for readers in OSNs to identify types of information based on argumentation in a sentence. Argumentation concepts can be applied to classify information more specifically, and that is also less generalized than the ones presented in Sect. 1. In this sense, we can say that the information that is propagated in OSNs can be: facts or data, opinions or value judgments, arguments, and counterarguments.

3 Exploratory Study

In this work, we have carried out an exploratory study on whether knowing the source of the information spread about a topic of general interest in a local area using Twitter influences the decisions of the participants when identifying types of information. In particular, the topic "delinquency in Quito" and a group of users from that city in Ecuador have been used as a case study. We think that this case could be representative of similar situations.

Thus, our null hypothesis (H0) is "Knowing the source of information does not influence people's decision when identifying types of information on Twitter." and, on the other hand, the alternative hypothesis (H1) is "Knowing the source of information does influence people's decision when identifying types of information on Twitter."

In this section we describe the tools and processes that allowed us to conduct the exploratory study that consisted in first, defining participants and their characteristics; second, gathering tweets; third, segmenting the tweets into ideas to make it easier for participants to identify the types of information; fourth, creating forms for participants to label information; and finally, fifth: conducting labelling process.

3.1 Participants

The experiment involved a group of 76 Spanish-speaking students living near Quito, 27 of them were women and 49 were men, and they are between 19 and 25 years old. All of them come from the third, fourth and fifth levels of the Computer Science Degree of "Universidad de las Fuerzas Armadas ESPE" in Ecuador. It is important to mention that none of the students who participated in the experiment posted tweets that were later labelled. The group of participants was divided into two subgroups:

- The first subgroup was made up of 41 participants. When labelling the information, these participants could see the user name who posted the tweets. That is, they knew the source (KS).
- The second subgroup was made up of 35 participants. When labelling the information, these participants could not see the user name who posted the tweets. That is, they did not know the source (NKS).

Both subgroups, KS Subgroup and NKS Subgroup are independent. This means that each participant belongs to only one of the two subgroups and is not related to the other.

3.2 Gathering

To conduct the experiment, we collected tweets in Spanish during a given period about a topic of interest, namely: "delinquency" near the City of Quito. To perform the data gathering, we use Twint[1], a tool written in Phyton that does not use the Twitter API but allows scraping tweets from Twitter profiles.

Twint defines objects to collect (both tweets and users) and for each of them the list of attributes to gather. In particular, we used the *twint.run.Search* function to search for

[1] https://github.com/twintproject/twint.

tweets through the tweet object and the attributes shown in Table 1. A random sample of 265 tweets was then taken from the tweets obtained during the period indicated, because of limitations on the number of records accepted by the application that will be used to perform the labelling process. Table 1 details the list of attributes we used to gather tweets using the Twint tool.

Table 1. Attributes of the tweet object used for the data gathering.

Attributes	Value	Description
search	delincuencia	Search for tweets containing a specified value (can be a word or a set of words)
near	Quito	Search for tweets near Quito
since	2022-08-09	Search for tweets since a date
until	2022-08-14	Search for tweets until a date
output	file.csv	Defines the output file format
count	True	Shows the number of tweets extracted
store_csv	True	Stores extracted tweets in a csv file

3.3 Data Segmentation

To obtain quality data for further analysis, we must consider that identifying facts, opinions, arguments and counterarguments in tweets is not an easy task because of the presence of, among others, a challenge: the separation between propositions. To tackle this challenge, we rely on the simple representation of a proposition within argumentation, that is, the sentence.

From the csv file obtained in the data gathering process, by simply using the "Text to columns" functionality of Microsoft Excel and taking the period (.) as a reference, we segment the tweets into sentences that from here on we will call ideas.

3.4 Form Development

After performing the segmentation process, we obtained an xlsx file with 786 ideas. We used the xlsx file and Google forms to create the following: for KS Subgroup, 9 forms and for NKS Subgroup, 12 forms, this is due to the limitations of Google forms to generate forms with a large amount of data and due to organizational issues of labelling sessions.

At this point, it is important to mention that the official language of the country where the experiment was carried out is Spanish. Therefore, the question type used in the forms to label the ideas was "The multiple-choice grid", where the questions were: "Hecho o dato" (in English, fact of data), "Opinión o juicio" (in English, Opinion or judgement), "Argumento" (in English, Argument) and "Contraargumento" (in English, Counterargument); and the answer options were a 5-point Likert-type scale (from 0 = Strongly Disagree to 5 = Strongly Agree) (see Fig. 1).

3.5 Labelling Process

Before starting the labelling process, which lasted five days, participants were briefed on some criteria related to the argumentation concepts to label the ideas. As mentioned in Sect. 3.1, KS Subgroup could see the name of the user who had posted the tweet, and NKS Subgroup could not see the name of the user who had posted the tweet. Also, as mentioned in Sect. 3.4, 9 forms were created for KS Subgroup and 12 forms were created for NKS Subgroup. According to these criteria, the forms were distributed for both, KS Subgroup and NKS Subgroup so that an idea was labelled by more than one participant and that a participant did not label the same idea twice.

4 Data Cleaning and Analysis of Results

Once the labelling process was complete, we generate two xlsx files for each subgroup with the participants' responses to perform the cleaning of the datasets first and then analyze them.

4.1 Data Cleaning

To obtain quality data sets that could be later analyzed, it was necessary to execute data cleaning for the two data sets. First, we define a threshold $\boldsymbol{\sigma}$ that consists of the sum of the squares of the deviations of the participants' responses for each idea to discard ideas for which there is no consensus among the participants who labelled them. That is when $\boldsymbol{\sigma > 0.7}$ it was considered that there was no consensus among the participants. On the other hand, when $\boldsymbol{\sigma < 0.7}$ we considered that there is consensus among the participants. Finally, the ideas that were labelled by a single participant were removed. The results of the data-cleaning process are shown in Fig. 1.

4.2 Analysis of Results

Once the data was cleaned out, we proceed to analyze the results shown in Fig. 1 and by observing the results obtained for the "Counterargument" label, we found an important phenomenon; there was a large difference between the amount of data for both KS Subgroup and for the NKS Subgroup, the value of the latter represents 4.07% of the total ideas before cleaning the data. This makes us think that, apparently, among the participants of the KS Subgroup there was a consensus when labelling the ideas as "Counterarguments". On the other hand, among the participants of the NKS Subgroup there was no consensus when labelling the ideas as "Counterarguments".

Continuing with the analysis process, we calculated the means of the scores given by the participants to the ideas to analyze them using statistical tests, which were carried out based on hypotheses statements, each leading to two possible outcomes:

- Null Hypothesis (H_0): Assumes no difference, association or relationship between the variables
- Alternative Hypothesis (H_1): Assumes a difference, association or relationship between the variables.

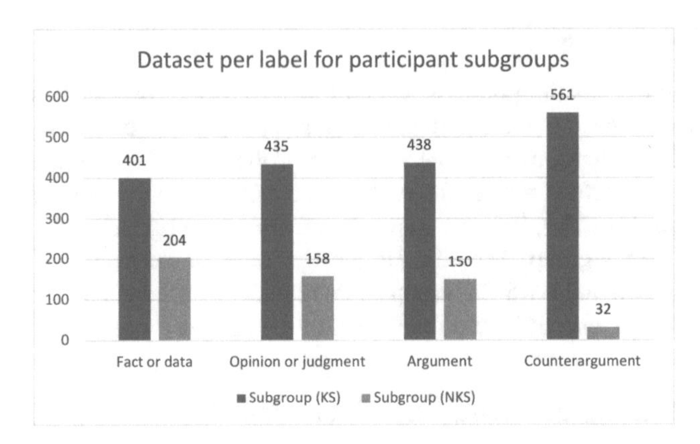

Fig. 1. Data cleaning process results.

The use of an appropriate statistical hypothesis test will allow us to decide between these two outcomes, by examining the probability (or p) value of the test statistic. For this analysis we consider that $p < 0.001$ is an adequate p-value to obtain statistically significant results (at 99.9%) in all tests [17]. We use R Commander to carry out the statistical tests.

The method defined for the analysis focuses on checking the significance of the performed process, applying the Student's t-test. To apply this test, we must first verify that three conditions are met:

1. The two groups must be independent. This means that each participant must belong to only one of the two groups and have no relationship with the participants in the other group.
2. The outcome variable must be continuous and follow a normal distribution in the two groups.
3. The homoscedasticity assumption must be met, i.e., equal variances in the two groups.

Before verifying the three conditions mentioned above, in Table 2 we showed the means per label for the KS Subgroup and for the NKS Subgroup.

Table 2. Means per label for the KS Subgroup and for the NKS Subgroup.

Labels	Fact or data		Opinion or judgment		Argument		Counterargument	
Subgroups	KS	NKS	KS	NKS	KS	NKS	KS	NKS
Mean	0.41	0.69	4.59	4.35	0.33	0.82	0.02	0.09

As we mentioned in Sect. 3.1, both subgroups of participants are independent. Next, it's important to review if the other conditions are met.

Normality Test. The assumption of normality of the variable in the two subgroups can be verified using the Kolmogorov-Smirnov test with the Lilliefors modification suitable for samples larger than 50 [18]. We determined the following hypotheses for the test:

- H_0: The data have a normal distribution.
- H_1: The data do not have a normal distribution.

The results obtained for each of the labels are shown in Table 3. We observe that the data have a normal distribution for neither of the labels since $p < 0.001$. Hence, we reject the H_0 hypothesis in all cases. Although the normality condition is not met, we continued with the analysis since according to [18] for these cases we can apply the Student's t test using Welch's correction.

Test for Homogeneity of Variance. Next, we tested the homoscedasticity assumption and we used Levene's test [18] for homogeneity of variance, which is recommended when the normality assumption is not met and so, determined the following hypotheses:

- H_0: Variances are equal in all groups.
- H_1: At least one group has a variance different from the rest.

The results obtained for each of the labels are shown in Table 3. We note that for the labels "Fact or data" and "Opinion or judgment" we accepted the null hypothesis since $p > 0.001$ and for the labels "Argument" and "Counterargument" we rejected the null hypothesis and accepted the alternative hypothesis since $p < 0.001$.

Student's t-Test. We carried out the Student's t-test, for the cases in which homoscedasticity exists; in the cases in which homoscedasticity does not exist we applied Welch's correction. We determined the following hypotheses:

- H_0: There are no significant differences between the means of KS and NKS Subgroups.
- H_1: There are significant differences between the means of KS and NKS Subgroups.

The results obtained for each of the labels are shown in Table 3. It's important to note that for the labels "Fact or data", "Opinion or judgment" and "Counterargument" we accepted the null hypothesis since $p > 0.001$ and for the label "Argument" we rejected the null hypothesis and accepted the alternative hypothesis since $p < 0.001$.

Finally, to complete the contrast test, we considered the use of a graphical method such as histograms (see Fig. 2).

Table 3. Statistical tests per label for the KS Subgroup and for the NKS Subgroup.

Labels	Subgroups	Lilliefors' test ($p < 1.0e-3$)	Levene's test ($p < 1.0e-3$)	t-test ($p < 1.0e-3$)
Fact or data	Ks & NKS	$p < 2.2e-16$	$p = 5.34e-3$	$p = 5.34e-3$
Opinion or judgment	KS & NKS	$p < 2.2e-16$	$p = 1.54e-2$	$p = 1.08e-2$
Argument	KS & NKS	$p < 2.2e-16$	$p = 7.79e-9$	$p = 2.63e-5$
Counterargument	KS & NKS	$p < 2.2e-16$	$p = 3.11e-4$	$p = 4.57e-2$

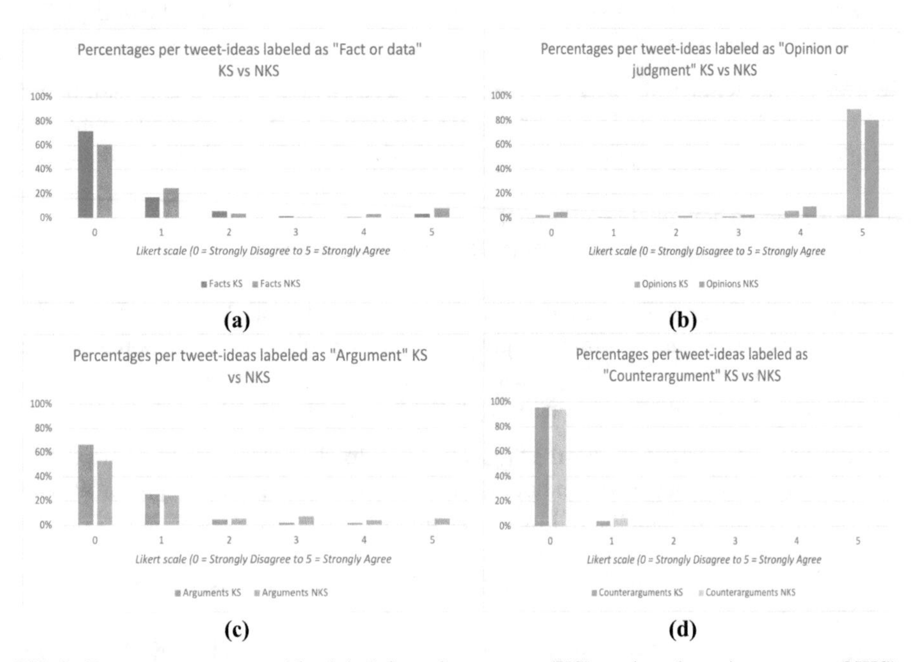

Fig. 2. Percentage per tweet-idea labels knowing sources (KS), and not knowing sources (NKS): (a) "Fact or data"; (b) "Opinion or judgment"; (c) "Argument"; (d) "Counterargument".

5 Conclusions, Limitations and Future Works

In this paper, we have detailed the exploratory study we conducted to check the influence of knowing the source of information on readers' perceptions. After carrying out the exploratory study, we must accept hypothesis H_0 for the types of information: fact or data, opinion or judgment, and counterargument, since it seems that knowing the source does not influence the identification of these types of information when spread on Twitter. On the other hand, for the type of information argument, we reject the null hypothesis H_0 and accept the alternative hypothesis H_1 defined for this study, since the results obtained show that knowing the source does influence the identification of this type of information on Twitter.

It was observed that for all cases, there are more ideas labelled in the subgroup that they knew the source, especially for the counterargument case, where a particular phenomenon occurs: the people who knew the source seem to have decided that the ideas they observed were not counterarguments. However, there was no consensus in labelling counterarguments among the people who did not know the source.

An important conclusion we can extract after the study is that, apparently, more opinions are disseminated on Twitter than facts, regardless of whether people know the source or not, and that the identification of both depends on the tweet, but for the identification of arguments and counterarguments, it is necessary to know the thread of the tweet.

After the results of the exploratory study, we must check whether the beacons can be extrapolated to other scenarios and types of information that could arise when analyzing the propagation of information in digital social networks to the analysis of topics that are not in the public domain such as those related to the political, economic or social situation of a country or a local area. Also, the sample size is relatively small and this should be increased for in-depth research.

Our future lines of research include the following: comparing the degree of discrepancy and the degree of consensus when identifying information between the two subgroups; identifying arguments and counterarguments in tweet threads; and analyzing tweets to try to predict whether they are facts or opinions by applying machine learning techniques. In addition, the study is intended to be replicated in other scenarios.

Acknowledgements. This research work has been co-funded by the Spanish National Research Agency (Agencia Estatal de Investigación) through the project Indigo! (PID2019-105951RB-I00), and the Regional Government of Madrid (Comunidad de Madrid) through the e- Madrid-CM project (P2018/TCS-4307). The e-Madrid-CM project is also supported by Structural Funds (ESF and ERDF).

References

1. DataReportal. Digital 2022 July Global Statshot report (2022). https://datareportal.com/reports/digital-2022-july-global-statshot. Accessed 14 Nov 2022
2. Peña, A., Herrera, L.: Indicadores de tecnología de la información y comunicación (2021). https://www.ecuadorencifras.gob.ec/documentos/web-inec/Estadisticas_Sociales/TIC/2020/202012_Boletin_Multiproposito_Tics.pdf. Accessed 14 Nov 2022
3. Ureña, R., Kou, G., Dong, Y., Chiclana, F., Herrera-Viedma, E.: A review on trust propagation and opinion dynamics in social networks and group decision making frameworks. Inf. Sci. **478**, 461–475 (2019). https://doi.org/10.1016/j.ins.2018.11.037
4. Lagnier, C., Gaussier, E., Kawala, F.: User-centered probabilistic models for content diffusion in the blogosphere. Online Soc. Networks Media **5**, 61–75 (2018). https://doi.org/10.1016/j.osnem.2018.01.001
5. Wang, Q., Jin, Y., Cheng, S., Yang, T.: ConformRank: a conformity-based rank for finding top-k influential users. Physica A **474**, 39–48 (2017). https://doi.org/10.1016/j.physa.2016.12.040
6. Zhang, W., Yang, J., Ding, X., Zou, X., Han, H., Zhao, Q.: Groups make nodes powerful: identifying influential nodes in social networks based on social conformity theory and community features. Expert Syst. Appl. **125**, 249–258 (2019). https://doi.org/10.1016/j.eswa.2019.02.007

7. Granese, F., Gorla, D., Palamidessi, C.: Enhanced models for privacy and utility in continuous-time diffusion networks. Int. J. Inf. Secur. **20**(5), 763–782 (2021)
8. Vosoughi, S., Roy, D., Aral, S.: The spread of true and false news online. Science **359**(6380), 1146–1151 (2018)
9. Brainard, J., Hunter, P.: Misinformation making a disease outbreak worse: outcomes compared for influenza, monkeypox, and norovirus. Simulation **96**(4), 365–374 (2019). https://doi.org/10.1177/0037549719885021
10. Shahi, G.K., Dirkson, A., Majchrzak, T.A.: An exploratory study of COVID-19 misinformation on Twitter. Online Soc. Networks Media **22**, 100104 (2021). https://doi.org/10.1016/j.osnem.2020.100104
11. Gutiérrez-Coba, L., Coba-Gutiérrez, P., Gómez-Díaz, J.A.: Las noticias falsas y desinformación sobre el Covid-19: análisis comparativo de seis países iberoamericanos. Rev. Lat. Comun. Soc. **78**, 237–264 (2020). https://doi.org/10.4185/RLCS-2020-1476
12. Rodríguez, J., Gómez, S.: La infodemia y su alcance en el área psicoemocional de las familias. Un aporte a la crisis de la salud a propósito del Covid-19. Revista Iberoamericana de Ciencia, Tecnología y Sociedad-CTS, vol. 16 (2021)
13. Pham, D.V., Nguyen, G.L., Nguyen, T.N., Pha, C.V., Nguyen, A.V.: Multi-topic misinformation blocking with budget constraint on online social networks. IEEE Access **8**, 78879–78889 (2020)
14. Kumar, K.P.K., Srivastava, A., Geethakumari, G.: A psychometric analysis of information propagation in online social networks using latent trait theory. Computing **98**(6), 583–607 (2016). https://doi.org/10.1007/s00607-015-0472-7
15. Palau, R.M., Moens, M.F.: Argumentation mining: the detection, classification and structure of arguments in text. In: Proceedings of the 12th International Conference on Artificial Intelligence and Law, pp. 98–107, June 2009. https://doi.org/10.1145/1568234.1568246
16. Van Eemeren, F.: Argumentation Theory: A Pragma-Dialectical Perspective. Springer, Cham (2018)
17. Samuels, P., Gilchrist, M.: Statistical hypothesis testing. Technical report (2014). http://www.statstutor.ac.uk/resources/uploaded/statisticalhypothesistesting2.pdf. Accessed 14 Nov 2022
18. Molina Arias, M., Ochoa Sangrador, C., Ortega Paez, E.: Comparación de dos medias. Pruebas de la t de Student. Evid Pediatr **16**(4), 51 (2020)

Knowledge Management: A New Business Success Factor in Companies in Southern Peru

Regis André Junior Fernández-Argandoña[1](✉), Alberto Miguel Alponte-Montoya[2], and Ben Yúsef Paul Yábar-Vega[3]

[1] Escuela de Posgrado Newman, Av. Bolognesi 789, Tacna, Perú
rfernandez@neumann.edu.pe
[2] Carver Research, 604 Courtland Street, Suite 131, Orlando, FL, USA
[3] Universidad Privada de Tacna, Av. Bolognesi 1177, Tacna, Perú

Abstract. This article entitled "Knowledge Management: A new business success factor for companies in southern Peru"; Its main objective is to determine the entrepreneur's valuation of aspects related to knowledge management and if this management is fostered through the creation of new knowledge and its transmission. The research approach was qualitative, of a correlational scope and of a basic or pure type. Regarding the data collection technique and instrument, the interview and the interview guide were used, respectively. Being a qualitative investigation, it was decided to interview six managers of recognized companies in the city of Tacna according to specified inclusion and exclusion criteria.

Keywords: Management · Knowledge · Organizational Climate · Corporate Identity

1 Introduction

At an international level, knowledge management has become a strategic resource, the creation, use and effective dissemination of knowledge is a key element for the success of organizations, allowing them to combine and classify resources in the best way, transforming them into new ones. And innovative, thus creating a differentiating value for customers (Rojas and Vera, 2016).

Likewise, it should be noted that knowledge management is seen as an organizational capacity, which develops a process that involves the exchange of information and that to be executed requires the participation of all members of the organization, this must be collected, transferred and managed. Appropriately to make the organization more profitable. According to Pereda and Berrocal (2001) "by carrying out adequate management, knowledge remains in the organization and is profitable, since it is shared with collaborators and superiors".

At the national level, Vilca (2013) and Pastor (2010) indicate that Peruvian companies tend to become more competitive as they transform their way of thinking and focus more on developing human resources (employees), that is, they seek the development of knowledge in the personnel they have in order to achieve the goals established by each organization.

A. Rocha et al. (Eds.): WorldCIST 2023, LNNS 800, pp. 63–69, 2024.
https://doi.org/10.1007/978-3-031-45645-9_5

The main objective of this scientific article is to determine the relevance and value that the entrepreneur gives to knowledge management, taking into consideration companies in the southern zone of Peru. Finally, regarding the relevance and pertinence of this research, we have to say that we are in a transition stage from a totally technological era to one where the assessment of the human factor is the most important thing.

2 Literature Review

Being the objective of this article, to determine the incidence of knowledge management in the success of companies in southern Peru; It is necessary to establish some concepts related to knowledge management. In Peru there are a large number of companies that have economic objectives, leaving aside intellectual capital or human knowledge (Tello and Armas, 2018).

Blas (2009) indicates that "employees are recognized as the source of innovation, although in some cases, these skills are not adequately developed because the employer does not provide adequate spaces for the creation and transmission of knowledge." The concept of knowledge management emerged in the 1990s and spread rapidly. This tool has created controversy, with many executives asking if there was really any advantage to it (Rosini and Palmisano, 2012).

The knowledge that an organization possesses can be a source of sustainable competitive advantage by implementing an effective knowledge management strategy that allows innovative actions to be taken to create products, services, management systems and processes to optimize the organization's resources and capabilities (Nofal, 2007).

Likewise, the authors Pérez (2016), Mejía (2008) and Nagles (2007), affirm that knowledge management is a discipline that provides a great focus on designing and implementing methods in order to identify and share knowledge with all the organization. Knowledge management can be understood as the process of transforming information and intellectual assets into lasting value (Calvo, 2017). That is why, faced with these changes, organizations can only create value through the effective management of their most valuable resources, knowledge and ideas to create a sustainable competitive advantage, unlike other organizations. (Godoy, Mora and Liberio, 2016).

On the other hand, it is also important to highlight that the success of a company is related to the organizational climate, which in turn is the perception of the members that make up the organization, so it must be taken into consideration to promote spaces for transmission of knowledge. (Figueroa and Sotelo, 2017; Hamidian, Torres, and Lamenta, 2018; Segredo, García, López, Cabrera, Perdomo, 2015).

Finally, it is necessary to analyze the companies and see how they implement knowledge management and if they maintain a good organizational climate to create a much more communicative environment with the members of the organization.

3 Method

This scientific article seeks to contribute to scientific knowledge, so we would be following a pure investigation, as well as with respect to our objective, focusing on the incidence of one variable on another, we are talking about a correlational study (Hernández, Fernández and Baptista, 2014).

Regarding the research approach, a qualitative approach was chosen since we collected the information through interview guides to be later analyzed in the discussion chapter.

Regarding the selection of businessmen to interview, as it is a qualitative investigation, inclusion and exclusion factors were taken into consideration in which we look for regionally recognized companies in southern Peru, which have at least 20 collaborators in total and also searched for choose one company per business to have a holistic understanding of the subject.

Once the managers of the companies were selected, an interview guide consisting of six (06) open questions was applied to them, for which they answered both in person and by telephone, the researcher later proceeding to transcribe their answers and be able to express them in the following (Tables 1, 2, 3, 4, 5, 6 and 7).

Table 1. List of companies interviewed

Business
Andina SAC Airports
Caja Tacna
JUMI Stores
Zofra Tacna
Calport E.I.R.L
Graficom EIRL

Note: Own elaboration

4 Analysis of Results

4.1 Interview Results

Table 2. Results ofiteminterview guide 01

Regarding question 1. Indicate in your opinion, what is knowledge management and what is its importance?	
Andina SAC Airports	Transferring knowledge, he also mentioned that the experiences in the printing press of each worker help the training
Caja Tacna	They are processes for the exchange of information within the organization It is necessary to improve performance
JUMI Stores	It is the effort that the employer makes to be able to motivate workers to innovate, it is very important since developing new skills in employees will help them achieve their goals and therefore the objectives of the company
Zofra Tacna	Share the work and studies carried out in a period
Calport E.I.R.L	When we transmit our knowledge achieved to all our staff
Graficom EIRL	It is an exchange of knowledge, information and experiences Important to improve results in the organization and/or projects

Note: Own elaboration based on the interview guide

Table 3. Results of item 02 of the interview guide

Regarding question 2. Do you consider that it is convenient to promote and disseminate knowledge in your organization? Why?	
Andina SAC Airports	Yes, for a better development of procedures
Caja Tacna	Of course, in order to improve under the same procedure and avoid gaps in companies as large as this
JUMI Stores	She stated that it is important, stating that the knowledge she manages to obtain is shared with her staff and group of officials, in order to promote and disseminate her experience
Zofra Tacna	Yes, due to the number of collaborators
Calport E.I.R.L	It is essential to achieve better capabilities
Graficom EIRL	Yes, to train new workers

Note: Own elaboration based on the interview guide

Table 4. Results of item 03 of the guideinterview

Regarding question 3. In your company, if a collaborator develops a better or faster way of carrying out an activity, is he valued in any way?	
Andina SAC Airports	Yes, and this motivates the other collaborators to innovate in their field of work
Tacna Box	It is valued with bonuses and public recognition
JUMI Stores	He mentioned that they do value their collaborators, providing them with productive bonuses (certificates of excellence and recognition in the monthly meetings that are held)
Zofra Tacna	Not because it is a state company with private management, incentives are very difficult
Calport E.I.R.L	Yes, you are given various incentives for finding new markets
Graficom EIRL	No, because they drive machines and they do everything

Note: Own elaboration based on the interview guide

Table 5. Results of item 04 of the interview guide

Regarding question 4. When a collaborator in your company acquires more experience and knowledge, does he pass it on to his colleagues?	
Andina SAC Airports	Yes, since everything is a joint effort, the knowledge has to be leveled for the proper functioning of the materials and logistics
Tacna Box	Depends on the worker, none is required unless new staff are assigned and must be trained
JUMI Stores	He stated that they train and monitor all their staff
Zofra Tacna	No, they are mostly jealous because of how old they are
Calport E.I.R.L	Only in the production area
Graficom EIRL	If, for example, in the handling of the colors that they develop for the impressions, tones must be mixed, then experience is needed and they help each other

Note: Own elaboration based on the interview guide

Table 6. Results of item 05 of the interview guide

Regarding question 5. If you would provide training to the staff, would you do so only to the most senior or would you train all the staff in general. Why?	
Andina SAC Airports	No, the trainings are for all staff
Tacna Box	Depending on the subject, since the number of areas of the company allows us to focus on them in a personalized way
JUMI Stores	He stated that he will do it to all the staff since it is important, since each collaborator must know how the company works
Zofra Tacna	To the new ones, and only in complex administrative processes
Calport E.I.R.L	It seeks to train each area independently
Graficom EIRL	We only train permanent staff, and we have a high staff turnover

Note: Own elaboration based on the interview guide

Table 7. Results of item 06 of the interview guide

Regarding question 6. In your opinion, does the creation and transmission of knowledge in your company influence the performance and satisfaction of your collaborators?	
Andina SAC Airports	Yes, to improve results and relieve possible inconveniences among workers
Tacna Box	It influences, but constant training is needed on the topics directed to the corresponding areas
JUMI Stores	He stated that, yes, it is important and currently it helps his collaborators a lot, since they acquire knowledge about sales, dispatches, understand the needs of customers and know the usefulness of each product
Zofra Tacna	It does influence, since when they are motivated their performance is higher
Calport E.I.R.L	Yes, because when the worker perceives a good environment he works better
Graficom EIRL	Of course they do, when they have confidence in their work team they are more motivated

Note: Own elaboration based on the interview guide

4.2 Discussion of Results

In general terms, after having obtained this information, it can be said that there is an assessment of knowledge management in the companies interviewed, although in some with marked limitations. The above is quite positive because the leaders of the organizations recognize the value of the creation and transmission of knowledge and the impact that it will have on their organizations, which is why they train, praise and recognize the efforts of their staff.

Regarding the first question, this was designed to define whether the interviewed managers knew about knowledge management, in this regard it was observed that all interviewees understand what knowledge management is and value its importance, understanding that the scope of the objectives depends on it. Business objectives.

Continuing with respect to item two, we also have a council among the interviewees to the extent that all of them see fit to promote and disseminate knowledge to strengthen work teams, improve processes, seek continuous improvement and company performance; This is positive to the extent that there is an assessment of the human factor in companies.

Now, regarding question three of the data collection instrument, four of the six interviewees refer to valuing a worker who has innovated in the way of carrying out their activities, where they differ is in terms of the way of valuation having some that grant verbal recognitions, diplomas but also others that indicate to offer economic bonuses; Something to take into account is that ZofraTacna sees this aspect limited as it is a state-owned company under a private regime and Graficom does not provide incentives because it understands that the process is carried out solely by machines, which in the investigator's opinion is not entirely real, since the human factor is needed to correctly manage these machines.

Regarding the fourth question of the interview guide, it was aimed at verifying whether there is a willingness to transmit knowledge within organizations, that is, to share what we know with those who still do not; In this regard, there were mostly encouraging responses regarding this good intention of sharing what we know, but as for ZofraTana, they stated that this courtesy is not given as the workers are jealous of what they know, which should set off the company's alarms to be able to execute actions that allow these collaborators to understand the importance of sharing knowledge.

Continuing, regarding the fifth question, this was aimed at defining whether managers trusted the transmission of knowledge by group leaders to their lower-ranking collaborators, finding very interesting answers since most companies understand that there may be generic training for all staff, but they also report that due to the size of their companies they opt for training focused on work areas.

Finally, regarding the sixth question of the interview guide, it was oriented to determine if the aspects related to the creation of knowledge influence the performance and satisfaction of the collaborators, in this regard it should be highlighted that all the interviewees recognized that a work environment in which camaraderie and the sharing of knowledge prevail, it influences work performance since collaborators are much more motivated.

5 Conclusions

As for our research topic, referred to knowledge management as a success factor in companies in southern Peru, it can be concluded that they consider and highlight aspects related to the creation and transmission of knowledge, this is observable since that the Management of these companies seeks to generate spaces for dialogue and fellowship among collaborators with the intention that there is greater trust and work experiences can be shared.

Bibliography

Blas, L.: Knowledge Management. Petrotechnics, pp. 12–26 (2009). https://www.petrotecnia.com.ar/junio09/gest.conoc_Blas.pdf

Calvo, O.: Knowledge management in organizations and regions. Magazine of the Faculty of Economic and Administrative Sciences **19**(1), 140–63 (2017). http://www.scielo.org.co/pdf/tend/v19n1/2539-0554-tend-19-01- 00140.pdf

Figueroa, E., Sotelo, J.: The organizational climate and its achievement with the quality of the service in an institution of higher secondary level education. RIDE Ibero-American Magazine for Educational Research and Development, 8(15) (2017). https://www.redalyc.org/pdf/4981/498154006021.pdf

Godoy, M., Mora, J., Liberio, F.: Knowledge management for the development of intelligent organizations. Publishing Magazine 3(9), 660–673 (2016). https://revistapublicando.org/revista/index.php/crv/article/view/393/pdf_245

Hamidian, B., Torres, K., Lamenta, P.: Organizational climate as knowledge management. Organizational Sapienza 5(9), 159–172 (2018). https://www.redalyc.org/articulo.oa?id=553056570008

Hernandez, R., Fernández, C., Baptista, M.: Investigation methodology, 6th Edition. Editorial McGraw-Hill, Mexico (2014)

Mejía, M.: Knowledge management model for companies, p 52 (2008). http://cybertesis.unmsm.edu.pe/bitstream/handle/20.500.12672/2135/Mejia_pm.pdf?sequence=1&isAllowed=y

Nagles, N.: Knowledge management as a source of innovation. EAN Magazine 77–88 (2007). https://journal.universidadean.edu.co/index.php/Revista/article/view/418/412

Nofal, N.: Knowledge management as a source of innovation. EAN Magazine 77–88 (2007)

Pastor Carrasco, C.: Strategic management for Peruvian companies. Faculty of accounting sciences 7(34), 199–208 (2010). https://revistasinvestigacion.unmsm.edu.pe/index.php/quipu/article/view/4734/3805

Perez, M.: Knowledge management: origins and evolution. Inf. Prof. **25**(4), 526–534 (2016). https://revista.profesionaldelainformacion.com/index.php/EPI/article/view/epi.2016.jul.02/31586

Pereda, S., Berrocal, F.: Management of human resources by competencies. Ramón Areces Study Center, Madrid (2001)

Rojas, G., Vera, M.: Organizational culture in knowledge management, Apuntes de Administración Colombia Magazine, Vol. 1(1), 50–59 (2016). https://revistas.ufps.edu.co/index.php/apadmin/article/view/993/940

Rosini, A., Palmisano, Â.: Administration of information systems and knowledge management, 2nd edn. Cengage Learning, São Paulo (2012)

Segredo, A. García, A. López, P. Cabrera, P., Perdomo, V.: Systemic approach to organizational climate and its application in public health, Cuban Journal of Public Health, 41(1), 115–129 (2015). https://www.scielosp.org/article/ssm/content/raw/?resource_ssm_path=/media/assets/rcsp/v41n1/spu10115.pdf

Tello, R., Armas, P.: Knowledge management and its relationship with the fulfillment of goals of Emapa San Martín sa district of Tarapoto, 2016. Thesis to opt for the Professional Title of Bachelor of Administration. National University of San Martin, Peru (2018). http://repositorio.unsm.edu.pe/bitstream/handle/11458/2965/ADMINISTRACION%2020Renato%20Tello%20Pi%C3%B1a%20%26%20Paola%20Armas%20Navarro.pdf?sequence=1&isAllowed=y

Vilca, Y.: Base model for knowledge management of Peruvian companies that carry out operational activities. Master's Thesis in Design, Management and Project Management. University of Piura Peru (2013). https://pirhua.udep.edu.pe/bitstream/handle/11042/1862/MAS_PRO_003.pdf?sequence=1

The Data Repositories in Agriculture. A Preliminary Investigation

Rubén Fernández Gascón[1], Jose Luis Aleixandre-Tudo[2,3](✉), Juan Carlos Valderrama Zurian[1], and Rafael Aleixandre Benavent[1,4]

[1] UISYS Research Unit, University of Valencia, Valencia, Spain
[2] Instituto de Ingeniería de Alimentos Para El Desarrollo (IIAD), Departamento de Tecnología de Alimentos, Universidad Politécnica de Valencia, Valencia, Spain
joaltu@upvnet.upv.es
[3] Department of Viticulture and Oenology, South African Grape and Wine Research Institute (SAGWRI), Stellenbosch University, Stellenbosch, South Africa
[4] Ingenio (CSIC-Valencia Polytechnic University), Valencia, Spain

Abstract. The primordial role of agriculture in the existence and future of humankind is clearly represented in the sustainable developmental goals (SDG) defined by the United Nations. A large amount of data is generated in agricultural research. However, it is still unknown if this data complies with the FAIR principles for scientific data management. This study aims at investigating the status quo of the data repositories in agriculture in terms of its data findability, accessibility, interoperability, and reusability. For this, a search strategy that includes the application of multiple filters has been executed to obtain a simplified list of data repositories. The most frequent keywords contained in data repositories within the main fields of agriculture were used to obtain a reduced list of repositories devoted the agricultural data.

Keywords: Open data · Data sharing · FAIR principles · Agriculture · Sustainable Development Goals

1 Introduction

The Oxford dictionary defines agriculture as the science or practice of farming. A more complete definition states that agriculture is "the practice of cultivating the soil, growing crops, or raising livestock for human use, including the production of food, fed, fiber, fuel, or other useful products". In addition, farming is given as a direct synonym. Agriculture is therefore a complex term that involves a wide variety of activities and processes [1]. Twelve out of seventeen of the United Nations Sustainable Developmental Goals (SDG) are directly or indirectly related to agriculture [2]. Among them the Zero Hunger goal (SDG 2) is directly related to agricultural practices. Specifically, attention is given to "sustainable agriculture" which is the practice of agriculture that considers the three dimensions of sustainable production. Sustainable agriculture is not only seen as the practice of farming in the pure sense of it as it also considers the environmental, economic,

A. Rocha et al. (Eds.): WorldCIST 2023, LNNS 800, pp. 70–74, 2024.
https://doi.org/10.1007/978-3-031-45645-9_6

and social dimensions of a sustainable activity [3]. Furthermore, the Food and Agriculture organization of the United Nations (FAO) main goal is also to defeat hunger at global level and ensure food security for all. This international organization promotes the regular access to high-quality food towards a population with active and healthy lives.

A big part of the scientific effort worldwide is placed on agriculture and related topics. This intense activity entails the generation of large amounts of relevant data. Making this data available will benefit scientific advancements by for example encouraging collaboration and improving decision making. A team of researchers from several institutions proposed "The FAIR guiding principle for scientific data management and stewardship" [4]. The FAIR principles provide guidelines to ensure that data is findable, accessible, interoperable, and reusable. Emphasis is placed on the interoperability of data due to the increasing dependence on computing systems. The final goal of the FAIR principles is to assure that maximum benefit is obtain for any kind of generated data. On the other hand, open data or open science is a movement that has been gaining traction lately. Open science was a concept introduced to solve accessibility restrictions to data by researchers, institutions or even the public [5]. Interestingly, intense debate is currently underway to understand if the proposed concept of open science is really solving the original accessibility problem [6]. The FAIR principles are not related to the specific case of open science and apply to the generation of any kind of data, either open or not. However, in the field of agriculture, initiatives to promote open data are in place. These initiatives claim open data as the way forward to make relevant information and knowledge freely available and accessible to everyone, therefore optimizing the decision-making process. One of these initiatives is the "Global Open Data for Agriculture and Nutrition (GODAN)". GODAN mission is to promote the open data initiative and advocate for cooperation and collaboration among stakeholders, enable data initiatives to place the focus on agriculture and nutrition, building high level policy, obtain support from public and private institutions and create awareness in open data matters in agriculture. Based on the above, this study aims at investigating the status of the available data repositories in agriculture. Repositories in other fields have shown varying degrees of FAIR principles compliance. However, to the authors knowledge no information on data repositories in agriculture is available in the literature. The final goal of this project will thus be to assess to what extent the FAIR principles are fulfilled in the data repositories devoted to open agricultural data or in other words if data related to agriculture is findable, accessible, interoperable, and reusable. However, before the data repositories can be analysis a reduced list of agriculture related data repositories needs to be obtained. For this, several filters have been applied to obtain an operational collection of data repositories in agriculture that will be later analyzed.

2 Materials and Methods

The registry of data repositories Re3data (www.re3data.org) was used to obtain a preliminary list of data repositories. The plain term agriculture was used. After the initial list was obtained, the results were filtered (filter 1) including only those repositories included in the subject areas of "Ecology and Agricultural Landscapes", "Agricultural economics and sociology" and "Agricultural and Food Processing Engineering". After the application of the first filter, a keyword analysis was performed. The most relevant keywords were then used to apply the second (filter 2) and third filter (filter 3). Repositories containing the first five most frequent keywords in the title field were selected. The same keywords were applied in this case to the description of the repository. Finally, a final filter (filter 4) was applied to obtain the institutions hosting repositories that contain the root "agri*" in the institution name. A single list of repositories, after eliminating the duplication of results was finally compiled.

3 Results and Discussion

Two hundred and seventy-five repositories were obtained from the initial search in the Re3data registry. After the application of filter 1, 69% of the repositories (190) were found within the three agriculture related areas mentioned above. The remaining 31% (85) of them were included in other thematic areas. A keyword analysis was later performed with the resulting 190 repositories. The most frequently keywords in the selected repositories were agriculture (39), biodiversity (24), multidisciplinary (20), ecology (20), ecosystem (18), health (17), environment (16), hydrology (12), economics (12), FAIR (11), forestry (11) and remote sensing (11).

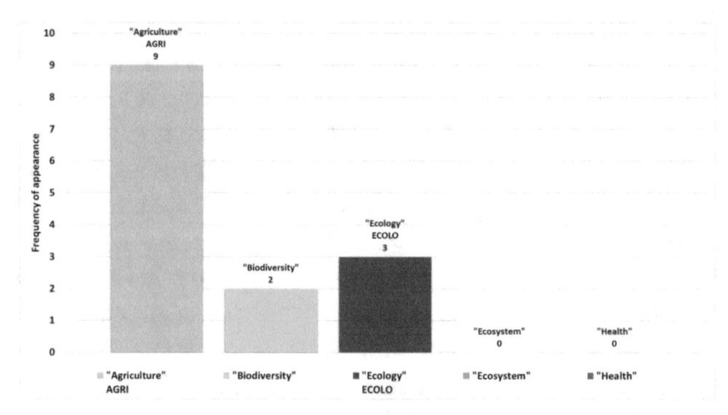

Fig. 1. Number of repositories containing the selected keywords in the title after the application of the first filter

From this list of keywords, the first five were selected and used for the application of filter 2 and 3. An exception was made with the keyword "multidisciplinary" as the authors consider that this term could be misleading as it does not directly reflect a topic specifically devoted to the field of agriculture. Figure 1 shows the number of repositories that contain the selected keywords (agriculture, biodiversity, ecology, ecosystem, and health). The roots agri* and ecolo* were used for the keywords agriculture and ecology, respectively. As expected, agriculture (agri*) was the most frequently keyword identified in the title of the selected repositories, followed by ecology and biodiversity. Surprisingly no repositories were identified containing the keywords ecosystem and health in the title field. Figure 2 shows the number of repositories containing the selected keywords in the description field. The same keywords and roots as in filter 2 were applied. After the application of filter 3, the keyword agriculture (agri*) also appeared as the most frequently used with 33 appearances in the description of the repositories, followed by ecology (18) and ecosystem (16) and health (12) and finally biodiversity (7). In this case, the keywords ecosystem and health were intensively represented as opposed to what was observed in the title field filter. After the combination of both sets and elimination of duplicated results 61 repositories were obtained. A final filter was also applied to ensure that some of the relevant repositories in agriculture were not omitted. In this this case, the keyword agriculture (agri*) was searched in the name of the institutions contain the repositories. This filter (filter 4) found 36 repositories whose hosting institutions contained the keyword agriculture. Finally, after the combination of all filters and elimination of duplicates 73 repositories were obtained.

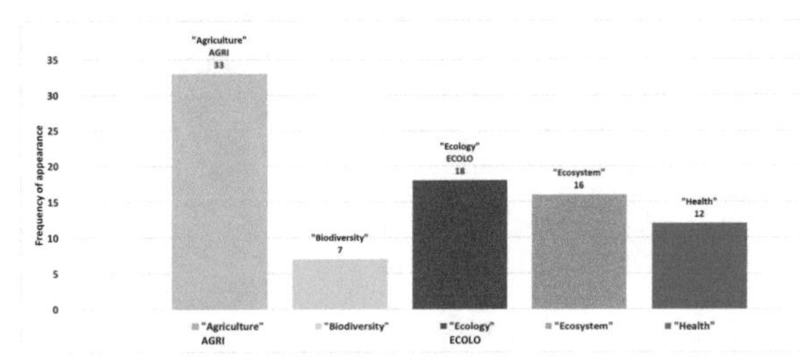

Fig. 2. Number of repositories containing the selected keywords in the description after the application of the first filter

4 Conclusion

The strategy followed in this investigation provided a simplified list of data repositories containing agricultural data. However, the final list should be individually inspected to ensure the validity of the data obtained that will be later used in more comprehensive work. Future work includes the evaluation of the repositories in compliance with the FAIR principles for scientific data management and stewardship.

References

1. van Huylenbroeck, G., Vandermeulen, V., Mettepenningen, E., Verspecht, A.: Multifunctionality of agriculture: a review of definitions, evidence and instruments. Living Rev. Landscape Res. **1**(3), 1–43 (2007)
2. Blesh, J., Hoey, L., Jones, A.D., Friedmann, H., Perfecto, I.: Development pathways toward "zero hunger." World Dev. **118**, 1–14 (2019)
3. Shahmohamadloo, R.S., Febria, C.M., Fraser, E.D., Sibley, P.K.: The sustainable agriculture imperative: a perspective on the need for an agrosystem approach to meet the United Nations sustainable development goals by 2030. Integr. Environ. Assess. Manag. **18**(5), 1199–1205 (2022)
4. Wilkinson, M.D., et al.: The FAIR guiding principles for scientific data management and stewardship. Sci. Data **3**(1), 1–9 (2016)

5. Attard, J., Orlandi, F., Scerri, S., Auer, S.: A systematic review of open government data initiatives. Gov. Inf. Q. **32**(4), 399–418 (2015)
6. Ramachandran, J., Bugbee, K., Murphy, K.: From open data to open science. Earth Space Sci. **8**(5), e2020EA001562 (2021). https://doi.org/10.1029/2020EA001562

Towards a KOS to Manage and Retrieve Legal Data

Bruno Oliveira[1(✉)] and Cristóvão Sousa[1,2]

[1] CIICESI/ESTG, Polytechnic of Porto, rua do Curral, Casa do Curral, 4610-156 Felgueiras, Portugal
bmo@estg.ipp.pt

[2] INESC TEC, Rua Dr. Roberto Frias, 4200-465 Porto, Portugal
cristovao.sousa@inesctec.pt

Abstract. Legislation is a technical domain characterized by highly specialized knowledge forming a large corpus where content is interdependent in nature, but the context is poorly formalized. Typically, the legal domain involves several document types that can be related. Amendments, past judicial interpretations, or new laws can refer to other legal documents to contextualize or support legal formulation. Lengthy and complex texts are frequently unstructured or in some cases semi-structured. Therefore, several problems arise since legal documents, articles, or specific constraints can be cited and referenced differently. Based on legal annotations from a real-world scenario, an architectural approach for modeling a Knowledge Organization System for classifying legal documents and the related legal objects is presented. Data is summarized and classified using a topic modeling approach, with a view toward the improvement of browsing and retrieval of main legal topics and associated terms.

Keywords: Legal Knowledge · Information Retrieval · Ontology · KOS

1 Introduction

In some domains, data is stored in an unstructured way and is composed of a complex corpus difficult to process and reuse, due to the ambiguity of the used terminology and its underlying meaning, which might lead to misinterpretations and confusion. The Legal domain is characterized by highly specialized knowledge reflected in a corpus with lots of particularities, both in terminology and coding techniques of each legal area. Furthermore, the complexity of the field means there are numerous references between documents that in many cases do not follow a standard structure, compromising the entire information retrieval process. Ambiguity is a compromising aspect of text interpretation by machines. Documents have different structures and are composed of several complex terms and arrangements applied in some specific contexts. Additionally, people who work in this domain tend not to recognize ambiguity in some text, since in their experience, those terms, framed with a given context, can only have one reasonable meaning. In fact, the use of information retrieval techniques over legal

A. Rocha et al. (Eds.): WorldCIST 2023, LNNS 800, pp. 75–84, 2024.
https://doi.org/10.1007/978-3-031-45645-9_7

documents is seen with criticism [4]. Theory and interpretation need to be used to identify the appropriate useful meaning of law in context. Meanings and uses of the law are difficult to capture or predict because legal data is continually evolving and context-dependent [4].

In Portugal, there have been several efforts to offer legal-related content digitally. Some initiatives are public like DRE[1] or DGSI[2], others private, like Lexit[3]. This last one furthers digitalization and offers a digital platform to support users by providing useful content for their legal practice activities. Beyond legislation, they offer their expertise by means of annotations, incorporated in the legal codes, which provide guidance to common interpretations of the law. Additionally, they also provide opinion articles and practical cases to add some more context for legal diplomas. Yet, the context is not digitally attached to content, which requires a high level of expertise to infer that. Accordingly, the following research question arises: How to put legal-related content into context, to support users in their digital journey toward searching and retrieving useful legal data for their legal practices? To answer this research question, a Knowledge Organization Systems (KOS) approach was followed and a knowledge model formalized using SKOS W3C recommendation was developed. Given the specificity of the legal domain, a KOS architecture is presented, embodying different technologies and concepts to describe, categorize and contextualize legal documents, metadata, amendments, and annotations in the legal domain, allowing the retrieval of information in a more precise way and adjusted to the users. needs. After the presentation of the related work in Sect. 2, a conceptual model for legal content management is presented in Sect. 3, demonstrating the main requirements to semantically describe a subset of legal domain components. Section 4 presents a case study used to categorize a legal corpus based on the presented conceptual primitives. Finally, the main conclusions and future work directions are presented in Sect. 5.

2 Related Work

Expressing and sharing legal knowledge from legal data is a challenging task. Due to domain specificities, produced documents are expressed in a natural language composed of complex terms and references embodying ambiguity and context-related data. In most cases, it is difficult to search for the right information. Traditional search engine algorithms use specific approaches to rank documents that do not always find the best match for domain-specific scenarios. Search results often return documents based on the search words occurrence and several results are hard to consume and use. For that reason, specific search engines are used, covering the domain's characteristics and improving accuracy and extra functionality not possible with general approaches [14]. For that, the knowledge representation, including semantics, properties, concepts, and their relationships,

[1] https://dre.pt/dre/home.
[2] http://www.dgsi.pt.
[3] https://informador.pt/legislacao/lexit/.

need to be expressed to contextualize data and their use unambiguously and expressively.

Legal documents are lengthy and don't have a fixed structure, syntax, or even interpretation that in some cases (e.g. jurisprudence) can be affected by external factors, including time. In [7], the authors refer to the search process as two steps: 1) searching for the right document and 2) understanding that document. Semantic networks are used to visualize and understand a document's content and support search capabilities to fulfill these needs. An entity recognition task using a NER Tagging tool is used to identify victims, places, or organizations as entities involved in the related legal case to produce nodes for the knowledge graph. To identify words their context and their relationship with associated words, Part-of-speech [12] tagging was used to identify edges (mainly verbs) between entities.

A legal cognitive assistant to analyze digital legal documents is presented in [11]. Machine learning and text mining are applied to several legal documents to create a Legal Question and Answer system. The authors explore Semantic Web technologies transforming lengthy texts into graph-based representations, allowing the creation of knowledge bases and the cross-referencing between related documents to support the extraction of similar legal entities and terms using deep learning techniques. The same problem is addressed in [5], in which authors purpose an ontology to provide reasoning for the relationships between texts. NLP techniques are used to extract textual information, mapping it to criminal law ontology-specific rules to determine inferences.

The approach presented in [3] connects the knowledge coming from different decisions and highlights similarities and differences between them. This research work uses the knowledge extracted from legal documents, metadata, and rules to define a semantic web framework to connect legal documents' raw text to the related semantics. The authors introduce JudO, an OWL2 ontology library of legal knowledge for modeling jurisprudence-related information and to enable reasoning on the resulting knowledge base to support legal argumentation on judicial decisions. JudO provides meaningful legal semantics while retaining a strong connection to source documents (fragments of legal texts). An OWL model to define legal norms is also used in [8] to develop expert systems for semiautomatic drafting, semantic retrieval, and legislation search.

Due to its semantics complexity, Semantic Web techniques are used for modeling legal information and reason about related data. Semantic layers are been widely explored [2] for representing relevant entities, their properties, and their relationship considering the legal domain [18]. Several other research works have been exploring the use of Semantic Web technologies to support several aspects of legal data representation. For example, for automating the semantic mapping between regulatory rules and organizational processes [16], for extracting domain-specific entity-relationships from judgments [17], or for classifying using Graph LSTM [13].

3 Conceptual Model for Legal Content Management

Considering the existing digital platforms and in particular, the Lexit platform, a set o requirements were gathered in the form o competency questions [6,9]. Within ontology engineering, competency questions(CQs) are a common way to specify the frame of context for a conceptualization process. The list of identified CQs are the following:

- Which legal practical acts are/were related to a specific document/content?
- Which is/was the legal doctrine associated to a specific document/content?
- Is/Was there any jurisprudence addressing a specific document/content?
- Is/Was there any provided interpretation for this specific document/content?
- When was this document consolidated?
- When was this document changed?
- When was this document Revoked?
- Which practical acts are/were based on a specific document/content?
- Which document/content are/were addressing a specific topic (legal subject)?

Notice that all the questions were posed either in the present and the past, considering the importance of time when dealing with legal issues. After the competency questions identification, the conceptualisation started, following a collaborative approach.

3.1 Conceptual Model

The "Informador Fiscal" (OIF), is a Portuguese entity that produces content related to several areas of legal knowledge, but with a particular focus on tax law. Currently, the OIF is using the basic elements of *wordpress*[4] to represent, digitally, a knowledge structure that includes: i) legal data (legal diplomas; legal codes and Jurisprudence); ii) cases and lessons learned; iii) articles, and; iv) annotations provided by a group of domain experts about specific aspects of the law. Moreover, a set of *ad-hoc* terms representing a single-person perspective, are being used to classify cases and articles, whereas the legal data has its own specific classification, based on a simple legal code's hierarchy. Yet, the interrelation of the available content is poorly addressed, which makes information recovery difficult and inaccurate. To overcome these issues there was the need to put the content in context, by following a collaborative process of domain conceptualization. The process, depicted in Fig. 1, comprises four main activities, wherein both the domain experts and knowledge engineers participated to jointly create a common view of the main domain entities and how they are related to each other.

Based on the typical content produced by OIF, the *know-how* of the domain experts, and the terminological-based approach, the main concepts were gathered (phase I) and a first draft of the conceptualization was built and revised (phase II). To enhance the data sharing capabilities for the developed model

[4] https://wordpress.org.

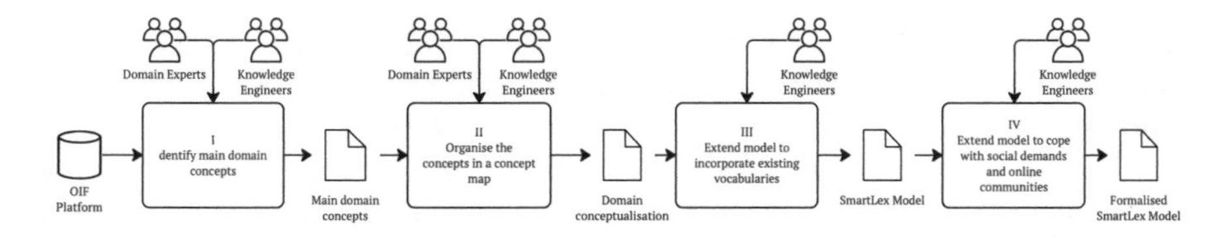

Fig. 1. Conceptualization Process

and towards a compliant FAIR Data model, the conceptualization result was extended to accommodate existing legal vocabularies namely the European Legislation Identifier (ELI)[5] (phase III). Lastly, the model was formalized according to its intention of use (phase IV) in which the domain model was enriched and mapped to *wordpress* main concepts.

Figure 2 (**A**), depicts the domain conceptualization by means of a concept map. Due to space constraints, we focused on tax law and avoided to represented all the identified types of entities. From the concept map we can read the following domain assertions:

- Tax law, which is a branch of the law, containing a list of legal codes and legal diplomas;
- The legal codes are based on specific legal diplomas, which approve the legal code. Additionally, legal codes provide regulation over a specific area of law;
- IRS (among others) is a type of legal code. the LGT, in turn, is a legal diploma related to tax law.
- Typically, a legal diploma provides information about some topics or themes, like Tax Benefits, Income Taxation, among others. These terms were revised and validates by means of NLP techniques (cf. Sect. 4)
- There are several types of legal diplomas, according to the legal action involved. Which, in turn, depends on the entity that produces the legal diploma;
- A Law or a Decree Law are two types of legal acts associated to legal diplomas;
- A legal diploma might be related to other types of content such as: i) practical cases; articles or jurisprudence.

The next step was to semantically enrich the conceptualization model increasing its interoperability capabilities. This was achieved by reusing and integrating the current artifact with other legal vocabularies, in particular the ELI ontology. ELI offers a standardized format to describe legal data and make legislation easy to access, share and reuse.

The schema represented by Fig. 2 (**B**) is the conceptual view on the integration result of the domain (local) concepts and the ELI entities, and to which we call SmartLex Model. The new semantic model has several *namespaces*: i) **sl**, which stands for SmartLex and refers to the domain concepts; ii) **eli**, for the ELI

[5] https://eur-lex.europa.eu/eli-register/about.html.

ontology entities and relations, and iii) **skos**, used to describe the domain knowledge in a formal way. Using the SKOS data model to translate the domain knowledge, it is possible to define each concept as an individual of the skos:Concept class and the skos:narrower and skos:broader properties, to construct the knowledge structure. In the cases we need to define different properties than the ones offered in the core of SKOS, to express extra information, we specialize the SKOS model. Following the recommendations of SKOS primer documentation[10] it is possible to specialize the SKOS model where an application designer is able to create new properties, based on existing vocabularies. For example:

eli:based_on rdfs:subPropertyOf skos:narrower

An example of a concept described using SKOS data model can be:

sl:Annotation rdf:type skos:Concept;
skos:prefLabel "Annotation"@en;
skos:prefLabel "Anotação"@pt;

In Fig. 2 (**B**), due to space constraints, only two sets of typologies are represented as SKOS concepts. Moreover, SKOS relations, such as **skos:narrower** and **skos:broader** were omitted for a better interpretation.

To the baseline domain model, the four classes of ELI ontology were added, namely: LegalResoure; ResourceType; LegalExpression, and LegalFormat. A **LegalResource** can represent a legal act or any component of a legal act, like an article. Legal resources can be linked together using properties defined in the model. A **LegalExpression** is the intellectual realization of a legal resource in the form of a"sequence of signs" (typically alpha-numeric characters in a legal context). **LegalFormat** is a physical embodiment of a legal expression, either on paper or in any electronic format.**ResourceType** is the type of a legal resource (e.g. "Directive", "Règlement grand ducal", "law", "règlement ministériel", "draft proposition", "Parliamentary act", etc.). Typically it corresponds to a local concept scheme. Beyond the main entities, the model was also extended with the ELI relations among concepts forming an enriched domain knowledge graph - the SmartLex Model.

Having this knowledge structure of the domain represented by means of a Smart Model (cf. 2 (B), the content should be putted in context. For this a terminological based approach was followed as described in Sect. 4.

4 Categorizing and Enriching Legal Corpus

Handling legal documents such as articles, jurisprudence, or regimentation among several legislative powers and organisms is not an easy task due to the amount of data that is still growing and the complexity of the legal corpus.

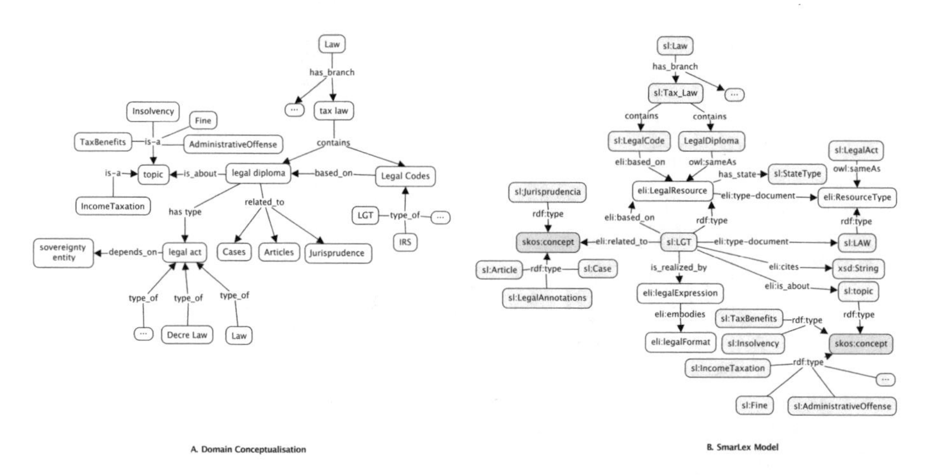

Fig. 2. Domain Conceptualization (**A**) and SmartLex Model (**B**)

Documents are composed of several terms associated with specific contexts and laws, creating ambiguity difficult to handle and index. Since legal documents are verbose and ambiguous, searching legal data can involve challenging procedures to identify the most relevant documents. Keywords or Key-phrases can be helpful because they resume the content to handle. Not only specific words can be identified but also groups of words or phrases can be used to identify if a specific term is addressed in one document. Natural Language Processing techniques need to be used for this task since representing a way to deal with unstructured text.

In our case study, the legal documents are composed of articles, practical cases of law application, jurisprudence, or annotated synthesis about a specific theme. Most documents include several references to sections or specific articles for specific codes. For that reason, performing a dummy search using a specific law code can return hundreds of documents, which may not be aligned with the purpose and search context. Moreover, based on the specific terms used, suggestions can be used to provide or suggest more context and provide new capabilities (for example, facet searches) to view and filter relevant documents.

The Fig. 3 presents the main phases involving the consumption of raw data and its processing to associate documents with topics. A Topic Modeling technique using Top2Vec model [1] is used and represents an unsupervised machine learning technique capable of detecting words and phrase patterns, and automatically generating clusters for grouping words into groups to characterize a set of documents. However, for generating topics, raw data need to be pre-processed. The Keyword Extraction Pipeline described in Fig. 1 starts by removing all HTML tags and special characters (such as "n") from each document's content. Next, a Lemmatization technique is applied to each word to reduce the word to its root, removing all inflections and identifying the lemma. This reduction results in a word that exists in the grammar. The grammatical class of the word

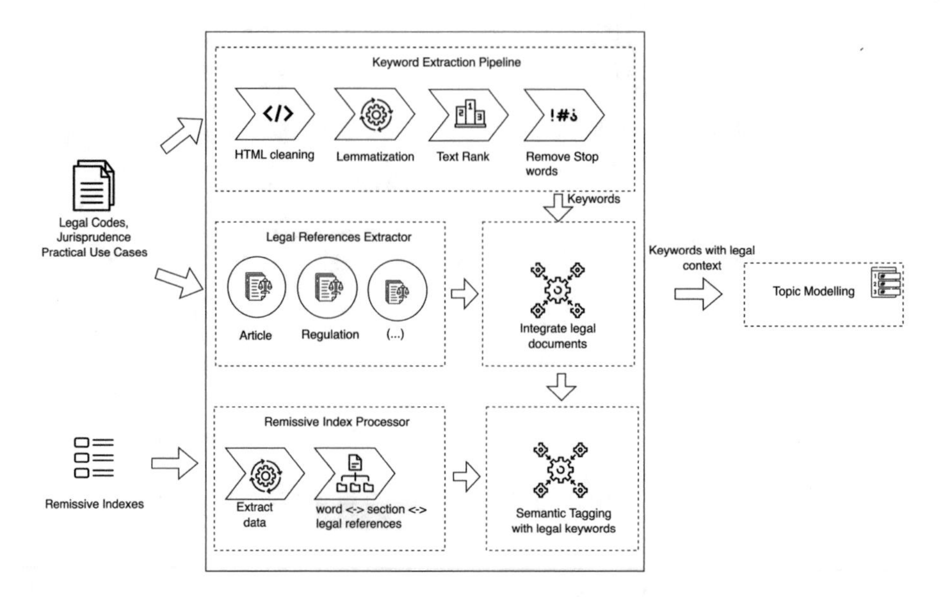

Fig. 3. Keyword extraction Pipeline Conceptual representation

will be considered to perform the reduction. We used spaCy[6] library to complete this task. After replacing words by their lemmas, a graph-based method for keyword extraction was used: Text-Rank [15]. The algorithm is applied on each input document separately, returning the words and multi-words based on the highest score. The final step for this pipeline removes stop words from single-word keywords, excluding possible nonrelevant terms for the selected keywords.

A Legal References Extractor is also applied to source documents. Based on specific regular expressing, law references are extracted. For example: "Art.–o 32.–o-A do EBF" is extracted and mapped to existing terms in a dictionary to identify the complete reference to the target legal document. The cross-reference indexes are also used for semantic tagging enrichment based on the law references. The cross-referenced index lists the terms and topics that are covered in a document, along with the pages on which they appear. These cross-reference indexes are manually created by domain specialists and serve as input to this process phase. That way, specific terms can be included (if not already by previous steps) based on their relationship with the section and the document they are related. After this semantic tagging procedure, the Topic Modelling component is performed to associate documents to keywords to form these unnamed topics. These topics are presented to domain experts that need to name them according to a set of pre-defined themes already identified by experts. The result of this interaction can be used in form of training models for machine learning techniques that can be used to regenerate and classify topics based on previous experts' interactions.

[6] https://spacy.io.

Following an NLP[7] approach we are able to extract some knowledge from the corpus, allowing us to populate the SmartLex model in an automatic way. This includes i) the automatic classification of legal diplomas according to their type, that is, legal act; ii) instantiating the different topics (legal subject), obtained from the analysis of cross-reference index; iii) tagging content as a result of a topic modeling approach, right after the results of topic modeling being validated according to existing cross-reference topic list and the *ad-hoc* categories baseline; iv) tagging specific articles of legal diplomas, based on the identified topics (legal subjects).

In practical terms, the result of this enrichment phase is the creation of a metadata layer over the existing content, providing the basis to discover new knowledge by means of inference mechanisms or advanced queries. A simple example consists of searching and retrieving all the content that shares the same topics (legal subject), allowing one to infer what practical cases are addressing a specific diploma that was retrieved by a user. Facet searching and advanced filtering are also new features that emerged from the KOS approach.

5 Conclusions and Future Work

In this paper, we explore a KOS architecture for dealing with the legal domain specificity. With this approach, existent documents can be described, categorized, and contextualized using keywords, legal metadata, and semantic tags allowing the information contextualization in a more useful way. A domain-specific conceptual model that provides the legal context and the possible exploration paths based on a specific term, law document, or legal framework. A case study based on unstructured data was explored and consumed by data pipelines that recurring to NLP techniques clean and normalize data to generate keywords based on regular words and legal documents references, being enriched with domain terms previously identified by cross-reference indexes created by domain experts. These processing stages produce keywords for each document that are used for clustering documents in topics, afterwards, it will be named by domain experts using an annotation tool. In future work, these inputs will be used to create training models that can help to reorganize documents in the respective topics as new legal documents are created and consumed by the system. Additionally, this knowledge representation will also be used to support a domain-specific search engine for a website, adapted to the particular legal domain needs and data semantic characteristics.

Acknowledgement. This work was supported by the Northern Regional Operational Program, Por- tugal 2020 and European Union, through European Regional Development Fund (ERDF) in the scope of project number 047223 - 17/SI/2019.

[7] Natural Language Processing.

References

1. Angelov, D.: Top2Vec: distributed representations of topics. Comput. Lang. (2020)
2. Casanovas, P., Palmirani, M., Peroni, S., Engers, T.V., Vitali, F.: Semantic web for the legal domain: the next step. Semantic Web **7**, 213–227 (3 2016)
3. Ceci, M., Gangemi, A.: An owl ontology library representing judicial interpretations. Semantic Web **7**, 229–253 (2016)
4. Devins, C., Felin, T., Kauffman, S., Koppl, R.: The law and big data. Cornell J. Law Public Policy **27**(2), 357–413 (2017)
5. Fawei, B., Pan, J.Z., Kollingbaum, M., Wyner, A.Z.: A semi-automated ontology construction for legal question answering. N. Gener. Comput. **37**, 453–478 (2019)
6. Gangemi, A., Presutti, V.: Ontology design patterns. In: Staab, S., Studer, R. (eds.) Handbook on Ontologies. IHIS, pp. 221–243. Springer, Heidelberg (2009). https://doi.org/10.1007/978-3-540-92673-3_10
7. Giri, R., Porwal, Y., Shukla, V., Chadha, P., Kaushal, R.: Approaches for information retrieval in legal documents. In: 2017 10th International Conference on Contemporary Computing, IC3 2017, Jan 2018, pp. 1–6 (2018)
8. Gostojić, S., Milosavljević, B., Konjović, Z.: Ontological model of legal norms for creating and using legislation. Comput. Sci. Inf. Syst. **10**, 151–171 (2013)
9. Grüninger, M., Fox, M.S.: The role of competency questions in enterprise engineering. In: Rolstadås, A. (ed.) Benchmarking — Theory and Practice. IAICT, pp. 22–31. Springer, Boston, MA (1995). https://doi.org/10.1007/978-0-387-34847-6_3
10. Isaac, A., Summers, E.: SKOS Simple Knowledge Organization System Primer, https://www.w3.org/TR/skos-primer/
11. Joshi, K.P., Gupta, A., Mittal, S., Pearce, C., Joshi, A., Finin, T.: ALDA: cognitive assistant for legal document analytics. In: AAAI Fall Symposium - Technical Report , vol. FS-16-01, pp. 149–152 (2016)
12. Koniaris, M., Anagnostopoulos, I., Vassiliou, Y.: Evaluation of diversification techniques for legal information retrieval. Algorithms **10**, 1–24 (2017)
13. Li, G., Wang, Z., Ma, Y.: Combining domain knowledge extraction with graph long short-term memory for learning classification of Chinese legal documents. IEEE Access **7**, 139316–139627 (2019)
14. McCallum, A., Nigam, K., Rennie, J., Seymore, K.: Building domain-specific search engines with machine learning techniques. In: Proceedings of AAAI 1999 Spring Symposium on Intelligent Agents in Cyberspace (1999)
15. Mihalcea, R., Tarau, P.: TextRank: bringing order into texts. In: Proceedings of EMNLP, vol. 2004, pp. 404–411. Association for Computational Linguistics (2004)
16. Sapkota, K., Aldea, A., Younas, M., Duce, D.A., Banares-Alcantara, R.: Automating the semantic mapping between regulatory guidelines and organizational processes. SOCA **10**, 365–389 (12 2016)
17. Thomas, A., Sangeetha, S.: A legal case ontology for extracting domain-specific entity-relationships from e-judgments (2017)
18. Winkels, R., Boer, A., Maat, E.D., Engers, T.V., Breebaart, M., Melger, H.: Constructing a semantic network for legal content. In: Belgian/Netherlands Artificial Intelligence Conference, pp. 405–406 (2005)

Supervised Machine Learning Based Anomaly Detection in Online Social Networks

Chi-Leng Che[1], Ting-Kai Hwang[2](✉), and Yung-Ming Li[1]

[1] Institute of Information Management, National Yang Ming Chiao Tung University, Hsinchu, Taiwan
yml@mail.nctu.edu.tw

[2] Department of Journalism, Ming Chuan University, Taoyuan, Taiwan
tkhwang@mail.mcu.edu.tw

Abstract. With the rapid development of online social networks (OSNs), a huge number of information provided by some entities around the world are well dispersed in OSNs every day. Most of those are useful but not all as anomalous entities utilize anomaly users to spread malicious content (like spam or rumors to achieve their pecuniary or political aims). In this paper, we propose a mechanism to detect such anomaly users according to the user profile and tweet content of each user. We design several features related to near-duplicate content (including lexical similarity and semantic similarity) to enhance the precision of detecting anomaly users. Utilizing the data by public honeypot dataset, the proposed approach deals with supervised learning approach to carry out the detection task.

Keywords: Online Social Network · Anomaly Detection · Near-Duplicate Computing · Supervised Machine Learning

1 Introduction

Due to the rapid explosion and popular use of online social networks (OSNs), everyone can share their personal statue with their friends and family, and publicly gives their point of views about a piece of news, a dish, or a person. To some extent, OSNs have morphed from a personal micro-blogging site to an information publishing platform. It becomes an important social tool for business promotion, customer service, political campaigning, and emergency communication.

However, malicious contents like spam and distribute malware and rumors threaten normal users and influence user experiment on the platform, such as Twitter, a large social network site on the world. A study said that up to 15% user on Twitter are bots [10] which shows the anomaly users still threaten on OSNs. Two existing policies on Twitter to recognize anomaly users, which are automatic anomaly user detection provided by Twitter and '@spam' function for user to report suspicious users. However, the permission of promoting product on Twitter makes figuring out malicious promoters be a hard nut to crack. In addition, there is no existing system for detecting rumors or report rumors in OSNs. Therefore, it is necessary to build a model for detecting such anomaly users in OSNs.

A. Rocha et al. (Eds.): WorldCIST 2023, LNNS 800, pp. 85–91, 2024.
https://doi.org/10.1007/978-3-031-45645-9_8

In this research, we focus on detecting anomaly users including spammers, content polluters, malicious promoters, and rumors. The proposed approach utilizes user information from Twitter, including user profile and posted tweets to build different fields and dimensions for system to distinguish anomaly users from benign users. The objection of this research is to provide more latent but useful factors from user tweets as a new measure to determine a user is good or not, which is according to plenty of previous studies by utilizing user profile features. In addition, user post attribute could also provide important message for defending anomaly users, like textual and semantic content which represent as an additional tool for measuring abnormality. Moreover, introduce innovatively new and useful user post features by utilizing near-duplicate computing which is good effectiveness of e-mail spam detection.

2 Literature Survey

2.1 User Profile Attribute Approach

Most of earlier studies leveraged basic profile information as features to construct detection model because such features are convenient to collect by using Twitter API. They figured out the difference of 'tweet count', 'number of friends/followers' and 'ratio of followers over friends' through statistical observations between human and non-human users. Some other studies [6–8] suggested more user attribute features which are discriminatory, such as frequency of tweeting, reputation of user, etc. Even though another research [11] showed anomaly user could evade such detection easily, such features performed stable, scalable and real-time detectability in large and modern scale dataset [12].

2.2 User Post Attribute Approach

Tweet can contain textual and non-textual content. Non-textual content includes URL, retweet, mention and hashtag which are defined as functional elements in this research. For functional elements analysis, many researches utilize whether the URLs contained in tweet are malicious or not to define such tweet, because malicious URL could infect victims. Das et al. [5] focused on detecting spam by leveraging embedded URLs in tweets and features of which are URL-oriented. In spite of the powerful discriminability of such models, which are only able to detect spam with URL. Chen et al. [3] show the external URL ratio of nonhuman (bot/cyborg) accounts are much higher than human. Previous research showed that anomaly users often mention other anomaly users. In addition, Twitter users can use hashtag function ('#topic') to tag a particular topic in a tweet [14].

For text content analysis, Cresci et al. [4] mentioned some websites offer several guideline for detecting fake users on OSNs which is related to tweeting near-duplicate post. Sandulescu & Ester [13] shows that they utilize semantic similarity, vectorial model and topic model similarity as an evaluation tool for classifying the relation between threshold of similarity and review spam detection performance. On the other hand, Ahmed [1] built the semantic metrics to detect opinion spam and fake news, which are separated into three components: word, text and order similarity and finally combine them as semantic similarity score between documents. Even though it cannot detect unique fake but it can help detect near-duplicate deceptive content which is similar to the other one labeled to spam comment.

3 The System Framework

In this section, the aim of detection mechanism is to detect anomaly users in OSNs, by crawling users' data and extract users profile information as well as the tweets generated by each user. In addition to leveraging functional elements analysis and textual near-duplicate analysis of textual similarity and semantic similarity as new features for enhancing accuracy of classification, the mechanism aims to avoid utilizing such tools as malicious URLs blacklist filter and spam-words dictionary that limit consequent and decline robustness of system. The flow of detection system framework is illustrated as in Fig. 1.

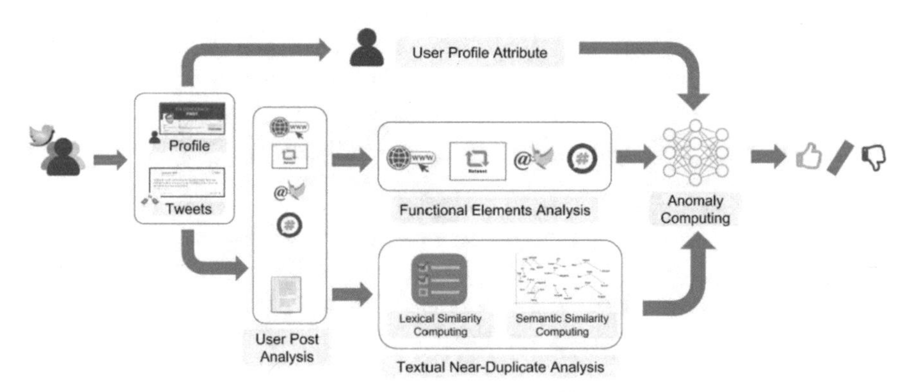

Fig. 1. Flow of Anomaly User Detection System

At the beginning, honeypot dataset from Lee et al. [9], which includes user information and tweet content on Twitter is pulled into our system for further analysis. The property of data is divided into two categories: profile attributes, tweet post based attributes. Profile attributes are related to basic information about a user on Twitter, such as age of account, number of tweet, etc. and it also includes social relationship features which record the social network of a user. For tweet post based attributes, they include the features which transform tweet content to be measurable.

The non-textual elements within a tweet include four functional elements: URL, retweet ('RT @' followed by a username), mention (symbol '@' followed by a username) and hashtag (words or phrases prefixed with a "#" symbol), which could be treated as spreading non-textual signals. Therefore, we will extract such elements to conduct specific experiments, which analyze the usage frequency about each type of functional elements, thereby being an evidence for the difference between benign users and anomaly users.

Based on the textual part of tweet, analyzing the probability as a duplication by lexical or semantic field, which is a tool to detect a promotional content and spam. Finally, we integrate all attributes which are statistically significant to explain detection model and use logistic regression and other machine learning method to classify benign users and anomaly users. Our system framework is shown in Fig. 2.

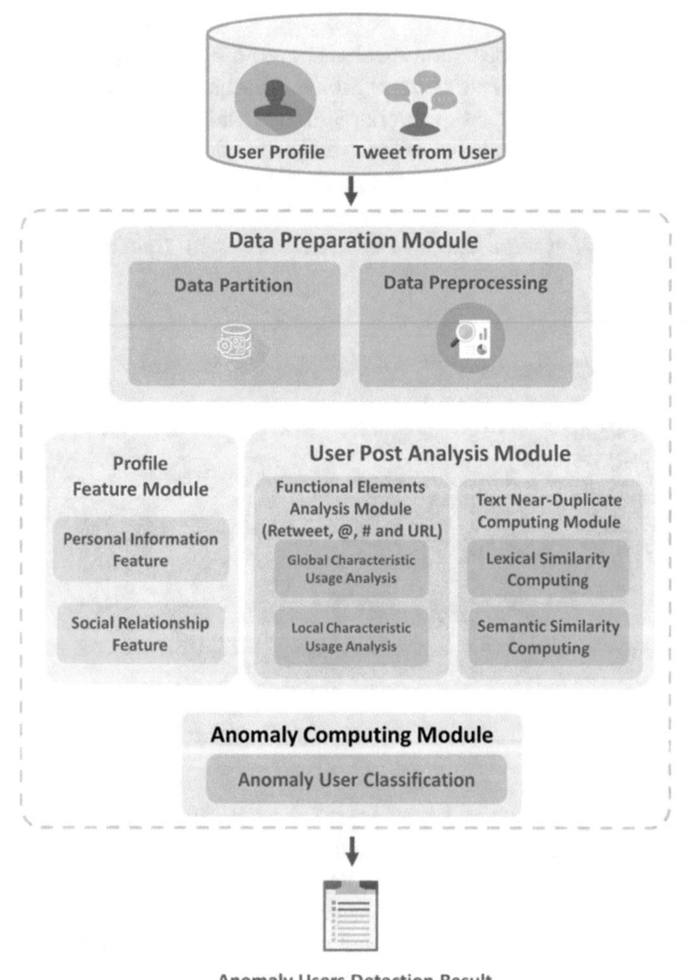

Fig. 2. System Framework

There are four main modules to build up our mechanism as follows.

1. Data Preparation Module: In this module, there are two procedures to prepare data for further analysis. The procedures include 'Data Partition and 'Data Preprocessing', which could segment date according to its attributes and remove the data noise.

2. Profile Attributes Module: This module aims to analyze basic information and social network of a user. Since lots of previous works on spammer and content polluter detection has proven such features like age of account, number of followers could be a measure for detecting anomaly users.
3. User Post Analysis Module: The property of a tweet can be separated into textual content and non-textual content (includes URL, retweet, mention and hashtag which are defined as functional elements in previous section). Because such contents reflect what kind of information the object user wants to spread to their audiences who are the followers of the object or the object is following, the tweet posted by an anomaly user is likely repeated and contains similar content due to malicious promotion. Based on such concept, considering the use frequency of specific functional elements and the duplicate level of textual content (includes the lexical similarity and the semantic similarity) could help us to figure out anomaly users.
4. Anomaly Computing Module: After collecting conductive features from previous modules, we utilize statistic models (e.g. logistic regression) and machine learning algorithms (e.g. support vector machine, SVM) to conduct an effective detection on anomaly user.

4 Analysis of Testing Results

4.1 Dataset

In this research, the crucial part affecting the accuracy of this model is the dataset which is a collection twitter users' data with label. As personal equation involves time-consuming of manual classification, we use a public social honeypot dataset [9]. In the dataset, there are 60 honeypot accounts to tempt content polluters by posting random sample tweets from Twitter public timeline and are designed to engage in none of the activities of legitimate users. It collected 22,223 spammers and 19,276 legitimate users profile data and their tweets.

4.2 Benchmark Approaches Comparison

We compare the performance of our proposed anomaly user detection model with two benchmark approaches (Table 1), which combined user profile attribute, functional elements analysis and tweet content analysis.

Table 1. Two benchmark approaches

Benchmark Approach	Modules
Approach 1 by Benevenuto et al. [2]	User profile attribute, Spam words detection
Approach 2 by El-Mawass & Alaboodi [6]	User profile attribute, Functional elements analysis, Tweet-duplicate computing but did not compute the lexical and semantic similarity between tweets

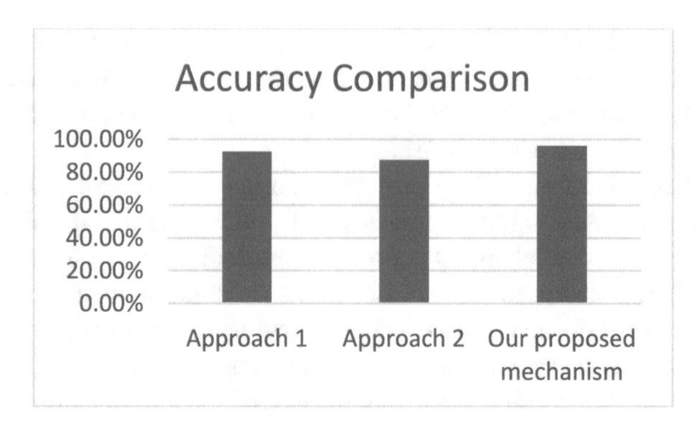

Fig. 3. Accuracy Comparison Between Benchmarks and Proposed Mechanism

According to Fig. 3, the comparison result shows that our proposed mechanism is more accurate than other two benchmarks. In spite of the same modules in these three mechanisms, there are some different feature selection contained in the mechanisms. For example, benchmark 1 did not consider the usage condition of different functional elements and only detected spam words in tweet content analysis. Although benchmark 2 has a similar feature selection with our approach, it did simple evaluation on replicate tweet computing, which did not consider semantic duplicate.

5 Conclusions

In this research, the approach by leveraging statistic model and machine learning approaches are adopted to detect anomaly users, including spammers, malicious promoters, content polluters, and even rumors. After we collect profile information and tweet data, the proposed mechanism utilizes the profile information to construct feature which is able to distinguish anomaly user and observe their statistic characters. On the other hand, we separate functional elements and textual content for near-duplication analysis as the foundation of determining anomaly degree of user.

In the functional element analysis module, we focus on the use behavior of those elements. According to the different usage of four types of elements for anomaly users, we design tailored features which represent the use condition of users and utilize statistical test to verify discriminate capacity of those features. For the analysis of textual content, we utilize the concept of near-duplicated detection, the lexical similarity, and semantic similarity of textual content as a tool to determine how similar is between tweets from one user and other tweets from other users, even the self-duplication. It can also seem like a tool for evaluating the anomaly condition of one user. Eventually, we integrate the features to build the final classification model. To verify our model, we use the initial label of dataset to compute the benchmark index of model.

References

1. Ahmed, H.: Detecting opinion spam and fake news using n-gram analysis and semantic similarity, Doctoral dissertation (2017)
2. Benevenuto, F., Magno, G., Rodrigues, T., Almeida, V.: Detecting spammers on twitter. In: Collaboration, electronic messaging, anti-abuse and spam conference (CEAS), vol. 6, p. 12 (2010)
3. Chen, C., Zhang, J., Chen, X., Xiang, Y., Zhou, W.: 6 million spam tweets: a large ground truth for timely Twitter spam detection. In: 2015 IEEE International Conference on Communications (ICC), pp. 7065–7070 (2015)
4. Cresci, S., Di Pietro, R., Petrocchi, M., Spognardi, A., Tesconi, M.: Fame for sale: efficient detection of fake Twitter followers. Decis. Support. Syst. **80**, 56–71 (2015)
5. Das, L., Ahuja, L., Pandey, A.: Analysis of twitter spam detection using machine learning approach. In: 2022 3rd International Conference on Intelligent Engineering and Management (ICIEM), pp. 764–769 (2022)
6. El-Mawass, N., Alaboodi, S.: Detecting Arabic spammers and content polluters on Twitter. In: 2016 Sixth International Conference on Digital Information Processing and Communications (ICDIPC), pp. 53–58 (2016)
7. Gilmary, R., Venkatesan, A., Vaiyapuri, G.: Discovering social bots on Twitter: a thematic review. Int. J. Internet Technol. Secured Trans. **11**(4), 369–395 (2021)
8. Karakaşlı, M.S., Aydin, M.A., Yarkan, S., Boyaci, A.: Dynamic feature selection for spam detection in Twitter. In: International Telecommunications Conference, pp. 239–250. Springer, Singapore (2019). https://doi.org/10.1007/978-981-13-0408-8_20
9. Lee, K., Eoff, B., Caverlee, J.: Seven months with the devils: a long-term study of content polluters on twitter. In: Proceedings of the International AAAI Conference on Web and Social Media, vol. 5, No. 1, pp. 185–192 (2011)
10. Newberg, M.: Nearly 48 million Twitter accounts could be bots, says study (2017). https://www.cnbc.com/2017/03/10/nearly-48-million-twitter-accounts-could-be-bots-says-study.html. Accessed 03 Nov 2022
11. Rao, S., Verma, A.K., Bhatia, T.: A review on social spam detection: challenges, open issues, and future directions. Expert Syst. Appl. **186**, 115742 (2021)
12. Samper-Escalante, L.D., Loyola-González, O., Monroy, R., Medina-Pérez, M.A.: Bot datasets on twitter: analysis and challenges. Appl. Sci. **11**(9), 4105 (2021)
13. Sandulescu, V., Ester, M.: Detecting singleton review spammers using semantic similarity. In: Proceedings of the 24th International Conference on World Wide Web, pp. 971–976 (2015)
14. Sedhai, S., Sun, A.: Hspam14: a collection of 14 million tweets for hashtag-oriented spam research. In: Proceedings of the 38th International ACM SIGIR Conference on Research and Development in Information Retrieval, pp. 223–232 (2015)

Leadership Effectiveness in Public Administration Remote Workers

Maria José Sousa[1](✉), Miguel Sousa[2], and Álvaro Rocha[3]

[1] Instituto Universitário de Lisboa, Business Research Unit, Lisbon, Portugal
maria.jose.sousa@iscte-iul.pt
[2] University of Essex, Colchester, England
[3] ISEG, Universidade de Lisboa, Lisbon, Portugal

Abstract. The main goal of this research is to identify the leadership styles that can be more effective in leading remote workers from Public Organizations. The research question can have a significant meaning because leadership is a significant issue for the effectiveness of the organizations, and in the current Pandemic situation, it has assumed even more importance. In this context, the research question is, "Which are the leadership styles that can unleash the potential of remote workers in Portuguese Public Organizations?" The methodology used was quantitatively based on a survey applied to a sample of 109 Public Organizations remote workers, and the results show that the leadership characteristics identified were grouped in sets – forming 5 leadership styles: a) Challenger – Transactional; b) Communicator – Transformational; c) Visionary – Democratic; d) Developer – Servant; and e) Innovator – Charismatic. These leadership styles emerge not separately but as a mix.

Keywords: Leadership · Leadership Styles · Remote Work · Public Organizations

1 Introduction

Leadership is a recurring theme in an academic and professional context, being considered by management literature a success factor in organizations, and that specific leadership style can lead to better performance and more innovative organizations. According to Levin, there are three major leadership styles (Lewin et al., 1939): "a) autocratic leaders make decisions without consulting their team members. This can be appropriate when decisions need to be made quickly, when there is no need for team participation, and when team agreement is not necessary for an outcome. However, this style can lead to high levels of absenteeism and staff turnover; b) democratic leaders make the final decision, but they include team members in the decision-making process". The third type of leader encourages creativity, and employees are often highly engaged in new activities. As a result, team members have high levels of job satisfaction and high productivity, develop their knowledge, their skills, and become self-motivated to do their work effectively in a remote work situation. This article intends to identify the leadership styles present in the contemporary Public Organizations in Portugal in the context of remote work, due to the social isolation provoked by the worldwide COVID-19 pandemic.

A. Rocha et al. (Eds.): WorldCIST 2023, LNNS 800, pp. 92–109, 2024.
https://doi.org/10.1007/978-3-031-45645-9_9

The research is important to the journal as it discusses a relevant issue that have emerged with the current situation caused by the Pandemic, pushing Public Administrations to implement new organizational practices related to remote work, which has been adopted by the Public Servants. In this context, the questions that Public Administrations face is how to lead a workforce that is disperse and using digital technologies to do their activities and achieve the defined goals. Moreover, the research has several managerial implications first the link between leadership, and remote work discussing the possibilities of leading a workforce that is doing digital work remotely; and, to give directions for Public Organizations that can be applied in practice, enlightening regarding different types of leaders and associated competencies to face that specific context. It also gives insights to theory, as it encompasses the elaboration of a literature review on themes leadership theories and also on remote work, a concept that is emerging in research and also at organizational level, giving this work a dimension of originality.

The article is structured as followed: first, the theoretical framework on leadership and remote work, second the methodology and results, and third the discussion and conclusions.

2 Theoretical Framework on Leadership

2.1 An Overview of the Leadership Concept

Through time many ideas of leadership have arisen to explain what leadership is. According to Avolio says that leadership is like a systematic relationship where no leader leads without followers, in this sense, for leadership effectiveness, the focus should be less on what the leader does and more on what the followers do or think. This one thinking is followed by situational and contingency theories, where we find the leadership most focused on the followers and the context in which it occurs. Regarding the theory of Bass's transformational leadership, one of the ways to distinguish management from leadership is in transactional and transformational leadership, respectively (Collins and Holton, 2004). According to Fiedler, leadership processes are highly complex, where leadership is an interaction between the leader and the leadership situation that translates into practice. "We cannot make leaders more intelligent or more creative, but we can design situations that allow leaders to use their intellectual abilities, expertise, and experience more effectively" (Fiedler, 1996, 249). According to the theory of transformational leadership, we find management associated with transactional leadership and leadership associated with transformational leadership.

Since 1985, new practices and leadership theories had emerged: Transformational Leadership (Bass, 1985), Team Leadership (Hackman and Walton, 1986; Larson and LaFasto, 1989), 360-degree feedback (Lepsinger & Lucia, 1997), and on-the-job experiences (McCauley & Brutus, 1998). In this leadership theories it is possible to verify that the learning methods have become dynamic, allowing to learn in more real and challenging scenarios. The focus of leadership development is different, as some organizations have integrated leadership development interventions with strategic and human resources objectives. Leadership development represents an important human resource development activity in organizations. In addition, given the dynamic and complex context in which organizations operate, these activities are considered vital to strengthening leadership capacity and the 'leadership pipeline'.

In short, defining leadership is a difficult task, hence the need for the various concepts found in the literature. Pfeffer (1997) considered the ambiguity of leadership in the face of difficulties in define and measure the concept. In a simplistic way, it can be said that leadership results from the reciprocal interaction between the leader and his followers.

2.2 Leadership Theories

In the last two decades, several theories on leadership have been established. Leadership theories aim to understand why confident leaders are successful when others are not and try to offer solutions for different scenarios. Following the initial theories of "Great Man," which originated in the 19th century, work on leadership has expanded significantly. Neo-charismatic ideas gained the most coverage from scholars between 2000 and 2012, with 294 publications (Dinh, Lord, Gardner, Meuser, Liden, & Hu, 2014). Nevertheless, the question still arises: what are the ingredients that are associated with a great leader? Bellow, we will share the main types of leadership theories:

The Great Man Theory and Trait Theory are based on the assumption that leaders were born with specific natural abilities and heroic traits that have influenced others to follow them. Those theories postulate that successful leaders are born in the position of leadership with some attributes that will make them hugely successful. Beginning throughout the first half of the twentieth century, both of these concepts became very similar as they were directed at defining individual attributes (e.g., age, height, and others), behavioral attributes (e.g., self-confidence, social competence, and others) and skills (e.g., verbal and writing fluency, and others) that can differentiate successful from unsuccessful leaders. Many traits have been defined, but no particular group of traits has been established as the standard for all circumstances. Stogdill (1948), analyzed 124 experiments and found that individuals did not automatically become leaders when they took advantage of a series of characteristics. Nonetheless, belief in leadership qualities persists to this day. Kouzes and Posner examined more than 1,500 executives (1988), investigated more than 1,500 managers and discovered that the four key traits of leadership success were: integrity, intuition, positive capacity, and competency.

The inability to define standard leadership characteristics prompted scholars to seek to discover what a leader is doing rather than what a leader in – the behavioral theories frame this. Such experiments seek to establish whether successful leaders differ from unsuccessful ones in their actions and are focused on the belief that leaders are created, not born. By shifting the analysis of leadership practices, this method broadened the horizons of awareness by examining leaders 'attitudes about followers in different contexts (Northouse, 2016). It analyzed not just how the leader acted against supporters, but also how this interacted with effectiveness. The behavioral hypotheses separated the task-oriented representatives, and those concerned with individuals.

Next, scholars started looking at theoretical and behavioral factors that had an impact on the success of leadership. In this sense, the theories of contingency suggest that there is no universal method of guiding, suggesting that leaders can evaluate the circumstance in which they work and adapt their actions to improve leadership effectiveness.

These theories argue that certain factors, such as collective characteristics and corporate climate, should be taken into account in order to recognize leadership. Significant situational factors are the followers 'characteristics, job setting characteristics and followers' activities, and the external situations.

Transactional or exchange theories rely on results and assess performance according to the system of reward and punishment in that organization. Throughout the quest to discover the elements that would contribute to optimum results, interactions that take place between leaders and followers were studied. Transactional leaders concentrate on supervisory positions, provide institutional control, elicit optimal success from followers by extrinsic encouragement, be rigid with policies, and establish a system of mutual support where individuals and organizational objectives are matched. Transactional theories promote a mutually advantageous partnership between leaders and followers, while people pursue pleasurable interactions and resist distasteful circumstances. The leader-member theory of exchange (LMX) (Dansereau, Graen, & Haga, 1975) is a transactional leadership method that directed researchers to the gaps that might occur between the leader and the followers. The earliest experiments were aimed at studying the essence of the vertical relations between the leader and each follower (vertical dyad linkage) (Northouse, 2016). Later research investigated the consistency of interactions between leading participants and the degree to which such interactions are correlated with excellent results for followers, parties, representatives, and the organization (Graen & Uhl-Bien, 1995). Transactional leadership factors, according to Bass (1985), are contingent reward and manage by exception. It implies the leader receives an incentive in which the follower retribute with an initiative, and the latter is correlated with performance indicators to measure success.

The transformational leadership started with Burns (1978), who studied the followers 'desires and motivations in order to fulfill leadership aims. Transformational leadership is a process that facilitates a link between leaders and followers. Thus the degree of engagement and morality in both (Northouse, 2016); Bass (1985) established a transformational leadership structure focused on four transformative leadership qualities: idealized influence, inspirational motivation, intellectual stimulation, and individualized consideration. Idealized influence means leaders are influencing the actions and dedication they want followers to imitate. Inspirational inspiration is linked to the sharing of a common goal and to encouraging followers by the clear presentation of essential intentions (Afsar et al., 2014). Intellectual stimulation is focused on inspiring followers to put their imagination, convictions, and ideas and work together to overcome problems. Finally, individualized consideration is when the leader listens carefully to the follower's desires and takes into account their increasing expectations.

The charismatic leadership theory explains the beneficial influence of leader attractiveness in follower excellent results because they become a good role model, highly trained, persuasive concepts with a spiritual overtone, have high aspirations for themselves and followers and are willing to encourage inspiration by becoming part of a more significant cause (Bass, 1985; Burns, 1978; Home, 1977; Weber, 1947). A meta-analytic analysis concluded that charismatic behaviors were related to the success of the leader and followers (DeRue, Nahrgang, Wellman, & Humphrey, 2011; Judge & Piccolo,

2004). Charismatic leaders express hope, passion, and trust, inspire followers to accomplish objectives and seek the leader's direction and send out an optimistic message to followers (Conger & Kanungo, 1998; Shamir, House, & Arthur, 1993). The charismatic leadership of the last 25 years received a great deal of consideration from the researchers. Nevertheless, not all leading scholars share the present interest in charisma because it may show a "dark side" (Hogan, Raskin, & Fazzini, 1990). Yukl (1999) concludes that charismatic leadership is not necessarily attractive since it is incompatible with mutual leadership and empowerment. Howell and Avolio (1992) note that unethical charismatic leaders operate by the coercion of followers to achieve personal objectives. Deluga (2001) provides a distinction between charismatic socialized leaders and responsive charismatic leaders. The former appears to be altruistic leaders who match their mission with the needs of the followers, and the latter appears to be exploitative leaders who use influence. Evidence has found that inspirational leaders influence their followers through the recognition and internalization of cognitive mechanisms (Gardner & Avolio, 1998; Howell, 1988). Still, Bass (1985) indicates the presence of intense emotional complexities that underlie the cycle of leadership.

Servant leadership is founded on the idea that leaders will be centered on building a connection with followers, empathizing their desires, taking care of and encouraging them, and providing a fertile environment for followers to grow and succeed (Greenleaf, 1970). Servant leadership combines self-interest with service to others. The model of servant leadership has three elements: antecedent conditions, servant leaders' behaviors, and outcomes (Cai et al., 2018; Wang et al., 2019). The behaviors of leaders that encourage servant leadership are: conceptualizing, emotional healing, putting followers first, helping followers grow and succeed, behaving ethically, empowering, and creating value for the community. These behaviors are likely to affect the individual, organizational, and societal levels (Northouse, 2016).

The contingency methods that aim to resolve shortcomings found with previous theories can be recognized in leadership theories (Kets de Vries, et al., 2004). The style of leadership is dependent in a situation according to these approaches (Den Hartog & Koopman, 2001). In the same line, Chemers (2000) considers that contingency theories understand the dynamic nature of leadership in the same line of thinking, as the leader's attributes and actions are analyzed in the light of situational parameters.

New theories were opened with the advancement of contingency methods in order to reconcile the contradictions of behavioral approaches. It appears that situational variables moderate the efficacy of different leadership behaviors through Fiedler's theories (1967) and House et al. (1971), leading some scholars to conclude that some competencies are universal (Bennis, 1999). The main models are as follows: - the Model Fiedler's Contingency (1967, 1996), the Hersey and Blanchard Situational Model (Hersey & Blanchard, 1974), the Vroom and Yetton Normative Decision-Making Model (Vroom & Yetton, 1973) and the Theory of Paths to House Objectives (House, et al. 1971; House, 1996; House et al. 1997). It appears that the Fiedler Contingency Model is based on an interaction of the leader's characteristics with the situational parameters by doing a brief study of the models. The inductive approach is used for this model and the criticism of it relates to the fact that the leader does not choose to be centered on either the mission or the relationship in the same situation (Chemers, 2000).

Because of the lack of theoretical base, the 1967 Fiedler model was criticized, giving rise to new developments. Fiedler and Garcia developed the theory of cognitive resources in 1987, where the success of the group depends on the interaction of two characteristics - intelligence and experience, a type of leader's actions - directive -, and two situational aspects - Interpersonal stress and group tasks in nature. The lack of empirical evidence for this theory was deemed (Den Hartog & Koopman, 2001, 169).

One of the most common is the Hersey and Blanchard situation model (1974), which has become the basis for leadership training (Den Hartog & Koopman, 2001). The conduct of the leader must conform to the maturity or level of growth of the teams as a whole and their individual members (Hersey & Blanchard, 1974). The essence of the model is that leaders change their behavioral style in accordance with the maturity of the subordinates. The understanding of the supporters is important for the modification of the leadership style (Hersey & Blanchard, 1974). Because of the reduced theoretical basis, and the uncertainty about the definition of maturity, this model has been criticized. Over time, the model has been improved by Blanchard, leading to the creation of Situational Leadership Model II, based on "the beliefs that people can and want to evolve and that there is no better leadership style to foster this growth. The style of leadership must be tailored to the situation" (Blanchard, 2010, 80).

The knowledge of the followers' perception of leaders and the followers' attitudes towards leaders affect the cognitive bases resulting from previous expectations and the processing of data (Chemers, 2000, 30). In this context, the Normative Model of Decision Making by Vroom and Yetton (1973), with the concept of the leader involving followers in various forms of decision making emerges. It uses a theoretical framework for deductibility and introduces itself as an important strategic decision model as it incorporates situational factors (Chemers, 2000, 30). Normative Decision Theory "identifies the decision procedures most likely to result in effective in a particular situation" (Yukl, 1989, 265). A revision of the model was proposed by Vroom in collaboration with Jago in 1988. The first model showed executives what not to do, but suggested an ideal decision-making mechanism (Yukl, 1989, 265). The model of Vroom-Yetton-Jago aimed to optimize the decision-making process for the leader. This model describes the situation and level of engagement in various leadership styles and recommends choosing a leadership style for group decision-making. Over time, Vroom tried to create a leadership style model where managers should engage team members in the decision-making process (Vroom, 2003). The model follows the authors' concept of leadership - "Leadership depends on the situation" (Vroom & Jago, 2007); therefore, the efficacy of decision-making procedures depends on many aspects of the situation, including the amount of knowledge available to the leader and supporters.

The Theory of Paths to House Goals (House et al., 1971) is considered to be the most influential and full approach to contingency, since it explains how leaders influence followers' motivation and satisfaction (House et al., 1971). According to this theory, productive leaders contribute to the atmosphere under which their subordinates' function, offering the cognitive clarifications required to ensure that they can achieve their work goals; the inherent happiness of the followers is evident through the experience of the objectives accomplished and the compensation for that. Initially, House described four types of leadership: Directive leadership, Positive leadership, Participatory leadership, and Effective leadership. The charismatic leadership theory of House (1977) and a reformulated theory of House arose from this theory in 1996. Over 25 years, more than 40 experiments were performed to test the proposals of the theory and reconcile the impact of leadership on tasks and the satisfaction and success of people (House, 1996). "A dyadic supervisory theory is the path-goal theory" (House, 1996, 325). The orientation towards tasks and individuals takes this into account, but the organisation as a whole is still concerned.

3 Remote Work

The concept of telework (remote work), created by Jack Nilles in 1973, is considered the work carried out by a teleworker, mainly in a place other than the traditional place of work, for an employer or a client, involving the use of advanced computer technologies as an essential element of the work (Niles, 1975, 1998).

The organization of remote work involves numerous variables, including the available resources, the necessary knowledge, the time needed to perform tasks, the location of the remote workers, customers, and equipment, among others (Gschwind, 2015). It also considers the establishment of an agreement between leaders and remote workers concerning the goals and expectations (Eurofound, 2015), considering factors such as quality, quantity, and deadlines (Gimenez-Nadal, 2018).

Remote work implies achieving objectives without having direct control of the procedures and the time of the workers, and this will need a climate of trust between the leader and the workers (Golden and Veiga, 2005). The leader can fell that there is a decrease in control and even a loss of his power over their subordinates (Dambrin, 2004). Many organizations see the control function as the first duty of leaders, and rewards and incentives are often based on factors related to it. Information and communication technologies (ICT) have a predominant role at this level, as they can increase control, remotely monitoring the use of equipment, the working hours of the machines, and the deadlines to be met (Lehdonvirta, 2018). However, the electronic surveillance process can cause problems in terms of confidentiality and information security.

Traditionally, the application of specific quantitative control techniques has been used to elaborate work programs; determine costs to serve as a basis for budgets; determine the performance of people and machines, and others. Although these techniques have so far proved to be a fundamental instrument for the decision-making process, with the adoption of new forms of work organization, management control must necessarily be replaced by work coordination (Eurofound and the International Labour Office, 2017).

The role of managers must be to lead workers and to be facilitating elements in the work development process. However, the culture that exists in most organizations is not open enough to allow such procedures by the management bodies, and management by objectives instead of direct performance supervision is a strategic option that many companies have not yet taken.

Leadership, in the case of remote work, is a crucial factor, being the ability to lead individuals to cooperate to share information and objectives (Song et al., 2019). However, there is a problem regarding the coordination of objectives, tasks, and behaviors so that the team members are in tune with the rules and ideas about the organization.

In remote work situations, the leadership becomes more complex, as the remote workers are isolated, and could be a tendency to create their own rules and to take a different approach to work (Dambrin, 2004). In order to not deviate from the objectives, procedures such as the creation of periodic reports with information on the projects being developed, the action plans, and the company policies, for example, must be developed. Teleworking thus requires the transition from spontaneous coordination in the workplace to more planned coordination, where ICT plays an important role.

The leader must be the mentor and the inspiration both at the individual level and at the level of teams. However, it is not very easy to exercise these influences in an organization with a network structure and where workers are not physically present (Dambrin, 2004). Methods such as "on the job training" will have to be replaced by others, such as - feedback to reinforce positive behaviors; - advice via communication tools; - contact the teleworker regularly to communicate promptly how the development of his work is going and the necessary changes; - establish a career plan, regularly reviewed with the worker to measure their progress.

The leader role in motivating the remote worker must include providing information about his work and the objectives achieved by the company itself, in the form of graphics, communications. According to Kanter (1989), the motivating factors that are important in a company with remote work situations are - mission (giving people something they believe in); - autonomy in the performance of work (feedback); - values (financial data, awards, public recognition); - learning (development of new knowledge); - reputation (opportunities to highlight each other's achievements). The leader of remote workers is responsible for integrating them, taking into account their different backgrounds, and providing networking, using empowerment instead of command and control from the hierarchy.

4 Methodology

The methodology approach was quantitative, based on a survey with 40 items representing leadership activities, behaviors, and capabilities (adapted from Sousa and Rocha, 2019), being the respondents Public Administration remote workers.

The first section of the questionnaire was to make a characterization of the remote workers (background information). The second section of the questionnaire (a scale, based on the work of Sousa and Rocha, 2019) integrates into the first question the leadership activities; on the second question of the questionnaire integrates the leader's behaviors; and, finally on the third question integrates items to investigate the perceptions of the respondents about the capabilities of the leader.

Remote workers were asked to rate the skills on a 4-point Likert scale ranging from 1-Not at all; 2-Sometimes; 3 – Fairly often; 4-Frequently, if not always. These measures helped to identify the leadership styles, and based on that information, a typology of leader's styles was created.

5 Data Analysis and Results

In order to answer the research question, a statistical analysis of data has been carried out using IBM SPSS 25.

5.1 Sample Characterization

Respondents were primarily from the male gender (n = 56), and secondly, from the female gender (n = 53), please see Table 1.

Table 1. Background information of Remote Workers - Gender

	N	%
MALE	56	51.4
FEMALE	53	48.6
TOTAL	109	100.0

Respondents characterized their jobs as Top management (n = 7), Middle management (n = 18), executive level (n = 58), and support staff (n = 26), please see Table 2.

Table 2. Background information of Remote Workers – Job Category

Job Characterization of respondents (n = 109)	n	%
Top Management	7	6,4
Middle management	18	16,5
Executive level	58	53,2
Support staff	26	23,9
Total	109	100,0

5.2 Leadership Styles for Remote Work Analysis

To analyze the leadership styles for Remote Work, as an initial set of measures, the Kaiser-Meyer-Olkin (KMO) was performed, and the result was 0.815, which is acceptable for proceeding with the subsequent analysis - factor analysis. It was decided to perform a Principal Component Analysis (PCA) to reduce the number of variables of the scale into factors. It is also important to refer that in the factor analysis process, the number of observations was 109, and five factors with eigenvalue > 1 were detected. They explain 63.1% of the total variance. The extraction method was iterated through principal factor analysis. The PCA has been selected to obtain a factor solution of a smaller set of variables, from the larger dataset we were working with, and that is reflected on the leading questions of the forums above mentioned. A varimax rotation method was used to spread variability more evenly amongst the variables. Many variables shared close similarities, as there were highly significant correlations.

Table 3. Factors Cronbach Alpha

	Cronbach Alpha
Factor 1	0,865
Factor 2	0,877
Factor 3	0,921
Factor 4	0,740
Factor 5	0,965

Concerning reliability, all variables were analyzed for internal consistency by using Alpha Cronbach which showed the reliability of 0.843, that can be considered very good, and as showed in Table 3, the factors Cronbach Alpha are also excellent:

The PCA has identified the patterns within the data and expressed it by highlighting the relevant similarities (and differences) in every component. The data has been compressed as it was reduced in some dimensions without much loss of information. The rationale for the data reduction process was to identify leadership styles in order to understand the reflections of the participants. Table 4 outlines the items taken into account, and that was considered for the identification of the factors.

The factor components were labelled following a cross-examination of the variables with the higher loadings. Typically, the variables with the highest correlation scores had mostly contributed towards the make-up of the respective component. The underlying scope of combining the variables by using component analysis was to reduce the data and make it readable allowing the leadership styles identification. A brief description of the extracted factor components is provided in the following topic of this paper.

6 Discussion and Conclusions

This study aimed to raise awareness of the importance of the leadership style for the effectiveness of remote workers of Public Organizations. Considering the results of the quantitative analysis it is possible to highlight two fundamental dimensions: a first dimension based on the characteristics and personality of leadership; a second dimension based on leadership or leadership style, which emerged from the study. Therefore, the result of the factor analysis was 5 factors: Challenger – Transactional; Communicator – Transformational; Visionary – Democratic; Developer – Servant; and Innovator – Charismatic:

The Challenger – Transactional is focused on the formal statements of organizational philosophy, values and goals, managing efficiently the resources. Defines objectively the criteria for reward, selection, promotion, and termination, but is also interested in the stories, legends and myths about key people and events of the organization. The critical incidents, crises and norms help them on the organizational design and structure, and also on the organizational systems and procedures, matching the remote workers goals with the organizational ones. He is also a role model that influences remote workers to

Table 4. Factor analysis – Leadership Characteristics to Potentiate Remote Work

	Factor 1	Factor 2	Factor 3	Factor 4	Factor 5
	Challenger - Transactional	Communicator - Transformational	Visionary - Democratic	Developer - Servant	Innovator - Charismatic
1. Promoted knowledge sharing helping to solve problems	0,041	0,055	0,127	**0,717**	0,166
2. Raised questions to bring out different viewpoints	−0,054	0,236	0,022	**0,587**	0,401
3. Guided discussions, but did not lead it	−0,066	0,152	0,489	**0,579**	0,075
4. Provided constructive criticism	−0,079	−0,246	**0,581**	0,264	−0,021
5. Understood the goals of the organization	−0,134	0,348	**0,692**	0,113	0,346
6. Kept the group on the agenda and moving forward	−0,266	0,143	**0,668**	0,091	0,225
7. Involved everyone in the organization activities	−0,087	0,234	**0,679**	−0,042	−0,029
8. Made sure that decisions were made democratically	−0,304	0,178	**0,701**	0,125	0,082
9. Identifying strengths and challenges	−0,238	−0,120	**0,755**	−0,078	0,065
10. Motivating and delegating	−0,043	0,256	**0,504**	0,374	0,441

(continued)

Table 4. (*continued*)

	Factor 1	Factor 2	Factor 3	Factor 4	Factor 5
11. Team building	−0,279	0,112	**0,678**	0,005	0,271
12. Providing feedback	−0,051	**0,543**	0,276	0,089	0,345
13. Resolve everyday challenges	0,067	**0,833**	0,098	0,089	0,087
14. Help employees become more self-aware	−0,058	**0,849**	0,031	0,007	−0,021
15. Change problematic behaviors	0,146	**0,913**	−0,045	0,069	0,095
16. Incentivate remote workers to seize opportunities to grow and improve	−0,093	**0,668**	0,114	0,067	0,278
17. Believe in remote worker's abilities	−0,145	**0,601**	0,074	0,265	0,198
18. Willing to invest time in the remote worker's development	0,016	**0,610**	0,245	0,178	0,255
19. Trust in remote worker's effectiveness	−0,165	**0,676**	0,294	0,021	0,184
20. Create and foster a vision of a new future	−0,068	0,067	0,265	0,298	**0,773**

(*continued*)

Table 4. (*continued*)

	Factor 1	Factor 2	Factor 3	Factor 4	Factor 5
21. Face up to behaviors, values and norms in current culture that must change	0,234	**0,657**	0,232	0,245	0,379
22. Initiate and lead the change	0,206	0,213	0,156	0,213	**0,678**
23. Create a willingness to separate from the past	−0,065	0,217	0,261	0,198	**0,765**
24. Build shared ownership through organization-wide participation	−0,165	0,267	**0,567**	0,242	0,284
25. Communicate the changes and new cultural messages	−0,190	0,365	−0,051	−0,154	**0,664**
26. Model the behavior that supports the new vision	−0,356	0,016	0,114	−0,237	**0,578**
27. Reward behavior which supports the new vision	−0,134	0,331	0,057	0,356	**0,635**
28. Maintain focus on the goal	**0,665**	0,067	−0,028	−0,476	−0,096

(*continued*)

Table 4. (*continued*)

	Factor 1	Factor 2	Factor 3	Factor 4	Factor 5
29. Bring in resources who uniquely add value to the change effort by modelling new ways to act, think and view things	**0,649**	−0,156	0,009	−0,456	0,075
30. The formal statements of philosophy, values and goals	**0,767**	−0,045	−0,328	−0,336	0,196
31. The criteria used for reward, selection, promotion, and termination	**0,589**	0,223	−0,413	−0,067	0,074
32. The stories, legends and myths about key people and events	**0,789**	−0,056	−0,345	−0,278	0,045
33. Critical incidents and crises and norms	**0,767**	−0,015	−0,367	−0,071	−0,090
34. Organizational design and structure	**0,890**	−0,267	−0,084	0,134	0,044
35. Organizational systems and procedures	**0,845**	−0,056	−0,167	0,076	−0,156

(*continued*)

Table 4. (*continued*)

	Factor 1	Factor 2	Factor 3	Factor 4	Factor 5
36. Matching the employee's goals with the organizational ones	**0,794**	0,112	−0,245	0,267	−0,156
37. Recognition and rewarding	**0,766**	−0,134	−0,245	−0,289	−0,178
38. Being a role model that influences remote workers to accomplish their goals	**0,823**	−0,015	−0,243	0,028	−0,258
39. Encouraging remote workers to get involved in organizational strategy	**0,865**	−0,068	−0,045	0,134	−0,279
40. Developing moral and team spirit	**0,834**	0,134	0,011	0,245	−0,067

Extraction method: principal component analysis: Varimax
Alpha = 0.843; KMO = 0.815

accomplish their goals, encouraging them to get involved in organizational strategies, and developing moral and team spirit.

The Communicator – Transformational base is leadership on providing feedback, resolving everyday challenges, and in the way helping remote workers become more self-aware, also helping to change problematic behaviors, and incentives remote workers to seize opportunities to grow and improve, believing in remote worker's abilities and development, as he trusts in remote worker's effectiveness.

The Visionary – Democratic involves every remote worker in the organization activities and provides constructive criticism. Makes sure that decisions are taken democratically, motivating and delegating, and consequently creating a strong team, building shared ownership through organization-wide participation. He also identifies strengths and challenges, understanding the goals of the organization, and keeping the group on the agenda and moving forward.

The Developer – Servant promotes knowledge sharing, helping to solve problems, raise questions to bring out different viewpoints, guiding the discussions in order to achieve the consensus regarding the strategies to achieve the organization's goals.

The Innovator – Charismatic creates and fosters a vision of a new future, initiating and leading the necessary changes, and creating a willingness to separate from the past. He communicates the changes and new cultural messages, and model and rewards the behavior of remote workers that supports the new vision.

All of these leadership styles have implications in remote workers of Public Organizations' performance, and it is possible to conclude that the leader assumes a reference role for the remote workers. On the one hand, the leader is a source of regulation, besides being an affective connection to the organization, revealed by his charisma. However, on the other hand, the leader in the inspirational motivation, based on a commitment to the organizational goals – is the emotional bonding factor, encouraging for development.

In the case of the more traditional Public Organizations, where it is possible to identify mainly transactional leaders profiles - the reward contingent, and the focus on resources for results is a primary framework for leader and follower relationships, as stated in the literature. However, a more flexible and innovative Public Organization requires changes in the leadership, and new emergent styles are appearing to face the digital economy, and the European Single Market challenges.

As the main limitations of this study are the nature and dimension of the sample, as because of the Global Pandemic of COVID-19, the sample was conditionate by the social isolation, and that fact is probably influencing the participants in the survey. For future research, it will be interesting to study the organizations operating in a global digital economy, with plenty of interactions occurring through virtual channels of communication, and it would also be essential to explore the role of new leadership styles in self-managed virtual teams and their participation in leveraging innovation in organizations.

References

Afsar, B., Badir, Y.F., Saeed, B.B.: Transformational leadership and innovative work behavior. Ind. Manag. Data Syst. **114**(8), 1270–1300 (2014)

Bass, B.M.: Bass & Stoghill's Handbook of Leadership: Theory, Research, and Managerial Applications. Free Press, New York, NY (1985)

Bennis, W.: The leadership advantage. Lead. Lead. **12**(Spring), 1–8 (1999)

Blanchard, K.: Leading at a Higher Level. Prentice-Hall, Upper Saddle River, NJ (2010)

Burns, J.M.: Leadership. Harper & Row, New York, NY (1978)

Cai, W., Lysova, E.I., Khapova, S.N., Bossink, B.A.: The effects of servant leadership, meaningful work, and job autonomy on innovative work behavior in Chinese high-tech firms: a moderated mediation model. Front. Psychol. **9**, 1767 (2018)

Chemers, M.M.: Leadership research and theory: a functional integration. Group Dyn. Theory Res. Pract. **4**(1), 27–43 (2000)

Collins, D., Holton, E.F., III.: The effectiveness of managerial leadership development programs: a meta-anaysis of studies from 1982 to 2001. Hum. Resour. Dev. Q. **15**(2), 217–248 (2004)

Conger, J., Kanungo, R.: Charismatic Leadership in Organizations. Sage, 2455 Teller Road, Thousand Oaks California 91320 United States (1998). https://doi.org/10.4135/9781452204932

Dambrin, C.: How does telework influence the manager-employee relationship? Int. J. Hum. Res. Dev. Manage. **4**(4), 358–374 (2004)

Dansereau, F., Jr., Graen, G., Haga, W.J.: A vertical dyad linkage approach to leadership within formal organizations: a longitudinal investigation of the role making process. Organ. Behav. Hum. Perform. **13**(1), 46–78 (1975)

Deluga, R.J.: American presidential machiavellianism: implications for charismatic leadership and rated performance. Leadersh. Q. **12**(3), 339–363 (2001)

den Hartog, D.N., Koopman, P.L.: Leadership in organizations. In: Anderson, N., Ones, D.S., Kepir Sinangil, H., Viswesvaran, C. (eds.), Handbook of Industrial, Work & Organizational Psychology, pp. 166–187. Sage (2001)

Derue, D.S., Nahrgang, J.D., Wellman, N.E.D., Humphrey, S.E.: Trait and behavioral theories of leadership: an integration and meta-analytic test of their relative validity. Pers. Psychol. **64**(1), 7–52 (2011)

Dinh, J.E., Lord, R.G., Gardner, W.L., Meuser, J.D., Liden, R.C., Hu, J.: Leadership theory and research in the new millennium: current theoretical trends and changing perspectives. Leadersh. Q. **25**(1), 36–62 (2014)

Eurofound and the International Labour Office. (2017). Working anytime, anywhere: The effects on the world of work. Geneva: Publications Office of the European Union, Luxembourg, and the International Labour Office

Eurofound: New Forms of Employment (Luxembourg, Publications Office of the European Union) (2015)

Fiedler, F.: Research on leadership selection and training: one view of the future. Adm. Sci. Q. **41**(2), 241–250 (1996)

Fiedler, F.: A Theory of Leadership Effectiveness. McGraw-Hill, New York (1967)

Gimenez-Nadal, J.I., Molina, J.A., Velilla, J.: Telework, the timing of work, and instantaneous well-being: evidence from time-use data. No. 11271. IZA discussion papers (2018)

Golden, T.D., Veiga, J.F.: The impact of extending of telecommuting on job satisfaction: resolving inconsistent findings. J. Manag. **31**(2), 301–318 (2005)

Graen, G.B., Uhl-Bien, M.: Relationship-based approach to leadership: development of leader-member exchange (LMX) theory of leadership over 25 years: applying a multi-level multi-domain perspective. Leadersh. Q. **6**(2), 219–247 (1995)

Greenleaf, R.K.: The servant as a Leader. Indianapolis: The Greenleaf Center (1970)

Gschwind, L.: Telework, New ICTs, and their Effects on Working Time and Work-Life Balance: Review of the Literature (Unpublished). International Labour Office, Geneva (2015)

Hackman, J.R., Walton, R.E.: Leading groups in organizations. In: Goodman e Associates, P.S. (eds.), Designing Effective Work Groups. Jossey Bass, São Francisco, pp. 72–119 (1986)

Hersey, P., Blanchard, K.: So you want to know your leadership style? Training & Development Journal, February, 22–37 (1974)

Hogan, R., Raskin, R., Fazzini, D.: The dark side of charisma. In Clark, K.E., et al. (eds.), Measures of leadership, pp. 343–354. West Orange, NJ: Leadership Library of America (1990)

House, R.J.: A 1976 Theory of Charismatic Leadership, Leadership: The Cutting Edge, pp. 189–207. Southern Illinois University Press, Carbondale (1977)

House, R., Aditya, R.: The social scientific study of leadership : quo vadis ? The social scientific study of leadership : quo vadis ? J. Manag. **23**(3), 409–473 (1997)

House, R.: Path-goal theory of leadership: Lessons, legacy, and a reformulated theory. Leadersh. Quart. **7**(3), 323–352 (1996)

House, R., Filley, A., Kerr, S.: Relation of leader consideration and initiating structure to R and D subordinates' satisfaction. Adm. Sci. Q. **16**(1), 19–30 (1971)

Howell, J.M., Avolio, B.J.: The ethics of charismatic leadership: submission or liberation? Acad. Manage. Perspect. **6**(2), 43–54 (1992). https://doi.org/10.5465/ame.1992.4274395

Judge, T.A., Piccolo, R.F.: Transformational and transactional leadership: a meta-analytic test of their relative validity. J. Appl. Psychol. **89**(5), 755 (2004)

Kanter, R.M.: Work and family in the United States: a critical review and agenda for research and policy. Fam. Bus. Rev. **2**(1), 77–114 (1989)

Kets de Vries, M., Vrignaud, P., Florent-Treacy, E.: The global leadership life inventory: development and psychometric properties of a 360-degree feedback instrument. Int. J. Hum. Res. Manage. **15**(3), 475–492 (2004)

Larson, C.E., LaFasto, F.M.J.: Teamwork: What must go right/what can go wrong. Sage series in interpersonal communication, Vol. 10. Sage (1989)

Lepsinger, R., Lucia, A.D.: The Art and Science of 360 Feedback. Pfeiffer Imp of Jossey-Bass Publishers, California, US (1997)

Lewin, K., Lippitt, R., White, R.K.: Patterns of aggressive behavior in experimentally created "social climates." J. Soc. Psychol. **10**(2), 269–299 (1939)

Lehdonvirta, V.: Flexibility in the gig economy: managing time on three online piecework platforms. N. Technol. Work. Employ. **33**(1), 13–29 (2018)

Liden, R.C., Wayne, S.J., Liao, C., Meuser, J.D.: Servant leadership and serving culture: influence on individual and unit performance. Acad. Manag. J. **57**(5), 1434–1452 (2014)

McCauley, C.D., Brutus, S.: Management Development Through Job Assignments: An Annotated Bibliography. Greensboro, NC: Center for Creative (1998)

Nilles, J.M.: Managing Telework: Strategies for Managing the Virtual Workforce (1998)

Nilles, J.M.: Telecommunications and organizational decentralization. IEEE Trans. Commun. **23**(10), 1142–1147 (1975)

Northouse, P.G.: Leadership: Theory and Practice, 7th edn. Sage, Thousand Oaks (2016)

Pfeffer, J.: New Directions for Organization Theory. Oxford University Press, New York (1997)

Posner, B.Z., Kouzes, J.M.: Relating leadership and credibility. Psychol. Rep. **63**(2), 527–530 (1988)

Shamir, B., House, R., Arthur, M.B.: The motivational effects of charismatic leadership: a self-concept based theory. Organ. Sci. **4**, 577–594 (1993)

Song, Y., Gao, J.: Does telework stress employees out? A study on working at home and subjective well-being for wage/salary workers. J. Happiness Stud. **21**, 2649–2668 (2019)

Stogdill, R.M.: Personal factors associated with leadership: a survey of the literature. J. Psychol. **25**(1), 35–71 (1948)

Uhl-Bien, M.: Relationship development as a key ingredient for leadership development. Future Leadersh. Dev. 129–147 (2003)

Vroom, V., Jago, A.: The role of the situation in leadership. Am. Psychol. **62**(1), 17–24 (2007)

Vroom, V.: Educating managers for decision making and leadership. Manag. Decis. **41**(10), 968–978 (2003)

Vroom, V., Yetton, P.: Leadership and Decision Making (Vol. 18) (1973)

Wang, Z., Meng, L., Cai, S.: Servant leadership and innovative behavior: a moderated mediation. J. Manag. Psychol. **34**(8), 505–518 (2019). https://doi.org/10.1108/JMP-11-2018-0499

Weber, M.: The Theory of Social and Economic Organization. Oxford University Press, New York, NY (1947)

White, R., Lippitt, R.: Autocracy and Democracy. Harper, New York (1960)

Yukl, G.: An evaluation of conceptual weaknesses in transformational and charismatic leadership theories. Leadersh. Q. **10**(2), 285–305 (1999)

Yukl, G.: Managerial leadership: a review of theory and research. J. Manag. **15**(2), 251–289 (1989)

Profiling the Motivations of Wellness Tourists Through Hotels Services and Travel Characteristics

Rashed Isam Ashqar[1,3](✉) and Célia M. Q. Ramos[2,3]

[1] Al Zaytona University of Science & Technology (ZUST), Salfit, Palestine
riashqar@ualg.pt
[2] ESGHT, University of Algarve, Faro, Portugal
cmramos@ualg.pt
[3] CinTurs, Faro, Portugal

Abstract. Recently, wellness tourism has grown motivated to improve physical, emotional and mental health, and well-being. The hotel industry needs to prepare for this trend in terms of offering services and products, it needs to identify the motivations and preferences of wellness tourists and define strategies according to the profile of this new type of guest. A questionnaire was prepared, which obtained 303 valid respondents, whose data were analyzed using descriptive statistics methods and clusters to identify the main segments. Two clusters associated with motivations were obtained: Holistic wellness (at the full-body level) and self-transformation (at the mental level). Moreover, the implications for the policymakers are presented to develop wellness programs on all the dimensions of physical and mental levels.

Keywords: Wellness Motivations · Wellness Tourism · Hospitality

1 Introduction

Medical tourism is a growing global phenomenon highly dependent on innovation and knowledge management [1]. The expanded interpretation of wellness tourism by Mueller and Kaufmann [2] includes "a state of health featuring harmony between the body, mind, and spirit, with self-responsibility, physical fitness/beauty care, healthy nutrition/diet, relaxation/meditation, mental activity/education, and environmental sensitivity/social contacts as fundamental elements" (see Fig. 1).

Previous research examined the several profiles of health and wellness tourists based on benefits sought, motivations, profitable hotel customers, different ages/belonging to other generations, socio-demographic characteristics of seniors, and customer satisfaction. For example, Dursun and Caber [3] focused on profiling profitable hotel customers

---

The original version of this chapter has been revised. the affiliation of the author "Célia M. Q. Ramos" has been corrected. A correction to this chapter can be found at https://doi.org/10.1007/978-3-031-45645-9_65

A. Rocha et al. (Eds.): WorldCIST 2023, LNNS 800, pp. 110–120, 2024.
https://doi.org/10.1007/978-3-031-45645-9_10

by Recency, Frequency, and Monetary (RFM) analysis and they found using data mining techniques for three five-star hotels operating in Antalya, Turkey that 369 profitable hotel customers were divided into eight groups: 'Loyal Customers', 'Loyal Summer Season Customers', 'Collective Buying Customers', 'Winter Season Customers', 'Lost Customers', 'High Potential Customers', 'New Customers', and 'Winter Season High Potential Customers. In addition, they inferred that most of the customers (36%) were positioned in the 'Lost Customers' segment, who stay for shorter periods, spend less than other groups, and tend to come to the hotels in the summer season.

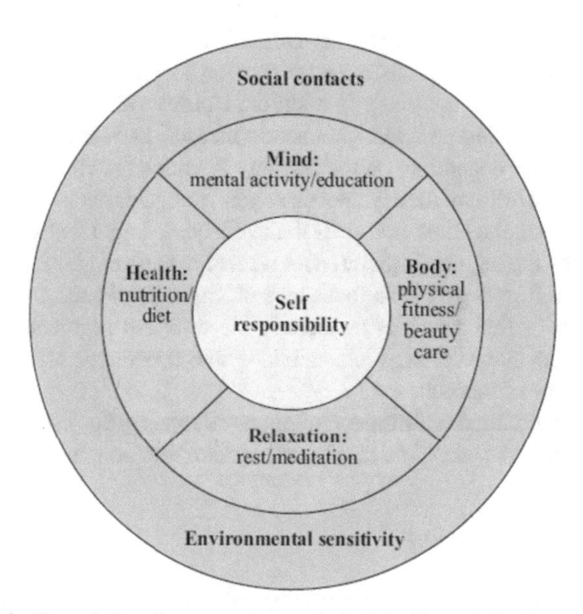

Fig. 1. Extended wellness tourism mode by Mueller and Kaufmann [2].

Moreover, et al. [4] explored the wellness orientations of consumers of different ages/belonging to different generations in the context of retailing namely Baby Boomers. They extracted six dimensions from a sample of 18–75-year-olds connected to different aspects of wellness are the following: 1) Responsible Consumption, 2) Recreational Shopping, 3) Appearance (self-responsibility), 4) Service(s) and quality, 5) Achievement, and 6) Store avoidance. In addition, they showed that the dimensions extracted among those aged 18–75 and 55−64 were fed into respective cluster analyses which both yielded seven segments are the following: 1) Health-oriented passive, 2) Passive leisure-oriented, 3) Reluctant shoppers, 4) Traditional, 5) Recreational shoppers, 6) Responsible proactive, and 7) For service-oriented hedonists.

On the other hand, Cain et al. [5] identified the spa customer sub-segments based on their motivations or benefits sought, and examine these segments in terms of their experience, subjective well-being, loyalty, and demographics. Consequently, Cain et al. [5] found that women were significantly more likely to be motivated to patronize the spa for reasons of rejuvenation and indulgence, whereas men were more likely to be motivated by socialization and self-esteem enhancement. Also, they concluded that spa customers were found to differ in their motivation to attend the spa based on age. Additionally, Alén et al. [6] determined the various profiles of senior tourism using socio-demographic variables, motivation, and characteristics of travel of seniors. They found five market segments were identified according to the behavioral variables analyzed as the following: (1) 'Women of advanced age who travel for health reasons, (2) 'Off-season holiday trips (IMSERSO)', (3) 'Visiting Friends and Relatives, staying with family and/or friends or in the second residence', (4) 'Short work-related travel' and (5) 'Travel for holiday and cultural purposes.

In a recent study, Chiny et al. [7] applied natural language processing (NLP) algorithms to examine 100,000 customer reviews left on the Airbnb platform to identify different dimensions that shape customer satisfaction according to each category studied (individuals, couples, and families). In addition, Chiny et al. [7] inferred that the customers are not equally interested in satisfaction metrics, and they noted the disparities for the same indicator depending on the category to which the client belongs.

In this sense, it is relevant for the hotel managers to acquire knowledge about wellness tourists and publicize the hotel's services to the adequate target audience, for each promotion or offer, by the segment's preferences where the main engine is their motivations, considered also as a strategy with a focus on service personalization.

The objective of this article is to identify the different segments associated with the profile of health and wellness tourists, taking into account their motivations for the practice of this type of tourism.

This paper is structured as follows. Section two describes the data and methodology. Section three reports the empirical results and section four concludes the paper.

2 Data and Methodology

Tourists today make consumptions of wellness products and services. Following trends, hotels started to define and implement marketing strategies and promotions to achieve more customers and increase their revenues [8].

A survey was done using a questionnaire addressed to all tourists aged 18 or more years old from different countries who have participated in a health and well-being activity, for example, hot springs, mineral springs, seawater, a comfortable climate, deep-sea water, or a spa in a hotel. The questionnaire was administered online through Microsoft Forms and prepared in two languages Portuguese and English, a fair pre-test was carried out by researchers and hoteliers to present suggestions and shared them on social networks.

The questionnaire was structured into three groups of questions: one related to profile response characterization and travel behavior characteristics, the second related to motivations [9], and the third with the importance attributed to the wellness services [10], as shown in Table 1. Overall, 303 questionnaires were completed by tourists.

Data analysis started with a descriptive statistical analysis of respondents. Then a cluster analysis was applied considering the Two steps method to classify wellness tourists' motivations profiles through hotel services and travel characteristics.

Table 1. Definition of the Variables.

Variable	Items	Ref.
Motivation	To find my inner self. To deepen my meditation experience. To learn how to meditate. To help better understand myself. To contemplate what is important to me. To be at peace with myself. To gain a sense of renewal. To think about what life means. To experience calmness. To enjoy an experience with all my senses. To maintain my physical fitness. To improve my physical fitness. To get exercise. To be active. To control my weight. To be more toned. To improve my appearance. To improve my health. To treat my body well. To overcome health problems. To relax. To escape the demands of everyday life. To be refreshed. To let go of my worries and problems. To reduce my stress levels. To get away from everything. To give me time and space for reflection. To be with friends To spend time with family members. To share my experience with people I am close to. To fulfil my curiosity. To tell others where I have been. To make a good impression on others. To experience something new. To gain more confidence about myself. To increase my self-esteem. To help recover from a major negative life event. To get away from other people. To spoil myself. To feel that I am the only person in the world.	[9]
Personnel Health Service	Service personnel are equipped with basic medical and emergency knowledge. Service personnel is compassionate toward customers.	[10]
Health promotion treatments	Massage Post-surgery recovery care Hot spring therapy guidance Medication consultation Traditional healing Aquatic workout guidance Fitness exercise guidance Weight programs Nutrition guidance Exfoliation services	[10]
Environment	Comprehensively planned walking trails Clean and hygienic hot spring bath environment With barrier-free facilities (accessibilities) Gym	[10]
Healthy diet	Use of non-toxic or detoxification food ingredients Health detoxifying meals Provide local ingredients-based cuisines	[10])
Relaxation	The atmosphere of relaxed tranquility provide landscape therapy (environment and psychological consultation) Meditation environment Provide music	[10]
Social activities	Provide recreation rooms for chatting and chess Provide family activities	[10]
Experience of unique tourism resources	Provide cultural custom experiences Local cultural celebration involvement Community development Provide private attractions Provide nocturnal exploration activities	[10]
Mental learning	Provide enlightenment lectures by resident religious and spiritual mentors. Provided relaxed learning atmosphere. Provide psychological consultation. Provide book clubs. Provide art exhibitions. Provide musical performances Provide group activities	[10]

3 Empirical Results

The sample characterization is shown in Table 2, there are 70.3% of our sample are females. There are 51.5% of the respondents are married, and 39.6% of the respondents have a master's degree. 69.3% of the respondents are Portuguese. There are 31.7% of the respondents are planning their trip in two weeks. Also, there are 44.6% of the respondents spend 1 to 5 days trip. In addition, there are 48.5% of the respondents are going on holiday and 30.7% of the respondents are taking information about the hotel from the internet.

Table 2. Sample characteristics.

Characteristic	%
Gender	
Male	29.7%
Female	70.3%
Age range	
18-25	14.9%
26-45	37.6%
46-65	35.6%
>65	11.9%
Monthly household income	
Up 500€	5.0%
From 501€ to 750€	14.9%
From 751€ to 1500€	23.8%
From 1501€ to 2250€	14,9%
From 2251€ to 3000€	17.8%
From 3001€ to 3750€	8.9%
From 3751€ to 4500€	14.9%
Marital status	
Single	39.6%
Married	51.5%
Divorced	8.9%
Highest level of education	
Basic	3.0%
High school	18.8%
Bachelor	22.8%
Master	39.6%
PhD	15.8%
Employment status	
Self-Employed	9.9%
Employed for someone else	56.4%
student	15.8%
Retired	11.9%
Unemployed	5.9%
Nationality	
Brazil	3.0%
Canada	3.0%
Iran	1.0%
Jordan	10.9%
Palestine	8.9%
Portugal	69.3%
Spain	3.0%
Syria	1.0%
Country of residence	
Angola	3.0%
Canada	3.0%
Jordan	9.9%
Palestine	5.9%
Portugal	69.3%
Spain	5.9%
The United Arab Emirates	3.0%
Trip planning	
Without planning	21.8%
Less than a week	10.9%
Between 1 and 2 weeks	31.7%
From 2 weeks to a month	17.8%
More than a month	17.8%
Length of trip	
1–5 days	44.6%
6–10 days	31.7%
11–15 days	8.9%
More than 15 days	14.9%
Information sources about the hotel	
Facebook	13.9%
Internet	48.5%
Referral from friends and family	37.6%
TripAdvisor	17.8%
Instagram	8.9%
Magazine Article	3.0%
Travel Agent	13.9%
Booking	5.9%
Motivation*	
Visiting friends or relatives	18.8%
Holiday	81.2%
Health	21.8%
Business	8.9%
Self-Knowledge	3.0%

*Multiple choices.

Moreover, we used a cluster analysis - Two steps method to classify wellness tourists' motivations profiles through hotel services and travel characteristics (see Fig. 2).

We defined the two clusters: the first one is Holistic wellness customers who have the motivation to practice wellness activities at both dimensions of physical and mental levels. On the other hand, the second cluster is self-transformation customers who have the motivation to practice wellness activities just on a mental level. Voigt et al. [9] showed that Self-transformation can be influenced by the benefit of transcendence which aims at self-awareness at the spiritual and psychological levels. Table 3 presents clusters profile by sociodemographic characteristics.

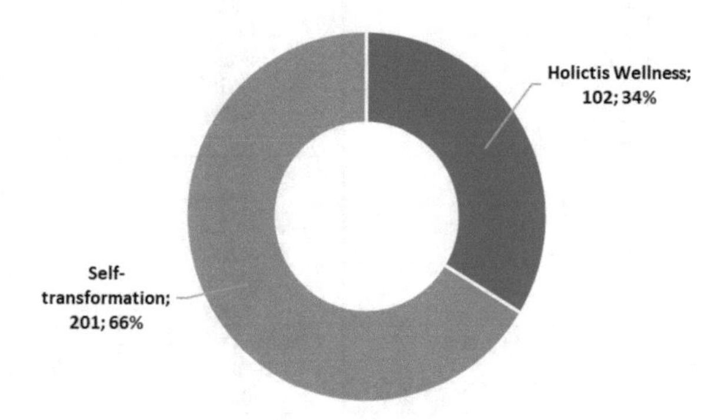

Fig. 2. Clusters Sizes.

As shown in Table 3, cluster two has 17.6% of old people aged more than 65 than cluster one. Also, there are 26.5% of people earn up to 4500 euros in cluster two more than in cluster one. In addition, there are 19.4% of people have PhD degrees in cluster one more than in cluster two.

Table 3. Clusters profile by sociodemographic characteristics.

Variables	Categories	Cluster 1 - Holistic wellness	Cluster 2 - Self-transformation
Gender	Male	34.3%	20.6%
	Female	65.7%	79.4%
Age range	18-25	11.9%	20.6%
	26-45	29.9%	52.9%
	46-65	49.3%	8.8%
	>65	9.0%	17.6%
Nationality	Brazil	0.0%	8.8%
	Canada	4.5%	0.0%
	Iran	1.5%	0.0%
	Jordan	11.9%	8.8%
	Palestine	9.0%	8.8%
	Portugal	67.2%	73.5%
	Spain	4.5%	0.0%
	Syria	1.5%	0.0%
Country of residence	Angola	4.5%	0.0%
	Canada	4.5%	0.0%
	Jordan	10.7%	8.8%
	Palestine	4.5%	8.8%
	Portugal	67.2%	73.5%
	Spain	4.5%	8.8%
Variables	**Categories**	**Cluster 1 - Holistic wellness**	**Cluster 2 - Self-transformation**
	The United Arab Emirates	4.5%	0.0%
Monthly household income	Up 500€	7.5%	0.0%
	From 501€ to 750€	13.4%	17.6%
	From 751€ to 1500€	29.9%	11.8%
	From 1501€ to 2250€	13.4%	17.6%
	From 2251€ to 3000€	17.9%	17.6%
	From 3001€ to 3750€	9.0%	8.8%
	From 3751€ to 4500€	9.0%	26.5%
Marital status	Single	40.3%	38.2%
	Married	55.2%	44.1%
	Divorced	4.5%	17.6%
Highest level of education	Basic	4.5%	0.0%
	High school	17.9%	20.6%
	Bachelor	25.4%	17.6%
	Master	32.8%	52.9%
	PhD	19.4%	8.8%
Employment status	Self-Employed	6.0%	17.6%
	Employed for someone else	62.7%	44.1%
	Student	13.4%	20.6%
	Retired	13.4%	8.8%
	Unemployed	4.5%	8.8%

Table 4 presents the cluster profile by travel behavior characteristics. As shown in Table 4, there is no big difference between the two clusters regarding travel behavior characteristics.

Table 4. Clusters profile by travel behavior characteristics.

Variables	Categories	Cluster 1 Holistic wellness	Cluster 2 Self-transformation
Trip planning	Without planning	23.9%	17.6%
	Less than a week	11.9%	8.8%
	Between 1 and 2 weeks	33.3%	28.4%
	From 2 weeks to a month	12.9%	27.5%
	More than a month	17.9%	17.6%
Length of trip	1–5 days	43.8%	46.1%
	6–10 days	32.3%	30.4%
	11–15 days	9.5%	7.8%
	More than 15 days	14.4%	15.7%
Information source about the hotel*	Facebook	9.6%	4.3%
	Internet	33.7%	14.9%
	Referrals from friends and family	23.4%	14.2%
	TripAdvisor	10.9%	6.9%
	Instagram	5.9%	3.0%
	Magazine Article	1.3%	1.7%
	Travel Agent	11.2%	2.6%
	Booking	3.3%	2.6%
Motivation*	Visiting friends or relatives	13.2%	5.6%
	Holiday	55.4%	25.7%
	Health	16.2%	5.6%
	Business	4.3%	4.6%
	Self-Knowledge	1.3%	1.7%

*Multiple choices.

Table 5 presents the cluster profiles by the importance of the services provided by the hotel. From Table 5, we can conclude that cluster two considered the services provided by the hotel to be more important than cluster one. According to the motivations of each cluster, in Table 5 it is verified that Cluster 2 highly values the activities characterized by "The atmosphere of relaxed tranquility", "Provide landscape therapy (environment and psychological consultation)", "Provide meditation environment", "Provide cultural custom experiences", "Provide relaxed learning atmosphere" and "Provide art exhibitions".

Table 5. Clusters profile by the importance of the wellness services provided by the hotel.

Services Offered		Cluster 1 (%)	Cluster 2 (%)	Total (%)
Service personnel equipped with medical and emergency knowledge	1	4.5	0.0	3.0
	2	0.0	8.8	3.0
	3	22.4	17.6	20.8
	4	44.8	20.6	36.6
	5	28.4	52.9	36.6
Service personnel is compassionate toward customers	1	4.5	0.0	3.0
	2	4.5	8.8	5.9
	3	7.5	26.5	13.9
	4	40.3	11.8	30.7
	5	43.3	52.9	46.5
Massage	2	1.5	0.0	1.0
	3	25.4	35.3	28.7
	4	31.3	8.8	23.8
	5	41.8	55.9	46.5
Post-surgery recovery care	1	9.0	0.0	5.9
	3	20.9	17.6	19.8
	4	37.3	17.6	30.7
	5	32.8	64.7	43.6
Hot spring therapy guidance	2	0.0	8.8	3.0
	3	40.3	8.8	29.7
	4	17.9	8.8	14.9
	5	41.8	73.5	52.5
Medication consultation	1	9.0	0.0	5.9
	2	6.0	0.0	4.0
	3	34.3	26.5	31.7
	4	31.3	17.6	26.7
	5	19.4	55.9	31.7
Traditional healing	1	4.5	0.0	3.0
	2	6.0	8.8	6.9
	3	38.8	26.5	34.7
	4	31.3	17.6	26.7
	5	19.4	47.1	28.7
Aquatic workout guidance	1	0.0	8.8	3.0
	2	1.5	0.0	1.0
	3	23.9	35.3	27.7
	4	32.8	17.6	27.7
	5	41.8	38.2	40.6
Fitness exercise guidance	2	1.5	0.0	1.0
	3	23.9	26.5	24.8
	4	37.3	17.6	30.7
	5	37.3	55.9	43.6
Weight programs	1	4.5	8.8	5.9
	2	0.0	8.8	3.0
	3	41.8	26.5	36.6
	4	29.9	17.6	25.7
	5	23.9	38.2	28.7
	1	4.5	0.0	3.0

Services Offered		Cluster 1 (%)	Cluster 2 (%)	Total (%)
Nutrition guidance	2	0.0	8.8	3.0
	3	34.3	26.5	31.7
	4	32.8	17.6	27.7
	5	28.4	47.1	34.7
Exfoliation services	2	14.9	17.6	15.8
	3	31.3	26.5	29.7
	4	26.9	26.5	26.7
	5	26.9	29.4	27.7
Comprehensively planned walking trails	2	1.5	0.0	1.0
	3	16.4	8.8	13.9
	4	64.2	44.1	57.4
	5	17.9	47.1	27.7
Clean and hygienic hot spring bath environment	2	4.5	0.0	3.0
	3	20.9	8.8	16.8
	4	20.9	17.6	19.8
	5	53.7	73.5	60.4
Barrier-free facilities (accessibilities)	3	26.9	8.8	20.8
	4	14.9	35.3	21.8
	5	58.2	55.9	57.4
Gym	2	4.5	8.8	5.9
	3	20.9	35.3	25.7
	4	32.8	8.8	24.8
	5	41.8	47.1	43.6
Use of non-toxic or detoxification food ingredients	1	4.5	0.0	3.0
	2	6.0	0.0	4.0
	3	7.5	8.8	7.9
	4	32.8	8.8	24.8
	5	49.3	82.4	60.4
Health detoxifying meals	3	25.4	8.8	19.8
	4	20.9	8.8	16.8
	5	53.7	82.4	63.4
Local ingredients-based cuisines	1	4.5	0.0	3.0
	3	7.5	0.0	5.0
	4	34.3	8.8	25.7
	5	53.7	91.2	66.3
Atmosphere of relaxed tranquility	1	4.5	0.0	3.0
	3	9.0	0.0	5.9
	4	23.9	0.0	15.8
	5	62.7	100.0	75.2
Environment and psychological consultation	2	4.5	0.0	3.0
	3	26.9	0.0	17.8
	4	37.3	8.8	27.7
	5	31.3	91.2	51.5
Provide meditation environment	3	29.9	0.0	19.8
	4	43.3	26.5	37.6
	5	26.9	73.5	42.6
Provide music	1	4.5	0.0	3.0

(*continued*)

Table 5. (*continued*)

Services Offered		Cluster 1 (%)	Cluster 2 (%)	Total (%)
	2	1.5	0.0	1.0
	3	16.4	17.6	16.8
	4	37.3	17.6	30.7
	5	40.3	64.7	48.5
Provide recreation rooms for chatting and chess	1	10.4	8.8	9.9
	2	4.5	26.5	11.9
	3	23.9	26.5	24.8
	4	47.8	8.8	34.7
	5	13.4	29.4	18.8
Provide family activities	2	4.5	8.8	5.9
	3	23.9	35.3	27.7
	4	31.3	17.6	26.7
	5	40.3	38.2	39.6
Provide cultural custom experiences	1	1.5	0.0	1.0
	2	4.5	0.0	3.0
	3	20.9	0.0	13.9
	4	46.3	26.5	39.6
	5	26.9	73.5	42.6
Local cultural celebration involvement	2	6.0	0.0	4.0
	3	19.4	17.6	18.8
	4	38.8	26.5	34.7
	5	35.8	55.9	42.6
Community development	2	0.0	8.8	3.0
	3	35.8	17.6	29.7
	4	37.3	26.5	33.7
	5	26.9	47.1	33.7
Provide private attractions	2	9.0	8.8	8.9
	3	38.8	8.8	28.7
	4	34.3	44.1	37.6
	5	17.9	38.2	24.8
Provide nocturnal exploration activities	1	4.5	26.5	11.9
	2	4.5	0.0	3.0
	3	31.3	8.8	23.8
	4	41.8	35.3	39.6
	5	17.9	29.4	21.8
Enlightenment lectures by resident religious and spiritual mentors	1	9.0	17.6	11.9
	2	6.0	8.8	6.9
	3	43.3	17.6	34.7
	4	32.8	8.8	24.8
	5	9.0	47.1	21.8
Provided relaxed learning atmosphere	2	9.0	0.0	5.9
	3	26.9	0.0	17.8
	4	41.8	44.1	42.6
	5	22.4	55.9	33.7
Provide psychological consultation	1	4.5	8.8	5.9
	2	4.5	26.5	11.9
	3	47.8	11.8	35.6
	4	29.9	17.6	25.7
	5	13.4	35.3	20.8
Provide book clubs	1	9.0	0.0	5.9
	2	9.0	11.8	9.9
	3	38.8	17.6	31.7
	4	29.9	44.1	34.7
	5	13.4	26.5	17.8
Provide art exhibitions	1	4.5	0.0	3.0
	2	6.0	8.8	6.9
	3	43.3	8.8	31.7
	4	28.4	29.4	28.7
	5	17.9	52.9	29.7
Provide musical performances	2	0.0	8.8	3.0
	3	35.8	17.6	29.7
	4	32.8	17.6	27.7
	5	31.3	55.9	39.6
Provide group activities	1	0.0	8.8	3.0
	2	4.5	0.0	3.0
	3	28.4	17.6	24.8
	4	38.8	35.3	37.6
	5	28.4	38.2	31.7

4 Conclusions

This paper investigates the classification of wellness tourists' motivation profiles through hotel services, sociodemographics, and travel characteristics. Two clusters were identified: the first one is Generic wellness people who have the motivation to practice wellness activities at both dimensions of physical and mental levels. The second cluster is self-transformational people who have the motivation to practice wellness activities just on the mental level. Moreover, we offer implications to the policymakers to develop wellness programs on all the dimensions of physical and mental levels.

The contribution of this research represents of the definition two segments within health and wellness tourists, where one is motivated and interested in activities related only to mental wellness, and the other with general wellness, i.e. living a healthy lifestyle that includes all the different dimensions associated with health. Thus, hoteliers can specialize in offering services only to a certain segment, as well as carry out specific promotions for a certain customer profile, which can help combat seasonality and increase revenues in periods of low activity.

For the academy, the identification of two segments can contribute to carrying out specific studies for each segment, with activities related to the mind becoming more sought after in today's society, which could be the forerunner of a new type of tourism - mental well-being, and holistic well-being (whole body). On the other hand, it will make it possible to establish bridges with activities considered non-traditional medicine, which can be sold as complementary services and included in innovative therapeutic packages.

This research suffers from limitations: short of the sample size of 303 tourists. Future research can extend to collecting more data from different groups of participants and be considered the technological environment provided by smart technologies in the hotel in a way to contribute to the wellness service personalization and consequently to customer satisfaction and loyalty.

Acknowledgment. This paper is financed by National Funds provided by FCT- Foundation for Science and Technology through project UIDB/04020/2020 and project Guest-IC I&DT nr. 047399 financed by CRESC ALGARVE2020, PORTUGAL2020 and FEDER.

References

1. Subasinghe, M., Magalage, D., Amadoru, N., Amarathunga, L., Bhanupriya, N., Wijekoon, J.: Effectiveness of artificial intelligence, decentralized and distributed systems for prediction and secure channelling for Medical Tourism. In: 11th IEEE Annual Information Technology, Electronics and Mobile Communication Conference, pp. 314–319 (2020)
2. Mueller, H., Kaufmann, E.: Wellness tourism: market analysis of a special health tourism segment and implications for the hotel industry. J. Vacat. Mark. **7**(1), 5–17 (2001)
3. Dursun, A., Caber, M.: Using data mining techniques for profiling profitable hotel customers: an application of RFM analysis. Tourism Manage. Perspect. **18**, 153–160 (2016)
4. Marjanen, H., Kohijoki, A.-M., Saastamoinen, K.: Profiling the ageing wellness consumers in the retailing context. Int. Rev. Retail, Distrib. Cons. Res. **26**(5), 477–501 (2016)
5. Cain, L., Busser, J., Baloglu, S.: Profiling the motivations and experiences of spa customers. Anatolia **27**(2), 262–264 (2016)
6. Alén, E., Losada, N., Carlos, P.: Profiling the segments of senior tourists throughout motivation and travel characteristics. Curr. Issue Tour. **20**(14), 1454–1469 (2017)
7. Chiny, M., Bencharef, O., Hadi, M., Chihab, Y.: A client-centric evaluation system to evaluate guest's satisfaction on AirBNB using machine learning and NLP. Appl. Comput. Intell. Soft Comput. (2021)
8. Park, H., Lee, M., Back, K.: Exploring the roles of hotel wellness attributes in customer satisfaction and dissatisfaction: application of Kano model through mixed methods. Int. J. Contemp. Hosp. Manag. **33**(1), 263–285 (2020)
9. Voigt, C., Brown, G., Howat, G.: Wellness tourists: in search of transformation. Tour. Rev. **66**(1/2), 16–30 (2011)
10. Chen, K.-H., Liu, H.-H., Chang, F.-H.: Essential customer service factors and the segmentation of older visitors within wellness tourism based on hot springs hotels. Int. J. Hospitality Manage. **35**, 122–132 (2013)

Decision Support System Based on Machine Learning Techniques to Diagnosis Heart Disease Using Four-Lead ECG Recordings

Mohamed Hosni[1](✉), Ibtissam Medarhri[2], Soufiane Touiti[3,4], Amal Mezalek Tazi[3,4], and Nabil Ngote[3]

[1] MOSI Team, M2S3 Laboratory, ENSAM, Moulay Ismail University of Meknes, Meknes, Morocco
m.hosni@umi.ac.ma
[2] MMCS Team, LMAID Laboratory, MINES-Rabat, Rabat, Morocco
medarhri@enim.ac.ma
[3] Abulcasis International University of Health Sciences, Rabat, Morocco
touiti.soufiane@uiass.ma, ngote@enim.ac.ma
[4] Cheikh Zaïd International University Hospital, Rabat, Morocco

Abstract. Cardiovascular disease is one of the leading causes of death in the world. Accurately and rapidly diagnosis of this disease remains an important challenge in research. Different decision support systems (DSS) implementing Machine Learning (ML) techniques have been proposed in the literature. In this study, we attempt to build a DSS that implements ML and try to classify patients according to their clinical characteristics and four-lead Electrocardiogram (ECG) recorded by mean of a smart watch. The principal goal of this study is to check whether the proposed DSS can play the same role as a standard ECG. Indeed, the obtained results suggest that this DSS can accurately screen and diagnose heart conditions and may be used as an alternative to a standard ECG. Moreover, among the four optimized ML techniques by grid search optimization technique, the Support Vector Machines achieved the highest accuracy score.

Keywords: ECG · Machine Learning · Cardiovascular Disease · Smart Watch · Withings ScanWatch

1 Introduction

Heart disease (HD) or cardiovascular disease is one of the most common diseases. In fact, there are several types of CVDs such as: coronary heart disease, coronary artery disease, cerebrovascular disease, and peripheral vascular disease, among others [1]. These CVDs may lead to serious events, such as heart attacks and strokes, which increase the risk of death and disability. It is one of the main causes

A. Rocha et al. (Eds.): WorldCIST 2023, LNNS 800, pp. 121–130, 2024.
https://doi.org/10.1007/978-3-031-45645-9_11

of mortality on the globe. Indeed, according to the latest report published by the World Health Organization (WHO), 17.9 million deaths globally in 2019-or 32% of all fatalities-were caused by CVDs [1,2]. It is therefore regarded as one of the top objectives for medical informatics research. Although the exact causes of CVDs remain unknown, several risk factors, including high blood pressure, smoking, high cholesterol, diabetes, inactivity, and family history, may make them more likely to occur [2].

Early diagnosis of CVD symptoms increases the probability of obtaining effective therapy, thereby increasing the chances of saving lives [2]. One of the most common tools used to monitor heart activity is the Electrocardiogram (ECG) machines [3] [4]. This device is considered as the most accurate method to discover anomalies in heart activity. The use of ECG consists of placing 10 electrodes in different locations of the patient body to capture the electrical activity of the heart, which results on 12 recordings leads in graph format. However, this process requires a skilled medical personal to use the machine and accurately interpret the results as well as some other important medical facilities, which is, therefore costly and time consuming [4].

In the last decade, the technological advances permitted the invention of several smart devices used to promote healthcare [4,5]. The smartwatches (SW) and smartphones are among these devices. Some recent generations of the SW have the ability of recording some important health parameters such as oxygen saturation (SpO2), heart rate, and heart electrical activity using the photoplethysmography (PPG) technique. In fact, these SWs are equipped with ECG sensor that can record one-lead ECG. These devices have optical heart rate monitors which are based on the flashings LEDs against the skin and detect and collect the smallest variations of light intensity related to blood flow. Afterward, the software starts processing the collected data to estimate the heart rate and inform the if any disorder is noticed. The collected data can be exported through other smartphone applications. Further consultation and examination by a cardiologist are required to decide upon on the heart condition using a standard ECG, which makes the process longer and costing.

Machine Learning (ML) approaches have been used during the past 30 years in different domains, including medicine, to create decision support systems (DSS) [2,6]. Several medical tasks were addressed successfully using these DSS. In fact, since medical datasets often contain a high number of entries, extracting knowledge from them manually is quite challenging and time-consuming. Therefore, applying ML [7] approaches to these datasets are advised for the quick and accurate extraction of relevant data that might aid physicians in their activity by giving them a second opinion. In fact, ML methods may be used to learn the hypotheses included in the medical datasets, which are then applied to make predictions about the health of the patients [8].

Within this context, this paper proposes to build a DSS using ML techniques to diagnosis heart activity to serve cardiologists to quickly screen and diagnosing patients. As relying on one-lead ECG is insufficient to screen the CVDs, this paper proposes the recording of four-lead ECG using smartwatch and build the ML models based those recordings and other clinical data.

In this study the Withings scanwatch was used to record the four-lead ECG, and the dataset consists of 140 patients recruited in Cardiology department in Sheikh Ziad hospital University in Morocco. The Sheikh Zaid University board gave its ethical approval, and all participants explicitly agreed to take part in the study.

To the best of authors knowledge, this is the first paper that examine the capability of ML techniques to diagnosis the heart conditions based on clinical indicators and four-lead ECG recorded by a smartwatch.

The main research question addressed in this study is as follow:

(RQ): Are Machine Learning able to diagnosis accurately the heart condition using the collected data as a standard ECG?

The main contributions of the current study are the following:

- Collecting a medical dataset that contains four-lead ECG recordings using a SW (i.e., Withings scanwatch), among other usual clinical data.
- Examining the potential of ML techniques to detect accurately the heart condition using the collected data.

The rest of this paper is structured as follows: Sect. 2 presents the material and methods used to conduct this study including dataset description, ML technique selected, and the experimental design. Section 3 reports and discusses the empirical results, and finally conclusions and further work are presented in the last Section.

2 Materials and Methods

This section presents the process followed to collect the dataset used to build the DSS, an overview of the four ML techniques used in this study, the experimental design followed to address our RQ which consists of describing the optimization techniques used to tune the hyperparameters of the four ML techniques and the performance measures used to assess the predictive capability of the developed DSS.

2.1 Dataset Description

For this study, 140 volunteers from across Morocco were recruited between 2021 and 2022 in the cardiology department and cardiology consultation of Sheikh Zaid University Hospital in Rabat, ranging in age from 18 to 95. A participant is included in this study if his heart conditions meet one of the two following criteria:

- Those who had healthy conditions as determined by their standard ECG recording.

- Those who had diagnosed with: Atrial fibrillation (AF), First degree atrioventricular block (AVB 1), left ventricular hypertrophy (LVH), ischemic disorders, premature ventricular complexes (PVCs).

The four-lead ECG recording process utilizing the Withings scanwatch (SW-ECG) began as soon as the normal ECG recording method was finished. Each participant in the cohort received some general instructions on how to utilize the smartwatch to capture the single lead ECG. The recording procedure was carried out under the identical circumstances as the prior session. A physician helped the participant perform the ECG recording. The four-lead ECG recordings consist of the three Wilson-like chest leads (i.e., VI, V3, and V6) and the Einthoven DI lead. Thereafter, the SW-ECG graphs were exported as PDF format using the Health Mate application for every patient. Afterward, the process of extracting features from these signals started manually. The focus in this study was on six parameters: P wave (amplitude (in millivolts), duration (in milliseconds), and morphology), PR interval (duration (in milliseconds)), QRS complex (amplitude (in millivolts), duration (in milliseconds), and morphology), ST segment morphology, T wave (amplitude (in millivolts), and morphology), and QT interval duration (in milliseconds). Some other clinical data that characterizes every patient was added to the dataset such as age, gender, diabetic, smoking, obesity among others, which result on 90 features describing each patient.

Finally, the labelling process is effectuated by attributing the class, either normal or anormal, to each patient according to the interpretation of the standard ECG graphs by a cardiologist.

2.2 Machine Learning: An Overview

This subsection presents a brief overview of the ML techniques selected to build the DSS. In fact, four ML techniques were selected in this study, namely: K-nearest neighbor (KNN), Multilayer Perceptron Neural Networks (MLP), Support Vector Machines (SVM), and Logistic Regression (LR). The excellent predictive performance demonstrated in several fields led to the selection of these techniques.

KNN: One of the simplest machine learning models is the KNN method, which is a form of instance-based learning [9]. This approach is used to deal with tasks in regression and classification. This method is based on similarity, it assumes that similar instances may have similar dependent variable. The process of KNN is made on three steps: identifying patients, computing the similarity between patients available in the dataset and the new patient, and the adaptation strategy where the class is computed.

MLP: The MLP neural networks are one architectural type of the artificial neural network [10]. These techniques can solve both classification and regression problems. The minimum number of layers that make up MLP is three, which are: the input layer; where the number of nodes is determined by the number of features in the input pattern, the hidden layer, and the output layer where the number of nodes is fixed by the type of the problem tackled. As in this study aims to classify the heart condition either normal or anormal, the number of nodes are two (i.e., binary classification). Several hyper-parameters should be tuned to achieve a good level of accuracy.

SVMs: are a class of machine learning models distinguished by the use of kernels. The SVM was introduced by Vapnik under the name of maximal margin classifier in 1992 [11]. SVMs have a strong theoretical foundation since they are implemented using statistical learning theory and structural risk minimization. These models were applied successfully in different fields. Three hyperparameters should be specified carefully; when using the SVM to achieve a good level of accuracy: the complexity parameter usually denoted with C, the round-off error denoted with Epsilon and the kernel with its parameters.

LR: one of the simple and widely used ML techniques for resolving a binary classification problem [12]. It is considered as a special case of linear regression. The technique searches the relationship between the independent attributes and the dependent attribute. It was applied successfully in different domain applications. By using the logit function, LR can forecast the probability that a binary event will occur. In this study, the norm of the penalty was varied.

2.3 Experimental Design

This subsection presents: the optimization technique used to tune the hyper-parameters of the four ML techniques selected to build our DSS, and the performance measures adopted to assess the predictive capability of the selected ML techniques.

Optimization Technique: The choices of ML methods' parameters, whose values vary from dataset to dataset, have a significant impact on their ability to predict the dependent variable [13]. This issue was revealed in several previous studies. They stated that to get accurate results, it is required to optimize the parameters settings of the ML models. In fact, this process remains as optimization problem. Unlike many papers conducted in the literature where this issue is not considered, this paper addresses this issue by using the Grid Search optimization technique. This technique has proved its effectiveness to identify the best model configurations in several fields.

Grid Search Technique (GS): runs an extensive search by testing every parameter setting over a predetermined range of parameter values, then chooses the configuration that produces the best classifier. Cross-validation is typically used to carry out the evaluation process.

Performance Measures: Four performance criteria were used to assess the predictive capability of the four ML selected in this paper, namely: accuracy, precision, recall, and the receiver operating characteristics curve (ROC Curve). The above-mentioned criteria are based on the confusion matrix. These performance metrics were the most used in the literature of medical field [2,14].

The ten-fold cross validation technique was used in all experiments performed in this study, as it is the most used validation method.

3 Results and Discussion

The empirical findings of the experiments raised in this study, which examined whether the chosen ML approaches may help cardiologists diagnose heart diseases effectively, are reported and discussed in the following paragraphs.

Toward this aim several software prototypes were developed based on Python libraries such as scikit-learn, Pandas, NumPy, using Python programming languages under Microsoft Environment.

As reported earlier our dataset was collected in department of cardiology in Sheikh Ziad Hospital University in Rabat, Morocco. In consist of 140 entries representing the population cohort recruited in this study. Each participant is described by 90 features. These features differ in their source as some are clinical and others derived from the ECG recording. In fact, the rationale behind adding clinical features that describes patients; such as diabetic, smoking, obesity, gender, SpO2, size, thoracique and so on, is due to the fact that the causes of CVDs are not well-known yet, therefore, these characteristics were added as they are considered as risk factors that may cause the heart diseases. Besides that, 60 features were extracted manually; using a standard ECG ruler, from the four-lead ECG recorded through the SW. For instance, several pieces of information were extracted from the recorded ECG for each lead according to the six parameters studied (See Sect. 2.1).

Before starting extracting knowledge from the dataset collected, a preprocessing step was mandatory to clean the data and check the consistency of all recorded values. Indeed, the dataset has several missing values and some inconsistent values with the type of data. As this study aims to check the predictive capability of ML techniques to accurately diagnosis heart conditions as the standard ECG, we procced to delete the rows containing missing values to provide a real recorded values as inputs to our ML techniques. To address the issue of values inconsistency noticed in some features, we double checked and recalculated these values.

Since the process of extracting values from the four-lead SW-ECG was done manually, and to ensure the correctness of the extracted values, two authors of the current study performed this process separately and thereafter they compared their results. If any mismatching occurred, another step of verification is performed by redonning the extraction.

Finally, and after performing the above-mentioned steps, our dataset remains with 48 entries described by 90 features. In fact, our dataset contains 203 missing values and as it well-known that the missing values significantly affect the performance of a given ML model, we choose to delete the rows containing these values. The distribution of classes was 60% for anormal cases and 40% for normal ones.

ML techniques are useful tools to extract hidden patterns present in a given dataset [7,8]. These techniques are used nowadays across different fields of research and practice, and they proved a high level of precision. In our study, these techniques are used to build a DSS, to help doctors to rapidly screen and diagnosis the heart conditions of a given patient, based on several clinical characteristics and four-lead ECG recorded by a SW. For that purpose, a comparison of four well-known ML methods is performed to verify; first, their ability to play the role of a standard ECG; second, to select the most accurate technique.

Before starting the learning process to build our ML models, an initial round of executions was performed to identify the best configuration of each selected ML technique using the GS optimization technique. The predefined range of values was determined based on the literature, Table 1. As known in the realm of ML, this step is critical and it should be effectuated for each model in every dataset, to guarantee the achievement of the best results. Table 1 lists the optimal parameters values identified by GS optimization technique.

Table 1. Search space for the GS optimization technique and the optimal configuration for the selected ML techniques

Technique	Search Space	Optimal configuration
KNN	K = {1 to 30} with increment of 2	K = 2
LR	Penalty = {'l1', 'l2', 'elasticnet', 'none'}	Penalty = 'l2'
SVM	Kernel = {'rbf', 'poly'} Gamma = {1e−3, 1e−4} C = {1, 10, 100, 1000}	Kernel = 'rbf' Gamma = 0.0001 C = 10
MLP	Hidden layer = {(10, 30,10), (20,)} Activation = {'relu', 'tanh'} Solver = {'sgd', 'adam'} Alpha = [0.0001, 0.05] Learning_rate = {'constant', 'adaptive'}	Hidden layer = {(20)} Activation = {'relu'} Solver = {'adam'} Alpha = 0.0001 Learning_rate = {'constant'}

Table 2 presents the performance accuracy of the four ML techniques selected in this study. The evaluation was performed using 10-fold cross validation technique and four performance criteria. Overall and as, it can be noticed from Table 2, all techniques achieved a good level of predictive capability. Indeed, according to accuracy criterion, all techniques achieved an acceptable level of performance, the SVM technique was ranked in the first position with 85%, followed by the remaining techniques with 74.5% each. Concerning the Precision criterion, the SVM technique occupied the first place with 82.5%, followed by the MLP technique with 73.08%, then the KNN method with 71.5% and finally the LR in the last position with 68.08%. As for recall measure, the best technique was the SVM with 83.40%, followed by the MLP technique (74.17%), the LR (71.67%) and finally the KNN technique with 70%.

Table 2. Performance accuracy of the selected ML techniques

Technique	Accuracy	Precision	Recall
SVM	0.85	0.825	0.8340
MLP	0.7450	0.7308	0.7417
LR	0.7450	0.6808	0.7167
KNN	0.7450	0.7150	0.7

Figure 1 achieved a good level shows the ROC curve of the four ML techniques along with their AUC measure. The ROC curve is generated by plotting the true positive rate against the false positive rate at various levels of threshold. The AUC provides an overall measure of performance across all possible classification thresholds. The ROC-AUC can be interpreted as the probability that a given model ranks a random positive instance more highly than a random negative instance. As it can be seen, the AUC ranked the SVM in the first place with 91% and a standard deviation (std) of 13%, followed by the LR technique (83% and a std of 22%), then the MLP technique (75% and 30%), and finally the KNN technique. In fact, the predictive ability of the SVM; according to the ROC-AUC and the std, is more stable than the remaining three techniques, which are characterized with high fluctuation in term of this performance measure (std > 30%).

Some conclusions can be drawn based on the empirical results reported in this study, which are as follow:

- The use of optimization technique to tune the models hyperparameters' is mandatory to achieve a good level of accuracy.
- All techniques achieved a good level of ability of detecting the actual heart condition.
- The SVM technique was the most accurate technique across different performance measures in this context and configuration (i.e., dataset characteristics, optimization technique, and validation technique), and its ability to differentiate between the normal and anormal cases was high and stable, compared to the other selected techniques.
- The obtained results are very promising and there is still room of improvement, especially by examining other ML models and investigating other optimization techniques.

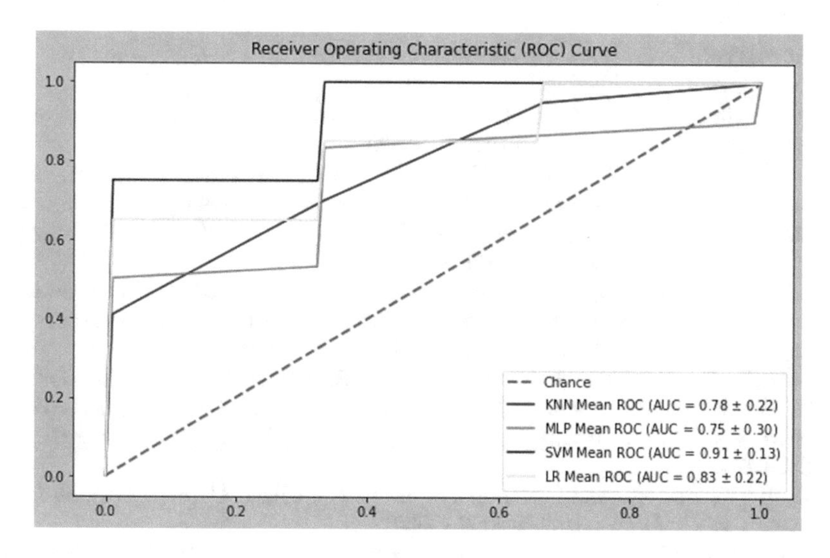

Fig. 1. ROC Cure and AUC for the selected ML techniques, and the ROC curve of a random classifier

4 Conclusion and Further Work

To assist cardiologists in rapidly screening and diagnosing patients, this study suggests developing a DSS that uses ML approaches to diagnose heart activity based on several clinical characteristics and four-lead ECG recorded by a SW. To determine if this DSS can successfully give the cardiologist a reliable assessment of the patient's heart status, four well-known ML approaches were chosen. These techniques were assessed using four performance criteria, using the 10-fold cross validation technique for the hyperparameters optimization, building and testing steps. We recall that the dataset used in this study was collect in Sheikh Ziad Hospital University in Morocco. The findings related to our RQ are as follows:

The four ML techniques were able to achieve a good level of ability to differentiate between normal and anormal cases. Furthermore, the SVM technique reached an excellent level of performance. These results remain very promising and will allow us to build an efficient DSS, that will help cardiologists to rapidly screen and diagnosis patient. The DSS will work based only on a few clinical information and four-lead ECG recorded by a SW that can be obtained automatically through a software. The results show the capability of the proposed DSS to serve as an alternative to the standard ECG, and with more enhancement it may be included in routine medical care.

Based on these promising results, several research directions arise to enhance the predictive capability of the proposed DSS. One of these directions is examining several ML techniques and other types (i.e., ensemble methods), and investigating other optimization techniques.

References

1. World Health Organization (WHO). https://www.who.int/en/news-room/fact-sheets/detail/cardiovascular-diseases-(cvds)
2. Hosni, M.: Carrillo de Gea JM, Idri A, el Bajta M, Fernández Alemán JL, García-Mateos G, Abnane I: A systematic mapping study for ensemble classification methods in cardiovascular disease. Artif. Intell. Rev. **54**(4), 2827–2861 (2021). https://doi.org/10.1007/s10462-020-09914-6
3. Ashley, E.A., Raxwal, V., Froelicher, V.: An evidence-based review of the resting electrocardiogram as a screening technique for heart disease. Prog Cardiovasc Dis **44**(1), 55–67 (2001). https://doi.org/10.1053/pcad.2001.24683
4. Behzadi, A., Shamloo, A.S., Mouratis, K., Hindricks, G., Arya, A., Bollmann, A.: Feasibility and reliability of smartwatch to obtain 3-lead electrocardiogram recordings. Sensors (Switzerland) **20**(18), 1–11 (2020). https://doi.org/10.3390/s20185074
5. Rk, K.: Technology and healthcare costs. Ann. Pediatr. Cardiol. **4**(1), 84 (2011). https://doi.org/10.4103/0974-2069.79634
6. Hosni M, Carrillo-De-Gea JM, Idri A, Fernandez-Aleman JL, Garcia-Berna JA: Using ensemble classification methods in lung cancer disease. In: Proceedings of the Annual International Conference of the IEEE Engineering in Medicine and Biology Society, EMBS (2020)
7. Faust O, Ng EYK: Computer aided diagnosis for cardiovascular diseases based on ECG signals: A survey. J Mech Med Biol 16 (2016)
8. Hijazi, S., Page, A., Kantarci, B., Soyata, T.: Machine Learning in Cardiac Health Monitoring and Decision Support. Computer (Long Beach Calif) **49**(11), 38–48 (2016). https://doi.org/10.1109/MC.2016.339
9. Hosni M, Idri A, Abran A: On the value of filter feature selection techniques in homogeneous ensembles effort estimation. Journal of Software: Evolution and Process 33(6). (2021) https://doi.org/10.1002/smr.2343
10. Simon H: Neural networks: a comprehensive foundation, 2nd edn.(1999) MacMillan Publishing
11. Vapnik V: Principles of risk minimization for learning theory. In: Advances in neural information processing systems. pp 831-838 (1992)
12. Kleinbaum, D.G., Klein, M.: Logistic Regression. Springer, New York, New York, NY (2010)
13. Idri A, Hosni M, Abnane I, Carrillo de Gea JM, Fernández Alemán JL: Impact of parameter tuning on machine learning based breast cancer classification. In: Advances in Intelligent Systems and Computing. Springer Verlag, pp 115-125 (2019)
14. Hosni, M., Abnane, I., Idri, A., Carrillo, J.M., Gea, D., Luis, J., Alemán, F.: Reviewing ensemble classification methods in breast cancer. Comput. Methods Programs Biomed. **177**, 89–112 (2019). https://doi.org/10.1016/j.cmpb.2019.05.019

Cognitive Computing in the Travel and Tourism Industry

Teresa Guarda[1,2,3](✉), Isabel Lopes[3,4], and Paula Odete Fernandes[4]

[1] Universidad Estatal Península de Santa Elena, La Libertad, Ecuador
tguarda@gmail.com
[2] CIST – Centro de Investigación en Sistemas y Telecomunicaciones, Universidad Estatal Península de Santa Elena, La Libertad, Ecuador
[3] Algoritmi Centre, Minho University, Guimarães, Portugal
isalopes@ipb.pt
[4] Applied Management Research Unit (UNIAG), Instituto Politécnico de Bragança (IPB), 5300-253 Bragança, Portugal
pof@ipb.pt

Abstract. Cognitive computing emerged from a mixture of cognitive science and computer science. Although cognitive computing is seen as a challenge for companies, it is an opportunity that raises the bar for the services provided, allowing companies that are committed to investing in innovation, and in this particular case in the travel and tourism sector, being able to stand out from the competition by adopting new technologies that make it possible to offer revolutionary and value-added experiences. Whether optimizing processes or developing new services, this technology is the key to innovation and competitiveness in the travel and tourism sector. With the user's history and previous search data, cognitive systems, even before the user realizes it, present specific options according to their profile\preferences. In this sense, the system has the ability to automatically restrict the entire travel package. The main objective of this work is to explore the area of Cognitive computing in the context of Travel and Tourism Industry.

Keywords: Cognitive computing · Travel and Tourism Industry · Natural Language Processing · Artificial Intelligence

1 Introduction

Throughout history computers and electronic devices have had to be told what to do usually programmed by software developers before they could complete a task. But these days the scenario is different: computers will start to learn how to perform different tasks based on experiences, which is very similar to the development of cognitive abilities in humans.

However, unlike humans, these intelligent computing systems will retain and remember everything that learn. With that in mind, imagine the possibilities for every human worker to have access to a computerized assistant who will always be ready to offer

A. Rocha et al. (Eds.): WorldCIST 2023, LNNS 800, pp. 131–138, 2024.
https://doi.org/10.1007/978-3-031-45645-9_12

useful information and guidance. This is the essence of the age of cognitive computing. It is very common to read or hear that the human brain is one of the most perfect and complex systems that exist. Not by chance, this is one of the bases for the development of cognitive computing.

What is cognitive computing in addition to its structure of neurons and synapses, there is another reason that places it in the focus of innovation: optimizing the performance of human beings in their activities [1].

Cognitive computing capabilities include natural language interfaces that enable users to interact with the application using common language [2] and the ability to monitor thousands of sensors embedded in the environment to understand changes in real time [3].

Less unique among cognitive applications is its ability to read unstructured data of all types, ranging from phone calls, emails, and scientific articles, to technical journals, with the aim of scanning databases to analyze and extract patterns.

Cognitive computing is available in the cloud and can be made available in almost any medium that people use. Thus, we can explore cognitive applications on smartphones, through automatic panels or digital assistants, and we can also make cognitive applications available through chips embedded in device, for example, an office machine that can talk to its user and help diagnose any problems encountered.

2 Cognitive Computing

Cognitive computing is a discipline that integrates concepts from neuroscience, cognitive psychology, information science, computational linguistic and artificial intelligence [4, 5].

Cognitive computing can be seen as the third era of computing [6]. This image has been built due to its development, which mixes the science of cognition and computer science to create technologies capable of simulating the human thought process.

The old computers needed to be commanded, which is no longer necessary nowadays, as they can use a self-learning cognitive algorithm, data mining, natural language processing and other elements, thus being able to imitate the human brain in the decision-making process.

In order to have a clearer idea of the evolution of computing, we can consider three eras: the tabulating era, the programming era, and the cognitive era [7] (Fig. 1).

Tabulating era (1900–1940s), was characterized by single-use electromechanical counter systems, and by the use of perforated cards, in which data and instructions for the machine were inserted and stored. In the case of the programming era (1950s–present), after the Second World War, due to scientific and military information needs, the evolution of digital computers was fast, becoming standard tools in companies and governments. The third era, the cognitive era (2011-) appeared as the natural evolution of programming era, to somehow overcome the limitations in the interaction with the human being, combining the human-machine forces [6–8].

AI can add a percentage point to the country's annual economic growth rate by 2035, and its results confirm this projection [9].

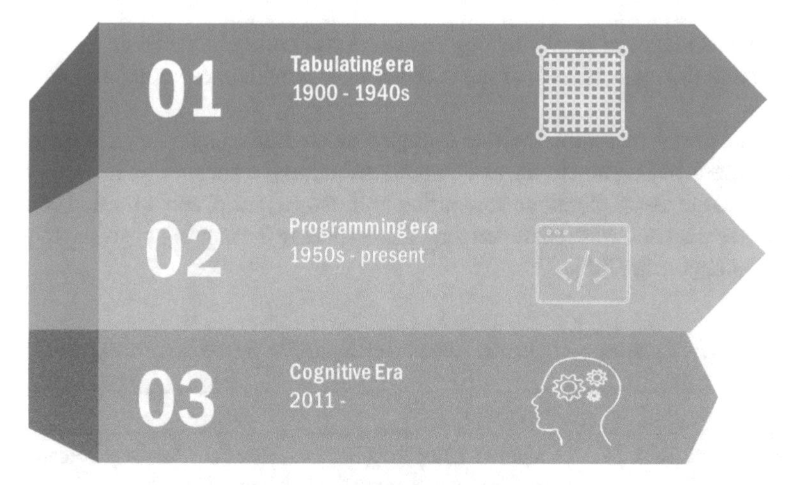

Fig. 1. Computing evolution eras.

AI has been developed to perform tasks of low decision-making and bureaucratic complexity, leaving human capital to focus its efforts on more relevant and strategic activities for the business.

In this mix of new technologies, there's still the machine learning (ML) category, which consists of improving the processing of data collected from different sources, allowing patterns and mathematical models to be transformed into forecasts, scenarios and trends.

The aim of cognitive computing applications, is the data interpretation, just like the human brain. Cognitive innovations go beyond the volume of data to be interconnected, considering structured and unstructured data for their analysis, such as text, audio and voice [1].

3 Advantages of Cognitive Computing

Systems that combine cognitive intelligence and AI allow the processing and analysis of data from different sources and formats, then where executed nothing is neglected (data, knowledge), allowing decision making to be more assertive and fast.

A professional in the accounting field, for example, can take months to understand the laws and rules of a particular country, quoted to host a new branch of the company. With cognitive computing, he could indicate that, after an analysis and comparison of all variables, the business would be unfeasible.

According to the IDC (International Data Corporation) study with predictions for the IT industry in 2020, it is projected that: by 2025, at least 90% of new enterprise applications will have Artificial Intelligence (AI) built in and that; and by 2024, more than 50% of user interaction interfaces will use AI-enabled computer vision, speech, natural language processing (NLP) and AR/VR [10].

Cognitive computing must be used to provide more resources to business leaders facing a time of great business impact [11]. And to succeed or even survive, they must reinvent their organizations with digital technologies.

Some advantages of implementing cognitive computing in the business, in addition to facilitating the management of the enterprise, is provide more efficiency for solving problems.

It is safe to say, therefore, in these examples of cognitive technology, that computers become experts in a certain topic, and their contributions can be unlimited for their users.

As in AI, the use of cognitive computing is already a reality and is used in companies in different segments, there are ten several factors where it can bring positive results with cognitive computing (Table 1).

Table 1. Cognitive computation factors with a positive impact on results [12–17].

Factors	Description
Processes Automation	Cognitive computing is an excellent ally for the automation of processes, since, through algorithms, it can help in the self-learning of these machines so that they perform activities in a more intelligent, agile way and with the reduction or elimination of repetitive tasks that before they were executed by people
Task complexity reduction	Cognitive technology is capable of simplifying processes that were previously performed manually. Through agile software, companies will be able to process large amounts of data to make the best possible decisions, which makes it possible to direct employees to other more strategic tasks for the business
Project management	When identifying patterns in the data, calculating scenarios is one of the most basic functions that cognitive computing can offer, as well as in project planning and tracking. There is also an improvement in the analysis of information and survey of trends that can offer advantages to the business, or allow them to be corrected within a safe limit for the success of the goals
Client management	Companies that offer different relationship channels have access to data in different. With cognitive computing and artificial intelligence, all the information generated from customer contacts can be used to customize the service and consumption experience. The history of individual contacts, evaluated with other customer data, ensures that the company can map behavior and consumption trends. From there, the company can create more efficient commercial strategies in attracting new leads and maintaining the relationship with its buyers

(*continued*)

Table 1. (*continued*)

Factors	Description
Increase in sales	Invest in this technology, where machines can be trained to approach consumers across multiple communication channels. Through the algorithms, a reading and evaluation of customer histories, their credit score and even predictions is performed. With this, it is possible to use the correct and more natural language to recommend products and offers, helping the company in its sales
Access by all media	Cognitive computing can be applied on any platform, as the execution of algorithms allows it to be accessible in any environment, while the results generated can be analyzed in an unlimited way
Reading unstructured data	A cognitive system is able to analyze both information called structured data, which are complete and clearly identified, and data that are apparently disconnected to generate conclusions, known as unstructured data, as they will not always be available or clear. Its reading becomes more complex for human readers, but not for machines. Thus, all types of information become input for the cognitive system to determine patterns and make connections, quickly and safely, adding positive results to your business
External sensor monitoring	With the growing use of Internet of Things applications, it has become necessary to implement smart sensors to track the data generated by these devices. The union of cognitive technology with IoT in external sensors makes it possible to capture more detailed information on the operation of equipment, monitor the data generated from operations and also seek solutions to improve their efficiency
Proactive security detection	Cognitive computing is widely used to control the security of company devices. You can monitor device performance data through cognitive analytics and check crash reports. Therefore, it is possible to identify the possible causes of malfunctions and suggest the best corrective actions to solve possible problems quickly and effectively

4 Cognitive Computing in Travel and Tourism Industry

The evolution of technology has led to new trends in the industry, and some of these trends have significantly affected the travel and tourism industry, and compared to other sectors, is at the forefront of digital transformation.

According to Statista [18], an increase in the global online travel sector was estimated in 2021 compared to the year 2020. The global online travel market is expected to be valued at around USD 433 billion in 2021, rising from around USD 396 billion in 2020, and by 2026 it is forecast to total approximately USD 691 billion (Fig. 2).

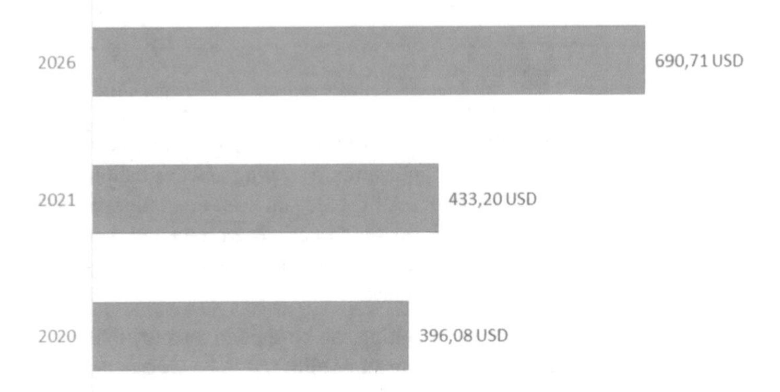

Fig. 2. Online travel market size worldwide in 2020 and 2021, with a forecast for 2026 in billion U.S. dollars [18].

Travel and tourism industry has always been supported by collaborative processes, since in each travel package there are different suppliers for each of the elements of the travel package. This collaboration allows us to offer customers the best possible options.

Cognitive computing in travel and tourism industry has a fundamental characteristic, which is the ability to understand and respond to human behavior in the event of unforeseen events or emergencies.

Customers often have to look for the package that meets their needs in different channels. It is in this sense that cognitive computing, supported by the customer's previous travel information regarding their preferences (duration, price), can book similar tickets and accommodation.

It will be possible to choose different travel options in case there is any unforeseen in terms of accommodation, and can also automatically modify itineraries making alternative reservations automatically, allowing the tourist to have a viable option without causing any inconvenience in their trip, facilitating automatic contact with their services supplier, so we have support in the solution reached, guaranteeing that the trip will not be affected by any unforeseen.

Cognitive computing can accompany the entire travel process from reservations (flights, accommodation, and transport), personalized recommendations during the trip, to travel assistance.

Cognitive computing can make transport and accommodation reservations autonomously, looking for the best offers, and selecting the most advantageous option for the user according to his profile, without his intervention being necessary.

It should be noted that one of the most important benefits of cognitive computing in travel and tourism is the ability to understand and respond to human behavior [19]. In this sense, in the event of an unforeseen occurrence, you can choose other travel options, and book alternative accommodation automatically, thus minimizing the inconvenience caused to the traveler.

5 Conclusions

Cognitive computing is based on self-learning systems that use techniques to intelligently perform specific human tasks.

Cognitive computing began to enhance business decisions and the performance of human thought processes and traditional analytics. Thus, it represents an approach to deploying software and solutions that encompass the use of artificial intelligence to mimic human thinking.

The growth of such innovations has been exponential as applications of the technology become more sophisticated, and the use of such technologies is taking root in places and industries with new applications being discovered almost daily.

Cognitive computing borrows its brain power from machine learning algorithms and artificial intelligence. After all, using these technologies, they continually learn from data received as input during the normal course of operations.

Pattern recognition systems, natural language processing, data mining increase the efficiency of these systems. With intelligence, which is continually built up, they are able to predict patterns and arrive at proactive decisions. That is, anticipating problems and deriving possible solutions. Over time, they become fully autonomous and can handle operations without human interference. In this way, the long-awaited future of complete automation becomes real.

To conclude, cognitive technology is going to enrich the interaction of tourists with the company or with the tourist destination where they are, as never before, in such a way that the visitor lives a much more pleasant experience. Cognitive systems will function as advisors with the capacity to dialogue with tourists, help them enjoy the tourist environment based on their preferences and even respond to their concerns. Cognitive computing ensures that travelers can make their trips and put aside worries about unforeseen events.

Acknowledgements. The authors are grateful to the UNIAG, R&D unit funded by the FCT – Portuguese Foundation for the Development of Science and Technology, Ministry of Science, Technology and Higher Education. "Project Code Reference: UIDB/04752/2020".

References

1. Garcez, A.D., et al.: Neural-symbolic learning and reasoning: a survey and interpretation. Neuro-Symbolic Artificial Intelligence: The State of the Art. IOS Press (2022)
2. Hurwitz, J.S., Kaufman, M., Bowles, A.: Cognitive Computing and Big Data Analytics. John Wiley & Sons, Hoboken (2015)
3. Hwang, K.C.: Big-Data Analytics for Cloud, IoT and Cognitive Computing. John Wiley & Sons, Hoboken (2017)
4. IDC: IDC FutureScape: Worldwide IT Industry. IDC. Obtenido de (2019). https://venturebeat.com/2020/07/15/mit-researchers-find-systematic-shortcomings-in-imagenet-data-set/
5. Jha, N., Prashar, D., Nagpal, A.: Combining artificial intelligence with robotic process automation—an intelligent automation approach. In: Ahmed, K.R., Hassanien, A.E. (eds.) Deep Learning and Big Data for Intelligent Transportation: Enabling Technologies and Future Trends, pp. 245–264. Springer International Publishing, Cham (2021). https://doi.org/10.1007/978-3-030-65661-4_12

6. Kaur, S., Gupta, S., Singh, S.K., Perano, M.: Organizational ambidexterity through global strategic partnerships: a cognitive computing perspective. Technol. Forecast. Soc. Chang. **145**, 43–54 (2019). https://doi.org/10.1016/j.techfore.2019.04.027
7. Kelly, J.E.: Computing, cognition and the future of knowing. Whitepaper. IBM Reseach. Obtenido de (2015). https://cloud.report/Resources/Whitepapers/e55108d4-92bd-428a-b432-64530b50c6b9_Computing_Cognition_WhitePaper.pdf
8. Laird, J.E., Lebiere, C., Rosenbloom, P.S.: A standard model of the mind: toward a common computational framework across artificial intelligence, cognitive science, neuroscience, and robotics. AI Mag. **38**(4), 13–26 (2017). Obtenido de https://ojs.aaai.org/index.php/aimagazine/article/view/2744/2671
9. Noor, A.K.: Potential of cognitive computing and cognitive systems. Open Eng. **5**(1), 75–88 (2014). https://doi.org/10.1515/eng-2015-0008
10. Ordóñez, M.D., et al.: IoT technologies and applications in tourism and travel industries. En Internet of Things–The Call of the Edge (págs. 341–360)). River publishers (2022)
11. Ploennigs, J., Ba, A., Barry, M.: Materializing the promises of cognitive IoT: How cognitive buildings are shaping the way. IEEE Internet Things J. **5**(4), 2367–2374 (2014). https://doi.org/10.1007/s11227-013-1021-9
12. Pramanik, P.K.D., Pal, S., Choudhury, P.: Beyond automation: the cognitive IoT. artificial intelligence brings sense to the Internet of Things. In: Sangaiah, A.K., Thangavelu, A., Sundaram, V.M. (eds.) Cognitive computing for big data systems over IoT, pp. 1–37. Springer International Publishing, Cham (2018). https://doi.org/10.1007/978-3-319-70688-7_1
13. PWC: The macroeconomic impact of artificial intelligence. PWC, Obtenido de (2018). https://www.pwc.co.uk/economic-services/assets/macroeconomic-impact-of-ai-technical-report-feb-18.pdf
14. Statista Research Department. Online travel market size worldwide 2020–2026 (2022)
15. Sumathi, D., Poongodi, T., Balamurugan, B., Ramasamy, L.K.: Cognitive intelligence and big data in healthcare. Scrivener Publishing LLC (2022). https://doi.org/10.1002/9781119771982
16. Tarafdar, M., Beath, C.M., Ross, J.W.: Using AI to enhance business operations. MIT Sloan Manage. Rev. **60**(4), 37–44 (2019). Obtenido de https://www.proquest.com/docview/2273705050/fulltextPDF/AB8DDA9F4C0443CPQ/1?accountid=39260
17. Wang, G.: DGCC: data-driven granular cognitive computing. Granular Comput. **2**(4), 343–355 (2017). https://doi.org/10.1007/s41066-017-0048-3
18. Xiang, Z., Magnini, V.P., Fesenmaier, D.R.: Information technology and consumer behavior in travel and tourism: Insights from travel planning using the internet. J. Retail. Consum. Serv.Consum. Serv. **22**, 244–249 (2015). https://doi.org/10.1016/j.jretconser.2014.08.005
19. Yao, Y.: Three-way decision and granular computing. Int. J. Approximate Reasoning **103**, 107–123 (2018). https://doi.org/10.1016/j.ijar.2018.09.005

Expectations and Knowledge Sharing by Stakeholders in a Bioproducts Community of Practice: An Exploratory Study

Leandro Oliveira[1](✉) and Eduardo Luís Cardoso[2]

[1] CBIOS - Universidade Lusófona's Research Center for Biosciences & Health Technologies, Campo Grande 376, 1749-024 Lisbon, Portugal
leandro.oliveira@ulusofona.pt

[2] Universidade Católica Portuguesa, CBQF - Centro de Biotecnologia e Química Fina – Laboratório Associado, Escola Superior de Biotecnologia, Rua de Diogo Botelho 1327, 4169-005 Porto, Portugal

Abstract. The Communities of Practice (CoP) are informal groups of individuals with common interests who come together to share experiences, discuss, and improve a given practice. This work aims to evaluate the perception of knowledge sharing in a community of practices in bioproducts by its participants and their expectations regarding added value, information and available resources/activities, and shared materials.

A questionnaire disseminated by CoP and social networks was developed between December 2020 and February 2021. Forty-three responses were collected. Most participants were female (67.4%), with a mean age of 32 years (SD = 9.1 years), and they had a degree (39.5%). Most participants have a very positive perception of knowledge sharing through a community of practice in bioproducts. Participants believe it is an added value to belong to a CoP due to the exchange of information and experiences and would like information on industry trends, scientific events, and innovations to be available. These results will make it possible to promote knowledge sharing at the CoP, adjusting the information, resources, and materials available there, to the expectations and preferences of the participants.

Keywords: Communities of Practice · Knowledge Sharing · Bioproducts · Attitudes · Expectations

1 Introduction

Bioeconomy is related to different policy areas, such as food, agriculture, energy, the environment, spatial planning, etc. Furthermore, there may be a demand by more parts for the same biological resources [1]. In this way, it is important to create environments that promote the exchange of experiences and ideas with the aim of promoting innovation and strategic partnerships for the development of solutions that correspond to the current challenges of the Bioeconomy. People's experience is a natural resource of knowledge for any organization and therefore it is important to find adequate ways to

A. Rocha et al. (Eds.): WorldCIST 2023, LNNS 800, pp. 139–146, 2024.
https://doi.org/10.1007/978-3-031-45645-9_13

extract it and transform it into communicable forms to be shared. However, sharing this knowledge implies a high motivation of each worker, a motivation built and sustained by a stimulating organizational culture, based on trust and respect [2].

Knowledge sharing has been identified as a focus area for knowledge management in organizations. The importance of this topic lies in the fact that knowledge only has value when it is transferred from individuals to the organization [3]. However, there are several barriers to knowledge sharing: lack of time, lack of trust, strength of relationships, lack of social networks, lack of structures for sharing, low motivation, competitiveness between different groups, among others [4]. If knowledge sharing has to overcome many barriers in organizations, in communities of practice it becomes almost a natural process [2].

Communities of practice are groups of voluntary participants who interact continuously around a shared concern. These provide an environment in which groups of people can share their experiences of practice, develop and discuss areas of interest and build a sense of community [5, 6].

The use of communities of practice to share knowledge brings some benefits, such as: facilitating collaboration and networking; provide opportunities for members to contribute; cross-sharing ideas, increasing opportunities for innovation; help people update their knowledge; develop professional skills to solve problems quickly; making efficient organizational decisions; allow the construction of an organizational memory and disseminate best practices; enable organizations to gain a competitive advantage [7]. Thus, it is important to study the perception of knowledge sharing in communities of practice to outline strategies capable of promoting it and improving organizational management and competitiveness.

This work aims to evaluate the perception of knowledge sharing in a community of practices in bioproducts by its participants. It also aims to study their expectations regarding the participation of the referred community of practices, especially regarding added value, information and available resources/activities, and shared materials.

2 Methodology

Based on a study [8], were constructed or adapted to the Portuguese language to assess specific aspects of knowledge sharing. The questionnaire contained 23 items answered on a 5-point Likert scale (1 – strongly disagree; 5 – strongly agree). The final score was obtained by adding the scores of the items that make up the scale (except for item 2, which is scored inversely). The total score, obtained by adding the scores of all items, varies between 23 and 115. In each item and in the total, higher scores correspond to more positive perceptions regarding knowledge sharing. The items consist of statements referring to aspects such as: attitudes (items 1, 2, 3, 4 and 5), perceived behavior (items 6, 7, 8 and 9), subjective norm (items 10, 11, 12, 13, 14 and 15), intention (items 16, 17, 18, and 19), and knowledge-sharing behavior (items 20, 21, 22, 23). The psychometric properties of this scale were previously published in another study [9].

An online questionnaire was developed and distributed between December 2020 and February 2021. This was disseminated in a community of practice in bioproducts (199 members) [10] on Facebook and shared again on the social networks of the Faculty of Biotechnology as a stakeholder in this area.

Questions were included for sociodemographic characterization, to assess attitudes, intentions and knowledge sharing in communities of practice. To assess the added value, information, materials, and resources of a community that practices bioproducts, multiple-choice questions were asked in which more than one option could be selected.

Participants were informed about the objectives and methods of the study, as well as their rights, in accordance with the Declaration of Helsinki and applicable legislation. Acceptance was given through online informed consent. Personal data was protected through an encryption system, which also guaranteed anonymity.

Statistical analysis was performed using IBM SPSS Statistics, version 26.0 for Windows. Descriptive statistics consisted of calculating the mean and standard deviation (SD) in the case of cardinal variables and calculating relative and absolute frequencies in the case of ordinal and nominal variables. To compare the mean orders of the independent samples, the Mann-Whitney test was used; and to compare proportions between qualitative variables, the chi-square test was used. Spearman's correlation coefficient (r) was calculated to assess the degree of association between pairs of continuous variables. A significance level (p) of 5% was considered in all analyzes [11].

3 Results

The sample of this study consisted of 42 Portuguese stakeholders (100.0%) who participated in communities of practice in bioproducts. Most participants were female (67.4%), with a mean age of 32 years (SD = 9.1 years), had a degree (39.5%), in food engineering, bioengineering or food biotechnology (30.2%), lived in the North region (83.7%) and were employed (48.8%).

Overall, participants scored well on the knowledge sharing scale, 84 (72; 94) out of 115 points. Regarding attitudes towards knowledge sharing in communities of practice, around 50% of participants agree that the knowledge they share with other members of the community of practice is useful. About 60% of participants strongly disagree/disagree that the knowledge they share with other members of the community of practice is of low quality. Participants agree/strongly agree that sharing their knowledge with other members of the community of practice is a pleasant experience (74.4%), important to them (53.5%), and something smart (60.5%).

Respecting perceived behavior in relation to knowledge sharing in communities of practice, around 50% of participants agree/strongly agree that they have the ability to control the knowledge they share (60.5%), the resources to share their knowledge (58.2%), and the knowledge to use (40.2%) the community of practice. Furthermore, around 50% of participants agree/strongly agree that, given the resources, opportunities, and knowledge, it would be easy for them to share their knowledge in the community of practice.

Concerning the subjective norm and normative beliefs about knowledge sharing, more than 50% of the participants agree/strongly agree that the director of the company where they work, or their supervisor/work coordinator or colleagues would consider it beneficial for them to share their knowledge with other members of the community of practice. Furthermore, about 50% of the participants agree/strongly agree that administrators, or members, or colleagues in the community of practice would find it beneficial to share their knowledge with other members of the community of practice.

Regarding the intention to share knowledge dimension, around 40% of the participants do not agree or disagree that they will share work reports/technical reports and official documents/legislation obtained within or outside the scope of their professional activity with other members of the community of practice.

About 50% and 60% of the participants agree/strongly agree that they will share the knowledge gained from their experience or working know-how with other members of the community of practice; and will share knowledge based on their academic or professional training with members of the community of practice, respectively.

Concerning the knowledge-sharing behavior dimension, around 40% of the participants do not agree or disagree that they often share work reports/technical reports and official documents/legislation obtained within or outside the scope of their professional activity with other members of the community of practice. About 35% of participants do not agree or disagree that they often share knowledge gained from their experience or working know-how with other members of the community of practice. And 39.5% of participants agree that they often share knowledge based on their academic or professional background with members of the community of practice.

No relationship was found between sex and the knowledge sharing scale or its dimensions. However, it was found that the older the age, the lower the score in the perceived behavior dimension (r: -0.337; p: 0.027).

Table 1 presents the results on the added value, information, materials, and resources of a community that practices bioproducts, according to the participants' responses. The exchange of information (19.1%) and the exchange of experiences (15.1%) are the added value that the participants consider existing because they belong to a community of practices in bioproducts. They would also like information about trends in the sector (20.8%), scientific events (19.4%) and innovations (18.1%) to be available. Regarding the materials, the participants would like to be shared in the community of practices in bioproducts, scientific articles (22.3%), articles and opinions (19.6%) and videos (16.9%) stand out. Finally, from the resources/activities that the participants would like to have available, the pages (25.4%), databases (16.4%) and URLs (12.3%) stand out.

Table 1. Added value, information, materials, and resources from a community of practice in bioproducts.

	n (%) *
What added value do you think there is in participating in a community of practice in bioproducts? (n = 152)	
Exchange of experiences	23 (15.1)
Exchange of information	29 (19.1)
Be up to date	17 (11.2)
Networking	15 (9.9)
Clarification of doubts/problem solving	17 (11.2)
Search for potential employees for the company	10 (6.6)
Find training materials	8 (5.3)
Stay up to date with events happening in the sector	18 (11.8)
Look for sources of inspiration for innovation	15 (9.9)
What information would you like to have available in a bioproducts community of practice? (n = 144)	
Scientific events	28 (19.4)
Sector events	9 (6.3)
Courses	18 (12.5)
Internship offers	8 (5.6)
Job offers	18 (12.5)
Job search	4 (2.8)
Industry trends	30 (20.8)
Innovations	26 (18.1)
What materials would you like to be shared in a bioproducts community of practice? (n = 148)	
Videos	25 (16.9)
Photos	21 (14.2)
Infographics	22 (14.9)
Scientific articles	33 (22.3)
Opinion articles	29 (19.6)
Leaflets	18 (12.2)
What resources/activities would you like to have available in the Bioproducts Community of Practice? (n = 122)	
File	16 (13.1)
Book	13 (10.7)
Page	31 (25.4)
Folder	9 (7.4)
Tag (label)	6 (4.9)
URL	15 (12.3)

(continued)

Table 1. (*continued*)

	n (%) *
Database	20 (16.4)
Chat	12 (9.8)

* Participants could choose more than one option.

4 Discussion

This study aimed to study the perception of knowledge sharing in a community of practice in bioproducts. It also intended to describe the expectations of belonging to a community of practice of this kind, namely regarding its added value, information and available resources/activities, and shared materials.

In the present study, it was found that the participants had a positive perception regarding the sharing of knowledge, however, a perception of sharing behavior was lower than the other dimensions, such as attitudes, beliefs, and intentions. This may be because participants in communities of practice with the ability to contribute knowledge can assess how other members behave and whether they can be trusted, which influences their decisions about helping other individuals by sharing. Your knowledge [12]. Another possible explanation for the low effective sharing of knowledge may be the lack of motivation, less motivated members do not actively participate in the exchange of information and, therefore, do not increase interactions within a community of practice [5]. Many individuals who participate in communities of practice are mainly motivated by the benefits they can have for their professional activity, for example participating in certain projects, improving their career prospects, facilitating their work and improving contact with colleagues [13].

Belonging to practical communities can bring added value as expected by the participants in this study. For example, work with students revealed that interactions resulting from participation in a community of practice indicated immediate learning and the creation of new insights (potential value), as well as the effective transfer of knowledge to academic practice, translated into significant improvement. of your performance [14]. In addition, another work in a business environment reports that knowledge-based human resource management facilitates knowledge sharing among co-workers; and co-workers' knowledge-sharing behaviors seem to facilitate their performance in creating innovation [15]. Another study in small and medium-sized companies revealed that knowledge sharing on social networks has a positive impact on innovation, with the transfer of knowledge through social networks between small and medium-sized companies and their international business partners helping them to have technological advances and reach new markets for their businesses [16]. This may also explain the demand for information about scientific events, courses, and innovations. In addition, the most mentioned resources/activities that participants would like to have available in the community of practice were pages (displays a page (WEB type) that can contain text, website/video links, images and other multimedia elements), the databases (collaboration tool, built by participants, which allows creating, updating, consulting and displaying a list of records on a given topic, in a pre-defined structure) and URL (provides a link to a page from

Internet). In other words, resources/activities that allow the sharing of information and experiences that can contribute to professional development.

To increase the sharing of knowledge in the practical community, a mechanism could be created in which voluntary contributions are effectively rewarded. For example, a system for evaluating members' publications, giving points for their quality. Another strategy to adopt would be the placement of more information according to the expectations of the participants. This may encourage members on their own initiative to also share information of interest.

This work has some limitations, namely its cross-sectional design that does not allow the extrapolation of the results, in addition the sample size is small. However, the results presented offer an overview of knowledge sharing in this specific community of practices and could serve to improve its dynamism, meeting the expectations of its members. In addition, it could serve as an exploratory study for larger ones and as a reference for other similar communities. As future works, it is suggested to relate knowledge sharing with trust in social networks, as well as carrying out a larger study and even in other practical communities in order to obtain more representative results.

5 Conclusion

Most participants have a very positive perception of knowledge sharing through a community of practice in bioproducts. Most agree/strongly agree that they intend to share knowledge in the community of practice, however they refer to having an intermediate position in the effective sharing of it. Participants believe it is an added value to belong to a practical community due to the exchange of information and experiences, and they would like to have information about trends in the sector, scientific events and innovations available. In addition, they would like the materials to be shared in the form of scientific and opinion articles and videos; and, that resources/activities such as pages, databases and URLs were available.

This study will promote the effective knowledge sharing in the community of practice under study, adjusting the information, resources and materials available in it, to the expectations and preferences of the participants.

Funding. This work was co-financed by the European Union - ERASMUS + Program, through the LEAD project (2021–1-HU01-KA220-VET-000033052) and Fundação para a Ciência e a Tecnologia (FCT) through projects UIDB/04567/2020 and UIDP/04567/2020 to CBIOS.

References

1. COWI A/S., Utrecht University: Environmental impact assessments of innovative bio-based product. Luxembourg: Publications Office of the European Union (2019)
2. Bratianu, C.: Knowledge sharing and communities of practice. In: Organizational Knowledge Dynamics: Managing Knowledge Creation, Acquisition, Sharing, and Transformation. edn. Hershey: IGI Global (2015)
3. Diab, Y.: The concept of knowledge sharing in organizations (Studying the Personal and Organizational Factors and Their Effect on Knowledge Management) (2021)

4. Nadason, S., Raj, S., Ahmi, A.: Knowledge Sharing and Barriers in Organizations: A Conceptual Paper on Knowledge-Management Strategy. Indian-Pacific Journal of Accounting and Finance, 1 (2017)
5. Wenger, E.: Communities of Practice: Learning, Meaning and Identity. Cambridge University Press, Cambridge (1998)
6. Wenger, E., McDermott, R., Snyder, W.M.: Cultivating Ccommunities of Practice. Harvard Business School Press, Boston, MA (2002)
7. Venkatraman, S., Venkatraman, R.: Communities of practice approach for knowledge management systems. Systems **6**(4), 36 (2018). https://doi.org/10.3390/systems6040036
8. Aziz, D.T., Bouazza, A., Jabur, N.H., Hassan, A.S., Al Aufi, A.: Development and validation of a knowledge management questionnaire. J. Inf. Stud. Technol. SLA-Arabian Gulf Chapter **1**(2) (2018)
9. Oliveira, L., Cardoso, E.L.: Psychometric properties of a scale to assess knowledge sharing in a community of practice. In: 2022 17th Iberian Conference on Information Systems and Technologies (CISTI), vol. 2022, pp. 1–7 (2022)
10. Oliveira, L., Cardoso, E.L.: Building a community of practice for engaging stackholders in bioproducts. In: 2021 16th Iberian Conference on Information Systems and Technologies (CISTI), vol. 2021, pp.1–4 (2021)
11. Marôco, J.: Análise Estatística com o SPSS Statistics, 7th edn. Pêro Pinheiro, Portugal: ReportNumber (2018)
12. Fang, Y.-H., Chiu, C.-M.: In justice we trust: exploring knowledge-sharing continuance intentions in virtual communities of practice. Comput. Hum. Behav. **26**(2), 235–246 (2010)
13. Zboralski, K.: Antecedents of knowledge sharing in communities of practice. J. Knowl. Manag. **13**(3), 90–101 (2009)
14. Mavri, A., Ioannou, A., Loizides, F.: Value creation and identity in cross-organizational communities of practice: a learner's perspective. Internet Higher Educ. **51**, 100822 (2021)
15. Singh, S.K., Mazzucchelli, A., Vessal, S.R., Solidoro, A.: Knowledge-based HRM practices and innovation performance: role of social capital and knowledge sharing. J. Int. Manag. **27**(1), 100830 (2021)
16. Ibidunni, A.S., Kolawole, A.I., Olokundun, M.A., Ogbari, M.E.: Knowledge transfer and innovation performance of small and medium enterprises (SMEs): an informal economy analysis. Heliyon **6**(8), e04740 (2020)

Gamification in Higher Education Assessment Through Kahoot

Geovanna Salazar-Vallejo(✉) and Diana Rivera-Rogel

Faculty of Social Sciences, Education and Humanities, Universidad Técnica Particular de Loja, Loja, CP 11-01-608, Ecuador
gesalazar2@utpl.edu.ec

Abstract. The approach of gamified assessments has significant implications in student learning and encourages educators to develop dynamic learning environments with permanent feedback, establishing a formative context that, through dynamic techniques, evaluates the content taught and measures the progress of learning. Responding to this background, the present research determines through a quantitative methodology, the potential of Kahoot in the university evaluation of Clinical Psychology students of the Universidad Técnica Particular de Loja (UTPL) and establishes a comparison between their average bimonthly evaluation score on the platform and the bimonthly written exam developed in the classroom. It is concluded that students improve their learning with the use of the Kahoot platform, presenting high levels of success in the topics raised throughout the academic course, which demonstrates its feasibility to reinforce the academic results obtained and process knowledge during professional training. In addition, it was established that the use of Kahoot is feasible in preparing students for a final written evaluation.

Keywords: Education · Gamification · Evaluation

1 Introduction

In recent years, gamification and other technologies are gradually incorporated into education, improving students' understanding through regular assessments [1]. This gamification executed through various applications is inserted in educational contexts to improve students' satisfaction, motivation and class attendance; within this context, Kahoot proves to be a promising, useful and advantageous tool as a formative evaluator [2, 3]. Kahoot encourages students to play, study, select the material to be studied and be aware of what they have learned in a subject, becoming a method to facilitate formative assessment and advance learning, demonstrating even greater efficiency than conventional techniques [4].

Games are a didactic tool in the classroom according to research on pedagogical resources, so implementing an online game encourages students to participate in class activities [5]. Students perceive Kahoot as a challenge for their study, besides consolidating as an entertaining and attractive tool that presents direct feedback to detect errors [6].

A. Rocha et al. (Eds.): WorldCIST 2023, LNNS 800, pp. 147–155, 2024.
https://doi.org/10.1007/978-3-031-45645-9_14

This platform is one of the main programs that can be used for gamification, formative assessment and peer-to-peer learning techniques, frequently used in the current educational landscape to introduce gamification practices in the teaching- learning process [7]. With the use of mobile devices, this platform allows students to answer online surveys created by teachers and quickly compare their answers with their classmates [8].

> Kahoot contributes to the teaching and learning of the curricular components of education by addressing the contents in a more engaging way for both teachers and students, providing topic review and knowledge consolidation, interaction and collaboration among participants, reasoning skills and immediate response, i.e., cognitive, social and motor experiences [9].

Those students who use Kahoot to learn the material of a course, obtain much better scores (in written evaluations), being feasible its inclusion within the different academic plans [10]. Therefore, the introduction of new material and the consolidation of information in a more interactive and visible way results in students' mastery of content and skills regarding the topics presented [11]. The creation of interactive presentations and games in which a variety of people can participate simultaneously creates a competitive incentive to compare knowledge with others [12].

Kahoot's direct feeback is particularly successful in rectifying errors and students feel that their learning momentum increases with respect to test preparation [6]. Studies show that this application helps eradicate learning problems; the use of gamification in educational settings increases the chances of long-term success and viability by resetting the student's attention clock and maintaining group learning as a method for memorizing material [7, 13].

In recent years, significant educational research efforts have been conducted in a variety of subject areas to evaluate the learning potential of Kahoot [14], highlighting the great acceptance by students, who consider this integrated learning approach as an exponent of new technologies, effective and precursor of teamwork and meaningful learning [15]. Authors highlight that Kahoot could improve students' extracurricular participation, as well as classroom interaction between students and teachers [16]. With the implementation of this game-based learning platform and digital technology, a high quality educational endeavor is supported, significantly improving the instructional and assessment process [17].

This maintains a level of attainment in education and ensures that students are better able to cope with challenges by turning potential failure into an opportunity to improve their studies, through fostering engagement, enjoyment and immersion in learning [20].

With this background, the present research aims to establish: 1) the influence and results of the Kahoot platform in the teaching-learning process at the university evaluation level; 2) establish a comparison with respect to the written evaluation developed by students at the university level.

2 Methodology

This research employs a quantitative methodology, based on the reports issued by the Kahoot learning platform. It uses a non-probabilistic sample, consisting of two parallels belonging to the career of Clinical Psychology of the Universidad Técnica Particular de Loja (UTPL): parallel 1 (n = 27) and parallel 2 (n = 38); these developed their on-site academic activities in the period April-August 2022 in the subject of Quantitative Research Methods.

Through Kahoot, two evaluations were made for each academic bimester: April- May (first bimester) and June-July (second bimester). We proceeded to establish a comparison between the general average of the course in the Kahoot evaluation and the general average obtained in the written bimestral exam. This approach responds to define the usefulness of the platform as a learning reinforcement with a view to a written evaluation, using a metric where 0 is the minimum score and 10 the maximum. For this, the scores obtained by the students in the indicated metric were converted with a rule of three and the respective average was calculated.

On the other hand, the percentage of successes/misses demonstrated by the students in the evaluations was established, demonstrating the effectiveness of the Kahoot platform in the assimilation of concepts. The evaluated topics respond to the syllabus of the subject, starting with general aspects until entering into specific issues that involve a higher level of complexity as stipulated in the academic structure of the subject.

3 Results

Kahoot as an evaluation method for the Clinical Psychology career issues the following reports: in the first academic bimester there is a total participation of students who complete all the questions. The level of correct answers is around 75% for the month of April and over 92% in May, in both analyzed parallels. In fact, from one month to the next, the level of students' correct answers rises, which suggests familiarity with the platform and efficiency in terms of identifying knowledge correctly (Tables 1 and 2).

Table 1. General data of the assessment developed in April through Kahoot - bimester 1

	Parallel 1	Parallel 2
Month/date	11 April 2022	14 April 2022
Players	27/27	38/38
Plays	9/9	9/9
Correct answers	75%	74.56%
Incorrect answers	25%	25.44%

Elaboration: Own.

In the second academic bimester, although the level of complexity of the topics increases, the percentage of correct answers by the students continues to be satisfactory.

Table 2. General data of the assessment developed in May through Kahoot - bimester 1

Month/date	Parallel 1	Parallel 2
Created by	16 May 2022	19 May 2022
Players	27/27	38/38
Plays	20/20	20/20
Correct answers	92.29%	93.72%
Incorrect answers	7.21%	6.28%

Elaboration: Own.

Parallel 1 and 2 present a level of response that in June exceeds 80% and in July 70%, being considered optimal to measure the learning achieved in the group of students. In addition, all students continue to participate in the proposed evaluation, which shows interest and motivation, not only in the contents of the subject, but also in the integrated platform to know the progress of learning (Tables 3 and 4).

Table 3. General data of the assessment developed in June through Kahoot - bimester 2

	Parallel 1	Parallel 2
Month/date	13 June 2022	16 June 2022
Players	27/27	38/38
Plays	20/20	20/20
Correct answers	80.58%	94.08%
Incorrect answers	19.42%	5.92%

Elaboration: Own.

Table 4. General data of the assessment developed in July through Kahoot - bimester 2

	Parallel 1	Parallel 2
Month/date	July 4, 2022	7 July 2022
Players	27/27	38/38
Plays	20/20	20/20
Correct answers	71.07%	96.84%
Incorrect answers	28.93.%	3.16%

Elaboration: Own.

In a second stage, the effectiveness of Kahoot can be measured by comparing the results obtained in the assessment through this platform and the results of a written assessment. In the first bimester (average of the April-May evaluations), the scores between both tests are compared with the results of the written test developed by the students, finding that there is no significant difference. A first piloting through Kahoot contributed to the fact that the average of the written evaluation was almost equal to the one obtained with this platform, reaching the following scores: in parallel 1, the evaluation in Kahoot reached a total average of 8.52, while the written exam reached 8.57; in parallel 2, the score increases slightly, with 8.68 in the evaluation in Kahoot and 8.71 for the written exam. The difference in scores between the two parallels is minimal (Fig. 1).

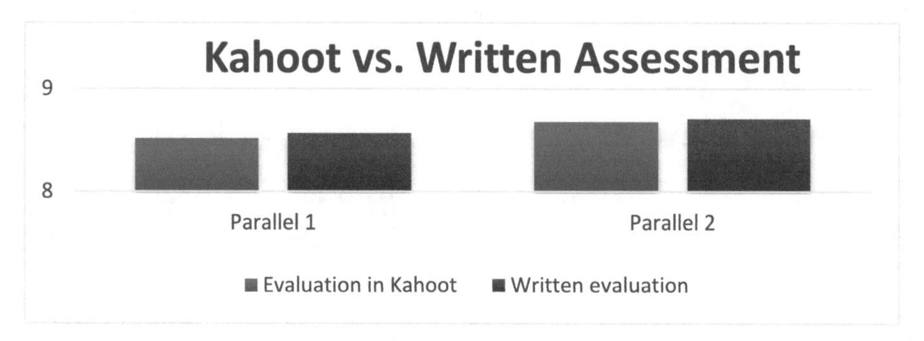

Fig. 1. Kahoot vs. Written Assessment - First bimester (Own Assessment)

Nevertheless, and as the complexity of the topics covered in class increases, the scores present some variations. In the second bimester, parallel 1 obtains in Kahoot an average of 8.32; however, in the written evaluation it reaches an 8. On the contrary, parallel 2 continues to present an almost exact similarity, both in Kahoot and in the written evaluation, with 8.68 and 8.71 respectively. This shows that, despite a greater complexity in the topics studied with the students, the previous reinforcement of Kahoot, not only as an evaluation but also as a review of contents, boosts the performance at the time of a printed evaluation (Fig. 2).

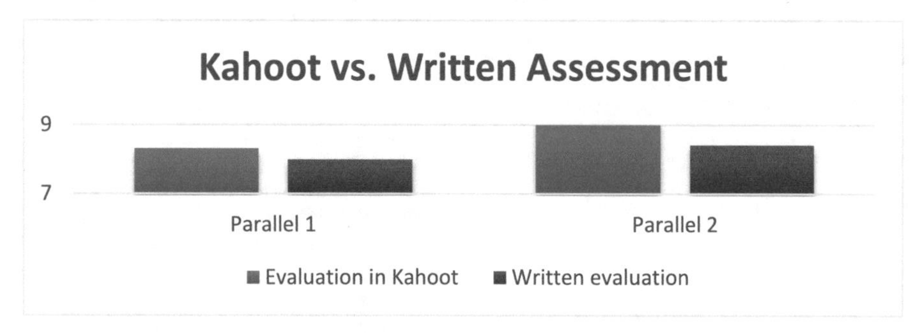

Fig. 2. Kahoot vs. Written Assessment - Second Bimester (Own Assessment).

As for the breakdown in the themes presented, the acceptance of Kahoot is evident in the percentages of successes obtained by the students. In the two evaluations that are developed in the first bimester, the students obtain the following scores with their participation: the first evaluation achieves in parallel 1 a minimum score of 4.17% (only in one question) and a maximum of 100%; the majority of the correct questions oscillates between 80% and 90%. Increasing these rates, parallel 2 presents a minimum score of

65.79% and a maximum of 92.11%, showing small deficiencies in two questions, which require reinforcement by the teacher.

In the second evaluation of the two-month period, parallels 1 and 2 maintain and even increase the scores obtained previously. The first one presents a minimum of 54.17% of correct questions and a maximum of 100%, nevertheless, most of the topics exceed 90% in correct answers. On the other hand, parallel 2 continues to mark a differentiating point with the other parallel, with a minimum correct score of 82.05% and a maximum of 100%, which implies a completely satisfactory performance on the part of the group of students (Table 5).

Table 5. Subjects evaluated and percentage of correct answers - First bimester

First evaluation	Parallel 1	Parallel 2	Second evaluation	Parallel 1	Parallel 2
Definition of research	83,33%	92,11%	Importance and need for a study	95,83%	89,74%
Approaches presented in the research	100%	89,47%	Evaluation of the importance of an investigation	54,17%	82,05%
Characteristic of the quantitative approach	95,83%	89,47%	Feasibility of a Research	91,67%	89,74%
Characteristic of the quantitative approach	79,17%	65,79%	Deficiencies in the knowledge of the problem	83,33%	97,44%
Characteristic of the quantitative approach	45,83%	68,42%	Consequences of the investigation	79,17%	84,62%
Characteristic of the quantitative approach	95,83%	92,11%	Quantitative approaches	100%	97,44%
Definition of subjective reality	83,33%	76,32%	Difficulty or frequent error in the problem statement	95,83%	94,87%
Definition of objective reality	87,50%	86,84%	Development of the theoretical perspective	83,33%	87,18%
Nature of the quantitative research	4,17%	10,54%	Development of the theoretical perspective	91,67%	89,74%
This first stage consists of an initial assessment of the competences of the students, with the aim of establish your knowledge of definitions basic knowledge related to the subject It also represents a premise for measuring the progress of students and make a comparison with future topics of study			Role of the theoretical perspective	100%	84,62%
			Function of the theoretical perspective	100%	94,87%
			Stages in development of the theoretical perspective	100%	97,44%
			Analytical review of the literature	100%	100%
			Link between the literature and the research problem	87,50%	97,44%
			Useful references	95,83%	92,31%
			Information gathered and logical criteria	100%	100%
			Index Methods	100%	97,44%
			Subpoena Process	95,83%	100%
			Clarity and structure of the theoretical framework	100%	97,44%
			Appeals for review of literature	91,67%	100%

Elaboration: Own.

However, in the second bimester, the list of issues raised in the subject of quantitative research methods obtains the expected responses and impact. In spite of the fact that the topics demand more effort from the students, the first evaluation of the second bimester obtains in parallel 1 a minimum of 53.85% (in a single question) and a maximum of 88.46%; however, the great majority of correct answers exceed 80%. In parallel 2, a minimum score of 78.95% and a maximum of 97.37% is obtained, demonstrating that, with the exception of a few questions, the majority of students demonstrate a mastery of knowledge.

Finally, the second evaluation (complex in terms of topics) maintains a good performance in the students: parallel 1 exceeds 70% in most of the questions and reaches 85.71% of correct answers as a maximum point; one of the questions posed presents less than 10%, which can be interpreted as an approach that was not clear for most of the students and needs to be reinforced.

Table 6. Subjects evaluated and percentage of correct answers - Second bimester

	Parallel 1	Parallel 2		Parallel 1	Parallel 2
Hypothesis in an investigation	61,54%	78,95%	Non-experimental studies	75%	92,11%
Hypothesis in an investigation quantitative research	84,62%	86,84%	Characteristic of non-experimental studies	67,86%	97,37%
How hypotheses emerge	84,62%	92,11%	Types of non- experimental designs	85,71%	97,37%
Types of assumptions	84,62%	94,74%	Exploratory Transectional Designs	50%	97,37%
Hypothesis testing	84,62%	97,37%	Trend and group evolution design	64,29%	94,74%
Hypothesis Definition	88,46%	97,37%	Characteristic of non-experimental studies	85,71%	92,11%
Variable definition	88,46%	97,37%	Non-experimental design and problem of research	35,71%	86,84%
Techniques for testing hypotheses	88,46%	92,11%	Populations	78,57%	97,37%
Design of research	88,46%	97,37%	Samples	85,71%	100%
Design of research	80,77%	97,37%	Delimitation of the population	85,71%	94,74%
Experimental design	88,46%	94,74%	Samples in the quantitative approach	75%	100%
Experimental design	84,62%	94,74%	Probabilistic sample	82,14%	94,74%
Pure experiments	84,62%	94,74%	Stratified sample	78,57%	97,37%
General Contexts of experiments	84,62%	97,37%	Case Study	78,57%	97,37%
Development of an instrument	76,92%	94,74%	Characteristic of the non-probability sample	85,71%	100%
Dependent/independent variable	69,23%	94,74%	Choice of elements	53,57%	100%
Handling of variable	53,85%	92,11%	Case Study	82,14%	100%
Independent variable	73,08%	94,74%	Sampling by marking telephone	85,71%	100%
Experimental group and control group	84,62%	97,37%	Census	78,57%	100%
Case study	76,92%	94,74%	Case Study	7,14%	97,37%

Elaboration: Own.

On the other hand, parallel 2 presents a minimum of 92.11% and a maximum of 100%, in fact, all the questions exceed 90% and show an ideal handling of the topics in the course (Table 6).

4 Discussion and Conclusions

The results demonstrate that Kahoot is a significant tool for gamifying assessment at the professional level, with results that show a significant improvement in the development of knowledge by students. This argument validates the fact that Kahoot currently becomes an ideal platform to enhance the learning process [3] and to evaluate the transmission of academic content.

The participation achieved through Kahoot is total in the groups studied, which demonstrates an intrinsic motivation in the students, allowing a fluid and dynamic integration that combines the game with the measurement of learning. Kahoot motivates and induces students to a permanent feedback [6], where gamification mediates an evaluative process that enhances knowledge and means progress for teaching-learning.

As some authors state, sometimes Kahoot demonstrates greater efficiency than traditional evaluation methods [4], however, the results show that the implementation of this platform in university classes reinforces the contents prior to a written evaluation. This means that teachers find in Kahoot an indispensable ally to review content, correct errors and deepen in topics where they detect shortcomings in the students; the results between the evaluations made in Kahoot and those written do not present a major difference in their scores.

A point that deserves to be highlighted is that the assessment dynamics with Kahoot are effective in both basic and more advanced knowledge, which means that students find in Kahoot a familiar platform and related to their interests, which uses technology and games to assess, contribute to meaningful learning and forge interactive processes between teachers and students [15–17].

Likewise, the gamification practices involved in Kahoot lead to a permanent use of audiovisuals as a connecting channel for knowledge and ideas. This is in line with presenting and assessing content more graphically and visually as a strategy for students to master the content [11].

References

1. Neureiter, D., Klieser, E., Neumayer, B., Winkelmann, P., Urbas, R., Kiesslich, T.: Feasibility of kahoot! as a real-time assessment tool in (histo-)pathology classroom teaching. Adv. Med. Educ. Pract. **11**, 695–705 (2020)
2. Martinez-Jimenez, R., Pedrosa-Ortega, C., Liceran-Gutierrez, A., Ruiz-Jimenez, M.C., Garcia-Marti, E.: Kahoot! as a tool to improve student academic performance in business management subjects. Sustainability **13**(5), 1–13 (2021)
3. Oz, G.O., Ordu, Y.: The effects of web based education and kahoot usage in evaluation of the knowledge and skills regarding intramuscular injection among nursing students. Nurse Educ. Today **103**, 1–6 (2021)
4. Ismal, M.A., et al.: Using kahoot! as a formative assessment tool in medical education: a phenomenological study. BMC Med. Educ. **19**, 1–8 (2019)

5. Kauppinen, A., Choudhary, A.I.: Gamification in entrepreneurship education: a concrete application of kahoot! Int. J. Manage. Educ. **19**(3), 1–14 (2021)
6. Iman, N., Ramli, M., Saridewi, N.: Kahoot as an assessment tools: students' perception of game-based learning platform. Jurnal Penelitian Dan Pembelajaran Ipa **7**(2), 245–259 (2021)
7. Uzunboylu, H., Galimova, E.G., Kurbanov, R.A., Belyalova, A.M., Deberdeeva, N.A., Timofeeva, M.: The views of the teacher candidates on the use of kahoot as A gaming tool. Int. J. Emerg. Technol. Learn. **15**(23), 158–168 (2020)
8. Curto Prieto, M., Orcos Palma, L., Blazquez Tobias, P.J., Molina Leon, F.J.: Student assessment of the use of kahoot in the learning process of science and mathematics. Educ. Sci. **9**(1), 1–13 (2019)
9. da Silva, A.M., Ferreira, D.P.C.: The use of the kahoot! platform as a gamification tool: a contribution to teaching and learning in basic education. Revista Edapeci-Educacao a Distancia E Praticas Educativas Comunicacionais E Interculturais **22**(2), 21–35 (2022)
10. Jankovic, A., Lambic, D.: The effect of game-based learning via kahoot and quizizz on the academic achievement of third grade primary school students. J. Baltic Sci. Educ. **21**(2), 224–231 (2022)
11. Kohnke, L., Moorhouse, B.L.: Using kahoot! to gamify learning in the language class-room. Relc J., 1–7 (2021)
12. Valles-Pereira, R.E., Mota-Villegas, D.J.: Kahoot aplicada en la evaluación sumativa en un curso de matemática discreta. Revista Científica **1**(37), 67–77 (2020). https://doi.org/10.14483/23448350.15236
13. Pertegal-Felices, L.M., Jimeno-Morenilla, A., Luis Sanchez-Romero, J., Mora-Mora, H.: Comparison of the effects of the kahoot tool on teacher training and computer engineering students for sustainable education. Sustainability **12**(11), 1–12 (2020)
14. Murciano-Calles, J.: Use of kahoot for assessment in chemistry education: a comparative study. J. Chem. Educ. **97**(11), 4209–4213 (2020)
15. Yelamos Guerra, M.S., Moreno Ortiz, A.J.: The use of ICT tools within the CLIL methodological approach in higher education (kahoot!, short films and BookTubes). Pixel-Bit- Revista De Medios Y Educacion **63**, 257–292 (2022)
16. Zhang, Q., Yu, Z.: A literature review on the influence of kahoot! on learning outcomes, interaction, and collaboration. Educ. Inf. Technol. **26**(4), 4507–4535 (2021)
17. Toma, F., Diaconu, D.C., Popescu, C.M.: The use of the kahoot! learning platform as a type of formative assessment in the context of pre-university education during the COVID-19 pandemic period. Educ. Sci. **11**(10), 1–18 (2021)

Education and Characteristics of Computational Thinking: A Systematic Literature Review

M. Juca-Aulestia(✉), E. Cabrera-Paucar, and V. Sánchez-Burneo

Universidad Técnica Particular de Loja, Loja, Ecuador
{jmjuca,ejcabrera3,vpsanchez}@utpl.edu.ec

Abstract. The purpose of this research paperwork is to learn about computational thinking (CT) in terms of its characteristics. How it influences teaching and learning through a systematic implementation of a literature review is also a part of this research. The implemented SLR methodology is oriented toward education and engineering, selecting 71 articles from the Scopus database. The results allow identifying the characteristics that influence computational thinking to generate skills and to solve problems with the application of software, block programming, and artifact design. On the other hand, how the results influence the teaching-learning of computational thinking for teachers and students to generate programming skills working in some environments such as Scratch or Snap, 3D technology to solve problems of industry 4.0 is also relevant. Thus, working on computer science principles become viable. Moreover, computational thinking allows students to decompose, generalize, and recognize patterns for the development of the industry. This is of utmost importance for problem-solving through digital tools, technology to produce ideas, and innovative and systematic solutions.

Keywords: computational thinking · CT · innovation · CT characteristics

1 Introduction

Computational thinking (CT) is related to education and technologies for problem-solving [1], so teachers must develop CT skills in students, using computer science principles to understand systems and verify human behavior [2], so it is essential to implement plans and curricula that are focused on CT, thus, all people must master these skills in the 21st century [3].

The research sought 71 studies related to computational thinking and solved the research questions posed in terms of characteristics, teaching, and learning.

We conducted the present study based on the methodology developed by [4], applying a conceptual mind fact; in the case of the searches, we undertook in the Scopus database. To know in depth the characteristics of how teaching-learning is carried out with computational thinking.

A. Rocha et al. (Eds.): WorldCIST 2023, LNNS 800, pp. 156–171, 2024.
https://doi.org/10.1007/978-3-031-45645-9_15

2 Literature Review

We conducted the research in three steps: planning, production, and presentation of the proposal taking into account the methodology applied for engineering and education [4].

2.1 Planning

Research questions
Computational thinking, is of significant importance since both teachers and students. They must have skills to work in the teaching-learning process. Furthermore, it is relevant to know the characteristics and influence of teaching-learning of computational thinking. Therefore, we considered two essential questions:

- RQ1: What characteristics influence computational thinking?
- RQ2: How does teaching-learning influence computational thinking?

Conceptual Mindset
The conceptual mindset proposed in Fig. 1 shows the initial steps of the systematic review concerning computational thinking. On the left side are the characteristics of the research approach. On the right side, we find the excluded parts of the research, and at the bottom are the keywords on which we worked the database.

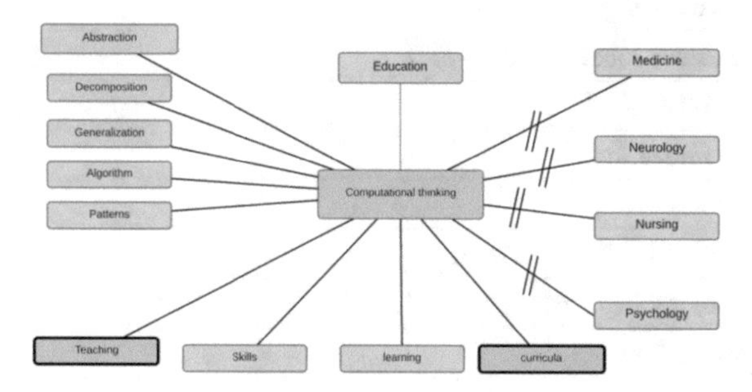

Fig. 1. Conceptual map according to [4].

Semantic Search Structure
We conducted research paperwork in four levels. The first focuses on computational thinking, the second is the characteristics that influence computational thinking, the third one is the teaching-learning in computational thinking, and the last level details the research questions of the semantic search.

2.2 Conducting the Review

Selection of Journal and Databases

The journals selected were from the Scopus database, and we show the most relevant journals in Tables 1, 2, 3 and 4.

Table 1. Review structure.

L1	Computational thinking	((("Computational thinking") AND ("steam education" OR "higher education" OR University*)))
L2	Characteristics	(Abstraction OR Decomposition OR Generalization OR Generalization OR Algorithm OR Patterns)
L3	Teaching-learning	(Skills OR learning OR teaching OR curricula)
L4	Questions	RQ1: What characteristics influence computational thinking? RQ2: How does teaching-learning influence computational thinking?

Table 2. Relevant journals where they have been published.

Journal	Q	IF
ACM Transactions on Computing Education	Q1	0.99
British Journal of Educational Technology	Q1	1.87
Communications of the ACM	Q1	1.56
Computer Science Education	Q1	1
Computers and Education	Q1	3.68
Education and Information Technologies	Q1	1.06
Educational Technology Research and Development	Q1	1.72
Frontiers in Computer Science	Q1	0.86
Informatics in Education	Q1	0.96
Information and Software Technology	Q1	1.45
Interactive Learning Environments	Q1	1.17
International Journal of Artificial Intelligence in Education	Q1	0.97
International Journal of Child-Computer Interaction	Q1	1.03
Journal of Computer-Assisted Learning	Q1	1.49
Journal of Educational Computing Research	Q1	1.28

(*continued*)

Table 2. (*continued*)

Journal	Q	IF
Journal of Research in Science Teaching	Q1	2.71
Proceedings of the ACM on Human-Computer Interaction	Q1	0.62
Science and Education	Q1	0.94
Social Media and Society	Q1	1.81
Studies in Science Education	Q1	2.06
Tec Trends	Q1	0.74
Theory and Research in Social Education	Q1	2.38
Thinking Skills and Creativity	Q1	1.16
ACM Transactions on Computer-Human Interaction	Q2	0.82
Computers	Q2	0.56
Education in the Knowledge Society	Q2	0.66
European Journal of Contemporary Education	Q2	0.62
Information (Switzerland)	Q2	0.62
International Journal on Software Tools for Technology Transfer	Q2	0.51
Journal of Computer Languages	Q2	0.63
Journal of Software: Evolution and Process	Q2	0.7
Natural Computing	Q2	0.58
ACM Inroads	Q3	0.27
International Journal of Advanced Computer Science and Applications	Q3	0.28
International Journal of Interactive Mobile Technologies	Q3	0.42

Table 3. Shows the papers according to RQ1: What characteristics influence computational thinking?

Variable	**Reference**	***f***
Abstraction	[5–29]	25
Algorithms	[5, 15, 16, 18, 23, 26, 30–33] [27, 34]	12
Patterns	[1, 5, 11, 12, 15, 27, 28, 35–38]	11
Decomposition	[12, 15, 23, 27, 39]	5
Generalization	[5, 40]. [21, 41]	4

Systematic Related Reviews

The present research performs systematic bibliographic research paperwork of the literature on impact and originality, which we did in the Scopus database. The difficulties

Table 4. Shows the papers according to RQ2: How does teaching and learning influence computational thinking?

Variable	Reference	*f*
Skills	[6, 10, 35, 40, 42–49]. [2, 18, 21, 32, 37, 41, 50–53]	22
Learning	[12, 20, 31, 36, 43, 54–60]. [49, 61, 62] [38, 45, 63, 64]	20
Teaching	[15, 26, 30, 32, 36, 59, 60, 65–72]	15
Curriculum	[11, 16, 54, 73]	4

presented were the adaptations of the proposed research questions. Thus, we detailed the following points in the review.

Definition of Inclusion and Exclusion Criteria
The research developed has some criteria that we have addressed for the resolution of the proposed research questions. We also aimed to fulfill the objective of the research work.

Specific Criteria:

- Studies that include the characteristics of computational thinking.
- Studies on the teaching and learning of computational thinking.
- Studies between the years 2018 -2022.

Exclusion Parameters:

- Studies of:
  - Psychology
  - Arts
  - Business
  - Energy
  - Medicine
  - Nursing
  - Economics

2.3 Review Report

Computational thinking combines different problem-solving abilities of students [5]; in the case of abstraction, it is a soft skill that allows solving problems of software applications and the design of artifacts that help teachers to have a better understanding of computational thinking [6]. We use abstractions in everyday life using programming language and we modeled figures and shapes [7]. Abstraction includes the generation of new concepts, actions, and characteristics of objects or processes [8]. Computational thinking is highly relevant in educational institutions, especially the use of abstraction and technological tools to allow the implementation of programs and curricula [9]. It increases the digital competence of students and teachers for problem-solving with the help of open educational resources [10]. Abstraction skills must be developed by the student performing tasks through block programming so that they can decompose, generalize and recognize patterns [11]. Likewise, for the implementation of specific practices in games by textual programming to develop analysis and abstraction in different contexts, especially that integrates various disciplines [12], students must know the different levels of abstraction for problem solving with attention to context [13]. The implementation of object-oriented programming activities and robots allow high performance in computational activities such as programming [14]. On the other hand, the generation of a model based on teacher-generated learning materials is necessary to improve curricula and integration to computational thinking [15]. This allows the application of strategies that enable generalizing from specific instances and patterns [16], as is the case of the constructionist approach [17], or also as one of the effective teaching methods is the learning environment with STEAM approach which allows developing abstraction [18]. It is also significant to make use of teaching approaches that enable using, modifying and creating different levels of abstraction to understand the programming code [19]. On the other hand the implementation of a model allows planning, implementation and documentation about behaviors and purposes of game levels in the case of analytical analysis [20].

A scratch is a tool that allows the development of logical thinking through the systematic processing of information [21]. The implementation of algorithmic solutions with the use of Scratch generates abstraction skills, explanations, and context insights for solutions [22]. Furthermore, automation and abstraction are relevant for the development of industry in terms of information processing being more efficient and reducing human error [23]. Design thinking helps projects using Scratch in Human-computer interaction design and refactoring of scenarios, activities, and tasks [24]. Patterns and rules of abstraction should be identified to operate a program to simplify practical procedures such as reading program codes [25]. It is essential to improve and understand the analysis of algorithms and didactic materials in a progressive way to conceive their function, order, and complexity [26]. Teachers should use creative ways for the connection of different topics with programming by integrating the opportunity to learn [27], and implement graphical patterns that will allow transformations through Structural Operational Semantics to address aspects of computation [28] Teachers should motivate students to the use of programs that allow the development of applications without the need for prior programming knowledge [29].

We consider the algorithm of a program that focuses on thinking and doing [30]. Furthermore, we should consider algorithms more than a set of order instructions [5]. The insertion of algorithms in academic plans as part of computational thinking in learning materials helps integration with content understanding [15]. We should use algorithms to create numerous games to improve students' performance in general skills [16]. The application of algorithms enables automation and innovation, which is significant for companies for information transmission [23]. We based the development of algorithmic thinking on new tools and didactic approaches [18] Participating in natural selections of algorithms to learn principles in different contexts and to change the student's perception of computing is necessary [31]. One option is text-based programming that allows generating problem-solving skills based on recurring patterns [32]. The stage-wise analysis of algorithms will enable students to reach different levels of understanding of abstraction [26]. We have observed that there is eliciting learning in students collaboratively when mining algorithms are applied [33] Within the software, it is relevant not to perform code cloning to maintain a straightforward compression in the projects [34]. The use of programming across disciplines integrate the way of learning computational thinking and other curricula at the same time, providing efficiency in learning [27].

Regarding patterns, we should note that gender and age do not influence students' computational thinking in pattern recognition. Instead, they generate generalization skills [35], 2022). It is relevant to integrate block-based programming teaching into the curricula as we intended to measure the abstraction of computational thinking in problem-solving through pattern recognition [11]. Teachers should integrate pattern-based learning materials into their classes to improve the self-efficacy of computational thinking integration [15], where we can implement structural operational semantics in patterns to enable flexible graphical language modeling [28]. Part of this is implementing games for computational practices in multidisciplinary contexts [12, 35]; programming through gaming is a pedagogical process that fosters interactivity and a deep understanding of programming concepts [36]. In itself, computational thinking is the outcome of several problem-solving skills, one of which is pattern recognition [5]. So, students recognize the use of computational thinking in different areas with a unique approach in analyzing data and solving problems that allow them to face complex puzzles. Through a structured and authentic interface [1], students can recognize patterns in the deployment of computational thinking and other disciplines [27]. Students make use of solution patterns in structures such as Scratch or Snap [37]. It is crucial to recognize patterns in block programming and to generate policies for programming education [38].

Decomposition and codification result in essential factors for the development of computational thinking; this characteristic potentiates to enhances the capacity for abstraction, for which it is advisable to design learning activities with a game-based approach [12]. In this framework of computational thinking, it is necessary to highlight automation as a factor that allows processing problems from a human resolution to then adapt them to computer simulators [23], faced with the requirement of integrating computational thinking in teacher management. The PRADA proposal marks a line of training that effectively brings TC closer to non-computer expert teachers, creating spaces that make educational actions with these characteristics visible [15]. In the framework of materializing TC in educational activities, it is necessary to highlight the main ideas of computing that give rise to linking learning objectives from a constructive path that involves the design of methods relevant to educational contexts [39]. In the purpose of incorporating TC into the schooled environment, a factor that can be substantial is to link it with artistic aspects. For this purpose, music and its compositional structure make it possible to analyze it in its parts through programming presenting students with creative ways to understand TC from a simple and practical application [27].

In generalization, it is necessary to identify the arrangement of the curricular programming with the aim of addressing its linkage to teaching practice. In this sense, we should highlight abstraction as a complex skill that requires its theoretical foundation [5]. One way to strengthen computational thinking in learning environments suggests the convergence of aspects related to learner dimensions. These are affectivity and motivation with the cognitive that binds elements in the training process based on pedagogical paradigms that articulate the academic with everyday [40]. To explain the influence of TC in the educational scenario, one must look at its origins in computational science and the exponential popularity with which it develops. This review shows the levels of progression and the dimensions that result in the consolidation of TC skills [41]. One of the pivotal factors in strengthening CT skills is Scratch, whose approach we based on basic programming competencies and whose result shows the generation of abstraction and logical thinking processes that, in synergy with other constructs, mark a path towards the concretization of CT [21].

Computational thinking is a skill that we all need to master. Particularly cognitive and learning skills [42], when applied at early ages it helps to develop essential literacy skills, cause/effect and design principles [43]. In computational thinking it is necessary to find solutions to develop some skills like motivation, social presence, security and some technical skills like coding, creativity [44], socioeconomic [40], skills like cognitive and relational [45] the same that should be linked to other skills such as numeracy, reading and writing [46], for the progression of development in educational institutions [35]. Therefore, it is important to develop problem-solving skills through digital tools [10], technology to produce innovative ideas and solutions [47], systematically through abstraction to apply to diverse needs and complex designs [6], as is the case of visual programming that helps a higher level of computational skill [48], and educational robotics to obtain more meaningful educational process [49]. To generate programming skills there are some programming environments such as Scratch or Snap [37], also the use of 3D technology allows training in computational thinking and solving problems of industry 4.0 [50]. Likewise it is of importance to use model-based pedagogy [2], the use of informal online communities to improve computational thinking [51], the development of computational thinking skills lies in academic performance in the use of information technologies and in the way students think [52], in addition to working on disciplinary ways of thinking and practicing computation [41]. In short, they should work on decomposition, formulation of algorithms with the help of patterns and code translation [32], abstraction, logical thinking, parallelism, control flow, data representation, patterns and systematic processing [21]Additionally, is also essential to work on projects for students to engage and generate skills such as generalization of patterns and communication of results [53], work on a STEAM approach to induce geography skills through resources [18].

The computational culture, in constant and undeniable growth, requires schools to provide training spaces for which we should structure in research parameters that pay tribute to the immersion of computer science at school [54]. To strengthen the TC work from the preschool stage, where logical thinking can be generated effectively, through exploration that in turn accounts for sequencing processes [43], also in principles (sequencing, repetition, and selection) and programming concepts (pattern matching, abstractions, and algorithms) through a creative pedagogy based on stories [55]. Thus, it is relevant to combine the study of natural selection and participation in the design of disconnected algorithmic explanations, allowing the integration of CT and science [31]. In mathematics learning, we have shown that integrating technology with both procedural and conceptual orientations through tutoring and exploratory learning environments guides an educational approach that brings knowledge efficiently closer [56]. Learning analytics represents a resource that allows reliably identifying problems with e-learning. Once identified, we made intervention proposals based on particular resources or OVA, exposed in stages, whose results are evident [57]. Teaching how to code to young students represents a significant step towards consolidating TC. Therefore, we must establish teaching strategies based on programming and computer literacy [58]. The TC approach must possess multidisciplinary characteristics in its academic training process for this purpose; interactive games stand out as elements that generate manipulation constitutes, bringing meaningful knowledge closer to the student in a meaningful way [12]. Designing learning activities with ludic aspects and assuming their rules to the educational environment means adapting the structure of the game; to the components that constitute the TC [20]. Gamification represents a component with pedagogical endowments that makes the teaching of programming feasible with interactive participation parameters and adjusted to a real context [36]. The approach of challenges as educational resources implies resorting to contests, which in the eagerness to solve them generates creative solutions with the consideration of competition as a prelude to cooperative learning [59]. In strengthening TC and OER as tools for the application of programming, we inserted them into the educational environment with a futuristic vision conceived as a necessity [60]. Learning how to program improves to the extent that students compare their solution with that of an instructor within a reflective process achieving a design trade-off between effort and impact on learning [61]. It is then relevant to work on algorithmic thinking sequences, abstraction patterns, and debugging) improvement through training programs in robotics and programming [49]. Computer-based activities such as automatic formative assessment and interactive feedback promote participation in Mathematics at the school level [62] is significant to implement visual block-based programming for different groups of students of different grades and gender [38], try to achieve distributed, automatic and efficient parallelization of familiar sequential Machine Learning (ML) code by making some mechanical changes, and we hid concurrency control details [63]. So machine learning is of importance when working with devices such as drones [64], to integrate experiences and different types of assessment that measures computational thinking [45].

We should implement teaching initiatives in a systemic and structured way for problem-solving [65]. Likewise, we should use digital technologies to enable the training of future specialist methodologists by creating personal trajectories based on self-learning [66], beginning teachers in the fundamentals of coding and computational thinking (sequences, algorithms, and loops) show increased growth in their self-efficacy by participating in continuous professional development [67]. Teachers consider that class discussion play an assertive role as an instructional method to provoke thinking about algorithms [30] they apply an effective methodology of teaching programming based on serious games that integrate gamification patterns, enabling deep learning of programming concepts [36], teachers should teach program design of algorithms with complex code fragments [68]. We have shown the use of programming in teaching to visualize technical and motivational aspects [59]. We should improve teaching in computational thinking through teaching materials and module evaluations [26]. Several teachers apply educational resources for programming to support their teaching process by facilitating the work of teachers through discovery, reuse, and sharing of them [60]. So, the student must know about parallel computing, handling theoretical concepts, and practical skills [69], the exploitation of tools such as CAS (Mathematica) for research allows for solving problems by finding the algorithm to obtain algorithmically correct results [70]. It is necessary to implement the Pattern Recognition, Abstraction, Decomposition, and Algorithms (PRADA) model that allows teachers to understand the central ideas of computational thinking, relate them to their curricula, and increase self-efficacy [15]. Likewise, the employment of the PRIMM (predict, execute, investigate, modify, do) method based on Vygotsky's sociocultural theory for teaching programming allows teachers to teach effectively in mixed-ability classes. And students to progress based on effective pedagogy by encouraging the use of language [71]. Computer science and technology students can interweave skills with explicit examples of decomposition, algorithm formulation, and translation to code-developed programming skills and problem-solving skills [32], which goes from simple to complex, supported by the elements of computational thinking (CT) (input, integration, output, and feedback) [72].

The implementations of computer science are significantly relevant, so we should integrate it into the curricula of both research and practice [54]. To teach computational thinking skills, especially that of abstraction for problem-solving [11], in an innovative way to implement patterns and generalization of instances [16]. So digital literacy with game creation is paramount and should be integrated into curricula [73].

3 Conclusions

Considering the review of the bibliography in the Scopus database, resulting in relevant documents for the research, and giving answers to the proposed research questions, we have the following: (Q1) abstraction is a soft skill that allows solving problems of software applications and artifact design. Abstraction skills should be developed through tasks through block programming so that it can decompose, generalize, and recognize patterns relevant for the development of the industry patterns. The rules of abstraction to operate a program to simplify practical procedures such as reading program codes should be identified. In addition, we can conclude that algorithms, more than being a set of instructions of order, are relevant to include the algorithms in the academic plans as part of computational thinking in learning materials. We should use algorithms to create different games for general skills, to improve the performance of students. The application of algorithms allows automation and innovation, which is paramount for companies through the transmission of information. Regarding patterns, we should note that gender and age do not influence the TC of students in pattern recognition to generate some generalization skills. Therefore, it is mandatory to integrate into the curricula teaching block-based programming. They intend to measure the abstraction of TC in problem-solving. Through pattern recognition, teachers should integrate pattern-based learning materials in their classes to improve the self-efficacy of TC integration, implement games for computational practices in multidisciplinary contexts, students make use of solution patterns in structures such as Scratch. Decomposition and codification are fundamental factors for the development of computational thinking. This element allows the development of the capacity for abstraction, for which it is advisable to design learning activities with a game-based approach. Concerning generalization, it is necessary to identify how to arrange the curricular programming to address its linkage to teaching practice. In this sense, we should highlight abstraction as a complex skill that requires its theoretical foundation. To explain the influence of CT in the educational setting, we must look at its origins in computational science and the exponential popularity with which it develops. One of the core factors in strengthening skills and (Q2) computational thinking is a skill that all of us must master. Particularly cognitive learning, essential literacy skills, cause/effect and design principles, motivation, social presence, safety, and some technical skills such as coding, creativity, and socioeconomic, so it is essential for problem-solving through digital technology tools to produce innovative and systematic ideas and solutions. When generating programming skills, there are some environments, such as Scratch or Snap, 3D technology that allow solving problems of industry 4.0. It is then worth combining the study of natural selection and participation in designing algorithmic considerations for integrating CT and science. In mathematics learning, we have shown that embedding technology with procedural and conceptual orientations manifests as an educational proposal that brings knowledge closer in an efficient way. Learning analytics represents a resource that allows us to identify reliability problems with e-learning. The teaching of coding to students represents a significant step towards the consolidation of TC. It is necessary to establish teaching strategies based on programming and literacy. The TC approach must have multidisciplinary characteristics in its academic training process, such as games, design learning activities with playful aspects, and assume its rules to the educational environment, such as gamification, which represents a component with pedagogical endowments that makes the teaching of programming viable. The approach of challenges as resources, the OER as tools for this implementation of programming. One must implement systemic and structured teaching initiatives for problem-solving using digital technologies that allow the formation of future methodologist specialists by creating personal trajectories based on self-learning. In addition, one must work with class discussions as they provide an increasingly important role as a method of instruction. To provoke reflection on algorithms and apply an effective methodology of teaching programming based on games that integrate gamification patterns, teachers should teach the design of algorithms programs with complex code fragments. Using programming in teaching allows visualizing technical and motivational aspects, improving teaching in TC through didactic materials and module evaluations, and applying educational resources for programming to support teaching through discovery, reuse, and sharing of these. It is necessary to implement PRADA towards the comprehension of the core ideas of TC and relate them to their curricula, computer science and technology students can intertwine skills with explicit examples of decomposition, algorithm formulation, and translation to code. The deployment of computer science is of utmost importance, so it should be integrated it into the curricula for both the research and the practice to teach TC skills; for easing the way to innovative patterns and generalization of instances.

References

1. Manfra, M.M., Hammond, T.C., Coven, R.M.: Assessing computational thinking in the social studies. Theory Res. Soc. Educ. (2021). https://doi.org/10.1080/00933104.2021.2003276
2. Ogegbo, A.A., Ramnarain, U.: A systematic review of computational thinking in science classrooms. Stud. Sci. Educ. (2021). https://doi.org/10.1080/03057267.2021.1963580
3. Xu, W., Geng, F., Wang, L.: Relations of computational thinking to reasoning ability and creative thinking in young children: mediating role of arithmetic fluency. Think. Ski. Creat. **44** (2022). https://doi.org/10.1016/j.tsc.2022.101041
4. Torres-Carrión, P.V., González-González, C.S., Aciar, S., Rodríguez-Morales, G.: Methodology for systematic literature review applied to engineering and education. In: 2018 IEEE Global Engineering Education Conference (EDUCON), pp. 1364–1373 (2018). https://doi.org/10.1109/EDUCON.2018.8363388
5. Oliveira, A.L.S., Andrade, W.L., Guerrero, D.D.S., Melo, M.R.A.: How do Bebras tasks explore algorithmic thinking skill in a computational thinking contest? In: Proceedings - Frontiers in Education Conference, FIE, vol. 2021 (2021). https://doi.org/10.1109/FIE49875.2021.9637151
6. Mirolo, C., Izu, C., Lonati, V., Scapin, E.: Abstraction in computer science education: an overview. Inform. Educ. **20**(4), 615–639 (2021). https://doi.org/10.15388/INFEDU.2021.27
7. Farris, A.V., Dickes, A.C., Sengupta, P.: Grounding computational abstractions in scientific experience. In: Computer-Supported Collaborative Learning Conference, CSCL, vol. 3, pp. 1333–1340 (2020). https://www.scopus.com/inward/record.uri?eid=2-s2.0-85102959650&partnerID=40&md5=49b18004f31686f8123e46548f996388
8. Gautam, A., Bortz, W., Tatar, D.: Abstraction through multiple representations in an integrated computational thinking environment. In: SIGCSE 2020 - Proceedings of the 51st ACM Technical Symposium on Computer Science Education, pp. 393–399 (2020). https://doi.org/10.1145/3328778.3366892
9. Chan, S.-W., Looi, C.-K., Ho, W.K., Kim, M.S.: Tools and approaches for integrating computational thinking and mathematics: a scoping review of current empirical studies. J. Educ. Comput. Res. (2022). https://doi.org/10.1177/07356331221098793
10. Morze, N., Barna, O., Boiko, M.: The relevance of training primary school teachers computational thinking. In: CEUR Workshop Proceedings, vol. 3104, pp. 141–153 (2022). https://www.scopus.com/inward/record.uri?eid=2-s2.0-85127399524&partnerID=40&md5=4b12103da187ad389516be3ae00eaf26
11. Çakiroğlu, Ü., Çevik, İ: A framework for measuring abstraction as a sub-skill of computational thinking in block-based programming environments. Educ. Inf. Technol. (2022). https://doi.org/10.1007/s10639-022-11019-2
12. Grizioti, M., Kynigos, C.: Code the mime: A 3D programmable charades game for computational thinking in MaLT2. Br. J. Educ. Technol. **52**(3), 1004–1023 (2021). https://doi.org/10.1111/bjet.13085
13. Rich, K.M., Yadav, A.: Applying levels of abstraction to mathematics word problems. TechTrends **64**(3), 395–403 (2020). https://doi.org/10.1007/s11528-020-00479-3
14. Çınar, M., Tüzün, H.: Comparison of object-oriented and robot programming activities: the effects of programming modality on student achievement, abstraction, problem-solving, and motivation. J. Comput. Assist. Learn. **37**(2), 370–386 (2021). https://doi.org/10.1111/jcal.12495

15. Dong, Y., et al.: Prada: a practical model for integrating computational thinking in K-12 education. In: SIGCSE 2019 - Proceedings of the 50th ACM Technical Symposium on Computer Science Education, pp. 906–912 (2019). https://doi.org/10.1145/3287324.3287431
16. Tseng, C.-Y., Doll, J., Varma, K.: Exploring evidence that board games can support computational thinking. In: Proceedings of International Conference on Computational Thinking Education, pp. 61–64 (2019). https://www.scopus.com/inward/record.uri?eid=2-s2.0-85093103248&partnerID=40&md5=d857c563e4439e4d15e04faf2d02a430
17. Lodi, M., Malchiodi, D., Monga, M., Morpurgo, A., Spieler, B.: Constructionist attempts at supporting the learning of computer programming: a survey. Olympiads Inform. **13**, 99–121 (2019). https://doi.org/10.15388/ioi.2019.07
18. Bedar, R.A.-H., Al-Shboul, M.: The effect of using STEAM approach on developing computational thinking skills among high school students in Jordan. Int. J. Interact. Mob. Technol. **14**(14), 80–94 (2020). https://doi.org/10.3991/IJIM.V14I14.14719
19. Law, R.: A pedagogical approach to teaching game programming: Using the PRIMM approach. In: Proceedings of the 14th International Conference on Game Based Learning, ECGBL 2020, pp. 816–819 (2020). https://doi.org/10.34190/GBL.20.071
20. Yunus, E., Zaibon, S.B.: Connecting computational thinking (CT) concept with the game-based learning (GBL) elements. Int. J. Interact. Mob. Technol. **15**(20), 50–67 (2021). https://doi.org/10.3991/ijim.v15i20.23739
21. Scullard, S., Tsibolane, P., Garbutt, M.: The role of scratch visual programming in the development of computational thinking of non-is majors. In: Proceedings of the 23rd Pacific Asia Conference on Information Systems: Secure ICT Platform for the 4th Industrial Revolution, PACIS 2019 (2019). https://www.scopus.com/inward/record.uri?eid=2-s2.0-85089224135&partnerID=40&md5=f59e657627ac8171d9f9cacb9c19dd9b
22. Statter, D., Armoni, M.: Teaching abstraction in computer science to 7th-grade students. ACM Trans. Comput. Educ. **20**(1), 8–837 (2020). https://doi.org/10.1145/3372143
23. Nuar, A.N.A., Rozan, M.Z.A.: Benefits of computational thinking in entrepreneurship. In: International Conference on Research and Innovation in Information Systems, ICRIIS, (2019). https://doi.org/10.1109/ICRIIS48246.2019.9073671
24. Lee, C.-S., Wong, K.D.: Comparing computational thinking in scratch and non-scratch web design projects: a meta-analysis on framing and refactoring. In: 29th International Conference on Computers in Education Conference, ICCE 2021 - Proceedings, vol. 2, pp. 457–462 (2021). https://www.scopus.com/inward/record.uri?eid=2-s2.0-85122920677&partnerID=40&md5=f1b60ee05ebc325f5c0a6b40eb395ef8
25. Ezeamuzie, N.O., Leung, J.S.C., Ting, F.S.T.: Unleashing the potential of abstraction from cloud of computational thinking: a systematic review of literature. J. Educ. Comput. Res. (2021). https://doi.org/10.1177/07356331211055379
26. Kay, A., Wong, S.H.S.: Discovering missing stages in the teaching of algorithm analysis: an APOS-based study (2018). https://doi.org/10.1145/3279720.3279738
27. Bell, J., Bell, T.: Integrating computational thinking with a music education context. Inform. Educ. **17**(2), 151–166 (2018). https://doi.org/10.15388/infedu.2018.09
28. Kopetzki, D., Lybecait, M., Naujokat, S., Steffen, B.: Towards language-to-language transformation. Int. J. Softw. Tools Technol. Transf. **23**(5), 655–677 (2021). https://doi.org/10.1007/s10009-021-00630-2
29. Martínez-Valdés, J.A., Martínez-Ijají, N.A.: An experience with the App Inventor in CS0 for the development of the STEM didactics. In: ACM International Conference Proceeding Series, pp. 51–56 (2018). https://doi.org/10.1145/3284179.3284189
30. Nijenhuis-Voogt, J., Bayram-Jacobs, D., Meijer, P.C., Barendsen, E.: Teaching algorithms in upper secondary education: a study of teachers' pedagogical content knowledge. Comput. Sci. Educ. (2021). https://doi.org/10.1080/08993408.2021.1935554

31. Peel, A., Sadler, T.D., Friedrichsen, P.: Learning natural selection through computational thinking: unplugged design of algorithmic explanations. J. Res. Sci. Teach. **56**(7), 983–1007 (2019). https://doi.org/10.1002/tea.21545
32. Piwek, P., Wermelinger, M., Laney, R., Walker, R.: Learning to program: From problems to code (2019). https://doi.org/10.1145/3294016.3294024
33. Emara, M., Rajendran, R., Biswas, G., Okasha, M., Elbanna, A.A.: Do students' learning behaviors differ when they collaborate in open-ended learning environments? In: Proceedings of the ACM Human-Computer Interaction, vol. 2, no. CSCW (2018). https://doi.org/10.1145/3274318
34. Perez-Castillo, R., Piattini, M.: An empirical study on how project context impacts on code cloning. J. Softw. Evol. Process **30**(12) (2018). https://doi.org/10.1002/smr.2115
35. Jiang, S., Wong, G.K.W.: Exploring age and gender differences of computational thinkers in primary school: a developmental perspective. J. Comput. Assist. Learn. **38**(1), 60–75 (2022). https://doi.org/10.1111/jcal.12591
36. Maskeliūnas, R., Kulikajevas, A., Blažauskas, T., Damaševičius, R., Swacha, J.: An interactive serious mobile game for supporting the learning of programming in javascript in the context of eco-friendly city management. Computers **9**(4), 1–18 (2020). https://doi.org/10.3390/computers9040102
37. Talbot, M., Geldreich, K., Sommer, J., Hubwieser, P.: Re-use of programming patterns or problem-solving?: Representation of scratch programs by TGraphs to support static code analysis. ACM Int. Conf. Proc. Ser. (2020). https://doi.org/10.1145/3421590.3421604
38. Kong, S.-C., Wang, Y.-Q.: Assessing programming concepts in the visual block-based programming course for primary school students. In: Proceedings of the European Conference on e-Learning, ECEL, vol. 2019-Novem, pp. 294–302 (2019). https://doi.org/10.34190/EEL.19.035
39. Rich, K.M., Binkowski, T.A., Strickland, C., Franklin, D.: Decomposition: A K-8 computational thinking learning trajectory. In: ICER 2018 - Proceedings of the 2018 ACM Conference on International Computing Education Research, pp. 124–132 (2018). https://doi.org/10.1145/3230977.3230979
40. Tsortanidou, X., Daradoumis, T., Barberá, E.: A K-6 computational thinking curricular framework: pedagogical implications for teaching practice. Interact. Learn. Environ. (2021). https://doi.org/10.1080/10494820.2021.1986725
41. Denning, P.J., Tedre, M.: Computational thinking: a disciplinary perspective. Inform. Educ. **20**(3), 361–390 (2021). https://doi.org/10.15388/infedu.2021.21
42. Xu, C., et al.: An automatic ordering method for streams in surface-water/groundwater interaction modeling [地表水和地下水相互作用模拟中河流的自动排序方法] [Une méthode d'ordonnancement automatique des cours d'eau dans la modélisation de l'interaction entre les eaux de surface et les eaux sout. Hydrogeol. J. **30**(6), 1789–1800 (2022). https://doi.org/10.1007/s10040-022-02531-3
43. McCormick, K.I., Hall, J.A.: Computational thinking learning experiences, outcomes, and research in preschool settings: a scoping review of literature. Educ. Inf. Technol. **27**(3), 3777–3812 (2022). https://doi.org/10.1007/s10639-021-10765-z
44. Robe, P., Kuttal, S.K.: Designing PairBuddy-a conversational agent for pair programming. ACM Trans. Comput. Interact. **29**(4) (2022). https://doi.org/10.1145/3498326
45. Basso, D., Fronza, I., Colombi, A., Pahl, C.: Improving assessment of computational thinking through a comprehensive framework (2018). https://doi.org/10.1145/3279720.3279735
46. Ezeamuzie, N.O., Leung, J.S.C.: Computational thinking through an empirical lens: a systematic review of literature. J. Educ. Comput. Res. **60**(2), 481–511 (2022). https://doi.org/10.1177/07356331211033158

47. Sun, M., Wang, M., Wegerif, R., Peng, J.: How do students generate ideas together in scientific creativity tasks through computer-based mind mapping? Comput. Educ. **176** (2022). https://doi.org/10.1016/j.compedu.2021.104359
48. Fanchamps, N.L.J.A., Slangen, L., Specht, M., Hennissen, P.: The impact of SRA-programming on computational thinking in a visual oriented programming environment. Educ. Inf. Technol. **26**(5), 6479–6498 (2021). https://doi.org/10.1007/s10639-021-10578-0
49. Caballero-Gonzáleza, Y.A., Garciá-Valcárcelb, A.: Learning with robotics in primary education? a means of stimulating computational thinking [¿aprender con robótica en educación primaria? un medio de estimular el pensamiento computacional]. Educ. Knowl. Soc. **21** (2020). https://doi.org/10.14201/EKS.22957
50. Bushmeleva, N.A., Isupova, N.I., Mamaeva, E.A., Kharunzheva, E.V.: Peculiarities of engineering thinking formation using 3D technology. Eur. J. Contemp. Educ. **9**(3), 529–545 (2020). https://doi.org/10.13187/ejced.2020.3.529
51. Xing, W.: Large-scale path modeling of remixing to computational thinking. Interact. Learn. Environ. **29**(3), 414–427 (2021). https://doi.org/10.1080/10494820.2019.1573199
52. Özgür, H.: Relationships between computational thinking skills, ways of thinking and demographic variables: a structural equation modeling. Int. J. Res. Educ. Sci. **6**(2), 299–314 (2020). https://www.scopus.com/inward/record.uri?eid=2-s2.0-85083390585&partnerID=40&md5=c8bfa8e81c88f29f2fe39738c6a32294
53. Sorensen, C., Mustafaraj, E.: Evaluating computational thinking in jupyter notebook data science projects. In: Proceedings of International Conference on Computational Thinking Education, pp. 115–120 (2018). https://www.scopus.com/inward/record.uri?eid=2-s2.0-85103846829&partnerID=40&md5=0ad9e8c92019d355ad332a73d555acf6
54. Novak, E., Khan, J.I.: A research-practice partnership approach for co-designing a culturally responsive computer science curriculum for upper elementary students. TechTrends **66**(3), 527–538 (2022). https://doi.org/10.1007/s11528-022-00730-z
55. Twigg, S., Blair, L., Winter, E.: Using children's literature to introduce computing principles and concepts in primary schools: work in progress. ACM Int. Conf. Proc. Ser. (2019). https://doi.org/10.1145/3361721.3362116
56. Mavrikis, M., Rummel, N., Wiedmann, M., Loibl, K., Holmes, W.: Combining exploratory learning with structured practice educational technologies to foster both conceptual and procedural fractions knowledge. Educ. Technol. Res. Dev. (2022). https://doi.org/10.1007/s11423-022-10104-0
57. Kew, S.N., Tasir, Z.: Developing a learning analytics intervention in e-learning to enhance students' learning performance: a case Study. Educ. Inf. Technol. (2022). https://doi.org/10.1007/s10639-022-10904-0
58. Relkin, E., de Ruiter, L.E., Bers, M.U.: Learning to code and the acquisition of computational thinking by young children. Comput. Educ. **169** (2021). https://doi.org/10.1016/j.compedu.2021.104222
59. Martins, G., Lopes De Souza, P.S., Jose Conte, D., Bruschi, S.M.: Learning parallel programming through programming challenges. In: Proceedings - Frontiers in Education Conference, FIE, vol. 2020 (2020). https://doi.org/10.1109/FIE44824.2020.9274009
60. De Deus, W.S., Barbosa, E.F.: An exploratory study on the availability of open educational resources to support the teaching and learning of programming. In: Proceedings - Frontiers in Education Conference, FIE, vol. 2020 (2020). https://doi.org/10.1109/FIE44824.2020.9274202
61. Price, T.W., Williams, J.J., Solyst, J., Marwan, S.: Engaging students with instructor solutions in online programming homework (2020). https://doi.org/10.1145/3313831.3376857
62. Barana, A., Marchisio, M., Rabellino, S.: Empowering engagement through automatic formative assessment. In: Proceedings - International Computer Software and Applications Conference, vol. 1, pp. 216–225 (2019). https://doi.org/10.1109/COMPSAC.2019.00040

63. Kim, J.K., Aghayev, A., Gibson, G.A., Xing, E.P.: STRADS-AP: simplifying distributed machine learning programming without introducing a new programming model. In: Proceedings of the 2019 USENIX Annual Technical Conference, USENIX ATC 2019, pp. 207–221 (2019). https://www.scopus.com/inward/record.uri?eid=2-s2.0-85077018135&partnerID=40&md5=10780cc333a4c544f84c208f187ce7e4
64. McQuillan, D.: People's councils for ethical machine learning. Soc. Media Soc. **4**(2) (2018). https://doi.org/10.1177/2056305118768303
65. Jansen, M., Kohen-Vacs, D., Otero, N., Milrad, M.: A complementary view for better understanding the term computational thinking. In: Proceedings of International Conference on Computational Thinking Education, pp. 2–7 (2018). https://www.scopus.com/inward/record.uri?eid=2-s2.0-85060020241&partnerID=40&md5=1d79c5482b9566029fa96668698d14cc
66. Soboleva, E.V., Suvorova, T.N., Bocharov, M.I., Bocharova, T.I.: Development of the personalized model of teaching mathematics by means of interactive short stories to improve the quality of educational results of schoolchildren. Eur. J. Contemp. Educ. **11**(1), 241–257 (2022). https://doi.org/10.13187/ejced.2022.1.241
67. Rich, P.J., Mason, S.L., O'Leary, J.: Measuring the effect of continuous professional development on elementary teachers' self-efficacy to teach coding and computational thinking. Comput. Educ. **168** (2021). https://doi.org/10.1016/j.compedu.2021.104196
68. Waite, J., Curzon, P., Marsh, W., Sentance, S.: Difficulties with design: the challenges of teaching design in K-5 programming. Comput. Educ. **150** (2020). https://doi.org/10.1016/j.compedu.2020.103838
69. De Jesus Oliveira Duraes, T., Sergio Lopes De Souza, P., Martins, G., Jose Conte, D., Garcia Bachiega, N., Mazzini Bruschi, S.: Research on parallel computing teaching: state of the art and future directions. In: Proceedings - Frontiers in Education Conference, FIE, vol. 2020 (2020) https://doi.org/10.1109/FIE44824.2020.9273914
70. Seidametova, Z.: Combining programming and mathematics through computer simulation problems. In: CEUR Workshop Proceedings, vol. 2732, pp. 869–880 (2020). https://www.scopus.com/inward/record.uri?eid=2-s2.0-85096101589&partnerID=40&md5=e771cfbc21faab03bfa1d6f2acbc860a
71. Sentance, S., Waite, J., Kallia, M.: Teaching computer programming with PRIMM: a sociocultural perspective. Comput. Sci. Educ. **29**(2–3), 136–176 (2019). https://doi.org/10.1080/08993408.2019.1608781
72. Christensen, D., Lombardi, D.: Understanding biological evolution through computational thinking: a K-12 learning progression. Sci. Educ. **29**(4), 1035–1077 (2020). https://doi.org/10.1007/s11191-020-00141-7
73. Troiano, G.M., et al.: Is My game ok Dr. scratch?: Exploring programming and computational thinking development via metrics in student-designed serious games for STEM. In: Proceedings of the 18th ACM International Conference on Interaction Design and Children, IDC 2019, pp. 208–219 (2019). https://doi.org/10.1145/3311927.3323152

Blockchain and Artificial Intelligence in a Business Context: A Bibliometric Analysis

Soraya González-Mendes(✉), Fernando García-Muiña, and Rocío González-Sánchez

Department of Business Administration (ADO), Applied Economics II and Fundaments of Economic Analysis, Universidad Rey Juan Carlos, Madrid, Spain
{soraya.gonzalez,fernando.muina,rocio.gonzalez}@urjc.es

Abstract. Speed and changes in the environment arise from the need to apply emerging technologies in business. The characteristics of Artificial Intelligence allowing the development of algorithms implanted in machines capable of imitating human behavior and the advantages of Blockchain technology being a distributed record of information stored in blocks in a decentralized, encrypted, confidential and secure way propel the generation of innovative business in companies. The article examines the relationship between Blockchain and Artificial Intelligence using a Bibliometric Analysis, examining data from 583 articles published between 2018 and October 2022 with the keywords 'Blockchain' and 'Artificial Intelligence' from the Web of Science database. To realize the Bibliometric Analysis, it has executed the VOSviewer program. The work has detected six research trends in the connection between these concepts.

Keywords: Blockchain · Artificial Intelligence · Business · Bibliometric

1 Introduction

The pandemic has forced many organizations to make a transformation in their business models and operations [3, 21]. Artificial Intelligence has an important role [26] and it has been applied to solve several challenging industrial problems in Industry 4.0 [9]. Blockchain is a disruptive technology capable of reshaping business operations and supply chain business models [24]. Blockchain goes beyond being just cryptocurrencies, being a decentralized technology and allowing the creation of Smart Contracts [19]. In recent years, the management of large amounts of data is industrializing in IoT playing an important role in Artificial Intelligence (AI), however, AI has some challenges such as centralized architecture, security and privacy and resource limitations that can be solved with Blockchain as a decentralized architecture that provides a secure exchange of data and resources to the various nodes of the IoT network, eliminates decentralization and AI challenges. The creation of an intelligent IoT architecture enabled for Blockchain with Artificial Intelligence efficiently provides the convergence between Blockchain and AI for IoT applying current techniques and applications [12]. The adoption of Blockchain has led to a paradigm shift towards a new digital ecosystem of smart city, where the

A. Rocha et al. (Eds.): WorldCIST 2023, LNNS 800, pp. 172–182, 2024.
https://doi.org/10.1007/978-3-031-45645-9_16

convergence between Artificial Intelligence (AI) and Blockchain allows to revolutionize the network architecture of smart cities allowing the generation of sustainable ecosystems [15].

This paper examines the following Research Questions (RQ's): 1) What is the evolution of the publications about Blockchain and Artificial Intelligence from 2018 to October of 2022?; 2) Which authors, institutions, countries and journals published the greatest number of documents and which were the most cited?; 3) Which areas are the most research in Blockchain and Artificial Intelligence?; 4) What are the top most articles related to this subject? The following parts contains the background of this article involving terms of Artificial Intelligence and Blockchain. Following the previous, Bibliometric Analysis will be realized belong the evolution of this topic, the most cited authors, institutions, countries and journals, the most studies in these fields and the topmost articles. To sum up, the work is finished with the major conclusions.

2 Methodology and Data Analysis

To conduct the Bibliometric Analysis, it used the Web of Science Database (WoS) [25]. A search was conducted with the words 'blockchain' AND 'artificial intelligence' from 2018 to October 2022 and it appeared 1312 investigations. It eliminated the year 2017. Then, it selected only articles in English. Later, it was collected and included only Science Citation Index Expanded (SCI-EXPANDED) and Social Sciences Citation Index (SSCI) databases. In total, the result obtained was 583 articles. This paper improves the description of the body of the work as did other authors [1].

2.1 The Growing Evolution

Because of Digital Transformation and Industry 4.0 experienced by cities, companies and society in general, the concerns of companies to obtain the greatest possible benefits and profitability based on innovative technology, concerns about war or future pandemics make possible the need to investigate more in this field where Blockchain technology is mixed with Artificial Intelligence. It can see how research on these concepts is increasing, especially from 2020, being the period of greatest growth in number of publications from 2020 to 2021 (see Fig. 1).

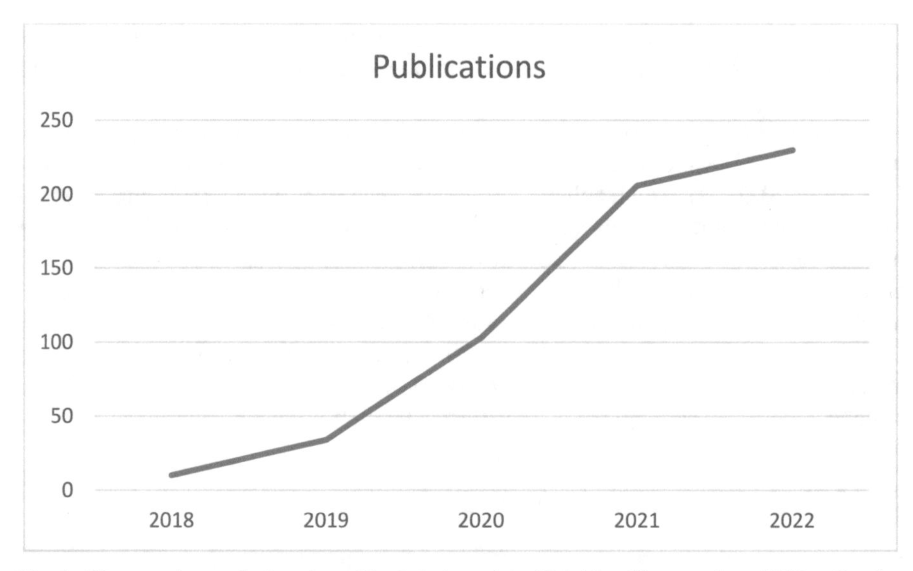

Fig. 1. The growing evolution about Blockchain and Artificial Intelligence since 2018 to October of 2022.

2.2 Co-authorship - Authors and Institutions

The following Table 1 illustrates the analysis of the authors ordered by the top of ten greatest number of citations. Of the 2170 authors, 18 meet the thresholds. The most cited author belongs of an institution of India and the author with the greatest number of documents and the greatest Total Link Strengths belong of an institution of India, too. Most of these authors who are part of the top ten in terms of number of citations, documents and Total Link Strengths are concerned, belong to institutions in India, China, Australia and the United States. This indicates that there is a strong geographical relationship between these institutions.

2.3 Co-authorship – Countries

It analyzes the influence and effect that countries have on Blockchain and Artificial Intelligence. In this analysis of 80 countries, 46 reached the thresholds generating 6 clusters. In this way, the correlation between these countries is examined by analyzing the strength of the links between documents and citations. The following Table 2 is ordered by Total Link Strength and it reveals the seven clusters that contain the most relevant countries. **Cluster 1** with 28 Links and 149 of Total Link Strength is the cluster that contain the most correlation belong countries, contains the node with the highest number of research related to Blockchain and Artificial Intelligence and has the highest correlation between the countries of China and United States.

Table 1. Most cited authors

Nr	Author	Institutions	D	C	TLS
1	Das, Ashok Kumar	Hyderabad (India)	6	198	2
2	Tanwar, Sudeep	Nirma University (India)	17	198	21
3	Gupta, Rajesh	San Diego (USA)	11	178	18
4	Nguyen, Dinh C	Purdue University (USA)	5	152	6
5	Pathirana, Pubudu N	Deakin University (Australia)	5	152	6
6	Li, Jianhua	Shangai University (China)	6	141	6
7	Wu, Jun	China	6	141	6
8	Kumar, Neeraj	Thapar Institute (India)	10	139	11
9	Wang, Fei-yue	Arizona (USA)	5	137	0
10	Srivastava, Gautam	Brandom University (USA)	6	129	2

Abbreviations: Nr = Number; D = Documents; C = Citations; TLS = Total Link Strengths

Table 2. Most relevant countries

C	Items	Links	TLS	D	Most Relevant Countries
1	10	28	149	166	Belgium, Germany, Iran, Japan, Netherlands, China, Poland, Russia, Switzerland, USA
2	7	24	65	49	Australia, Bangladesh, Hungary, Malaysia, Taiwan, Tunisia, Vietnam
3	7	27	120	62	Canada, Egypt, Kuwait, Pakistan, Qatar, Saudi Arabia, U Arab Emirates
4	7	38	157	107	France, India, Mexico, Morocco, Singapore, South Africa, Spain
5	6	17	31	20	England, Greece, Iraq, Italy, Scotland, Turkey
6	6	15	29	12	Brazil, Denmark, Norway, Portugal, Sweden, Romania
7	3	25	67	48	Finland, Ireland, South Korea

Abbreviations: C = Clusters; D = Documents

2.4 Citation - Journals

The Journal with the greatest number of documents, citations and Total Link Strength related to these topics is IEEE Access. Other relevant journals that contain a good relationship with IEEE Access are IEEE Internet of Things Journal, IEEE Communications Surveys and Tutorials, IEEE Transactions on Industrial Informatics and IEEE Transactions on Network Science and Engineering. In Table 3, all the 10 topmost journals are about investigations of technology and there is one about sustainability.

Table 3. Most important journals

N	Source	D	C	TLS
1	IEEE Access	52	1047	44
2	IEEE Internet of Things Journal	24	446	28
3	Transactions On Emerging Telecommunications Technologies	7	112	19
4	IEEE Communications Surveys and Tutorials	8	526	15
5	Sustainability	25	186	15
6	IEEE Transactions on Industrial Informatics	7	448	14
7	IEEE Network	20	295	12
8	Annals of Operations Research	5	107	12
9	IEEE Transactions on Network Science and Engineering	6	72	11
10	Future Generation Computer Systems-the International Journal of eScience	5	249	10

Abbreviations: N = Number; D = Documents; C = Citations; TLS = Total Link Strength

2.5 Co-citation Analysis – Keywords

It has performed an analysis of co-occurrences of keywords. It has found 2329 keywords from the 583 articles analyzed and it has designated a minimum of 5 occurrences and 185 keywords met the threshold. In this manner, it was examined areas indicated by research domains, hottest topics and emerging trends. It was realized a search on the WoS Database and it was analyzed the parameters in the VOSviewer program. Keywords such as AI is grouped in Artificial Intelligence and terms as Internet of Things is grouped in IoT. In Table 4, we show the 10 keywords sorted by Total Link Strength. Words that have a strong relationship such as Blockchain, Artificial Intelligence, Internet, Security, Big Data and IoT.

Table 4. 10 topmost keywords

Nr	Keyword	Occurrences	Total Link Strength
1	Blockchain	353	1705
2	Artificial Intelligence	197	985
3	Internet	118	814
4	Security	93	626
5	Big Data	88	545
6	IoT	75	512
7	Challenges	71	466
8	Management	62	414
9	Framework	62	397
10	Things	48	352

Abbreviations: Nr = Number

In the following Table 5 appears the 185 results classified in clusters considering the relation of links between keywords and documents. **Cluster 1 (Emergent Innovative Circular Technologies Management)** analyzes the relationship between Artificial Intelligence and other emerging technologies that allow generating innovation and digital transformation in the management of companies supporting the circular economy through prediction analysis. **Cluster 2 (Healthy Cryptoassets)** examines that the application of Blockchain technology, the use of cryptocurrencies, augmented reality and Machine Learning allow the design of new payment models and the creation of new systems that allow improving health and facing new pandemic situations. **Cluster 3 (Security Network)** explores the importance of increasing security in Internet of Things devices, in 5G and 6G networks, and in the creation of new computational and consensus models. **Cluster 4 (Safe Identification)** establishes the need to create new protocols of authentication to prevent the intrusion of attacks in social networks and use security sensors using drones. **Cluster 5 (Efficient Smart Cities)** indicates the need to develop intelligent and efficient algorithms and models based on distributed ledger in Smart Cities. **Cluster 6 (Intrusion Detection Cloud)** discover the need to create intrusion detection models, which solve privacy problems improving access control through the creation of new Industrial Internet of Things devices that allow storing information in Cloud Computing. The outcomes permit us to recognize the *'state of the art'* and can be employed for future researchers. The article shows important lines of research that focus principally on how Blockchain with Artificial Intelligence allows to improve the management of companies supporting the circular economy, the confidence the security of Smart Cities.

VOSviewer software permits us to realize an analysis by occurrences of the keywords and visualizes the interpretation of data in a word map. It can be seen the most explored areas related to Blockchain and Artificial Intelligence. Six clusters allow grouping 185 words and the intensity and lines that connect the research areas are observed (see

Table 5. Clusters with relevant keywords

C	C and N	Most Relevant Keywords	AA
1	Emergent Innovative Circular Technologies	Artificial Intelligence; Big Data; Management; Framework; Innovation; Impact; Networks; Optimization; Performance; Industry 4.0; Future; Digital Transformation; Logistics; Circular Economy; Analytics; Resilience; Predictive Analytics	Emergent Technologies Management Circular Economy
2	Healthy Cryptoassets	Blockchain; Machine Learning; System; Technology; Design; Deep Learning; Augmented Reality; Bitcoin; Ethereum; Evolution; Trust; Governance; Healthcare; Pandemics	Crypto assets Payments Cryptographic
3	Security Network	5G; 6G; Security; IoT; Servers; Federated Learning; Architecture; Informatics; Consensus; Network; Computational Modeling; Training	Networks Security
4	Safe Identification	Authentication; Attacks; Collaboration; Drones; Internet; Protocols; Robotics; Safety, Sensors; Social Media; Traceability	Identification
5	Efficient Smart Cities	Algorithm; Cities; Distributed Ledger; Efficient; Intelligence; Models, Smart Cities	Smart Cities
6	Intrusion Detection Cloud	Access Control; Intrusion Detection; Cloud Computing; IIoT; Issues; Privacy	Cloud

Abbreviations: C = Cluster; C and N = Color and Name; AA = Application Area

Fig. 2). To aid the interpretation of the data, the numbers and words that are larger are more relevant in a field, closer and with a stronger relationship to it. Colors allow to distinguish the various areas of research, emerging topics and trends [27].

There are associations between terms related to management and payment with Cryptoassets, the improving of authentication security in networks, the storage in the cloud that is more secure when encrypted, and the management of Smart Cities.

3 The Topmost Cited Articles

In relation to the 20 articles most cited, most of these authors considers that: Industry 4.0 is based on a series of technological paradigms [18]. Through a qualitative study, it is shown that companies are increasingly aware of the development of digital transformation using new digital technologies, Artificial Intelligence, Blockchain and Internet of Things that improve the experience and satisfaction of users, optimize operations and create new business models [2]. Cities are increasingly turning to technologies to embroider problems that concern society such as ecology and, therefore, smart cities integrate sensors and Big Data through Internet of Things (IoT) devices [4]. The Digital Twin has become a concept and emergent as an effortless integration of data between a physical and virtual machine [5]. The integration of Big Data together with the Internet of Things in the field of health generates an immense advance allowing to collect a large amount of data [6]. The new emerging technologies of Industry 4.0 such as Robotics, Internet of Things, Artificial Intelligence and Blockchain allow to change the way work is done showing more competitive operations in cost, with greater flexibility, speed and quality [7]. Through an empirical study using the Technology, Organization and Environment Framework in the application of Blockchain it is discovered that competitive pressure, complexity and cost have significant effects on the intention to use Blockchain [8]. The use by many companies of new technologies such as Robotics, Drones, Internet of Things, Artificial Intelligence or Blockchain allows the development of cyber-physical systems that can change the landscape of competition, creating economic, environmental

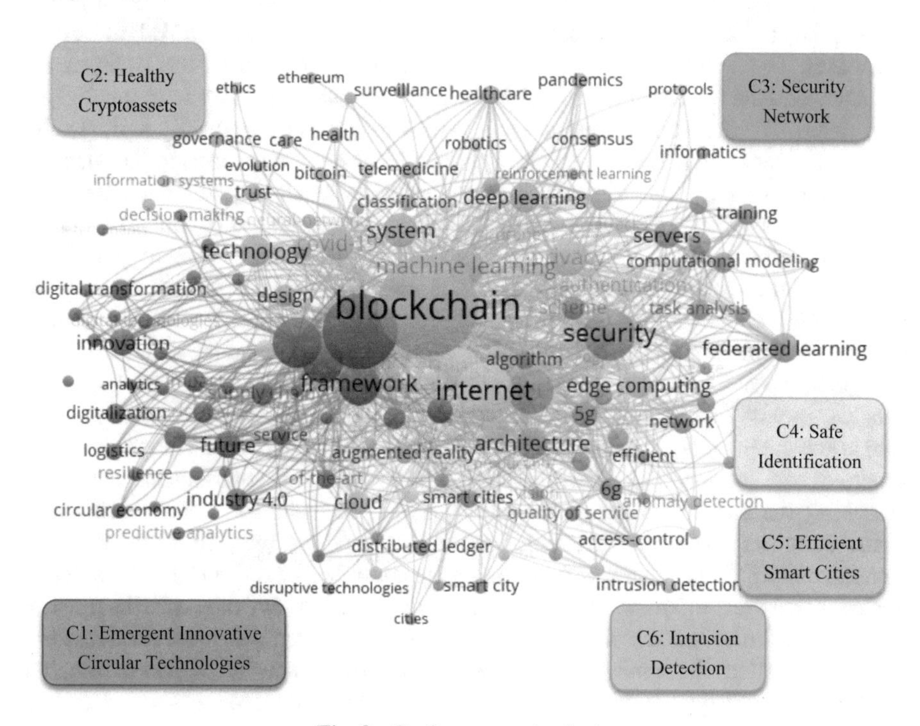

Fig. 2. Co-Ocurrence Analysis

and social value [10]. The use of Internet of Things (IoT) applications in traditional systems is leading to the Industrial Development of the Internet of Things (IIoT) automating intelligent objects allowing greater productivity and operational efficiency [11, 22]. Edge Intelligence is a key enabler for IoT along with Blockchain and Edge Intelligence for flexible and secure edge service management [13]. Smart technologies make it possible to change the organizational lineal [14]. IoT devices are increasingly used by agricultural entrepreneurs [16]. Emerging technologies such as Internet of Things, Robotics, Artificial Intelligence, Big Data Analytics and Blockchain would create an improvement of industrialized agricultural production processes and the industrialized agri-food supply chain [20]. The integration of various technologies such as Artificial Intelligence (AI) and Internet of Things (IoT) leads to the emergence of intelligent ecosystems in aspects such as transportation, agriculture, logistics, education, healthcare, which can be automated, controlled and accessed remotely through smart devices [17]. Industry 5.0 is expected to be the next industrial revolution with applications in smart healthcare, supply chain or manufacturing [23].

4 Conclusions

The major objective of this paper is responds to the Research Questions (RQ's) associated to Blockchain and Artificial Intelligence evaluating the evolution in the scientific literature (RQ1), the most cited authors, institutions, countries and journals (RQ2), the most areas studies (RQ3) and the most cited articles (RQ4). To elaborate this investigation, we searched the keywords 'Blockchain' and 'Artificial Intelligence' in Web of Science Databases. The VOSviewer program is used to analyze the Bibliometric Analysis and to understand the impact of utilizing Blockchain technology and Artificial Intelligence in business. The evolution about this topic is growing (RQ1), specifically in the United States and China and regions that share a geographical situation such as India, Australia and certain European countries (RQ2). The results designate research tendencies, the first focused on emergent technologies, management and circular economies; the second related to payment with Cryptoassets and the use of cryptographic technology; the third focused on the security of the network; the fourth examines the identification of the user on the social media; the fifth investigates the improves in Smart Cities with these technologies and the last trend explores the privacy of the data on the cloud (RQ3). In relation to the most cited articles (RQ4) the greatest of these authors contemplate that Blockchain with Intelligence Artificial permits to create new business models in companies and the integration with other technologies such as Internet of Things permits to generate more sustainability in Smart Cities. This investigation contributes theoretically examining the association between Artificial Intelligence and Blockchain technology, contributing to recognize the growing trend of these notions, recognizing the principal authors, countries, and journals with the greatest number of documents and citations related to these terms. This work permits you to examine the keywords with the research domains and the hottest keywords. The paper helps to go ahead with theoretical research in the field of Blockchain and Artificial Intelligence showing new future lines of research. Related to the practical implications, more studies related to these areas are needed. In relation to the limitations, only articles from the WoS database are analysed. The study can be completed by incorporating articles included in other relevant databases such as Scopus. Finally, Blockchain has problems to solve, so it is essential to examine about its issues and challenges for the companies such as trust and security on the networks. It is necessary to create new mechanism of consensus more ecological for the environment creating new business models with human resources specializes in Blockchain.

References

1. Bernardino, C., Costa, C.J., Aparício, M.: Digital evolution: blockchain field research. In: 2022 17th Iberian Conference on Information Systems and Technologies (CISTI), pp. 1–6 (2022). https://doi.org/10.23919/CISTI54924.2022.9820035
2. Warner, K.S.R., Waeger, M.: Building dynamic capabilities for digital transformation: an ongoing process of strategic renewal. Long Range Plan. **52**(3), 326–349 (2019). https://doi.org/10.1016/j.lrp.2018.12.001
3. Dwivedi, Y.K., et al.: Impact of COVID-19 pandemic on information management research and practice: transforming education, work and life. Int. J. Inf. Manage **55**, 102211 (2020). https://doi.org/10.1016/j.ijinfomgt.2020.102211
4. Allam, Z., Dhunny, Z.A.: On big data, artificial intelligence and smart cities. Cities **89**, 80–91 (2019). https://doi.org/10.1016/j.cities.2019.01.032
5. Fuller, A., Fan, Z., Day, C., Barlow, C.: Digital twin: enabling technologies, challenges and open research. IEEE Access **8**, 108952–108971 (2020). https://doi.org/10.1109/ACCESS.2020.2998358
6. Qadri, Y.A., Nauman, A., Bin Zikria, Y., Vasilakos, A.V., Kim, S.W.: The future of healthcare internet of things: a survey of emerging technologies. IEEE Commun. Surv. Tutor. **22**(2), 1121–1167 (2020). https://doi.org/10.1109/COMST.2020.2973314
7. Olsen, T.L., Tomlin, B.: Industry 4.0: opportunities and challenges for operations management. M&Som-Manuf. Serv. Oper. Manag. **22**(1), 113–122 (2020). https://doi.org/10.1287/msom.2019.0796
8. Wong, L., Leong, L., Hew, J., Tan, G.W., Ooi, K.: Time to seize the digital evolution: adoption of blockchain in operations and supply chain management among Malaysian SMEs. Int. J. Inf. Manag. **52**, 101997 (2020). https://doi.org/10.1016/j.ijinfomgt.2019.08.005
9. Hao, M., Li, H., Luo, X., Xu, G., Yang, H., Liu, S.: Efficient and privacy-enhanced federated learning for industrial artificial intelligence. IEEE Trans. Industr. Inf. **16**(10), 6532–6542 (2020). https://doi.org/10.1109/TII.2019.2945367
10. Tang, C.S., Veelenturf, L.P.: The strategic role of logistics in the industry 4.0 era. Transp. Res E-Log **129**, 1–11 (2019). https://doi.org/10.1016/j.tre.2019.06.004
11. Khan, W.Z., Rehman, M.H., Zangoti, H.M., Afzal, M.K., Armi, N., Salah, K.: Industrial internet of things: recent advances, enabling technologies and open challenges. Comput. Electr. Eng. **81**, 106522 (2020). https://doi.org/10.1016/j.compeleceng.2019.106522
12. Singh, S.K., Rathore, S., Park, J.H.: BlockIoTIntelligence: a blockchain-enabled intelligent IoT architecture with artificial intelligence. Future Gener. Comput. Syst. Int. J. eSci. **110**, 721–743 (2020). https://doi.org/10.1016/j.future.2019.09.002
13. Zhang, K., Zhu, Y., Maharjan, S., Zhang, Y.: Edge intelligence and blockchain empowered 5G beyond for the industrial internet of things. IEEE Netw. **33**(5), 12–19 (2019). https://doi.org/10.1109/MNET.001.1800526
14. De Keyser, A., Koecher, S., Alkire (nee Nasr), L., Verbeeck, C., Kandampully, J.: Frontline service technology infusion: conceptual archetypes and future research directions. J. Serv. Manag. **30**(1), 156–183 (2019). https://doi.org/10.1108/JOSM-03-2018-0082

15. Singh, S., Sharma, P.K., Yoon, B., Shojafar, M., Cho, G.H., Ra, I.: Convergence of blockchain and artificial intelligence in IoT network for the sustainable smart city. Sustain. Cities Soc. **63**, 102364 (2020). https://doi.org/10.1016/j.scs.2020.102364
16. Gupta, M., Abdelsalam, M., Khorsandroo, S., Mittal, S.: Security and privacy in smart farming: challenges and opportunities. IEEE Access **8**, 34564–34584 (2020). https://doi.org/10.1109/ACCESS.2020.2975142
17. Ahad, M.A., Paiva, S., Tripathi, G., Feroz, N.: Enabling technologies and sustainable smart cities. Sustain. Cities Soc. **61**, 102301 (2020). https://doi.org/10.1016/j.scs.2020.102301
18. Aceto, G., Persico, V., Pescape, A.: A survey on information and communication technologies for industry 4.0: state-of-the-art, taxonomies, perspectives, and challenges. IEEE Commun. Surv. Tutor. **21**(4), 3467–3501 (2019). https://doi.org/10.1109/COMST.2019.2938259
19. Angelis, J., da Silva, E.R.: Blockchain adoption: a value driver perspective. Bus. Horiz. **62**(3), 307–314 (2019). https://doi.org/10.1016/j.bushor.2018.12.001
20. Liu, Y., Ma, X., Shu, L., Hancke, G.P., Abu-Mahfouz, A.M.: From industry 4.0 to agriculture 4.0: current status, enabling technologies, and research challenges. IEEE Trans. Ind. Inform. **17**(6), 4322–4334 (2021). https://doi.org/10.1109/TII.2020.300391
21. He, W., Zhang, Z., Li, W.: Information technology solutions, challenges, and suggestions for tackling the COVID-19 pandemic. Int. J. Inf. Manag. **57**, 102287 (2021). https://doi.org/10.1016/j.ijinfomgt.2020.102287
22. Tange, K., De Donno, M., Fafoutis, X., Dragoni, N.: A systematic survey of industrial internet of things security: requirements and fog computing opportunities. IEEE Commun. Surv. Tutor. **22**(4), 2489–2520 (2020). https://doi.org/10.1109/COMST.2020.3011208
23. Maddikunta, P.K.R., Quoc-Viet, Pham, P., Deepa, N., Dev, K., Gadekallu, T.R., Liyanage, M.: Industry 5.0: a survey on enabling technologies and potential applications. J. Ind. Inf. Integr. **26**, 100257 (2022). https://doi.org/10.1016/j.jii.2021.100257
24. Wamba, S.F., Queiroz, M.M.: Blockchain in the operations and supply chain management: benefits, challenges and future research opportunities. Int. J. Inf. Manag. **52**, 102064 (2020). https://doi.org/10.1016/j.ijinfomgt.2019.102064
25. Zupic, I., Cater, T.: Bibliometric methods in management and organization. Organ. Res. Methods **2015**(18), 429–472 (2015). https://doi.org/10.1177/1094428114562629
26. Aparicio, S., Aparicio, J.T., Costa, C.J.: Data science and AI: trends analysis. In: 2019 14th Iberian Conference on Information Systems and Technologies (CISTI), pp. 1–6 (2019). https://doi.org/10.23919/CISTI.2019.8760820
27. Van Eck, N.J., Waltman, L.: Software survey: VOSviewer, a computer program for bibliometric mapping. Scientometrics **84**(2), 523–538 (2010). https://doi.org/10.1007/s11192-009-0146-3

Preliminary Research to Propose a Master Data Management Framework Aimed at Triggering Data Governance Maturity

Leonardo Guerreiro[1], Maria do Rosário Bernardo[2], José Martins[3,4,5], Ramiro Gonçalves[1,4,5], and Frederico Branco[1,5](✉)

[1] Universidade de Trás-os-Montes e Alto Douro, Vila Real, Portugal
leonardo@guerreiro.com, {ramiro,fbranco}@utad.pt
[2] Universidade Aberta, Lisbon, Portugal
maria.bernardo@uab.pt
[3] Instituto Politécnico de Bragança, Bragança, Portugal
jose.martins@aquavalor.pt
[4] AquaValor – Centro de Valorização e Transferência de Tecnologia da Água, Chaves, Portugal
[5] INESC TEC, Porto, Portugal

Abstract. Data management solutions became highly expensive and ineffective mainly due to the lack of transparent processes and procedures to measure and provide clear guidance on the steps needed to implement them. The organizations and specialists agree that the only manner solve the data management issues requires the implementation of data governance. Many of those attempts had failed previously because they were based only on IT, with rigid processes and activities frequently split by systems or the areas supported by systems and their data. It shows that Data governance has been acquiring significant relevance. However, a consensus or even a holist approach was not achieved so far. This paper that is part of an ongoing thesis research that aims to identify the main gaps and opportunities by summarizing and study the literature consistently and as result at the end of the research it will propose a standard framework for data governance measuring its impact on the Data Governance maturity level before and after its implementation and thus as contribute to the community by trying to mitigate the problems found.

Keywords: Data Governance · Framework · DAMA · DMBOK2

1 Introduction

Data governance is a subset of IT governance and, despite of the focus on data assets the Data Governance and IT Governance are not so distinct each other. IT governance makes decisions about IT investments, the IT application portfolio, and the IT project portfolio. IT governance aligns the IT strategies and investments with enterprise goals and strategies.

There are several definitions to data governance. Newman and Logan [1] say that data governance is the collection of decision rights, processes, standards, policies and

A. Rocha et al. (Eds.): WorldCIST 2023, LNNS 800, pp. 183–189, 2024.
https://doi.org/10.1007/978-3-031-45645-9_17

technologies required to manage, maintain and exploit information as an enterprise resource. Thomas [2] describes data governance referring to the organizational bodies, rules, decision rights, and accountabilities of people and information systems as they perform information-related processes. It goes on to state data governance sets the rules of engagement that management will follow as the organization uses data.

Data management (DM) is the business function of planning for, controlling and delivering data and information assets [3]. Data governance refers to the exercise of authority and control over the data management. The purpose of data governance is to increase the value of data and minimize data-related cost and risk also preserving the transparency and accountability during the data management, the main characteristics of governance [4].

Although data governance has recently gained significant relevance, the opinion regarding this matter remains fragmented and scattered. Articles and publications approach separately or join only a few subjects, such as data quality, security, and lifecycle, to address data governance. Even the frameworks currently proposed are either conceptual or focused on one specific organization or business branch [4–8].

The frameworks provide practical standards to help organizations implement multiple processes and procedures [9] and being the foreseen artefact of the research. Using the design science research methodology (DSRM) as a guide [10], in a near future the research will lead to a framework based on DAMA concepts [3], that will be applied as a case of study using CMMI [11] as maturity level and tool of measurement.

1.1 Motivation and Objectives

In the ongoing preliminary research, it was raised that even with this growing relevance of data governance, there isn't a centralized vision addressing all the strands necessary to implement a generic and adaptable framework. The lack of parameters to measure the level of data governance maturity that organizations find before and after implementing a systematic data governance process. While data governance was once a differential, this is currently a subject of extreme importance and relevance in the organizations [12].

To overcome these limitations, the ongoing research has been synthesising some publications and articles in a systematic and structured way trying to figure out how to create a standard framework that could increase the performance, satisfaction, and level of data governance maturity control of the initial and final evaluation of the maturity index in data management. Whose could provide a comprehensive framework that helps organisations achieve their IT and corporate governance goals, creating added value and maintaining balance, optimising resources and mitigating risks [13].

By assembling concepts from various existing theories is generally associated with theory building rather than theory testing. A conceptual framework explains narratively and graphically the main ideas and the hypothesized or proposed relationship among such concepts. As such, it is essential. The frameworks can fit in both quantitative and qualitative research. In a quantitative study, the research wants to get down to variables because the objective is measuring things and do quantitative work with them. In qualitative study, the research tends to stay within a conceptual level. However, the research should be carefully defined the terms, including the research problem, paradigm, aim and objectives and literature review at the beginning of the study. The conceptual framework can be developed if you do an excellent job on these [14]. However only if applied, tested and produces significant results can be an artefact and innovative contributor that aggregates value to the community and organizations what is precisely the aim of the ongoing research which this paper is based on, and the result intended in the near future at the end of the research.

As such this paper is a preliminary report status of ongoing research that as contextualized will propose a standard data management framework and apply it in a case study that will diagnose the data government maturity and analyze the scenario before and after the application of this framework that should cover the main domains of DAMA-DMBoK2$_{TM}$ [1] and answer the main question of the research: *Is it possible to create a standard data governance framework that positively impacts organizations and contribute to reduce and mitigate the gaps found in the literature?*

2 Related Work and Choose of Methodologies and Models

The ongoing research which this paper is based on following a systematic literature review through a state-of-the-art will led to understand data governance, the tools we can use to accomplish its implementation, and the gaps we need to fill. As already mentioned before, the focus will be on creating a standard framework to guide us on this path and then measuring the effects caused after this implementation.

Several academic reviews accept the definition of Weill and Ross [15] and define data governance as specifying the framework for decision rights and accountabilities to encourage desirable behaviour in the use of data [16–18]. DAMA [1] disagrees with this general concept and defines data governance as more than a simple framework specification unless this can be practised. According to Otto [7], important formal goals of data governance are: 1. to enable better decision making, 2. to ensure compliance, 3. to increase business efficiency and effectiveness, 4. to support business integration.

Researchers have been proposing conceptual or initial frameworks regarding data governance [17, 19] and have analysed influencing factors [20] as well as the morphology of data governance [7]. The lack of consensus can be seen in articles and publications approaching separately or joining only a few subject's scopes [5–8]. The frameworks currently proposed are either conceptual or were not applied in a practical context, mainly presented in a specific [4].

The design science research methodology (DSRM) was selected to guide this study because of the structure composed of practices and procedures needed to conduct the research and aim the following objectives: helping on a research model, being consistent with existing literature and finding a mental model for evaluation and demonstrate research regarding Data governance [10].

The big challenge of the research guided by DSRM is to contribute by developing and applying a framework in compliance with the scientific community and up to date with the most used concepts of the market by using DAMA [1]. Address the main pillars described by DAMA: organisational, data, domain, data governance items and finally, precedents and results by using a framework in a practical scope and measuring the effects gained after this application, comparing against the precedents and be able to rate the data governance maturity level before and after as a tool take leverage of one of the most used maturity level models the CCMI [11].

The DAMA (data management association international) promotes understanding, development, and practice of managing data. All this information and best practices widely used in the global market are compiled in a manual named DAMA-DMBoK2 where the most recent version is the 2nd [1]. There are a few reasons why DAMA was selected and considered best suited for use as a guide for this study. First, this framework is concise, consisting of only a few components to evaluate. This makes it less time-consuming for staff in a research function to access the data they manage. Second, the components in this framework are general enough to apply to most data being used. Given that much of the data being used in research will have varying formats and relatively broad components that can apply broadly, this framework is flexible. Finally, this model is a standardised way of conveying the same concepts researchers talk about in an unstructured way. The components in this framework reflect many statistical and practical considerations often discussed in analysis about the pros and cons of data use without veering too far into technological or business use concepts. In brief, this framework is a concise, flexible standardisation of common research concepts that does not delve too deeply into tangential concerns more appropriate for another business function [21].

Initially based on software development industry and administrated by ISACA (Information Systems Audit and Control Association), the Capacity Maturity Model Integration, or simply CMMI is a model of collections containing the best practices that help to improve the processes of an organization allowing measurement of the maturity distributed in 5 crescent levels [22]. The CMMI is the most used and known model, there are over 10,000 organizations and businesses that use CMMI models from over 101 countries, moreover, 11 governments invest in CMMI to support economic development in their countries, CMMI models have been translated into ten languages to support governments and organizations [23]. The CMMI is the maturity model selected on this research because this relevance and because it has provided a method from various disciplines that can be used for process improvement activities. Each maturity level of CMMI contained a set of process areas. The process area has activities an organization must do to improve its business processes [11] and this can be easily adapted to the data governance principles as well as DAMA.

3 Visual Conceptual Idea of Process in a Near Future

At this point a simple attempt of drawing of the framework would be vague and disconnected whereas a glimpse of it tied to a complete process clearly show the importance of this tool that this research intends to propose. Hence, from left to right we have: 1. The Antecedents: initial assessment based on CMMI adapted to DAMA, 2. The impacts over current day by day activities, 3. The actors composed by the relationship between human resource and information systems, 4. Application of the designed framework to be developed, 5. That should result in gains, 6. Consequences: The final assessment based on CMMI to effectively check the impacts and gains. This is how the process intend to look like when ready:

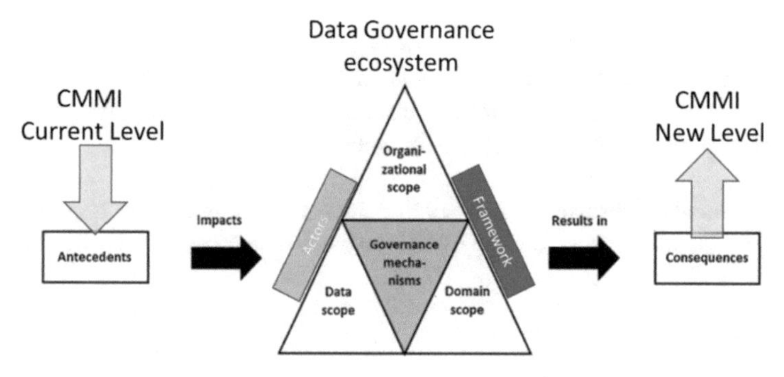

Fig. 1. Conceptual Data Governance process using the future proposed framework

In detail the process illustrated in Fig. 1, would represent conceptually all the efforts and aims of the ongoing research which this paper is based on. According to Abraham et al. [4], the organisational scope determines the organisational expansiveness of data governance and roughly corresponds to the unit of analysis. There is a differentiate between the intra-organizational and the inter-organizational range. The data scope pertains to the data asset an organisation needs to govern. The domain scope covers the data decision domains that apply to governance mechanisms, and here is precisely where the framework would act and help. They comprise data quality, data security, data architecture, data lifecycle, metadata, and data storage, infrastructure and all domains mentioned in DAMA [1]. Antecedents cover the contingency factors that impact data governance adoption and implementation, opening space to understand what the current maturity level of the organization is. Distinguishing between internal and external ancestors. Finally, consequences contain the effects of data governance, and this can result or not in an improvement of the maturity level.

4 Final Considerations and Next Steps

Even with the efforts to find out a comprehensive and deep analysis of data governance in the preliminary literature review of the ongoing research, the results were so far limited. They mostly have revealed definitions, disconnected concepts, and a lack of consensus, almost leading to data management instead of data governance. Regarding the data governance framework, the results almost lead to conceptual or very particular frameworks conceived for a few specific business branches. They have been much more related to best practices than a standard framework.

In the claim to overcome these limitations and gaps as next steps the research will be focused in keeping deepening into the systematic review of the literature raising new challenges, concepts and theories concerning the data governance eco-system and how frameworks can impact positively in the maturity level. All this knowledge is fundamental to support and provide the prototype of the mentioned framework, however as mentioned by Taheri [3] this framework can only break the barrier of the conceptual if it can be tested and therefore applied. In our case it will proceed through a case of study and baselines of measurements with the continuous communication and evolution of the research, aiming to contribute to the community at the end of research trying to fill the gaps that exists between data management and data governance and helping to guide the organizations to pursuit or improve data governance by applying the proposed framework and measure effectively the impacts in the maturity level.

Acknowledgements. This work is financed by National Funds through the Portuguese funding agency, FCT – Fundação para a Ciência e a Tecnologia, within project LA/P/0063/2020.

References

1. Newman, D., Logan, D.: Governance Is an Essential Building Block for Enterprise Information Management. Presented at the , Stamford, CT (2006)
2. Thomas, G.: Alpha males and data disasters: the case for data governance. Brass Cannon Press, Orlando, FL (2006)
3. DAMA International: DMBOK - Data Management Body of Knowledge. Technics Publications LLC (2012)
4. Abraham, R., Schneider, J., vom Brocke, J.: Data governance: A conceptual framework, structured review, and research agenda. Int. J. Inf. Manage. **49**, 424–438 (2019). https://doi.org/10.1016/j.ijinfomgt.2019.07.008
5. Donaldson, A., Walker, P.: Information governance—a view from the NHS. Int. J. Med. Informatics **73**, 281–284 (2004). https://doi.org/10.1016/j.ijmedinf.2003.11.009
6. Ballard, C., et al.: Information Governance Principles and Practices for a Big Data Landscape. IBM Redbooks (2014)
7. Otto, B.: A morphology of the organisation of data governance. In: ECIS 2011 Proceedings (2011)
8. Tallon, P.P., Ramirez, R.V., Short, J.E.: The information artifact in IT governance: toward a theory of information governance. J. Manag. Inf. Syst. **30**, 141–178 (2013). https://doi.org/10.2753/MIS0742-1222300306
9. Smits, D., Hillegersberg, J.V.: The continuing mismatch between IT governance theory and practice: results from a systematic literature review and a delphi study with cio's. J. Manag. Syst. **24** (2014)
10. Peffers, K., Tuunanen, T., Rothenberger, M.A., Chatterjee, S.: A design science research methodology for information systems research. J. Manag. Inf. Syst. **24**, 45–77 (2007). https://doi.org/10.2753/MIS0742-1222240302
11. Ahern, D.M., Clouse, A., Turner, R.: CMMI distilled: a practical introduction to integrated process improvement. Addison-Wesley, Boston (2004)
12. Haneem, F., Kama, N., Taskin, N., Pauleen, D., Abu Bakar, N.A.: Determinants of master data management adoption by local government organizations: An empirical study. Int. J. Inf. Manage. **45**, 25–43 (2019). https://doi.org/10.1016/j.ijinfomgt.2018.10.007
13. Kerr, D.S., Murthy, U.S.: The importance of the CobiT framework IT processes for effective internal control over financial reporting in organizations: An international survey. Inform. Manag. **50**, 590–597 (2013). https://doi.org/10.1016/j.im.2013.07.012
14. Taheri, B.: 10 tips on developing a conceptual framework in quantitative studies. Getting Published, The final years (2017)
15. Weill, P., Ross, J.W.: IT governance: how top performers manage IT decision rights for superior results. Harvard Business School Press, Boston (2004)
16. Otto, B., Weber, K.: Data governance. In: Hildebrand, K., Gebauer, M., Hinrichs, H., and Mielke, M. (eds.) Daten- und Informationsqualität. pp. 277–295. Vieweg+Teubner, Wiesbaden (2011). https://doi.org/10.1007/978-3-8348-9953-8_16

17. Khatri, V., Brown, C.V.: Designing data governance. Commun. ACM **53**, 148–152 (2010). https://doi.org/10.1145/1629175.1629210
18. Wende, K., Otto, B.: A contingency approach to data governance. In: 12th International Conference on Information Quality, Cambridge, USA (2007)
19. Otto, B.: Organizing data governance: findings from the telecommunications Industry and consequences for large service providers. CAIS **29** (2011). https://doi.org/10.17705/1CAIS.02903
20. Weber, K., Otto, B., Österle, H.: One Size Does not fit all—a contingency approach to data governance. J. Data and Inform. Quality. **1**, 1–27 (2009). https://doi.org/10.1145/1515693.1515696
21. Becker, T.: Discussing Data: Evaluating Data Quality TB (2019). https://doi.org/10.18651/TB/TB1903
22. Chrissis, M.B., Konrad, M., Shrum, S.: CMMI: guidelines for process integration and product improvement. Addison-Wesley, Upper Saddle River, NJ (2007)
23. Alsawalqah, H., Alshamaileh, Y., Alshboul, B., Shorman, A., Sleit, A.: Factors impacting on CMMI acceptance among software development firms: a qualitative assessment. MAS. **13**, 170 (2019). https://doi.org/10.5539/mas.v13n3p170
24. Taheri, B.: 10 tips on developing a conceptual framework in quantitative studies (2016). https://researchportal.hw.ac.uk/en/publications/10-tips-on-developing-a-conceptual-framework-in-quantitative-stud,

Adoption of GDPR – In the Intermunicipal Community of Terras de Trás-os-Montes, Portugal

Pascoal Padrão[1], Isabel Lopes[1,2,3](✉), and Maria Isabel Ribeiro[1,4]

[1] Instituto Politécnico de Bragança, Campus de Santa Apolónia, 5300-253 Bragança, Portugal
{isalopes,xilote}@ipb.pt

[2] UNIAG, Instituto Politécnico de Bragança, Campus de Santa Apolónia, 5300-253 Bragança, Portugal

[3] Algoritmi, Universidade do Minho, Largo do Paço, 4704-553 Braga, Portugal

[4] Centro de Investigação de Montanha (CIMO), Laboratório Associado Para a Sustentabilidade e Tecnologia em Regiões de Montanha (SusTEC), Instituto Politécnico de Bragança, Campus de Santa Apolónia, 5300-253 Bragança, Portugal

Abstract. In the face of rapid globalization and rapid technological evolution, new challenges have emerged in terms of personal data protection, demanding a solid and more coherent protection framework in the European Union (EU). Thus, the European Union created a Regulation, in 2016, with the aim of protecting the privacy of personal data of citizens of the European Union. Thus, in May 2018, Regulation (EU) 2016/679 - General Data Protection Regulation (GDPR) entered into force. Thus, after 6 years since its creation and four since its entry into force, with this research work, we intend, through a survey, to assess the state of implementation and adoption of the GDPR in Local Public Administration in Portugal, focusing on Municipalities inserted in the Intermunicipal Community of Lands of Trás-os-Montes (CIM-TTM). The results are discussed in the light of the literature and future work is identified with the aim of evaluating the implementation of such an important regulation with regard to secrecy, privacy and preservation of personal data.

Keywords: General Data Protection Regulation · Municipalities · Data Protection Officer · Information Systems · Security

1 Introduction

According to Jornal Económico, "the protection of natural persons regarding the processing of Personal Data is a fundamental right. Rapid technological evolution and globalization have created new challenges in terms of Personal Data protection, demanding a solid and more coherent protection framework in the European Union. Thus, arises the need for a Regulation that introduces important changes on the protection of natural persons regarding the processing of Personal Data, imposing new obligations on citizens, companies and other private and public organizations" [1].

A. Rocha et al. (Eds.): WorldCIST 2023, LNNS 800, pp. 190–197, 2024.
https://doi.org/10.1007/978-3-031-45645-9_18

During the entry into force of the GDPR, the Local Public Administration was obliged, in the same way as other companies and organizations, to implement the said regulation, given that the GDPR is a legal imperative of the EU. This represents a real revolution in the rules applicable to the processing of information relating to identified or identifiable natural persons, as such and due to the fact that the Local Public Administration exercises a great proximity relationship with the citizens, it is essential that it prioritizes the privacy of the data from citizens, hence this investigation focuses on the implementation of the GDPR in Portuguese municipalities.

As the universe of study for the 308 Portuguese municipalities would become too time-consuming, we limited our investigation to the municipalities belonging to the Intermunicipal Community of Terras de Trás-os-Montes (CIM-TTM), and in the future it is intended to cover all Portuguese municipalities.

This article, in terms of structure, begins with this brief introduction to the topic that the research work addresses, followed by a review of the literature on the general data protection regulation, as well as its implementation. Section 3 addresses the research methodology, identifying the research universe and the research structure. Section 4 presents the results of the questionnaire, followed by the conclusions of this research work. In this same Sect. 5, the limitations encountered in the course of this work are presented and, therefore, future works are presented.

2 General Data Protection Regulation

The GDPR entered into force on May 25, 2018. That date marked a turning point in all European Union member states with regard to data protection. Since then, the GDPR has been the applicable personal data protection legislation in all 27 EU member states, replacing the 1995 Data Protection Directive (95/46/CE of October 24 1995).

The implementation of the GDPR made it possible to unify data protection in the European Union, creating a clear and unique regulation that, in addition to strengthening individual rights, also establishes clear responsibilities for the organizations that collect, that have this data in their possession and that, in addition, also process this data. It reflects the growing concern for privacy and the need for stricter regulations in the face of the increasingly evolving digital landscape.

Since the GDPR deals with personal data, it is important to know first how it defines it. The GDPR defines personal data as "any information relating to an identified or identifiable natural person [2].

This includes any information that can directly or indirectly identify a specific person. Some examples of personal data include names, addresses, identification numbers, contact details, financial information, medical records, photographs, IP addresses, among others" [3].

The GDPR also includes a special category of personal data called "sensitive personal data" or "sensitive data". This data refers to information such as racial or ethnic origin, political opinions, religious or philosophical beliefs, trade union membership, genetic data, biometric data for unique identification of a person, health data or data relating to sex life or sexual orientation. It should be noted that this type of "sensitive data" requires different treatment.

As of May 25, 2018, companies are required to demonstrate that they have taken appropriate steps to ensure GDPR compliance. Among these measures, the following stand out [4]:

- "Adopt personal data security mechanisms;
- Clarify and provide training to employees on GDPR rules;
- Assess the need/mandatory appointment of a Data Protection Officer (DPO) or Data Protection Officer, who should be responsible for managing the compliance process within the company;
- Evaluate the need to prepare a Privacy Impact Assessment (PIA), that is, a document that assesses the impact on the processing of personal data, and respective monitoring. This document should contain a risk assessment on the technology and on the processes that support the processing of personal information in the organization, as well as the identification of measures to be adopted to minimize possible risks;
- Mapping and categorization of personal data collected and processed;
- Creation of automatisms that simplify compliance with the Regulation;
- Communicate to the regulatory authorities and the respective data subjects the occurrence of data breach incidents, within 72 h, after a security breach is known".

It is important that all companies, regardless of their size, are fully informed about the requirements of the GDPR and are ready to apply them, as sanctions and the protection of this data are increasingly pervasive [5].

3 Research Strategy

The term "research method" or "research strategy" is used when you want to address "a set of procedures, techniques and systematic approaches used to research data in a given study. The research method describes the general strategy that a researcher adopts to answer a series of questions that lead him to reach the proposed objective with the research work [6].

The research method adopted in this study involved the application of a survey and document analysis. The survey was developed with the objective of collecting data directly from the participants, while the document analysis was carried out to examine and extract relevant information from important and pertinent documents that were against the subject under study. This combined approach allowed a comprehensive understanding of the investigated aspects, providing quantitative data through the questionnaire and additional and contextualized information through the analysis of the documents [7, 8].

The survey involved the municipalities under study, that is, the 9 Municipal Councils that make up the CIM-TTM: Bragança, Vimioso, Mirandela, Macedo de Cavaleiros, Miranda do Douro, Vila Flor, Alfândega da Fé and Mogadouro.

The survey's main objective was to provide a clear and direct knowledge of the municipalities belonging to this intermunicipal community, regarding the acceptance and implementation of the RGPD.

As for the document analysis, it was very important as a complement and basis for all the research work, having been transversal to the entire study.

The survey was carried out in two months: August and September 2022. As for the response rate, 8 of the 9 possible responses were obtained, which corresponds to a response rate of 84%.

4 Results

After collecting and processing the surveys, it was possible for us to draw up the final graphs, carrying out the quantitative analysis of the responses, and presenting the results obtained.

Always bearing in mind the fact that the GDPR entered into force four years ago, more precisely on May 25, 2018, and its implementation being mandatory in both the public and private sectors, the municipalities surveyed at the CIM- TTM of the total of 8 municipalities that responded to the questionnaire, only one of them has not yet implemented the GDPR, which corresponds to only 11% (see Fig. 1).

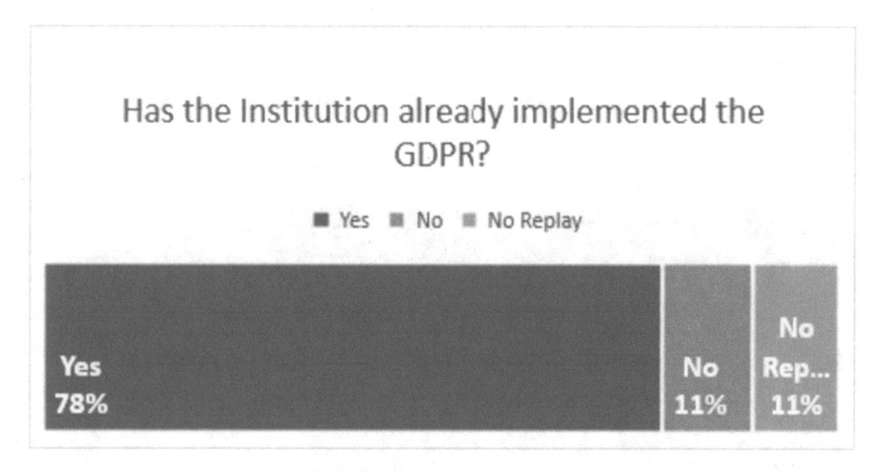

Fig. 1. Adoption of GDPR.

Regarding the CIM-TTM, most of the municipalities that implemented the GDPR did so in an initial phase (see Fig. 2) and 22% did so in the last year, due to the fact that the implementation in the first phase went wrong.

Directly related to the previous data are the data shown in Fig. 3, as we found that the processes triggered by external entities were implemented at an early stage, while those triggered by the municipalities themselves were only implemented this last year.

The reason for adopting the GDPR was another question raised. In this way, there would be three possible reasons, namely through a Legal Imperative, for reasons of Institution Certification or initiative of the municipality itself. As you can see below in Fig. 4, of those who responded, they all implemented the GDPR.

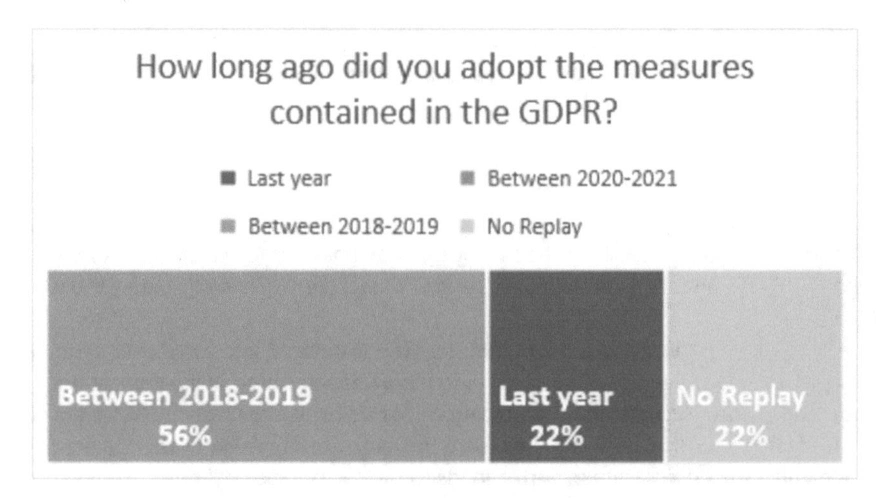

Fig. 2. Time in compliance with the regulation.

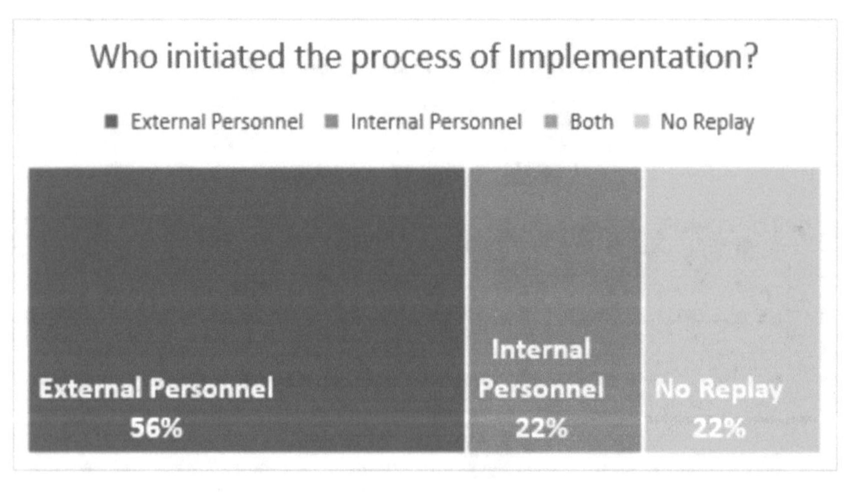

Fig. 3. Who initiated the process of Implementation.

Figure 5 that follows tells us whether there was adequate training or not, so that in most municipalities there was no adequate training, which leads us to deduce that when implementing the GDPR, the municipalities lacked the preparation of an adequate training plan.

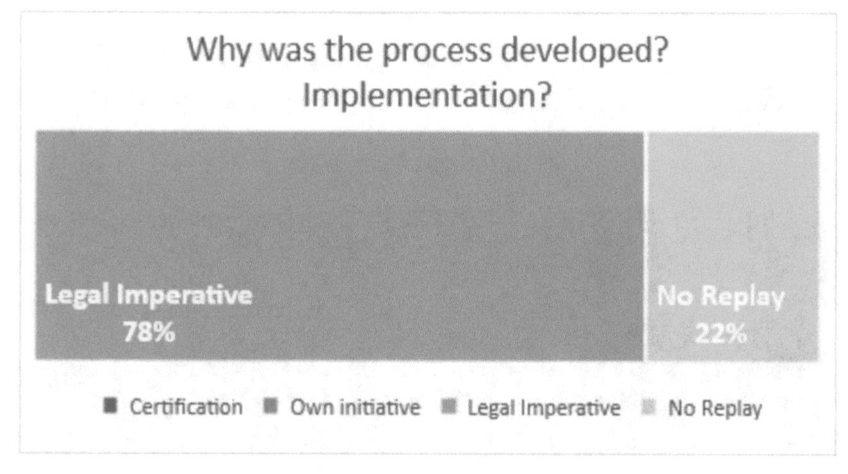

Fig. 4. Why was the implementation process developed.

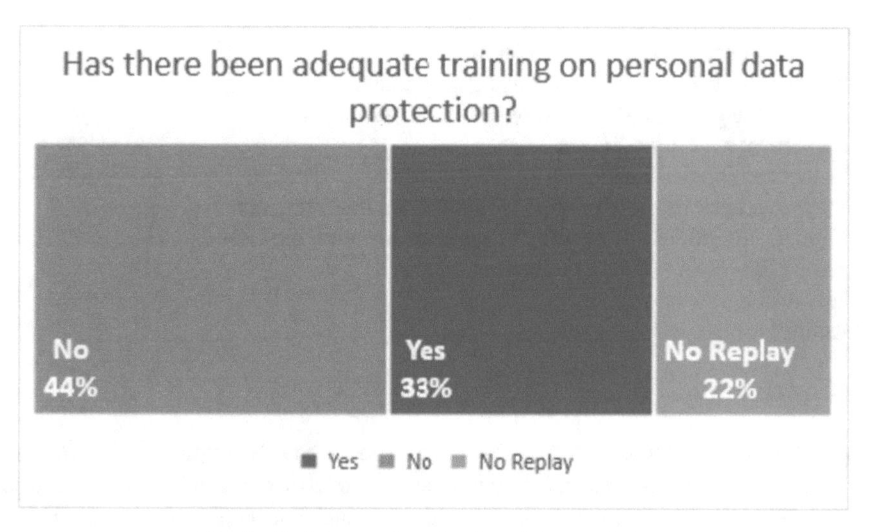

Fig. 5. Training on the adoption of the regulation.

The existing correlation between GDPR and Security is very evident in Fig. 6, as the majority of respondents consider that with the adoption of the GDPR, the data in their institution became more secure, demonstrating the importance that the GDPR attaches to the security of personal data.

This study demonstrated that the Municipalities belonging to the CIM-TTM are concerned with the security of personal data, as the GDPR is implemented in large numbers. It is an excellent finding, but it should be noted that it is not enough to implement the regulation, this was the first step, other aspects must deserve the attention of this CIM from now on, which are:

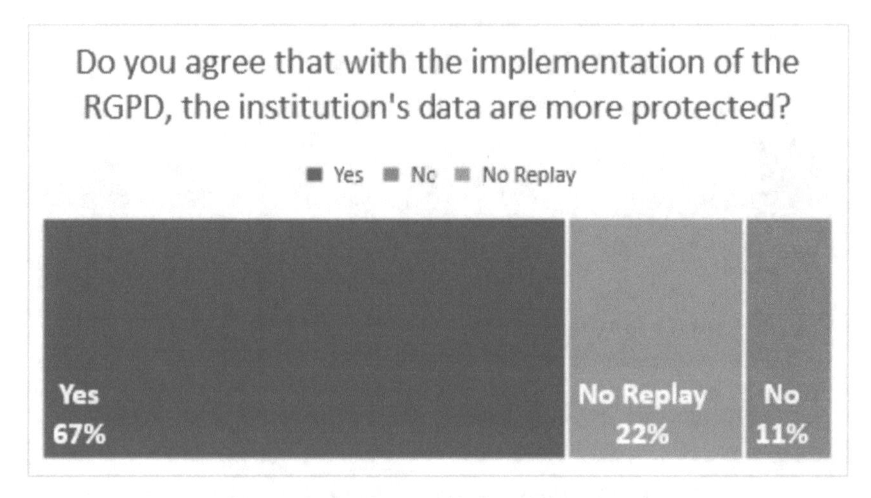

Fig. 6. With the adoption of GDPR, data is more protected.

- Update security and privacy policies;
- Adapt data collection and processing processes;
- Identify the personal data held by the organization;
- Implement security policies;
- Detect data breach incidents;
- Prepare processes to respond to requests from data subjects;
- Validate compliance of existing IT applications with the GDPR;
- Assess the impact of data protection;
- Monitoring;
- Audits.

5 Conclusions

The Public Administration must be aware that it needs to fully implement the GDPR.

With regard to the security and protection of personal data, the Public Administration has the obligation to ensure compliance with technical and organizational measures in order to protect these same data, however, for this purpose, it may subcontract an external organization designated as subcontractor, although he has to give guarantees that he applies and complies with the rules set out in the GDPR.

When the topic is personal data, everyone without exception must comply with the provisions of the GDPR. The sanctions for non-compliance with the regulation are high and it is personal and often sensitive data that we are talking about. It is true that the personal data that Municipalities collect from citizens are necessary for the necessary processing that must be carried out within their competences, but it is also true that more and more citizens have or are starting to fall into reality in relation to the disclosure of your data in the most diverse sectors and that this starts to be worrying. Thus, the Local Public Administration and other companies and organizations must have direct support, in order to be able to properly implement the RGPD and thus manage to protect the data of their citizens as best as possible.

According to Oliveira as cited in Lambelho & Mendes, 20019), "In addition, it is important to bear in mind that Local Authorities must combine the obligations arising from the GDPR, with the positions of Commission for Access to Administrative Documents and the National Data Protection Commission, without losing sight of the Code of Administrative Procedure" [9].

It should be noted that one of the positive impacts that the GDPR added was undoubtedly greater legal uniformity, as well as new rights were granted to the holder, namely the right to be forgotten and portability.

After completing this investigation, we concluded that most municipalities have already implemented the GDPR, despite the fact that there are still municipalities that have not done so.

Given the sample size limit of 9 municipalities, we can consider this to be a limitation of this study, but given that the responses to the survey were 84%, this limitation is thus more diluted and not as impactful.

However, and as is evident in these research works, there is always more to be achieved and the whole is never studied, so, as future work and since this study focused on the implementation of the RGPD, it was interesting to know if these municipalities had already had to communicate to the competent authorities any violation of the stipulations in the regulation.

Acknowledgements. The authors are grateful to the Foundation for Science and Technology (FCT, Portugal) for financial support through national funds FCT/MCTES (PIDDAC) to CIMO (UIDB/00690/2020 and UIDP/00690/2020) and SusTEC (LA/P/0007/2020).

References

1. Jornal Económico. A importância do regulamento geral de proteção de dados, May 4 2017. https://jornaleconomico.pt/noticias/a-importancia-do-regulamento-geral-de-protecao-de-dados-154198
2. European Parliament and Council. Regulation (EU) 2016/679 of the European Parliament and of the Counciel of 27 April 2016, Official Journal of the European Union 2016 (2016)
3. Ryz, L., Grest, L.: A new era in data protection. Comput. Fraud Secur. **2016**(3), 18–20 (2016)
4. Primavera. Regulamento Geral de Proteção de Dados, Saiba como adaptar a sua empresa! (2022). https://pt.primaverabss.com/pt/tudo-que-precisa-saber-sobre-o-rgpd/. Accessed 14 Nov 2022
5. Lopes, I.M., Oliveira, P.: Implementation of the general data protection regulation: a survey in health clinics. In: 13ª Iberian Conference on Information Systems and Technologies, vol. 2018-June, pp. 1–6 (2018)
6. Hudson, L., Ozanne, J.: Alternative ways of seeking knowledge in consumer research. J. Consum. Res. **14**(4), 508–521 (1988)
7. Martins Junior, J.: Trabalhos de conclusão de curso: instruções para planejar e montar, desenvolver e concluir, redigir e apresentar trabalhos monográficos e artigos (2008)
8. Stake, R.: The Art of Case Study Research. Sage Publications, Thousand Oaks (1995)
9. Lambelho, A., Mendes, J.B.: O RGPD e o impacto nas organizações: 6 meses depois. Atas (IPL, Ed.; pp. 1–154). X Congresso Internacional de Ciências Jurídico-Empresariais (2019). www.cicje.ipleiria.pt

Drivers of Long-Term Work from Home Adoption in Small and Medium-Sized Enterprises: A Managerial Perspective

Ann-Kathrin Röpke and Mijail Naranjo-Zolotov(✉)

NOVA Information Management School (NOVA IMS), Campus de Campolide, 1070-312 Lisbon, Portugal
{m20201109,mijail.naranjo}@novaims.unl.pt

Abstract. The pandemic shifted the labor market resulting in an increased demand for work from home (WFH), leading companies to be pressured into implementing it to stay competitive and attractive to employees. Nevertheless, the managers make the decision regarding employees' requests to conduct their tasks from home and therefore play a crucial part in the adoption of WFH in enterprises. This study explores the factors that influence its long-term adoption from a managerial perspective. We propose and qualitatively evaluate a model, guided by the task-technology fit theory, to examine the role of the individual, technological, organizational, and task characteristics for long-term adoption. The findings show that individual factors are weighted highest by managers for their decision-making. Those factors include their experiences with working from home and the corresponding trust toward employees, which are prerequisites for the implementation and shape the attitude of managers. Technological, organizational, and task characteristics are key enablers for WFH and are the building stock for its effective long-term adoption. If trust and positive experiences exist, the interviewed managers indicated that they strive to introduce a hybrid model, in the long term, to cope with the changes in the labor market by staying competitive and attractive to qualified employees.

Keywords: Work from home adoption · Managerial perspective · Small and medium-sized enterprises · Task-Technology Fit

1 Introduction

With the unexpected outbreak of the pandemic in early 2020, companies had to rethink their business models. They had to quickly adapt to emerging technologies to stay competitive and ensure business continuity. Consequently, the outbreak of the pandemic pushed digitalization within enterprises and led to new ways of working which might shape the future of the labor market, also known as the "new normal". Flexible scheduling and hybrid work are expected to be in demand and may even become the "new normal" in the post-pandemic era [1]. Therefore, investments in new collaboration tools had to be made, and new flexible work arrangements were introduced, such as work

A. Rocha et al. (Eds.): WorldCIST 2023, LNNS 800, pp. 198–207, 2024.
https://doi.org/10.1007/978-3-031-45645-9_19

from home (WFH). Several digital platforms have been adopted to share and encourage communication and collaboration during the pandemic, for instance, Zoom, Skype, Slack, or Dropbox [2]. A sustainable WFH environment can be achieved through different lenses of tools and management support to increase the potential benefits and to reduce drawbacks.

Due to the expected shift in the labor market regarding WFH in the post-pandemic era, there is a need for managers to adapt to the "new normal" [3]. To date, there is still little research on pivotal factors for managers to consider when deciding whether to permit or prohibit WFH for their employees, especially in the area of SMEs. Numerous analytical studies have examined managerial and organizational difficulties related to the implementation of WFH in large businesses as due to growing interest in its adoption, but little research has been done on smaller businesses [4]. As the pandemic continues to surge, the long-term effect of the pandemic deserves further attention and needs more research in this area [5]. Most of the recent literature mainly focuses on crucial factors from an employee's perspective, neglecting the research on the managerial perspective. Therefore, this study explores the factors that ensure a long-term adoption post-pandemic of WFH within SMEs from a managerial perspective.

2 Background

2.1 Impact of WFH on the Organizations

WFH might have a positive impact within and outside of an organization, e.g., the decrease of real-estate costs for employers, better work-family balance for employees because they can spend more time at home, as well as a reduction of air pollution and traffic since more people work from home instead of traveling to the physical office [6]. One of the most reported outcomes is increased job satisfaction [7]. Additionally, further literature states stress reduction, employee satisfaction, and lower turnover rate as a benefit, to name a few [7].

Moreover, due to the ongoing globalization of businesses, companies require higher mobility of ICT to keep up with the competition and changing needs within and outside the organization. Those developments impact the traditional way of working and shift it towards a more strategic organizational innovation resulting in a geographically working independent trend. ICT allows individuals to communicate and collaborate with colleagues, clients, and other corporate stakeholders, even though they are geographically separated. Therefore, advancements in ICT, organizational shifts towards WFH, and innovations make the concept of WFH more relevant for companies in the future and make it even easier, more accessible, and more effective for employees and employers in the long-term. In the process of adopting WFH, managers may need to reexamine the traditional methodologies in supervision and evaluation of job performance.

2.2 Adapting to WFH

Managers play a crucial part in the implementation of WFH in their enterprises since they may have the ultimate say when it comes to employee requests for telework arrangements. This is because remote work can be a disorienting experience for everyone if managers turn to a traditional set of leadership behavior, which can cause problems with task completion, performance reviews, and employee engagement [8]. In those situations, employees look to managers to offer direction, consolation, hope, and truthful information [8]. Thus, managers should ensure that proper tools (IT) are applied within the organization and that the skill level of employees is sufficient to ensure that employees can be shifted fast and effectively to remote working. Using remote work provides employees with autonomy and flexibility, and leaves organizations with increased human resource potential and savings in direct expenses [9].

Several companies now provide working packages containing the freedom to pursue WFH at any time requiring the continuation of work performance and related factors. This implies a cultural change within the organization combined with the continuous utilization of ICT and the corresponding digitalization. Digitalization is being more understood to be essential for post-pandemic company survival and prosperity [10]. However, small and medium-sized enterprises (SMEs) frequently struggle with acquiring digital skills and integrating digital processes. Therefore, SMEs tended to be more vulnerable during the pandemic based on, for instance, the lack of financial resources and specialized knowledge [11]. Identifying potential factors relevant for managers from literature makes it possible to support prospective future academic studies into other, yet unexplored subjects. Table 1 presents a summary of the influential factors relevant for managers for WFH adoption identified in recent studies.

Table 1. Influential factors for manager's decision-making about WFH adoption

Authors	Study Objective	Method used	Influential factors
[12]	Challenges of employing older workers as home-based teleworkers	Questionnaire	Trust, Reliability, ability to work independently, time-management ability, adaptability, technology skills
[13]	How institutional context affects how managers form their attitudes towards teleworking	Large-scale survey	Mimetic pressures: productivity, numbers of previous adopters
[14]	How managerial telework permit decisions in German firms were affected by person-related, task-related, and organizational context factors	Vignette study	Loss of control and authority, high-quality relationship between manager and employee, WFH experience of the manager, self-management skills of employees

(*continued*)

Table 1. (*continued*)

Authors	Study Objective	Method used	Influential factors
[15]	Potential indirect impacts on worker performance through organizational commitment and job satisfaction	Survey	Difficulties in managing employees' performance
[16]	Factors relevant to the application of new management practices "telework" when determining managers' ambition to innovate	Online Questionnaire	Attitude toward their view of organizational support for innovation, control over innovative behaviours
[17]	What factors affect middle managers' attitudes toward the implementation of telework	Questionnaire	Managerial practices, self-efficacy of employees, information security tools improvement in organization, ease of use
[18]	Managers' allowance decisions on WFH on organizational, group, and individual level	Survey	Organizational level: organizational policy Group level: trust, reduction in social learning and innovation based on lack of face-to-face interactions Individual level: Productivity, trust, psychological contract, performance
[5]	Decisions made during a pandemic and their impact on objectives deemed essential for survival by managers	Survey	Decisions based on survival and continuity
[19]	Determining advantages of and obstacles to WFH from an organizational standpoint	Survey	Control of effects on work
[20]	Preferences of managers for potential organizational implementation strategies: 1) totally remote, 2) totally office, 3) hybrid	Questionnaire	Type of work performed, average age of client and themselves, ability to manage people from a distance

3 WFH Adoption: A TTF Perspective

This study examines the WFH adoption from the task-technology fit (TTF) [21] model lens. The TTF provides deeper insights into the extent a technology aids a person in their range of tasks which in turn affects the user's ability to do tasks effectively and the likelihood that the technology will be favorably accepted and used [21]. In general, it aims to comprehend why and how current information technologies that enable technologically connected tasks improve performance. It addresses the need to go beyond only examining people's intentions to use technology and focuses on how doing so improves performance at the individual level [21].

The model in this study builds upon TTF and adds one more characteristic, the personal view and experience of the manager regarding WFH, in this sense we evaluate a model with four pillars: individual, technological, organizational, and task characteristics. Individual characteristics contain general information about the common understanding of WFH, how managers have experienced the pandemic in the workplace, and demographic characteristics. Technological characteristics include factors that explain the nature and accessibility of various resources needed to complete activities e.g., the adopted Software and Hardware. Organizational characteristics elaborate further on the changes in policies and contracts due to the new work modality and managers' expectations of the outcome of implementing WFH in the future. Task characteristics portray the current performance as well as the degree of different tasks that can be performed from home. Figure 1 demonstrates the main factors.

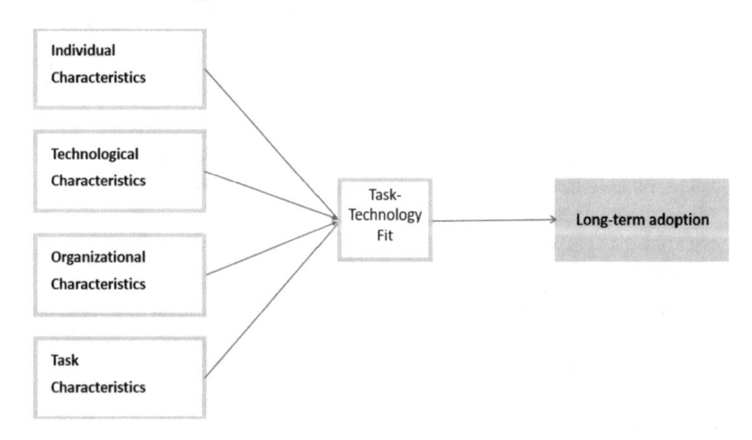

Fig. 1. Research Framework

4 Research Methodology

This study follows a qualitative semi-structured interview procedure to identify the main factors that establish a WFH environment from a managerial perspective and validate the proposed framework. Four middle-managers from different SMEs located in Austria participated in the interviews. The questionnaire includes open-ended as well as closed-ended questions. Each formulated question is related to one of the main constructs of the developed model, namely individual, technological, task, and organizational characteristics. The semi-structured interview approach started with a broad discussion of the general understanding of the concept of WFH and their personal experience during self-isolation. After, the interview shifted towards more specific areas of WFH, for instance, their perception of the usefulness of the software and hardware for communication and collaboration purposes, to gain deeper insights into relevant factors for the managers regarding a long-term implementation of WFH in their organizations. The interviews were carried out within a four-week timespan between August 2022 and September 2022 and were performed via online communication tools. All interviewees agreed for the interviews to be recorded. Table 2 shows the basic profile of the interviewees.

Table 2. Participants in the interview

ID	Gender	Type of industry	Age
I1	male	IT sector	42
I2	male	Agriculture sector	35
I3	male	Manufacturing sector	60
I4	male	Watch and jewelry sector	30

5 Results

The semi-structured interview with four different CEOs identified crucial factors relevant to the long-term implementation of WFH. Additionally, their perceptions of the concept of WFH and their own experience regarding being constantly connected to work during the period of self-isolation have been analyzed. Table 3 summarizes the main findings from the interviews by highlighting the factors relevant for long-term adoption categorized by the characteristics and the type of industry the interviewee operates in. Additionally, the corresponding interview outcomes and the related impact on TTF for each interviewee has been highlighted.

Table 3. Main findings extracted from interviews

Framework dimension	Type of industry	Relevant factors for long-term adoption
Individual	IT sector	Experience, attitude, trust
	Agriculture sector	Experience, attitude, trust
	Manufacturing sector	Experience, attitude, trust
	Watch and jewelry sector	Experience, attitude, trust
Technological	IT sector	Quality of communication tools, broadband connection
	Agriculture sector	Quality of communication tools, broadband connection, ease of use, monitoring tools
	Manufacturing sector	Broadband connection, ease of use
	Watch and jewelry sector	Quality of communication tools, broadband connection, ease of use
Organizational	IT sector	Trust, IT Infrastructure, IT Security, family support, training, tools, and equipment, employees' performance
	Agriculture sector	Trust, communication, collaboration, IT infrastructure, IT security, policies, training, tools, and equipment; employees' performance
	Manufacturing sector	Trust, communication, collaboration, IT infrastructure, employees' performance
	Watch and jewelry sector	Trust, communication, collaboration, IT infrastructure, IT security, policies, training; employees' performance
Task	IT sector	Access to information, type of task, collaboration
	Agriculture sector	Access to information, type of task, collaboration
	Manufacturing sector	Type of task, collaboration
	Watch and jewelry sector	Access to information, type of task, collaboration

6 Discussion and Conclusions

The TTF models provided a valuable lens for WFH since it gives the ability to explain complicated and interrelated information clearly and logically in this context. Individual characteristics have the highest impact on managers' decision-making toward WFH adaption in the long-term. Whereas, technological, organizational, and task characteristics served as key enablers for the effective implementation of WFH and thus only have a moderate impact on TTF.

The most crucial factors that ensures the implementation of WFH are managers' experiences with remote work (prior to and during the pandemic) and trust toward their employees regarding how well and reliable they perform their tasks, which relate to individual characteristics. Trust is built by establishing a positive experience [22]. In general, a person who trusts another person is convinced that this person will fulfill his or her expectation [22]. Thus, experiences shape managers' attitudes toward the concept of WFH. One interviewed manager stated that he had negative experiences with WFH during the pandemic, e.g., long response times of their employees or the manager was not able to reach them, resulting in a lack of trust and the decision of not implementing WFH anytime soon.

Technological, organizational, and task characteristic factors are considered key enablers for WFH. They build the building blocks for effective implementation of WFH in the long-term. Crucial technological factors include good broadband connection, quality of communication and collaboration tools, and the perceived usefulness of IT from managers. The perceived usefulness of technologies depends on experiences and was more favorably viewed by those managers with relevant positive prior experiences on WFH. Those factors are a prerequisite to providing effective WFH and can increase productivity on both an individual and team level [23]. When switching to WFH during the pandemic, the managers who did not offer WFH prior to the pandemic offered instruction on how to carry out tasks and how the different channels have to be used. In the post-pandemic era, training might only be relevant to new employees to familiarize them with the corresponding technology. Lastly, the type of tasks and access to relevant information serve as enabling factors within the task characteristics. Thus, creative, and innovative processes which need a great amount of information exchange are preferred by managers to be performed at the physical location. In general, these enablers must be ensured to successfully adopt WFH. The findings of Thulin et al. (2019) confirm that routine and standardized tasks are gradually becoming workable from home because they do not demand high collaboration with team members. As stated by the interviewees, relevant business information might get lost when communicating virtually and the relationship with the employees may suffer.

Implementing or approving WFH depends heavily on trust [25]. The essential elements of a manager-employee relationship include trust since it directly affects the level of commitment, performance, and corresponding job satisfaction [26]. Managers who implement WFH strategies show a stronger level of employee support and trust [27]. Nevertheless, key enablers must be confident to effectively adapt and perform WFH in the long-term. Therefore, individual characteristics are more weighted as relevant for the implementation from a managerial perspective than technological, organizational, and task characteristics which serve as WFH enablers. To stay competitive and attractive in the labor market, companies nowadays are feeling pressure to offer WFH and the associated flexibility since more employees are now demanding it. Most managers acknowledged the benefits of WFH and want to exploit them post-pandemic. To overcome potential disadvantages of WFH, the hybrid solution was implemented by some managers. Nevertheless, there is no "one-size-fits-all" answer, therefore, businesses have to proactively design a long-term WFH, or hybrid work strategy based on their individual needs [1]. In the post-pandemic era, WFH is a significant trend, still there are numerous pressing concerns regarding the well-being of employees, regulations, and cyber-security that necessitate monitoring and further solutions. To develop a more sustainable model for "new normal" work habits, relevant parties at all levels of society, including governments and corporations, must collaborate [1].

7 Limitations

This study focuses only on a few selected managers, all males, from different industries. Future research may include also female CEOs to remove a possible bias caused by CEOs gender. The opinions toward WFH adaption might differ between managers in a company and at the same time, even between industries. For instance, managers who already established WFH prior to the pandemic or are mainly active in the IT industry are more likely to adopt WFH than other managers. Additionally, external factors may influence the decisions of managers for implementing WFH, for instance, governmental regulations and policies. Future research may explore other components not considered in this study.

References

1. Vyas, L.: "New normal" at work in a post-COVID world: work–life balance and labor markets. Policy Soc. **41**, 155–167 (2022). https://doi.org/10.1093/polsoc/puab011
2. Tønnessen, Ø., Dhir, A., Flåten, B.T.: Digital knowledge sharing and creative performance: work from home during the COVID-19 pandemic. Technol. Forecast. Soc. Change **170**, 120866 (2021). https://doi.org/10.1016/j.techfore.2021.120866
3. Appelgren, E.: Media management during COVID-19: behavior of Swedish media leaders in times of crisis. J. Stud. **23**, 722–739 (2022). https://doi.org/10.1080/1461670X.2021.1939106
4. Dickson, K., Clear, F.: Management issues in the adoption of telework amongst SMEs in Europe. In: Portland International Conference on Management of Engineering and Technology, pp. 1703–1708 (2006). https://doi.org/10.1109/PICMET.2006.296745
5. Chudziński, P., Cyfert, S., Dyduch, W., Zastempowski, M.: Leadership decisions for company SurVIRval: evidence from organizations in Poland during the first Covid-19 lockdown. J. Organ. Chang. Manag. **35**, 79–102 (2022). https://doi.org/10.1108/JOCM-09-2021-0289
6. Bailey, D.E., Kurland, N.B.: A review of telework research: findings, new directions, and lessons for the study of modern work. J. Organ. Behav. **23**, 383–400 (2002). https://doi.org/10.1002/job.144
7. Manroop, L., Petrovski, D.: Exploring layers of context-related work-from-home demands during COVID-19. Pers. Rev. (2022). https://doi.org/10.1108/PR-06-2021-0459
8. Lagowska, U., Sobral, F., Furtado, L.M.G.P.: Leadership under crises: a research agenda for the post-covid-19 era (2020). https://doi.org/10.1590/1807-7692bar2020200062
9. Harpaz, I.: Advantages and disadvantages of telecommuting for the individual, organization and society. Work Study **51**, 74–80 (2002). https://doi.org/10.1108/00438020210418791
10. Doerr, S., Erdem, M., Franco, G., Gambacorta, L., Illes, A.: BIS Working Papers No 965 technological capacity and firms' recovery from Covid-19 (2021)
11. Klein, V.B., Todesco, J.L.: COVID-19 crisis and SMEs responses: the role of digital transformation. Knowl. Process. Manag. **28**, 117–133 (2021). https://doi.org/10.1002/kpm.1660

12. Sharit, J., Czaja, S.J., Hernandez, M.A., Nair, S.N.: The employability of older workers as teleworkers: an appraisal of issues and an empirical study. Hum. Factors Ergon. Manuf. **19**, 457–477 (2009). https://doi.org/10.1002/hfm.20138
13. Peters, P., Heusinkveld, S.: Institutional explanations for managers' attitudes towards telehomeworking. Human Relations **63**, 107–135 (2010). https://doi.org/10.1177/0018726709336025
14. Beham, B., Baierl, A., Poelmans, S.: Managerial telework allowance decisions – a vignette study among German managers. Int. J. Hum. Resour. Manag. **26**, 1385–1406 (2015). https://doi.org/10.1080/09585192.2014.934894
15. de Menezes, L.M., Kelliher, C.: Flexible working, individual performance, and employee attitudes: comparing formal and informal arrangements. Hum. Resour. Manage. **56**, 1051–1070 (2017). https://doi.org/10.1002/hrm.21822
16. Massu, J., Caroff, X., Souciet, H., Lubart, T.I.: Managers' intention to innovate in a change context: examining the role of attitudes, control and support. Creat. Res. J. **30**, 329–338 (2018). https://doi.org/10.1080/10400419.2018.1530532
17. Silva-C, A., Montoya R, I.A., Valencia A, J.A.: The attitude of managers toward telework, why is it so difficult to adopt it in organizations? Technol. Soc. **59**, 101133 (2019). https://doi.org/10.1016/j.techsoc.2019.04.009
18. Williamson, S., Colley, L., Foley, M.: Public servants working from home: exploring managers' changing allowance decisions in a COVID-19 context. Econ. Labour Relat. Rev. **33**, 37–55 (2022). https://doi.org/10.1177/10353046211055526
19. Urbaniec, M., Małkowska, A., Włodarkiewicz-Klimek, H.: The impact of technological developments on remote working: insights from the Polish managers' perspective. Sustainability (Switzerland) **14**, 552 (2022). https://doi.org/10.3390/su14010552
20. Kis, I.A., Jansen, A., Bogheanu, C.D., Deaconu, A.: Research upon the evolution of the preference regarding the way of working in COVID-19 crisis time. Proc. Int. Conf. Bus. Excellence **16**, 965–980 (2022). https://doi.org/10.2478/picbe-2022-0090
21. Goodhue, D.L., Thompson, R.L.: Task-technology fit and individual performance. MIS Q. **19**, 213 (1995)
22. Buchner, D., Schmelzer, J.A.: Vertrauen. In: Führen und Coachen, pp. 27–46. Gabler Verlag, Wiesbaden (2003). https://doi.org/10.1007/978-3-322-87014-8_4
23. Bosua, R., Gloet, M., Kurnia, S., Mendoza, A., Yong, J.: Telework, productivity and wellbeing: an Australian perspective (2012)
24. Thulin, E., Vilhelmson, B., Johansson, M.: New telework, time pressure, and time use control in everyday life. Sustainability (Switzerland) **11**, 3067 (2019). https://doi.org/10.3390/su11113067
25. Stout, M.S., Awad, G., Guzmán, M.: Exploring managers' attitudes toward work/family programs in the private sector. Psychol.-Manag. J. **16**, 176–195 (2013). https://doi.org/10.1037/mgr0000005
26. Golden, T.D.: Applying technology to work: toward a better understanding of telework. Organ. Manag. J. **6**, 241–250 (2009). https://doi.org/10.1057/omj.2009.33
27. Golden, T.D., Fromen, A.: Does it matter where your manager works? Comparing managerial work mode (traditional, telework, virtual) across subordinate work experiences and outcomes. Human Relations **64**, 1451–1475 (2011). https://doi.org/10.1177/0018726711418387

Study of Digital Maturity Models Considering the European Digital Innovation Hubs Guidelines: A Critical Overview

Daniel Babo[1], Carla Pereira[2], and Davide Carneiro[1](✉)

[1] CIICESI, ESTG, Instituto Politécnico do Porto, Rua do Curral, 4610-156 Margaride, Felgueiras, Portugal
{dfrb,dcarneiro}@estg.ipp.pt, davide.r.carneiro@inesctec.pt

[2] INESCTEC, ESTG, Instituto Politécnico do Porto, Rua do Curral, 4610-156 Margaride, Felgueiras, Portugal
carla.pereira@inesctec.pt

Abstract. Nowadays the concept of digitalization and Industry 4.0 is more and more important, and organizations must improve and adapt their processes and systems in order to keep up to date with the latest paradigm. In this context, there are multiple developed Maturity Models (MMs) to help companies on the processes of evaluating their digital maturity and defining a roadmap to achieve their full potential. However, this is a subject in constant evolution and most of the available MMs don't fill all the needs that a company might have in its transformation process. Thus, European Digital Innovation Hubs (EDIH) arose to support companies on the process of responding to digital challenges and becoming more competitive. Supported by the European Commission and the Digital Transformation Accelerator, they use tools to measure the digital maturity progress of their customers. This paper analyzes several MMs publicly available and compares them to the guidelines provided to the EDIH.

Keywords: Maturity Model · Digital Maturity · European Digital Innovation Hub (EDIH)

1 Introduction

Digital transformation is a priority in most organizations. It is a complex task that tackles the changes digital technologies can impose [1]. Maturity Models (MMs) have proved to be an important instrument as they allow for a better positioning of the organization and help find better solutions [2]. To ascertain and measure dedicated aspects of social and technical systems' maturity, a wide range of assessment models have been developed by both practitioners and academics [3]. However, some enterprises might lack technical expertise, experimental capabilities, and specialist knowledge. To lower barriers, especially for SMEs, and to realize the potential of growing autonomy in Cyber-Physical Systems, competence centers and, with a broader perspective, (regional/pan-EU) Digital Innovation Hubs (DIH) are arising [4].

A. Rocha et al. (Eds.): WorldCIST 2023, LNNS 800, pp. 208–217, 2024.
https://doi.org/10.1007/978-3-031-45645-9_20

MMs are a prospering approach to improve a company's processes and business process management (BPM) capabilities [5]. MMs enable users to identify the need for change and to derive the necessary measures to accompany the change process [6]. They monitor the company's performance, finding information that can be used in the development of strategies to achieve the company's full potential, improving its processes and operations. Most MMs conduct an assessment based on 4–5 evolutionary maturity levels. While some models use status levels describing the internal digital penetration, other models use archetypes or clusters of companies that describe common characteristics of these [7].

2 Maturity Models Analysis

This research featured 17 MMs, out of which the first four were developed by consultancy institutions, one by the World Economic Forum in collaboration with Singapore Economic Development Board (Smart Industry Readiness Index [14]) and the remaining 12 are scientific/academic MMs.

Smart Grid Maturity Model (SGMM): [8] The SGMM is a management tool that an organization can use to appraise, guide, and improve its smart grid transformation, developed by IBM and the Global Intelligent Utility Network Coalition with the assistance of APQC. It contains 8 domains (Strategy, management and regulatory; Organization; Technology; Societal and environment; Grid operations; Work and asset management; Customer management and experience; Value chain integration) described at five levels of maturity (exploring and initiating, functional investing, integrating, optimizing enterprise-wide and innovating the next wave of improvements).

PwC Maturity Model: [9] This MM, developed by PwC, aims to speed the process of digitizing a business, helping to understand what strengths can already be built on, and which systems/processes may be needed to integrate future solutions. It consists of seven dimensions (Digital business models and customer access; Digitization of product and service offerings; Digitization and integration of vertical and horizontal value chains; Data & Analytics as core capability; Agile IT architecture; Compliance, security, legal & tax and Organization, employees and digital culture) and four stages (Digital novice, Vertical integrator, Horizontal collaborator and Digital champion).

Digital Maturity Model (DMM): [10] DMM is the first industry-standard digital maturity assessment tool, created to provide guidelines for a clear path throughout the digital transformation journey. It evaluates digital capability across five clearly defined business dimensions (Customer, Strategy, Technology, Operations and Organization & Culture). Divided into 28 sub-dimensions, which breakdown into 179 individual criteria. There is no information about the maturity levels of this MM.

Digital Readiness Assessment (DRA): [11] DRA, developed by KPMG, analyzes the state and quality of relevant areas of a company, incorporating two different perspectives for digital solutions: management (transformation intensity) and pervasiveness (operational effectiveness). It consists of four dimensions (Development and purchasing, Production, Marketing and Sales), categorized in four readiness levels (reactive participant, digital operator, ambitious transformer and smart digitalist).

Digital Industry Survey (DIS): [12] DIS is a digital maturity diagnosis, working as a tool for reflection and autonomous goals definition regarding the digital transformation of a manufacturing company. This MM evaluates the digital readiness through three analysis dimensions (Intelligent products, Intelligent processes and Strategy & organization), subdivided into 21 indicators, rated on a scale from 0 to 5 (absence, first signals, existence, survivability, comfort and maturity). Currently, the author of this paper is working on this MM as a developer.

IBM Big Data Model: [13] This model's objective is to determine the importance of large-scale investments to support planned business projects. It contains six dimensions (Business strategy, Information, Analytics, Culture & execution, Architecture and Governance) and five maturity levels (Ad hoc, Foundational, Competitive, Differentiating and Breakaway).

Smart Industry Readiness Index (SIRI): [14] Comprises a suite of frameworks and tools to help manufacturers start, scale and sustain their digital transformation journeys. It identifies three fundamental building blocks of Industry 4.0 on its topmost layer, which are broken down into eight pillars in the second layer, representing critical aspects that companies must focus on. A third layer comprises 16 dimensions that should be referenced when evaluating the current maturity levels of manufacturing facilities. These dimensions are evaluated on a scale from 0 to 5.

System Integration Maturity Model Industry 4.0 (SIMMI 4.0): [15] SIMMI 4.0 enables companies to classify its IT system scenery with focus on Industry 4.0 requirements. It contains four dimensions (Vertical integration, Horizontal integration, Digital product development and Cross-sectional technology criteria) and five stages (Basic digitization level, Cross-departmental digitization, Horizontal and vertical digitization, Full digitization and Optimized full digitization).

Maturity Model for Digitalization within the Manufacturing Industry's Supply Chain: [16] This MM splits the process of digital transformation into two different perspectives depicting "smart product realization" and "smart product application". Both assess nine dimensions (Strategy development, Offering to the customer, "Smart" product/factory, Complementary IT system, Cooperation, Structural organization, Process organization, Competencies and Innovation culture) through five levels of maturity (Digitalization awareness, Smart networked products, The service-oriented enterprise, Thinking in service systems and The data-driven enterprise). Both perspectives display certain similarities, as activities on certain paths may not diverge from another.

Supply Chain Systems Maturing Towards the Internet-of-Things [17]: This framework investigates layers of ICT deployment, enabling the identification of possibilities for improvement and helping in identifying gaps where IoT can strengthen future applications. It depicts four dimensions (Business, Application, Information and Technical Infrastructure), named as layers, through four maturity levels, labeled by the main technological driver (ERP, ERP 2.0, SOA/SAAS, IoT).

Industry 4.0-MM: [18] Industry 4.0-MM's goal is to assess the establishment of companies' Industry 4.0 technology and guide them towards maximizing the economic benefits of Industry 4.0. This MM defines five dimensions (Asset management, Data governance, Application management, Process transformation, Organizational alignment), and six maturity stages (incomplete, performed, managed, established, predictable and optimizing).

Digital Readiness Assessment Maturity Model (DREAMY): [19] DREAMY's goal is to help manufacturing companies and researchers understating the digital readiness level in the state of practice. It contains four dimensions (Process, Monitoring & Control, Technology and Organization) and five maturity levels (initial, managed, defined, integrated & interoperable and digital-oriented).

Industry 4.0 Maturity Model: [20] This model assesses the Industry 4.0 maturity of industrial enterprises in the domain of discrete manufacturing. It defines nine dimensions (Strategy, Leadership, Customers, Products, Operations, Culture, People, Governance, Technology), divided into 62 items, and five maturity levels. Nevertheless, the author only describes the first level (complete lack of attributes supporting the concepts of Industry 4.0) and the fifth level (state-of-the-art of required attributes).

Industry 4.0 Maturity Model: [21] This MM was created to assess the SMEs digital readiness. It contains 5 dimensions (Business and organization strategy, Manufacturing and operations, Technology driven process, Digital support and People capability) divided into 43 sub-dimensions, and 5 maturity levels (not relevant, relevant but not implemented, implemented in some area, mostly implemented, full implementation).

Process Model towards Industry 4.0: [22] This model is a framework to guide and train companies to identify opportunities for diversification within Industry 4.0. It depicts three dimensions (Vision, Roadmap and Projects) and five maturity levels (Initial, Managed, Defined, Transform and Detailed BM).

Maturity Model for Data-Driven Manufacturing (M2DDM): [23] M2DDM addresses companies' problems with their IT architectures, guiding them in the adoption process of Industry 4.0. The model only defined the IT systems, thus being it the only dimension of this MM, assessed through six maturity levels (nonexistent IT integration, data and system integration, integration of cross-life-cycle data, service-orientation, digital twin and self-optimizing factory).

IoT Technological Maturity Model: [24] This MM assists manufacturers in adopting IoT-technologies. Since it is limited to look only at these technologies, this is also a one-dimension maturity model. The model contains eight maturity levels (3.0 maturity, initial to 4.0 maturity, connected, enhanced, innovating, integrated, extensive and 4.0 maturity).

In the end of this first analysis, given the diversity of dimensions and different names used to label them and with the purpose of comparing the MMs, all the dimensions were grouped into four areas: Processes (Data analysis, production, digitalization of data), Technology (IT systems, analytics tools), Organization (workforce training, culture, investment, projects) and Strategy (vertical and horizontal integrations, marketing, sales). Table 1 compares all the analyzed MMs within these areas.

Table 1. Areas assessed by the Maturity Models (P – Process, T – Technology, O – Organization, S - Strategy)

	P	T	O	S
Smart Grid Maturity Model	X	X	X	X
PwC Maturity Model	X	X	X	X
Digital Maturity Model	X	X	X	X
Digital Readiness Assessment	X	X		X
Digital Industry Survey	X	X	X	X
IBM Big Data Model	X	X	X	
Smart Industry Readiness Index	X	X	X	X
System Integration Maturity Model Industry 4.0	X	X		X
MM for Digitalization within the Manufacturing Industry's Supply Chain	X	X	X	X
Supply Chain Systems Maturing Towards the Internet-of-Things	X	X	X	X
Industry 4.0-MM	X	X	X	X
Digital Readiness Assessment Maturity Model	X	X	X	
Industry 4.0 Maturity Model	X	X	X	X
Process Model towards Industry 4.0	X	X	X	X
Maturity Model for Data-Driven Manufacturing		X		
IoT Technological Maturity Model		X		

Although consultancy and scientific/academic MMs share the same goal, there's a slight difference in the way they are presented, mostly on PwC Maturity Model [9] and Digital Maturity Model [10]. Since they are not presented in scientific papers, the way information is displayed, with multiple graphics and boards, facilitate the interpretation of their objectives through statistics and data related to the dimensions and companies that already have assessed their maturity. Smart Industry Readiness Index [14] also fits in this alternate way of presenting the MM, being the most extensive and detailed document, alongside PwC Maturity Model [9]. Thus, it would be important to have some sort of guidelines on the creation of a MM, assuring that all the most important issues are covered. On that basis, European Digital Innovation Hubs appeared.

3 European Digital Innovation Hubs

European Digital Innovation Hubs (EDIH) are, together with Partnerships & Platforms, Skills & Jobs and Regulatory Framework, one of the four key elements of the Digitizing European Industry (DEI) strategy, launched by the European Platform of national initiatives on digitizing industry [25].

EDIH support companies to improve business/production processes, products, or services using digital technologies by providing access to technical expertise and testing, as well as the possibility to "test before invest". They also provide innovation services, such as financing advice, training, and skills development that are central to successful digital transformation. Environmental issues are also taken into account, in particular with regard to the use of digital technologies for sustainability and circularity [26].

Besides that, since they are located in a certain region, EDIH are able to follow closely the companies' progress, as they know the environment and local aspects of the company. As this is a European project, the best practices are exchanged easily between hubs all over Europe, as well as the provision of certain services from other hubs when they are not locally available. Furthermore, they will act as a multiplier and widely diffuse the use of all the digital capacities built up under the different specific objectives of the Digital Europe Programme on High Performance Computing, Artificial Intelligence, Cybersecurity, Advanced Digital Skills and Accelerating the best use of technologies [27].

The EDIH network is supported by the Digital Transformation Accelerator (DTA), which supports the European Commission in building a community of hubs and other stakeholders. Furthermore, the DTA assesses the performance of the EDIH network, measuring the impact that EDIH have on the digital maturity of the organizations they support. To this end, the Joint Research Center provides a guideline in how to build a digital maturity assessment for companies that are interested in receiving EDIH support, which contains two different modules and also a scoring criteria to calculate the final result of the assessment. The first module gathers basic information about the enterprise to analyze how it compares itself with others in the same sector, size category, region and country. The second module aims to measure the digital maturity of the enterprise through the dimensions represented in the Fig. 1, collecting information that will help to characterize the enterprise's departing point on the digital transformation journey, identifying areas where it might need EDIH support [28].

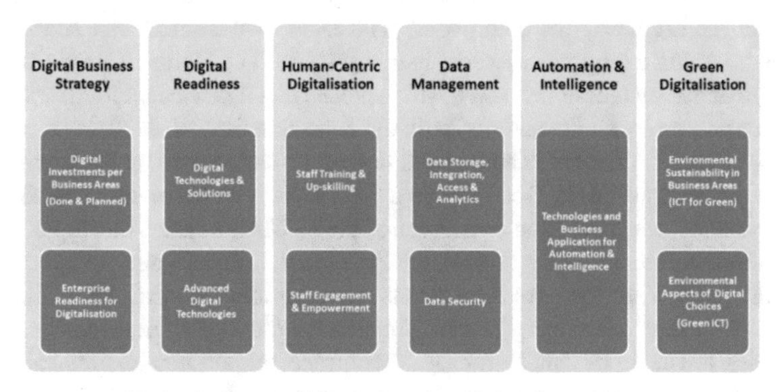

Fig. 1. Assessed Dimensions on the EDIH model [28]

3.1 The Existing Maturity Models and the EDIH Model Guidelines

The next step of this research was comparing the analyzed MMs with the guidelines provided by the European commission. The scientific reports and documents about the

MMs are not extensive and some only contain a poor description of how the digital maturity assessment of an enterprise occurs. Also, some MMs are created to provide support to a specific niche, as it's the case of Digital Industry Survey, designed to help manufacturing companies on the Portuguese region of Tâmega e Sousa. Thus, there might be a topic that is approached by a subdimension and it is not possible to deduct that only from the dimensions' name and brief explanation, without getting into details.

The criteria used for the comparison is whether the documentation of the MM addresses or not the topics assessed in the questions present in the guideline [28]. The first dimension, Digital Business Strategy, seeks to determine the overall status of the digitalization strategy from a business perspective. The first set of questions address if the enterprise already invested or pretends to invest in certain business areas (e.g., product design, marketing and sales, project planning) and the second set evaluates the company's readiness to digitalization, which might require organizational and economic efforts not yet foreseen. The Digital Readiness dimension provides an evaluation of the current uptake of digital technologies, divided into mainstream technologies (e.g., cloud services, e-commerce sales, remote tools) and advanced technologies (e.g., virtual/augmented reality, Internet of Things, Blockchain). Human-centric digitalization looks at how staff are skilled, engaged and empowered with and by digital technologies, and their working conditions improved, with a view to increase their productivity and wellbeing. Data Management and connectedness assesses how data is managed within the enterprise (e.g., if it is stored digitally, if it is accessible in real time), considering the data is sufficiently protected via cybersecurity schemes. Automation & Intelligence evaluates the level of integration of both aspects in the business processes (e.g., image recognition, autonomous devices, recommendation systems). Finally, Green Digitalization evaluates how the enterprise uses digital technologies to contribute to environmental sustainability.

Through Table 2 we conclude that the main focus of the MMs is the preparation and adoption of the digital by the enterprises, with Data Management and Automation & Intelligence having a preponderant role in that process. The Human-Centric Digitalization is approached by half of the MMs, which means that it is relevant for the process but still not considered by all the authors, with the consultancies' MMs giving it more relevance than the scientific ones. There is still little to no concern about the environmental aspects of the digitalization. Deep research for the keywords "green", "environment" and "sustainable" was made and only 2 out of the 17 MMs address this topic.

As noticed before, it is clear that Digital Readiness, Data Management and Automation & Intelligence are the core dimensions when building a MM. These dimensions represent the base of a company's transformation to the digital, evolving/changing its technologies and using its data to improve their processes, automating them at the same time. The structural aspects of the organizations come right after that, with the investments plans on the digital and work force qualification. Green Digitalization is still a rudimental aspect, but certainly something that will be addressed often by the MM developed in the next years. We should consider that this table does not qualify how the dimensions are assessed, the same topic can be addressed by two MMs but one of them explores much more how a company deals with it on its digitalization process. For instance, Smart Grid Maturity Model [8] has a dimension directed exclusively to green

Table 2. Comparison between the MMs and EDIH dimensions (DBS - Digital Business Strategy, DR - Digital Readiness, HCD - Human-Centric Digitalization, DM - Data Management, A&I - Automation & Intelligence, GD - Green Digitalization)

	DBS	DR	HCD	DM	A&I	GD
Smart Grid Maturity Model	X	X	X	X	X	X
PwC Maturity Model	X	X	X	X	X	
Digital Maturity Model	X	X	X	X	X	
Digital Readiness Assessment	X	X	X	X	X	
Digital Industry Survey	X	X		X	X	
IBM Big Data Model	X	X		X	X	
Smart Industry Readiness Index	X	X	X	X	X	
SIMMI 4.0	X	X		X	X	
Maturity Model for Digitalization within the Manufacturing Industry's Supply Chain	X	X	X	X	X	
Supply Chain Systems Maturing Towards the Internet-of-Things	X	X		X	X	
Industry 4.0-MM	X	X	X	X	X	
Digital Readiness Assessment Maturity Model	X	X		X	X	
Industry 4.0 Maturity Model	X	X	X	X	X	
Industry 4.0 Maturity Model	X	X	X	X	X	X
Process Model towards Industry 4.0	X	X				
MM for Data-Driven Manufacturing		X		X	X	
IoT Technological Maturity Model		X		X	X	

and environmental issues, with 4 subdimensions, while Industry 4.0 Maturity Model [21] addresses this topic only as a subdimension of Business and organization strategy. With the growth of the EDIH most of these models will become obsolete due to the arising of models that comply with the guidelines, being in advance more sturdy and complete. Although, all the existing models will continue to serve the purpose of research with their ideas and methods being used on the development of more advanced models.

4 Conclusions

This paper aims to review several MMs publicly available and compare them to the most recent guidelines provided by the European Commission. The 17 chosen MMs were analyzed one by one, and their dimensions categorized for a better understanding of the dimensions assessed. There might be some flaws on this analysis due to the information available on the respective research papers not being fully immersive, making it impossible to analyze deeply the topics that the MM assesses. Although, it is enough to have a general overview of the addressed topics and if they match with the desired by the EDIH.

The obtained results are somehow expected, since most of the MMs were developed several years ago they don't address the most recent concerns so frequently, especially green and environmental ones, because the focus was the adoption of the digital on the companies' processes and improvement of their technologies to become more competitive and stay on track. It is expected that the future development of MMs includes environmental issues and address this topic more often, since it is a main concern of our society nowadays and companies play a big role in it. The fact that we are currently in the decade of climate awareness will boost this aspect and raise its importance, probably getting to the point that it will surpass the currently most important issues and become one of the top concerns.

Another important point is that the way digital maturity is assessed on the existing MMs might become obsolete with the tool provided by EDIH, but those models who provide a roadmap will keep their importance. The methodologies on their strongest assessed areas are still going to be used by companies to improve their digital maturity and become more competitive. With the technology paradigm shifting, new aspects such as green digitalization and sustainability are gaining their importance in the development of MMs, but the core elements of digitalization will still be technology, data management and automation of the processes, followed by the companies' strategies and workforce management.

Acknowledgements. This work is financed by National Funds through the Portuguese funding agency, FCT - Fundação para a Ciência e a Tecnologia, within project LA/P/0063/2020.

References

1. Brown, N., Brown, I.: From Digital Business Strategy to Digital Transformation – How? A Systematic Literature Review (SAICSIT 2019), 17–18 September 2019, Skukuza (2019)
2. Becker, J., Kanckstedt, R., Pöppelbuß, J.: Developing Maturity Models for IT Management – A Procedure Model and Its Application (2009)
3. Mettler, T.: Maturity assessment models: a design science research approach. Int. J. Soc. Syst. Sci. **3**(½) (2011)
4. Sassanelli, C., et al.: Towards a reference model for configuring services portfolio of digital innovation hubs: the ETBSD model. In: Camarinha-Matos, L.M. et al. (eds.) PRO-VE 2020, IFIP AICT 598, pp. 597–607 (2020)
5. Röglinger, M., Pöppelbuß, J., Becker, J.: Maturity models in business process management. Bus. Process Manag. J. **18**, 2 (2012)
6. Felch, V., Asdecker, B., Sucky, E.: Maturity models in the age of Industry 4.0 – do the available models correspond to the needs of business practice? In: Proceedings of the 52nd Hawaii International Conference on System Sciences, pp. 5165–5174 (2019)
7. Chanias, S., Hess, T.: How digital are we? Maturity models for the assessment of a company's status in the digital transformation. Munich (2016)
8. CMU. Cert's podcasts: security for business leaders: show notes (2010). https://resources.sei.cmu.edu/asset_files/Podcast/2010_016_102_67772.pdf

9. PwC. Industry 4.0: Building the Digital Enterprise (2016). https://www.pwc.com/gx/en/industries/industries-4.0/landing-page/industry-4.0-building-your-digital-enterprise-april-2016.pdf
10. Deloitte: Digital Maturity Model. (2018). https://www2.deloitte.com/content/dam/Deloitte/global/Documents/Technology-Media-Telecommunications/deloitte-digital-maturity-model.pdf
11. KPMG: Digital Readiness Assessment. (2016). https://assets.kpmg/content/dam/kpmg/pdf/2016/04/ch-digital-readiness-assessment-en.pdf
12. Duarte, N., Pereira, C., Carneiro, D.: Digital maturity: an overview applied to the manufacturing industry in the region of Tâmega e Sousa, Portugal. In: 12th International Scientific Conference Business and Management. Vilnius, Lithuania (2022)
13. Nda, R., Tasmin, R., Hamid, A.: Assessment of big data analytics maturity models: an overview. In: Proceedings of the 5th NA International Conference on Industrial Engineering and Operations Management Detroit, Michigan, 10–14 August 2020 (2020)
14. World Economic Forum. The Global Smart Industry Readiness Index Initiative: Manufacturing Transformation Insights Report 2022 (2022). https://www.edb.gov.sg/en/business-insights/market-and-industry-reports/manufacturing-transformation-insights-report-2022.html
15. Leyh, C., et al.: SIMMI 4.0 – a maturity model for classifying the enterprise-wide IT and software landscape focusing on Industry 4.0. In: Proceedings of the Federated Conference on Computer Science and Information Systems, pp. 1297–1302 (2016)
16. Klötzer, C., Pflaum, A.: Toward the development of a maturity model for digitalization within the manufacturing industry's supply chain. In: Proceedings of the 50th Hawaii International Conference on System Sciences, pp. 4210–4219 (2017)
17. Katsma, C., Moonen, H., Hillegersberg, J.: Supply chain systems maturing towards the internet-of-things: a framework. In: 24th Bled eConference eFuture: Creating Solutions for the Individual, Organisations and Society, Bled, 12–15 June 2011 (2011)
18. Gökalp, E., Şener, U., Eren, P.: Development of an Assessment Model for Industry 4.0: Industry 4.0-MM (2017)
19. De Carolis, A., et al.: A maturity model for assessing the digital readiness of manufacturing companies (2017)
20. Schumacher, A., Erol, S., Sihn, W.: A maturity model for assessing Industry 4.0 readiness and maturity of manufacturing enterprises (2016)
21. Chonsawat, N., Sopadang, A.: The development of the maturity model to evaluate the smart SMEs 4.0 readiness. In: Proceedings of the International Conference on Industrial Engineering and Operations Management Bangkok, 5–7 March 2019 (2019)
22. Ganzarain, J., Errasti, N.: Three stage maturity model in SME's toward Industry 4.0. J. Indust. Eng. Manag. OmniaScience, Barcelona **9**(5), 1119–1128 (2016). ISSN 2013-0953
23. Weber, C., et al.: M2DDM – a maturity model for data-driven manufacturing. In: The 50th CIRP Conference on Manufacturing Systems, pp. 173–178 (2017)
24. Jæger, B., Halse, L.: The IoT Technological Maturity Assessment Scorecard: A Case Study of Norwegian Manufacturing Companies (2017)
25. European Commission. Digitising European Industry Reaping the Full Benefits of a Digital Single Market. (2016). https://eur-lex.europa.eu/legal-content/EN/TXT/?uri=CELEX%3A52016DC0180
26. European Digital Innovation Hubs. https://digital-strategy.ec.europa.eu/en/activities/edihs. Accessed 2022/11/18
27. European Commission. Annex to the Commission Implementing Decision on the Financing of the Digital Europe Programme and Adoption of the Multiannual Work Programme – European Digital Innovation Hubs for 2021–2023 (2021)
28. European Commission. Digital Maturity Assessment for EDIH Customers. https://ec.europa.eu/newsroom/dae/redirection/document/82255

A Comparison Between the Most Used Process Mining Tools in the Market and in Academia: Identifying the Main Features Based on a Qualitative Analysis

Gyslla Santos de Vasconcelos(✉), Flavia Bernardini, and José Viterbo

Universidade Federal Fluminense, Computing Institute, Rua Passo da Pátria 156, Niterói, Brazil
{gysllav,fcbernardini,viterbo}@id.uff.br

Abstract. There are several tools focused on process mining with different functionalities and purposes of use, ranging from the most intuitive to the extremely complex, configuring a difficult task to choose the tool that best applies to your work. With this in mind, a comparative study of the process mining tools used in the commercial and academic environments was developed. The work was developed from a bibliographic research of the business process management area, process mining, event logs definition, the XES standard, of the process mining tools, studies advanced about the process mining tools and identification of the most cited tools. Differences were verified between tools with academic proposals of the commercial ones, based on functionality, availability, customization possibility and support. In addition, the tools that simultaneously attend to these different kinds of audiences (commercial and academic) has been identified, were Celonis and ProM 6, respectively.

Keywords: Process mining · Event logs · XES standard · Commercial tools

1 Introduction

With the advancement of technology, the interconnection of information systems, and the immensity of data generated daily, the management of information produced in real time, ensuring reliability and efficiency in the production of results, has motivated several research in the area of process mining, which had origin in artificial intelligence and data mining. With this, the area of process mining has retrieved attention of professionals and students, with different focuses and to meet their specific needs. Currently, there are several Process Mining tools on the market, but with different applications. Some are intuitive and easy to use, others are extremely complex and focused on research. With this large possibility of available tools, and not very clear differences, it can be a difficult task for a potential user to choose the most appropriate tool or algorithm, to fulfil yours necessities, given the infinite functionalities.

We could find a few works in literature that also compare process mining tools. Devi [1] performed a comparative study of the tools ProM, Disco, and Celonis, with a

A. Rocha et al. (Eds.): WorldCIST 2023, LNNS 800, pp. 218–228, 2024.
https://doi.org/10.1007/978-3-031-45645-9_21

contextualization of process mining, presenting benefits as well as disadvantages of these tools. Among other aspects, the author compared the tools in the context of importing and exporting records from an event log and checking the conformity between the log and the process model [2]. Bru and Claes [3] focused on end users, applying a questionnaire aimed at identifying the most used tools, similar to the research of Ailenei [2], searching for the functionality that matters most and the most used environment as web or computer platforms. We could observe in these works that there is a lack of a unified methodology for comparing the process mining tools from the user perspective.

This study aims to describe the main process mining tools that meet the expectations of both, researchers in academia and professionals working in the market, with analysis of commercial tools. This work was elaborated based on bibliographical research and, the choice of the commercial tools, was based on user ratings. In this way, for the development of this study, we formulated the following Research Questions (RQs): [Q1] What are the main uses and functionalities of process mining tools? [Q2] Which process mining tools are best for different users? [Q3] Are there tools capable of serving the academic and the commercial environment? Answering these RQs turned possible to structure a framework from the bibliographic research, contextualized within the area of business process management, defining what is process mining and its importance, what are the process mining techniques used. Besides that, the importance of event logs and the XES standard, highly significant to process mining tools in the context of standardization, and the possibility of adding numerous information to the log. From this, a comparison between the most used process mining tools on the market and academia was developed. The main functionalities that arouse interest and satisfy their users were also identified. So, evaluation tables were developed with qualitative and quantitative criteria, and the most used, most cited, and highest ranked were identified.

This work is organized as follow: Initially, Sect. 2 discusses process mining and its importance, as well as process mining techniques, clarifying the use of event logs, the XES standard, and the process mining tools. Afterwards, Sect. 3 presents our methodology used for conducting this study and the results found with the comparative research, were addressed. Finally, the conclusions and future work.

2 Background

2.1 Process Mining and Its Importance

The word "process" comes from the Latin word *processus*, meaning method or set of actions with some objective, suggesting changes with sequential steps. Process can be defined as a set of tasks conducted by people or systems to reach a goal [4]. It is a logical, correlated and sequential set of activities, from an input, adds value and produces an output [5]. Business refers to the personas that interact in the execution of a sequence of activities, adding value with the aim of generating return. Business covers the various types of enterprises private or public organizations, in any segment of activity [6]. According to Dumas [7], events correspond to things that happen in a unique way, meaning that they have no specific duration. As an example, we can cite the arrival of a piece of equipment at a shipyard. However, this event can trigger a series of activities that will have a duration. Still in the example, due to the arrival of the equipment to the shipyard, the responsible engineer will need to inspect it. This inspection is an activity.

The Business Process Management (BPM) is a system of performance management of an enterprise from beginning to end, starting from the creation of the process [7]. According to the specialized literature, there are several BPM life cycle models as described by [7, 8] and [9], with similar aspects, but different divisions of the activities, being that, the main objective is to manage the operational processes of a company, separating BPM in stages. The BPM CBOK [6] defines that no matter the amount of steps and descriptions in a process life cycle, they will always a PDCA cycle for continuous improvement. The Deming Cycle or PDCA Cycle focuses on the process design, and as stated before, on the sequential actions of plan, do, check and act. Although such representation is simple, it represents a rupture in the way that companies were managed in the past, with trial and error, now requiring planning and investments from beginning to end of the process [10].

Therefore, in real situations and over time, several improvements may be needed in business processes already implemented, for various reasons such as no longer meeting expectations, and constant monitoring is required. Processes without monitoring tend to degradation [7]. According to Hammer [11], all good processes can become bad, unless they are continuously evaluated and adapted to changing objectives, new business requirements, technology innovations as well as new customer expectations.

The company Gartner, a market reference specialized in research and consulting Information Technology, until the year 2010 disclosed in its reports the class of BPM support systems as BPMS [12]. However, realizing the nature of the changes and the market evolution, in 2012 it published the first Gartner Magic Quadrant for intelligent business process management suites (iBPMS), positioning the systems focused on real-time agility, process optimization and other features as intelligent business operations [13].

So, Process mining is the extraction of important information related to the processes recorded in an event log of the organizations information systems. From these data, it can be generated models consistent with the active behavior observed in the log. Thus, instead of the starting point be the ideal process modeling, was traditionally done, it starts dynamically from information about the way these processes are executed in practice, in the order which they are executed, because any information system can record these events [14].

2.2 Process Mining Techniques

Process mining techniques can be classified from action types or use cases: (i) automated process discovery (*discovery*), (ii) conformance checking (*conformance*), (iii) performance mining (*enhancement*), and (iv) variant analysis [7, 15]. Aalst [15] described as three types of actions, comprising items (i), (ii) and (iii), and Dumas detailed as four use cases. For Aalst [15], (iv) is part of (ii).

Automated process discovery techniques have as input an event log and produce as output a business process model corresponding with the behavior observed in the event log either directly with the logged or implicitly from the paths traversed. The techniques of compliance verification have as input, besides the event log, a business process model, and as output, they provide a list of differences between the model and the event log. Performance mining techniques also use a process model and event log as input, but the output is different in that they present an improved process model, identifying bottlenecks and delays along the process. This input model can be provided by the analyst or from an automated process. Finally, the techniques of variant analysis have as input two event logs, corresponding to two variants, that is, different cases of the same process. As output, they produce a list with the differences between these two logs. Basically these two logs describe differently ended cases, one with a positive outcome (for example a satisfied customer or process completed successfully) and the other ending with a negative outcome (dissatisfied customer or process completed late). This verifies how important the event logs are for the application of process mining techniques, and in the next section they will be detailed in their basic structure necessary for information processing.

Event logs: When a process is executed, the responsible system is usually in charge of coordinating the individual cases and informing users of the tasks they need to perform. All access management and job complexities are hidden to the user, but are present in the execution scope. If there is a defined process model, each work item corresponds to a task in the model. As an example, we can cite a task related to an Order-to-cash process that would be "Confirm the order", and the work items would be the orders from different customers. All these events can be registered in a log file [7]. So, formally defining, an event log is a collection of records with date and time (timestamp), which deal with the execution of a work item, consequently a task in a process. An event log records when the task started, if it was completed and even if it occurred within a particular process. This information refers to the cases or instances of a process, which is the data regarding which the activities or tasks occur. As an example, we can cite the case of an employee confirming the order of client A, within the process of Request for Payment [7, 16]. Logs can also describe a set of traces of the life cycle of a particular case, which in turn is composed of events. The simplest event logs are represented as tables and stored in Comma-Separated-Values (CSV) format. They should contain the minimum of attributes presented. If they have several other attributes, the format needs to be more versatile, capable of handling more complex data, such as the eXtensible Event Stream (XES) format.

The **XES Standard** defines a grammar for a tag-based language. Its goal is to provide a unified and extensible methodology to capture the behavior of systems through event logs defined in the standard. This standard includes definitions of validation rules (XML Schema), describing the structure of an event recorded by an information system log [18]. This is a more versatile format for storing event logs and has been standardized by the IEEE Task Force on process mining [17]. The structure of a XES file is based on a data model, where each XES file represents an event log, which contains several traces. Each trace can correspond to several events and these can contain different attributes. Due to standardization, most of the process mining tools work with event logs in XES format [15, 18].

Process Mining Tools. In recent years, several researches are being developed on new techniques, new algorithms as well as new process mining software aimed at the academic and commercial area. In addition, studies of existing tools to have the understanding of the operation of each of them, checking their functionality and differences [2]. There are several tools capable of performing process mining. In this work, we analyzed 42 process mining tools, as described later.

3 Our Methodology

In order to identify the tools to be comparable, we conducted two types of research: (i) an extensive literature research, with an academic focus, seeking information in various types of publications such as scientific articles, books, dissertations and theses; and (ii) a monitoring of market trends, with an overview of the highest ranked commercial process mining tools from May of 2021 to April of 2022. For conducting a data analysis from the academic perspective, we developed a qualitative analysis methodology based on the main criteria already identified in other scientific researches [1–3]. The main criteria are: market availability, accessibility as to the license, academic license, existence of tutorials, different types of input format, multi-platform tools, usability, support to process mining techniques, academic popularity and number of citations. For conducting a data analysis from the market perspective, we used Gartner Peer Insights platform, promoted by Gartner, Inc., which is a reference in the corporate environment with individual insights on companies and technology tools, providing a practical, independent, and objective view of numerous tools with their features, favoring the decision making and performance increase in companies. The primary methodology used by the Gartner Peer Insights team is to recognize the best companies by the customers that use them. So, we proposed items regarding data collected as a high-level synthesis, selecting the vendor software most valued by information technology professionals. Table 1 presents all our evaluation criteria items, used for conducting our study.

For classifying the process mining tools, we scored the tools based on two criteria: (i) academic data analysis and (ii) market data analysis. For the academic data analysis, we summed up the values of items 4 to 21 from Table 1 and normalized the total values for each tool. For the market data analysis, we scored the tools according to their relevance. The relevance of the tools was calculated based on the number of reviews and the score. So, *Relevance of the tool = number of reviews / Total of reviews* and *Ranking = Score X relevance of the tools*. We calculated the relevance of the tools according to the user experience based on several requirements, such as evaluation and hiring, integration and deployment service and support and product capability.

Table 1. Items for analyzing Process Mining Tools. Column I/C/G refers to the data type of each item, where I means Informative data, collected from the tools website; C means Collected data, qualitatively collected by us; and G means summary data collected from Gartner Inc.

Item n	I/C/G	Item description	Domain of collected data
1	I	Responsible for research and development	Text
2	I	Web address	Text
3	I	Current version	Text
4	C	Available on the market	Yes or no (1 or 0)
5	C	License	Accessibility: Open-Source = 1 Open-Source/Commercial = 1 Commercial = 0.5
6	C	Academic license	Yes or no (1 or 0)
7	C	Tutorials	Yes or no (1 or 0)
8	C	Input format (import)	(a) Multiformat: Scored based on the tool with the most formats (b) Standardized: If it has XES = 1 otherwise 0
9	C	Output format (export)	Scored by number of output models (up to 1)
10	C	Capacity of the imported log	Maximum capacity of file sizes (in GB) - only in commercial tools
11	C	Execution platform	More than 3 platforms = 1 Two platform = 0.8 One platform = 0.5
12	C	Data filtering	Yes or no (1 or 0)
13	C	Process discovery	Yes or no (1 or 0)
14	C	Conformance check	Yes or no (1 or 0)
15	C	Social network mining	Yes or no (1 or 0)
16	C	Decision rule mining	Yes or no (1 or 0)
17	C	Process view	Yes or no (1 or 0)
18	C	Performance analysis	Yes or no (1 or 0)
19	C	Usability	Easy = 1 Intermediate = 0.5 Hard = 0
20	C	View quality	Excelent = 1 Good = 0.5 Poor = 0
21	C	Academic popularity	1 or 0
22	C	Amount of citations	Number of citations in Google Scholar search with a string equal to "process mining" AND " < tool name >" (normalized)
23	G	Number of reviews	Integer
24	G	Percentage of 5-star assignments	Percentage
25	G	Product Capabilities	Grade calculated by Gartner, which varies from 0.0 to 5.0
26	G	Relevance of the tool	Grade calculated by Gartner, varying from 0.0 to 1.0
27	G	Ranking	Ranking calculated by Gartner

4 Results and Analysis

A total of 42 process mining tools were analyzed, found from various publications and market research. For aims of replicability of our work, we turned available all our collected data in tables in an online repository [19]. The available tables are: (i) Tools202204.xlsx, containing the name of the tool and the company responsible for its development and maintenance; (ii) Rank202204.xlsx, containing the results of our ranking process according to the academic perspective of data analysis (items 4 to 22 from Table 1); (iii) GartnerRank202204.xlsx, containing the collected data from Gartner Peer Insights as well as the ranking values calculates by us; (iv) Previous-ToolsCatalog202204.xlsx, containing diverse information of discontinued tools; and (v) Local202204.xlsx, containing data regarding the countries where the headquarters of the companies responsible for the tools are located. Of this total, 30 tools are currently available for open-source or commercial use, 17 of which have an academic license. Most of the tools support CSV, XES, and XLSX file types for importing event log data. As for the execution platform, the main ones are On-Premises (17) and SaaS (13). For execution on the user's machine (stand-alone) only 5 support it. Considering these 30 tools, we also analyzed where the companies responsible for them are located. Most of them (20) are located in Europe (Germany, Netherlands and France) and in the USA (Fig. 1).

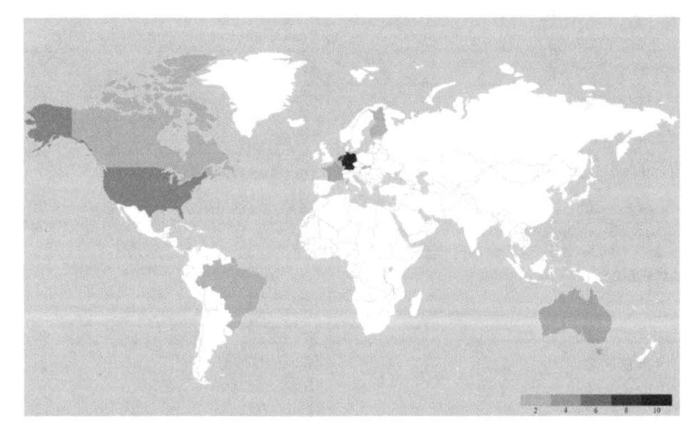

Fig. 1. Location of top companies with process mining tools.

Regarding to importing format, we discovered that only the following tools support XES standard: Disco (Fluxicon), Celonis Process Mining, ProcessGold Enterprise Platform, Minit, myInvenio, Signavio Process Intelligence, QPR ProcessAnalyzer, LANA Process Mining, Rialto Process, Icris Process Mining Factory, Worksoft Analyze & Process Mining for SAP, SNP Business Process Analysis, webMethods Process Performance Manager and Perceptive Process Mining. Regarding academic popularity (Item 22 – Table 1), the following 5 tools are the most frequently cited (with number of citations): ProM 6 (5710); Disco (1930); Aris Process Performance Manager (1550); Celonis Process Mining (714); EMiT process mining tool (525).

We consider that an important analysis is related to the process mining groups of techniques, which we discussed in Sect. 2.2. More specifically, we were interested in discovering which of the groups of techniques (i), (ii), (iii) and (iv) are supported by the tools. Analyzing the data, we verified that only 9 of the 42 tools support all the four groups, which are: Apromore, ARIS, bupaR, Celonis, Disco, EverFlow, Minit, ProM 6 and QPR. In our research, the techniques were separated in groups according to Dumas et al. [7], but in practice they can be part of each other. For instance, the group (iv) can be seen as part of (ii), or they can be displayed as graphs and run in plugins, that can be installed. For example, in the Apromore tool, the discovery capability is encapsulated in a Process Discoverer plugin, and the enhancement (iii) can be shown to the user in a form of charts, via dashboards, or in a process model automatically discovered from the log. In Celonis tool, all techniques are within the workspace in Process Analytics.

Scoring the tools according to the academic data analysis, relative to items 4 to 22 from Table 1, we scored the tools and we present in Table 2 the 5 tools with more than 50% of compliance with our adopted criteria and their respective scores. We could observe that the tools with scores higher than 60% of compliance with our adopted criteria were ProM 6, Disco by Fluxicon, Apromore and Celonis' tool.

Table 2. Ranked tools according to the academic perspective with more than 50% of compliance according to our evaluation instrument (items 4 to 22 from Table 1).

Tools	Research and Development	Scores
ProM 6	TU/e, EIT, STW, NWO, and Beta	89%
Disco	Fluxicon	71%
Apromore	Apromore Pty Ltd	66%
Celonis Process Mining	Celonis GmbH	60%
Signavio	Signavio Process Intelligence	57%

Scoring the tools according market data analysis (items 23 to 27 from Table 1), the top 10 tools were selected by Gartner's assessment, with the highest scores in evaluation & contracting, integration & deployment, and service & support requirements (Table 3). We presented all the tools which have at least 2 reviews in the Gartner website. We verified that the tools with higher scores, considering those with user contributions higher than 30 were the Celonis tool, IBM and UiPath (ProcessGold). Other tools were also evaluated by their users, but the amount of contributions was almost unitary, such as Minit, EverFlow, Puzzle Data and ProM.

Table 3. Overview of process mining tool evaluations according to the market perspective.

Process Mining Ratings Overview	Reviews	5-star assignments	Product Capabilities	Relevance of the tool	Ranking
Celonis Process Mining	215	60%	4.7	0,676101	7,7
IBM Process Mining	37	68%	4.8	0,116352	5,2
UiPath Process Mining	34	62%	4.6	0,106918	5,1
InVerbis	3	100%	5.0	0,009434	5,0
Livejourney	3	100%	4.3	0,009434	5,0
Fluxicon Disco	2	100%	4.5	0,006289	5,0
Scout Platform	7	86%	4.4	0,022013	5,0
ARIS Process Mining 10	8	75%	4.9	0,025157	4,9
SAP Signavio Process Intelligence	6	67%	4.3	0,018868	4,8
Kofax Insight	3	0%	3.0	0,009434	3,0

5 Conclusions and Future Work

The study was guided as from an extensive bibliographic research based on the works already carried out on the process mining tools as scientific articles, books and several other relevant publications in the area. In addition, an analysis was conducted based on Gartner Peer Insights studies, through the recognition of the best tools and their companies, ranked by the customers themselves. As of this, it was verified the difference between the most academically used tools and the commercial ones. Those that have academic support and are more evaluated as the ProM 6, Fluxicon's Disco and Apromore, that present great possibility of contributions and implementations. Another tool in this context is the Celonis, that is very well evaluated by its users and has a huge support and investment for researchers and students.

Another important analysis was about the process mining techniques, discussed in Sect. 2.2, items (i), (ii), (iii) and (iv), it was identified that only 9 tools support all process mining techniques. The best ranked commercial tools were Celonis, IBM and UiPath (ProcessGold) according to the market perspective. All of them present an easy to use layout and a very well evaluated user support. Thus, each type of tool is classified according to how expectations and different needs are met, as is the case of the academic and commercial versions. It is worth observing that, from the academic perspective, the focus of the researchers is having tools more flexible to be adapted for new experiments. However, from the market perspective, it is more important to a tool having more functionalities and being more easy to use. Celonis has applied many efforts to attend both target audiences and that is why it appears on both ranking tables. Therefore, Celonis is able to meet both audiences quite satisfactorily. On the other hand, ProM 6 is a very good tool for academics, but it does not appear in the ranking list from market perspective as it is difficult to use.

As future work from this comparative research, it would be the construction of conceptual framework to support process mining, with general mining activities and with the possibility of extension to various domains. A framework attending to data patterns, types of input and output data, models, and which algorithms can be executed for each type of activity.

References

1. Devi, T.A., Kumudavalli, M.V., Sudhamani: An informative and comparative study of process mining tools. Int. J. Sci. Eng. Res. **8**(5) (2017)
2. Ailenei, I., Rozinat, A.A.: Process mining tools: a comparative analysis (2011). PhD Thesis
3. Bru, F., Claes, J.: The perceived quality of process discovery tools (2018). https://arxiv.org/abs/1808.06475
4. Teixeira, L.M.D., Aganette, E.C.: Document management associated with business process modeling. Brazilian J. Inf. Sci. **13**(1), 33–44 (2019)
5. Harrington, J.H., Essling, E.K.C., Nimwegen, H.: Business process improvement workbook: documentation, analysis, design, and management of business process improvement. Quality Progress **31**(6) (1997)
6. ABPMP: Guide to the business process management common body of knowledge (BPM CBOK). In: BPM CBOK Version 3.0: Guide to the Business Process Management Common Body of Knowledge (2013)
7. Dumas, M., Rosa, L., Mendling, J., Reijers, H.A.: Fundamentals of Business Process Management, 2nd edn. Springer-Verlag, Berlin Heidelberg (2018)
8. vom Brocke, J., M.R.: Handbook of BPM: Business Process Management. Bookman Publisher (2013)
9. van der Aalst, W.M.P., Günther, C.W.: Finding structure in unstructured processes: the case for process mining. In: Proceedings - 7th International Conference on Application of Concurrency to System Design, ACSD (2007)
10. Deming, W.E.: Out of the crisis: centre for advanced engineering study. In: Quality and Reliability Engineering International (Issue 1). MIT (1986)
11. Hammer, M.: What is business process management? In: Handbook on Business Process Management 1 (2010)
12. Sinur, J., Hill, J.: Gartner magic quadrant for business process management suites. In Gartner Information Technology Research. https://www.gartner.com/en/documents/1453527/magic-quadrant-for-business-process-management-suites (2010)
13. Sinur, J., Schulte, W. R., Hill, J., & Jones, T.: Gartner Magic Quadrant for Intelligent Business Process Management Suites. https://www.gartner.com/en/documents/2179715/magic-quadrant-for-intelligent-business-process-management-suites (2012)
14. Lopes, N. C. de S.: Modelo de Gestão por Processos Baseado em Mineração (2017)
15. van der Aalst, W.M.P.: Process mining: discovery, conformance and enhancement of business processes. In: Media, vol. 136, Issue 2 (2011)
16. van Eck, M.L., Lu, X., Leemans, S.J.J., van der Aalst, W.M.P.: PM2: a process mining project methodology, 9097 (2015)

17. IEEE: IEEE 1849–2016 XES Standard. IEEE (2021). https://xes-standard.org/
18. IEEE Comp. Intel. Soc.: Standard for eXtensible Event Stream (XES) for Achieving Interoperability in Event Logs and Event Streams. In: Proceedings of the IEEE (2016)
19. Vasconcelos, G.S., Bernardini, F., Viterbo, J.: Online repository of the collected data: Comparison of process mining tools on the market and in academic environment (2022). With an academic focus, seeking information in various types of publications such as scientific articles, books, dissertations and theses. https://doi.org/10.6084/m9.figshare.21804198

Anonymisation Methods for Complex Data Based on Privacy Models

Michael Boch[1(✉)], Emmanouil Adamakis[2], Stefan Gindl[1], George Margetis[2], and Constantine Stephanidis[2,3]

[1] Research Studio Data Science, RSA FG, 1090 Vienna, Austria
{michael.boch,stefan.gindl}@researchstudio.at
[2] Institute of Computer Science, Foundation for Research and Technology Hellas, Heraklion, Greece
{madamakis,gmarget,cs}@ics.forth.gr
[3] Department of Computer Science, University of Crete, Heraklion, Greece

Abstract. As the demand for personal data rises, so does the risk of potential de-anonymisation attacks. De-anomymisation is a breach of privacy where an attacker can identify an individual in a published dataset, despite the removal of personal identifying information. A risk analysis based on privacy models can be applied to assess the de-anonymisation risks. The challenge lies in finding appropriate anonymisation methods based on the results. So far, a large number of privacy enhancing methods of non-complex data types based on privacy models was already proposed, but only a few for high-dimensional and complex data types. Therefore, this study focuses on identifying methods for the anonymisation of such data types based on privacy models. In order to evaluate possible approaches and to assess the associated challenges a total of 9 prototypes was developed for 5 different data types. The result was a guideline to determine which method is suitable for each of the data types. Here, the data controller has to decide between more labour-intensive manual methods with more accurate anonymisation results and thus a lower privacy-utility trade-off or the faster, automatic methods with a decrease in utility. It was also shown that even after applying privacy enhancing methods, it is still important that the anonymised datasets are again subjected to risk analysis. The presented methods and guidelines support data controllers in complying with privacy regulations such as GDPR.

Keywords: De-anonymisation · Data Privacy · Privacy Models

1 Introduction

The demand for personal data is constantly growing, both in academia and in industry. Previously the focus lied on tabular and statistical data, now the demand for highly personal data (microdata) is rapidly increasing [25]. Particularly with the abundance of big data analysis and the proliferation of sophisticated relevant tools [7,9,19,24,30], data privacy protection and transparency

A. Rocha et al. (Eds.): WorldCIST 2023, LNNS 800, pp. 229–237, 2024.
https://doi.org/10.1007/978-3-031-45645-9_22

are of paramount importance. Characteristics to identify individuals are called personal identifying information (PII) and include information such as an individual's full name, phone number, and social security number [4]. However, this does not mean that a dataset with PIIs removed is safe from de-anonymisation. There might still be other clues in the dataset to identify individuals [25]. For example, based on the three paramaters *gender*, *date of birth* and *postcode*, which are not PIIs, 87% of the US population could already be identified [28]. These types of parameters are called quasi-identifiers (QID). Independently, they cannot be used to infer individuals, but when combined they may result in a privacy breach. Datasets can therefore only be anonymous if both PIIs and QIDs are removed [16]. However, full removal of QIDs lowers the utility, since important features in the dataset might get lost. Consequently, it is crucial to keep the trade-off between privacy and utility as low as possible while maintaining the highest level of privacy.

A dataset is high-dimensional when the amount of features of a table exceeds the number of observations [6]. From a de-anonymisation viewpoint, this inevitably results in a high number of QIDs and complicates the risk analysis and the subsequent anonymisation. Other complex data types include fields which are not based on tabular principles, such as textual or spatiotemporal data, where privacy models cannot be applied without further consideration. Complementing [28], by showing that 3 QIDs are enough to identify most of the US population, [8] showed that similar results can be achieved with complex data types. Already 4 spatiotemporal data points were enough to successfully de-anonymise 95% of the 1.5 million individuals in their dataset.

The goal behind these theoretical de-anonymisation attacks is to identify individuals with the help of QIDs to assess the risk of a potential privacy breach. While this threat is apparent when working with PIIs, it is not so clear when it comes to QIDs. In these cases, privacy models like k-anonymity [25,28] and l-diversity [17] can be applied. They were originally invented for data in tabular format and applied by grouping the values according to k or l, but appropriate adaptations are needed for high-dimensional and complex data.

2 Related Work

The increasing demand also raises the challenge that datasets become larger and more complex. There are already many options for non-complex datasets, but existing state-of-the-art tools like ARX[1] only offer limited options for complex and multidimensional datasets [23]. However, interest in de-anonymisation of data without PIIs has been growing recently and there is an increasing effort in academia to examine different types of datasets for possible de-anonymisation risks. This includes all sorts of different data types, such as official statistics published by state authorities [12], language models [29] and also clickstream data [22]. There are also efforts for further development of ARX like an extension for the anonymization of high dimensional biomedical datasets [18].

[1] https://arx.deidentifier.org/.

Since k-anonymity is the most popular privacy model [13], we focus on the identification of anonymisation strategies for high-dimensional and complex datasets to achieve it. For both k-anonymity and l-diversity, QIDs are considered. With k-anonymity, records with the same values within the QIDs are grouped together. A dataset is k-anonymous as soon as each of these groups, so-called equivalence classes, are consisting of at least k values. l-diversity [17] goes one step further and also considers sensitive attributes. Sensitive attributes are parameters where it should not be possible to associate them with an invidual, for example medical information. A dataset is considered l-diverse if the sensitive attributes are also sufficiently distributed within the equivalence classes based on l. [2] has already successfully demonstrated methods to visualize the de-anonymisation risks of selected high-dimensional and complex data types. Building on this, we will present corresponding methods for anonymising these data types.

3 Methodology

The structure of this study is based on the design science research (DSR) process [20] (see Fig. 1) and is divided into 2 parts. Starting with a (1) *literature review* in order to identify suitable anonymisation methods, followed by the (2) *implementation of prototypes* for the individual anonymisation methods.

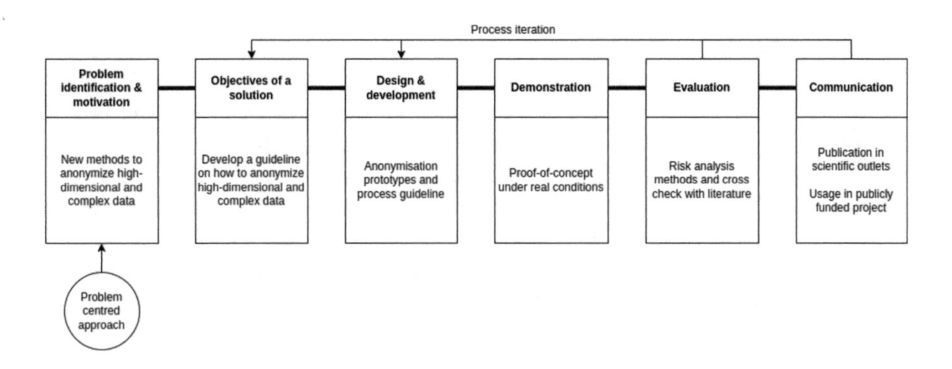

Fig. 1. The applied design science research process derived from [20].

The research question of this work is to find ways to achieve k-anonymity with high dimensional and complex datasets. Based on the data types identified by [2], corresponding anonymisation methods are assessed. This includes anonymisation methods for spatiotemporal, textual, financial transaction, and aggregation based datasets. Types of anonymisation are also collected for tabular datasets where a simple grouping and splitting into value ranges is not possible due to, for example, a high number of identical values within a column.

3.1 Anonymisation Methods

In this paper, we will address anonymisation methods for high-dimensional and complex data, building on the k-anonymity [28] and l-diversity [17] privacy models. A risk analysis based on privacy models will serve as an input and evaluation method for the anonymisation process. The goal is to achieve k-anonymity with high-dimensional and complex data.

[3] are describing 3 different anonymisation strategies with k-anonymity: (1) *Bucketisation*, (2) *microaggregation*, and (3) *generalisation and suppression*. *Bucketisation* uses statistical metrics to divide the data into buckets of equal size. In *microaggregation*, data are grouped into small aggregates. To determine the number of aggregates, k can be used as a guideline [10]. *Generalisation and suppression* is often carried out in the form of value generalisation hierarchy (VGH). For this purpose, the data is anonymised using predefined hierarchies. A typical example are postcodes, where one digit less is displayed in each step. However, numerical values can also be grouped together into value ranges like age into age ranges. Another option when working with categorial values is to group them semantically with their hypernyms, like dogs and cats to animals or cars and trains to vehicles. The last step in the hierarchy is always depicted as "*" and represents full suppression of the value [3]. Generalisation can be distinguished between global and local recoding. With global recoding, the generalisation is applied to all values based on the hierarchy [25]. With local encoding, it is only applied to values that require anonymisation based on the selected privacy model [4].

In the following sections we will describe the five different types of data that were examined for this study. For tabular data, we used the Adult dataset [11] provided by the University of California, Irvine (UCI) with 30,162 records and 9 QIDs. For textual data, we used the Amazon dataset [14] containing a total of 142.8 million records where we used a subset of around 1.000 records including the QIDs *reviewer id*, *product id* and *review text*. The other datasets for aggregation, spatiotemporal, and financial transaction data were provided within the TRUSTS[2] project. The aggregation-based dataset consisted of 19 QIDs with multiple aggregated values. The financial transactions dataset included next to the obligatory fields *customer id* and *timestamp* also invoice relevant QIDs like *revenue*. The spatiotemporal dataset included *userid timestamp*, and *latitude* and *longitude*.

Tabular Data. For numerical values in tabular data anonymisation with k-anonymity can be achieved in a straightforward way. You can group values together in categorical range objects with group sizes based on the desired k or l value. However, this does not work in every possible scenario e.g. if the identical value occurs too often in the dataset grouping by k or l is not possible. The higher the k-value, the more the data quality decreases and thus also the utility of the dataset for training machine learning models. However, the quality

[2] TRUSTS - trusted secure data sharing space. https://www.trusts-data.eu/.

varies depending on the dataset and the associated anonymisation method [26]. For this reason, it is important to choose the best possible method with the help of different anonymisation approaches including evaluation based on risk analyses. For tabular data, all of the proposed anonymisation methods can be used. We recommend a generalisation based on a hierarchy, because this allows to adjust values precisely and thus to keep the privacy-utility trade-off low. Especially in the scenario presented, this enables anonymisation where perhaps the automatic methods would not perform effectively.

Textual Data. Textual data can be seen as any text that is generated by individuals containing personal information. There is a wide range of options, i.e. blogposts, internet comments, shop reviews, etc. [27]. Here we cannot apply k-anonymity or l-diversity, because every text is different purely from the point of view of characters. However, with the help of Jaccard Similarity, the uniqueness of text can be calculated. For this, a score between 0 and 1 is calculated (the closer to 1 the higher the syntactic similarity). According to [16] there are two different approaches for automated anonymisation of textual data submitted: (1) *natural language processing* and (2) *privacy-preserving data publishing.*

We propose sentiment analysis as another anonymisation option. Automatic assessment of texts with the aim of identifying the sentiment/opinion of a text as positive or negative. Instead of the whole text, only the information about positivity and negativity remains in the dataset. To evaluate the anonymisation method, we will use the risk analysis approach by [2]. Here, texts are compared based on their score value on a heat map.

Aggregation-Based Data. This data type contains all datasets with attributes on which an aggregation operation has been performed, e.g., sum, count or average. The risk is that if an aggregated value is only generated from a supposedly small number of values, a privacy breach is possible. We propose two methods for anonymisation: (1) *Generalisation with hierarchies* on the one hand and (2) *bucketisation* on the other hand. In the case of generalisation with hierarchies, the data controller can use the results of a risk analysis to convert compromised records into a range of values or remove them completely. With bucketisation, the data is automatically split into so-called buckets of similar size. It is important again that the number of records within a bucket is greater than the value of k.

Spatiotemporal Data. This data type combines individuals' information from two dimensions: the (1) spatial information of an individual, i.e. the location, and the (2) temporal/time level. If only the time dimension existed here, the points in time could be grouped together similar to our approaches with aggregated or financial transaction data.

We propose cluster analysis as an anonymisation method for this data type. Clustering combines similar values and partitions them into so-called clusters. The values in different clusters thus differ greatly from each other. In principle, a variety of clustering methods are suitable for this purpose. The key factor is that the sizes of the clusters must be able to follow the value k [15].

Financial Transactions Data. This data type contains two types of information, the temporal information, when the transaction took place and the amount of the payment within the transaction.

For this data type, four of the presented anonymisation methods can be considered. *Generalisation* with hierarchies where the data controller determines to what extent the times of the transactions are summarised or form value ranges for the amount of the transactions. In addition to this manual classification, automated methods can be applied. *Microaggregation*, where the time dimension can be aggregated to higher levels and the transaction amounts are aggregated accordingly. *Bucketisation*, where the amount of the transaction is automatically divided into value ranges to enable a splitting into groups with values higher than k. *Cluster analysis*, as already described for spatiotemporal data. Instead of the geographical dimension, the temporal dimension is considered for financial transaction data. Rather than dividing the date values into value ranges or grouping them at a higher level, similar time series are identified and grouped into clusters. The prototype was implemented using time series cluster methods adapted from [21].

4 Evaluation

For each of the selected data types, anonymisation methods are identified, and respective prototypes are developed and compared. We use the risk analysis

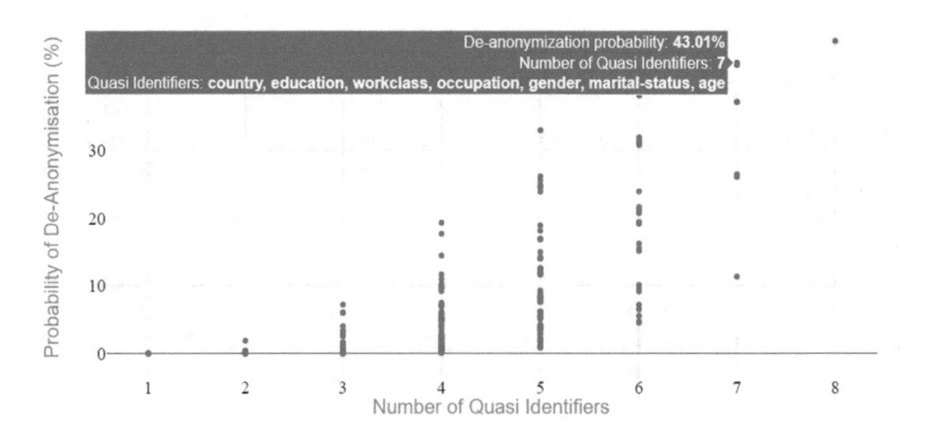

Fig. 2. 2-anonymity risk analysis on the UCI Adult dataset [11]

methods as presented in [1,2] to evaluate the de-anonymisation risks of data before and after the anonymisation methods are applied. To conduct an anonymisation procedure, we must first identify the de-anonymisation risks of the data being analyzed. Then, based on the findings of the risk analysis, we determine the necessary anonymisation measures and configure the appropriate anonymisation method in accordance with our privacy goals. Figure 2 illustrates the risk analysis results of a tabular dataset. As we can see, the likelihood of de-anonymisation increases in direct proportion to an adversary's number of QIs possession. To mitigate these risks, we must employ an appropriate anonymisation method. From the methods available for tabular data, we choose to apply hierarchy generalization (see Sect. 3.1). The resulting anonymised dataset is then re-evaluated using a 2-anonymity risk analysis, as shown in Fig. 3. The results show that the risks of de-anonymisation are now considerably lower (under this configuration).

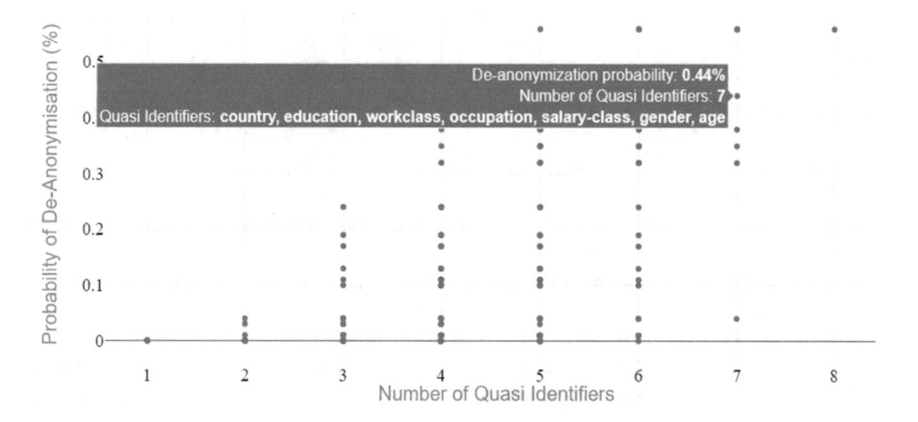

Fig. 3. 2-anonymity risk analysis on the anonymised UCI Adult dataset [11]

5 Conclusion and Future Work

In this paper, we have presented anonymisation methods that allow complex and high-dimensional data to be anonymised based on privacy models. Manual and automatic methods were proposed for the different data types. Manual generalisation with hierarchies can be targeted, thereby minimising the trade-off between privacy and utility. However, this involves considerable overhead and requires a level of data understanding by the data controller. Automatic methods are less labour-intensive but can lead to a greater privacy-utility trade-off. We also showed that it is still important to evaluate the results of anonymisation processes before publication, as the results can vary depending on the anonymisation method and data type. The presented methods should assist data controllers to comply with GDPR. As part of the demonstration phase of the DSR process (see Fig. 1) the developed prototypes will be the starting point for the risk analysis and anonymisation component developed for the TRUSTS data sharing platform. However, the anonymisation methods presented can be used in a wide range of applications. Another possible application are data management platforms (DMP). And there is a large number of open-source platforms that can be extended [5]. For future work we will explore other NLP-based anonymisation methods, e.g. topic modelling, as an alternative to sentiment analysis.

Acknowledgements. The research leading to these results has received funding from the European Union's Horizon 2020 Research and Innovation Programme, under Grant Agreement no 871481 - Trusted Secure Data Sharing Space (TRUSTS), from the H2020-ICT-2018-20/H2020-ICT-2019-2 Call.

References

1. Adamakis, E., Boch, M., Bampoulidis, A., Margetis, G., Gindl, S., Stephanidis, C.: Darav: a tool for visualizing de-anonymization risks. In: 2023 IEEE 39rd International Conference on Data Engineering (ICDE). IEEE (2023)
2. Adamakis, E., Boch, M., Bampoulidis, A., Margetis, G., Gindl, S., Stephanidis, C.: Visualizing the risks of de-anonymization in high-dimensional data. In: Rocha, Á., Ferrás, C., Ibarra, W. (eds.) ICITS 2023. LNNS, vol. 691, pp. 27–37. Springer, Cham (2023). https://doi.org/10.1007/978-3-031-33258-6_3
3. Ayala-Rivera, V., McDonagh, P., Cerqueus, T., Murphy, L., et al.: A systematic comparison and evaluation of k-anonymization algorithms for practitioners. Trans. Data Priv. **7**(3), 337–370 (2014)
4. Bampoulidis, A., Markopoulos, I., Lupu, M.: Prioprivacy: a local recoding k-anonymity tool for prioritised quasi-identifiers. In: IEEE/WIC/ACM International Conference on Web Intelligence-Companion Volume, pp. 314–317 (2019)
5. Boch, M., et al.: A systematic review of data management platforms. In: Rocha, A., Adeli, H., Dzemyda, G., Moreira, F. (eds.) WorldCIST 2022. LNNS, vol. 469, pp. 15–24. Springer, Cham (2022). https://doi.org/10.1007/978-3-031-04819-7_2
6. Bühlmann, P., Van De Geer, S.: Statistics for High-Dimensional Data: Methods, Theory and Applications. Springer, Heidelberg (2011). https://doi.org/10.1007/978-3-642-20192-9
7. Carlisle, S.: Software: tableau and microsoft power BI. Technol. Architect. Des. **2**(2), 256–259 (2018)
8. De Montjoye, Y.A., Hidalgo, C.A., Verleysen, M., Blondel, V.D.: Unique in the crowd: the privacy bounds of human mobility. Sci. Rep. **3**(1), 1–5 (2013)
9. Divya Zion, G., Tripathy, B.K.: Comparative analysis of tools for big data visualization and challenges. In: Anouncia, S., Gohel, H., Vairamuthu, S. (eds.) Data Visualization, pp. 33–52. Springer, Singapore (2020). https://doi.org/10.1007/978-981-15-2282-6_3
10. Domingo-Ferrer, J., Mateo-Sanz, J.M.: Practical data-oriented microaggregation for statistical disclosure control. IEEE Trans. Knowl. Data Eng. **14**(1), 189–201 (2002)
11. Dua, D., Graff, C.: UCI machine learning repository (2017). http://archive.ics.uci.edu/ml
12. Favato, D.F., Coutinho, G., Alvim, M.S., Fernandes, N.: A novel reconstruction attack on foreign-trade official statistics, with a Brazilian case study. arXiv preprint arXiv:2206.06493 (2022)

13. Gkoulalas-Divanis, A., Loukides, G., Sun, J.: Publishing data from electronic health records while preserving privacy: a survey of algorithms. J. Biomed. Inform. **50**, 4–19 (2014). https://www.sciencedirect.com/science/article/pii/S1532046414001403. Special Issue on Informatics Methods in Medical Privacy
14. He, R., McAuley, J.: Ups and downs: modeling the visual evolution of fashion trends with one-class collaborative filtering. In: Proceedings of the 25th International Conference on World Wide Web, pp. 507–517 (2016)
15. Kabir, M., Wang, H., Bertino, E., et al.: Efficient systematic clustering method for k-anonymization. Acta Informatica **48**(1), 51–66 (2011)
16. Lison, P., Pilán, I., Sánchez, D., Batet, M., Øvrelid, L.: Anonymisation models for text data: state of the art, challenges and future directions. In: Proceedings of the 59th Annual Meeting of the Association for Computational Linguistics and the 11th International Joint Conference on Natural Language Processing (Volume 1: Long Papers), pp. 4188–4203 (2021)
17. Machanavajjhala, A., Kifer, D., Gehrke, J., Venkitasubramaniam, M.: l-diversity: privacy beyond k-anonymity. ACM Trans. Knowl. Discov. Data (TKDD) **1**(1), 3-es (2007)
18. Meurers, T., Bild, R., Do, K.M., Prasser, F.: A scalable software solution for anonymizing high-dimensional biomedical data. GigaScience **10**(10) (2021). https://doi.org/10.1093/gigascience/giab068
19. Murray, D.G.: Tableau Your Data!: Fast and Easy Visual Analysis with Tableau Software. Wiley, Hoboken (2013)
20. Peffers, K., Tuunanen, T., Rothenberger, M.A., Chatterjee, S.: A design science research methodology for information systems research. J. Manag. Inf. Syst. **24**(3), 45–77 (2007)
21. Petitjean, F., Ketterlin, A., Gançarski, P.: A global averaging method for dynamic time warping, with applications to clustering. Pattern Recogn. **44**(3), 678–693 (2011)
22. Plant, R., Giuffrida, V., Gkatzia, D.: You are what you write: preserving privacy in the era of large language models. arXiv preprint arXiv:2204.09391 (2022)
23. Prasser, F., Eicher, J., Spengler, H., Bild, R., Kuhn, K.A.: Flexible data anonymization using ARX-current status and challenges ahead. Softw. Pract. Exp. **50**(7), 1277–1304 (2020)
24. Quezada-Sarmiento, P.A., Ramirez-Coronel, R.L.: Develop, research and analysis of applications for optimal consumption and visualization of linked data. In: 2017 12th Iberian Conference on Information Systems and Technologies (CISTI). IEEE (2017)
25. Samarati, P.: Protecting respondents identities in microdata release. IEEE Trans. Knowl. Data Eng. **13**(6), 1010–1027 (2001)
26. Slijepčević, D., Henzl, M., Klausner, L.D., Dam, T., Kieseberg, P., Zeppelzauer, M.: k-anonymity in practice: how generalisation and suppression affect machine learning classifiers. Comput. Secur. **111**, 102488 (2021)
27. Sousa, S., Kern, R.: How to keep text private? A systematic review of deep learning methods for privacy-preserving natural language processing. Artif. Intell. Rev. **56**, 1–66 (2022)
28. Sweeney, L.: k-anonymity: a model for protecting privacy. Int. J. Uncertain. Fuzziness Knowl.-Based Syst. **10**(05), 557–570 (2002)
29. Vamosi, S., Platzer, M., Reutterer, T.: AI-based re-identification of behavioral clickstream data. arXiv preprint arXiv:2201.10351 (2022)
30. Vitsaxaki, K., Ntoa, S., Margetis, G., Spyratos, N.: Interactive visual exploration of big relational datasets. Int. J. Hum.-Comput. Interact. **39**, 1–15 (2022)

A Mobile Application for Wooden House Fire Risk Notifications Based on Edge Computing

Ruben D. Strand[1](✉), Lars M. Kristensen[2], Thorbjørn Svendal[2], Emilie H. Fisketjøn[2], and Abu T. Hussain[2]

[1] Department of Safety, Chemistry and Biomedical Laboratory Sciences, Western Norway University of Applied Sciences, Haugesund, Norway
ruben.dobler.strand@hvl.no

[2] Department of Computer Science, Electrical Engineering and Mathematical Sciences, Western Norway University of Applied Sciences, Bergen, Norway
lars.michael.kristensen@hvl.no

Abstract. We investigate how recent advances in computational modelling of fire risk for single-unit wooden houses can be used as a foundation for a mobile application capable of notifying users of in-home fire risk. Our approach relies on publicly available cloud services providing historic weather data and weather forecast data to compute fire risks for geographical locations on a mobile (edge) device. We undertake an iterative usability study focusing on the user interface to determine how fire risk must be conveyed in a form suitable for end-users. Furthermore, we evaluate CPU and memory consumption to assess the computational feasibility of implementing the fire risk computations directly on mobile devices. Our contribution is a validated design of a user interface for a mobile fire risk application, and a demonstration of fire risk computations being feasible on mobile devices.

1 Introduction

Information about present and near-future fire risk is traditionally provided to the general public and authorities via centralised cloud computing services in conjunction with weather forecasts. These services typically express fire risks originating in wild-land conditions and aggregate larger geographical areas despite potentially significant local variation, e.g., due to large variation in vegetation, building construction material, and building density. In particular, limited attention is given to the quantification of the fire risk associated with buildings, such as the in-home fire risk that may develop with outdoor weather conditions. A concrete example of this shortcoming is the severe fire in Lærdalsøyri in Norway in the winter of 2014, where low wildfire risk was predicted by the general weather forecast service, but where more than 40 wooden buildings ended up being lost partly due to very dry indoor and outdoor wooden claddings [5].

Recently, increased efforts have been undertaken in the field of building fire safety through fire risk quantification and assessment for structures and

A. Rocha et al. (Eds.): WorldCIST 2023, LNNS 800, pp. 238–248, 2024.
https://doi.org/10.1007/978-3-031-45645-9_23

wildland-urban interface (WUI) areas [2,13]. In particular, ambient dew point has been suggested as an explanation for the increased fire frequencies during the winter in cold climate regions [11,19] such as in Scandinavia, Canada, and Japan. A recent study [14] outlined the possibilities for a structural fire danger rating system, accumulating into a computational model that predicts a potential time to flashover (TTF) for an enclosure fire. Combining the model with cloud services providing access to weather data sources enables data-driven applications supporting dynamic fire risk assessments and predictions.

The above developments motivated us to conduct research into the feasibility of exploiting edge computing [20] in the sense that fire risk is computed on a mobile (edge) device and presented to the user in a mobile application. The potential advantage of an approach based on edge computing is that it can contribute towards more rigorous, location-tailored, and integrated risk assessments to support decision making and control at a community scale. As examples, it can be used for location-allocation of resources at high risk regions, such as densely built wooden home areas, or wooden homes located in the WUI where structure-to-vegetation fire spread is a concern. Specifically, we consider two research questions: 1) is it feasible in terms of CPU and memory usage to compute fire risk on a mobile device; 2) how should fire risk be presented to different kinds of end-users (e.g., private citizens, fire brigade personnel).

The contribution of this paper is the design, development, and evaluation of a mobile application prototype for fire risk computations and notification using edge computing. The application implements the wooden home fire risk indication model [12] and computes fire risk indications based on historic and predicted weather data obtained via cloud-services. The application regularly requests weather data, computes fire risk indications for the given locations, updates the local database and notifies the user if fire risk increases.

This paper is organised as follows. Section 2 provides a first glimpse of the mobile application, summarises the basic principle of the underlying fire risk model, and introduces the cloud weather data sources being used. Section 3 presents the design and implementation of the mobile application. Section 4 presents the usability study, and Sect. 5 presents the results from the performance evaluation of the mobile application. Finally in Sect. 6 we sum up the conclusions, provide a further discussion of related work, and discuss future work.

2 Fire Risk Model and Cloud Data Sources

The fire risk indication model [12] that serves as the computational foundation of the mobile application uses outdoor relative humidity and outdoor temperature to compute indoor relative humidity and FMC (Fuel Moisture Content) of indoor wooden panels. The FMC is then used to compute a TTF [8], which resembles how quickly the fire would evolve in case of an established ignition. Currently, the model indicates a fire risk via TTF within the first house catching fire, but when combined with wind-data it may express a potential fire spread risk

[14]. The ability for this model to provide reliable fire risk predictions with low computational overhead was demonstrated in [22]. The model is especially suited for wooden homes, where the interior surfaces are covered by hygroscopic materials, such as wood. It was developed for the densely built wooden town areas of Norway, but is generally applicable to cold climate regions as found in Scandinavia, Canada, Japan, and China [22].

The weather data required to compute a single fire risk indication is fetched by the mobile application from two REST APIs provided by the Norwegian Meteorological Institute (MET). The MET API [18] provides weather forecast data which is needed for the fire risk prediction. The predicted weather data is a result of sophisticated modelling based on a network of weather data measurement stations. The spatial resolution follows from the modelled weather data, and is given on a interpolated grid consisting of 1×1 km cells. The temporal resolution is every hour for the upcoming 72 h, then every sixth hour up until 9 days of forecast data. The FROST API [17] provides measurements (historic data) from MET measurement stations which are needed for 3–5 days into the past in order to adapt the computational fire risk model. The FROST API is based on specifying a measurement station identifier or by stating the longitude and latitude of a desired location. If specifying a location, the service will return weather data from the nearest station.

Figure 1 provides a first glimpse of the main screens of the mobile application. We discuss the user interface in detail in Sect. 4. At the left is the first time startup screen with a description on how to proceed. When a MET ID (user identifier required to access the MET cloud data services) has been registered, the user can start to add locations using longitude and latitude coordinates or by the *Get my location* functionality. A summary of all locations currently being monitored by the user is provided in the middle screen, here showing four cities in Norway. The rightmost screen shows the expanded detailed view of fire risk for the city of Stavanger.

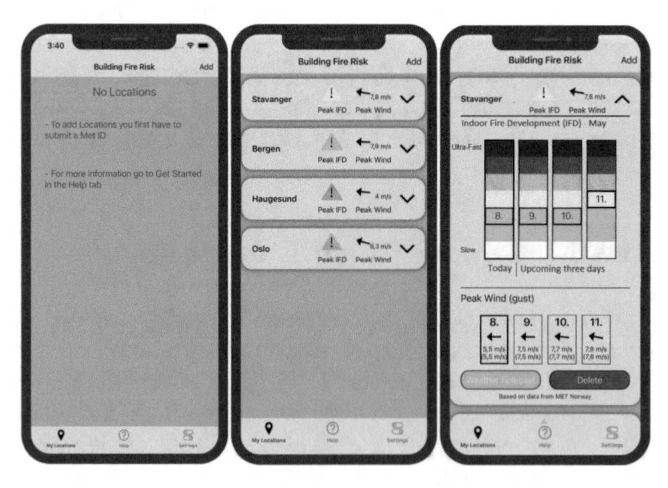

Fig. 1. Mobile application: startup, location summary, and expanded location screens.

3 Internal Application Design and Implementation

The mobile application has been implemented using the Xamarin [15] framework in order to make the application available across the Android and iOS platforms using a common codebase. Figure 2 shows the high-level application software architecture which has been based on the Model-View-ViewModel (MVVM) design pattern. In addition, the *Services* component provides an integration point for the *Weather Data Sources* and *Fire Risk Model* software components.

The *View* and *ViewModel* components are coupled, as the *View* trigger commands in *ViewModel*. The *ViewModel* can store and update content in the *Local database* and uses *Services* to retrieve new fire risk computations. The *Services* component requests weather data (measurement and forecasts) from *Weather Data Sources* and uses *Model* to format the received data. The *Fire Risk Model* receives data from *Services* and performs fire risk computations in accordance with the fire risk model.

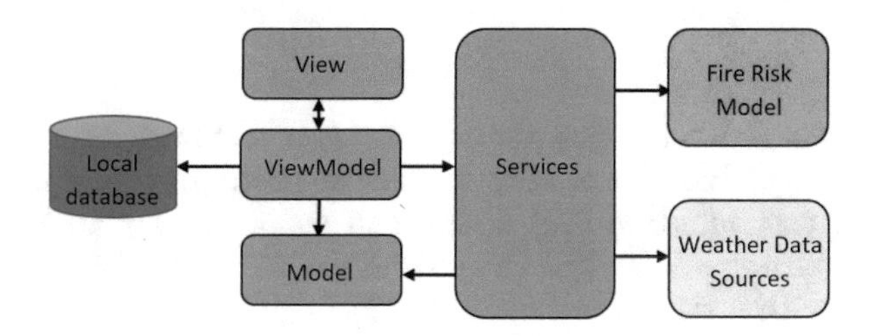

Fig. 2. High-level internal software architecture for the mobile application.

The sequence diagram in Fig. 3 illustrates the interaction between the application components in the case of receiving a fire risk notification (push). The notification is expected to be triggered by a timer that determine when to update fire risks for all monitored locations. As can be seen from the figure, *ViewModel* requests all locations within the database and for each location uses *Services* to retrieve the new fire risk computations. The *Services* then sends API requests to obtain weather data from the FROST and MET APIs, represented here by *Weather Data Sources*. Received data is in JSON format and must be converted into .NET objects by the *Model* prior to being provided to the *Fire Risk Model* for fire risk computation. When receiving the new fire risk, *Services* passes the result to *ViewModel*, which updates the *Local database*. If changes in indicated fire risks are detected, the user receives a notification. The interaction between the components for other use cases is similar.

4 User Interface Development and Usability Evaluation

For the software development methodology, we used an incremental and iterative approach similar to [4], but started by addressing the visual design and use

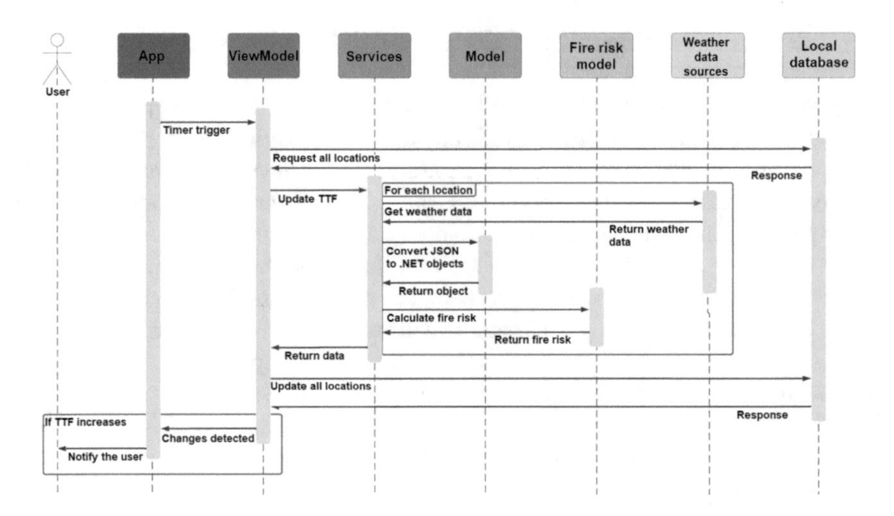

Fig. 3. Component interactions for the push notification use case.

cases through an extended horizontal prototype. Figure 4 depicts the overall methodology for arriving at a fully functional prototype. The left-hand part represents the user interface (UI) design and usability evaluation (as parts of the horizontal prototype development) while the right-hand part represents the full prototype with vertical functionality implemented for all use cases and user interface elements. The user interface design was the outcome of an iterative test-driven design process, including usability evaluation with three groups of end-users (private citizens, fire safety experts, and fire brigade personnel). The

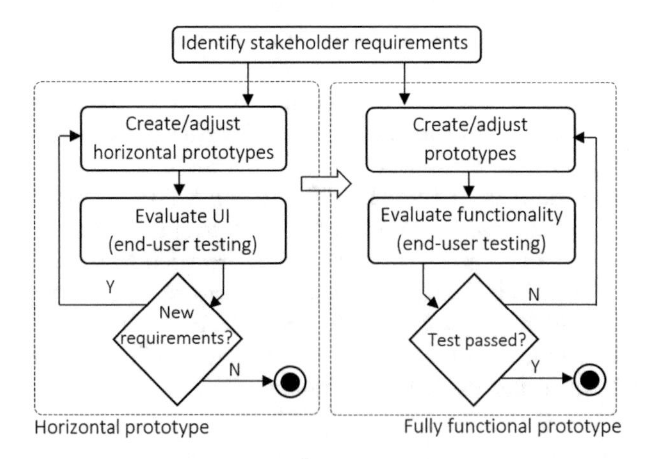

Fig. 4. Software design methodology: from initial horizontal prototype (left) to fully functional vertical prototype (right).

horisontal prototype was developed and evaluated both prior to- and during the development of the fully functional prototype.

The elicitation of initial stakeholder requirements was a process primarily based on a national questionnaire among Norwegian fire brigades, as well as dialogue between three participating fire brigades and fire safety expert opinions. The questionnaire was carried out as part of a greater mapping on the technological status and available risk based tools within Norwegian fire departments. It aided in providing general guidelines on how to design the visual presentation in the UI of the current and near-future risk situation based on the preferences from the individual fire brigades. A second iteration included requirements from all stakeholders, including private citizens and fire safety experts.

The extended horizontal prototype simulated vertical functionality by use of screenshots, allowing end-users to test and evaluate UI design and functionalities not only for a single risk scenario, but investigate the prototype through a set of different risk scenarios. This allowed for a comprehensive testing of the risk communication concept, being the most important design feature requiring feedback. When developing the UI, a framework for iterative development was used, supporting continual redesign based on usability input from end-users [6]. The framework was used in conjunction with the cognitive, physical, sensory and functional affordances of interaction design [7].

Figure 5 presents the evolution of the user interface as driven by the feedback from the iterative evaluation process. As can be seen from the figure, the UI started off quite simple (center left screen) and developed throughout, becoming more descriptive and extending on the content. The initial design was based on identified requirements and the inclusion of well-known concepts to depict an intuitive communication of the risk scenario. The design had a calendar view as the basis, where time (days) progresses from left to right. The risk level is then presented per day by use of vertical colour-bars to immediately give some

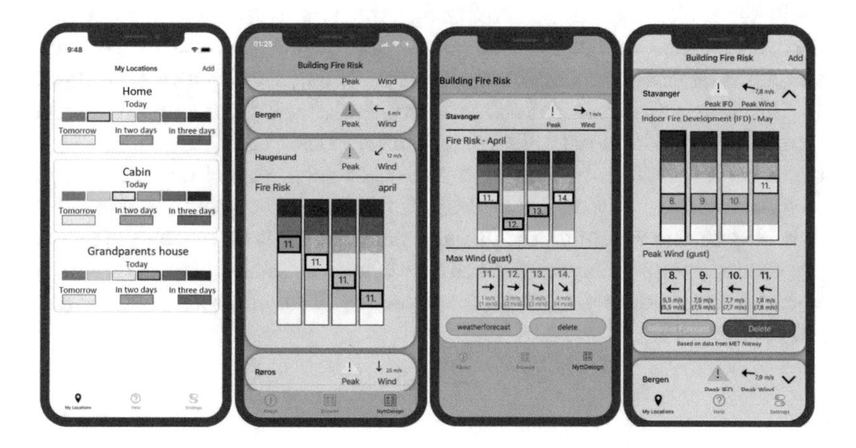

Fig. 5. User interface design evolution (left to right) resulting from the iterative redesign based on usability evaluations by end-user groups.

reference to maximum and minimum risk values. The colours chosen correspond to the colours used in the Norwegian Forest Fire Risk Index [16], as to avoid confusion for stakeholders making use of both systems.

The most important feedback related to the readability of the risk communicating colourbars. In general, this concerned how to best communicate the progress in time, and how to distinguish between the current and predicted risk. More specifically, it involved increasing the highlighting of risk levels (black rectangles) and text size and making clear which days the vertical colourbars represented. The last iteration was presented in Fig. 1. Furthermore, the preferred terminology would vary between user groups without any fire safety knowledge compared to e.g., the local fire department or municipality. As can be seen from Fig. 5 (right most screen), the chosen presentation in terms of indoor fire development (IFD) is a result from the evaluation process. The terminology is chosen aimed at being intuitive to people without in-depth knowledge on fire safety, i.e., the *private citizens* group. It replaces the term *Time to flashover* (used in earlier iterations) as a less intuitive quantity as it would require the user to be knowledgeable of the fire-safety domain-concept of *flashover*. The term indoor fire development is related to whether it is *Ultra-Fast* or *Slow* and appears easier to grasp with for private citizens.

The usability tests followed an evaluation plan, describing the progress and guidelines of how the application was going to be tested. It included a test use case to make sure that users tested specific functionality as well as considered the communication of different risk scenarios. The evaluation involved feedback on the risk communication design, as well as general usability, i.e., navigation and functionality. Usability testing was undertaken by (1) *private citizens* with no background knowledge from fire safety; (2) people with background knowledge from safety sciences with emphasis on *fire safety*, including risk assessments; and (3) the *fire brigade*, consisting of three fire brigades represented by people within management or the analytical department.

Feedback took place post completing the test use cases. It was done by detailed discussions and walk-through of the application and a standardised questionnaire designed to assess perceived usability, the System Usability Scale (SUS) [10]. The questionnaire relates to general usability and involves ten questions concerning: complexity, learning curve, consistency, functions, and how functions are integrated. Table 1 summarises selected data from the anonymous online questionnaire. It can be seen that the two user groups *private citizens* and *fire brigade* have similar results and a larger variation, when compared to the *fire safety* group. The results may indicate an initial biased design as the first version, while complying to stakeholder requirements, was developed by fire safety experts.

Table 1. Survey results for end-user groups. Scores are given according to SUS (0-100) scale together with average score (Avg) and standard deviation (StD).

User group	Participants	Min score	Max score	Avg score	StD
Private Citizens	11	38	80	66	14.5
Fire Safety	5	70	95	84	10.1
Fire Brigade	6	43	85	64	15.7

5 Application Performance Evaluation

A key aspect of our work is to undertake the computation of fire risk and retrieval of weather data on a mobile device. To validate the feasibility of such an edge computing-based approach, we have evaluated the performance of the mobile application in terms of CPU and memory/storage usage, battery consumption, and data communication usage. The evaluation was performed on an iPhone XS running iOS 15.4.1 and a Samsung Galaxy S10 running Android 12 using platform-associated profiling tools for the measurements.

Data Usage (Bandwidth). The application uses data communication in order to retrieve weather data from the MET cloud services. We considered the monitoring of between 1 and 10 locations for which the fire risk was recomputed four times per day. The measurements showed that data usage per location ranged from 155 Kb to 168 Kb per re-computation per location with an average of 162 Kb. In particular, the total data usage grows linearly with the number of locations being monitored, demonstrating the scalability of the application with respect to tracked locations.

CPU Usage and Memory Usage. Figure 6 shows the CPU usage of the application on the Iphone when performing three sets (numbered) of fire risk re-computation for three locations per set. Each location re-computation can be seen to reach 20–25% of CPU usage on a single core. This amount to a peak of approximately 4% when considering all six cores of the mobile phone. The memory footprint of the application ranged between 120 Kb and 131 Kb. The performance evaluation on the Android platform showed a similar modest CPU and memory consumption.

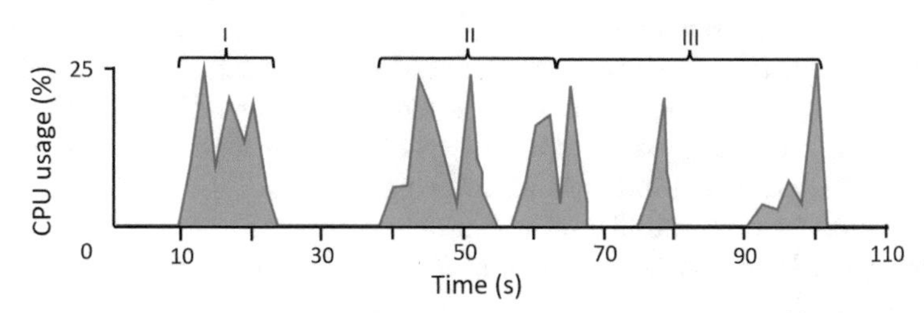

Fig. 6. CPU usage (IPhone) when performing three sets (numbered) of fire risk re-computation for three locations per set. Given percentage is for a single core.

Energy Consumption. The energy (battery) consumption of the mobile application was assessed by running the application for a period of 10 days for three locations, with re-computation of fire risks taking place four times per day. The total energy consumption of the application was at 2% for the period. The profiling on the Android platform also showed a similar low energy consumption.

6 Conclusions and Future Work

We have proposed, implemented, and experimentally evaluated a mobile phone application for fire risk prediction in single-unit wooden houses. We have emphasised single-unit houses due to its relevance in the context of the Scandinavian tradition for wooden homes. However, subject to adjustments, the model is also suitable for other types of wooden homes and buildings [12].

Our performance evaluation showed that recently developed fire risk models are sufficiently light-weight for the computation to be conducted on a mobile device without having to rely on high performance cloud services. Our usability evaluation with different users groups indicates that our application is well-suited for conveying fire risk for people within safety engineering, but appears less suited for private citizens and the fire brigade. Our usability study hence emphasises the challenges in developing a single UI to communicate fire risk across disciplines. A threat to the validity of this latter finding, is the low number of participants in the user groups. Furthermore, the UI evaluation was performed on a horizontal prototype and consequently the encountered scenarios were predefined.

Edge computing is a candidate technology to tackle existing technical limitations of centralized cloud data centers. While different surveys has been undertaken [1,9], addressing definitions, advantages, requirements, challenges, architectures and application areas, we bring forth some recent work affecting the private citizen and other stakeholders, similar to our study. In [21], an energy recommendation system was developed and implemented using an edge-based architecture. Results and recommendations were communicated through their developed Home-Assistant UI, available on personal computer and mobile devices. In the feasibility study of [3], edge computing was used for AI-based in-home processing of video data to reduce the number of unwanted notifications to the end-user. The two related studies both used a PC as a local server, in contrast to our suggested smartphone solution. Another promising study is the work of [23], where a cloud-fog architecture was used in combination with wireless sensor networks for data acquisition. The system indicated wildfire risk through the chandler burn index and communicated risk-levels through their web-based UI called Fog-assisted Environmental Monitoring System.

The results of our usability study prompts further research into finding a UI design suited for professional use of the application, e.g., for fire brigades. The ability to choose between UIs may serve as a practical solution within the application. Our mobile application may form the basis of a fully distributed solution where fire risk are computed only locally. Following this direction, it may then be investigated how the use of proxy servers can reduce the load on the cloud weather data services in case several users are tracking fire risks for the same location. Taking this idea further, it is relevant to explore a hybrid architecture in which the mobile application pushes back fire risks into a cloud service which could then be retrieved by other users tracking the same location.

Acknowledgements. This study was partly funded by the Research Council of Norway, grant no 298993, Reducing fire disaster risk through *dynamic risk assessment and management* (DYNAMIC). The study was also supported by Haugaland Kraft Nett, Norwegian Directorate for Cultural Heritage and Stavanger municipality.

References

1. Abbas, N., et al.: Mobile edge computing: a survey. IEEE Internet Things J. **5**(1), 450–465 (2018). https://doi.org/10.1109/JIOT.2017.2750180
2. El Ezz, A.A., et al.: Framework for spatial incident-level wildfire risk modelling to residential structures at the wildland urban interface. Fire Saf. J. **131**, 103625 (2022)
3. Ayuningsih, T., et al.: Feasibility study of artificial intelligence technology for home video surveillance system. In: 2022 1st International Conference on Information System & Information Technology (ICISIT), pp. 210–215 (2022)
4. Chàvez, A., et al.: Design and evaluation of a mobile application for monitoring patients with Alzheimer's disease: a day center case study. Int. J. Med. Inform. **131**, 103972 (2019). https://doi.org/10.1016/j.ijmedinf.2019.103972
5. DSB: Brannene i Lærdal, Flatanger og på Frøya vinteren 2014. Technical report, Norwegian Directorate for Civil Protection (2014). (in Norwegian)
6. Gould, J.D., Lewis, C.: Designing for usability: key principles and what designers think. Comms. ACM **28**(3), 300–311 (1985). https://doi.org/10.1145/3166.3170
7. Hartson, R.: Cognitive, physical, sensory, and functional affordances in interaction design. Behav. Inf. Technol. **22**(5), 315–338 (2003)
8. Kraaijeveld, A., Gunnarshaug, A., Schei, B., Log, T.: Burning rate and time to flashover in wooden 1/4 scale compartments as a function of fuel moisture content. In: 8th International Fire Science Engineering Conference, Interflam, pp. 553–558 (2016)
9. Khan, W.Z., et al.: Edge computing: a survey. Future Gener. Comput. Syst. **97**, 219–235 (2019)
10. Lewis, J., Sauro, J.: Item benchmarks for the system usability scale. J. User Exp. **13**, 158–167 (2018)
11. Log, T.: Cold climate fire risk; a case study of the Lærdalsøyri fire. J. Fire Technol. **52**, 1815–1843 (2014)
12. Log, T.: Modeling indoor relative humidity and wood moisture content as a proxy for wooden home fire risk. Sensors **19**(22) (2019). https://doi.org/10.3390/s19225050
13. Papathoma-Köhle, M., et al.: A wildfire vulnerability index for buildings. Sci. Rep. **12**(1), 6378 (2022)
14. Metallinou, M.M., Log, T.: Cold climate structural fire danger rating system? Challenges **9**, 12 (2018). https://doi.org/10.3390/challe9010012
15. Microsoft: What is Xamarin. https://learn.microsoft.com/en-us/xamarin/get-started/what-is-xamarin

16. Norwegian Meteorological Institute: Norwegian Fire Risk Index. https://skogbrannfare.met.no/
17. Norwegian Meteorologisk Institute: Historical Weather Data. http://frost.met.no/
18. Norwegian Meteorologisk Institute: Weather Forecast Data. http://api.met.no/
19. Pirsko, A., Fons, W.: Frequency of urban building fires as related to daily weather conditions. Technical report, 866, US Department of Agriculture (1956)
20. Satyanarayanan, M.: The emergence of edge computing. Computer **50**, 30–39 (2017)
21. Sayed, A., et al.: Intelligent edge-based recommender system for internet of energy applications. IEEE Syst. J. **16**(3), 5001–5010 (2022)
22. Strand, R.D., Stokkenes, S., Kristensen, L.M., Log, T.: Fire risk prediction using cloud-based weather data services. J. Ubiquit. Syst. Pervasive Netw. **16**(1), 37–47 (2021). https://doi.org/10.5383/JUSPN.16.01.005
23. Tsipis, A., et al.: An alertness-adjustable cloud/fog IoT solution for timely environmental monitoring based on wildfire risk forecasting. Energies **13**(14), 3693 (2020)

Diagnosis Model for Detection of e-threats Against Soft-Targets

Sónia M. A. Morgado[1,2,3](✉), Margarida Carvalho[1,2], and Sérgio Felgueiras[1,2,4]

[1] Research Center (ICPOL), Lisbon, Portugal
{smmorgado,srfelgueiras}@psp.pt
[2] Instituto Superior de Ciências Policiais e Segurança Interna, Lisbon, Portugal
[3] CIEQV - Life Quality Research Center, Santarém, Portugal
[4] Centre for Public Administration and Public Policies, Institute of Social and Political Sciences, Universidade de Lisboa, Lisbon, Portugal

Abstract. Threats online - internet, social media platforms, blogs - has risen with development of cybernetics tools of communication. Since the threats against soft targets are recurrent on cyberspace, the focus becomes an e-threat. This paper aims to contribute to construct a roadmap to apprehend e-threats against soft targets, considering the triangle ground of coexistence – normative, empirical and theoretical. An all-inclusive guide amalgamates the concepts of e-threats detections, risk, IA, soft targets, is developed to understand processes. The awareness of the added-value and the limitations of the integration of this technology in policing, is an initiative whose enforcement is legitimized and stranded in a simple concept of security. Despite the IA advances, and its efficiency for detection, the process is complex, because deceptive behaviour has a multidimensional nature. Our analysis reveals that the disruptive potential of e-threats of soft targets is tackled and some orientations are given to guarantee the safety and security of crowds.

Keywords: e-threat · police intelligence · soft-target · cyberspace

1 Introduction

The notion of threats seems self-evident: a person or a group of persons sets a scenario of intentions and can set offence to soft targets.

The key issues revolve around the "who", "what", and "how". In the digital era (internet, IoT, big data and artificial intelligence), there are new mechanisms to define how communication is processed, who is being targeted in terms of frequency, intensity and vulnerability, and what tools (human, technical, weaponry) to achieve the purpose.

The digital and technological revolution impacted the definition of threats when the process goes around the internet, establishing a new way of design thinking about the problem. These mechanisms focus on cyberspace behaviour, which expands the concept from threats to e-threats and how to assess risk.

The 19th century was the baseline for discussing threat assessment [1] and has evolved with the works of other authors during the 20th century [2–5] till now [6, 7].

A. Rocha et al. (Eds.): WorldCIST 2023, LNNS 800, pp. 249–262, 2024.
https://doi.org/10.1007/978-3-031-45645-9_24

The central characteristic of all the concepts is the need to consider warning behaviours that indicate increasing or hastening the threat risk [7].

Regardless of the specific model, traditional – threat - or digital – e-threat – each one will involve a continuing dynamic of design and implementation of a risk diagnosis model.

In the contemporary world, cyberspace is critical, and the undercover actions used to perpetrate crimes or terrorist attacks require new models of determining the e-threats [8].

Cyberspace or e-space is today a new dimension of public space that reshapes security needs. Technologies, as sophisticated decentralised means used in favour of the State, organisations or individuals, may help mitigate the negative impact of newly emerging forms of threats. On the one hand, they translate knowledge and benefits in assisting security forces, services, and other security providers. On the other hand, their permanently changeable, ambiguous and multifaceted nature represents a possible pathway for the practice of various forms of crime in a vast universe of deviant behaviours, which translate into threats and quantify into risks. In this international arena, there is a trend of migration of numerous threats to the cyber network, "causing the displacement of the battlefield to cyberspace" (p. 32) [9]. Which inevitably means exponentiating the risk of cybercrime, according to the World Economic Forum [10]: "as society continues to migrate into the digital world, the threat of cybercrime looms large" (p.46).

The threats that emerge in the network are produced by actors whose appearance takes on different contours. They are diverse, unidentified actors who present themselves in disguise. They can now reach a geography that extends to every corner of the world, going beyond the borders physically imposed by States. In this universe, the criminal can access many potential victims, hide his identity, engage in illicit activities and develop crime on a larger scale than ever before. Europol [11] warns that this form of crime will skyrocket in the coming years - "In light of these developments, the market for criminal goods and services is booming" – so an objective and joint response is needed to combat this new form of threat, which starts in its detection (p.8).

The research philosophy used for this theoretical paper is grounded in peer-reviewed literature. Besides the scientific literature highlighting was given to grey literature such as: i) guidelines 07/2020 European Data Protection Board (EDPB) (2020); ii) preliminary opinion on data protection and scientific research of the European Data Protection Supervisor (EDPS) (2020); iii) documents from the European Commission, and iv) United Nations Interregional Crime and Justice Research Institute (UNICRI). The peer-reviewed literature considered the most well-known and comprehensive databases: EBSCO, Scopus, Science-Direct, Google Scholar, B-on, and RCAAP. The exclusion criteria for not contemplating some documents that might compromise the integrity of the paper were: i) no added-value of the document; ii) the document was no connection to the main topic; iii) poor quality of the documents; iv) working paper, posters, doctoral dissertations or master thesis, abstracts.

2 Theoretical Conceptualization

National security is one of the most fundamental tasks of the government. Security and safety, because they are inexhaustible and non-exclusive, they are considered public goods. Law enforcement agencies are vital in protecting society against anarchy, anomy and social unrest. Despite the various entities with competencies in the field of social regulation, to achieve these ends, proactive and predicting policing must be efficient in many areas, including soft targets in the era of cyberspace.

Soft-Targets

The vulnerabilities intrinsic to soft targets have mobilised the security community to ensure their protection. A European Commission considers a soft target as locations that "are vulnerable and difficult to protect and are also characterised by a high probability of mass casualties in the event of an attack" (p. 4) [12].

Despite multiple contributions to the definition of a soft target, there are some difficulties in getting a common approach which identifies the soft target with the proposed concept. A clear idea is needed considering the soft target definition can refer to another kind of target. In this sense, the critical argument against the actual meaning of a soft target, as Zeman [13] points out, "definitions are very common, which does not allow their use in the process of soft targets identification" (p. 109).

As Kalvach et al. [14] state, "there is no official definition for the term "soft targets" "(p. 6). In security circles, however, the term denotes places with a high concentration of people and a low degree of security against assault, which creates an attractive target, especially for terrorists. Beňová et al. [15] have the same approach when they state "soft targets are characterised as places with a high concentration of people and low level of security" (p. 2).

Beyond the theoretical discussion on the concept of a soft target, we can see some common features in the different proposals. In these terms for this communication, we use the idea presented by Schmid [16], who considers that "soft targets are undefended and/or under-protected open access locations with low or no security provided where civilians can be found in significant numbers" (p. 818).

The trend for the protection of soft targets has been to increase the level of security, which ends up deconfiguring the target itself. This transformation process is a sign of fear. This sign can be seen as a victory for terrorists because of a change in the space's normality. However, hardening soft targets are not feasible because no country and no city can transform all soft targets into hard targets. Even if this strategy were possible, the societal consequences would introduce profound changes in people's lives, particularly regarding exercising their fundamental rights.

A terrorist threat continues to manifest in Europe through the perpetration of attacks on the territory of several member states. The Executive Director of Europol, Catherine De Bolle, states in TE-SAT [17] that

> terrorism remains a key threat to the internal security of the EU. Fifteen completed, foiled and failed terrorist attacks were recorded in the EU in 2021. The four completed attacks included three jihadist terrorist attacks and one left-wing terrorist attack." (p. 3).

In the terrorist attacks that have taken place on European soil, it is vital to understand the facilitating role of cyberspace. According to TE-SAT [17].

> the online environment plays a key role in this as it facilitates (self-)radicalisation and the spread of terrorist propaganda. Europol has significantly scaled up its capabilities in identifying terrorist and violent extremist content, and is working with online providers to remove it (p. 3).

e-Threats

The last few years of the global pandemic have led to an increase in the use of cyberspace, and there is evidence of deviant activities, particularly among young people. The TE-SAT [17] report testified to the pandemic's role:

> the combination of social isolation and more time spent online during the pandemic has exacerbated the risks posed by violent extremist propaganda and terrorist content online, particularly among younger people and minors. Gaming platforms and services are increasingly used by right-wing terrorists to channel terrorist propaganda targeting a younger generation of users. Pandemic-linked restrictions also have the potential to exacerbate pre-existing mental health issues potentially prompting violent acts that resemble terrorist or violent extremist attacks (p. 5).

According to TE-SAT reports, the link between cyberspace and terrorist attacks seems to be a reality. Terrorist attacks are preceded by cyberspace activities for recruitment purposes and training, self-indoctrination, promotion of radicalisation processes, and acquisition of weapons, equipment and explosives (TE-SAT, 2022) [17]. Preparatory activities triggered by terrorists in cyberspace can be characterised by recruiting and connecting new activists to affiliated communities and groups, facilitating training and acquiring resources to perpetrate attacks. On this approach, TE-SAT 2022 states that

> "Online community building often plays a key role, as it connects peers virtually on a global scale. This drives radicalisation and provides access to terrorist propaganda, instructional material and opportunities for procurement of weapons and explosives precursors" (p. 30) [17].

Identifying e-threats is one of the keys to preventing attacks against soft targets. What do we know about the detection of e-threats?

To this end, it is essential, firstly, to understand what an e-threat is. According to a United Nations [18] report on security, a threat is understood today as "any event or process that leads to large-scale death or lessening of life chances and undermines States as the basic unit of the international system (…)" (p.25). In the words of Creppel [19], "threat is compared to a house on fire, where the urgency of danger is palpable" (p.451), divided into three entities "physical threats, economic threats and cyber threats" (p.156) [20]. This idea means that the emergence of cyberspace adapts the concept of threat to its dimension, called cyber threat or e-threat (although its definition varies). In this paper, e-threat refers to any factor, event or action that occurs in cyberspace with the potential to "cause damage to the integrity of people, beings or things" (p. 1168)

[21], be it material or moral damage, of varied nature (social, economic, political). E-threats may take the form of: "social intervention ("cyber-activism", "cyber-hacktivism", "cyber-vandalism", or "cyber-graffiti"); the form of criminal actions (hacking, cracking, cybercrime or cyber-terrorism), or even the form of acts of war (cyber-warfare)" [22], whose motivations are essential "egoism, anarchy, money, destruction and power" [20] (p. 159).

The concept may also be perceived as the intention to provoke an outcome and the ability to execute it. However, it is crucial to mention that an e-threat results from a social construction that may differ according to the actor who defines (or not) an issue as an existential threat.

The uncertainty and non-knowledge [23] that characterise crime in the modern world require innovative approaches that prevent the realisation of increasingly challenging threats: the case of e- threats. A study by Eachus et al. [24] targeting hostile intent in crowded public spaces considered stress a crucial element in detecting individuals intending to consummate a terrorist attack. The stress likely felt, considered a symptom that could denounce a terrorist threat will manifest itself in different ways, analysed and categorized - biological, physiological, behavioural and psychological factors - as integral to hostile intent. However, in e-space, this is not a symptom we can identify. The criminal appears as an invisible face that does not physically exist at the moment of crime prevention, which means that, based on the study guidelines, as mentioned earlier, the biological and physiological components will not be identifiable. For the diagnosis of e- threats, it remains for us to follow the conduct of the user, through the analysis and collection of information, based on their speeches and activities in the environment, which may denounce an e-threat.

As an example the simple keyboarding on social networks. Social networks have originated a new dimension of communication, allowing, in real-time and freely, the diffusion of ideals, opinions and knowledge. In the network society [25], globalised and massified, there is an increase in the number of users of social networks. As a natural cause, an increase in the volume of offensive speech: "In fact, social media by definition can be viewed as a potential accelerator for hate speech in uncontrolled contexts" (p. 161) [26]. According to Khan et al. [27] "hate speech are words that are generally targeted to be hurtful towards a person or a group of people on the basis of race, caste, gender, colour, ethnicity and also support and promote violence against the targeted groups" (p. 688), e.g. against soft targets. In order to create a safe online environment, it will be necessary to diagnose and block this type of content. The approach of Mussiraliyeva et al. [28], aimed at detecting online radicalisation and extremism on the Twitter platform, allows us to identify linguistic variables - vocabulary and specific semantic patterns - as "informative predictors of the user's fundamental interests" (p.2) [28]. It is a beneficial use of classification algorithms that vary according to the type of words used, the frequency of their use, the frequency of punctuation marks, and emotions experienced by the author of the text, among others. Also, a study by Aziz et al. [29], which aims at detecting tweets with racist content through keywords, is an example of the importance of cyber communication for detecting hostile user interests. This circumstance means it may be possible to recognise e-threats successfully by employing verbal language.

In cyberspace, in addition to speech, digital activity can also be a means of diagnosis of e-threats. We take as an example the consumption of content and e-commerce in platforms such as Youtube, Google, Yahoo, and Amazon, among the best known. Actions of this type make up the so-called "digital footprint" to which platforms have access and the opportunity to monitor. According to Bujlow et al. [30], "Web services make continuous efforts to obtain as much information as they can about the things we search, the sites we visit, the people with whom we contact, and the products we buy", adding that their objectives are usually economic as "tracking is usually performed for commercial purposes" (p. 1) [30]. The authorities can access this information, which is essential for identifying deviant behaviour according to the type of goods and services purchased or the consumption of internet users' content that may represent a hostile intention.

The modus operandi of some acts leaves digital footprint, for instance "the planning and preparation of attacks were frequently accompanied by auxiliary offences such as self-training, including the exchange of tactical information online" (p. 26) [17], which is the preliminary screening "during pre-crime phase" (p. 491) [31]. It means that the digital footprint is fundamental to detecting e- threats. In the cyber dimension, we are reduced to behavioural analysis - speech and activities that denounce conduct - and, once limited to this aspect, the issue of e-threat detection becomes even more complex.

3 Perspectives and Directions

Reflection on the protection of soft targets should focus on the importance of anticipating threat identification processes, in general, and e- threats, in particular. The existing relationship between e-threats and soft-target attacks is the preferred area of work for detecting early signs that evidence the intention to perpetrate an attack, with a particular focus on identifying preparatory acts.

The difficulty in identifying e-threats stems from a number of aspects that need to be carefully analysed. This process is like looking for a needle in a haystack.

The first barrier is related to the immensity of events that occur daily in soft targets. The limited resources associated with bounded rationality make an exhaustive analysis of each event unfeasible.

In cyberspace, there is a high probability that the information elements associated with an e-threat are fragments that require an in-depth study to build an intelligible meaning for security managers. Moreover, these fragments are often encrypted, which facilitates their camouflage.

A third aspect is the complexity of a terrorist attack, which means that operations are initiated through preparatory acts. Detecting the preparatory acts may not lead to subsequent actions, and, for this reason, the processes of analysing data, fragments and information are often doomed to failure.

The use of digital tools to identify e-threats represents a significant advance in the process. The exponential increase in the amount of data analysed, the decrease in research time, the use of models that allow for data triangulation, the constitution of databases with the historical elements constituting the e-threats, the identification of the relations established between e-threats and terrorist attacks unleashed in the past, boost the entire intelligence effort because it diminishes the limitations of bounded rationality. Despite the stated advantages, societal, ethical, legal, and privacy constraints must be considered.

The construction of analysis models made up of criteria and rules that enable the identification of technical indications of an e-threat should be a challenge for States, international organisations and the international community. The digitalisation of these analysis models is a priority and a multidisciplinary effort involving technical experts, legal advisors and security specialists. The confidentiality associated with constructing e-threat detection tools is a further barrier to cooperation processes.

To analyse the preliminary considerations for constructing the diagnostic model for detecting e-threats against soft targets, we should present the reflection according to three plans: the theoretical, the empirical and the normative.

Delivering intelligence through analysing elements gathered with a surfeit of hi-tech tools allows us to understand the ubiquity of social media as an open source of information [32, 33]. Considering that the world has shortened because of globalisation, geographic barriers, cultural, social, and economic proximity [34], as well as criminal, have evolved to almost non-existent, the leverage of applying different technologies in terrorism, as well in the e-threat detection is a concern.

The recognition of the need for e-threat detection to develop intelligence is embracing technology in objective-driven police is a vital resource for every security force [35]. In this challenging and changing world of geo-globalisation, geo-terrorism, and geo-e-threats, the protection of critical assets requires the application of the 4P (Prevention, Proactivity, Predictively, Patterns) [36].

The 4 P's are only manageable if law enforcement agencies and secret services can apply the latest technology (Body worn-cameras - BWC, augmented reality, blockchain, human sensors, AI and the Internet of Things - IoT, Advanced Computation, Facial Recognition Technology - FRT, machine learning), which means altering the landscape of Police action. It countenances the possibility in a chaotic environment to integrate into law enforcement and secret organisations to provide the means to predict, analyse, recognise, explore and communicate, increasing their accountability and transparency [37].

However, this technological road entails the triangulation of the algorithms and the interoperability of every stakeholder to enhance the benefits, and mitigate the risks, that Rizzi and Pera underscore: i) Fundamental Human Rights; ii) criminal and terrorist groups use; iii) jeopardising human survival [38].

Exploring the architecture of the different dimensions of the challenges for e-threats detection may allow a better understanding of the all.

Harvesting up the normative basis upheavals questions such as i) Fundamental Human Rights; ii) freedom of speech; iii) protection of personal data; iv) the principle of privacy, and iv) ethics [39–43].

The intricate balance between the convergence of law and regulation and technologies affects human rights because human the sacredness of human life is the core idea [44]. The interloping of technology for e-threats detection can be acknowledged in privacy and data protection. Articles 7° and 8° of the Convention for the Protection of Human Rights and Fundamental Freedoms state the right to respect private life and to protect personal data [45]. This consideration aims to allow the growth of individual characteristics and liberty of thought to ensure the construction of critical thinking.

Whether it is through no interference in private life or the protection of personal data whenever it is processed is essential, even though respecting these elements is not applied to security forces or security organisations.

However public bodies are not subjected to it, the intertwined relationship with private security makes it necessary to highlight this to avoid the violation of fundamental rights by the private sector [46] (Wakefield, 2002) when using technological tools [47] (Benson, 1998).

Being said that the fundamental postulate of a modern democratic society is inextricably linked with respect and protection of guaranteed human rights, a question arises. The protection of privacy personal data in digital life has a new ecosystem.

The EU Directive 95/46/EC [48] codifies the principles for managing personal data and the free movement of data. The protection of personal data is endorsed in processing, letting the concepts of data use and data reuse out of that perception [49]. According to Article 2, processing is the set of operations performed on personal data. Those operations are "collection, recording, recording, organisation, storage, adaption or alteration, retrieval, consultation, use, disclosure by transmission, dissemination or otherwise making available, alignment or combination, blocking, erasure or destruction" (p. 4) [49].

The combination of articles Article 6.1(b), 5, and 9(2) determine that the collection of personal data must be specific and for a legitimate purpose. It also requires its lawful processing and consolidates the exemption for competent authorities with the Police Directive [50] and the governance of data treatment under GDPR.

The ultimate challenge of processing personal data is to avoid embedded unfairness in the algorithmic automated decision-making process and the bias in the human decision, which can be conscious, unconscious or metabolism driven.

Using an algorithm and AI requires considering three main aspects that convey it as trustworthy, which are lawful, ethical, and robust [51]. Ethical behaviour in the era of AI needs to comply with the principles of respect for human autonomy; prevention of harm; fairness; explicability [52] as to provide and develop a set of requirements (human agency and supervision; technical robustness and safety; privacy and data governance; transparency; diversity, non-discrimination and fairness; environmental and societal well-being and, accountability) that helps to overcome negative externalities of using AI, and extrapolating its positive benefits.

Nevertheless, considering the thin line between positive and negative effects, from bias, excess use of algorithms, AI, and IoT, the European Commission, considering the work of HLEG, provides a checklist of ethical requirements that must be applied in the phase of design, developing and in the phase of use (deploying) [53].

When using data for the risk diagnosis model, the challenge of the systems is coping with and handling human independence, self-determination, and self-esteem, while respecting their right to privacy and personal data protection. Also, in the form of technology (CCTV, augmented reality, drones), or human sensors (apps), the data recollected must be representative, truthful, vital, and non-biased to allow a non-discriminatory agenda. The ultimate purpose of these reflections is, on the one hand, to avoid negative impact in individual, social or environmental set-ups, and on the other hand, to promote transparency and accountability to their stakeholders and end-users. If this is not achieved rather than democratising human rights exercise, high-tech advances are expected to be skewed for inequity [54]. Therefore, establishing preventive measures and a normative framework is fundamental to ensure that the technology used for e-threat detection does not compromise any right of the individual and his data while guaranteeing the security and safety of the soft targets.

Empirically concerns about technology are raised in terms of what its defaults are. The roadmap of the progress of the IA is undoubtedly strong and has been achieving positive results and evolved from machine learning to deep learning models.

Though, the methodology applied to detect threats/e-threats and the methods still depend on hyper-parameter configurations to be effective [55, 56]. Those metrics help to construct diagnosis models bridging the data recollected with the decision-making process for the analysis of an e-threat.

Mainstreaming technology and big data recollected in e-threats detection somehow thought-provoking the efficiency of the technology and the ineffectiveness of pondering individuality [57].

The increase in technology dependency, the way the data is obtained, the high error susceptibility, the misuse that can lead to some scale of devious analysis, and the interpretation of results [58, 59], the bias (beliefs, values, religion, professional and personal experiences) of the person that creates the metrics of the algorithm, the mismatch from reading the same criteria in different contexts, the increase of technology dependency, up brings the natural disadvantages of this process.

However, globalisation is an active stakeholder of complexity and interdependence. In fact, as a form of economic, cultural, social and criminal spatial reduction, with timeless constraints, it induces the spread of technologies, even the dangerous ones, the dissemination of new forms of terrorism, and unveils the creation of new and unthinkable threats [60], and e-threats.

It is clear that the Holy Grail of technology also entails inquiries on police legitimacy and public recognition [57].

The transition from the traditional degree of threat that a crowd represents to public order to dynamic risk assessment, an external threat which may fall on the crowd [61], is the natural theoretical evolution for the e-threat against soft-targets.

Even though the positive side effects of machine learning, AI, deep machine learning, and all the technological paraphernalia at the disposal of law enforcement agencies and secret services, special attention to their disruptive effects should be taken in place.

Siding with the idea of Beridze [62], this entails that governance, and appropriate international frameworks, must be up to date with technological evolution to safeguard all the above mentioned aspects without undermining public trust in authorities and avoiding adverse impacts on the lives of individuals.

The triangulation of these three approaches is essential to perform and avoid false positives or negatives, the fake news, and to overcome problems in the analysis of the discourse (for instance, the use of metaphors – language, imagery – code and belittling language, figurative linguistic expressions, numbers, acronyms, meaningful places).

The interoperability influences the possibility of conglomerating stratagems to serve a single-mindedness: minimising the risks of the technology for decision-making [63].

In light of the aforementioned a first approach for a diagnosis model is possible to draw as shown by the Fig. 1.

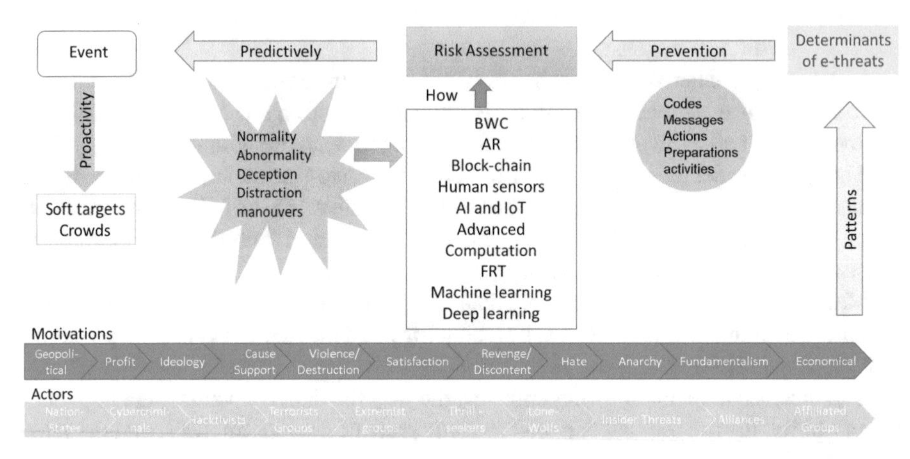

Fig. 1. Diagnosis model for e-threats

4 Conclusion and Discussion

The desire and expectation, which forms the intention, the resources and knowledge that build the capacity – manifested in cyberspace – conveys the new formulation of threat, the e-threat.

In the digital era, critical attention to deviant discourse to diminish the negative externalities of information processing to augment the efficiency of the process and consequently in the process decision-making and reducing the risk. Even though the mathematical algorithms are improving, human interaction is necessary when evaluating the information obtained from them for security and safety. The efficiency and equity of the information are managed through the triangulation of the empirical, theoretical and normative framework, which is only possible with the law enforcement officers' intervention. It increases the validity of the information, guarantees good governance, and builds proactive and predictive policing strategies adequate to the defined risk.

Integrating risk assessment of e-threats with technology and different techniques to face crime and terrorism against soft targets requires a multilevel enactment when interacting with citizens. Also, safeguarding human rights and social, ethical and legal matters must be possible in predictive and proactive policing, self-policing, and threats and danger awareness.

References

1. Laschi, M., Lombroso, C.: Le délit politique. Actes du Premier Congrès International d'Anthropologie Criminelle, Rome 1885, pp. 379–389, Bocca Fréres, Turin (1886)
2. Calhoun, T., Weston, S.: Contemporary threat management. San Diego: Specialized Training Services (2003)
3. Dietz, P.E., Martell, D.: Mentally disordered offenders in pursuit of celebrities and politicians. National Institute of Justice, Washington, DC (1989)
4. Fein, R., Vossekuil, B.: Assassination in the United States: an operational study of recent assassins, attackers, and near-lethal approachers. J. Forensic Sci. **44**, 321–333 (1999)
5. Fein, R., Vossekuil, B., Holden, G.: Threat assessment: An approach to prevent targeted violence (NCJ 155000). Washington, DC: U.S. Dept. of Justice, Office of Justice Programs, National Institute of Justice (1995)
6. Dietz, P.E., Martell, D.: Commentary: approaching and stalking public figures—a prerequisite to attack. J. Am. Acad. Psychiatry Law **38**, 341–348 (2010)
7. Meloy, J.R., Hoffmann, J., Guldimann, A., James, D.: The role of warning behaviors in threat assessment: an exploration and suggested typology. Behav. Sci. Law **30**, 256–279 (2012). https://doi.org/10.1002/bsl.999
8. Morgado, S.M.A., Felgueiras, S.: E-threats detection against public spaces and events. In: Yang, X.S., Sherratt, S., Dey, N., Joshi, A. (eds.) Proceedings of Eighth International Congress on Information and Communication Technology. Lecture Notes in Networks and Systems (Forthcoming). Springer, Singapore (2023)
9. Martins, M.: Ciberespaço: uma Nova Realidade para a Segurança Internacional 5(133), 32–49 (2012). https://www.idn.gov.pt/pt/publicacoes/nacao/Documents/NeD133/NeD133.pdf
10. World Economic Forum: The Global Risks Report 2022: 17h Edition (2022). https://wef.ch/risks22
11. Europol: Internet Organised Crime Threat Assessment (IOCTA) 2021. Luxembourg: Publications Office of the European Union (2021). https://doi.org/10.2813/113799
12. European Comission: COM/2017/041 final. Fourth progress report towards an effective and genuine Security Union. Luxembourg, Publications Office of the EU (2017)
13. Zeman, T.: Soft targets: definition and identification. Acad. Appl. Res. Mil. Publ. Manag. Sci. **19**(1), 109–119 (2020). https://doi.org/10.32565/aarms.2020.1.10
14. Kalvach, Z.: Basics of soft targets protection–guidelines. Soft Targets Protection Institute, Prague (2016)
15. Beňová, P., Hošková-Mayerová, Š, Navrátil, J.: Terrorist attacks on selected soft targets. J. Secur. Sustain. Issues **8**(3), 453–471 (2019)
16. Schmid, A.: Layers of preventive measures for soft target protection against terrorist attacks. In A. Schmid (ed.), Handbook of terrorism prevention and preparedness. The Hague, NL: ICCT Press (2020). https://doi.org/10.19165/2020.6.01
17. Europol: EU Terrorism situation & Trend report (TE-SAT) (2022). https://www.europol.europa.eu/publications-events/main-reports/tesat-report
18. United Nations: A more secure world: Our shared responsibility. Report of the High-level Panel on Threats, Challenges and Change (2004). http://providus.lv/article_files/931/original/HLP_report_en.pdf?1326375616
19. Creppell, I.: The concept of normative threat. Int. Theor. **3**(3), 450–487 (2011). https://doi.org/10.1017/S1752971911000170
20. Urinov, B.N.: Theoretical aspects of organizational behavior and corporate culture. Econ. Innovative Technol. **2020**(2), 1–8 (2020)
21. Granjo, P.: Quando o conceito de «risco» se torna perigoso. Análise Social XL **I**(181), 1167–1179 (2006)

22. Nunes, P.: Mundos virtuais, riscos reais: fundamentos para a definição de uma Estratégia da Informação Nacional. Revista Militar (2506), 1169–1198 (2010). https://www.revistamilitar.pt/artigo/608
23. Beck, U.: Risk society towards a new modernity, 2nd edn. Sage Publications, Newbury Park (1992)
24. Eachus, P., Stedmon, A., Baillie, L.: Hostile intent in public crowded spaces: a field study. Appl. Ergonomic **44**, 703–709 (2013). https://doi.org/10.1016/j.apergo.2012.05.009
25. Castells, M.: The rise of the network society, 2nd edn. Blackwell Publishing Ltd, New Jersey (2009)
26. Alotaibi, A., Hasanat, M.H.A.: Racism detection in Twitter using deep learning and text mining techniques for the Arabic language. In: 2020 First International Conference of Smart Systems and Emerging Technologies (SMARTTECH), pp. 161–164. IEEE, Riyadh (2020)
27. Khan H., Yu F., Sinha A., Gokhale S.: A Parsimonious and Practical Approach to Detecting Offensive Speech. In: 2021 International Conference on Computing, Communication, and Intelligent Systems (ICCCIS), pp. 688–695. IEEE, Greater Noida (2021). https://doi.org/10.1109/ICCCIS51004.2021.9397140
28. Mussiraliyeva, S., Bolatbek, M., Omarov, B., Medetbek, Z., Baispay, G., Ospanov, R.: On detecting online radicalization and extremism using natural language processing. In: 21st International Arab Conference on Information Technology (ACIT). IEEE, Giza (2020). https://doi.org/10.1109/ACIT50332.2020.9300086
29. Aziz, N., Maarof M., Zainal, A.: Hate speech and offensive language detection: a new feature set with filter-embedded combining feature selection. In: The 3rd International Cyber Resilience Conference (CRC) IEEE, Ghent (2021). https://doi.org/10.1109/CRC50527.2021.9392486
30. Bujlow, T., Español C.V., Pareta S.J., Ros B.P.: Web Tracking: Mechanisms, Implications, and Defenses. ARXIV.ORG Digital Library (2015). https://www.researchgate.net/publication/280590332_Web_Tracking_Mechanisms_Implications_and_Defenses
31. Weert, A., Eijkman, Q.: Early detection of extremism? The local security professional on assessment of potential threats pose by youth. Crime Law Soc. Chang. **73**, 491–507 (2019). https://doi.org/10.1007/s10611-019-09877-y
32. Morgado, S.M.A., Ferraz, R.: Análise de conteúdo e pesquisa em ciências policiais: Contextualização em ambiente virtual, facebook, e o planeamento de grandes eventos. In Costa, A. P., Tuzzo, S., Ruano, L., Silva, C. T., Souza, F. N, Souza, D. N. (eds.) CIAQ2016 – Investigação Qualitativa em Ciências Sociais, vol 3, pp. 600–608, CIAQ, Lisboa (2016a)
33. Morgado, S.M.A., R. Ferraz, R.: Social media: facebook e a gestão policial no planeamento de grandes eventos. International Journal of Marketing Communication and New Media (Special Number 1—QRMCNM), 27–47 (2016b)
34. Morgado, S: Going Global: Health organizations and networking – information society and social media. In: Proceedings in Scientific Conference 2013, pp 47–51. Thomson, Slovakia (2013)
35. Morgado, S., Alfaro, R.: Technology embracement in objective driven police: UAV's in public security police in Portugal. In N. S. Teixeira, N. S., Oliveira, C. S. Lopes, M., Sardinha, B., Santos, A., Macedo, M. (eds.), International Conference on Risks, Security and Citizens: Proceedings/Atas, pp. 298–305. Setúbal: Município de Setúbal (2017)
36. Mendes, S., Morgado, S.: Intelligence services intervention: Constraints in portuguese democratic state. In: N. S. Teixeira, N. S., Oliveira, C. S. Lopes, M., Sardinha, B., Santos, A., Macedo, M. (eds.), International Conference on Risks, Security and Citizens: Proceedings/Atas, pp. 285–297. Setúbal: Município de Setúbal (2017)
37. Campbell, T.A.: Artificial intelligence: an overview of state initiatives. Future Grasp, LLC, Evergreen (2019)

38. Risi, F.T., Pera, A.: Artificial Intelligence in the field of criminal law. Special collection: Artificial Intelligence, pp. 6–17. UNICRI, Torino (2020)
39. Morgado, S.M.A., Felgueiras, S.: Technological Policing: Big Data versus Real Data. Politeia – Revista Portuguesa de Ciências Policiais XIX, 139–151 (2022). https://doi.org/10.57776/hkcb-br21
40. Bernal, P.: Data gathering, surveillance and human rights: recasting the debate. J. Cyber Policy **1**(2), 243–264 (2016). https://doi.org/10.1080/23738871.2016.1228990
41. Bridges, L.E.: Digital failure: Unbecoming the "good" data subject through entropic, fugitive, and queer data. Big Data Soc. **8**(1), 2053951720977882 (2021). https://doi.org/10.1177/2053951720977882
42. Sanders, C.B., Sheptycki, J.: Policing, crime and 'big data'; towards a critique of the moral economy of stochastic governance. Crime Law Soc. Chang. **68**(1), 1–15 (2017). https://doi.org/10.1007/s10611-016-9678-7
43. UNICRI: Strategic programme framework. UNICRI, Torino (2020)
44. Risse, M.: Human rights and artificial intelligence: an urgently needed agenda. Hum. Rights Q. **41**(1), 1–16 (2019). https://doi.org/10.1353/hrq.2019.0000
45. Council of Europe: CETS 005 - Convention for the Protection of Human Rights and Fundamental Freedoms (coe.int) (2019). https://rm.coe.int/1680a2353d
46. Wakefield, A.: The public surveillance functions of private security. Surveill. Soc. **2**(4) (2002). https://doi.org/10.24908/ss.v2i4.3362
47. Benson, B.L.: Crime control through private enterprise. Independent Rev. **2**(3), 341–371 (1998). http://www.jstor.org/stable/24561016
48. European Parliament, Council of the European Union: EU Directive 95/46/EC - on the protection of individuals with regard to the processing of personal data and on the free movement of such data (1995). https://eur-lex.europa.eu/legal-content/EN/TXT/PDF/?uri=CELEX:31995L0046
49. Custers, B., Uršič, H.: Big data and data reuse: a taxonomy of data reuse for balancing big data benefits and personal data protection. Int. Data Privacy Law **6**(1), 4–15 (2016). https://doi.org/10.1093/idpl/ipv028
50. Risse, M.: Human rights and artificial intelligence: an urgently needed agenda. Hum. Rights Q. **41**(1), 1–16 (2019). https://doi.org/10.1353/hrq.2019.0000
51. Council of Europe: Council Framework Decision 2008/977/JHA - On the protection of personal data processed in the framework of police and judicial cooperation in criminal matters (2008). https://eur-lex.europa.eu/legal-content/EN/TXT/?uri=celex%3A32008F0977
52. European Commission. Directorate-General for Communications Networks, Content and Technology: Ethics guidelines for trustworthy AI, Publications Office (2019a). https://doi.org/10.2759/346720
53. European Commission. Directorate-General for Communications Networks, Content and Technology: Assessment List for Trustworthy Artificial Intelligence (ALTAI) for self-assessment. Publications Office (2020). https://digital-strategy.ec.europa.eu/en/library/assessment-list-trustworthy-artificial-intelligence-altai-self-assessment
54. European Commission: Ethics by design and ethics of use approaches for artificial intelligence (2021). https://ec.europa.eu/info/funding-tenders/opportunities/docs/2021-2027/horizon/guidance/ethics-by-design-and-ethics-of-use-approaches-for-artificial-intelligence_he_en.pdf
55. Land, M.K., Aronson, J.D.: Human rights and technology: new challenges for justice and accountability. Ann. Rev. Law Soc. Sci. (Forthcoming) **16**, 223–240 (2020)
56. Lei, Y., Yang, B., Jiang, X., Jia, F., Li, N., Nandi, A.K.: Applications of machine learning to machine fault diagnosis: a review and roadmap. Mech. Syst. Signal Process. **138**, 106587 (2020). https://doi.org/10.1016/j.ymssp.2019.106587

57. Pfisterer, F., van Rijn, J.N., Probst, P., Müller, A.C., Bischl, B.: Learning multiple defaults for machine learning algorithms. In: Proceedings of the Genetic and Evolutionary Computation Conference Companion, pp. 241–242. GECO, Kyoto (2021)
58. Morgado, S.M.A., Felgueiras, S.: Big Data in Policing: Profiling, patterns, and out of the box thinking. In: Rocha, Á., Adeli, H., Dzemyda, G., Moreira, F., Ramalho Correia, A.M. (eds.) WorldCIST 2021. AISC, vol. 1365, pp. 217–226. Springer, Cham (2021). https://doi.org/10.1007/978-3-030-72657-7_21
59. Khanzode, K.C.A., Sarode, R.D.: Advantages and disadvantages of artificial intelligence and machine learning: a literature review. Int. J. Libr. Inf. Sci. (IJLIS) **9**(1), 30–36 (2020)
60. Ray, S.: A quick review of machine learning algorithms. In: 2019 International Conference on Machine Learning, Big Data, Cloud and Parallel Computing (COMITCon), pp. 35–39. IEEE, India (2019). https://doi.org/10.1109/COMITCon.2019.8862451
61. Boin, A., Comfort, L.K., Demchak, C.C.: The rise of resilience. Designing Resilience Preparing Extreme Events **1**, 1–12 (2010)
62. Pais, L.G., Felgueiras, S., Rodrigues, A., Santos, J., Varela, T.: Protesto político e atividade policial: a perceção dos" media". Análise Social **216**(3º), 494–517 (2015)
63. Beridze, I.: Foreword. Special collection: Artificial Intelligence. UNICRI, Torino (2020)
64. Felgueiras, S., Pais, L.G., Morgado, S.M.A.: Interoperability: Diagnosing a novel assess model. European Law Enforcement Research Bulletin, Special Conference Edition: Innovations in law enforcement: Implications for practice. Educ. Civ. Soc. (4), 255–260 (2018). https://bulletin.cepol.europa.eu/index.php/bulletin/article/view/339/301

x2OMSAC - An Ontology Population Framework for the Ontology of Microservices Architecture Concepts

Gabriel Morais[1(✉)], Mehdi Adda[1], Hiba Hadder[1], and Dominik Bork[1,2]

[1] Université du Québec à Rimouski, 1595, boulevard Alphonse-Desjardins, Lévis, QC G6V 0A6, Canada
{gabrielglauber.morais,mehdi_adda,hiba.hadder}@uqar.ca
[2] TU Wien, Business Informatics Group, Favoritenstrasse 11, Vienna, Austria
dominik.bork@tuwien.ac.at

Abstract. Applying the Ontology of Microservices Architecture Concepts (OMSAC) as a modelling language calls users to have expertise in ontology engineering. However, ontology practice remains restricted to a limited pool of practitioners, leading to a barrier to widely adopting such a modelling approach. Here, we present x2OMSAC, an ontology population framework that enhances the modelling of microservices architectures using OMSAC. We instantiate our framework by FOD2OMSAC, which limits modellers' manual tasks to data selection, cleaning, and validation of created models, thereby eliminating the need for ontology expertise and, consequently, expanding the potential of OMSAC adopters for modelling microservices architectures.

Keywords: OMSAC · ontology population · microservices · conceptual modelling · machine learning · OpenAPI · Docker-compose · feature modelling

1 Introduction

Microservices Architecture (MSA) is a recent software engineering paradigm based on a compositional approach. Systems built using this paradigm are arrangements of microservices providing limited functionalities, which are put together to form complex systems [6]. It has been widely adopted by industry [2] in different domains to address various challenges, from modernizing monolithic financial applications to large IoT systems.

The Ontology of Microservices Architecture Concepts (OMSAC) [11] is the MSA domain ontology formalized in the Web Ontology Language (OWL2 [18]) using the semantics of the Description Logic (DL). It supports modelling, exploration, understanding, and knowledge sharing of MSA concepts, and MSA-based systems representation. For instance, modellers can apply the OMSAC's terminology component (TBox) as a modelling language [12]. This application of

A. Rocha et al. (Eds.): WorldCIST 2023, LNNS 800, pp. 263–274, 2024.
https://doi.org/10.1007/978-3-031-45645-9_25

OMSAC allows modellers to link heterogeneous concepts and viewpoints simultaneously, bringing the capability to explore MSAs holistically and produce different system views according to stakeholders' information needs throughout semantic queries. The resulting models are instances of OMSAC's concepts, i.e., an assertion component (ABox) that, with the OMSAC's TBox, make up a knowledge base of microservices systems.

However, OMSAC – as any ontology – lacks an established process to handle *ontology population*, which is creating concept instances related to an existing TBox, i.e., creating an ABox. The ontology population process comprises two essential components: The TBox and an instance extraction engine [15]. Manually executing this process requires "tremendous effort and time" [9] and calls for specialized expertise and specific ontology engineering and exploration tools. Consequently, having an information extraction system that automates this process is highly desirable [9].

Accordingly, this paper presents x2OMSAC an ontology population framework tailored for OMSAC. It aims to address the automation challenge by facilitating existing knowledge to automate the creation of OMSAC ABoxes. We also present FOD2OMSAC, an implementation of the x2OMSAC framework based on a semi-automatic approach which populates OMSAC knowledge bases from a restrained number of inputs: *F*eature models, *O*penAPI, and *D*ocker-compose files. This implementation applies Sentence Transformer (SBERT [16]), a Natural Language Processing (NLP) machine learning (ML) model, to linkage proposes. We evaluate x2OMSAC using FOD2OMSAC on two open-source microservices systems. The source code and the evaluation kit are available at this paper's code companion repository [13].

The remainder of this paper is as follows. In Sect. 2, we present x2OMSAC followed by FOD2OMSAC in Sect. 3. In Sect. 4, we present the background in SBERT and details concerning the semantic matching mechanism implemented in the framework instance. Then, we detail the adopted evaluation process (Sect. 5). We discuss our framework in Sect. 6 and close this article with a short conclusion and feature work perspectives in Sect. 7.

2 x2OMSAC

We designed x2OMSAC inspired by the ontology population process presented in Petasis et al. [15], which comprises four components: Input (TBox and knowledge resources), instance extraction engine, population process, and validation (consistency check) [9,15]. Hence, x2OMSAC contains four steps: Knowledge resources selection, model extraction, creation, and validation.

Knowledge Resource Selection. In the knowledge resources selection, one must identify appropriate resources containing knowledge about the microservice's concept to model. The resource may vary according to the perspective of the microservice system to be modelled. For instance, modelling microservices' functional perspective would require information sources containing knowledge related to the microservice's functionalities (e.g., requirements, conditions, constraints). In contrast, technical perspectives would call for non-functional information sources. This knowledge should be cleaned and organized to be correctly used in the following steps.

Model Extraction and Creation. The model's extraction and creation steps compose the x2OMSAC's instance extraction engine and ontology population. The selected knowledge sources are explored to identify instances of OMSAC concepts and their relations. The challenge here is linking instances representing different perspectives which may be extracted from distinct knowledge sources in various processing tasks or at other points in time. Indeed, these relations could be implicit and dictated by the modelling goal.

For instance, when modelling a given microservice, it is possible to describe the domain functionalities it implements and provide its specific implementation details. These are two perspectives in OMSAC that are explicitly linked at the TBox level, but doing the linkage at the instance level is challenging because it calls for pairwise analysis of instances' properties, and the knowledge resources selected could be unlikely to provide any clue to unveil them. Thus, one must implement mechanisms to identify these linkages regardless of the processed knowledge resource.

Therefore, the model creation step should handle the instance creation and linkage using the available information about existing instances in the ontology and the information extracted in the previous step as knowledge input.

Validation. The validation step comprises the consistency check and linkage review tasks. The former should be conducted by assessing the created ABox consistency concerning the TBox rules, and the latter should verify that suitable linkages between different perspectives have been created. The linkage validation should focus on detecting instance linkages that are optional or hidden at the conceptual level, i.e., liking individuals from different perspectives.

3 FOD2OMSAC

This section presents FOD2OMSAC, which stands for *F*eature model, *O*penAPI and *D*ocker compose *to* OMSAC. It implements the x2OMSAC framework using a semi-automatic approach for extracting models covering three microservices' modelling perspectives: Functional, implementation and deployment. It uses three knowledge resources: Feature model descriptions to handle the functional perspective, OpenAPI (specification version 3) files for the implementation perspective, and docker-compose for the deployment perspective. We provide a detailed view of FOD2OMSAC in Fig. 1.

FOD2OMSAC implements x2OMSAC's knowledge resource selection, extraction, creation and validation steps. It is a hybrid ontology population [9] approach using rule-based and ML-based techniques. It comprises manual tasks for knowledge resource selection and validation steps, and Python scripts implementing automatic instance extraction and creation steps. The automated

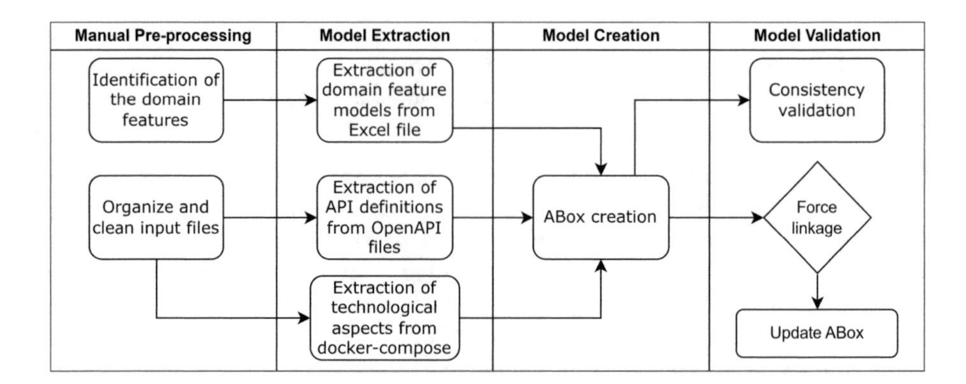

Fig. 1. FOD2OMSAC: Semi-automatic Approach for Creating OMSAC models

```
usage: micro_extractor.py [-h] -i INPUT (-fm | -sm) -n MODEL_NAME [-o OUTPUT_DIR]


Create an OMSAC-based model

options:
  -h, --help            show this help message and exit
  -i INPUT, --input INPUT
                        The file (for functional domain models) or folder(s) (for
 systems models) from where the model will be extracted
  -fm, --functional_model
                        Creates a functional domain model
  -sm, --system_model   Creates a microservices system model
  -n MODEL_NAME, --model_name MODEL_NAME
  -o OUTPUT_DIR, --output_dir OUTPUT_DIR
```

Fig. 2. Command line interface of the OMSAC ABox creator tool.

tasks can be invoked by a script providing a command-line interface (CLI), the *micro_extractor*, which offers commands handling different inputs, allowing the user to create specific models using specific files. It is possible to create models from a unique system or create various systems simultaneously using batch mode. Figure 2 provides a screenshot of the CLI of the OMSAC ABox creator tool. In the following, we present each framework step in detail.

3.1 Manual Pre-processing

The manual pre-processing prepares the different files to be processed by the *micro_extractor* scripts. The first task is defining the feature model input file, which must be a CSV file containing ten columns corresponding to concepts in the OMSC TBox. Table 1 provides the column names and meanings.

In the second task, the user verifies that the OpenAPI files are defined using version 3 of the OpenAPI specification. Otherwise, a conversion is needed and can be achieved manually using the Swagger Editor conversion facility.

Each microservice-based system must be organized in different folders if the batch mode is invoked. OpenAPI files must be named differently (e.g., the microservice name) to prevent overwriting.

Table 1. Description of the columns required by the CSV file to be processed.

Column	Content
goal	The goal's name, it is used as the goal ID
goal_desc	A textual description of the goal
requirement	The requirement's name, it is used as the requirement ID
type	The requirement's type, it accepts functional or technical as values
requirement_desc	A textual description of the requirement
condition	The condition's name, it is used as the condition ID
condition_desc	A textual description of the condition
constraint	The constraint's name is used as the condition ID
constraint_desc	A textual description of the constraint
depends_on	The IDs of goals, requirements, conditions or constraints that have a dependency on each other

3.2 Model Extraction

In this step, FOD2OMSAC extracts only mandatory and highly informative optional fields to guarantee higher compatibility. The extraction order is fixed; thus, the user must start with files containing the feature models, then the OpenAPI files and finally, the Docker-compose file. When using the batch mode, these three files must be in each system folder.

When processing the OpenAPI files, the script first injects all the relative content to obtain an inline file. For doing so, it relies on the *dereference* function of the JSON Schema $Ref Parser[1], which enhances the extraction of the inputs and outputs of each endpoint. Then, the script extracts from each endpoint: The path, the operation, inputs and outputs, including input and output types and whether they are mandatory.

The extraction of information from the Docker-compose file identifies the type of services, including platform-provided services used by the microservices (e.g., database, messaging, API gateway, and aggregator), the Docker image used for each service and deployment dependencies.

Finally, the script merges all the information extracted from both files and stores the extracted models used in the model's creation step.

[1] https://apitools.dev/json-schema-ref-parsert/.

3.3 Model Creation

In this step, the script creates the OMSAC ABox instances and links them using the extracted data from the previous step. It relies on the OWLReady2 [8] Python library and proceeds in the same order as the extraction step, starting with creating the functional perspective and then the implementation and deployment.

We also implemented the inter-perspective instance linkage mechanism. In other words, we link individuals from the OMSAC's *Feature* class to those of the OMSAC's *Requirement* class throughout the OMSAC's *fulfills* object property based on the instance names. It compares a short text similarity to a 0.7 threshold and automatically links individuals accordingly. The short text similarity is established using the *msmarco-distilbert-cos-v5* pre-trained SBERT model. The script generates a report of unmatched individuals (i.e., unmatched_features.csv) to be handled manually in the validation step.

We detailed in Sect. 4 the process of the SBERT model's choice and the identification of the suitable threshold and provided background about SBERT NPL models.

3.4 Model Validation

In this step, we validate the created models (i.e., the OMSAC's ABoxes). First, the user reviews the unmatched individuals and makes the necessary changes for complete linkage. Then, she validates the model consistency using Protégé [14].

To accomplish the first task, the user relies on the provided report of unmatched features (*unmatched_features.csv*), which contains two columns: Feature and requirement. The first column contains the Feature class individuals' names, and the second column is the name of the closest individual of the Requirement class that has been found. The user can then confirm the linkage, change it, or delete the line. Once the report is reviewed, the user can submit the modified report using the *force_linkage* command that will force the linkage.

To assess the ABox's consistency, first, the user must open the ontology in Protégé (Fig. 3a). A pop-up appears asking her to provide the path to the OMSAC TBox (Fig. 3b). Finally, the user must start one of Protégé's provided reasoners, such as Pellet (Fig. 3c). If the ABox is inconsistent, the reasoner will show an error message, and the user can ask Protégé to explain the causes of the inconsistency. In such a case, the user should identify the cause of the inconsistency and correct it manually.

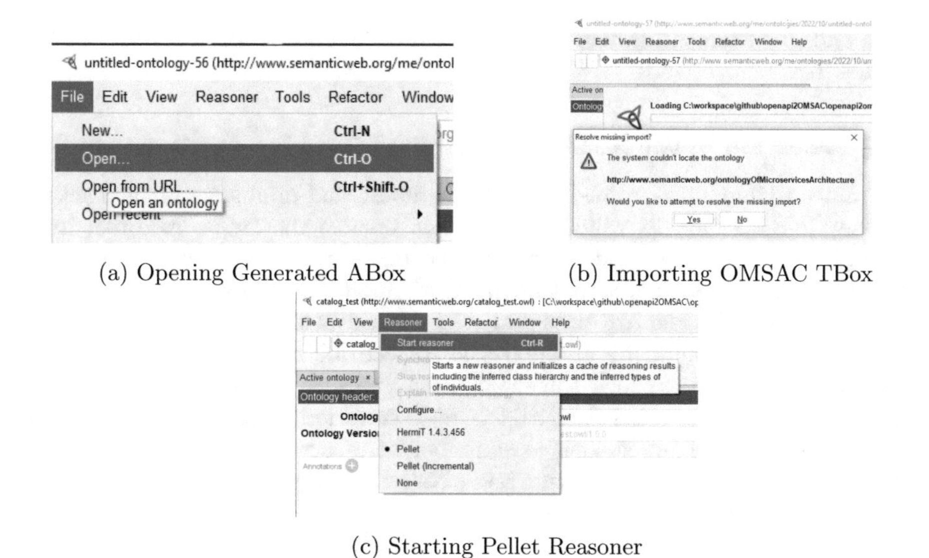

(a) Opening Generated ABox (b) Importing OMSAC TBox

(c) Starting Pellet Reasoner

Fig. 3. Manual ABox Consistency Validation with Protégé

4 The Semantic Comparison Mechanism

Using a word or character-based similarity measure may not consider that different words could represent a similar concept. NPL techniques have coped with this issue [3], and it seemed natural for us to rely on one of them for handling text comparison in FOD2OMSAC. This makes NPL machine learning techniques highly suitable for handling semantic similarity in heterogeneous knowledge source contexts [3]. Among these techniques, Bidirectional Encoder Representations from Transformers (BERT) [5] has become a state-of-the-art technique [3].

Sentence-BERT (SBERT) [16] are ML models applied for the NLP of short texts. SBERT fine-tunes BERT pre-trained models using siamese and triplet network architectures to "derive semantically meaningful sentence embeddings that can be compared using cosine-similarity." [16] SBERT has exceeded the performance of previous BERT models for establishing semantic similarity [7]. In addition, they need little effort to implement, as we can rely on a Python framework and a set of pre-trained SBERT ML models[2].

There is no ML model explicitly trained to process API descriptions based on the OpenAPI specification and able to match them to other natural language sources. Training such a model is out of the scope of this paper. Consequently, we decided to rely on existing pre-trained SBERT models used in the semantic search for automating concept linkages when building OMSAC ABoxes. Specifically, for linking the functional and implementation perspectives, i.e., Fea-

[2] https://www.sbert.net/.

ture and Requirement individuals, extracted from the OpenAPI and the feature model files, respectively.

4.1 Selection of the SBERT Pre-trained Machine Learning Model

We aimed to find a pre-trained SBERT ML model that minimizes manual actions in the OMSAC models' validation step and creates the fewest incorrect links. In addition, we need to couple the selected model to a mechanism to intercept unsuitable linkages that would be handled manually. We relied on a threshold mechanism to identify potential mismatches before effectuating the linkage so that they could be automatically placed for manual processing. Thus, we aimed to identify which threshold most limits this risk. Consequently, we assessed selected pre-trained SBERT models considering as criteria the number of required manual processing, the number of incorrect undetectable linkages (i.e., mismatches with a similarity rate above the threshold) and their accuracy (using the accuracy and F1 measures).

First, we built an evaluation dataset by extracting 47 API endpoints from three microservices-based systems proposed by Assunçao et al. [1]: *EshopOnContainers*, *Hipstershop*, and *Socksshop*, and manually mapping them to a requirement extracted from the feature model of the e-shopping domain [10]. This evaluation dataset contains the expected matches the different pre-trained SBERT models should find.

Table 2. Assessment of the Pre-trained SBERT Models

Criteria	Required Manual Actions						Incorrect Undetected Link						Accuracy						F1					
Model /Threshold	0	0.5	0.6	0.7	0.75	0.8	0	0.5	0.6	0.7	0.75	0.8	0	0.5	0.6	0.7	0.75	0.8	0	0.5	0.6	0.7	0.75	0.8
msmarco-MiniLM-L6-v3	23	23	24	22	29	31	23	18	14	8	6	**1**	0.511	0.574	0.617	0.723	0.617	0.660	0.676	0.688	0.690	0.745	0.571	0.529
multi-qa-mpnet-base-cos-v1	**19**	**19**	23	26	28	32	19	12	8	6	2	**1**	0.596	0.660	0.617	0.596	0.617	0.553	0.747	**0.750**	0.667	0.612	0.571	0.432
msmarco-distilbert-cos-v5	**19**	**19**	20	23	30	31	19	15	9	3	2	2	0.596	0.617	0.681	0.702	0.574	0.553	0.747	0.735	0.746	0.708	0.500	0.462
msmarco-MiniLM-L12-cos-v5	24	23	24	25	26	30	24	19	15	11	4	2	0.489	0.574	0.617	0.638	**0.745**	0.681	0.657	0.688	0.690	0.667	0.714	0.571
multi-qa-MiniLM-L6-cos-v1	21	20	23	24	26	29	21	16	13	5	5	**1**	0.553	0.617	0.596	0.681	0.638	0.660	0.712	0.727	0.678	0.681	0.622	0.579
multi-qa-distilbert-cos-v1	21	20	20	24	25	29	21	14	9	6	2	2	0.553	0.638	0.702	0.660	0.723	0.638	0.712	0.730	**0.750**	0.667	0.698	0.564

Table 3. Selection of the Pre-trained SBERT Models

Model	Threshold	Accuracy	F1	Required Manual Action	Undetected Incorrect Links	Criteria Met
msmarco-distilbert-cos-v5	0.7	0.70	0.71	23	3	**2/4**
multi-qa-distilbert-cos-v1	0.75	0.72	0.70	25	2	0/4
multi-qa-MiniLM-L6-cos-v1	0.8	0.66	0.58	29	1	1/4

Following, we pre-selected six SBERT pre-trained models according to recent measurements of their performance rate in handling semantic search in asynchronous contexts [4,17]. We assessed their similarity score regarding our evaluation criteria using six thresholds: 0, 0.50, 0.60, 0.70, 0.75 and 0.80 (see Table 2 for the results).

To select the appropriate pre-trained SBERT model, we first identified the model and threshold pairs with the fewest undetected incorrect links and the smallest number of required manual actions. The pair multi-qa-MiniLM-L6-cos-v1 and 0.80 had the best scores (one undetected incorrect link and 29 required manual actions). Then, we recursively selected the next best score for undetected incorrect links having fewer manual actions than the previously selected models and kept the best pair model threshold. Finally, we looked at each pair and their performance regarding the four criteria announced above and kept the combinations that met the most criteria. Table 3 summarises the set considered and the selection rounds. From this analysis, we identified the *msmarco-distilbert-cos-v5* pre-trained SBERT model and the threshold of 0.7 as the most appropriate for our purpose.

5 Evaluation

We evaluated x2OMSAC's capacity to enhance the use of OMSAC as a modelling language. As it is an abstraction, we assessed its capabilities by evaluating the FOD2OMSAC implementation. For doing so, we relied on a comparison of the ABox generated by a junior developer with no ontology engineering knowledge using FOD2OMSAC to one developed manually by an ontology practitioner with three years of experience with ontology engineering and more than ten years of experience in modelling data structures (e.g., relational and no-relation databases) supported by a domain expert.

We relied on two open-source [1] microservices-based systems from the e-shopping domain as use cases: *EshopOnContainers* and *Socksshop*, and on the e-shopping domain feature model proposed by Mendonça et al. [10]. The domain expert reviewed the proposed feature model and conducted the input files' organization and cleaning. This input was then used by both manual and automatic ABox creation. Table 4 summarizes the evaluation results of each model covering a particular perspective.

Table 4. Accuracy of FOD2OMSAC generated and manually created models.

Model	FOD2OMSAC Accuracy		
	Systems		Average
	eShopOnConatiners	Sockshop	
Functional model	100%	100%	100%
Implementation model	100%	100%	100%
Deployment model	53%	67%	60%
Inter-perspective linkage	67%	67%	67%

The evaluation results showed that the functional and implementation models created by FOD2OMSAC were similar to those made by the ontology practitioner. Concerning the inter-perspective linkage, we observed that both approaches had the same output for the Socksshop system and a difference of three links for the eShopOnContainers system. Nevertheless, FOD2OMSAC performed inaccurately when extracting deployment concept instances.

Regardless, when comparing the performance of FOD2OMSAC concerning the amount of data correctly processed, we observed that the distance between it and the manual approach is reduced. Indeed, the implementation perspective was built from seven OpenAPI JSON files representing more than three thousand lines of code. In contrast, the deployment perspective, which performed the worst, was built from two Docker-compose files representing 340 lines of code, ten times fewer lines.

6 Discussion

We understand that FOD2OMSAC improved the OMSAC ABox creation dramatically and validated the aim of the x2OMSAC framework for reducing the complexity of creating OMSAC models. The x2OMSAC met the need for technological openness mandatory to handle technological heterogeneity observed in the Microservices domain. Indeed, the implementation of the x2OMSAC presented in this paper handled only a limited set of technologies used for representing aspects of microservices-based architectures. By enabling FOD2OMSAC to process widely adopted standards, we ensured it covers many microservice-based systems, building a heuristic solution for creating OMSAC ABoxes that avoids manual-intensive tasks and limits the need for users' ontology knowledge. As observed by the user of the FOD2OMSAC, the CLI provided the information needed to pass through the extraction and creation steps, even if the user had no previous experience with Protégé.

We noted that the instance extraction from Docker-composed files performed poorly because the script covered a limited number of kinds of services, which were extracted based on comparing the service name and a set of words. For instance, if a service contained the string *data* or *db*, the script identified it as a *Database* instance. However, these files contain limited information, and the gap between potential and actual instances extracted could be filled by other knowledge sources, i.e., the OpenAPI files.

This work comes with limitations. The most obvious is the limited number of inputs used in the evaluation. We limited the experiment to an amount of information meeting human cognitive capabilities. Indeed, a larger set of inputs would make it impossible to compare the output of FOD2OMSAC to models created manually. Besides, the quality of the input used can impact the extraction accuracy and, consequently, the quality of the final models. Indeed, using resources extracted from source-code repositories may be prone to error because we have no guarantee that they are accurate and follow the standards of the state of the practice.

Similarly, the performance of the pre-trained S-BERT ML models we applied could impact the quality of the inter-perspective linkage and create undetectable unsuitable links when used in a different dataset. To handle this risk, we applied a threshold mechanism that can be tuned to ensure a higher intercept of undetectable mismatches. However, we expect the accuracy of the inter-perspective linkage mechanism to decrease if the input document uses a language other than English because, in such a case, the document will contain English (i.e. OpenAPI and docker-compose keywords) and non-English words likely to impact the semantic similarity performance, as the selected ML model was not trained with a multi-language dataset.

7 Conclusion

This paper presented the x2OMSAC, an ontology population framework tailored for the OMSAC ontology. We demonstrate how implementing this framework (FOD2OMSAC) based on a semi-automatic approach could handle existing knowledge sources to build models representing microservices architectures, which limits manual tasks, abstracts ontology engineering complexity for non-ontology practitioners, and effectively enhances the creation of OMSAC ABoxes. In future work, we plan to explore other machine learning models applied to NPL to increase the linkage mechanism's accuracy and expand its capabilities. Besides, we ambition to implement instances of x2OMSAC to handle other formalisms used in the microservices domain, including Proto Buffers and Kubernetes files.

Acknowledgements. We acknowledge the support of the Natural Sciences and Engineering Research Council of Canada (NSERC) grant number 06351, and Desjardins.

References

1. Assunção, W.K., Krüger, J., Mendonça, W.D.: Variability management meets microservices: six challenges of re-engineering microservice-based webshops. In: Proceedings of the SPLC (A), pp. 22.1–22.6 (2020)
2. Bogner, J., Fritzsch, J., Wagner, S., Zimmermann, A.: Microservices in industry: insights into technologies, characteristics, and software quality. In: IEEE International Conference on Software Architecture Companion, pp. 187–195 (2019)
3. Chandrasekaran, D., Mago, V.: Evolution of semantic similarity-a survey. ACM Comput. Surv. (CSUR) **54**(2), 1–37 (2021)
4. Craswell, N., Mitra, B., Yilmaz, E., Campos, D., Voorhees, E.M.: Overview of the TREC 2019 deep learning track. arXiv preprint arXiv:2003.07820 (2020)
5. Devlin, J., Chang, M.W., Lee, K., Toutanova, K.: Bert: pre-training of deep bidirectional transformers for language understanding. arXiv preprint arXiv:1810.04805 (2018)
6. Dragoni, N., Giallorenzo, S., Lafuente, A.L., Mazzara, M., Montesi, F., Mustafin, R., Safina, L.: Microservices: yesterday, today, and tomorrow. In: Mazzara, M., Meyer, B. (eds.) Present and Ulterior Software Engineering, pp. 195–216. Springer, Cham (2017). https://doi.org/10.1007/978-3-319-67425-4_12

7. Han, M., Zhang, X., Yuan, X., Jiang, J., Yun, W., Gao, C.: A survey on the techniques, applications, and performance of short text semantic similarity. Concurr. Comput. Pract. Exp. **33**(5), e5971 (2021)
8. Lamy, J.B.: Owlready: ontology-oriented programming in python with automatic classification and high level constructs for biomedical ontologies. Artif. Intell. Med. **80**, 11–28 (2017)
9. Lubani, M., Noah, S.A.M., Mahmud, R.: Ontology population: approaches and design aspects. J. Inf. Sci. **45**(4), 502–515 (2019)
10. Mendonça, W.D., Assunção, W.K., Estanislau, L.V., Vergilio, S.R., Garcia, A.: Towards a microservices-based product line with multi-objective evolutionary algorithms. In: 2020 IEEE Congress on Evolutionary Computation, pp. 1–8 (2020)
11. Morais, G., Adda, M.: OMSAC-ontology of microservices architecture concepts. In: 2020 11th IEEE Annual Information Technology, Electronics and Mobile Communication Conference (IEMCON), pp. 0293–0301. IEEE (2020)
12. Morais, G., Bork, D., Adda, M.: Towards an ontology-driven approach to model and analyze microservices architectures. In: Proceedings of the 13th International Conference on Management of Digital EcoSystems, pp. 79–86 (2021)
13. Morais, G., Bork, D., Adda, M., Hadder, H.: Companion source code repository (2022). https://github.com/UQAR-TUW/fod2OMSAC
14. Musen, M.A.: The protégé project: a look back and a look forward. AI Matters **1**(4), 4–12 (2015). http://protege.stanford.edu/
15. Petasis, G., Karkaletsis, V., Paliouras, G., Krithara, A., Zavitsanos, E.: Ontology population and enrichment: state of the art. In: Paliouras, G., Spyropoulos, C.D., Tsatsaronis, G. (eds.) Knowledge-Driven Multimedia Information Extraction and Ontology Evolution. LNCS, vol. 6050, pp. 134–166. Springer, Heidelberg (2011). https://doi.org/10.1007/978-3-642-20795-2_6
16. Reimers, N., Gurevych, I.: Sentence-BERT: sentence embeddings using Siamese BERT-networks. In: Proceedings of the 2019 Conference on Empirical Methods in Natural Language Processing. Association for Computational Linguistics (2019). http://arxiv.org/abs/1908.10084
17. Semantic BERT: Pretrained models. https://www.sbert.net/docs/pretrained_models.html
18. W3C OWL Working Group: Owl 2 web ontology language document overview (second edition) (2012). https://www.w3.org/TR/owl2-overview/

Calibration Techniques and Analyzing the Website Design with Eye Tracking Glasses

Zirije Hasani(✉), Samedin Krrabaj, Nedim Faiku, Shaban Zejneli, and Valon Ibraimi

Faculty of Computer Science, University "Ukshin Hoti" Prizren, Prizren, Kosovo
{zirije.hasani,samedin.krrabaj,200306052.b,200306096.b, 200306031.b}@uni-prizren.com

Abstract. Web site design is a key factor for a successful website. The number of websites is very large for this reason a successful website needs to fulfill this key element. The website design affects also the Universities world-wide because it affects the number of student applications. For this reason, in this study, we have analyzed the design and easy of use of all websites of public universities in our country. For analyzing the website design, we have used Eye tracking Pupil Labs which is tested with different calibration techniques and then its equipment and the result is compared. To analyze the design different parameters are measured, such as: time for applying as a new student, how much time consumes the navigation bar, how much time consumes photo shower as slider, time which is taken by news and events, footer is analyzed and video presentation how much time it takes. All these parameters are measured for each University and the smallest time taken with these actions result in better design from others. Also, a questionnaire is used for collecting data related to the website's design from participants in the eye tracking analyzing.

Keywords: Eye Tracking · Pupil Labs · calibration · design evaluation · websites

1 Introduction

Analyzing the web site design is very important for website performance. There are different techniques for analyzing the website design. In our research we have used Pupil Labs for analyzing the public Universities websites in Kosovo, especially the applying for study at each University. In applications where eye tracking is essential, especially in the critical application of the position where the viewing positions are highly influential in decision making, it is necessary to have an accurate system for better results. Applications like controlling mouse clicks through eye movement, maneuvering mouse points on the screen using eye movement, controlling an RC controller using eye movement, games and much more, require precise locations to make decisions safe [1, 2]. On the other hand, there are less critical applications regarding positions where the decision made may be based on ambiguous positions, based on a range or perhaps the direction of the view itself. Applications such as rotating a surveillance device in the direction of sight, using the eye tracker in web usability search, product packaging analysis and so on. While all of these applications use one or the other hardware device for eye tracking, the physiology of the human eye is not uniform across the population.

A. Rocha et al. (Eds.): WorldCIST 2023, LNNS 800, pp. 275–284, 2024.
https://doi.org/10.1007/978-3-031-45645-9_26

The eye tracker has a work in hand to calibrate manually or by itself before use in the user physiology, to achieve the desired results. We can say that calibration is essential for any eye tracking system and should be considered thoroughly. In this paper, we present the techniques for calibrating the eye tracking system of Pupil labs and also the performance of Universities' website is measured. We have created a group of people who have tested the application for study at each University where the time for doing the application form is measured and based on this recommendations are given where the problems are and how to modify the website design.

The aim of this paper is to compare the calibration techniques for eye tracking glasses and then to use them to analyze the website designs of public Universities in Kosovo. All this is done to find what the design problems are in order to be improved in the future.

The following in second chapter is related work in this particular topic compared to what we have done in this research. In the third chapter the Pupil Labs equipment are described which are used for research and the calibration techniques of this eye tracking equipment. Also, comparison of different calibration techniques is given. Then in fourth chapter the methodology used for this research is explained. Next is discussion of the results and in the end is conclusion.

2 Related Work

Through the virtual glasses Pupil Labs which we have used for the realization of our work in order to analyze the websites of public universities in Kosovo, many others have used them for different purposes.

As discussed in [3], both webcam eye trackers and wearing eye trackers can be suitable candidates for visual-based communication.

In [4] an approach to automatically and dynamically compile viewing data into a fixed reference image is also proposed, facilitating the use of this methodology in experimental contexts. They addressed this issue by developing a method to evaluate the accuracy of wearing eye trackers and tested this method against 3 existing eye wear tracker models and from those tested models realized that Pupil labs Binocular 120Hz glasses display the error with small accuracy. Pupil Labs 120Hz and SMI ETG 2.6 displayed similarly precise performance overall, with both showing stronger accuracy than the Tobii Pro Glasses 2.

In the following research it is proposed [5] to use a stationary eye tracker and a set of reproductions and modified works of art, and by this we can attempt to quantify how subjects view different works of art. They found that subject fixations in a controlled experiment followed some general principles in common (e.g., withdrawal from prominent regions) but with a great deal of variability for figurative paintings, according to the subject's assessment and personal knowledge.

Next in [6] a summary of modern student detection algorithms for outdoor application is proposed. The focus was mainly on the robustness of these algorithms in relation to lighting conditions that change frequently and rapidly, the position of the camera off-axis, and other noise sources. Six modern approaches were evaluated in over 200,000 images with true terrestrial notes collected with various eye-tracking devices in a variety of different everyday environments.

Also, in the following research [7] a new test battery is discussed that allows the analysis of a range of eye tracking measures. They used this test battery to evaluate two well-known eye trackers and to compare their performance.

Next we discussed in [8] how through using the eye-tracking technology we can investigate visual attention (i.e., counting of fixation and residence time) on the patient monitor during simulated anesthesia inductions with and without critical incidents occurring. They assumed that when managing an emergency situation, anesthesia staff would pay more visual attention to the patient monitor than when dealing with a non-critical situation.

In the following research the authors propose [9] that we use eye tracking equipment for research we have to know about:

(1) Oculomotor component: if the eye moves slowly or rapidly;
(2) Functional component: what purposes does eye movement serve; the coordinate system used: in relation to what the eye moves;
(3) Computational definition: how the event is presented in the eye tracker signal.

This should enable researchers of eye movement from different fields and the use of different eye trackers (fixed in the world or with the head) to discuss without misunderstanding.

The proposal in [10] is a new solution for the analysis of cellular eye tracking data. The proposed method combines sight-measured behavior from mobile eye-tracking glasses with the emergence of a recent DL-based object recognition and segmentation algorithm (R-CNN Mask) and enables fully automatic region and objects observed in nature as navigation evidence.

There are several researches but none of them have evaluated the Universities website especially Universities in Kosovo. Also, we use several calibration techniques and we use different measurement techniques for testing the usability of these websites.

3 Pupil Labs' Eye Tracking System

Pupil Labs[1] 'eye tracking system [11] is an open-source eye tracking device that comes with open-source software to calibrate and evaluate vision. The tracker uses images of the eyes and the pupil's position on the perimeter of the eye head to determine the user's gaze. This image-based eye tracking system consists of three cameras as shown in Fig. 1 below.

Two cameras point to the eye and one away from the eye mounted on the head to give the user a field of view perspective (Field Of View - FoV). The camera pointing outwards away from the eye in the direction of the user's gaze is also called the Global Camera. It is a high-resolution camera that can record images at 1920x1080 at 30 frames per second. The global camera's FoV is 90 degrees with focus lens. It is also able to record audio with integrated stereo microphone. However, the microphone is less used in eye tracking.

[1] The equipment is bought by project: "The development and implementation of a PhD Program in ICT for the Kosovo Education System", co-funded by Erasmus+ Programme of the European Union under the reference number: 609990-EPP-1-2019-1-SE-EPPKA2-CBHE-JP.

The eye camera or camera that tracks the eyes comes in two commercial versions. Eye cameras can be monocular or binoculars depending on whether both eyes are considered for tracking or just one. The eye cameras are able to record videos with a resolution of 640x480 and with a frequency of 120 frames per second.

Also, these cameras are IR Cameras (Infrared) which are able to illuminate the pupil for eye tracking. There is a known problem with the Pupil Labs eye tracker. The eye camera tends to heat up quickly due to its IR brightness and high degree of capture. So, it makes it very impossible to use the eye tracker for any application that requires prolonged use of the eye tracker. However, for academic purposes and for prototyping purposes, this device is the cheapest and best option currently available.

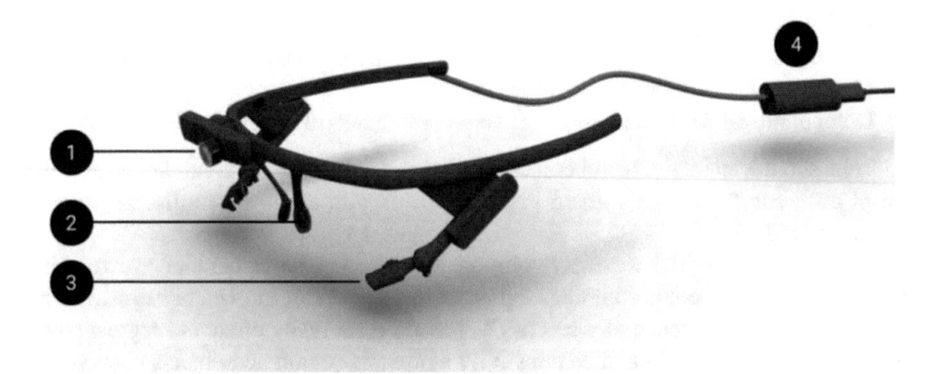

Fig. 1. Binocular Eye Tracker

3.1 Pupil Labs' Real-Life Application

The Pupil Labs eye tracker comes with its Pupil Capture app, which features object detection techniques to detect the eye and pupil transmitted by eye cameras. The pupil's position is designed in a 640 × 480 frame by normalizing the image to give an effective value of the pupil's position in the image relative to the position of the eye. It is also assumed that the diameter of a pupil is 12 mm for an average person.

The position of the pupil is a primary parameter on which the entire assessment of tracking and viewing depends. The gaze is nothing but the intersection of the visual axis of the two eyes in a plane normalized in the coordinate system of the world camera. However, the viewing values are not the same for everyone who uses Pupil Labs' eye tracker. The application is equipped with various calibration functions to activate the tracker and correct errors in the visual assessment. Also, the application software is equipped with recording sessions and reviewing them as needed and when required. Pupil Labs APIs are exposed to extract real-time viewing data in any personalized application.

Calibrating the pupil tracker using the Pupil Capture software is the essence of this article. The difference in the accuracy of the visual data is noted with a scenario considered for experimentation.

3.1.1 Calibration Techniques

There are three main techniques used for calibration. Screen calibration, manual calibration and calibration of natural features which are widely used in the Pupil Labs' eye tracker. The application software is equipped with functionalities to perform the required calibration. It should be noted that calibration can only be performed after the eye tracker is connected to the system via a serial cable. Also, calibration must be performed before the data retrieval event. The calculated calibration is estimated and provided by the device or application software.

3.1.2 Screen Calibration

Used mainly when the focus is close to the eye. In other words, the viewing points that are expected to be tracked are at close range of the user. This calibration is optimal in applications such as mouse movement using viewing, analysis of web application usage and similar types. Signs are lit by on-screen application software in 9 different locations to capture the user's FoV. During this phase, the movement of the head is limited to ensure that compensation is made only to the movement of the pupil and the values of the point of view. The points can be of any number based on the user's choice, there is marker calibration with 3 points, 5 points, 9, 13 and so on as shown in Fig. 2.

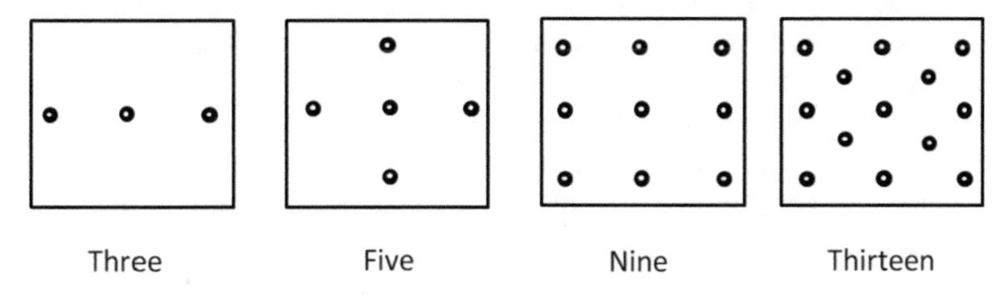

Fig. 2. Calibration points used to calibrate the eye tracker

However, it should be noted that screen calibration is only used for screen-specific FoV, which is very narrow compared to physiologically possible FoV with a predefined resolution. The marker image is as shown in Fig. 3 shows multi-dot markers used for screen calibration.

Fig. 3. Calibration marker

3.1.3 Manual Marker Calibration

While this technique is like screen calibration, markers can be placed elsewhere in the environment. It is mainly used for tracking and evaluating mid-range vision. The calibration technique requires a fixed head while the markers are strategically placed in the environment to cover the user's field of vision. Once again, any change in the field of view does not qualify the tracker's confidence in the assessment of the view.

3.1.4 Natural Calibration

Mainly used for remote eye tracking with a large FoV, it is not useful for many applications unless the tracker is used for moving vision tracking. However, even this technique uses a fixed field of view to calibrate selected points in the vicinity, as displayed by the world-wide on-screen camera with trackers connected to the system. Prominent features within FoV have been selected for FoV finalization. One can select as many points as possible within the FoV of the world camera and calibrate the tracker.

4 Methodology

To analyze the web site design are many techniques but the most important are they which are based on the people eye because is the main factor to like or dislike the design of one web site. Based on this we use in our research eye tracking glasses to do the research. The research is conducted by two type of methodologies, questionnaire and eye tracking glasses which test directly the design on people.

In this research, we used Pupil Labs glasses to analyze the Universities web site design, especially is measured the time needed to apply for study at the respective Universities.

The experiments are done by a group of people and then the data collected from them is analyzed and a conclusion is given. The selection of people is based on their experience in using websites because they will give more correct data related to what we are interested in.

A group of seven students and 10 high school students are conducted. Each of them is given the same task applying for study at a respective University. For each of them the eye tracking through the web site is recorded. Then the time needed for this process is measured.

During this research, we used some of the calibration techniques mentioned above, such as: screen calibration, manual calibration and calibration of natural features, based on which each of us has done the analysis of the design of the websites of the universities of Kosovo. Every student recorded their experience of all websites of public universities in Kosovo.

These recordings have been done in order to evaluate these websites, how easy they are to access them and to make the application process at the respective university online.

For each student while browsing the web pages, the browsing time is classified for some of the main elements (such as: nav-bar, slider, footer, news/events) of the respective web-site. The videos are saved in the Pupil-Player application and after reviewing them the data are placed in an Excel file manually. Is important to mention that every person who is involved in doing the experiments have done the same task "applying online for study at public university" at each six web sites of respective universities and then a comparison between results we get from different people give us approval that the testing is done good and that there aren't any problems related to the person who is being tested. This form of approving the correctness of data collected is used because there may be a person who could not be honest about a specific task for some reason and this may affect the result of the study.

We have used two methods for evaluating the website design: questionaire with people involved in evaluation process where they are asked to give a mark to the overall website design from 1 to 5 and also the eye tracking methodology for evaluating the time spent in certain parts of the web page. In the following section the results for this research are given.

5 Result Discussion

As we also mentioned in the methodology of this work, the experiments for each university websites are repeated with different people in order to collect correct data. The result collected from the research is compared and for each university an approximate value is given based on the data collected.

Based on the statistics we have obtained; we have concluded some data that make a difference in the design of websites. As seen in the diagram below, we notice that student feedback on the universities of Prizren and Gjakova have resulted in higher scores, as a result of the shorter time required for online application. While with the lowest evaluation are the University of Prishtina, Peja and Mitrovica and with an average evaluation is the University of Gjilan. In the figure below we have presented the grade for website designed (maximum grade 5). The grading form is done with marks from 1 to 5 where each person who was involved in experiments is asked to give a mark for the design of respective website.

After this, for each university an approximate mark of all involved people in experiment. The results are shown in the following chart below (Fig. 4).

The second form of testing the web site design used is time spent in certain parts of the web page as: time for application, time spend in navigation bar, time spend for slide show of photos, also are measured the time consumed to read about news, then time spent in footer and in the end the total amount of time for respective university is shown.

The following table shows the time of analysis of the main elements from the websites of the above-mentioned Universities. Different measurement is done as time taken for application, and then the navigation bar is analyzed on how much time is needed to find the needed information. Also, the photo presented by slide how much time it requires, then news and events presented in particular side of the web site; the time consumed by footer. The measurements are done for each person involved in this study for all respective web sites, and then an approximate time is taken for each specific task (Fig. 5).

Based on this table we can say that the best design and easiest application procedure has the University of Gjilan with time 1 min in total for the online application procedure followed by University of Prizren by 1:14 min for the same task. But with the largest time

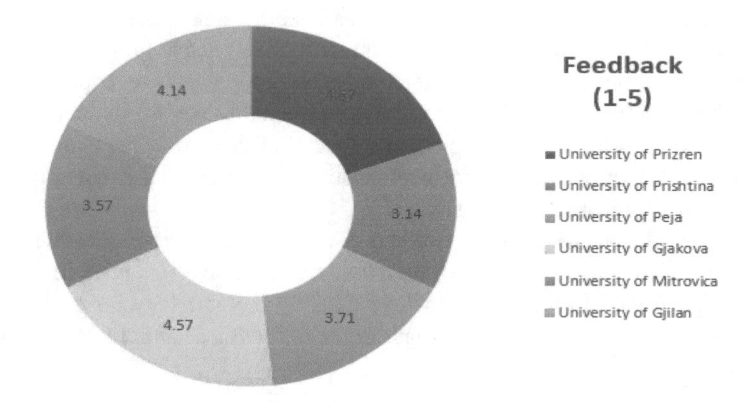

Fig. 4. Grade of design evaluation by students

Universitetet	Time for application	Nav Bar	Slide/Photo	News/Events	Footer	Video
Univeristy of Prishtina	03:02	00:57	00:09	00:26	00:12	03:40
Univerisy of Gjilan	00:34	00:10	00:06	00:20	00:06	01:00
University of Peja	00:35	00:25	00:06	00:14	00:06	01:27
University of Gjakova	01:13	00:37	00:08	00:36	00:11	01:59
Univeristy of Mitrovica	02:31	00:41	00:13	00:56	00:11	03:03
University of Prizren	01:10	00:13	00:05	00:45	00:07	01:14

Fig. 5. Time data for application to universities

for applying is University of Prishtina 3:40 min and they need to improve their website design because compared to others universities they have the worst design. The same situation is with the grading form shown in the chart above where University of Prishtina gets the lowest grade 3.14 and the highest grade is given to University of Prizren and University of Gjakova 4.57.

6 Conclusion

Due to the heavy use of the internet recently, many processes have been digitized including application to universities. In this case, the purpose of our research was to analyze the usability of the online application on the websites of universities in our country.

The aim of this research is to find the problems that the universities have with their web site design and if it affects the number of new students applying for study at respective university. Six public web site Universities are used for doing experiments and a group of 17 people (university students and high school students) are involved into the evaluation procedure.

Also, different calibration techniques are used for this research as: screen calibration, manual calibration and calibration of natural features which are widely used in the Pupil Labs' eye tracker.

From results we got, the highest mark for website design gets the University of Prizren and University of Gjakova 4.57 whereas the lowest mark is given to University of Prishtina 3.14 which means they need to improve the design of their web site because visually is not liked by the students. Also related to eye tracking time measurement for the usability of the respective web site the fastest online application has University of Gjilan, by 1 min, and the longest application time also has University of Prishtina with 3.40 min.

After browsing these websites through Pupil Labs virtual glasses, we came to the conclusion that some of the websites have an easier application method and some more difficult.

Based on the result we get and the time we measured for each part of the website, the respective institution can take action to modify the design in order to decrease the time needed for online application to the respective University.

In the future we are going to analyze if the results we found here affects the number of students who apply to respective Universities, because many studies that Universities do worldwide in recent years are especially focused on how the design of the University Web Site affects the number of new students applying for study. It is a very important aspect for each university.

References

1. Tripathi, S., Guenter, B.: A statistical approach to continuous self-calibrating eye gaze tracking for head-mounted virtual reality systems. In: 2017 IEEE Winter Conference on Applications of Computer Vision (WACV), Santa Rosa, CA, SHBA, pp. 862–870 (2017)
2. Jackson, L.A., Eye, A.V., Barbatsis, G., Biocca, F., Fitzgerald, E.H., Zhao, Y.: The social impact of Internet use on the other side of the digital divide. ACM **47**(7), 43–47 (2004)
3. Arslan, K.S., Acartürk, C.: Eye Tracking for Interactive Accessibility: A Usability Analysis of Alternative Communication Interfaces. Neuroergonomic Conference, München, Germany (2021)
4. Macinnes, J.J., Iqbal, S., Pearson, J., Johnson, E.N.: Wearable Eye-tracking for Research: Automated dynamic gaze mapping and accuracy/precision comparisons across devices, bioRxiv 299925. https://doi.org/10.1101/299925
5. Quian Q.R., Pedreira, C.: How do we see art: an eye-tracker study. Front. Hum. Neurosci. **5**, 98 (2011). https://doi.org/10.3389/fnhum.2011.00098
6. Fuhl, W., Tonsen, M., Bulling, A., et al.: Pupil detection for head-mounted eye tracking in the wild: an evaluation of the state of the art. Mach. Vis. Appl. **27**, 1275–1288 (2016). https://doi.org/10.1007/s00138-016-0776-4
7. Ehinger, B.V., Groß, K., Ibs, I., König, P.: A new comprehensive eye-tracking test battery concurrently evaluating the Pupil Labs glasses and the EyeLink 1000 (2019). https://doi.org/10.7717/peerj.7086
8. Roche, T.R., Maas, E.J.C., Said, S., et al.: Anesthesia personnel's visual attention regarding patient monitoring in simulated non-critical and critical situations, an eye-tracking study. BMC Anesthesiol. **22**, 167 (2022). https://doi.org/10.1186/s12871-022-01705-6
9. Hessels, R.S., Hooge, I.T.C.: Eye tracking in developmental cognitive neuroscience—the good, the bad and the ugly. Dev. Cogn. Neurosci. **40**, 100710 (2019). https://doi.org/10.1016/j.dcn.2019.100710
10. Deane, O., Toth, E., Yeo, S.H.: Deep-SAGA: a deep-learning-based system for automatic gaze annotation from eye-tracking data. Behav. Res. **55**, 1372–1391 (2022). https://doi.org/10.3758/s13428-022-01833-4

11. Pupil Labs' eye tracking system. https://pupil-labs.com/. Accessed 20 Apr 2022
12. Barba-Guaman, L., Quezada-Sarmiento, P.A., Calderon-Cordova, C., Lopez, J.P.O.: Detection of the characters from the license plates by cascade classifiers method. In: Paper Presented at the FTC 2016 - Proceedings of Future Technologies Conference, pp. 560–566 (2017). https://doi.org/10.1109/FTC.2016.7821662

HealthCare Fake News Detection: A New Approach Using Feature Selection Concept

Mostafa R. Kaseb[1(✉)], Saad M. Darwish[2], and Ahmed E. El-Toukhy[1,3]

[1] Faculty of Computers and Artificial Intelligence, Fayoum University, Fayoum, Egypt
mrk00@fayoum.edu.eg

[2] Institute of Graduate Studies and Research, Alexandria University, Alexandria, Egypt
saad.darwish@alexu.edu.eg

[3] Department of Computer Science, Higher Institute of Management and Information Technology, Kafr Elsheikh, Egypt

Abstract. Life is now lot easier than it was before the development of Internet-based technologies. Social media platforms have become exponentially more popular, which has improved human communication as well as the ability to link individuals in distant locations. However, social networking sites have also occasionally been used for immoral and illegal actions. People's mental and physical health has gotten worse as a result of bogus news being spread on social media during the COVID-19 epidemic. Thus, a number of research have been conducted to automatically detect the false information about COVID-19 using a variety of intelligent algorithms in order to control the flow of false information about the novel coronavirus. However, the performance of the forecasting models has varied according to investigations. In this research, we offer an intelligent model for the automatic identification of fake news about healthcare utilizing machine learning and deep learning principles. Simulation and evaluation results on real-life healthcare datasets reveal the efficiency of the suggested model in differentiating between fake news and real news. It achieves an improvement of 6% , for accuracy contrasted with traditional systems.

Keywords: HealthCare · Fake News · Classification · Machine Learning

1 Introduction

Social media usage is growing with time. In 2021, there are over 4.26 billion users on social media worldwide, and by 2027, it is expected that there will be almost 6 billion users [1]. Social media has brought us many benefits like faster and easier communication, brand promotions, customer feedback, etc.; however, it also has several disadvantages, and one of the prominent ones being fake news. Fake news is unarguably a threat to society [2] and it has become a challenging problem for social media users and researchers alike. WHO Director-General Tedros Adhanom Ghebreyesus recently said: "We are not just fighting an epidemic; we are fighting an infodemic" [3]. Fake news in the medical field is a far greater issue since it may lead individuals to believe that the

A. Rocha et al. (Eds.): WorldCIST 2023, LNNS 800, pp. 285–295, 2024.
https://doi.org/10.1007/978-3-031-45645-9_27

news is accurate, which might lead them to act in extreme ways. For instance, ('Alcohol is a cure for COVID-19') led to many deaths and hospitalizations in Iran [4]. This shows how vulnerable we are to fake news in these hard times and how severe the outcome can be if we ignore them. The first step towards tackling fake news is to identify it. We primarily restrict our investigation of the social media content to the topic of the medical news.

Fake news detection is a cumbersome task for two reasons: firstly, it requires a specially trained human to make a clear distinction between fake news and real news, and secondly, a huge velocity, veracity, and diversity of fake news are available on various social media platforms. Due to the lack of efficient skills and human resources, an automatic tool for fake news detection is required specially in medical domain. The variety and velocity of fake news keep changing, so the existing methods fail to detect misinformation consistently [5].

Fake news has existed for a very long time, nearly the same amount of time as news began to circulate widely after the printing press was invented in 1439 [6]. However, there is no agreed definition of the term "fake news". A more specific definition of misinformation "Fake News "is news articles that purposely and demonstrably mislead readers [7]. This definition's authenticity and aim are its two main components. First, incorrect information that can be verified as such is a component of fake news. Second, false news is produced with the malicious goal to deceive consumers. For the purposes of this proposal, false news is defined strictly. Formally speaking, we define (FAKE NEWS) as a news piece that is blatantly and demonstrably untrue.

As we know, In the age where the internet is frequently the main source of information, news audiences are at higher risk than ever of encountering and sharing fake news. Every day, consumers all over the world read, watch, or listen to the news for updates on everything from their favorite celebrity to their preferred political candidate, and often take for granted that what they find is truthful and reliable. Identifying fake news is made more difficult because it is becoming an industry, with individuals paid to write sensationalist stories and create clickbait content to increase site traffic. So, "when it comes to the spread of fake news, social media is the main culprit".

As we know, 2020 and 2021 were marked by the coronavirus pandemic and have changed everyday life in various ways, one of which is, without a doubt, the way people use the internet. There was a significant increase in the average time U.S. users spent on social media in 2020: 65 min daily, compared to 54 min and 56 min the years before. The amount of time spent on social networking is expected to remain stable in the upcoming years [8].

Many tasks can be performed on social platforms and users resort to them for different reasons such as keeping track of news and information about the pandemic. Where, updates on the virus itself seemed to be in high demand at the start of 2020, as of January 2021, information about vaccines was the most sought piece of information. A survey held between October 2020 and September 2021 revealed that 72% were most interested in reading information about a vaccine, and COVID-19 [8].

The remaining of this work is organized as follows. In Sect. 2, review some related works. Section 3 presents the proposed model. In Sect. 4, we report some empirical results. Finally, the conclusion and future work is presented in Sect. 5.

2 Related Work

Many efforts have recently been completed to address the issue of identifying fake news, rumors, misinformation, or disinformation in social media networks. The majority of these works may be classified as supervised or unsupervised learning methodologies. In this paper, we will overview only the supervised learning approaches.

For the supervised approach, a system based on machine learning techniques for detecting fake news or rumors from social media during the COVID- 19 pandemic is presented in [9]. The authors used Twitter's Streaming API to collect one million Arabic tweets. The tweets were evaluated by identifying the subjects mentioned throughout the epidemic, recognizing rumors, and predicting the source. A sample of 2,000 tweets was manually classified as incorrect information, true information, or irrelevant. Support Vector Machine, Logistic Regression, and Naïve Bayes were among the machine learning classifiers used. They were able to identify rumors with 84% accuracy.

Also, a supervised learning strategy for detecting Twitter trustworthiness is proposed in [10]. Five machine learning classifiers were trained using a collection of features that included both content-based and source-based information. When applied with a combined set of features, the Random Forests classifier outperformed the other classifiers. A total of 3,830 English tweets were manually classified as reliable or untrustworthy. The influence of textual characteristics on believability detection was not investigated. Another supervised machine learning strategy to detecting rumors from business evaluations was suggested in [11]. Rumor identification studies were carried out using a publicly accessible dataset. To categorize business evaluations, several supervised learning classifiers were utilized. The experimental findings revealed that the Nave Bayes classifier attained the maximum accuracy and outperformed three other classifiers, namely the Support Vector Classifier, K-Nearest Neighbors, and Logistic Regression. The drawback of this research is the modest size of the dataset utilized to train machine learning classifiers.

Advocated detecting bogus news using n-gram analysis and machine learning algorithms [12]. Two alternative feature extraction approaches and six machine learning algorithms were explored and evaluated based on a dataset from political articles that were acquired from Reuters.com and kaggle.com for real and false news. [13] authors proposed spotting bogus news via social media. They utilized multiple pre-processing methods on the textual data before employing 23 supervised classifiers with TF weighting. The integrated text pre-processing and supervised classifiers were evaluated on three real-world English datasets: BuzzFeed Political News, Random Political News, and ISOT Fake News.

A large corpus for struggle against the COVID-19 infodemic on social media has been proposed in [14]. The authors created a schema that includes numerous categories such as advise, cure, call to action, and asking a question. They saw such categories as beneficial to journalists, legislators, and even the community as a whole. Tweets in Arabic and English were captured in the dataset. Classification tests were carried out utilizing three classifiers and three input representations: word-based, Fast-Text, and BERT. Only 210 of the classified tweets were made public by the authors.

There are various publicly available corpora for fake news identification in different languages to combat the spread of misleading information. [15] describes a multilingual cross-domain fact-checking news collection for COVID-19. The gathered dataset covers 40 languages and is carefully annotated using fact-checked articles from 92 distinct fact-checking websites. The dataset may be found on GitHub. [16] provided another freely accessible dataset named "TweetsCOV19." More than eight million English tweets regarding the COVID-19 epidemic are included in this dataset. The dataset is freely available online and may be used to train and test a wide range of NLP and machine learning approaches. [17] presents a unique Twitter dataset that was created to define COVID-19 misinformation networks. The tweets were divided into 17 categories by the authors, including fake cure, fake treatment, and fake information or prevention. They carried out many tasks on the produced dataset, such as community identification, network analysis, bot detection, socio-linguistic analysis, and vaccination attitude. The limitations of this study include that just one individual did annotation, the results are correlational rather than causal, and the data obtained was only for three weeks. MM-COVID is a multilingual and multidimensional library of false news data [18]. The collection includes 3981 fake news stories and 7192 authentic news stories in English, Spanish, Portuguese, Hindi, French, and Italian. The authors examined the obtained material from many angles, including social activities and social media user profiles.

3 Proposed Model

The proposed model, as shown in Fig. 1, consists of two main steps. Firstly, preprocessing step to discard any noise from the text in the purpose of enhance the accuracy. Secondly, the feature extraction step to extract the features from the text that helps in detecting the healthcare fake news. The workflow of the proposed model is presented in detail below.

3.1 Preprocessing

Most social media datasets contain many unnecessary words such as stop words, misspellings, etc. All of these words affect system performance. In this section, we show how to handle these words and how to clean the datasets.

3.1.1 Tokenization

Tokenization is a pre-processing method that breaks a stream of text into words, phrases, symbols, or other meaningful elements called tokens [19, 20]. The main goal of this step is the investigation of the words in a sentence [21]. For example: **Alcohol is a cuere for #COVID-19!!!**

In this case, the tokens are as follows:

{"Alcohol" "is" "a" "cuere" "for" "#COVID-19" "!!!"}

3.1.2 Spelling Correction

Spelling correction is an optional step in the pre-processing process. Typographical errors are widespread in texts and documents, particularly in social media text collections. In NLP, several algorithms, approaches, and methodologies have been developed to address this issue [22]. For researchers, several strategies and methodologies are accessible, including hashing-based and context-sensitive spelling correction systems [23]. For example, the correction of the word "cuere" is "cure".

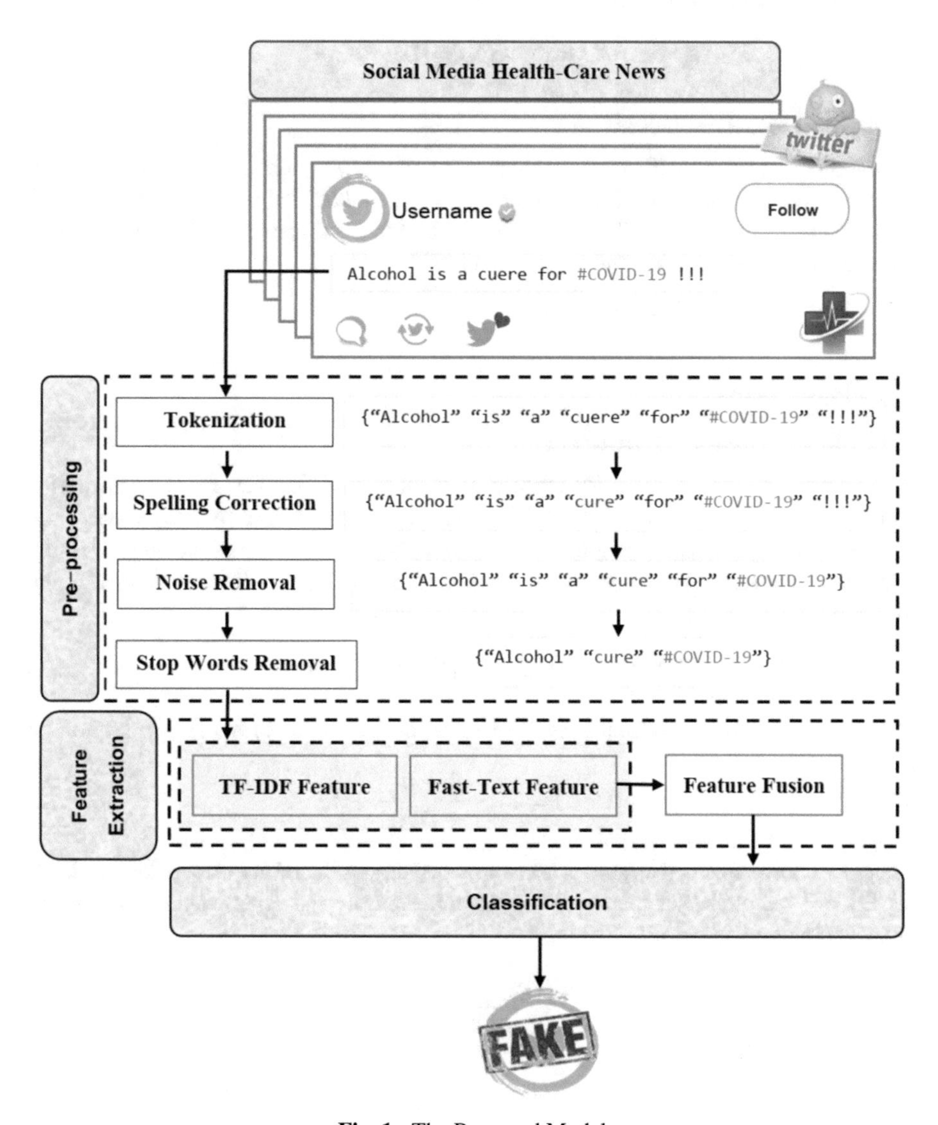

Fig. 1. The Proposed Model

{"Alcohol" "is" "a" "cure" "for" "#COVID-19" "!!!"}

3.1.3 Noise Removal

Most text, document, or tweet datasets contain a lot of extraneous characters like punctuation and special characters. Although critical punctuation and special characters are necessary for human comprehension of papers or tweets, they can be detrimental for classification algorithms [24].

{"Alcohol" "is" "a" "cure" "for" "#COVID-19"}

3.1.4 Stop Words Removal

Text and document classification includes many words which do not contain important to be used in classification algorithms, such as {"a", "about", "above", "across", "after", "afterwards", "again",...}. The most common technique to deal with these words is to remove them from the texts and documents [25].

{"Alcohol" "cure" "#COVID-19"}

3.2 Feature Extraction

3.2.1 TF-IDF Feature Extraction

We will use a bag of word (BoW) representation method to convert each tweet into the corresponding feature vector. The BoW representation consists of two steps: counting, and normalizing. First, we weight each tokenized word "the output of the preprocessing step" using the term frequency-inverse document frequency (TF-IDF) as defined in the following equation:

$$\mathrm{TF}-\mathrm{IDF}(t,d)=\mathrm{TF}(t,d)\times\mathrm{IDF}(t), \tag{1}$$

where TF (*t, d*) is the term frequency of token (*t*) in document (*d*) and IDF is defined as:

$$\mathrm{IDF}(t)=\log\frac{1+n}{1+\mathrm{DF}(t)}+1, \tag{2}$$

where DF(t) represents the number of tweets containing the term t on the dataset. Last, the TF-IDF vector (V) for each tweet document is normalized using L2-norm as defined in the following equation:

$$V=\frac{v}{\sqrt{v_1^2+v_2^2+\ldots v_n^2}} \tag{3}$$

3.2.2 Word Embedding Feature Extraction

Word embedding is a technique of representing a word into a fixed-size vector with the help of contextual information. They preserve the contextual information of each token, unlike the TF-IDF-based method that is purely based on the frequency of words rather than their contexts. The widely used word embedding vectors for English languages are word2vec [26], GloVe [27], and Fast-Text [28]. Herein, we choose Fast-Text-based word embedding in our work because it is an open-source deep learning model pretrained on large Wikipedia corpus on various languages and a recent study on Tweets dataset shows that Fast-Text-based feature extraction method is promising for the classification of Health-Care related tweets [26]. It produces the vector of size 300-D for each word/token. As a result, a matrix of size $n \times 300$ is obtained for each tweet, where n is the total number of tokens present in each input tweet.

$$W = \text{fastText}(d), \tag{4}$$

where W denotes the word embedding matrix ($n \times 300$) obtained from Fast-Text-based embedding (fs) for tweet dataset d.

3.2.3 Feature Fusion

Similar to the role of content and context features in scene image representation more accurately as in [30], the role of syntactic and contextual information is also complementary to each other to represent the tweets more accurately. The TF-IDF method captures the syntactic information of tokens, whereas the Fast-Text-based method captures the contextual information. Given the efficacy of both kinds of information to better represent each tweet, we propose to combine them as suggested by the authors in [31, 32], for the performance improvement as shown in (5).

$$\mathrm{H}_{ij} = \Sigma_{k=1}^{n} V_{ik} W_{kj}, \tag{5}$$

where H_{ij} is the final feature matrix, V_{ik} is TF-IDF tweet matrix ($m \times n$), and W_{kj} is a Fast-Text-based word embedding matrix ($n \times 300$). Note that m and n represent the number of tweets and number of tokens, respectively. The computational complexity of our hybrid feature is mainly based on feature fusion procedure, which is determined by the multiplication cost of two matrices (V_{ik} of size $m \times n$ and W_{kj} of size $n \times 300$). Hence, the total time complexity for feature fusion is $O(m \times n \times 300)$.

3.3 Classification

The existing studies have found that the Deep Learning models (DL) is performing better as compared to the traditional Machine Learning models (ML) [33, 34]. So, in this paper we used convolutional neural network (CNN) which is a discriminative supervised DL model that automatically extracts features from the input data [35]. The CNN is an enhancement to the conventional artificial neural networks. Each of the layers of the CNN considers optimal hyper parameter settings to produce acceptable results, while reducing the complexity of the model. The CNN also supports a mechanism to handle the overfitting of the model which may not be found in traditional neural networks [36].

4 Experimental Results

The evaluation metrics are derived from the four parameters given in Table 1, based on the predicted class label and actual class label. We used Four evaluation metrics to evaluate the output of the proposed models. A brief detail of the evaluation metrics is discussed in Table 1. Accuracy is the fraction of the correct predictions made by the classifier. It is given by Eq. (6). Precision is used to evaluate the effectiveness of the classifiers. It is given by Eq. (7). Recall is known as sensitivity/TP rate, is the ratio between the TP predictions to the total number of TP and FN predictions made by the classifier. It is given by the Eq. (8). f-Measure is the harmonic mean of precision and recall; thus, it is a combination of two metrics in a single measure. It indicates the performance of the classifiers in the case of imbalanced datasets. It is given by the Eq. (9).

$$Accuracy = \frac{TP + TN}{TP + TN + FP + FN} \tag{6}$$

$$Precision = \frac{TP}{TP + FP} \tag{7}$$

$$Recall = \frac{TP}{TP + FN} \tag{8}$$

$$f - Measure = 2 * \frac{Precision * Recall}{Precision + Recall} \tag{9}$$

Table 1. Detail of the Evaluation Metrics.

Parameter	Actual Class Label	Predicted Class Label
True Positive (TP)	True	True
True Negative (TN)	Fake	Fake
False Positive (FP)	Fake	True
False Negative (FN)	True	Fake

Table 2. Shows the performance of the classifiers on the Covid-19 Kaggle Fake News dataset in terms of Accuracy, precision, recall and f-measure. It is found that the proposed model achieves more stable and accurate results in comparison with the aforementioned classifiers. As shown in Fig. 2, the proposed model achieved higher accuracy and precision than other approaches.

Table 2. Classification performance of different models

	Naïve Bayes	KNN	Proposed Model
Accuracy	75%	76%	81%
Precision	82%	77%	83%
Recall	79%	90%	77%
f-Measure	81%	83%	80%

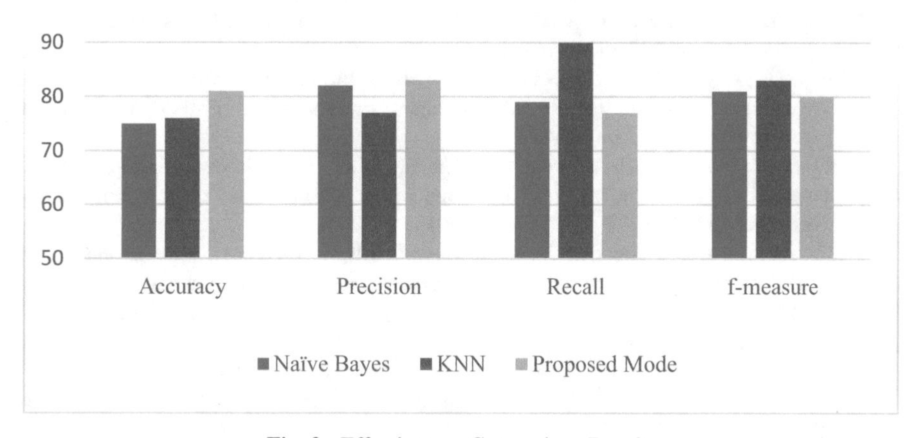

Fig. 2. Effectiveness Comparison Results

5 Conclusion and Future Work

In this paper, we have proposed an approach for the detection of Healthcare Fake News. We have used the Covid-19 Kaggle's Fake News dataset. We used TF-IDF and Fast-Text Features for extracting features after the preprocessing step, then we used CNN to detect if the news is fake or not. Finally, the proposed model achieved higher accuracy and precision than other approaches. In future work, we will try to enhance the sensitivity of our model besides we expect to enhance the accuracy of the model by adding a feature selection step to choose only the core and powerful features that can help in the detection of fake news. Finally, it should be noted that the proposed model is not exclusive only to the healthcare dataset, it also can be used in detecting fake news in various domains, which opens the area to other researchers to do their best in this field.

References

1. Cement, J.: Number of social media users 2027. https://www.statista.com/statistics/278414/number-of-worldwide-social-network-users/. Accessed 06 Dec 2022
2. Panke, S.: social media and fake news. AACE (2020). https://www.aace.org/review/social-media-and-fake-news/

3. Mesquita, et al.: Infodemia, fake news and medicine: science and the quest for truth. Int. J. Cardiovasc. Sci. **33**(3), 203–205 (2020)
4. Karimi, N., Gambrell, J.: Hundreds die of poisoning in Iran as fake news suggests methanol cure for virus. Times of Israel (2020). https://www.timesofisrael.com/hundreds-die-of-poisoning-in-iran-as-fake-news-suggests-methanol-cureforvirus/
5. Kishore, S.G., Nandini, D.: FakeCovid--A multilingual cross-domai-fact check news dataset for COVID-19 (2020)
6. http://www.politico.com/magazine/story/2016/12/fakenews-history-long-violent-214535
7. Kai, S., et al.: Fake news detection on social media: a data mining perspective. ACM SIGKDD Explor. Newslett. **19**(1), 22–36 (2017)
8. Social media use during COVID-19 worldwide - statistics & facts, Published by Statista Research Department, May 19, 2021. https://www.statista.com/topics/7863/socialmedia-use-during-coronavirus-covid-19-worldwide/
9. Alsudias, L., Rayson, P.: Covid-19 and arabic twitter: How can arab world governments and public health organizations learn from social media? In: Proceedings of the 1st Workshop on NLP for COVID-19 at ACL (2020)
10. Hassan, N.Y., Gomaa, W.H., Khoriba, G.A., Haggag, M.H.: Supervised learning approach for twitter credibility detection. In: 2018 13th International Conference on Computer Engineering and Systems (ICCES). IEEE, pp. 196–201 (2018)
11. Habib, A., Akbar, S., Asghar, M.Z., Khattak, A.M., Ali, R., Batool, U.: Rumor detection in business reviews using supervised machine learning. In: 2018 5th International Conference on Behavioral, Economic, and Socio-Cultural Computing. IEEE, pp. 233–237 (2018)
12. Ahmed, H., Traore, I., Saad, S.: Detection of online fake news using n-gram analysis and machine learning techniques. In: Traore, I., Woungang, I., Awad, A. (eds.) ISDDC 2017. LNCS, vol. 10618, pp. 127–138. Springer, Cham (2017). https://doi.org/10.1007/978-3-319-69155-8_9
13. Ozbay, F.A., Alatas, B.: Fake news detection within online social media using supervised artificial intelligence algorithms. Physica A **540**, 123174 (2020)
14. Alam, F., et al.: Fighting the covid-19 infodemic in social media: a holistic perspective and a call to arms (2020). arXiv preprint arXiv:2007.07996
15. Shahi, G.K., Nandini, D.: Fakecovid–a multilingual cross-domain fact check news dataset for covid-19 (2020). arXiv preprint arXiv:2006.11343
16. Dimitrov, D., et al.: Tweetscov19-a knowledge base of semantically annotated tweets about the covid-19 pandemic. In: Proceedings of the 29th ACM International Conference on Information & Knowledge Management, pp. 2991–2998 (2020)
17. Memon, S.A., Carley, K.M.: Characterizing covid-19 misinformation communities using a novel twitter dataset (2020). arXiv preprint arXiv:2008.00791
18. Li, Y., Jiang, B., Shu, K., Liu, H.: Mm-covid: A multilingual and multidimensional data repository for combatingcovid-19 fake new (2020). arXiv preprint arXiv:2011.04088
19. Gupta, G.; Malhotra, S. Text Document Tokenization for Word Frequency Count using Rapid Miner (Taking Resume as an Example). Int. J. Comput. Appl. (2015)
20. Verma, T., Renu, R., Gaur, D.: Tokenization, and filtering process in RapidMiner. Int. J. Appl. Inf. Syst. **7**, 16–18 (2014)
21. Aggarwal, C.C.: Machine Learning for Text. Springer, Berlin/Heidelberg, Germany (2018).https://doi.org/10.1007/978-3-319-73531-3
22. Mawardi, V.C., Susanto, N., Naga, D.S.: Spelling Correction for Text Documents in Bahasa Indonesia Using Finite State Automata and Levinshtein Distance Method. EDP Sci. (2018)
23. Dziadek, J., Henriksson, A., Duneld, M.: Improving terminology mapping in clinical text with context-sensitive spelling correction. In: Informatics for Health: Connected Citizen-Led Wellness and Population Health; IOS Press, Netherlands, vol. 235, pp. 241–245 (2017)

24. Pahwa, B., Taruna, S., Kasliwal, N.: Sentiment analysis-strategy for text pre-processing. Int. J. Comput. Appl. **180**, 15–18 (2018)
25. Saif, H., Fernández, M., He, Y., Alani, H.: On stopwords, filtering and data sparsity for sentiment analysis of twitter. In: Proceedings of the 9th International Conference on Language Resources and Evaluation, Iceland (2014)
26. Mikolov, T., Chen, K., Corrado, G., Dean, J.: Efficient estimation of word representations in vector space. In: Proceedings of the Workshop at ICLR, USA (2013)
27. Pennington, J., Socher, R., Manning, C.D.: GloVe: global vectors for word representation. In: Proceedings of the 2014 Conference on Empirical Methods in Natural Language, Processing, pp. 1532–1543, Qatar (2014)
28. Bojanowski, P., Mikolov, T.: Enriching word vectors with subword information. Trans. Assoc. Comput. Linguist. **5**, 135–146 (2017)
29. Sitaula, C., Basnet, A., Mainali, A., Shahi, T.B.: Deep learning-based methods for sentiment analysis on Nepali covid-19-related tweets. Comput. Intell. Neurosci. **2021**, 2158184 (2021)
30. Sitaula, C., Xiang, Y., Aryal, S., Lu, X.: Scene image representation by foreground, background and hybrid features. Expert Syst. Appl. **182**, 115285 (2021)
31. De Boom, C., Van Canneyt, S., Demeester, T., Dhoedt, B.: Representation learning for very short texts using weighted word embedding aggregation. Pattern Recogn. Lett. **80**, 150–156 (2016)
32. Onan, A., Toçoglu, M.A.: Weighted word embeddings and clustering-based identification of question topics in MOOC discussion forum posts. Comput. Appl. Eng. Educ. **29**(4), 675–689 (2020)
33. LeCun, Y., Bengio, Y., Hinton, G.: Deep learning. Nature **521**(7553), 436–444 (2015)
34. Mikolov, T., Deoras, A., Povey, D., Burget, L., Cernocky, J.: Strategies for training large scale neural network language models. In: Proceedings of the Automatic Speech Recognition and Understanding, USA (2011)
35. LeCun, Y., Bottou, L., Bengio, Y., Haffner, P.: Gradient-based learning applied to document recognition. Proc. IEEE **86**(11), 2278–2324 (1998)
36. Aurélien, G.: Hands-on machine learning with Scikit-Learn, Keras, and TensorFlow. O'Reilly Media, Inc., (2022)

Buildings and Vehicles Discrimination in Drone Images

António J. R. Neves[1(✉)], Rita Amante[1], Daniel Canedo[1], and José Silvestre Silva[2]

[1] DETI and IEETA, University of Aveiro, Aveiro, Portugal
{an,rita.amante,danielduartecanedo}@ua.pt
[2] Portuguese Military Academy and CINAMIL and LIBPhys-UC, Lisboa, Portugal
jose.silva@academiamilitar.pt

Abstract. The aim of this work is to analyze the performance of Deep Learning algorithms for detecting vehicles and buildings in aerial images. Two algorithms, Faster R-CNN and YOLO, are compared to identify the best performance. The results showed that there is a considerable discrepancy between the two algorithms, both in terms of performance and speed. Faster R-CNN only proved to be superior in terms of training speed, but YOLO achieved the best results.

Keywords: Deep Learning · Transfer Learning · Computer Vision · Object Detection

1 Introduction

The facility with which Unmanned Aerial Vehicle (UAV) access hard-to-reach places, in a mobile way, has boosted their use in different application areas, both in a civil and military context. At the civil level, they have several benefits in areas such as surveillance, public and private security, disaster assistance, agriculture and the environment, among others [1]. They are of enormous practical interest in a military context, especially for the defense sector, as they help to predict enemy movements and plan preventive measures, saving time and effort [2].

Object detection in aerial images is a recent topic, with a very significant growth outlook, making its study complex and interesting [3]. Vehicle detection, specifically, can provide several real-world applications, such as road traffic monitoring, parking management and control, road accidents, screening for illegal activities, operations to support government agencies and surveillance of enemy troops. Building detection can be useful for urban planning, roof monitoring and control of illegal constructions.

There is an increasing interest in the study of vehicle and building detection in aerial images obtained by UAV. Two Deep Learning (DL) algorithms, well known in the literature for object detection, are studied: Faster Region-based Convolutional Neural Network (Faster R-CNN), a two-stage algorithm, and You Only Look Once (YOLO), a one-stage algorithm. Regarding the YOLO algorithm, two of its versions will be studied, YOLOv3, which is one of the most mentioned algorithms in literature, and YOLOv5l, which is the most recent version.

A. Rocha et al. (Eds.): WorldCIST 2023, LNNS 800, pp. 296–303, 2024.
https://doi.org/10.1007/978-3-031-45645-9_28

2 Literature Review

In the last 20 years, progress has been made in object detection, with two important periods being highlighted: traditional object detection period (before 2014) and deep learning-based detection period (after 2014) [4]. During the first period, the features base used in object detection algorithms was artisanal, where the image features were manually extracted (color, texture, contours, among others) and then the classifier was trained. The second period can be divided into two-stage and one-stage.

Faster R-CNN Algorithm

Since the R-CNN performs excessive calculations on numerous proposals, resulting in a slow detection speed, the selective search algorithm is not flexible, and training is time-consuming. The CNN is responsible for extracting the characteristics of all the RoI created. Later, to overcome some shortcomings of the R-CNN, the Fast Region-based Convolutional Neural Network (Fast R-CNN) algorithm appeared, where, unlike the R-CNN, the full image is provided to the CNN to create a feature map [5].

However, Fast R-CNN has several limitations such as a high calculation time and a complicated and time-consuming approach [6]. Then came the Faster R-CNN, proposed to overcome the limitations of the previous algorithms. The latter's main contribution was the introduction of the Region Proposal Network (RPN) [4]. In Faster R-CNN, the full input image is provided to the CNN, which learns and identifies the RoI, combined with a parallel network [7]. The Faster R-CNN detection process follows several steps. Firstly, it extracts a feature map from images; secondly, it generates hypotheses from the feature map, determining the approximate coordinates; thirdly, it compares the hypothesis coordinates with the feature map, using RoI and, finally, it classifies the hypotheses, updating the coordinates [8].

YOLO Algorithm

YOLO uses a fully convolutional neural network which performs the object recognition and localization step at the same time and returns the limiting box position and its class directly in the output layer [9]. Initially, YOLO resizes the input image to a specific size, which is then sent to the CNN system, and finally, the network prediction results are processed to detect the target [10].

The third version, YOLOv3 uses Darknet-53 as backbone for resource extraction and has 53 convolutional layers and 23 residual layers. The novelty of its architecture is that it does not contain a pooling layer and uses three-scale feature maps to predict the position of the object [9]. YOLOv3 has two main problems. The first is due to the detector not using rewritten limiting boxes for the training process. The second happens when the centers of two limiting boxes are present in the same cell, and one can be replaced by the other [10].

The latest version, YOLOv5 was developed from the open-source PyTorch library and is considered the lightest version of the YOLO algorithms. It uses CSPDarknet53 as a backbone and its structure replaced the first three layers of the YOLOv3 algorithm, reducing Compute Unified Device Architecture (CUDA) memory needed and increasing direct propagation and backpropagation [11].

3 Methodology

The choice of "deep learning based detection period" algorithms was included in this first step, once that traditional methods are slow and inefficient compared to the more recent methods based on CNNs. Of these, a two-stage (Faster R-CNN) and a one-stage (YOLO) algorithm were chosen as the most representative algorithms used in this type of problem. Regarding YOLO, two versions were used, YOLOv3 and YOLOv5l.

The steps taken to achieve the final results are shown in Fig. 1.

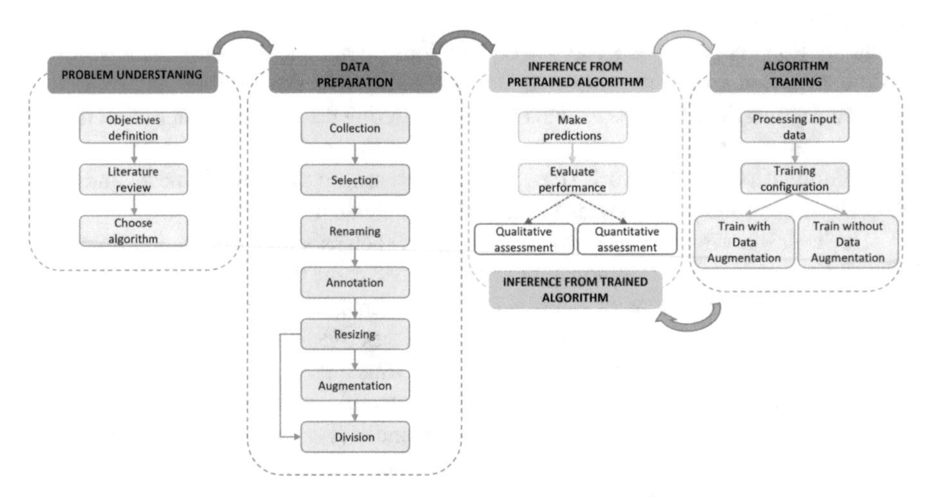

Fig. 1. Diagram of the methodology used to detect vehicles and buildings in aerial images.

The second step was based on the preparation of data. It started by selecting the images, annotated and resized from a private dataset made by Portuguese Military Academy. Subsequently, two datasets were created, with and without data augmentation, called Portuguese Military Academy (PMA) and Portuguese Military Academy with Data Augmentation (PMA-DA), respectively. Finally, both datasets were divided into three subsets: training, validation and testing.

The third step included the inference process of the pretrained algorithms, using the pretrained network as a classifier approach of TL. The performance of the algorithms was assessed qualitatively, based on the observation of the predictions obtained for a new image, and quantitatively, where the detection times were analyzed.

In the fourth step, the input data were processed, some parameters were configured and the algorithms were trained. The following were the adjusted training parameters: batch size, epochs, steps, learning rate. The Faster R-CNN algorithms are configured with the step parameter, while YOLO is configured with the epoch parameter.

To compare the algorithms, metrics were computed to assess their quality. Among the various existing metrics, the following were used to evaluate the algorithms: IoU, Precision, Recall and mean Average Precision (mAP). Precision and Recall were evaluated for just a single 50% IoU value for each class, named P@0.5 and R@0.5, respectively. The mAP was evaluated for a single value of 50% IoU, named by mAP@0.5.

4 Results and Discussion

The implementation of the Faster R-CNN algorithm was made with the TensorFlow 2 Object Detection API. For the implementation of the YOLO algorithm, for both version 3 and version 5, the open source developed by Ultralytics was used. The chosen algorithms were pretrained on the MS COCO dataset.

4.1 Dataset

The images used was provided by the Portuguese Military Academy, consisting of two sets of private images obtained by UAV, with 146 RGB images and 227 IRG images, for a total of 373 images with a size of 3000 $\times$ 4000 per image. To standardize the input data set for the implemented algorithms, the images were resized to 640 $\times$ 640 pixels.

To create a more effective dataset, the two sets of RGB and IRG images were joined, resulting in the PMA Dataset. Of the 373 images provided, 10 images were discarded, since they were excessively unfocused.

The tool LabelImg, written in Python, was used for the annotation of the images.

Two classes of objects were created: vehicle and building. The number of object instances of each class, from a total of 11945 instances, resulted in 9500 (79.53%) of vehicles and 2445 (20.47%) of buildings. The application of data augmentation resulted in a new dataset, named PMA-DA, which is three times larger than the PMA dataset.

The division was executed with 80% of the images of the dataset were used for training, 10% were used for validation and the remaining 10% were used for testing.

4.2 Inference from Trained Algorithms

The Faster R-CNN algorithm was trained for 14500 and 29000 steps, equivalent to 100 and 200 epochs, respectively, with the PMA dataset; also trained for 43500 steps, equivalent to 100 epochs with the PMA-DA dataset. The analysis of Table 1 revealed that the in both classes, vehicles and buildings, the Precision value was high (above 91%) and the Recall value was low (less than 50%), which indicates that most objects were correctly detected, but with detection losses. Comparing the results of 14500 steps PMA and 43500 steps PMA-DA, it was observed that without data augmentation the algorithm obtained a superior performance.

Table 1. Faster R-CNN algorithm training results.

		PMA Dataset		PMA-DA Dataset
		14500 steps	29000 steps	43500 steps
Training time	(hours)	8	16	25
mAP	(all)	58.1%	56.2%	52.5%
Precision	(vehicle)	91.9%	93.7%	**94.8%**
Recall		50.4%	47.9%	58.2%
Precision	(building)	**98.9%**	97.7%	97.2%
Recall		47.0%	**44.5%**	50.1%

The results of training the YOLOv3 algorithm, for 100 and 200 epochs with the PMA dataset and for 100 epochs with the PMA-DA dataset. The results shown in Table 2 revealed that the algorithm training time was slow for any epoch value. Precision and Recall values, for both classes, were high, above 91% and 76%, respectively, which indicates most objects were correctly detected. Training the algorithm with the PMA-DA dataset, compared to the PMA dataset, presented worse performance. The best results of the YOLOv3 algorithm were achieved in the training of 200 epochs.

Table 2. YOLOv3 algorithm training results.

		PMA Dataset		PMA-DA Dataset
		100 steps	200 steps	100 steps
Training time	(hours)	30	60	127
mAP	(all)	91.4%	**92.2%**	88.7%
Precision	(vehicle)	**94.1%**	93.2%	91.9%
Recall		91.1%	92.3%	88.9%
Precision	(building)	93.7%	94.5%	**97.1%**
Recall		82.3%	85.4%	**76.0%**

The YOLOv5l algorithm, it was also trained for 100 and 200 epochs with the PMA dataset and for 100 epochs with the PMA-DA dataset. The results presented in Table 3, show that in both classes, the Precision and Recall values were high, above 92% and above 81%, respectively, which means that most objects were correctly detected. When comparing the training performance of 100 epochs for the PMA and PMA-DA datasets, it was noticed that the algorithm performed better without data augmentation. The best mAP value obtained was 93.3% for the training of 100 epochs with the PMA dataset.

Table 3. YOLOv5l algorithm training results.

		PMA Dataset		PMA-DA Dataset
		100 steps	200 steps	100 steps
Training time	(hours)	24	48	73
mAP	(all)	**93.3%**	91.8%	91.4%
Precision	(vehicle)	**95.4%**	92.8%	93.4%
Recall		90.8%	91.6%	91.4%
Precision	(building)	96.0%	**97.2%**	96.4%
Recall		85.4%	91.5%	81.6%

Figure 2 illustrates the detection results of two images, RGB on the left and IRG on the right for each algorithm. To visualize the results, vehicles were identified in blue and buildings in yellow. Through a direct observation of the images, relative to the Faster R-CNN, the presence of several false negatives and the existence of limiting boxes with a localization deficit were obvious. For example, some boxes did not include the object in its entirety. The YOLOv3 algorithm presented better detections compared to Faster R-CNN, being able to detect practically all target objects. However, it still had some problems with false positives and negatives. As for the YOLOv5l algorithm, it presented the best detection results, mainly due to the reduction of false negatives.

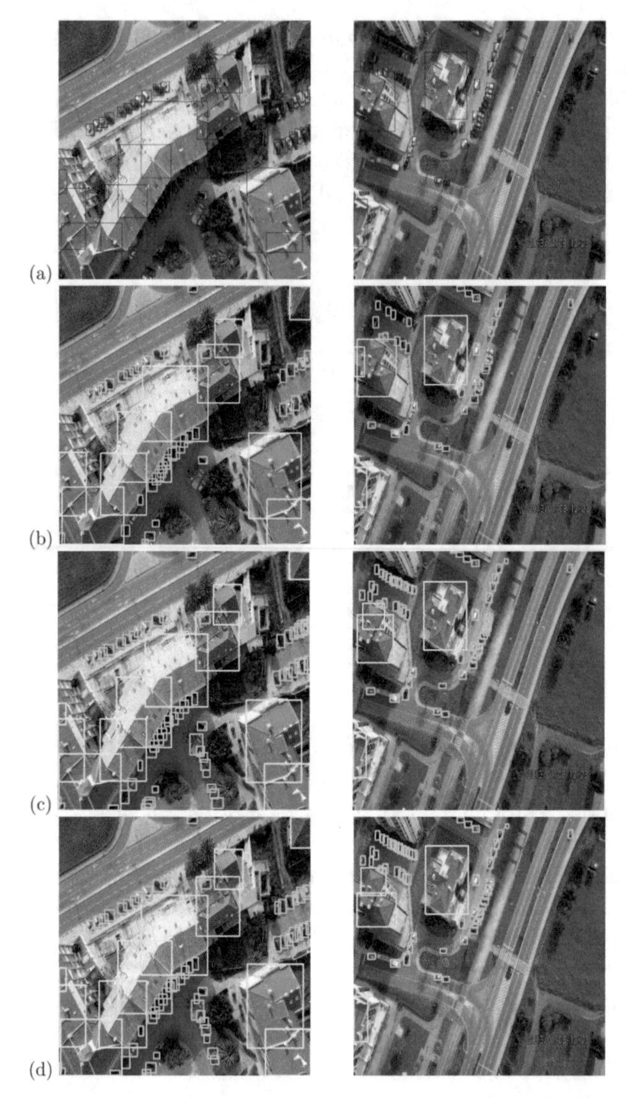

Fig. 2. Detection results of the trained algorithms, for an RGB image (left) and an IRG image (right): (a) original annotations; (b) Faster R-CNN; (c) YOLOv3 and (d) YOLOv5l.

5 Conclusions

The adopted methodology included data preparation, annotation, resizing, data augmentation and division of the PMA and PMA-DA datasets. Also the inference of pretrained algorithms with the MS COCO dataset, and performance evaluations of the algorithms using training time, evaluation metrics, and detection time.

Based on the results, it was found that the data augmentation technique did not produce improvements in the quality of the algorithms, possibly because there are new added difficulties and the algorithms would need more training time to learn.

The number of object instances in an image and the number of epochs or steps did not influence the detection time; the building class showed a higher Precision than the vehicle class, indicating that the algorithms are more accurate in detecting larger objects. The building class showed a lower Recall than the vehicle class, probably because there are fewer instances of this class in the PMA and PMA-DA data sets.

The YOLOv3 algorithm, despite presenting a relatively fast detection time and a performance exceeding that of Faster R-CNN, could not surpass YOLOv5l; the YOLOv5l algorithm was the one which achieved the best results, with the best mAP of 93.3%, the highest percentage of Precision and Recall, as well as standing out for having a faster detection time. Thus, it is concluded that the YOLOv5l, presented the best performance for the detection of vehicles and buildings in aerial images.

References

1. Srivastava, S., Narayan, S., Mittal, S.: A survey of deep learning techniques for vehicle detection from UAV images. J. Syst. Architect. **117**, 102152 (2021)
2. Kamran, F., Shahzad, M., Shafait, F.: Automated military vehicle detection from low-altitude aerial images. In: 2018 Digital Image Computing: Techniques and Applications (DICTA), pp. 1–8. IEEE (2018)
3. Budiharto, W., et al.: Fast object detection for quadcopter drone using deep learning. In: 2018 3rd International Conference on Computer and Communication Systems (ICCCS), pp. 192–195. IEEE (2018)
4. Zou, Z., Shi, Z., Guo, Y., Ye, J.: Object detection in 20 years: A survey (2019). arXiv preprint arXiv:1905.05055
5. Girshick, R.: Fast R-CNN. In: Proceedings of the IEEE International Conference on Computer Vision, pp. 1440–1448 (2015)
6. Shetty, A.K., et al.: A review: object detection models. In: 2021 6th International Conference for Convergence in Technology (I2CT), pp. 1–8. IEEE (2021)
7. Ren, S., He, K., Girshick, R., Sun, J.: Faster R-CNN: towards real-time object detection with region proposal networks. In: Advances in Neural Information Processing Systems, vol. **28** (2015)
8. Saetchnikov, I.V., Tcherniavskaia, E.A., Skakun, V.V.: Object detection for unmanned aerial vehicle camera via convolutional neural networks. IEEE J. Miniaturization Air Space Syst. **2**(2), 98–103 (2020)
9. Lin, F., Zheng, X., Wu, Q.: Small object detection in aerial view based on improved Yolo v3 neural network. In: 2020 IEEE International Conference on Advances in Electrical Engineering and Computer Applications (AEECA), pp. 522–525. IEEE (2020)
10. Ding, W., Zhang, L.: Building detection in remote sensing image based on improved YOLOV5. In: 2021 17th International Conference on Computational Intelligence and Security (CIS), pp. 133–136. IEEE (2021)
11. Nepal, U., Eslamiat, H.: Comparing YOLOv3, YOLOv4 and YOLOv5 for autonomous landing spot detection in faulty UAVs. Sensors **22**(2), 464 (2022)

Information Technologies in Education

AI Chatbot for Teachers for an LMS Platform

Camilo Rivillas Cardona[1,2], Gabriel M. Ramirez V[3], Jaime Diaz[4], and Fernando Moreira[5,6](✉)

[1] Escuela de Ingeniería, Universidad Internacional de la Rioja, Av. de la Paz, 137, 26006 Logroño, La Rioja, Spain
[2] Nusoft, Transversal 65A #32-56, Rionegro, Antioquia, Colombia
camilorivillas@nusoft.com.co
[3] Facultad de Ingeniería, Universidad de Medellín, Cra. 87 #30-65, Medellín, Colombia
gramirez@udemedellin.edu.co
[4] Departamento Cs. De la Computación e Informática, Universidad de la Frontera, Temuco, Chile
jaimeignacio.diaz@ufrontera.cl
[5] REMIT. IJP, Universidade Portucalense, Porto, Portugal
fmoreira@uportu.pt
[6] IEETA, Universidade de Aveiro, Aveiro, Portugal

Abstract. This work consists of creating a Chatbot agent using RASA, Python, and PHP to automate the teachers' assistance in the learning and education management platform Ciudad Educativa, which is focused on primary and secondary education. A knowledge base has been created in RASA; this knowledge base has 159 intentions. It comprises 131 frequently asked questions and 28 management and control intentions to train the machine learning model. Primary and secondary schools significantly vary in parameterization as well as pedagogical and educational methodologies. This variability is identified according to the contexts and intent combined to provide adequate assistance to the user. The Chatbot has been integrated with the chat module of the same platform. It was evaluated by a sample of teachers who rated 82.9% of the answers presented by the Chatbot to the questions as applicable.

Keywords: chatbot · education · learning platforms · artificial intelligence · interactive techniques

1 Introduction

Nowadays, organizations that provide educational services commonly have computer systems that allow them to manage their processes: Teaching, learning, content, methodologies, official records, resources, and students, among others. A modality of use of this type of system is the software as a service (SaaS) through the Internet.

Ciudad Educativa (CE) is one of these platforms; CE offers school and learning management. This platform is manufactured by the company Núcleo Software [1, 2]. Its clients are mainly official and private primary and secondary schools in Colombia.

A. Rocha et al. (Eds.): WorldCIST 2023, LNNS 800, pp. 307–317, 2024.
https://doi.org/10.1007/978-3-031-45645-9_29

In the present work, we have explored the various market options for implementing a chatbot that can assist CE teachers by solving frequently asked questions. After defining the platform's requirements, RASA [3], in its open-source version, has been selected as the framework for constructing the Chatbot; the Chatbot is called EdyBot.

Edybot has been implemented as a microservice connected to CE. Teachers can access it from the contact list of the 'personal messaging system' (PMS) already in CE.

During the development, some difficulties presented by the framework related to the management of intentions, the parameterization of certainty and ambiguity thresholds, the prediction of intentions when grouped into categories, and the use of logical operators in conditional responses have been solved.

Finally, tests have been conducted with a sample of teachers to verify whether such a solution is adequate to assist them in school management tasks. The model had an accuracy of 95.44%. The participants marked 82.9% of the answers received as valuable, and 87% of them stated that they would recommend using EdyBot to other teachers.

2 Background

2.1 Educational Management

The platform has most of its clients in Colombia [1]. According the Colombian Ministry of National Education (MEN), a teacher must diagnose, plan, communicate, execute and evaluate teaching processes as part of their pedagogical duties. These steps are framed in their institution's educational project. At the end of each cycle, the teacher must analyze and communicate the results of his students [4].

The teacher is in charge of a certain number of courses and, in addition to their academic responsibilities, there are other tasks to be performed, such as vocational guidance, participation in community spaces, dialogue with parents, professional updating, and cultural and sports activities, among others [5].

The actual functions of a teacher can be much more numerous. Still, the development of EdyBot, has only considered the activities of planning, evaluation, communication, and results management [6].

These processes are manifested in two scenarios: The learning management and the administration of the educational organization [7]. For both, different systematization tools are needed. Platforms for education cover a vast spectrum of products [8–14], among which are 'Learning Management Systems (LMS), 'School Administration Systems' (SIS) and 'Learning Record Stores (LRS). CE can categorize as an LMS platform that includes the functionalities of SIS and LRS.

2.2 Chatbot

Chatbots (also called conversational agents, virtual conversational assistants, etc.) are computer programs that attempt to simulate human conversations via text or voice and supported by Artificial Intelligence (AI) and Natural Language Processing (NLP) [15, 16]. The simplest type of chatbot would be one able to execute an action or issue a response after user's utterances or questions.

In 2015 the University of Montreal used a neural network to overcome previous models' inability to take into account information from interactions before the last one [17]. Detailed research sought an algorithm to generate context-sensitive smooth responses, but the model was found to have the problem of lack of consistency [18].

In 2016, with the emergence of RNN (recurrent neural networks), LSTM (long short-term memory), and reinforcement learning in general, some chatbots are able, to some extent, to maintain a fluid conversation, taking into account the data previously provided in that conversation [19], giving rise to experiments that at the time constituted a breakthrough such as Kuki.ai [20] formerly known as Mitsuku from Pandora Bots [21].

When studying the offerings of NLP tools providers' vendors like Microsoft Luis [22], Amazon Lex [23], and Google DialogFlow [24], it is interesting to note that, despite accelerated advances, the benefits obtained through APIs for natural language understanding and generation commercially viable for use in chatbots are not very different from those achieved with statistical keyword analysis and conversation trees. That is, there are not yet commercial chatbots capable of having a smooth and sensible conversation that is not constrained by strict input rules.

2.3 Chatbot Learning

While using CE and getting to understand its tools, a teacher is a learner. Therefore, in this section it will be necessary to consider the teacher as a student.

In 2001, the term "Technology mediated learning" (TML) was defined, which consists of environments in which the learner interacts with learning resources, content, teachers, and other learners using information systems [25].

The TML concept is refined in the "Chatbot-mediated learning" (CML) specification, a learning method in which the learner communicates with one or more virtual agents to consume educational content [26].

Several studies present the impact obtained with the use of CML considering the characteristics of the student. However, regardless of these characteristics, if the student feels an affinity with the use of chatbots, chatbots help to improve self-confidence, as they provide greater control over learning [26, 27].

Some analyses have been conducted to identify how the medium in which educational chatbots are deployed predicts success or failure. For a chatbot to be successful in the academic environment, it is essential that it is accessible and fast and that the student does not have to access multiple applications. Chatbots for education should be available in the media that the student commonly uses, such as social networks and messaging systems, or directly in the LMS, as this is the site where students perform their educational activities [28–31].

2.4 LMS and Chatbot

Some platforms were found to have implemented chatbots in their service offerings designed to provide answers to frequently asked questions from their users [32–34]. For example, Canvas has a proprietary development integrated into its support system, which is currently in the testing stage [35]. PowerSchool uses an external AI provider to implement its knowledge-based chatbot [36]. Blackboard, on the other hand, offers its customers to be the ones to implement the FAQs, and thus, the institutions can provide virtual assistance to their users just as the LMS platform does [37].

In the case of Moodle, the most widely used open-source LMS in the world, multiple sources were found that explain how integrate chatbots. One example is by Farhan Karmali, who demonstrates the implementation of Google DialogFlow in Moodle [38].

3 Chatbot Development

3.1 General Description

EdyBot was designed to answer the most common doubts of CE teachers. For example, it is expected to help the user with questions/expressions such as 'Why can't I create lessons?', 'What does it mean that an activity is associated with a test?', among others. Expressions with the same purpose make up a group called 'Intent.'

The provided answer may depend on the context of the teacher. Therefore, an intention may be associated with several solutions. To deliver the most accurate answer, the PMS sends EdyBot that context as a vector which includes information about current module, course, evaluation term, general system configuration, user role and resources.

EdyBot has been trained using 2037 common user utterances obtained from messages received by CE's technical support department, which have been classified into 159 different intentions, some of them with more than one response.

In order to facilitate the training of the model, a module for managing intentions, has been developed within CE, which allows adding records as users express themselves in ways that the artificial intelligence model does not yet know or ask unknown questions.

3.2 Components of EdyBot

The following diagram presents the components of the solution: CE connects with the PMS and this, in turn, interacts with EdyBot. In this diagram, EdyBot represents the front end of the conversational agent, while Chatbot represents the back end (Fig. 1).

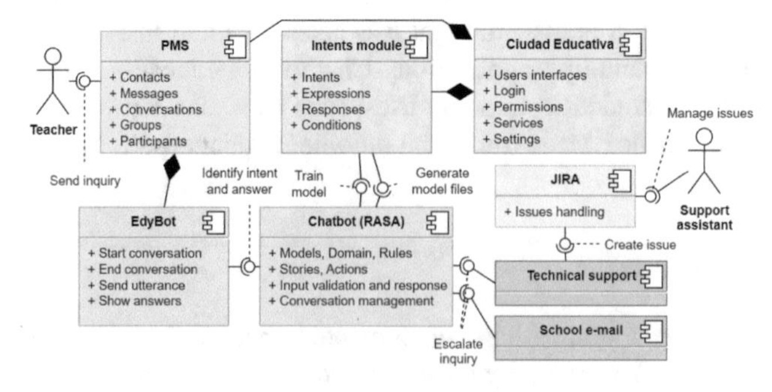

Fig. 1. Components of EdyBot

A diagram showing the EdyBot components is presented. The components related to the chatbot are shown in green and those associated with CE are shown in blue.

The solution consists of the chatbot built in RASA [3], a module for intention management, and EdyBot, which will interface the chatbot and the teacher. In the RASA code, some complementary classes perform operations such as sending e-mails, recording conversations, receiving the context of the current user, and saving the effectiveness of the responses, among others.

EdyBot gives the teacher the option to escalate its inquiry in case the chatbot is not able to deliver a satisfactory answer for which, the JIRA support system [39] receives e-mail messages that will then be classified as incidents.

3.3 Improvements in the Definition of Confidence and Identification of Intentions

RASA establishes the sequence of steps (pipeline) that must be executed each time an expression is received from the user. Within this pipeline a parameter called 'natural language understanding threshold' or simply 'threshold' defines the minimum level of confidence to issue a response to a user's expression.

RASA also includes the parameter 'ambiguity_threshold' which defines the minimum difference between the two most likely detected intentions to determine whether the first can be selected with sufficient confidence [40]. If the detected intention has a confidence below 'threshold', an alternative course of action can be defined for example, to indicate the user his message was not understood.

RASA allows similar intentions to be grouped into categories. EdyBot used the categories 'chitchat' and 'FAQ' to group casual conversation and frequently asked question's intentions. The advantage of doing this is defining what to do upon detecting an event or a sequence of events without the need of creating a rule for each intention. Another set of non-clustered intentions was made up of 'control intentions,' which include greeting, saying goodbye, thank you, affirming, denying, among others.

However, grouping into categories brought with it the problem that thresholds only apply to the categories and not to the intentions within them. Therefore, when the user uttered an expression outside the domain, for example, 'How do I live a hundred years?' it was found that RASA identified such expression as a frequent question with a probability above 98% since, although the question is outside the EdyBot domain, its composition made it very similar to the expressions in the 'FAQ' group.

That is, many of the frequently asked questions start with very similar expressions, for example: 'How do I (...)?', 'How can I (...)?', 'What is (...)?', which can also be standard ways of initiating questions outside the domain.

Because RASA calculates questions outside the domain as frequently asked questions, it proceeded to give the user the answer with the highest probability within the 'faq' group. The highest likelihood within the group always yielded values below 20%. Thus, for the question 'How do I live a hundred years?', the model would give an incorrect answer such as 'To add more columns you must...' or any other.

A second problem arose from the fact that the school management context has questions that are often too similar. The difference between probabilities of the first options tended to be very low, so the ambiguity threshold needed to be small (less than 7%).

For example, the questions 'How to create a gradable activity?' and 'How to grade an activity?' are similar, but their intentions are different. In contrast, the intention groups of control, casual conversation, and frequently asked questions tend to have very different expressions, so the ambiguity threshold should be higher.

To solve the above, we implemented a control procedure. This receives the user's expression, identifies the category and then the probability of the intention within the category. This class implements different thresholds for FAQ and casual conversation and different thresholds for intentions.

Use of Logical Operators in Conditional Responses

RASA allows defining several responses for the same intention. In absence of conditions, RASA will randomly select one of the responses. If, on the contrary, any response has conditions and these conditions are met, RASA selects that response [41].

However, it is only possible to compare if a variable (slot) is defined or if it has a certain value. At the time of the study, it was not possible to create a condition that compares, for example, if a variable is 'greater' than a certain value.

We designed an extended domain file which includes the logical operators that RASA does not support (greater than, lesser than, greater than or equal to, lesser than or equal to). Next, the conditions are analyzed before returning a response.

EdyBot in the PMS

EdyBot was designed as a contact in the PMS. When starting the conversation, the PMS sends the user's context variables to the back-end. The next screenshot shows one interaction between the user and EdyBot. The chatbot has not understood the user's intention and therefore attempts to remove ambiguity by returning similar questions (Fig. 2).

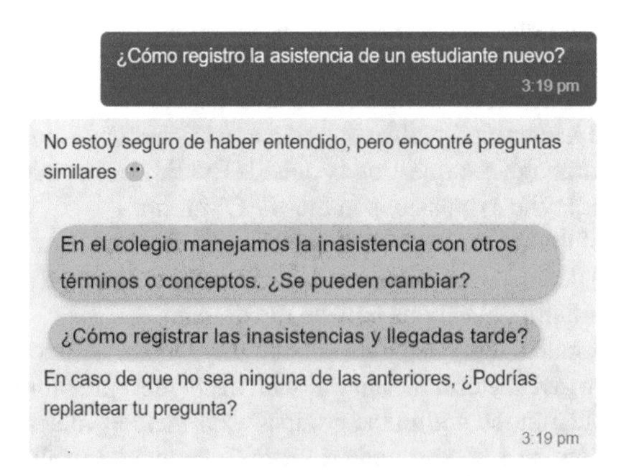

Fig. 2. PMS, where EdyBot receives a question, it does not understand and presents the user with similar questions.

4 Evaluation and Results

The following is the description of the evaluation and the results obtained from developing the Chatbot for the CE platform.

4.1 Evaluation 1

The first is an evaluation of the artificial intelligence model where the algorithm's accuracy and the ability to generalize unknown examples are measured. Cross-validation has been used by dividing the data into five segments.

The distribution of the intents per group is not even. EdyBot is a FAQs chatbot and therefore, that group is wider. Some intents had more expressions than others, because the data was collected from the CE support team conversations and some doubts are more common than others. Minimum of expression per intent was 11 and maximum was 38 (Table 1).

Table 1. Metrics of the evaluation. Weighed data.

Precision	Recall	F1-Score	Support	Accuracy
95.44%	95.44%	95.37%	2041	95.44%

The metrics obtained by performing a cross-validation test with k-fold = 5 is presented. For a volume of 2041 sample data, an accuracy of 95.44% was obtained (Table 2).

Table 2. Distribution of the dataset.

	FAQ	Casual conversation	Conversation control
Intents	133	18	10
Expressions	1528	323	190

The distribution of intents and expression is presented. The FAQ group had 82% of total intents.

4.2 Evaluation 2

An evaluation was also conducted to find out how helpful EdyBot was according to the opinion of a group of 103 teachers (0.43% of the total) from 7 educational institutions invited to test the tool.

A total of 218 questions were recorded. After each answer, the agent sent the following message to the users: 'I hope I was helpful. Did I solve your concern?'. An answer was considered helpful if after the first utterance of the user or subsequent utterances for a given intent, the user replied 'Yes' to the aforementioned question. It is worth to note that EdyBot is not processing information from previous interactions.

The written feedback from each user was automatically recorded. EdyBot was programmed to try to find similar questions or, failing that, given the user the option to escalate their concern to a human being.

Figure 3 shows that 82.91% of the participants considered that the response provided by the chatbot was useful.

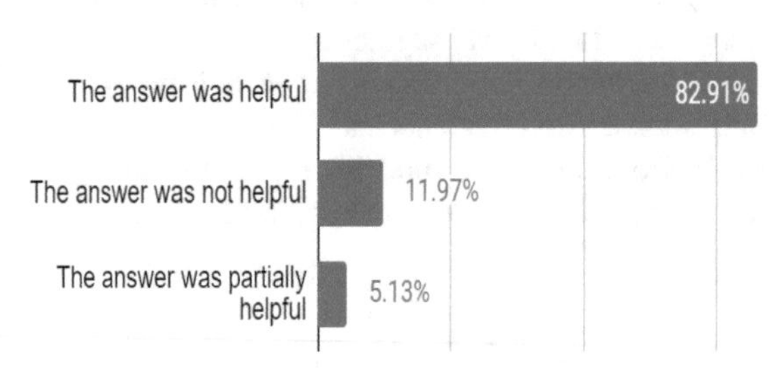

Fig. 3. Participants' opinion of the usefulness

Figure 4 shows that in 46.15% of the events in which the chatbot did not understand the question, and then shows similar questions, some of those were marked as helpful.

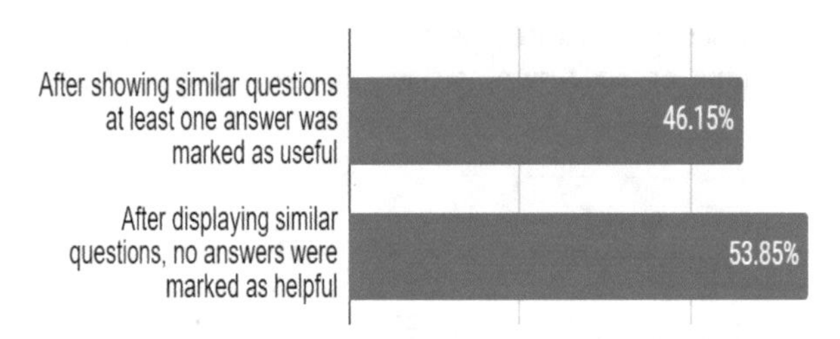

Fig. 4. Helpful answers displaying

4.3 Discussions

The evaluation metrics were obtained using the RASA command 'rasa test'. This instruction does not consider the source code modifications mentioned above but only evaluates the model's efficiency against the yml files containing the training data. This means that results could vary positively since EdyBot performs subsequent validations.

However, it is essential to remember that unlike other models (e.g., a spam classifier), the categories used to classify the input data are not static but increase and change over time.

In the second evaluation, it was evident that EdyBot could not understand some questions. For EdyBot to be able to attend to a more significant number of intentions, it is necessary to have it implemented in a production environment.

When participants stated that an answer was not helpful, it may have been because it was not easy to understand or because they disagreed with the response. In other words, the answer may be correct but may not be to a user's liking.

Those answers that were not useful should be improved. Even so, it is possible that there may be cases of users who are definitely not comfortable using a chatbot.

5 Conclusions and Future Work

A functional chatbot for CE, called EdyBot, has been developed using the RASA framework in Python 3.8. This chatbot has been integrated into the internal messaging system of the above-mentioned management platform, which has allowed teachers to receive instant help for frequently asked questions.

A cross-validation test was performed on the training data. The test yielded a 95.44% model accuracy on the weighted average metrics.

The developed tool was evaluated with the participation of 103 teachers to measure the usefulness of the answers returned by EdyBot. Regarding the intentions in the agent's domain, in 82.9% of the cases, the participants considered that the response received was helpful. In 5.1%, it was partially valuable, and in 12%, it was not helpful. EdyBot was also able to disambiguate 46.2% of non-understood user's intentions.

EdyBot has proven to be a suitable tool for teacher assistance, and it is expected to reach a higher level of approval after going into production and expanding its knowledge base.

EdyBot can represent an excellent utility for users and in the future could implement new NLP elements to improve the accuracy of its answers. EdyBot can also be extended to service other platforms users.

References

1. Ciudad Educativa. Plataforma de gestión escolar. https://www.ciudadeducativa.com. Accessed 10 Jun 2022
2. Núcleo Software. Núcleo Software. https://www.nusoft.com.co. Accessed 10 Jun 2022
3. RASA. Rasa Open Source. https://rasa.com/open-source/. Accessed 8 Apr 2022
4. Decreto 1278 de 2002. [Ministerio de Educación Nacional de Colombia]. Por el cual se expide el Estatuto de Profesionalización Docente
5. Plan nacional decenal de educación [PNDE], 2006. Ministerio de Educación Nacional. https://www.mineducacion.gov.co/1780/articles-392871_recurso_1.pdf. Accessed 14 Jun 2022
6. Resolución 15683 de 2006 [Ministerio de Educación Nacional de Colombia]. Por la cual se subroga el anexo I de la Resolución 9317 de 2016 que adoptó el anual de Funciones, Requisitos y Competencias para los cargos de directivos docentes y docentes del sistema especial de Carrera Docente. 1 de agosto del 2016. https://www.mineducacion.gov.co/portal/normativa/Resoluciones/357769:Resolucion-N-15683-01-de-Agosto-de-2016. Accessed 15 Jun 2022
7. Plan nacional decenal de educación [PNDE], 2016. Ministerio de Educación Nacional. https://www.mineducacion.gov.co/1621/articles-312490_archivo_pdf_plan_decenal.pdf. Accessed 14 Jun 2022
8. Cybercraft. Types of elearning platforms. Medium - Cybercraft. https://cybercraft.medium.com/types-of-elearning-platforms-1dffa14152e. Accessed 15 Jun 2022

9. Feffer, M. LXP vs. LMS: What are the differences?". TechTarget. https://www.techtarget.com/searchhrsoftware/tip/LXP-vs-LMS-What-are-the-differences. Accessed 15 Jun 2022
10. Fortune Business Insights. Learning management system market size: LMS industry trends 2029. Fortune Business Insights. https://www.fortunebusinessinsights.com/industry-reports/learning-management-system-market-101376. Accessed 15 Jun 2022
11. Mooc.org. Learn about moocs - massive open online courses: An EDX site. MOOC Organization. https://www.mooc.org/about-moocs. Accessed 15 Jun 2022
12. OpenSesame. LMS VS LCMS vs CMS...changing one letter makes a big difference. OpenSesame. https://www.opensesame.com/site/blog/lms-vs-lcms-vs-cmschanging-one-letter-makes-big-difference. Accessed 15 Jun 2022
13. Whittemore, B. 6 types of online learning platforms. Extension Engine Blog. https://blog.extensionengine.com/six-types-of-online-learning-platforms. Accessed 15 Jun 2022
14. XAPI. What is an LRS? learn more about learning record stores. xAPI.com. https://xapi.com/learning-record-store/. Accessed 15 Jun 2022
15. Gartner. Definition of Chatbot". IT Glossary - Gartner. Gartner Inc. https://www.gartner.com/en/information-technology/glossary/chatbot. Accessed 12 Apr 2022
16. Finwin. What are the different types of chatbots?. Finwin Technologies. https://medium.com/finwintech/what-are-the-different-types-of-chatbots-c99cdd6b3248. Accessed 15 Jun 2022
17. Sordoni, A., et al.: A Neural Network Approach to Context-Sensitive Generation of Conversational Responses. https://arxiv.org/pdf/1506.06714.pdf. Accessed 4 Apr 2022
18. Vinyals, O., Le, Q.: A neural conversational model. In: Proceedings of the International Conference on Machine Learning. https://arxiv.org/pdf/1506.05869.pdf. Accessed 29 Mar 2022
19. Li, J., Monroe, W., Ritter, A., Galley, M., Gao, J., Jurafsky, D.: Deep Reinforcement Learning for Dialogue Generation. https://arxiv.org/abs/1606.01541. Accessed 27 Mar 2022
20. Kuki. Kuki - Chat with me Kuki. Kuki AI. https://chat.kuki.ai/chat. Accessed 28 Apr 2022
21. Pandorabots. Developers Home Page. https://developer.pandorabots.com/home.html. Accessed 28 Apr 2022
22. Microsoft, Inc. Language Understanding (LUIS). https://www.luis.ai. Accessed 10 Jun 2022
23. Amazon Web Services, Inc. Conversational AI and Chatbots - Amazon Lex. Amazon Web Services. https://aws.amazon.com/lex/. Accessed 10 Jun 2022
24. Alphabeth, Inc. Lifelike conversational AI with state-of-the-art virtual agents. Dialogflow - Google Cloud. https://cloud.google.com/dialogflow. Accessed 10 Jun 2022
25. Alavi, M., Leidner, D.: Research commentary: technology-mediated learning—a call for greater depth and breadth of research. Inf. Syst. Res. **12**(1), 1–10 (2001)
26. Winkler, R., Söllner, M.: Unleashing the Potential of Chatbots in Education: A State-Of-The-Art Analysis. Academy of Management Annual Meeting (AOM). Chicago, USA. https://www.researchgate.net/publication/326281264_Unleashing_the_Potential_of_Chatbots_in_Education_A_State-Of-The-Art_Analysis. Accessed 16 Apr 2022
27. Hattie, J.: Visible Learning for Teachers: Maximising Impact on Learning. Routledge/Taylor & Francis Group ISBN: 9780415690140 (2011)
28. Pereira, J.: Leveraging chatbots to improve self-guided learning through conversational quizzes. In: F. J. García-Peñalvo (Ed.), Proceedings of the Fourth International Conference on Technological Ecosystems for Enhancing Multiculturality - TEEM 2016, pp. 911–918. New York, New York, USA: ACM Press (2016)
29. Gonczi, A., Maeng, J., Bell, R.: Elementary teachers' simulation adoption and inquiry-based use following professional development. J. Technol. Teach. Educ. **25**(2), 155–184 (2017)
30. Luebeck, J., Roscoe, M., Cobbs, G., Diemert, K., Scott, L.: Re-envisioning professional learning in mathematics: teachers' performance, perceptions, and practices in blended professional development. J. Technol. Teach. Educ. **25**(3), 273–299 (2017)

31. Ellis, J., Polizzi, S.J., Roehrig, G., Rushton, G.: Teachers as leaders: the impact of teacher leadership supports for beginning teachers in an online induction program. J. Technol. Teach. Educ. **25**(3), 245–272 (2017)
32. Sadler, M.: Corporate LMS Software Market Update (2021). Trust Radius. https://www.trustradius.com/vendor-blog/corporate-learning-management-system-LMS-software-market. Accessed 27 Jul 2022
33. Hill, P.: State of Higher Ed LMS Market for US and Canada: Year-End 2021 Edition. PhilOnEdTech. https://philonedtech.com/state-of-higher-ed-lms-market-for-us-and-canada-year-end-2021-edition/. Accessed 27 Jul 2022
34. Verified Market Research. Learning Management Systems (LMS) Market size worth $ 76.18 Billion, Globally, by 2030 at 19.1% CAGR. Verified Market https://www.prnewswire.com/news-releases/learning-management-systems-lms-market-size-worth--76-18-billion-globally-by-2030-at-19-1-cagr-verified-market-research-301563911.html. Accessed 27 Jul 2022
35. Canvas. How do I use the Canvas Virtual Assistant?". Instructure community. https://community.canvaslms.com/t5/Canvas-Virtual-Assistant/How-do-I-use-the-Canvas-Virtual-Assistant/ta-p/493505. Accessed 27 Jul 2022
36. Khoros. Best in Class Community: PowerSchool. Khoros. https://khoros.com/resources/best-in-class-community-powerschool. Accessed 27 Jul 2022
37. Blackboard. How chatbot works. Blackboard. https://www.blackboard.com/services/student-success-services/blackboard-chatbot/how-chatbot-works. Accessed 27 Jul 2022
38. Karmali. F. Creating a Chatbot in Moodle. Moodle. https://www.youtube.com/watch?v=5_ecxwJxK5U. Accessed 27 Jul 2022
39. JIRA. Issue and project tracking software". Atlassian Inc. https://www.atlassian.com/software/jira. Accessed 27 Jul 2022
40. RASA. Rasa core policies fallback. Rasa Docs. RASA Technologies Inc. https://rasa.com/docs/rasa/2.x/reference/rasa/core/policies/fallback/. Accessed 17 Jun 2022
41. RASA. Responses. Rasa Docs. RASA Technologies Inc. https://rasa.com/docs/r. Accessed 14 Apr 2022

Chatbot to Assist the Learning Process of Programming in Python

Miguel Sánchez[1], Gabriel M. Ramirez V[2], Jeferson Arango-López[3], and Fernando Moreira[4,5](✉)

[1] Escuela de Ingeniería, Universidad Internacional de La Rioja, Av. de la Paz, 137, 26006 Logroño, La Rioja, Spain
[2] Facultad de Ingeniería, Universidad de Medellín, Cra. 87 #30-65, Medellín, Colombia
gramirez@udemedellin.edu.co
[3] Depto. de Sistemas e Informática, Universidad de Caldas, Manizales, Colombia
jeferson.arango@ucaldas.edu.co
[4] REMIT. IJP, Universidade Portucalense, Porto, Portugal
fmoreira@uportu.pt
[5] IEETA, Universidade de Aveiro, Aveiro, Portugal

Abstract. As the technology of conversational assistants' advances, new uses and needs arise in multiple sectors such as training and education. In this work, a chatbot is developed with Artificial Intelligence technology to help and support doubts and queries when studying the Python programming language. This chatbot will be able to respond in Spanish language and in written form to questions related to programming so that it can speed up the resolution of doubts to beginner users. To create this chatbot library and the free and open-source Rasa framework will be used to enable the processing of natural language and neural networks to provide answers to users' questions. In addition, the chatbot will be integrated into Telegram by accessing a specific account created for this purpose.

Keywords: Artificial Intelligence · Chatbot · Programming · Python · Rasa

1 Introduction

This paper focuses on creating a text chatbot in the Spanish language to help beginners' users learn the syntax and features and understand the main structures used in the Python 3.9 programming language. Beginner users can find support with the chatbot and improve their skills, as well as they can learn the possible errors that may occur during its execution [1]. To reach this goal, data collection from one or more reliable sources is needed since the quality of these sources will depend on the quality of the answers that will be obtained. In this case, it will be obtained from the Python Language Reference Manual [1], and from the Covantec web repository of manuals for Basic Python Programming [2], where the syntax, semantics, and basic structure of the Python programming language in the Spanish language are described.

The conversational agent (chatbot) is presented as the entry and exit point with the user, implemented with the Rasa framework [3], providing a fluid, comfortable and fast

A. Rocha et al. (Eds.): WorldCIST 2023, LNNS 800, pp. 318–328, 2024.
https://doi.org/10.1007/978-3-031-45645-9_30

interface that allows a good user experience, which causes a greater predisposition to use the system. In turn, the chatbot is integrated with the Telegram messaging application [4], so that the flow of dialogue with the chatbot will be done from the Telegram interface itself, accessing a specific account created for this purpose, with the advantages of accessibility, security, and availability.

2 Background

Chatbots are increasingly used and can be defined by different names: intelligent conversational agents, intelligent conversational assistants, and intelligent conversational bots, among others, but they all refer to computer programs that interact and maintain a conversation in natural language with the user. They are usually used to request information or carry out specific procedures, through different communication channels, from any place and at any time of the day or night and 365 days a year.

In 1950 Alan Turing experimented called the Turing test [5]. It was the first-time people began to wonder if machines could think for themselves. Later ELIZA [6], created by Joseph Weizenbaum, appeared. The first computer program simulated a natural language conversation with the user. In 1972, Kenneth Colby created PARRY [7]. It consisted of a computer program based on natural language processing that simulated the behavior of a patient with paranoid schizophrenia. In the 1990s, Dr. Sbaitso [8] Creative Labs, 1991 was created, which emulated giving answers to the user as if he were a real psychologist. Then, in 1995, Richard Wallace created ALICE or Alicebot [9]. Later, in 1997, JABBERWACKY was launched. It can be considered the first chatbot that imitated the human voice. In 2001 SmartChild [10] the precursor of Siri [11], appeared. In 2008 Cleverbot was created from JABBERWACKY, and in 2009 the company WeChat created a very advanced chatbot. Finally, in 2010, Siri appeared, creating the concept of virtual assistants, rivaling the virtual assistants created by companies such as Alexa, Google Assistant, and Cortana.

Specific platforms such as Xenioo [12] or Voiceflow [13] can be used for small chatbots with simple structure logic. However, using specific development frameworks for more complex chatbots is more enjoyable. Therefore, the most important ones are described, considering development, deployment and understanding: Microsoft Bot Framework [14]; DialogFlow [15]; Amazon Lex [16]; Facebook Messenger [17]; IBM Watson [18]; Pandorabots [19]; Wit.ai [20]; Botsify [21]; SAP Conversational AI [22]; Botfuel [23]; Rasa [3] and others.

3 Method

The chatbot allows interaction between the programmer and itself to answer queries about Python programming with phrases to interpret the programmer's answers better.

The main goal is to develop a conversational agent, a chatbot, that allows beginners users to learn the Python programming language. In this way, the user will be able to interact with the chatbot and ask questions, doubts, or queries that may occur during the Python 3.9 programs development.

This work involves software development; a methodology with a quantitative approach of an applied and technological type is followed. This type of methodology allows, from an objective analysis of reality, to find strategies that address specific problems and generate knowledge that is put into practice in the productive sector, implementing the proposed approach. It follows a sequential software development model designed cascade, in which a series of stages are executed one after the other. In the last step, it is verified all the components implemented work correctly and perform the function for which they were designed. In addition, at this stage, integrated tests are performed with an environment like the final one, testing the integration of the chatbot in the corresponding messaging application and being able to find defects or bugs, thus increasing the software quality.

4 Contribution

The tool developed is a chatbot that interacts in text format and in Spanish to guide beginners in the first steps of Python 3.9 programming language so that the user asks questions about the syntax, sentences, and structures, and the chatbot responds with the corresponding help information.

4.1 Knowledge Base Extraction

The knowledge base of this work, which is used to train the chatbot, is extracted from two web pages corresponding to reference manuals of the Python language. For this purpose, web scraping techniques are used, which extract the information from the web page in text format, converting it later to a format that the chatbot will then understand.

For example, this extraction process is shown in the following figure (Fig. 1):

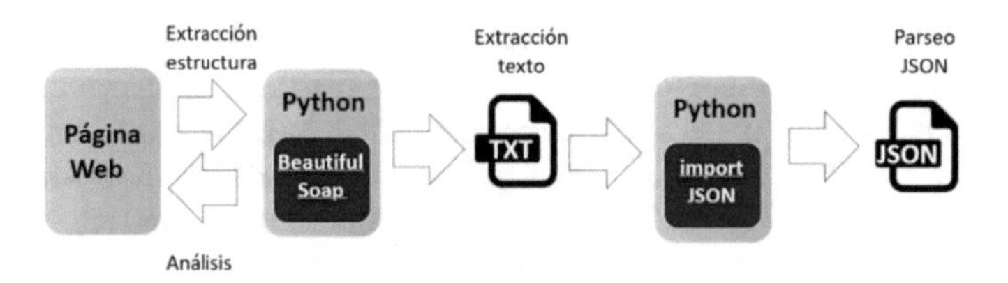

Fig. 1. Knowledge base extraction process.

The first web page for information extraction is the Python reference manual [1], where the Python programming language's syntax, semantics, and basic structure. In addition, the web page allows select the language to translate the manual and select the version of Python to which it refers; in the case of the present work, Python 3.9.5.

The second web page is the web repository of manuals for Covantec's Basic Python Programming [2]. It also contains the Python programming language's syntax, semantics, and base structure. Both web page's structure the information in the form of a tree, facilitating the extraction of information in an orderly fashion. The information extracted from the two web pages corresponds to each of the titles of the sections and subsections. In addition, to the explanatory text of each of them and the examples that may exist in each subsection. On the other hand, the knowledge base also refers to multimedia content or web pages to explain the related content. This information is not extracted from the web pages mentioned above; it is searched in other sources such as Youtube [24] and added manually to the extracted text file.

A Python library called Beautiful Soup [25], which uses web scraping techniques, is used to extract the information from the web page. The tool generates a tree from the elements of the page's source code, allowing it to extract the content of certain elements. It is only necessary to indicate the URL of the web page, and the library extracts the content of the HTML file into plain text.

Once all the information has been extracted, it is necessary to give it a specific structure and convert it to a format that the chatbot can consult. In this work, the JSON format is used to store and structure the knowledge base. The structure has a primary or parent key for each object type. Each object type corresponds to each of the groupings of objects that the user can ask about, e.g., Arithmetic Operators, Loops, Delimiters, Structured Programming, and About Python. In turn, each object type is associated with a list of objects corresponding to each element belonging to the corresponding object.

For each of these objects, a series of attributes correspond to the chatbot's response when a question related to the given object and attribute is asked. Five attributes are defined for each object: *id*; *name*; *description*; *example*; *url-media.*

4.2 RASA Based Architecture

For the present work and after studying several frameworks in the market such as DialogFlow [15], IBM Watson [18], ManyChat [26] or Chatfuel [27], Rasa 2.0 was chosen. The following is a description of the main features that lead to decide for the Rasa 2.0 tool, compared to other frameworks:

- It does not run-on Provider servers; it runs on the user's machine.
- The data is stored on the user's machine. In this way, providers cannot make use of the data.
- Free, free, open and generous tool in terms of access to software and information. The user can add or modify programming code to give more functionality and flexibility to the chatbot.
- It has a very powerful and active community that gives support to doubts, errors and collaborates in the improvement of different projects, through specific websites or specialized forums.
- It has official documentation of sufficient quality and quantity to cover the basic needs for the resolution of doubts and queries.

The architecture designed for this chatbot (Fig. 2) is divided into two main parts: the first, the external or visible part of the chatbot, the Front-end used as an interface for communication between the user and the chatbot. The second part, the Backend or internal part of the chatbot, which performs the interpretation of the questions, using interpreters and controllers to obtain and classify the intentions of the question and be able to give a satisfactory answer to the user. The Backend is divided into:

- Training module. Rasa NLU identifies the intent and extracts the question features. The minimum percentage confidence level of the prediction is defined. Rasa NLU makes use of Natural Language Processing (NLP) and Machine Learning (ML) technologies, forming what is called the Rasa Backend.
- Bot module. Rasa Core decides how to proceed once the intention has been identified and the features have been extracted, obtaining several possible answers, choosing the one with the highest degree of coincidence. In this module the actions are executed.
- Actions module. Where actions are executed, being able to access external systems to search for information. In this case, the JSON knowledge base.

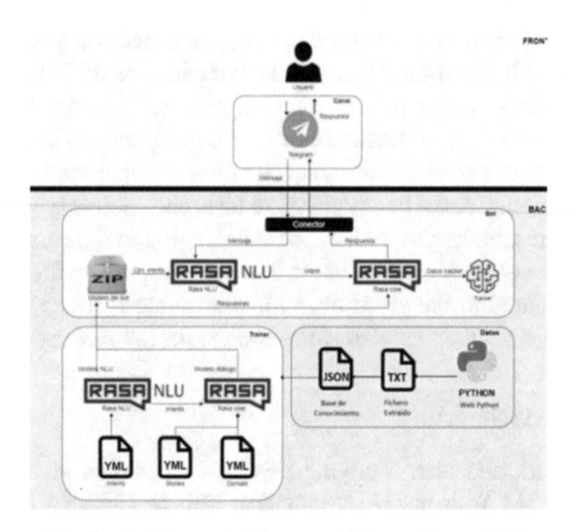

Fig. 2. Architecture Diagram (Adapted from [28]).

4.3 Questions and Answers

Below are the different types of questions that the user can ask and the answers that can be received from chatbot, depending on the type of data being asked about, as shown in Table 1.

Table 1. Questions and answers.

User questions	Chatbot answers
Direct questions defined in terms of intent	Defined responses of greeting, farewell, or emotion
Questions were out of context or in other languages or incorrect	Defined responses due to inability to find matches
Questions to get menu options	All entities of an intention
Questions from part of an entity's information	Description or examples of video of the entity
Questions of all information of an entity	Description, examples, and video of the entity

5 Evaluation and Results

In the development of software products, it is common to use expressions, or terms such as User Experience and Usability [29] or Accessibility [30] to refer to certain aspects of the quality of the products developed [31].

Once the test was developed, the chatbot was evaluated with three people with different profiles to provide different points of view according to their experience with programming languages. Two of them are professionals in developing informative applications, and a third one is a student with basic knowledge of computer science in general.

5.1 Evaluation

User 1: Professional user, an expert in programming languages such as Java, but without the knowledge or professional experience in Python (Table 2).

Table 2. User 1 – Evaluation.

User 1 – Professional (Java)	Satisfaction grade
Activities and factors of measurement	
Degree of incomplete or inappropriate responses from other contexts	High
Amount of information in the python knowledge base	Medium
Level of complete and accurate answers to specific questions about python	Medium
Conversation latency	Medium
Level of accessibility	High

User 2: Computer science student without professional experience with programming languages (Table 3).

User 3: Computer science student with professional experience with programming languages (Table 4).

Table 3. User 2 – Evaluation.

User 2 – Student	Satisfaction grade
Activities and factors of measurement	
Degree of incomplete or inappropriate responses from other contexts	High
Amount of information in the python knowledge base	Medium
Level of complete and accurate answers to specific questions about python	Medium
Conversation latency	Medium
Level of accessibility	High

Table 4. User 3 – Evaluation.

User 2 – Student	Satisfaction grade
Activities and factors of measurement	
Degree of incomplete or inappropriate responses from other contexts	High
Amount of information in the python knowledge base	High
Level of complete and accurate answers to specific questions about python	High
Conversation latency	Medium
Level of accessibility	High

In terms of accessibility, the degree of satisfaction is excellent, among other reasons. Furthermore, the display interface used is that of the Telegram messaging application, allowing to enjoy all its advantages of security, accessibility, fluidity, user interface, and availability on the most essential and current platforms.

Telegram has a security layer for private chats, an incognito mode keyboard option, messages with automatic deletion, protection against screenshots, and the creation of groups of up to 200,000 members. Another essential feature is that it does not require sharing a phone number to communicate with other people; it is enough for the Telegram user.

Due to the above features, Telegram is considered a good tool for Mobile Learning. The study on the pedagogical value of Telegram as a complement to Mobile Learning [32] relates the use of Telegram with learning models in a very positive way due to the ease of sharing theoretical knowledge and acquiring different practical skills through audiovisual content from different platforms and devices or through interaction between students or students and teachers and all this in a safe, private, easy and fast way.

Once the Telegram features and their suitability for the current project had been reviewed, different tests were performed by users. The tests are grouped by different usability and performance criteria. It should be noted that users have tested the chatbot without providing any feedback or information that could influence or affect the tests performed.

5.2 Results

The different users are introduced in written form to the chatbot and some cases. Then, the chatbot is asked about its origin, identity, or functionality. The chatbot, in all cases, behaves correctly, giving coherent answers in Spanish, responding to greetings, and asking about the user's needs. It will be present only two figures due to number of pages' limit. The Fig. 3 shows how the chatbot greets and identifies itself, returning to answer the same response at the user's insistence.

Fig. 3. Welcome Dialog.

Once the chatbot greets, it waits for a response from the user requesting information, support, or some help on how Python works. In these cases, the chatbot always responds by displaying the main menu (Fig. 4) to start the menu-driven support. Even if the user asks for other menus or other information about Python, the chatbot will show the main menu again.

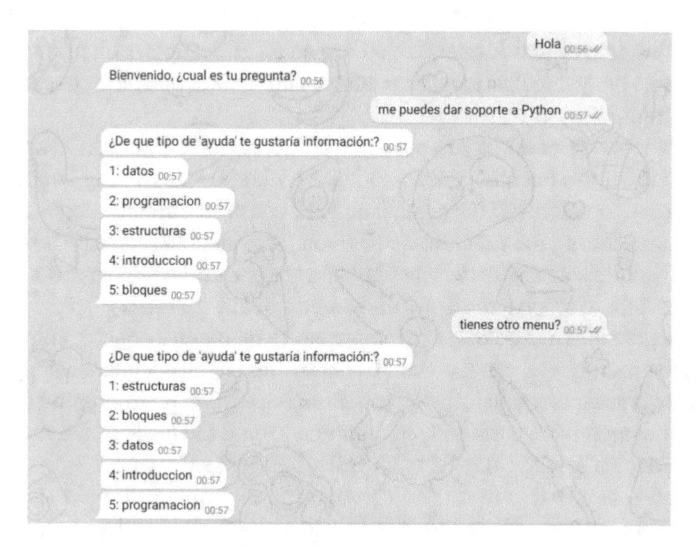

Fig. 4. Chatbot menu interaction.

5.3 Discussions

In general terms and according to the measurements and tests performed, it is concluded that the chatbot behaves correctly when answering questions out of context, inappropriate questions, questions with special characters, unannounced questions, or those related to the knowledge of Python language. In addition, the latency time between users' questions and the chatbot's answers is adequate, with no significant time difference between obtaining directly configured answers and those obtained by consulting the knowledge base.

It should be considered that more experienced users ask more advanced and complex questions when evaluating the chatbot's behavior in different situations and when consulting the information in the knowledge base. In the latter case, it may be that more experienced users seek more advanced information that is not included in the knowledge base.

In terms of accessibility, the degree of satisfaction is excellent, among other reasons, because the display interface used is that of the Telegram messaging application, allowing users to enjoy all the advantages of security, accessibility, fluidity, and user interface.

6 Conclusions and Future Works

In an increasingly digitized, and globalized world, where the answers to any day-to-day question require more and more speed, reliability, security, and availability at any time of the day, it confirms that solutions of this type are becoming more and more valuable and necessary. In addition, since the appearance of COVID-19, there have been many social and economic effects on the training sector, which have caused a change in the way of teaching to a digital format, which has enabled the approach of a large number of people to this type of technology, providing training tools and new channels to enable on-line training.

In conclusion, it is necessary to convey the importance of this educational software solution, which allows beginners users to learn Python programming language. In this way, the user can interact with the chatbot and ask questions, doubts, or queries that may occur to the user during the execution of Python programs. In this way, it is possible to increase and facilitate access to the teaching of a programming language such as Python, making it possible to reach a more significant number of users.

The main goal is achieved because it manages to maintain a coherent flow of conversation with the user, giving clear, concrete answers and in an adequate response time. In addition, answers with enough information about the Python programming language are obtained, allowing novice users to learn and understand the basis of the main concepts and features, accessing the contents in an agile, fast, and safe way.

Therefore, it is possible to offer a solution to the problem of help and support to beginner users in Python programming. It is also remarkable that the implementation of the technological solution is optimal and efficient, based on a reliable and complete data reference domain. Regarding the validation by users of different profiles, the objective is partially achieved since even though the functional and technical validation by users of different professional profiles was achieved remarkably, the corresponding tests were not completed to determine the degree of accessibility for all types of users, including those with different types of disabilities or special conditions.

Another improvement would be to increase the knowledge base, add more advanced information about Python programming language concepts and features, or learn about previous and new versions of Python [33]. It would increase the number of potential users with a more advanced profile. It could also include information on other programming languages, such as Java [34], C#, and Ruby, and give the option to provide two answers for the same question, one for each programming language.

References

1. Python: Referencia del Lenguaje Python. https://docs.python.org/es/3.9/reference/index.html. Accessed 21 Apr 2022
2. Covantec: Repositorio de manuales y recursos del entrenamiento Programación en Python - Nivel básico. https://github.com/Covantec. Accessed 21 Apr 2022
3. Rasa: Rasa Chatbots. https://rasa.com. Accessed 23 Apr 2022
4. Telegram: Telegram Web. https://web.telegram.org. Accessed 21 Apr 2022
5. Turing, M.A.: Computing machinery and intelligence. Mind **49**, 433–460 (1950)
6. Weizenbaum, J.: Symmetric list processor. Commun. ACM **6**, 524–544 (1963)
7. Colby, K.M.: Turing-like indistinguishability test for the validation. Artif. Intell. **3**, 199–221 (1972)
8. Creative Labs: Dr. Sbaitso, artificial intelligence speech synthesis. https://archive.ph/20130111132657/http://www.x-entertainment.com/articles/0952/. Accessed 21 Apr 2022
9. Wallace, R.: The elements of AIML style. http://www.Alicebot.org/. Accessed 15 May 2022
10. Matthew, B.: Alexa, Siri, Cortana, and More: an introduction to voice assistants. Med. Ref. Serv. Q. **37**(1), 81–88 (2018)
11. Epstein, J., Klinkenberg, W.D.: From Eliza to Internet: a brief history of computerized assessment. Comput. Hum. Beahv. **17**, 295–314 (2001). https://doi.org/10.1016/S0747-5632(01)00004-8
12. Xenioo: Xenioo. Omnichannel Chatbot Platform. https://www.xenioo.com/. Accessed 20 May 2022
13. Voiceflow: Voiceflow. Design, prototype & launch voice & chat apps, https://www.voiceflow.com. Accessed 25 May 2022
14. Microsoft: Microsoft Visual. https://code.visualstudio.com/. Accessed 10 June 2022
15. Google: Dialogflow. https://dialogflow.cloud.google.com. Accessed 17 May 2022
16. Amazon: Amazon Lex. https://aws.amazon.com/es/lex. Accessed 10 June 2022
17. Facebook: Facebook Messenger. https://www.facebook.com/messenger/. Accessed 21 May 2022
18. IBM: IBM Watson. https://www.ibm.com/es-es/watson. Accessed 14 June 2022
19. Pandorabots: Pandorabots. https://home.pandorabots.com. Accessed 13 Apr 2022
20. Wit.ai: Wit.ai. https://wit.ai. Accessed 13 June 2022
21. Botsify: Botsify - A Fully Managed Chatbot Platform To Build AI-Chatbot. https://botsify.com. Accessed 21 May 2022
22. Sap: Sap Conversational. https://www.sap.com. Accessed 9 May 2022
23. Botfuel: Botfuel. https://www.botfuel.io. Accessed 9 May 2022
24. Youtube: Youtube. https://www.youtube.com/. Accessed 4 June 2022
25. BeautifulSoup: BeautifulSoup. https://www.crummy.com/software/BeautifulSoup. Accessed 5 June 2022

26. Manychat: Manychat. https://manychat.com/. Accessed 5 June 2022
27. Chatfuel: Chatfuel. https://chatfuel.com/. Accessed 11 June 2022
28. Castillo, A.: RASA Framework Análisis e Implementación de un Chatbot
29. Madrid, N.: Métricas de usabilidad y experiencia de usuario. https://www.nachomadrid.com/2020/01/metricas-de-usabilidad-y-experiencia-de-usuario/. Accessed 17 May 2022
30. Carreras, O.: Validadores y herramientas para consultorías de accesibilidad y usabilidad. https://www.usableyaccesible.com/recurso_misvalidadores.php. Accessed 3 Apr 2022
31. Antevenio: Chatbots efectivos para captar clientes: los mejores ejemplos. https://www.antevenio.com/blog/2019/07/chatbots-efectivos-para-captar-clientes/. Accessed 18 Apr 2022
32. Rios, J.: El valor pedagógico de Telegram como complemento del mobile learning en la formación en finanzas: aplicación práctica a un caso de estudio
33. Vega, D.: Programación con Lenguaje Python 3.X basado en Microsoft Bot Framework. Universidad Privada Telesup
34. Weizenbaum, J.: ELIZA, a computer program for the study of natural language communication between man and machine. Commun. ACM **9**(1), 36–45 (1966)

Formulating a Group Decision Support Systems (GDSS) Model for Accreditation: An Early Childhood Institution Perspective

Abdul Kadir[2], Syed Nasirin[1(✉)], Esmadi A. A. Seman[1], Tamrin Amboala[1], Suddin Lada[1], and Azlin A. P. Kinjawan[3]

[1] Universiti Malaysia Sabah, 87000 Labuan, Malaysia
snasirin@ums.edu.my
[2] Universitas Sari Mulia, Banjarmasin City, Indonesia
[3] Ministry of Education, Putrajaya, Malaysia
azlin.azlan@moe.gov.my

Abstract. This research introduces an innovative approach to accrediting early childhood education institutions in South Kalimantan, Indonesia, aiming to address the labour-intensive nature of traditional accreditation processes. By integrating PCA-AHP-TOPSIS-VIKOR algorithms, the study proposes a hybrid model that streamlines accreditation, overcoming existing limitations and ensuring a more precise assessment of institution quality. Emphasizing the necessity of a comprehensive approach to accreditation, the study underscores the need for improved methodologies. Through the amalgamation of multiple algorithms, the hybrid model enhances the evaluation process, offering a robust means of assessing early childhood education institutions. Two models are presented: one without PCA and another incorporating PCA. The PCA method aids in summarizing assessment criteria, thereby enhancing the efficiency of the evaluation process.

Keywords: DSS · GDSS · Accreditation · Early Childhood · Childhood Education

1 Introduction

A well-accredited early childhood education institution symbolises a quality early childhood educational institution [1]. With the accreditation, early childhood education institutions form a prosperous future generation as acceptable accreditation classification are proven quality education programs [2, 3]. In general, it is a formal recognition given to an institution. Therefore, it is a vital asset to determine the position of an educational institution [13]. In addition, good accreditation results from an accreditation board recognize the eligibility rating [14].

In school accreditation, it can be interpreted as an activity to assess the feasibility of a school based on criteria set by the school. Accreditation management aims to motivate schools to improve and systematically develop their program with better quality [15]. This development of an early childhood education accreditation system is one of the

A. Rocha et al. (Eds.): WorldCIST 2023, LNNS 800, pp. 329–337, 2024.
https://doi.org/10.1007/978-3-031-45645-9_31

steps made by the government to guarantee the implementation of early childhood education [4, 5] and [6]. In other words, accreditation is a process of assessing and assessing the quality of an institution carried out by an assessment team based on predetermined quality standards [7, 8], under the direction of an independent accreditation body or an institution outside the institution concerned [9, 10]. Accreditation status guarantees quality graduates from educational institutions with well-controlled management processes [11]. This process involves several agency decision-making processes that carry the accreditation [12]. Accreditation of early childhood education institutions involves various parties and requires accurate data for filling out accreditation forms.

In Indonesia, the National Accreditation Board for Early Childhood Education and Non-Formal Education has completed the accreditation process. Since the inception of the accreditation programs, the process has been carried out manually. This manual process requires numerous administrators tasked with receiving application forms, checking them, and reporting the assessment outcomes, including a report on the actual conditions of the institution [17]. The activities carried out by these officers usually lead to compassion fatigue which may cause some serious assessment errors [18, 19].

Given the current stand-alone condition in South Kalimantan in carrying out the above series of processes, it is necessary to have an application tool for the group decision-making process. The need to track down the ideal choice of criterion is thus a must from every possible option. However, the human criteria cannot be isolated [20]. Therefore, this study proposes a GDSS-based approach that can be used with a typical accreditation assignment [21, 22]. The approach is appropriate for streamlining and dissecting complex and interrelated accreditation criteria. Using this approach, tough spots can be rearranged without extinguishing the fundamental expectations of accreditation [23]. The model also applies to decision-making by exchanging ideas or preferences to increase group member participation in providing opinions and decision choices on an accreditation problem [24–26].

2 Review of the Literature

Accreditation is a formal commendation of an educational institution. Therefore, it is an important asset to determine the institution's position at the rank of the competition [27]. Furthermore, the value of accreditation is a benchmark for an educational institution that ensures the quality of graduates born from a well-managed institution that meets the standard accreditation criteria [28]. This optimum decision-making process is thus crucially needed to allow an accurate classification to be awarded to the institution applying for accreditation [29–31]. In addition, many models were developed to solve these accreditation problems [32].

In Indonesia, similar models were developed to support decision-makers in sorting out their accreditation assignments. Several studies have shown that AHP is a method that is appropriate in helping to complete such assignments (Nurcahyo, Gabriel, Ivan, and Sari, 2018). The AHP method solves complex decision-making problems by making a hierarchical arrangement for each criterion. In practice, the method has provided much support in the dynamics of various decision-making related to the accreditation process of an institution. These models could accommodate the dynamic nature of the accreditation criteria [33, 34, 48]. For example, a study by [45] describes the TOPSIS method for accreditation of universities in Morocco. It selects and prioritizes action for effective decisions. It was also applied to determine the accreditation of universities in Yemen [46]. In addition, fuzzy AHP algorithms further assisted King Khalid University in helping the Accreditation Board for Engineering and Technology (ABET) determine the ranking relative importance [47].

Furthermore, AHP–TOPSIS helped the Philippine Government Accreditation Agency overcome the complexity of decision-making for ISO 9001 certification [17]. The algorithm provides a comparison value against the requirements throughout the accreditation process [50]. At the same time, other studies have employed similar algorithms to determine the accreditation process criteria at the Tehran Social Security Hospital [51].

The research findings prove that the AHP method can rank the importance of success factors [47]. Therefore, many universities use this model to reference their academic programs evaluation to pass ABET accreditation optimally. Nonetheless, these studies have impediments: The choice measures might change under certain circumstances. The AHP method is also developed based on decision-makers' input, which increases the tendency to bias the results. Hence, future studies need to add a process that validates the results. One might nonetheless conclude that the AHP method is still competent in providing the criteria to support decision-making in the accreditation process. However, using the AHP method with other models is necessary to improve its outcome.

Another common method used for accreditation assessment is the Order Preference by Similarity to Ideal Solution (TOPSIS). It is a method that is more straightforward as it uses criteria indicators and alternative variables to determine decisions. In addition, several studies related to the accreditation process have shown that the model has better computational calculations [45]. The TOPSIS method was applied to avoid closing academic programs at universities in Morocco. More studies on fuzzy TOPSIS prioritize the criteria outlined by accredited universities in South India [55]. There were eight required criteria and 40 sub-requirements utilised as parameter assessments by the nation's accreditation board. However, TOPSIS was only applied to analyze agency ranking parameters, so additional methods are needed to provide institutions' ranking values. Hence, a combination of TOPSIS and methods is welcome to assist in decision-making.

On the other hand, VIKOR (Vase Kriterijumska Optimizacija I Kompromisno Resenje) uses a compromise ranking technique for multi-criteria decision-making. The model was applied to make a multi-criterion ranking of an institution's accreditation [58]. It may also be considered to avoid conflicting criteria in the rankings. The basic concept of VIKOR is to sort the existing samples by looking at the utility and regret values of each sample. Therefore, the VIKOR method can provide the best identification of each criterion. However, this study has limitations, as it has not considered the criteria that influence the effectiveness from the point of view of other stakeholders. In addition, the VIKOR method makes it possible to calculate additional criteria such as reliability and safety. For example, the VIKOR method was employed to calculate different criteria weight in an airport rating [28]. However, this study is limited to only using green practice criteria. The method was also employed and reliable in evaluating e-government websites. However, the study did not consider the interdependence criteria between the websites, so it is necessary to add other methods for the evaluation process. In short, combining the VIKOR method with other methods is thus needed for a more comparative evaluation [28]. As a result of the weaknesses inherited in the individual models, hybrid models started to be developed.

The combination of AHP and TOPSIS methods improved the study comparison matrix. At the same time, the TOPSIS method's basic concept is simple and able to calculate the performance of decision choices. This method was used to identify the most and least important dimensions of the government agencies in the Philippines. The combined method's basic strength was designing the GDSS concept evaluations under uncertain environments [65]. Although the combined AHP-TOPSIS method has a positive contribution, it has some limitations due to its subjective evaluation of risk and hazard parameters. More objective assessment is thus needed needs weighing opinions among experts, using risk parameters with different weights.

Moreover, some studies argued that the mixture between VIKOR and TOPSIS techniques is more reasonable as they possess better computational abilities [67]. In contrast, others have promoted combining the AHP and VIKOR methods as they complemented each other so well through their computational abilities. The combination can support the accreditation assessment decisions. The combination of three major algorithms has also been proposed in which all AHP-TOPSIS-VIKOR algorithms were utilised to support group decision making. However, the combination of the AHP, TOPSIS and VIKOR methods is weak when different multi-criteria are considered. The recent combination of a hybrid model includes the PCA. The principal component analysis (PCA) is so effective in determining in reducing a huge number of criteria (i.e., 50 or more items) into a small number of constructs [35–37, 40–44]. Nonetheless, we strongly believe that PCA should come before applying the AHP and other methods. PAC may help identify the hidden dimensions or constructs of the criteria, which may or may not be apparent from a direct analysis.

3 Experimental Set-Up: Modelling Early Childhood Education Accreditation Institutions GDSS

3.1 The Model Without PCA

In In the given scenario, the Accreditation Board has outlined 60 criteria that need to be evaluated. The evaluation process involves assigning weights to each criterion based on its perceived level of importance, and the assessment is carried out by a team of assessors acting as the decision-maker (DM). The DM can use an ordinal scale to assess each criterion and assign weights accordingly. Criteria that are deemed less significant will receive lower weights, while those that are considered more critical will receive higher weights.

Once the weights have been assigned to the criteria, the next step is to rank the decision alternatives using the TOPSIS method. This involves a voting process in which the decision alternatives are evaluated based on their relative strengths and weaknesses. The alternatives are then ranked based on their overall performance, considering the assigned weights for each criterion.

By following this approach, the assessors can arrive at a well-informed and objective decision regarding the accreditation of the entity being evaluated. The use of both AHP and TOPSIS methods enables the assessors to consider all relevant factors and make an informed decision that is based on the collective wisdom of the assessor team.

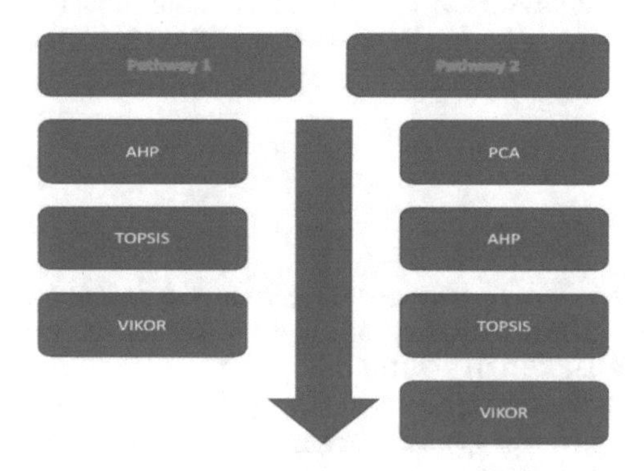

Fig. 1. Proposed Experimental Pathways

3.2 The Model with PCA

In the second formulation, the PCA method summarizes all the criteria, i.e., 60 assessment criteria (Fig. 1). Once this is done, the formalization continues with AHP-TOPSIS and VIKOR algorithms. Then, two tests will be carried out, with construct validity and user feedback (Fig. 2). Construct validity is an instrument designed to measure certain constructs of a model system. Construct validity measures validity by testing whether a model system or GDSS model is as expected.

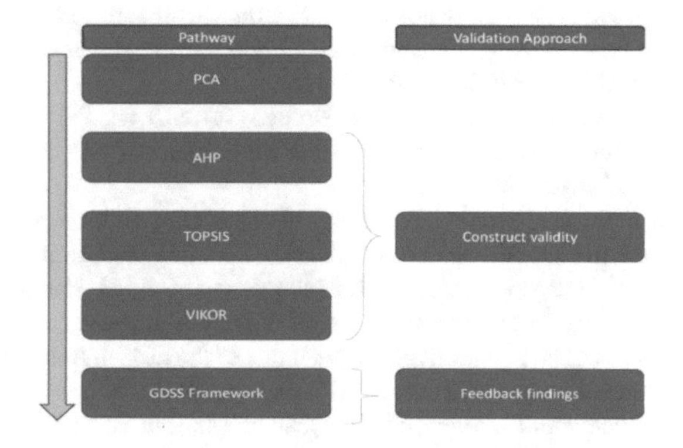

Fig. 2. Proposed Validity Approaches

4 Conclusion

Many accreditation models have been developed for early childhood education institutions, including Indonesia. Nonetheless, these models could not identify the hidden dimensions or constructs of the variables, which may or may not be apparent from a direct analysis. The proposed PCA model is much more robust than developed before. It can identify the groups of inter-related variables to see how they relate.

References

1. Ngoc-Tan, N., Gregar, A.: Knowledge management and its impacts on organisational performance: An empirical research in public higher education institutions of Vietnam. J. Inf. Knowl Manage **18**(02), 1950015 (2019)
2. Abdel-Basset, M., Zhou, Y., Mohamed, M., Chang, V.: A group decision-making framework based on neutrosophic Vikor approach for e-government website evaluation. J. Intell. Fuzzy Syst. **34**(6), 4213–4224 (2018)
3. Ahmad, N., and Qahmash, A. "Implementing Fuzzy AHP and FUCOM to evaluate critical success factors for sustained academic quality assurance and ABET accreditation," PLoS ONE, 2020, 15(9 September), 1–30
4. Ak, M. F., and Gul, M. "AHP-TOPSIS integration extended with Pythagorean fuzzy sets for information security risk analysis," Complex & Intelligent Systems, 2019, 5(2), 113–126. Springer International Publishing
5. Akgün, İ, Erdal, H.: Solving an ammunition distribution network design problem using multi-objective mathematical modelling, combined AHP-TOPSIS, and GIS. Comput. Ind. Eng.. Ind. Eng. **129**(January), 512–528 (2019)
6. Alqadasi, R.M., Rassam, M.A., Ghaleb, M.: An Integrated academic accreditation program (IAAP). Inter. J. Quality Control Stand. Sci. Eng. **7**(1), 42–67 (2020)
7. Alvarenga, A. D., Salgado, E. G., and Mendes, G. H. de S. "Ranking criteria for selecting certification bodies for ISO 9001 through the Analytic Hierarchy Process (AHP).," International Journal of Quality and Reliability Management, 2018, 35(7), 1321–1342
8. Bahadori, M., Teymourzadeh, E., Bagejan, F.F.: Factors affecting the effectiveness of quality control circles in a hospital using a combination of fuzzy VIKOR and grey relational analysis. Proc. Singapore Healthcare **27**(3), 180–186 (2018)
9. Bakioglu, G., and Atahan, A. O. "AHP integrated TOPSIS and VIKOR methods with Pythagorean fuzzy sets to prioritise risks in self-driving vehicles," Applied Soft Computing, 2021, 99, 106948, Elsevier BV
10. Bakken, L., Brown, N., and Downing, B. "Early childhood education: the long-term benefits," Journal of Research in Childhood Education, 2017, 31(2), 255–269, Routledge
11. Bedregal-Alpaca, N., Delgado-Barra, L., Baluarte-Araya, C., and Sharhorodska, O. "Reflections on the teaching body criterion in an accreditation process: proposal for teaching evaluation from the student perspective," Proceedings of the 2019 International Symposium on Engineering Accreditation and Education, 2019, ICACIT 2019
12. Benmoussa, N., Elyamami, A., Mansouri, K., Qbadou, M., Illoussamen, E.: A multi-criteria decision-making approach for enhancing university accreditation process. Eng. Technol. Appli. Sci. Res. **9**(1), 3726–3733 (2019)

13. Bera, B., Shit, P. K., Sengupta, N., Saha, S., and Bhattacharjee, S. "Susceptibility of deforestation hotspots in Terai-Dooars belt of Himalayan Foothills: A comparative analysis of VIKOR and TOPSIS models," Journal of King Saud University – Computer and Information Sciences, 2021
14. Bogren, M., Banu, A., Parvin, S., Chowdhury, M., and Erlandsson, K. "Findings from a context-specific accreditation assessment at 38 public midwifery education institutions in Bangladesh," Women and Birth, 2021, 34(1), e76-e83. Australian College of Midwives
15. Bordelois, M.I.: The career self evaluation in improving the quality of higher education. Inter. J. Soc. Sci. Human. (IJSSH) **2**(1), 84–91 (2018)
16. Boyd-Swan, C., and Herbst, C. M. "Influence of quality credentialing programs on teacher characteristics in centre-based early care and education settings," Early Childhood Research Quarterly, 2020, 51, 352–365, Elsevier Inc.
17. Carneiro, J., Alves, P., Marreiros, G., Novais, P.: Group decision support systems for current times: Overcoming the challenges of dispersed group decision-making. Neurocomputing **423**, 735–746 (2021)
18. Carneiro, J., Saraiva, P., Martinho, D., Marreiros, G., Novais, P.: Representing decision-makers using behaviour styles: An approach designed for group decision support systems. Cogn. Syst. Res.. Syst. Res. **47**, 109–132 (2018)
19. Chatterjee, P., Chakraborty, S.: A comparative analysis of the VIKOR method and its variants. Decision Sci. Lett. **5**(4), 469–486 (2016)
20. Dewi, K.H.S., Kusnendar, J., Wahyudin, A.: Sistem Pendukung Keputusan Penyusunan Prioritas Perbaikan Standar Akreditasi Program Studi Menggunakan Metode AHP dan. Teori san Aplikasi Ilmu Komputer **1**(1), 45–54 (2018)
21. Ebrahim, Z., Nasiripour, A.A., Raeissi, P.: Identifying and prioritising the effective factors on establishing accreditation system in tehran hospitals affiliated with the social security organization in Tehran. J. Health Manag. Inform. **5**(April), 3–5 (2019)
22. Fenech, M., Sumsion, J., Goodfellow, J.: The regulatory environment in long day care: a 'double-edged sword' for early childhood professional practice. Australas. J. Early Childhood. J. Early Childhood **31**(3), 49–58 (2006)
23. Fitrisari, A., Harapan, E., Wahidy, A.: The school heads' strategy for planning to boost the standard of education. JPGI (Jurnal Penelitian Guru Indonesia) **6**(1), 155 (2021)
24. Ghadami, L., Masoudi Asl, I., and Hessam, S. "Developing hospital accreditation standards: Applying fuzzy DEMATEL," International Journal of Healthcare Management, 2019, 0(0), 1–9, Taylor & Francis
25. Gul, M., and Yucesan, M. "Performance evaluation of Turkish Universities by an integrated Bayesian BWM-TOPSIS model," Socio-Economic Planning Sciences, 2021, October, 101173, Elsevier Ltd.
26. Gupta, H. "Evaluating service quality of airline industry using a hybrid best worst method and VIKOR," Journal of Air Transport Management, 2018, 68, 35–47, Elsevier Ltd.
27. Hakim, A.R., Suharto, N.: The role of accreditation in medical education. Educ. Humanities Res. **60**(8), 211–212 (2018)
28. Indrawan, I.: Pelaksanaan Kebijakan Akreditasi PAUD. Jurnal Pendidikan dan Konseling **03**(01), 46–54 (2020)
29. Indrayani, R., and Pardiyono, R. "Decision Support System to Choose Private Higher Education Based on Service Quality Model Criteria in Indonesia," Journal of Physics: Conference Series, 2019, 1179(1)
30. Ishikawa, S.I., Chen, J., Fujisawa, D., Tanaka, T.: The development, progress, and current status of cognitive behaviour therapy in Japan. Aust. Psychol. **55**(6), 598–605 (2020)
31. Jaberipour, M., and Khorram, E. "Two improved harmony search algorithms for solving engineering optimisation problems," Communications in Nonlinear Science and Numerical Simulation, 2010, 15(11), 3316–3331, Elsevier BV

32. Jacqmin, J., and Lefebvre, M. "The effect of international accreditations on students' revealed preferences: Evidence from French Business schools," 2021, 85(October), 102192, Elsevier Ltd.
33. Khairunnisa, Farmadi, A., and Candra, H. K. "Penerapan Metode Ahp Topsis Pada Sistem Pendukung Keputusan Penentuan Taman," KLIK: Kumpulan Jurnal Ilmu Komputer, 2015, 02(01), 1–10
34. Khan, S., Khan, M. I., and Haleem. "Prioritising the risks in Halal food supply chain: an MCDM approach," Journal of Islamic Marketing, 2019
35. Khojah, A., Shousha, A.: Academic accreditation process of english language institute: challenges and rewards. High. Educ. Stud. **10**(2), 176 (2020)
36. Kholifah, U., Dini, R. H., Wibawa, A. P., and Nugraha, E. "Comparing the characteristics of the undergraduate program of information systems in public and private universities," Proceeding -2017 3rd International Conference on Science in Information Technology: Theory and Application of IT for Education, Industry and Society in BigData Era, ICSITech 2017, 2017, 2018-January, 483–487
37. Khumaidi, A., Purwanto, Y. A., Sukoco, H., Wijaya, S. H., and Darmawan, R. "Design a smart system for fruit packinghouse management in the supply chain," 2020 International Conference on Computer Science and Its Application in Agriculture, ICOSICA 2020
38. Koltharkar, P., Eldhose, K. K., and Sridharan, R. "Application of fuzzy TOPSIS for the prioritisation of students' requirements in higher education institutions: A case study: i-criteria decision-making approach," 2020 International Conference on Systems, Computation, Automation and Networking, ICSCAN 2020. 2020
39. Kumar, A., A, A., and Gupta, H. "Evaluating the green performance of the airports using hybrid BWM and VIKOR methodology," Tourism Management, 2020, 76(June 2019), 103941, Elsevier
40. Kumar, P., Shukla, B., Passey, D.: Impact of accreditation on quality and excellence of higher education institutions. Investigacion Operacional **41**(2), 151–167 (2020)
41. Kurniah, N., Andreswari, D., and Kusumah, R. G. T. "Achievement of Development on Early Childhood Based on National Education Standard," 295(ICETeP 2018), 351–354
42. Listyaningsih, V., Utami, E.: Decision support system performance-based evaluation of village government using AHP and TOPSIS methods: Secang sub-district of Magelang regency as a case study. Inter. J. Intell. Syst. Appl. **10**(4), 18–28 (2018)
43. Maharani, S., Hatta, H. R., Anzhari, A. N., and Khairina, D. M. "Paskibraka member selection using a combination of AHP and TOPSIS methods on the office of youth and sports of kutai kartanegara regency," E3S Web of Conferences, 2018, 31
44. Mondal, S. D., Ghosh, D. N., and Dey, P. K. "Prediction of NAAC Grades for Affiliated Institute with the help of Statistical Multi-Criteria Decision Analysis," 2021, 1(2), 116–126
45. Mufazzal, S., and Muzakkir, S. M. "A new multi-criterion decision making (MCDM) method based on proximity indexed value for minimising rank reversals," Computers and Industrial Engineering, 2018, 119(MCDM), 427–438
46. Nasution, R. H. S., Khadijah, K., and Sinaga, A. I. "Implementation of PAUD Unit accreditation in south padangsidimpuan City," Budapest International Research and Critics in Linguistics and Education (BirLE) Journal, 2020, 3(2), 981–987
47. Nurcahyo, R., Gabriel, D. S., Ivan, E., and Sari, I. P. "ISO / IEC 17025 Implementation at Testing Laboratory in Indonesia," 2018, January 22–23, IEEE
48. Ocampo, L., Alinsub, J., and Casul, R. A. "Public service quality evaluation with SERVQUAL and AHP-TOPSIS: A case of Philippine government agencies. Socio-Economic Planning Sciences, 2019, 68, Elsevier Ltd.
49. Olszak, C. M., and Kisielnicki, J. "A conceptual framework of information systems for organisational creativity support. Lessons from empirical investigations," Information Systems Management, 2018, 35(1), 29–48. Taylor & Francis

50. Ortiz, L., and Hallo, M. "Analytical Data Mart for the Monitoring of University Accreditation Indicators," EDUNINE 2019 – 3rd IEEE World Engineering Education Conference: Modern Educational Paradigms for Computer and Engineering Career, 2019, Proceeding IEEE
51. Pamučar, D., Stević, Ž, Zavadskas, E.K.: Integration of interval rough AHP and interval rough MABAC methods for evaluating university web pages. Appl. Soft Comput. J. **676**, 141–163 (2018)
52. Payne, J. W., and Schimmack, U. "Construct validity of global life-satisfaction judgments: A look into the black box of self–informant agreement," Journal of Research in Personality, 2020, 89, 104041, Elsevier Inc.
53. Reinke, S., Peters, L., Castner, D.: Critically engaging discourses on quality improvement: Political and pedagogical futures in early childhood education. Policy Futures Educ. **17**(2), 189–204 (2019)
54. Riaz, M., and Tehrim, S. T. "A robust extension of the VIKOR method for bipolar fuzzy sets use SPA theory-based metric spaces connection numbers," Artificial Intelligence Review, 2021, (vol.54). Springer Netherlands
55. Rochefort, C., Baldwin, A. S., and Chmielewski, M. "Experiential Avoidance: An Examination of the Construct Validity of the AAQ-II and MEAQ," Behaviour Therapy, 2018, 49(3), 435–449, Elsevier Ltd.

Considerations in the Design of Pervasive Game-Based Systems for the Older Adult Population

Johnny Salazar Cardona[1], Jeferson Arango-Lopez[2], Francisco Luis Gutiérrez Vela[1], and Fernando Moreira[3,4](✉)

[1] Departamento de Lenguajes y Sistemas Informáticos, ETSI Informática, Universidad de Granada, 18071 Granada, Spain
jasalazar@correo.ugr.es, fgutierr@ugr.es
[2] Departamento de Sistemas e Informática, Facultad de Ingenierías, Universidad de Caldas, Calle 65 # 26-10, Edificio del Parque, Manizales, Caldas, Colombia
jeferson.arango@ucaldas.edu.co
[3] REMIT, IJP, Universidade Portucalense, Rua Dr. António Bernardino Almeida, 541-619, 4200-072 Porto, Portugal
fmoreira@upt.pt
[4] IEETA, Universidade de Aveiro, Aveiro, Portugal

Abstract. Currently, the different technological advances and developments have allowed the generation of new computing paradigms, one of these being pervasive systems. This type of systems has been used to generate pervasive game experiences, to which new devices have been incorporated in the contexts of game-based systems (GBS). These advances have allowed the emergence of new enriched experiences through the integration of the virtual and real worlds within the game world. Although these new paradigms offer different benefits, they also generate new challenges for their design, being more complex when they are intended to be used in specific populations such as the elderly, which is one of the main objectives of this research work. These experiences can be oriented in different ways, some of them being entertainment, leisure, learning or obtaining benefits. In order to achieve such pervasive experiences, the existing technological gap and the consequences of human aging, such as physical and cognitive abilities and the impact on their emotional level, must be considered. For this reason, this article identifies the different aspects that should be considered for the design of pervasive GBS in order to adjust to the particularities of this population.

Keywords: Pervasive · Game Experience · Older adults · Game Based Systems · Player Experience

1 Introduction

The existing technological gap among older adults leads to a lack of optimal use of the different technological developments available for this group of people. This is reflected in the different experiences and technological systems focused on entertainment, leisure,

A. Rocha et al. (Eds.): WorldCIST 2023, LNNS 800, pp. 338–347, 2024.
https://doi.org/10.1007/978-3-031-45645-9_32

learning and generation of wellbeing, where the different GBS oriented to these purposes are not adequately used, as they are not attractive to this population because these experiences are not adjusted to their particularities.

In this respect, in previous works [1], some important considerations have been evidenced, such as the fact that game experiences offered through direct input, natural and immersive technological devices are accepted, to a greater extent, by the older adult population. These types of experiences, such as those offered by "pervasive games," are an emerging genre that has a long way to go as a mechanism for generating positive experiences and emotions for older adults. These pervasive experiences can offer different benefits to this population by incorporating them into their daily lives, promoting the improvement of their health through physical activity, cognitive training and socialization.

Technology-based products are designed and evaluated based on usability principles, focused on user interaction processes. This is applied to information systems, web pages, mobile applications or similar, to measure the user experience (UX). However, in the context of games, UX evaluation falls short because, being a GBS, it should not only consider interaction processes, but also subjective elements such as emotion, motivation and fun. For this reason, the evaluation of GBS should be oriented on the playability and the Player eXperience (PX), thus evaluating not only how easy and efficient is the interaction process with the game, but also how these can be structured to generate subjective feelings such as positive emotions, satisfaction in the experience with the game system and engagement [1], oriented mainly to entertainment and fun.

To understand what is gameplay and PX, two points of view must be considered. First, the game is a software product that must be analyzed to determine its quality. The "Playability" property has been used for this because of its ability to adapt usability to GBS and because it is a more accurate measure of how much fun a game is. Second, "Player Experience", is directly related to the concept of UX, but, when referring to the context of games, it is oriented to more subjective and personal measures such as "emotion", "pleasure" or "engagement" or to measures more specific to game systems such as "immersion", "frustration" or "reaction", which are key to describe and improve the interactive experience that humans enjoy while playing [1, 2].

This article addresses the particularities that should be considered when offering pervasive gaming experiences focused on the older adult population, in order to obtain a higher degree of acceptance of the use of technology in this type of experiences and to make the best possible use of them. The document is organized as follows: Sect. 2 gives a brief description of the definition of pervasiveness, its relationship with older adults and related previous works; Sect. 3 explains the elements that should be considered to design a system based on pervasive gaming oriented to this population; finally, Sect. 4 presents the discussion, conclusions and future lines of work.

2 Background

The concept of "pervasiveness" is relatively new, but it has generated a lot of expectation and this is evidenced by the different experiences that can be found in the market today. Some technological developments that have facilitated this pervasive growth have been virtual assistants, motion and location sensors such as GPS, accelerometer and gyroscope, immersive devices such as virtual reality (VR), augmented reality (AR) and mixed reality (AR), haptic devices, and many others.

The pervasive experiences have had a great reception in the leisure and entertainment sector, creating the called "pervasive games", by providing an enriched experience, breaking the limits of the game world [3], by not being played always in a certain place, nor during a certain time, nor with a certain number of people, thus achieving an extension of the limits of conventional games with respect to what is traditionally known as "the magic circle" [4], from its spatial, social, contextual and time limits. Some uses of pervasive games are georeferenced games such as "Pokemon GO", tangible interaction systems, socially based recommendation systems such as "Foursquare" or serious games such as "Foldit" for protein folding.

Pervasive experiences in older adults are currently used as a means to achieve rehabilitation [5], physical-cognitive training [6, 7], fall prevention [8], socialization and generation of positive emotions [9]. When designing these experiences, it is possible to make use of a wide variety of elements offered by the different expansions of pervasivity, both spatially, socially, temporally and contextually, allowing, as a whole, to offer game experiences that integrate both the virtual and the real world, thus generating motivating experiences in different situations.

Based on the results of previous research [10], we have proposed a model called "pervasiveness pyramid", in order to determine in different degrees, the level of pervasiveness of a technology-based game experience, according to the different elements that are integrated in its design and in the experience it offers. This was proposed because, although a game makes use of technologies that allow pervasiveness, it does not necessarily mean that it is pervasive. This proposal also involved a set of specific transversal elements, which must be considered when designing this type of pervasive experiences. Previously this model was used and adjusted for its specific application in serious games [11], which are only one type of GBS. The proposal presented here contemplates all types of GBS, being a more complete proposal than the one presented previously.

3 Transversal Elements in Pervasive GBS

In the proposal of the "pervasiveness pyramid" we highlight 6 transversal aspects [10]: technology, narrative, aesthetics, purpose, rules and ethics. These transversal elements should be considered when designing a pervasive experience oriented to older adults, in order to offer an adequate experience to the player. Although this proposal arose as a means to determine the degree of pervasiveness of a game experience oriented to older adults, it can be applied to any type of pervasive experience oriented to any type of target audience, adjusting the respective transversal aspects. For this reason, the necessary elements for each transversal aspect will be specified below, adjusting them to the needs and particularities of the older adult population, in order to generate more motivating, fun and attractive game experiences.

3.1 Aesthetics

Digital games are symbolic and cultural tools that provide different resources and game elements such as characters, narratives and game mechanics for the interaction process [12]. All this must be accompanied by an aesthetic element that enhances the player's experience, offering a pleasant environment not only visually, but in its interaction process. Aesthetics in a pervasive game, refers to all the visual and sound elements that are presented in the game experience, as well as its usability. Here 4 basic characteristics stand out: visual, sound, interaction and support elements.

Visual Characteristics. In pervasive games, screens can be used to facilitate the process of access to information, states and game elements, such as smartphones, tablets, televisions, projectors or VR glasses. The graphic and visual level used in this type of technologies, should generate a high contrast between the text and the background of the game, being this of a size equal to or greater than 12 points [13], applying simple interfaces, with short, clear and concise text content, in bright and warm environments that are familiar to players [14]. On-screen elements should offer smooth transitions, therefore, images and texts that are presented with abrupt and fast movements that generate dizziness should be avoided [15], mainly in virtual reality environments where this symptom is easy to generate (Fig. 1).

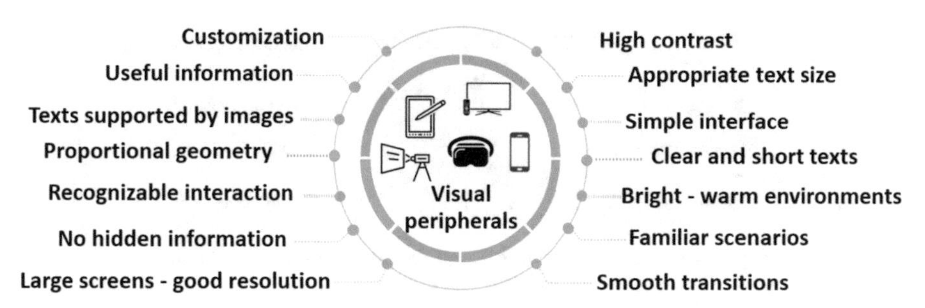

Fig. 1. Visual characteristics of the pervasive game aesthetics.

Sound Characteristics. Pervasive games can make use of different technologies that offer sound reproduction, such as virtual assistants, tablets, smartphones, televisions, projectors and VR glasses. Game sounds should be oriented to provide game status information, feedback, support, recommendations, recognition and confirmation of actions performed in the game. It is important that the sounds are easily recognizable in a soft, clear and simplified way so as not to generate discomfort in the older adult, using familiar sounds that generate confidence and relaxation [14]. These sounds should also generate immersion in the game in such a way that they enrich the experience, but should also be used as a complementary form to the text. In addition, they should be used in conjunction with colors and images, in order to provide different ways of accessing the information provided by the game [16] (Fig. 2).

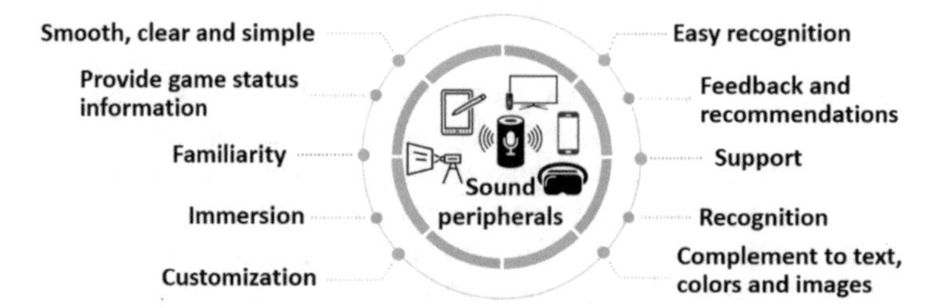

Fig. 2. Sound characteristics of the pervasive game aesthetics.

Interaction Characteristics. In pervasive gaming experiences designed for older adults, different technologies oriented to the interaction of the experience can be integrated, such as: motion sensors, touch screens, virtual or augmented reality, tabletops, virtual assistants, accelerometers, gyroscopes or GPS. All the elements described above encourage interaction with the game in a natural way through movement or physical activity, and it is necessary to provide a safe environment for the older adult so that they do not hurt themselves [17]. Some of the important elements to consider during the design of the interaction can be seen in Fig. 3.

Fig. 3. Interaction characteristics of the pervasive game aesthetics.

Supporting Characteristics. Regardless of whether the game makes use of images, sounds, or specific interaction processes, constant support elements should be provided to guide and instruct the older adult during the gaming experience. Depending on the technology used, feedback should be as complete as possible being visual, audio and even haptic to be aware of all important events in the game [18]. Contextual help, tutorials and practice spaces should be simple and clear, showing them at any time if difficulty of achievement by the older adult is detected [19] (Fig. 4).

3.2 Purpose

The purpose is the reason why a player decides to play, investing his time to live the game experiences offered. These can range from having fun and passing the time, to deeper motives such as physical or cognitive training or overcoming personal challenges. Older adults are a very heterogeneous group and their play purposes may vary according to

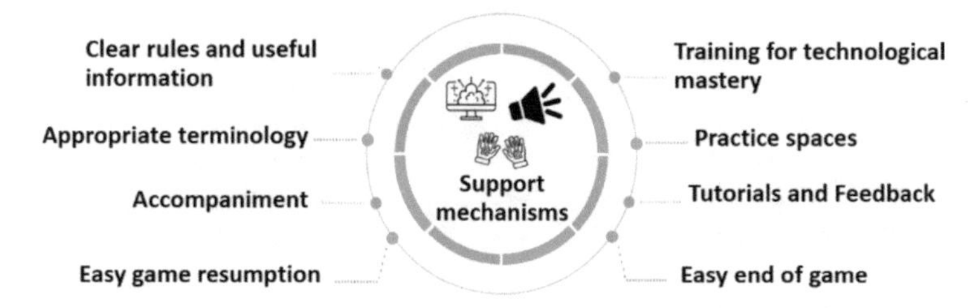

Fig. 4. Supporting characteristics of the pervasive game aesthetics.

motivators such as: perceived benefits, participation, keeping busy, social contact, enjoyment of more casual games, learning, obtaining some benefit, or generating emotional, social or health wellbeing [20] (Fig. 5).

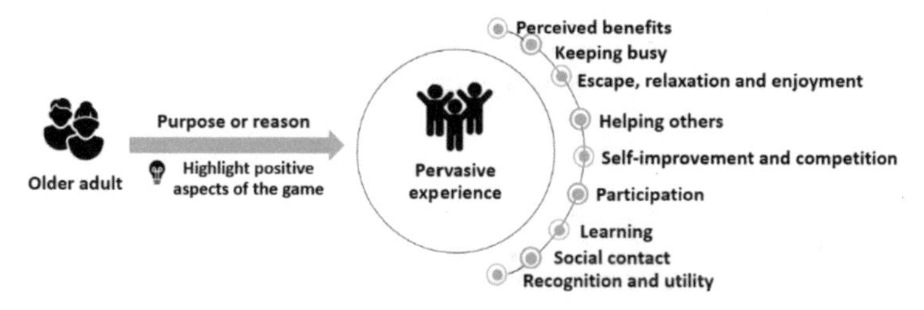

Fig. 5. The purpose for interacting with a pervasive gaming experience.

3.3 Rules

Although older adults' preferences for games are very varied, they share certain characteristics that provide an opportunity to identify in a general way the types of games that could be focused on for this population. Game rules that attract attention in older adults are games that are simple, familiar environment, slow paced, can be played for a short period of time, involve some type of cognitive challenge that is not too demanding, remove any possible barriers such as complex learning curves and difficult to learn rules, are easy to play, do not require high commitment or high skills, are visually appealing, do not punish their mistakes in the game, and can be played with a counterpart in person rather than online. It is notable that games that require fast response times are not well received by older adults, such as shooting games or games with violent themes [20] (Fig. 6).

All game elements with respect to their dynamics, mechanics and game experiences should be oriented to offer the player the reach of the flow zone, understood as the balance between the challenges of the game and the skills of the participant [21]. A pleasant game flow is achieved through experiences that allow the player to feel sensations of achievement and victory, offering a balanced difficulty in the challenges with respect to their skill, thus generating a psychological immersion [22].

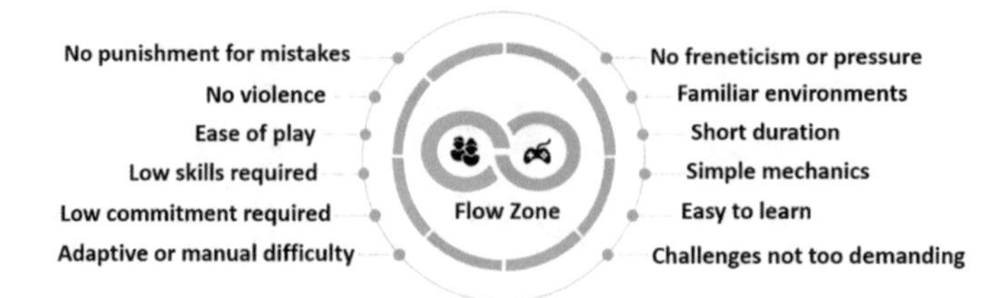

Fig. 6. Elements that enable the flow zone in older adults.

3.4 Technology

The cognitive and motor skills of older adults are directly affected by the set of interaction peripherals used in pervasive play. The muscle changes and responses of this population impact actions such as pointing and picking up items in the game. The interaction technologies provided to older adults should be those that offer few buttons in order not to generate psychological stress. In addition, this population prefers portable, direct and natural input devices such as touch screens that allow simple actions, transparently and without much effort [23] (Fig. 7).

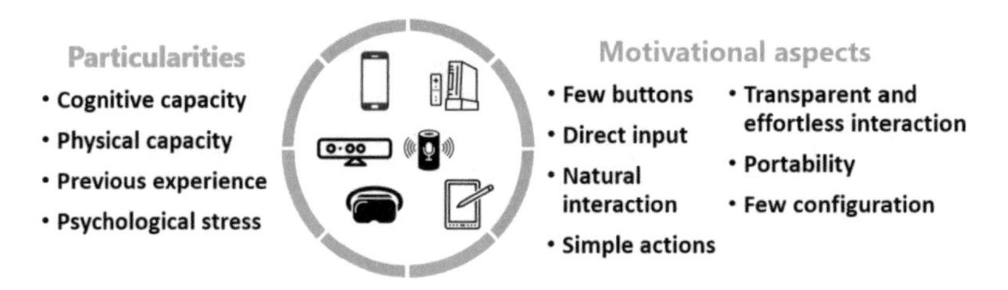

Fig. 7. Elder-friendly devices.

3.5 Narrative

The narrative of the games is responsible for generating a large part of the interaction processes of the game, because logical actions must be performed and challenges must be met through interaction with the elements offered during the story (characters, objects, scenarios). In addition, it allows interaction with other players who are involved in the story, which is a facilitating aspect to achieve pervasiveness [24]. When narrative and pervasiveness are integrated, a narrative evolution or extended narrative must be generated, where the objects, characters and all the components of the story that will be transmitted in the game experience through the narrative in the real or virtual world are defined, and can evolve from the player's behavior or the game world [11] (Fig. 8).

3.6 Ethics

Older adults are a vulnerable and influenceable population, so ethics in both the design and the game itself should be a fundamental principle that should always be incorporated

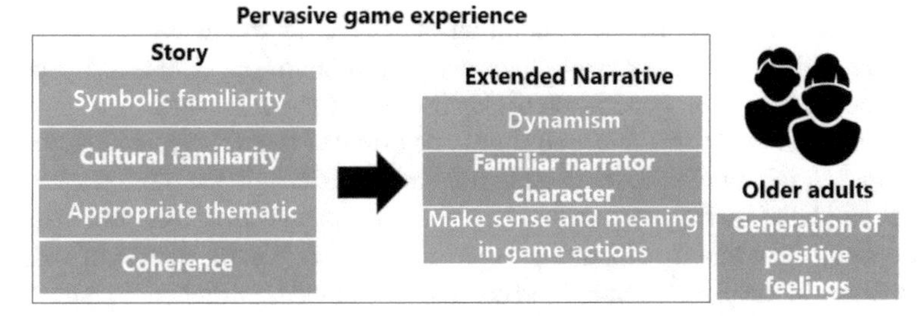

Fig. 8. History, narrative and motivations of older adults

in order to ensure their wellbeing, respect and non-misuse of the game for commercial purposes. With this in mind, the following ethical guidelines are proposed to be integrated in any type of pervasive game in order not to break the thin line of ethical and unethical in these gaming experiences: ensure physical and cognitive safety, avoid commercial uses, capture data not allowed, consider the culture and regional practices where the game will be used, promoting respect and avoiding the promotion of depression, sadness or isolation (Fig. 9).

Fig. 9. Ethical aspects to consider.

3.7 Conclusions and Future Work

The implementation of pervasive games oriented to the elderly population is not an easy task, since different considerations of the end users must be involved. Due to the benefits and good acceptance offered by this GBS approach, efforts should be focused on the construction of this type of solutions, integrating the real and virtual worlds, providing a new world of possibilities and new experiences to the elderly population.

The proposed model based on the "pyramid of pervasiveness" and its transversal elements were adequately adjusted to the particularities of the older adult population. These elements considered should offer deeper experiences to the participants, taking the game from the virtual world to the real one, with a theme of interest, a deep narrative and allowing a comfortable and safe interaction with the game experience.

The above approaches can be applied to different types of GBS, such as serious games, simulation games, gamified experiences, ludification and digital games. In the specific case of older adults, the generation of wellbeing and active aging is encouraged through leisure, entertainment, learning, promotion of physical activity, cognitive training, socialization and generation of positive emotions in general.

Future extensions of this research should be oriented to a formal validation process of the proposal, taking the model to a tangible process through the definition of a set of heuristics validated by experts that can be used as tools both in design processes and in the evaluation of the quality of gaming experiences. With this, the degree of satisfaction and enjoyment in the elderly population could be evidenced, further encouraging the use of these new technologies.

Acknowledgements. This work was supported by the FCT – Fundação para a Ciência e a Tecnologia, I.P. [Project UIDB/05105/2020].

References

1. Salazar Cardona, J., Gutiérrez Vela, F.L., Lopez Arango, J., Gallardo, J.: Game-based systems: towards a new proposal for playability analysis. CEUR Workshop Proc. **3082**, 47–56 (2021)
2. González, J.L., Gutierrez, F.L.: Caracterización de la experiencia del jugador en videojuegos (2010)
3. Arango-López, J., Gutiérrez Vela, F.L., Collazos, C.A., Gallardo, J., Moreira, F.: GeoPGD: methodology for the design and development of geolocated pervasive games. Univers. Access Inf. Soc. (2020). https://doi.org/10.1007/s10209-020-00769-w
4. Huizinga, J.: Homo Ludens: A Study of the Play-Element in Culture. Eur. Early Child. Educ. Res. J. **19**, 1–24 (1938). https://doi.org/10.1177/0907568202009004005
5. Lauze, M., et al.: Feasibility, Acceptability and Effects of a Home-Based Exercise Program (2018)
6. Loos, E.: Exergaming: meaningful play for older adults? In: Zhou, J., Salvendy, G. (eds.) ITAP 2017. LNCS, vol. 10298, pp. 254–265. Springer, Cham (2017). https://doi.org/10.1007/978-3-319-58536-9_21
7. Loos, E., Zonneveld, A.: Silver gaming: serious fun for seniors? In: Zhou, J., Salvendy, G. (eds.) ITAP 2016. LNCS, vol. 9755, pp. 330–341. Springer, Cham (2016). https://doi.org/10.1007/978-3-319-39949-2_32
8. Guimarães, V., Pereira, A., Oliveira, E., Carvalho, A., Peixoto, R.: Design and evaluation of an exergame for motor-cognitive training and fall prevention in older adults. ACM Int. Conf. Proc. Ser. 202–207 (2018). https://doi.org/10.1145/3284869.3284918
9. Pereira, F., Bermudez I Badia, S., Jorge, C., Da Silva Cameirao, M.: Impact of game mode on engagement and social involvement in multi-user serious games with stroke patients. In: International Conference on Virtual Rehabilitation, ICVR, July 2019, pp. 1–14 (2019). https://doi.org/10.1109/ICVR46560.2019.8994505
10. Salazar, J., López, J., Gutiérrez, F., Trillo, J.: Design of technology-based pervasive gaming experiences: properties and degrees of pervasiveness. Congr. Español Videojuegos (2022)
11. Salazar, J., Arango, J., Gutierrez, F.: Pervasiveness for learning in serious games applied to older adults. In: García-Peñalvo, F.J., García-Holgado, A. (eds.) Proceedings TEEM 2022: Tenth International Conference on Technological Ecosystems for Enhancing Multiculturality. TEEM 2022. LNET. Springer, Cham (2022). https://doi.org/10.1007/978-981-99-0942-1_65
12. Zhang, F., Hausknecht, S., Schell, R., Kaufman, D.: Factors affecting the gaming experience of older adults in community and senior centres. Commun. Comput. Inf. Sci. **739**, 464–475 (2017). https://doi.org/10.1007/978-3-319-63184-4_24

13. López-Martínez, Á., Santiago-Ramajo, S., Caracuel, A., Valls-Serrano, C., Hornos, M.J., Rodríguez-Fórtiz, M.J.: Game of gifts purchase: computer-based training of executive functions for the elderly. In: 2011 IEEE 1st International Conference on Serious Games and Applications for Health, SeGAH 2011 (2011). https://doi.org/10.1109/SeGAH.2011.6165448
14. Yang, Y.C.: Role-play in virtual reality game for the senior. ACM Int. Conf. Proceeding Ser. Part **F1483**, 31–35 (2019)
15. Xu, W., Liang, H.N., Yu, K., Baghaei, N.: Effect of gameplay uncertainty, display type, and age on virtual reality Exergames. In: CHI: Conference on Human Factors in Computing Systems (2021). https://doi.org/10.1145/3411764.3445801
16. Acosta, C.O., Palacio, R.R., Cortez, J., Echeverría, S.B., Rodríguez-Fórtiz, M.J.: Effects of a cognitive stimulation software on attention, memory, and activities of daily living in Mexican older adults. Univers. Access Inf. Soc. (2020). https://doi.org/10.1007/s10209-020-00742-7
17. Tahmosybayat, R., Baker, K., Godfrey, A., Caplan, N., Barry, G.: Move well: design deficits in postural based Exergames. What are we missing? In: 2018 IEEE Games, Entertainment & Media Conference, GEM 2018, pp. 24–27 (2018). https://doi.org/10.1109/GEM.2018.8516516
18. Borrego, G., Morán, A.L., Meza, V., Orihuela-Espina, F., Sucar, L.E.: Key factors that influence the UX of a dual-player game for the cognitive stimulation and motor rehabilitation of older adults. Univers. Access Inf. Soc. (2020). https://doi.org/10.1007/s10209-020-00746-3
19. Lee, S., Oh, H., Shi, C.K., Doh, Y.Y.: Mobile game design guide to improve gaming experience for the middle-aged and older adult population: User-centered design approach. JMIR Serious Games **9**, 1–18 (2021)
20. Salazar, J., Gutièrrez, F., Arango, J., Paderewski, P.: Older adults and types of players in game-based systems: classification based on their motivations. In: ACM International Conference Proceeding Series. Association for Computing Machinery (2022). https://doi.org/10.1145/3549865.3549900
21. Csikszentmihalyi, M.: Flow: the psychology of optimal experience. Acad. Manag. Rev. **16**, 636 (1991). https://doi.org/10.2307/258925
22. Tabak, M., De Vette, F., Van DIjk, H., Vollenbroek-Hutten, M.: A game-based, physical activity coaching application for older adults: design approach and user experience in daily life. Games Health J. **9**, 215–226 (2020). https://doi.org/10.1089/g4h.2018.0163
23. Doroudian, A., Loos, E., Ter Vrugt, A., Kaufman, D.: Designing an online escape game for older adults: the implications of playability testing sessions with a variety of Dutch players. In: Gao, Q., Zhou, J. (eds.) Human Aspects of IT for the Aged Population. Healthy and Active Aging. HCII 2020. LNCS, vol. 12208, pp. 589–608. Springer, Cham (2020). https://doi.org/10.1007/978-3-030-50249-2_42
24. Arango López, J.: GeoPGD: Metodología para la Implementación de Juegos Pervasivos Georreferenciados Apoyados en Linked Open Data (2019). http://repositorio.unicauca.edu.co:8080/xmlui/handle/123456789/1784

The Effect of Using the Chatbot to Improve Digital Literacy Skill of Thai Elderly

Kanyarat Sriwisathiyakun(✉)

School of Industrial Education and Technology, King Mongkut's Institute of Technology Ladkrabang, Bangkok 10520, Thailand
kanyarat.sr@kmitl.ac.th

Abstract. This research aims to demonstrate the learning success and contentment of elderly people in Thailand using a chatbot innovation for improving digital literacy. Three parts of the research were carried out: Phase 1 of the research focused on developing the chatbot and any relevant educational digital media; Phase 2 engaged in pre-experimental by experts' validation; and Phase 3 involved an experiment where the chatbot was deployed with elderly people. Samples were collected from 33 elderly people. The information was gathered through expert interviews, pretests, and posttests on chatbot usage, and satisfaction surveys. The data was then analyzed using the dependent t-test, percentage, mean, standard deviation, and content analysis. Results showed that our Senior See Net chatbot was simple to use and navigate, with an appropriate artistic appearance. Findings demonstrate that this chatbot was simple to use and access, had an appropriate creative composition, and was made up of sufficiently thorough media and content on digital literacy. Furthermore, the statistical analysis revealed that after using the chatbot, the elderly's understanding of digital literacy improved statistically significant at the 0.05 level. Furthermore, their overall satisfaction was rated high.

Keywords: Digital literacy · Chatbot · Elderly · Learning outcome · User satisfaction

1 Introduction

Digital literacy skills are practically a requirement in this day and age, especially for the elderly. According to statistics, more elderly people were using digital platforms, and this increase in senior users in social networks was noticeable. Hence, the elderly seems to be playing a more vital role in society's news circulation, and as a part of the cycle, the elderly could be exposed to positive and negative physical and mental impacts. On this note, the elderly should be protected from victimization caused by currently problematic online scams and equipped with fundamental skills and experiences that would help them further autonomously learn and solve problems as they arise from an online platform, which includes digital and information competencies [1, 2]. Promoting digital literacy for the elderly is a form of social immunization. Digitally competent

A. Rocha et al. (Eds.): WorldCIST 2023, LNNS 800, pp. 348–358, 2024.
https://doi.org/10.1007/978-3-031-45645-9_33

seniors are projected to have sufficient skills to protect personal information, choose and use digital tools, and understand how to access information online securely. To build an active aging society, literacy training for the elderly is perceived as a crucial milestone [3], and chatbots are viewed as a beneficial supplement to reinforce such development. Chatbots are conversational AI agents that allow users to interact with them by exchanging textual and audio inputs and outputs. They are perceived as a convenient and personalized solution for users to access services [4]. Additionally, chatbots have the ability to make informal connections with users, allowing them to be used to support ubiquity, self-directed learning, and personal and professional development [5].

This study developed a chatbot innovation and investigated elderly learning outcomes and satisfaction after using the chatbot to improve digital literacy. The chatbot was planned to be used as a communication tool to improve digital literacy among Thai elderly, who can learn and develop their digital literacy abilities by conversing with chatbots that offer intelligent services. This study was aimed at contributing to national policies to prepare Thailand for a complete transition to an aging society. Research questions were studied as follows: 1) Will the elderly's digital literacy significantly improve after experiencing the chatbot? 2) Is it satisfactory for the elderly to learn about digital literacy through chatbots? 3) What are the elderly' thoughts on utilizing the chatbot to improve their digital literacy?

2 Literature Review

2.1 Chatbot Development

Chatbots are AI assistants and computer applications that utilize natural language processing (NLP) to virtualize convincing human-machine conversational interactions [6, 7]. As a technological solution, chatbots have been widely implemented in several fields and services, ranging from online customer support to education. Chatbots will improve learning engagement by being available around the clock, giving prompt and consistent responses, and making the overall learning process more engaging for the participants. Evidence of chatbots in educational services includes self-service learning support and critical thinking practices, where humans autonomously converse with AI agents for skill development following specific goals and paces [8].

Chatbots are most commonly used to help students strengthen their skills, according to the present state of chatbot use in education. Chatbots are being utilized in pedagogy to enhance learning and teaching activities, which helps students learn new skills and become more motivated [9]. Additionally, it was discovered that meaningful learning was considerably improved by a chatbot that could provide scaffolds via voice and text when necessary during an online video course. This type of voice based chatbot has been successful in motivating elderly people to exercise for better health. Elderly users of the "FitChat" AI chatbot on mobile were evaluated, and it was discovered that they prefer voice-based chat over text notifications or free text entry and that voice is an effective method of boosting motivation [10]. The impact of AI chatbot on education is notable. In November 2022, OpenAI released ChatGPT (Generative Pre-trained Transformer), an artificial intelligence chatbot that is as easy to use as a search engine but can generate almost any text in a conversational style upon request in response to a human query. As a result, it can provide students and educators with personalized, real-time responses to their questions and needs. Imagine being able to swiftly scan through many academic papers by utilizing ChatGPT to generate summaries or abstracts. This could save learners hours of time and effort, allowing them to focus on the most relevant and important information [11].

Chatbots engage with users through conversational user interfaces (CUI) or integration with current messaging platforms rather than graphical user interfaces (GUI) [12]. The chatbot's job is to interpret user requests expressed in messages and respond appropriately. In general, a chatbot design has a user interface component for receiving the user's request through a text- or speech-based application, such as a messaging program like the Line application, Facebook Messenger, WhatsApp, WeChat, etc. The user messages will subsequently be sent to the user message analysis component. In order to extract the user's goal, which is the intent, and the entities from the user request, which are the parameters of intent, the user's language is identified and translated into the language of the chatbot's NLU (Natural Language Understanding). A Dialog Management Component that controls and updates the conversational context and anticipates the next course of action is also part of the chatbot architecture. The "next action" may involve returning to the user with a suitable message, asking for further context information to complete any missing entities, or it may involve locating the data required to carry out the user's intent through external APIs calls or database queries. Response Generation Component uses pre-defined templates with placeholders to generate response messages for users. After the user sees the right message, the chatbot enters a waiting mode and waits for the next input of data [13, 14].

Using the Software Development Life Cycle (SDLC) concept, chatbot software is created. The SDLC is a framework that describes the various activities and tasks that need to be carried out during the software development process. The waterfall model is the longest-standing and best-known SDLC paradigm. This paradigm is often used by both big businesses and government organizations. Its distinguishing characteristic is the model's sequential steps. This model's steps don't cross over, so each step starts and ends before moving on to the next, creating a waterfall effect [15].

In this study, we have developed the chatbot for the elderly in Thailand based on the chatbot development framework from the previous research sequence as shown in Fig. 1 [16].The framework uses Line messaging and API.AI (Dialogflow) as the main platform. The planning, analysis, design, implementation, and deployment phases of the waterfall SDLC model served as the framework for constructing a chatbot.

2.2 Digital Literacy for Elderly

Digital literacy can refer to abilities gained from knowledge and skills to use digital devices (such as smartphones, tablets, and personal computers) as well as an Information and Communications Technology (ICT) skill set, which is crucial in today's computerized world [17, 18]. Because of lacking adequate digital literacy, many seniors today are overwhelmed by the information that is available online, including both truths and lies. Despite the benefits that information technology (IT) offers, even while IT has the potential to improve the elderly's quality of life, those who lack digital literacy are missing out on a number of enriching opportunities. Thais would be wise to equip themselves with digital literacy skills since the internet has become a fundamental force in Thai society. These skills should include using digital media in digital formats like social media, email, online transactions, and web sites, as well as securely accessing other forms of media like videos and chatting. In light of this, it is essential to discover engaging learning techniques that will aid in their development of cybersecurity and digital literacy skills in order to reduce potential online risks [19–21]. In order to increase Thai students' knowledge, [22] used chatbot technology as a digital learning tool. It was discovered that the use of chatbot in educational settings to enhance students' research literacy produced beneficial learning outcomes and helped create a more personalized learning environment for students. Because of this, chatbots are one of the learning tools that researchers are interested in adopting to assist the elderly in developing digital literacy skill.

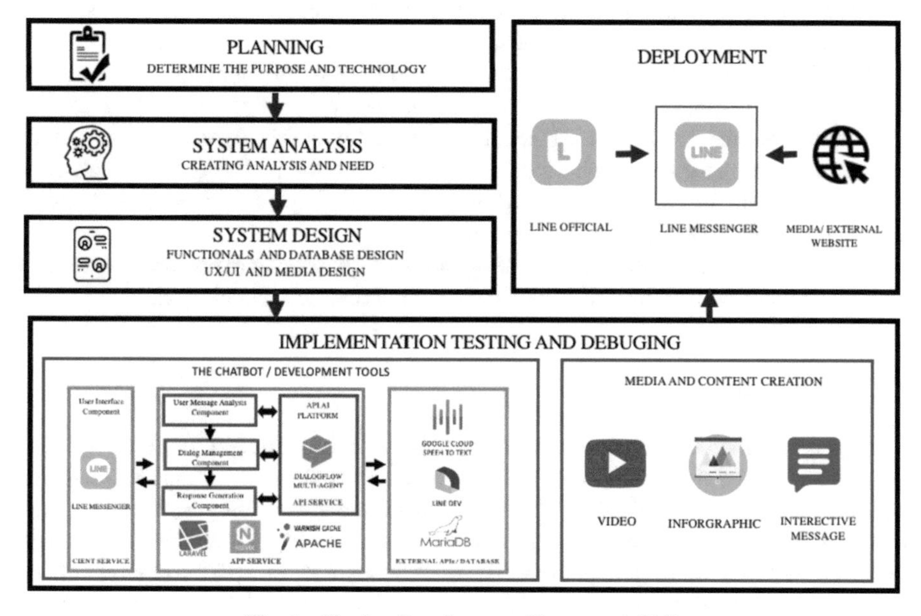

Fig. 1. Chatbot Development Framework [16]

3 Methodology

This study used a research and development methodology to experiment and analyze the chatbot. The investigation was carried out in three phases: 1) Chatbot, instructional resources, and content were created and delivered to professionals for approval; 2) Before the chatbot was made available to elderly people, it was revised by using the advice and comments of experts;3) The chatbot was launched for experimentation, and dimensions of outcomes were investigated, including knowledge of digital literacy, user satisfaction, and questionnaire suggestions.

3.1 Participants

The data for this study was gathered through the completion of questionnaires with 33 Thai elderly people.

3.2 Data Collection and Analysis

The data collected from the elderly people and the experts were separately analyzed in four sections as follows:

Section 1: Qualitative analysis was employed to process experts' feedback and suggestions, and results were incorporated to revise the chatbot before the experimentation with the participants.

Section 2: Knowledge of digital literacy were compared before (pretest) and after (posttest) the chatbot experimentation. The data in this section were analyzed by mean, standard deviation, and dependent t-test.

Section 3: The satisfaction data obtained from the elderly who utilized the chatbot were statistically analyzed in percentage, mean, and standard deviation. Five experts validated response items in the satisfaction survey for Item-Objective Congruence (IOC) and confidence via the Cronbach's alpha coefficient. The survey had gone through a try-out with ten samples before it was used with the 33 participants. The satisfaction data were reported in mean and standard deviation.

Section 4: The data gathered from the open-ended questionnaire was processed using qualitative analysis. The overall results were summarized in a conclusion.

4 Results

This study designed and constructed a chatbot, called "Senior See Net", within the SDLC framework in five steps using Chatbot Development Framework in Fig. 1. To design and construct the Senior See Net Chatbot, relevant tools, technologies, and data were extracted from research papers, journals, academic articles. The chatbot was designed to work with the Line messaging platform because the majority of the Thai elderly utilized Line for communicating. Dialogflow (dialog-flow.cloud.google. Com) was used to build the chatbot's initial prototype system, which was then tested for responsiveness and utility. Media and content were created in interactive messages, videos, and infographics with functions that are most friendly to senior users. After the Senior See Net Chatbot had been fully developed, it was sent to the five experts for a pre-experimental assessment.

Section 1: Feedback and suggestions from the five experts, two specialized in Information Technology, one in Educational Technology, and two in the Thai elderly were summarized in Table 1.

Table 1. A summary of feedback and suggestions of the five experts on the chatbot

Issues	Experts	Comments/suggestions
Accessibility	Information Technology	It was simple to use and access the chatbot. Since the menu and its functions were composed of both visuals and text, they were simple to use. The elderly key menu items were gathered in one location
	Educational Technology	Pre-use instructions were provided, and the elderly might find them beneficial when learning to use and access the chatbot
	Elderly	The chatbot was simple to operate and not overly complex for the elderly
Artistic composition	Information Technology	The layout and color scheme were acceptable and straightforward. Given the common eyesight problems that seniors may have, the font size seemed appropriate. The chatbot menu was aligned in the sense that it was friendly to the target group
	Educational Technology	The chatbot was designed with appropriate fonts and font sizes. The colors were plain and friendly to elderly' eyes
	Elderly	The fonts were extra-large. The colors were toned to comfort the eyes. The chatbot was suitable for elderly
Media and content	Information Technology	The information was sufficiently comprehensive to teach elderly digital literacy
	Educational Technology	The information was properly organized. Elderly should be able to understand media and content in order to interpret, verify, and take action on them
	Elderly	The content was well-diversified since it included options like infographics and videos that might aid elderly in their learning processes
Functionality	Information Technology	The menu options included all of the functions that elderly would need. For instance, the schedule reminder was regarded as the chatbot's best feature. The elderly would typically have all of the necessary features

(*continued*)

Table 1. (*continued*)

Issues	Experts	Comments/suggestions
	Educational Technology	Games were put into place to help elderly review their knowledge. Links to relevant external content were present
	Elderly	To enable instant communication between elderly and agents, the agent contact feature should be revised

Section 2: The elderly digital literacy comparison of pre- and post- Senior See Net Chatbot experiment results in Table 2.

Table 2. The pretest and posttest results on digital literacy n = 33

Tests	n	$\overline{X}$	SD	t	Sig
Pretest	33	13.09	2.42		
				4.84*	0.0000
Posttest	33	14.76	2.32		

* At the 0.05 significance level

Table 2 revealed that the elderly yielded higher posttest scores after learning with chatbot than their pretest scores at the 0.05 level.

Section 3: Most of the elderly respondents was female (81.82%), in the age range of 60–70 years (81.82%), followed by 71–80 years (15.15%) and lower than 60 years (3.03%), respectively. Their highest education was lower than the bachelor's level (69.70%), followed by the bachelor's level (24.24%), and the master's level (6.06%), respectively. The data in Table 3 analyzed in mean, and standard deviation indicated that the respondents were moderately satisfied overall ($\bar{x} = 3.57$, SD = 0.67). Dimensionally, the satisfaction report is broken down by item as follows:

Table 3. The elderly satisfaction in Senior See Net Chatbot, n = 33.

Assessment	$\overline{X}$	SD	Result
1. Perceived benefits from the Senior See Net Chatbot	4.14	0.61	High
2. Accessibility and ease of use	3.97	0.61	High
3. Content, design, and functionality of the chatbot	4.41	0.57	High
4. Overall user satisfaction	4.15	0.87	High

1) Perceived benefits from the Senior See Net Chatbot: The respondents' satisfaction in the chatbot on perceived benefits was rated high ($\bar{x} = 4.14$). Specifically, the most satisfying aspect was that "the chatbot proved to be a helpful assistant in daily life", followed by "a useful assistant in educating digital-platform security", and "helped enhance digital literacy skills", respectively.
2) Accessibility and ease of use: The respondents' satisfaction in the Senior See Net Chatbot on the ease of use was rated high ($\bar{x} = 3.97$). Specifically, the most satisfying aspect was that "the menu and functions within the chatbot were easy to use", followed by "the user was able to seamlessly and continuously use the chatbot", and "the user was able to autonomously learn to operate and operate the chatbot", respectively.
3) Content, design, and functionality of the chatbot: The respondents' satisfaction in the Senior See Net Chatbot on the content, design, and functionality of the chatbot was rated high ($\bar{x} = 4.14$). Specifically, the most satisfying aspect was that "the chatbot's design was suitable, beautiful, and modern", followed by "the media and content presented via the chatbot were suitable and appealing", and "the functions were practical for everyday use", respectively.
4) Overall satisfaction: The respondents were overall highly satisfied with the Senior See Net Chatbot ($\bar{x} = 4.15$).

Section 4: The data obtained from open-ended questions was analyzed and revealed that most senior citizen respondents expressed the need for extra training activities introducing ways to operate the Senior See Net Chatbot and a user manual for later consultation. The reasoning was that training would accelerate their learning of how to use the chatbot when compared to self-training. At least, the training might be provided for senior leaders in the group to promote more effective use of the Senior See Net Chatbot, as these trained representatives would be able to pass on knowledge to other elderly in a broader scope and in a shorter time, such as in other elderly networks in Thailand.

5 Discussion

In terms of chatbots, [23] viewed them as a form of AI that could be used to interact and converse with the elderly, who might have age-related challenges. Specifically for this study, Senior See Net Chatbot has been created to help Thai elderly improve their digital literacy. The Chatbot was described by the experts as easy to access and utilize, designed with a suitable artistic composition, and containing adequately comprehensive content on digital literacy education for the elderly. To answer the first research question, the test results of digital literacy understanding before (pretest) and after (posttest) the chatbot experimentation were compared. At a significance level of 0.05, it was found that the elderly citizens had significantly greater levels of digital literacy after using the Senior See Net Chatbot. The results of this study support [22]'s findings that using chatbot as a digital assistant tool significantly increases learners' knowledge.

Regarding the second research question, overall satisfaction with the Senior See Net Chatbot was rated high. This finding is in line with [22], who also reported that the chatbot had a high level of learner satisfaction. Chatbot provided knowledge in a fun, interesting, and innovative way, and it was effective when used as a digital learning tool to provide personalized learning support. Engaging with the elderly through a chatbot application designed and developed exclusively for them is more satisfying and likely to help them achieve the chatbot's goals. Correspondingly, [24]'s research revealed that older adults saw a chatbot, a form of virtual companion, based on an embodied conversational agent as a helpful tool that may enhance their everyday life. They expressed satisfaction with the application, expressed enjoyment with the features and activities offered, and indicated that they would suggest it to friends.

The third research question was answered in the fourth section of the above results. The elderly's needs are in line with [25]'s research findings, which concluded that the elderly's learning is made easier with the assistance of family and friends. The most optimal learning format was determined to be a blend of traditional and digital learning formats. The main challenges were a lack of qualified, motivated, and trained educators.

6 Conclusion and Suggestions

The results obtained from the development of the chatbot for digital literacy improvement were projected to introduce new experiences to the elderly in Thailand. The Senior See Net Chatbot could work as a virtual assistant or a communication representative who interacts with users via the Line Application Platform, using diverse content formats, i.e., short videos, images, infographics, and conversational interactions. Furthermore, the chatbot was also intended to be distributed to large elderly groups via group leaders, who would then share knowledge with elderly in other networks to increase the number of users. The mentioned plan would be driven by the Education Center for the Thai Elderly, Ministry of Social Development and Human Security to position the chatbot innovation as a policy driver for the safer use of information technology among the elderly in Thailand.

References

1. Loureiro A., & Barbas M.: Active Ageing – Enhancing digital literacies in elderly citizens. In: Zaphiris P., Ioannou A. (eds) Learning and Collaboration Technologies. Technology-Rich Environments for Learning and Collaboration. LCT 2014. Lecture Notes in Computer Science, vol 8524. Springer, Cham. (2014). https://doi.org/10.1007/978-3-319-07485-6_44
2. Chaichuay, W.: Elderlies' experience in using LINE application: a phenomenological study. Veridian E-Journal, Silpakorn University, Thai edition in the field of humanities: social sciences: andarts. 10,1 (January-April 2017) : 905–918 (2017)
3. Pipatpen, M.: Education for the elderly: a new educational dimension for the learning society of Thailand. International Journal of Behavioral Science Copyright 2017, Behavioral Science Research Institute 2017, Vol. 12, Issue 1, 43–54 (2017)
4. Brandtzaeg, PB., Følstad, A.: Chatbots: Changing user needs and motivations. interactions 25(5):38–43 (2018). https://doi.org/10.1145/3236669
5. Garrett, M.: Learning and educational applications of chatbot technologies: Cinglevue International. (2017). [Online]. Available: https://cinglevue.com/learning-educational-applications-chatbot-technologies/. [Accessed September 26, 2021]

6. Almansor, E.H., Hussain, F.K.: Survey on intelligent chatbots: state-of-the-art and future research directions. In: Barolli, L., Hussain, F.K., Ikeda, M. (eds.) Complex, Intelligent, and Software Intensive Systems: Proceedings of the 13th International Conference on Complex, Intelligent, and Software Intensive Systems (CISIS-2019), pp. 534–543. Springer International Publishing, Cham (2020). https://doi.org/10.1007/978-3-030-22354-0_47
7. Shevat, A.: Designing bots: Creating conversational experiences. O'Reilly Media, Inc. (2017)
8. Goda, Y., Yamada, M., Matsukawa, H., Hata, K., Yasunami, S.: Conversation with a chatbot before an online EFL group discussion and the effects on critical thinking. J. Inf. Syst. Educ. **13**(1), 1–7 (2014)
9. Wollny, S., Schneider, J., Di Mitri, D., Weidlich, J., Rittberger, M., Drachsler, H.: Are we there yet? - a systematic literature review on chatbots in education. Front. Artif. Intell. **4**, 654924 (2021). https://doi.org/10.3389/frai.2021.654924
10. Wiratunga, N., Cooper, K., Wijekoon, A., Palihawadana, C., Mendham, V., Reiter, E., Martin, K.: FitChat: Conversational Artificial Intelligence Interventions for Encouraging Physical Activity in Older Adults. ArXiv, abs/2004.14067. (2020)
11. Metzler, K., ChatGPT.: How ChatGPT Could Transform Higher Education. (2022). [Online]. Available: https://www.socialsciencespace.com/2022/12/how-chatgpt-could-transform-higher-education/. [Accessed December 30, 2022]
12. Rollande, R., Grundspenkis, J., Mislevics, A.: Systematic approach to implementing chatbots in organization - RTU Leo showcase. BIR Workshops. (2018)
13. Kompella, R.: Conversational AI chatbot Architecture overview. [Online]. (2018). Available: https://towardsdatascience.com/architecture-overview-of-a-conversational-ai-chat-bot-4ef3dfefd52e. [Accessed September 26, 2021]
14. Adamopoulou, E., Moussiades, L.: Chatbots: History, technology, and applications. Machine Learning with Applications. Vol 2. Dec. 2020. (2020). https://doi.org/10.1016/j.mlwa.2020.100006
15. Rather, M.A., Bhatnagar, V.: A comparative study of softwaredevelopment life cycle models. International Journal of Application or Innovation in Engineering & Management (IJAIEM).Volume 4, Issue 10, October 2015. (2015)
16. Sriwisathiyakun, K., Dhamanitayakul, C.: Enhancing digital literacy with an intelligent conversational agent for senior citizens in Thailand. Educ. Inf. Technol. **27**, 6251–6271 (2022). https://doi.org/10.1007/s10639-021-10862-z
17. Friemel, T.N.: The digital divide has grown old: determinants of a digital divide among seniors. New Media Soc. **18**(2), 313–331 (2016). https://doi.org/10.1177/1461444814538648
18. Deursen, A. J.A.M V., Helsper, E.J., Eynon, R.: Development and validation of the Internet Skills Scale (ISS), Information, Communication & Society, 19:6, 804–823 (2016). DOI: https://doi.org/10.1080/1369118X.2015.1078834
19. Tayati, P., Disathaporn, C., Onming, R.: The Model of Thai Elderly Learning Management for Information and Communication Technology Literacy. Veridian E-Journal, Silpakorn University, Vol.10 No. 3 September-December 2017. (2017)
20. Takagi, H., Kosugi, A., Ishihara, T., Fukuda, K.: Remote IT education for elderly. In Proceedings of the 11th Web for All Conference (W4A '14). Association for Computing Machinery, New York, NY, USA, Article 41, 1–4 (2014). DOI:https://doi.org/10.1145/2596695.2596714
21. Blackwood-Brown, C.G.: An Empirical Assessment of Elderly' Cybersecurity Awareness, Computer Self-Efficacy, Perceived Risk of Identity Theft, Attitude, and Motivation to Acquire Cybersecurity Skills. Doctoral dissertation. Nova Southeastern University. Retrieved from NSUWorks, College of Engineering and Computing. (1047). (2018)
22. Vanichvasin, P.: Chatbot Development as a Digital Learning Tool to Increase Students' Research Knowledge. International Education Studies. 14. 44. https://doi.org/10.5539/ies.v14n2p44. (2021)

23. Tascini, G.: AI-Chatbot Using Deep Learning to Assist the Elderl. In book: Systemics of Incompleteness and Quasi-Systems. 303–315 (2019). https://doi.org/10.1007/978-3-030-15277-2_24
24. Jegundo, A.L., Dantas, C., Quintas, J., Dutra, J., Almeida, A.L., Caravau, H., Rosa, A.F., Martins, A.I., Rocha, N.P.: Perceived Usefulness, Satisfaction, Ease of Use and Potential of a Virtual Companion to Support the Care Provision for Older Adults. Technologies. 8. 42. (2020). https://doi.org/10.3390/technologies8030042
25. Prodromou, M., Lavranos, G.: Identifying latent needs in elderly digital literacy: the PROADAS study. European Journal of Public Health. 29. (2019). https://doi.org/10.1093/eurpub/ckz186.092

Rethinking the Approach to Multi-step Word Problems Resolution

Abdelhafid Chadli[1(✉)], Erwan Tranvouez[2], Abdelkader Ouared[1], Mohamed Goismi[1], and Abdelkader Chenine[1]

[1] University Ibn Khaldoun of Tiaret, Tiaret, Algeria
chadli68@yahoo.fr
[2] LIS–Aix Marseille Université, Marseille, France

Abstract. Mathematical multi-step word problems (MSTWP) have received a lot of research attention from academics and researchers in the past, and a variety of solutions have been offered. The main reason why students commonly fail with MSTWP is because of how the many components of an MSTWP interact to one another. According to the literature, the stage of the problem-solving process that involves creating the solution plane is the most challenging. Students require assistance in recognizing the semantic relationships between the problem's components and information organization to facilitate problem translation and solution in order to solve this issue.

In order to help students learn how to resolve conflicts, we suggest in this work to rethink the MSTWP resolution technique by using a goal-driven approach during the making plan stage of the resolution process to assist student in learning how to effectively solve MSTWP.

Keywords: Goal Driven Approach · Polya's Problem Solving Strategy · Mathematical Multi-step Word Problems

1 Introduction

The difficulty with word problems is that so many pupils still struggle to solve them! One of the most challenging and irritating tasks for teachers, according to their reports, is helping pupils improve their word problem-solving skills [1].

In fact, solving word problems is a difficult multi-step process that requires students to read the problem, comprehend the statement it makes, recognize the key details in the text, create an abstract mental model of the problem, decide on the solution steps, and then implement those steps to solve the problem [2]. Additionally, a number of studies have demonstrated that word problem solving is frequently a challenging and complicated activity for many pupils, see for instance [3].

The method that math is currently taught is geared toward helping students learn and retain procedures and facts. This approach does not promote critical thinking or problem-solving skills. It is possible to develop mathematical word problem solving abilities by employing complicated thinking through systematic and successful self-regulation, as

A. Rocha et al. (Eds.): WorldCIST 2023, LNNS 800, pp. 359–365, 2024.
https://doi.org/10.1007/978-3-031-45645-9_34

well as by providing effective instruction [4]. Additionally, techniques and strategies for problem-solving should be thorough and incorporate both doing and thinking activities in addition to being directed toward doing activities.

Students need to understand how the many components of a multi-step word problem relate to one another in particular. This demonstrates that in order to properly solve multi-step word problems in mathematics, it is essential to first comprehend the narrative before focusing on the semantic intricacies of the relationships between the various parts of the solution.

In this article, we suggest using a mobile application called "KHOTWA" to help students solve MSTWP by using a goal-oriented approach. The four stages of Polya's strategy for problem-solving provide the foundation for the creation of the mobile app [5].

2 Related Work

A mathematical multi-step word problem is a circumstance that occurs in the real world where numerical numbers are provided, the values of a number of quantities are available, but the values of several additional quantities are unknown.

The keywords method, the multi-step based problem-solving strategy, and the schema-based approach are three often taught problem-solving techniques.

2.1 Keywords Method

The "keyword" method, in which students are taught key words that serve as cues for what arithmetic operation to employ when solving problems, is one typical instructional strategy. Students discover, for instance, that the terms completely, in all, and sum denote the application of the addition operation, whereas left denotes subtraction [6]. The words times and among both suggest the necessity for multiplication and division, respectively. Some authors [6, 7] asserted that this training leads to students reacting to cue words at the superficial level of analysis rather than engaging in a deep-understanding analysis of the relationships between the word and its surroundings.

2.2 Multi-step Based Problem Solving Strategy

Students are given a list of steps to take in order to solve a problem using a multi-step method to mathematical problem solving. The reasoning behind this is consistent with how long division is taught in school. The first publication of George Polya's book "How to Solve It" in 1945, a mathematician at Stanford University, is credited as the beginning of the multi-step problem-solving approach [5].

The use of Polya's model, according to some scholars, encourages various misconceptions. Griffin and Jitendra [8] observed, for instance, that the plan step involves adopting a broad strategy for the problem-solving activity. Despite all of its detractors, Polya's paradigm is still widely employed in scientific research projects.

2.3 Computer and Mobile Assisted Instruction

The teaching of mathematics can benefit significantly from the use of technology [9]. Particularly, educational technology like laptops, tablets, and other applications has gained popularity. It has been claimed that mobile-assisted instruction is beneficial in assisting children's education regardless of their aptitudes and can assess language, reading, motivation, and ability to solve problems [10]. For instance, Sheriff and Boon [11] investigated the effects of computer-based graphic organizers when solving one-step word problems using Kidspiration 3 software. Results showed that when using computer-based graphic organizers instead of more conventional instructional methods, all students' ability to solve one-step word problems increased.

3 System Layout

For the purpose of addressing mathematical problems, a wide range of intriguing software products have been created [12]. For instance, the computerized problem-solving tool created by Chang et al. [13] uses solution trees and schema representations. However, employing the resolution tree method fosters at least two myths. The learner first chooses the built-in steps of the solving process in the plan stage of Polya's model based on his preferred sequence of problem-solving. In this sense, the plan is restricted to a list of sequential built-in procedures and may not necessarily emphasize the significance of outlining the relationships between the problem's many steps that must be solved in order to solve it successfully [14]. Second, the resolution tree is created during the planning step rather than at the execution stage, when it would be more advantageous for students. As a result, we use the goal-driven approach to assist students in creating solution plans that are displayed as goal-driven plan-trees.

3.1 Goal-Driven Plan-Tree

The relationships between the steps in our software system are represented using a schema. The goal-driven plan-tree is a diagram that shows a collection of interconnected goals that have been assembled to give a precise explanation of the problem-solving procedure. The KAOS technique [15], a goal-oriented approach to capturing software requirements for requirements engineering, is the first fundamental principle underlying the goal-driven plan-tree.

By analogy, our system helps students identify the highest level goal (primary issue of the problem) in the building plan phase of Polya's model by posing the question, "Why do we want that?" Additionally, students must ask themselves questions like "how shall we achieve that objective?" in order to find more detailed subgoals. The cycle repeats itself until the lowest goals can no longer be broken down into lower goals. The learner must specify the arithmetic operator to use between subgoals of the same level in order to reach the goal of the immediately higher level after recognizing the primary goals and all subgoals. An example of a goal-driven plan-tree is shown in Fig. 1.

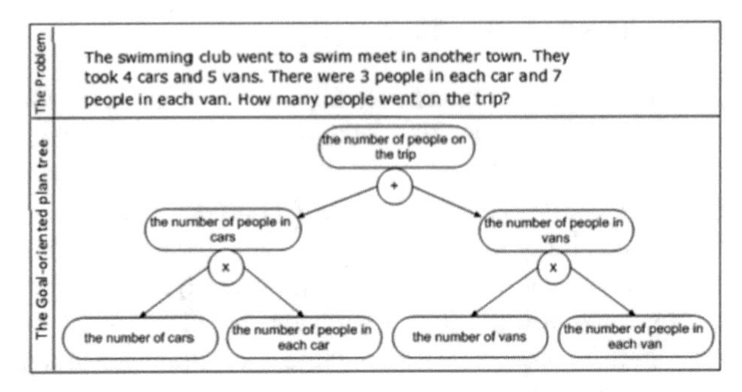

Fig. 1. An illustration of a goal-driven plan-tree.

3.2 The Guidance Process of Problem-Solving

Figure 2 displays the staged guidance process for students' problem-solving. In order to lighten the stress on users learning how to operate the system, a straightforward interface was used to assist students in the problem-solving process. The student learning-path controller pulls the student's information from the student information repository after they have logged in, and any notifications of newly posted problems are then displayed. The student has the option of selecting a problem from those that have been submitted by teachers or letting the system choose a problem for him from the database of study materials. Next, the four support modules help the student through each step of Polya's problem-solving process: (1) Understanding the problem, (2) Making a plan, (3) Executing the plan, and (4) Revising the solution. In the end, a module for assessment evaluates and assesses the students' performance, and feedback is produced. The following describes each stage of the app's advice process.

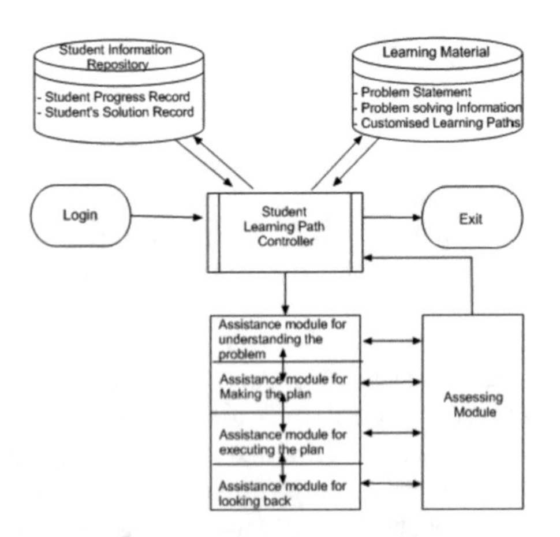

Fig. 2. Students' problem-solving guidance process.

Understand the Problem. The system first shows the student the problem statement (see Fig. 3 for an example). The goal of this stage is for students to comprehend the purpose of the problem statement, identify the known elements of the problem, the unknowns, and the relationship between them. The system provides several boxes at this stage that contain information on all of the problem's components, some of which are extraneous and aren't relevant to the solution. The initial screen (see Fig. 3-a) asks the learner to choose the primary objective from the list of suggested elements. The student then provides information regarding knowns and unknowns of the problem by selecting various colors for the boxes enclosing elements in the following screen, as shown in Fig. 3-b.

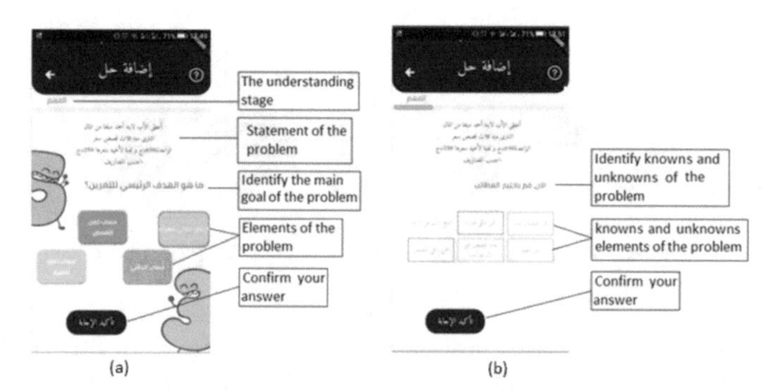

Fig. 3. (a)-The first stage's screenshot (understand the problem). (b)- The first stage's screenshot (detecting the problem's knowns and unknowns)

Plan Design. The heart of our contribution is this stage. The main purpose of this stage is to give the student some built-in potential goals and subgoals for creating the plan, from which the student chooses the highest level goal and then subgoals of each level that help meet the higher level goal (parent goal). There are also goals that are unsuccessful in helping find a solution. When the learner touches the high-level goal icon (see Fig. 4), built-in objectives already recorded in the system are offered as potential candidates for his selection as he arranges the order of goals by level of hierarchy to solve the problem. By picking the subgoals at each level of the hierarchy, the learner increases the size of the goal-driven plan-tree. The system establishes the relationship between a subgoal and its parent goal automatically after the learner touches and selects the subgoal from the provided list of subgoals. The student must indicate the mathematical operation to be used between each subgoal and the parent goal (aside from the leaves) after creating the goal-driven plan-tree.

Carrying out the Plan. The execution frame and the calculation frame are the two frames that are currently provided by the app. The system modifies the goal-driven plan-tree somewhat and shows it in the execution frame. Each goal has two attributes: label and value, which, respectively, indicate the goal's meaning and its numerical worth. The execution of the plan starts with the leave-goals. The student describes the value of each leave-goal and determines the value of the corresponding parent-goal either manually or with the use of the provided calculator.

Fig. 4. The second stage's screenshots (Making the plan).

Looking Back and Reflect. In the last stage, the system suggests an alternative solution to the problem that might contain some errors in order to support the answer provided in earlier stages. The student is then asked to confirm or deny the system's assertions. When this stage is finished, the system examines the outcomes, and messages display to show whether any errors were committed. The system then shows a four-row assessment table as its final output. Each row represents a step in the solution process. The total grade takes into account how many times a student failed to arrive at the right answer.

4 Conclusion

This study used Polya's solution model and a goal-driven approach created for the building plan stage to help children improve their multi-step word problem solving abilities and math accomplishment. The idea of developing a plan in the making plan stage using a goal-driven approach was inspired by the software requirements capturing approach in requirements engineering, where analysts create new system goals in the form of a graph so that each goal is refined as a group of subgoals describing how the refined goal can be reached. As a result of such instruction, students become more focused on the semantics of the connections between all the parts of solutions in order to properly solve issues.

Goal-driven plan-tree approach has the potential to be a helpful tool to aid students in solving multi-step word problems, as the idea in this paper implies. To improve students' mathematical skills to answer multi-step problems when it is used in the classroom, further research into the usefulness of the mobile app is required.

References

1. Charles, R.: Solving word problems: developing quantitative reasoning. Res. Pract. Math. (2011)
2. Thevenot, C., Oakhill, J.: The strategic use of alternative representations in arithmetic word problem solving. Q. J. Experimental Psychol. Sect. A **58**(7), 1311–1334 (2005)
3. Sweller, J.: Cognitive load theory, learning difficulty, and instructional design. Learn. Instr. **4**(4), 295–312 (1994)
4. Montague, M.: Self-regulation strategies to improve mathematical problem solving for students with learning disabilities. Learn. Disabil. Q. **31**(1), 37–44 (2008)
5. Polya, G.: How to solve it. Princeton University Press, Princeton, New Jersey (1945)
6. Oakhill, J., Cain, K., &Elbro, C.: Understanding and teaching reading comprehension: a handbook. Routledge (2014)
7. Reeves, D.B.: The learning leader: How to focus school improvement for better results. ASCD (2006)
8. Griffin, C.C., Jitendra, A.K.: Word problem-solving instruction in inclusive third-grade mathematics classrooms. J. Educ. Res. **102**(3), 187–202 (2009)
9. Cuoco, A.A., Goldenberg, E.P.: A role for technology in mathematics education. J. Educ. **178**(2), 15–32 (1996)
10. Borba, M.C., Askar, P., Engelbrecht, J., Gadanidis, G., Llinares, S., Aguilar, M.S.: Blended learning, e-learning and mobile learning in mathematics education. ZDM Math. Educ. **48**, 589–610 (2016)
11. Sheriff, K.A., Boon, R.T.: Effects of computer-based graphic organizers to solve one-step word problems for middle school students with mild intellectual disability: a preliminary study. Res. Dev. Disabil. **35**(8), 1828–1837 (2014)
12. Chadli, A., Tranvouez, E., Ouared, A.: A goal oriented approach for solving mathematical multi-step word problems. In: Proceedings of the 10th International Conference on Information Systems and Technologies, pp. 1–7 (2020)
13. Chang, K.E., Sung, Y.T., Lin, S.F.: Computer-assisted learning for mathematical problem solving. Comput. Educ. **46**(2), 140–151 (2006)
14. Van Garderen, D., Montague, M.: Visual-spatial representation, mathematical problem solving, and students of varying abilities. Learn. Disabil. Res. Pract. **18**(4), 246–254 (2003)
15. Dardenne, A., Van Lamsweerde, A., Fickas, S.: Goal-directed requirements acquisition. Sci. Comput. Program. **20**(1–2), 3–50 (1993)

Data Science Analysis Curricula: Academy Offers vs Professionals Learning Needs in Costa Rica

Julio C. Guzmán Benavides[1](✉), Andrea Chacón Páez[1,2], and Gustavo López Herrera[1]

[1] University of Costa Rica, San José, Costa Rica
{julio.guzman,andrea.chaconpaez,gustavo.lopezherrera}@ucr.ac.cr
[2] State Distance University, San José, Costa Rica

Abstract. The analysis of large volumes of information has become necessary for decision-making in companies, creating new roles and capacities that must be covered. To meet this demand, institutes and universities have created curricular programs that satisfy this need; however, the coverage of these programs is non-standard and leaves important topics uncovered. This paper compares a set of academic curricula inside and outside of Costa Rica based on the report Computing Competencies for Undergraduate Data Science Curricula - ACM Data Science Task Force.

Likewise, the results of a survey with professionals from different industries to identify the topics related to data science that they learned in formal education, work or that they would need to know, providing an overview of the issues that should strengthen or add within the academic curricula analyzed.

Keywords: Education · Data Science · Curricula · ACM Data Science

1 Introduction

Information has become a company's most valuable asset for decision-making. Managing the exponential increase of data has required the development of new capabilities and roles within companies. Consequently, there is a rapidly increasing demand for data scientists and related jobs. Research conducted by Mckinsey and IBM forecasts the need for hundreds to thousands of data science jobs and associated roles in the next decade [1, 2].

To meet the high demand for jobs related to data science, institutes and universities have seen the opportunity to create programs to train students and fill the growing number of jobs. For example, a 2019 study identified 581 programs in data science, data analytics, and other related fields at more than 200 universities worldwide; including Ph.D., Master, Bachelor, and Certificate programs [3].

Data science is an evolving field that has generated new roles in companies, causing the emergence of study programs that meet that demand in the market. Different terms with similar meanings are used in the study programs, such as Big Data Analytics, Data & Analytics, Business Intelligence & Analytics, and Business Analytics [4]. Unfortunately,

A. Rocha et al. (Eds.): WorldCIST 2023, LNNS 800, pp. 366–375, 2024.
https://doi.org/10.1007/978-3-031-45645-9_35

these terms have been used as synonyms both within the academic context and in the industry, creating confusion in study programs or job profile descriptions.

Costa Rica has also seen the need to prepare professionals in this role, providing scholarships from the government in data science programs to encourage professionals from different areas to specialize in this subject [5]. Considering the high demand for Data Science specialists, assessing study programs in this field is critical. This paper analyzes the Costa Rican context of Data Science education. First, we provide a revision of several curricula offered in the country. Second, we created a survey and described the results. Finally, we aim to answer the following research questions:

RQ1. To what extent do the revised curricula cover the areas of knowledge proposed by the report Computing Competencies for Undergraduate Data Science Curricula - ACM Data Science Task Force?

RQ2. How much discrepancy is there between the areas covered by academic offerings and the educational needs of data science professionals?

2 Background and Related Work

Data Science is a term that became a "catch-all." By the nature of the concept, it tends to be misunderstood, or confused by others. In this section, we will describe the different perspectives encompassing the concept. In addition to understanding the concept, it is also relevant to delve into how it has been taught.

In 1985, C.F. Jeff Wu was the first to use the term data science in a paper. Later, Wu emphasized the need for computing power and interdisciplinary training. Raj et al. [6] reported the evolution of the concept from textbooks, the job market, current degree programs, and online course content. The authors mentioned how the concepts of data mining and informatics dropped once the concept of Data Science started to grow.

In the case of Wu, they make the difference between data science and data engineering. Data science is a multidisciplinary field that integrates approaches from mathematics, statistics, machine learning, data mining, and data management. On the other hand, data engineering is a process driven by the construction of data infrastructure.

In a study by Berkeley School of Information, data science has five stages on a life cycle: Capture, Maintain, Process, Analyze, and Communicate. Also, a data scientist is a professional skilled in organizing and analyzing massive amounts of data [7].

In their work, Bonnell et al. [8] define data science as a discipline that aims to develop analytical research and facilitate decision-making. However, data science is not a single discipline. It is an interdisciplinary effort between mathematics, statistics, computer science, communication, and relevant application domains. Furthermore, a professional in data science needs a set of skills (programming language skills, calculus, linear algebra, discrete mathematics, data structures and algorithms, statistical analysis, machine learning, databases, and data visualization). Finally, the authors define a group of non-linear seven steps of the data science lifecycle.

As indicated Bonnell et al. there are many efforts to close the supply-demand gap, like massive online open courses and degree/certificate programs in Data Science. Still, each type has problems depending on the context. There is no "one-size-fits-all" solution.

Some prestigious university programs aim to teach students fundamental knowledge, give exposure to domain-specific, and provide opportunities through internships.

For Urs and Minhaj [9], Data Science emerged as a heavenly field of study across schools/departments of computer science, business, and others. While statisticians argue that data science is the new avatar of statistics, some consider it an attractive new term in computer science. Authors address Data Science as a significant opportunity for transformation from traditional statistics and computing research to cross-disciplinary data-driven discovery. For the authors, beyond the hype of the topic, many transformative factors have influenced the attention of everyone and turned data science into a sought-after educational program. They used curriculum analysis to draw the landscape of Data Science using a text-mining approach. They scanned and analyzed the websites of 122 information schools and manually collected data on course titles of the data science educational programs.

Hassan et al. [10, 11] address Data Science as a fast-evolving subject with increasing demand. However, names of similar programs change from data science to data analytics, Business Analytics, or Business intelligence. According to the authors, the three foundation areas are mathematics, statistics, and computer science. They support their study in the ACM Taskforce on Data Science Education. They analyzed 122 data science degrees in the U.S they focused on 101 programs, where 90% included introductory computer programming, the database coverage was 74.3%, and the calculus and linear algebra was over 74%.

Wu et al. [12] planted the question of the competencies required for data scientists and broke down disciplinary boundaries from related disciplines. The authors reviewed existing data science courses to figure out what knowledge topics are covered. They used a manual review procedure. In the end, they collected 96 data science graduate programs. In their study, they obtained a wide range of competencies and skills such as data analysis, data management, data processing, solving practical problems, supporting decision-making, understanding requirements, Python, SQL, and many other fields of knowledge. This study also proposes a human-centered data science graduate curriculum model. In their proposal, the curricula should match students' requirements for developing data science competencies, contrary to the thinking algorithmically and statistically offered.

For Demchenko et al. [13], Data Science education needs an effective combination of theoretical, practical, and workplace skills. They used the EDISON Data Science Framework as a tool to understand the educational environment and created a new approach to building effective curricula in Cloud Computing, Big Data, and Data Science. In their study, the curricula work as facilitate digital transformation.

Raj et al. [6] report the evolution of data science education, analyzing programs of degree level. They used information from 315 practicing analytics professionals. In their conclusions, they offer suggestions that may be useful to developing a data science education program, like following the data lifecycle (acquire, clean, use/reuse, publish and preserve/destroy) and closing the gap between what the professionals' skills need and educational programs.

3 Methods

This paper aims to analyze the Costa Rican context of Data Science education. To answer the research questions, we performed a revision of a set of curricula in this area.

The curricula were analyzed under three criteria: first, the duration of the program, we search for curricula with a minimum of twelve months. The second criteria were the availability to study the program in Costa Rica, either in person or virtually. And third, complete access to the agenda of the curricula.

In some cases, the whole program was available on the Internet. Meanwhile, in other cases, we needed to contact the institutions responsible for the curricula. Under these conditions, we examined six curricula and their depth in Data Science topics using expert judgment. In addition, we found other curricula, but in some cases, only some information was available, the program was not available for Costa Rica, or the duration was less than a year, so those were out of the study.

Our intention is not to judge, nor to create a position on which curriculum is better over another, so we have decided to keep the names of the universities and programs analyzed anonymously.

There have been several efforts to define a Data Science curriculum; nevertheless, we based on the Computing Competencies for Undergraduate Data Science Curricula - ACM Data Science Task Force report [14] because this project looks at data science from the perspective of the computing disciplines but recognizes that other views contribute to the whole picture, for example, The EDISON Data Science Framework a project with the purpose to accelerate the creation of the Data Science profession [15].

We used the Computing Competencies for Undergraduate Data Science Curricula - ACM Data Science Task Force report as a starting base to analyze the most common elements of the different curricula. This report, offered by ACM, has eleven areas of knowledge and a set of subtopics for each.

(AP) Analysis and Presentation: Visualization can be used to provide easily understandable summaries but can also help guide activities such as grouping and sorting.

(AI) Artificial Intelligence: The concepts and methods developed to build AI systems are helpful in data science. Professionals skilled in artificial intelligence can apply different techniques in a data science context.

(BDS) Big Data Systems: The computational problems associated with managing and processing large amounts of data tend to increase as the number of data increases.

(CCF) Computer Fundamentals: Data Scientists should understand – at least at a high level – the structure of operating systems, file systems, compilers, and networks, as well as security issues related to them.

(DG) Data Acquisition, Management, and Governance: A Data Scientist must understand concepts and approaches of data acquisition and governance, including data shaping, information extraction, information integration, data reduction and compression, data transformation as well as data cleaning.

(DM) Data Mining: Basic types of analysis include clustering, classification, regression, pattern mining, prediction, association, time series, and web data.

(DP) Privacy, Security, Integrity, and Analysis for Data Security: Data scientists must be able to consider data privacy concerns and related challenges when acquiring, processing, and producing data.

(ML) Machine Learning: Also known as statistical learning, it refers to a broad set of algorithms for identifying patterns in data to build models that could then be and possibly be produced.

(PR) Professionalism: In their technical activities, data scientists must conduct themselves responsibly, giving credit to the profession.

(PDA) Programming, Data Structures, and Algorithms: Data scientists must be able to implement and understand algorithms for data collection and analysis and integrate them with existing software or tools.

(SDM) Software Development and Maintenance: Data scientists can be expected to build (or help build) deployable systems for data analytics purposes or to put data analytics results to work.

Each area of knowledge has a group of subtopics that can be checked in the curricula.

Continuing with the work of this study, for each of the six selected academic curricula, we studied to what extent they cover the areas of knowledge proposed by the report Computing Competencies for Undergraduate Data Science Curricula - ACM Data Science Task Force. In addition, we created a survey instrument on the Data Science educational needs. The survey instrument was used to understand the discrepancy between the areas covered by academic offerings and the educational needs of data science professionals, in the same way using the report Computing Competencies for Undergraduate Data Science Curricula - ACM Data Science Task Force. The analysis instrument comprises 24 elements to evaluate the areas of knowledge and subtopics that are considered relevant for the study:

- Basic Programming Skills (PDA)
- Algorithms and Data Structures (PDA)
- Statistics (AP)
- Probability (AP)
- Machine learning (ML)
- Deep Learning (ML)
- Data Mining (DM)
- Decision-making Modeling (BDS)
- Algorithm Analysis (CCF)
- Linear algebra (AP)
- Data Management (AP)
- Data Visualization (AP)
- Big Data (DP)
- Software Engineering (SDM)
- Discrete Mathematics (PDA)
- Ethics (PR)
- Operations Research (DM)
- Calculus (PDA)
- Cybersecurity (DP)
- Artificial Intelligence (AI)
- Soft Skills (PR)
- NoSQL Databases (BDS)
- Relational Databases (BDS)
- Parallel programming (CCF)

4 Execution

We started to search curricula in Costa Rica under established standards and found three programs from Costa Rican academies and three from the United States. Each curricula was crumbled into the topics of the ACM Data Science Task Force report. As every topic has subtopics, we looked for the subtopics in the curricula to define the coverage grade.

We follow a linear execution process for the survey and getting the answers. After the design and review of the survey, we contact different Costa Rican institutions. The condition to complete the study was that the participants had to work in Data Science or a related topic. Professionals in analysis and statistics from the University of Costa Rica helped in the creation and validation of the survey.

We started gathering answers using an online survey site and launched it for two months. After that, the invitation was opened to anyone who fits the profile. We get involved in the process National IT Business Association Chamber of Costa Rica (Cámara Costarricense de Tecnologías de Información y Comunicación CAMTIC), College of Professionals in Informatics and Computing (CPIC), the list of graduates of the postgraduate program in computing and informatics of the University of Costa Rica, Facebook groups on Data Science in Costa Rica, and some personal contacts.

Once the survey was closed, the respective analysis was carried out. The professionals who completed the survey come from different areas such as engineering, mathematics, statistics, administration, among others. The common feature they share is that they all work in data science. Also, they all work in Costa Rica. More than 30 companies are represented in the results. For a total of 104 responses. 72 of them men, 30 women and 2 preferred not to identify with any gender. In addition, the majority, 70%, are between the ages of 30 and 49.

For each element, participants were asked if they acquired knowledge in an educational process, through their work experience, or if they wished to receive training in the subject (regardless of whether they have previous knowledge).

5 Findings and Discussion

5.1 Academy Offers

We reviewed six curricula, of which three are offered by Costa Rican entities and three are offered by international entities, but it has virtual programs available in the country. For each curriculum, it is determined the percentage of coverage on each topic. Table1 shows the rate of each curricula.

Table 1. Coverage of the reviewed curricula on the data science areas of knowledge.

	Curricula CR - 1	Curricula CR - 2	Curricula CR - 3	Curricula INT - 1	Curricula INT - 2	Curricula INT - 3
Analysis and Presentation	80%	20%	40%	40%	40%	40%
Artificial Intelligence	0%	0%	0%	0%	0%	0%
Big Data Systems	44%	44%	44%	0%	11%	0%
Computer Fundamentals	0%	0%	0%	14%	0%	0%
Data Acquisition, Management	25%	25%	88%	50%	63%	25%
Data Mining	90%	80%	30%	70%	80%	60%
Data Privacy, Security, Integration	0%	0%	0%	0%	0%	0%
Machine Learning	100%	80%	20%	80%	100%	80%
Professionalism	0%	0%	11%	0%	11%	0%
Data structures, and Algorithms	100%	33%	33%	50%	50%	17%
Software Development	0%	0%	0%	0%	50%	0%
Total rate	**40%**	**26%**	**24%**	**28%**	**37%**	**20%**

All curriculas have a percentage of coverage in the topics of Analysis and Presentation, Data Mining, Machine Learning, Programming, data structures, and Algorithms, and Data Acquisition, Management and Governance.

The Big Dada topic is in only four curriculas of the six curricula analyzed. Likewise, professionalism is only in two curriculas of the six curricula analyzed.

Regarding Computing and Computer Fundamentals, and Software Development and Maintenance, it was only added by one curricula each topic.

Two areas of knowledge were not addressed by any of the six curricula analyzed: artificial intelligence and privacy, security, integrity, and analysis for data security.

A total percentage by curricula was also obtained, which is between 20% and 40% of compliance according to what is indicated by the Computing Competencies for Undergraduate Data Science Curricula.

5.2 Learning Needs of Professionals in Costa Rica

Figure 1 shows the information provided by 100 participants, ordered by the knowledge most needed by the professionals.

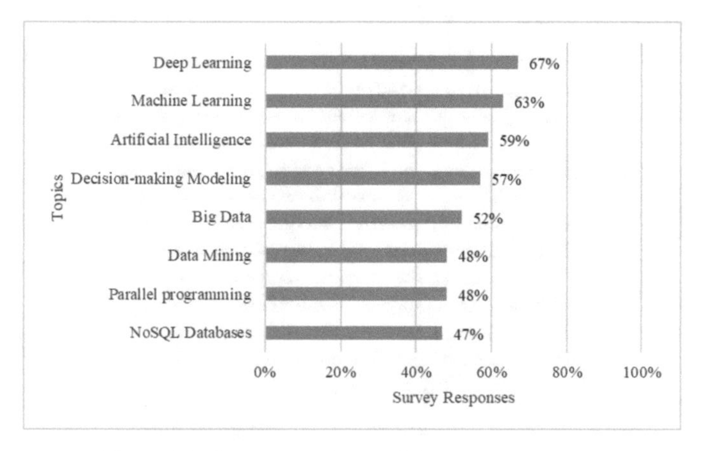

Fig. 1. The knowledge that participants need to learn.

Figure 2 shows the knowledge that is acquired in formal education, also information provided by the same 100 participants. According to the results is calculated with 83% of people who indicate learning the subject in formal education, 73% in Linear Algebra, 69% in Basic programming skills, 64% in Statistics, 63% algorithms and data structures, 60% probability, 58% discrete mathematics, 56% relational databases, 55% ethics, 53% operations research, 51% algorithm analysis, 49% systems engineering.

The field of knowledge most needed by professionals is deep learning; 67% of the participants indicated that they would need to learn about this topic, followed by 63% on machine learning, 59% artificial intelligence, 57% modeling for decision making, 52% Big Data, 48% parallel programming, 48% data mining and 47% NoSQL databases.

Regarding the knowledge acquired at work, 55% of the participants report soft skills and 48% data visualization.

There are two topics in which knowledge is sometimes acquired at work and other times in formal education, and a significant percentage wants to learn about the topic. These topics are data management principles and techniques, 35% learn it in formal education, 32% at work, and 22% want to know. The second topic is safety, with 37% learning at work, 31% in formal education, and 27% wanting to know about the subject.

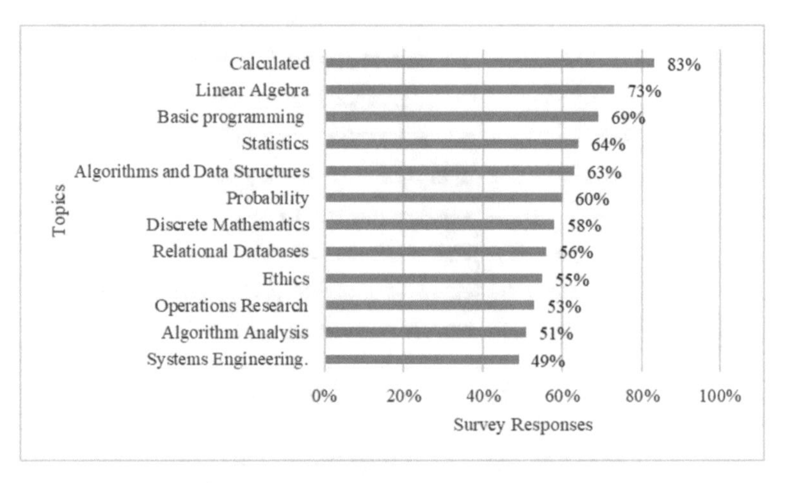

Fig. 2. Knowledge acquired in formal education.

6 Conclusions

About the knowledge needed by the participants, it concluded that there is a strong need for training in topics related to data science that people have not acquired in formal education or at work. According to table 2, people want to learn about deep learning (67%), machine learning (63%), artificial intelligence (59%), modeling for decision-making (57%), Big Data (52%), parallel programming (48%), data mining (48%) and NoSQL databases (47%).

On the other hand, topics taught in formal education, but that can be strengthened to have low coverage are data management (35%), security (31%), data visualization (24%), and soft skills (16%).

The discrepancy between the areas covered by the academic offerings and the educational needs of data science professionals is found in Artificial Intelligence and Security because no curricula include these topics. There are also discrepancies in some of the curricula that do not have Big Data within their curriculum. Yet, it is one of the topics with the most significant educational need among Data Science professionals.

The comparison between the ACM proposal for data science and the curricula shows that the curricula do not include Artificial Intelligence, security, and data privacy. In addition, only some curricula included computer fundamentals, big data, professionalism, and software maintenance. There are many curricula from different countries that could accompany this study, for example, taking longer programs into consideration can expand the research. Also, it would be interesting to ask professionals outside of Costa Rica for their position.

References

1. M. Analytics, The Age of Analytics: Competing in a Data-Driven World. McKinsey Global Institute Research (2016)
2. Miller, S., Hughes, D.: The quant crunch: how the demand for data science skills is disrupting the job market, Burning Glass Technologies (2017)
3. Hassan, I.B., Liu, J.: Data science academic programs in the US. J. Comput. Sci. Coll. **34**(7), 56–63 (2019)
4. Pedro, J., Brown, I., Hart, M.: Capabilities and readiness for big data analytics. Procedia Comput. Sci. **164**, 3–10 (2019)
5. M. de C. Presidencia de la República de Costa Rica, 'Micitt otorgara 250.000 en becas para el programa de data science impartido por el TEC', Gobierno de Costa Rica, January 2019
6. Raj, R.K., et al.: An empirical approach to understanding data science and engineering education. In: The Proceedings of the Working Group Reports on Innovation and Technology in Computer Science Education, Aberdeen, Scotland UK, pp. 73–87 (2019)
7. 'What is Data Science? [Digital edition]. https://ischoolonline.berkeley.edu/data-science/what-is-data-science-2/. Accessed 6 Oct 2022
8. Bonnell, J., Ogihara, M., Yesha, Y.: Challenges and issues in data science education. Computer **55**(02), 63–66 (2022)
9. Urs, S.R., Minhaj, M.: Evolution of data science and its education in iSchools: an impressionistic study using curriculum analysis. J. Assoc. Inf. Sci. Technol. **74**, 606–622 (2023)
10. Hassan, I.B., et al.: Data science curriculum design: a case study. In: The Proceedings of the 52nd ACM Technical Symposium on Computer Science Education, Virtual Event, USA, pp. 529–534 (2021)
11. Hassan, I.B., Liu, J.: Embedding data science into computer science education. In: The 2019 IEEE International Conference on Electro Information Technology (EIT), pp. 367–372 (2019)
12. Wu, D., Xu, H., Sun, Y., Lv, S.: What should we teach? A human-centered data science graduate curriculum model design for iField schools. J. Assoc. Inf. Sci. Technol. (2022)
13. Demchenko, Y., José, C.G.J., Brewer, S., Wiktorski, T.: EDISON Data Science Framework (EDSF): addressing demand for data science and analytics competences for the data driven digital economy. In: The 2021 IEEE Global Engineering Education Conference (EDUCON), pp. 1682–1687 (2021)
14. Danyluk, A., et al.: Computing competencies for undergraduate data science programs: an ACM task force final report. In: The Proceedings of the 52nd ACM Technical Symposium on Computer Science Education, pp. 1119–1120 (2021)
15. Demchenko, Y., Belloum, A., Wiktorski, T.: EDISON Data Science Framework: Part 1. Data Science Competence Framework (CF-DS), Release, vol. 2 (2016)

On-the-Job Training: From Actual Processes to Training Scenarios, a Methodology

Helisoa Randrianasolo, Ulysse Rosselet(✉), and Cédric Gaspoz

University of Applied Sciences Western Switzerland (HES-SO), HEG Arc, Neuchâtel, Switzerland

{ulysse.rosselet,cedric.gaspoz}@he-arc.ch

Abstract. Given the increased competitiveness of the business environment, organizations strive to constantly adapt their business processes and supporting tools while maintaining the skill level of their workforce. On-the-job training is an important tool for this purpose. Following a design science research approach, we present in this paper a methodology for transforming business process activities into simulation based serious games that can be embedded in real ERP software and used for effective on-the-job training. In order to create a verisimilar environment for the learners, the key principle in our methodology is the encapsulation of the training activity inside a sub-process where all the relevant aspects of the process can be simulated and controlled without impacting the overall business process. Implementing this approach in our own simulation software integrated to the Odoo ERP software enabled us to use this training approach in a very effective way.

Keywords: enterprise information system · user assistance system · on-the-job training · serious game · simulation

1 Introduction

The application of information and communication technologies has led to increased competitiveness in the business environment (Haračić et al., 2018), forcing companies to be more responsive to changes. Therefore, they are constantly striving to streamline their operations and to adapt their business processes while balancing costs and quality considerations (Goel et al., 2021). As regards business processes and related tools, effective computer training is viewed as a major contributor to organizational performance. The rapid pace of change of business processes and ICT tools make it hard for companies to maintain a sufficient level of training, resulting in lower productivity and higher technostress caused by perceived work overload and job insecurity (Ayyagari et al., 2011; Shu et al., 2011). On-the-job training is a valuable tool for the existing workforce or for integrating new employees. Our research aims at developing innovative methods and solutions that leverage enterprise applications to enable immersive on-the-job training addressing both the business process and the support software level.

A. Rocha et al. (Eds.): WorldCIST 2023, LNNS 800, pp. 376–394, 2024.
https://doi.org/10.1007/978-3-031-45645-9_36

This paper presents a methodology for transforming business process activities into self-contained user tasks in a serious game for on-the-job training, which is running inside the company's business software. First, we present the current state of theories on user assistance systems, educational games, business process management and on-the-job training. Then we derive from these theories a methodology for creating valid serious game scenarios, i.e. scenarios that help the employee acquire the knowledge and skills required to perform his or her activities in the context of more global business processes. We demonstrate the application of our methodology with one business process, and evaluate the resulting scenarios through a live experiment with more than two hundred learners. Finally, we discuss our findings and outline future research directions.

2 Literature Review

On-the-Job Training

Traditional training methods have difficulties in supporting the on-demand development of skills and intellectual abilities. Existing research recognizes the need for identifying effective and efficient ways of enabling and supporting employee-centric training and development (Cascio & Montealegre, 2016). On-the-job training methods have been generally regarded as effective in addressing this issue (Nguyen et al., 2021; Sree et al., 2019). On-the-job training is a practical approach that is based on the acquisition of skills, attitudes and behaviors needed to perform a job (Campbell, 1990) in a real or near-real work environment (Sree et al., 2019). On-the-job training requires a significant investment by the company in the design of the content and the integration of the learning environment (Sree et al., 2019). Developing similarity between the company's business processes and the training provided requires serious analysis and requires the skills of managers and business staff, which are not necessarily available (Jacobs, 2014). Furthermore, this complexity can be reflected at the technical level. When training is carried out on the company's own information system, integration efforts are to be expected (Sree et al., 2019).

Studies have shown that on-the-job training benefits the firm and the employees (Nguyen et al., 2021). Having qualified and proficient human capital improves productivity, which results in a stronger competitive position (Bilal et al., 2021). On-the-job training of employees significantly optimizes their potential to do jobs effectively and efficiently (Bilal et al., 2021), thus reducing the perceived stress in their performance (Sree et al., 2019).

Furthermore, by developing employees' skills, knowledge and competences to perform well, on-the-job training also has an impact on commitment, motivation to work and job satisfaction (Bilal et al., 2021; Niati et al., 2021; Sree et al., 2019).

Finally, employees become familiar with the tools and know-how much more quickly thanks to the practical nature of on-the-job training (Sree et al., 2019). As this training is carried out in a similar or real environment, the stress perceived by the learner is reduced (Sree et al., 2019).

Business Process Management

A process is defined as a "[...] set of interrelated or interacting activities that uses inputs to produce a created result" (ISO 9000, 2015). Caseau (2011) elaborates on this definition by stating that a process is a tool that enables actors to cooperate, with the aim of delivering a product or service to the customer. A business process is therefore an ordered sequence of activities performed within a given economic and technical framework (Weske, 2012). Companies operate business processes whose outputs create value for their stakeholders. Following the evolution of ICT, business processes are now highly integrated with enterprise software that supports its operational, decision-making and reporting needs. This integration of processes by information systems is reflected in a strategic approach known as business process management (BPM), often considered a key lever for reaching business objectives (Baporikar, 2016). BPM is a comprehensive system that includes process identification, modeling, analysis, redesign, automation and monitoring in order to better manage and transform organizational operations (Fischer et al., 2020; Rosemann & Brocke, 2015).

In order to increase efficiency and adapt to their environment, companies have to improve their processes (Tsagkani & Tsalgatidou, 2022; Weske, 2012). To do this, they optimize resources, standardize activities or put in place technologies to support them. These modifications will have a direct impact on the employees who are actors in these processes. The improvements generate changes in the way they work and involve new skills (Baporikar, 2016). In this sense, for the process to be effective, the employee must master its various components. He or she must acquire the necessary skills to carry out the new activities and be supported to integrate the new processes (Hrabal et al., 2020; Eicker et al., 2008).

Game Based Learning

Learning through games is a concept recognized by the educational field (Krath et al., 2021; de Freitas & Oliver, 2006; Gee, 2005). Thus, educational game design is considered a cornerstone of achieving cognitive learning effectiveness. Consequently, studies emphasize the need for a more conscious and goal-oriented game design if educational goals are to be met (Arnab et al., 2015; Gunter et al., 2008). Gunter et al. (2008) argue that educational games "*must be evaluated in terms of how well they immerse academic content within the game's fantasy context and how tightly the game designers couple gameplay with other fundamentally sound instructional strategies*". In order to cope with the lack of effectiveness of most simulation games, the authors developed the RETAIN (Relevance Embedding Translation Adaptation Immersion & Naturalization) model, which is a design and evaluation model for educational games. The RETAIN model builds on existing and well-established learning and instructional theories and principles: Keller's (1987) ARCS model, which addresses motivational aspects; Gagne's (1985) events of instruction for a successful game; Piaget's (1970) adaptation theory; and Bloom's (1956) taxonomy of educational objectives. The RETAIN model proposes six criteria that evaluate both the cognitive and the motivational aspects of the game through the environment it creates.

The resulting taxonomy is a two-dimensional evaluation of educational games, which helps designers of educational games to build effective and motivational games. The first dimension assumes a bottom-up hierarchy of the criteria, largely inspired by Bloom and others, where one criterion builds upon the previous ones. The second dimension evaluates each criterion on four maturity levels (from 0 to 3), similar to many maturity models that we are familiar with in our own field. The authors claim that the RETAIN model has potential predictive capabilities for determining the eventual success of a game intended for use in educational settings.

Simulation Training and Serious Games

In business, it has been found that the use of game-based learning approaches positively impacts engagement and provides better performance than traditional training approaches (Baxter et al., 2017; Hamari et al., 2016). When the learning material is too technical or boring, the game aspect helps to make the training more accessible (Azadegan et al., 2012). These tools offer engaging and contextually appropriate scenarios with achievable objectives. Employees can carry out tasks, experience failures and make repeated attempts (Roh et al., 2016). In this experiential learning approach, the learner also receives feedback that not only validates his or her journey but also 'also enables them to solve the challenge. In this sense, they build skills (de Freitas & Oliver, 2006). Furthermore, the playful aspect allows a paradigm shift by transforming the passive employee into an active player (Michael & Chen, 2006). Well-designed serious games make learning fun, challenging and rewarding. Learners immersed in the game do not realize that they are in a learning process. They achieve objectives without realizing it, so focused are they on achieving goals, competing with others and enjoying the game (Donovan, 2012). In addition to these empirical contributions, researchers such as Azadegan et al. (2012) have conducted surveys within companies to identify the real impact of serious games. Their research on more than 200 entities has highlighted that the use of game principles really allows the company to be more efficient, to develop competences and to be more agile within a changing environment.

Serious games often comprise elements of simulation where the participants are operating in a simulated environment. These environments can include only the abstract concepts of the real world (simplified), reproduce real-world situations with reduced interactions (verisimilar) or include all aspects of reality as best as possible (realistic) (Stainton et al., 2010). Botte et al. (2009) argue that simulation is a representation of the real world, allowing participants to evaluate and predict the dynamic development of a set of events predefined by conditions. Thus, a simulation is first and foremost a model of how a system works (Greenblatt, 1987), which allows the participant to operate in an environment with some aspects of reality and low error cost (Crookall et al., 1987). These environments offer the possibility of engaging the learner in situations or events that would be too costly, tedious, dangerous or unethical in real life (Garris et al., 2002). In addition, the coaching and support provided in these environments puts the learner in a stressful yet positive situation which is conducive to learning (Clarke, 2009). Finally, among the benefits of simulation, it is emphasized that designs based on real-life scenarios, such as business scenarios, help learners to integrate the business concept, understand their role in a process and adhere to the resulting objectives (Abdullah et al., 2013). Simulation – particularly in a real-life environment – is therefore a tool with huge potential both for learning new skills and for engaging the learner in the performance of their activities.

User Assistance Systems

User assistance systems (UAS) are systems that enrich the information systems of a company and help individuals to better understand their tasks and increase their user experience (Acar & Tekinerdogan, 2020). These systems can be categorized by their degree of interactivity and intelligence (Maedche et al., 2016). Given the different functionalities that UAS offer, the literature agrees on the fact that they can make a real contribution to the process of accompanying users for a given task and mastering the technical elements involved (Acar & Tekinerdogan, 2020; Diederich et al., 2019; Morana et al., 2020). Indeed, these systems not only produce advice, but are also able to detect activities performed in real time by the user, or changes in the environment, and thus provide contextualized hints and assistance (Maedche et al., 2019). In this sense, they are levers of skill acquisition since through feedback and personalization they promote the user's interaction with the system and maintain their motivation during the performance of their activities (Gee, 2005).

With these elements of the literature identified, we can now develop a methodology for transforming business process activities into autonomous user tasks in a serious game for on-the-job training that runs in the company's enterprise software.

3 Research Methodology

Our research follows the design science research perspective, which consists of the following ontology, epistemology, methodology and axiology. The ontology postulates different context-dependent alternative realities/state of the world, made possible by socio-technological factors (Vaishnavi & Kuechler, 2015). The epistemology is based on the creation of knowledge through 'doing', i.e. the objectively constrained construction of artifacts in a given context, and where the iterative circumscription of phenomena reveals their meaning (Vaishnavi & Kuechler, 2015). The methodology is oriented towards the development and measurement of the impacts of the artifact on the system studied (Vaishnavi & Kuechler, 2015). In terms of values and ethical aims the axiology of the design science approach is oriented towards the control, creation, progress (improvement) and understanding of socio-technological phenomena and artifacts.

A particular feature of the design science approach is that it places central importance on the relevance of the research to the environment: its results – embodied in the produced artifact – must be applicable in a predefined socio-technical context (Brocke et al., 2020). The activities that result from this constraint constitute the relevance loop (vom Brocke et al., 2020; Hevner & Chatterjee, 2010; Hevner, 2007). For such an approach to be differentiated from design and to qualify as research, it must simultaneously draw on and contribute to the knowledge base. In order to ensure the rigor of the research, researchers must select the appropriate scientific theories and methods for both the design and evaluation of their artifact (Peffers et al., 2012). The fact that the construction of the artifact and its evaluation are guided by scientific principles allows researchers to make valid contributions to the knowledge base. This constitutes the rigor loop. The activity of designing the artifact itself follows an iterative approach and is referred to as the design loop (vom Brocke et al., 2020).

Our research, aimed at providing effective tools for on-the-job training of workers, corresponds to a problem-centered initiation (Gregor et al., 2020; Peffers et al., 2007). In this paper, we present two artifacts, the first being a methodology for creating serious games tasks from business processes for on-the-job training. Using Gregor's taxonomy of information systems theory (2006, 2020), this methodology can be considered as a theory for design and action. Then secondly we present an instantiation resulting from the application of our methodology. This instantiation consists of a serious game for on-the-job training and onboarding in a service company. It is composed of simulated business cases (gamified realistic work scenarios) that run directly inside the ERP software Odoo. As recommended by Gregor & Hevner (2013), the next section will present the principles of form, function and methods and will explain the relations with the justificatory theoretical knowledge presented in the literature review section that is used in the development of serious games for on-the-job training. The instantiation provides the basis for the evaluation of our methodology (Peffers et al., 2012).

4 Methodology for On-the-Job Training Scenarios from Actual Business Processes

Our methodology requires that the target business processes for training have been modeled. We rely on the business process modeling and notation (BPMN) standard which uses events, activities and branching as key concepts to represent the way work is carried out by actors in an organization. Activities, as they are being performed by the employees, are the focus of our methodology which consists in encapsulating the worker's activity in a verisimilar simulation context. Figure 1 shows a generic BPMN diagram example that will be used as the illustration for our methodology. In order to encapsulate an activity into a serious game scenario we have applied the principles of the RETAIN model proposed by Gunter et al. (2008).

First, we advocate the setting up of a verisimilar environment. This will facilitate the acquisition of skills and their application in another context, in our case real work (Gunter et al., 2008). The employee, whom we will call the learner from now on, can carry out the activity in conditions close to their daily work. Moreover, they should be more involved and committed since they see the relevance of what they are doing, i.e. the importance of these activities in real life. We emphasize that this environment must be less complex than the real world. Actions that do not add value to learning or tasks that are too complex or even too time-consuming should be performed by the simulator. This way, the learner can focus on the main actions and the skills they underpin (Keller, 1987). We emphasize that the allocation of tasks to the simulator does not remove the learner's freedom of action. As Piaget's (1970) theory of adaptation suggests, the learner constructs their knowledge through their own actions. It is therefore essential to let the learner carry out the actions required by the activity on their own (Gunter et al., 2008). The simulator only carries out the tasks of initializing the different states, simulating other actors' actions or validating the tasks.

Furthermore, we stress the need for information. Before the start of the game, and at each new stage, in order to reduce negative states such as the anxiety perceived during the training, we suggest informing the learner of the scope of their work and the objectives required of them (Gagne, 1985). In this way, they are aware of the work context, the actions to be carried out and the resources at their disposal (Gunter et al., 2008).

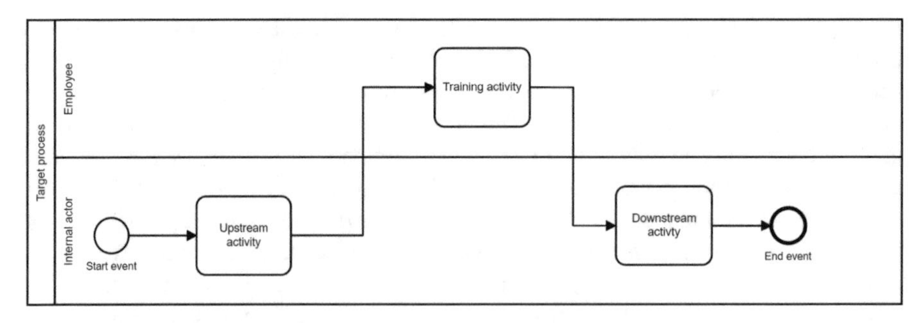

Fig. 1. Generic target process for on-the-job serious game simulation training

We also retain, in this approach, the importance of learning through experimentation. The learner can test and make mistakes without this having any impact on real life. Through this process, they can develop automatisms in problem solving (Gunter et al., 2008). Exception loops should be provided to allow the learner to perform the task again and correct their mistakes themselves. At the same time, as mentioned, clear instructions will be given to ensure that the objective is achieved (Gagne, 1985). They may also ask for help or advice if they cannot find a way to solve the problem. As Piaget points out (1970), this information accompanies the learner in the process of creating skills. Since they are faced with a new difficulty, the instructions help them to adapt their approach and return to the assimilation of the expected skills.

We also stress the need for evaluation of skills. In order to maintain the learner's motivation in the training process, successes and errors will be accounted for in their performance (Gagne, 1985). These elements can take the form of points that are attributed based on the learner's performance at each stage of the scenario. The attribution of points should reflect the difficulty of the task that is asked of the learner. They can thus have confidence in the learning process they are undertaking (Gunter et al., 2008).

Finally, as Gunter et al. (2008) suggest, the learner must be aware of the journey they are taking and be satisfied with the new knowledge. Feedback is important, to inform the learner of their performance (Gagne, 1985). This allows explicit validation of skills and helps maintain a positive motivational flow (Keller, 1987).

Our methodology consists in eight steps:

1. **Model the business process**. To create the game scenario for a specific BPMN activity, it is essential to define the business process in which it occurs. Identifying and modeling this process enables the training activity to be targeted and its sequence with other activities to be understood. On the other hand, this approach allows the definition of the scope of the work and the expected business objective. This information will be passed on to the learner so that they understand the relevance of the training approach.
2. **Target the training activity**. Once the process has been identified, it is necessary to define the activity with which the training process is concerned. This is the element that contains the learning scenario during which the learner will progress. The choice of the activity can be made on the basis of the competences that the learner must acquire during the training.

3. **Identify the participating actors**. The definition of the process and the activity makes it possible to identify precisely the actors concerned. At the same time, it is possible to define the existing interactions with actors external to the process and which nevertheless influence the activity. The actions of these external actors are attributed to the simulator.
4. **Frame the task**. Define the task that will be performed by the learner in the context of the learning process. This task should be meaningful and should produce verifiable results that can be assessed by the simulator. This is usually a process task for which it is desired that the learner acquire the skills to perform it independently. This step requires defining the success criteria for the task and defining the work instructions. To engage the learner in the interaction we provide the information necessary to carry out the activity. The learner must be able to access this information at the appropriate time. Thus we suggest setting up a textual introduction phase indicating to the learner the expected objectives and the operating rules of the action to be carried out. The same applies to the provision of advice in the event of a deadlock on a given problem.
5. **Provide incentives**. To stimulate motivation and allow the learner to validate their progress and achievements, we provide incentives. These could include positive verbal feedback from the simulator, objective KPIs relevant to the process and game-related indicators such as the score achieved and the learner's ranking.
6. **Manage exceptions**. To support the experiential model, learning goes through iterative phases of understanding and testing. In order to provide feedback to help the learner enter a reflexive process, we support two feedback mechanisms. First, after a predefined timeout, the learner receives a reminder to perform the task. Secondly, the simulator checks for correct execution of the training activity according to the defined success criteria. In case of errors, the learner receives remedial instructions and is asked to correct the task.
7. **Check data objects**. The definition of data objects allows, on the one hand, the identification of the elements that the learner manipulates in the activity. On the other hand, they are validation components and can include the expected result of the activity.
8. **Encapsulate the training activity**. In order to embed the training activity in the serious game, we encapsulate the training activity within a sub-process. This allows us to isolate the main task, to initialize the required master data, to control the triggering element, to provide work instructions and assistance to the user, to check the correctness of the task and to provide corrective instructions without impacting the overall business process. The activities of the process that are not carried out by the learner doing the training need to be automated in order to form a verisimilar environment for the training activity. This includes performing all actions necessary for the learner to have a similar experience to what would happen during a real execution of the process. Figure 2 shows a BPMN metamodel of the training activity encapsulated in a sub-process containing all the serious game logic and simulation elements.

As we can see, this methodology respects the initial BPMN model and follows the recommendations for designing effective serious games. In terms of integration into existing processes and enterprise software, it also has the advantage of only impacting the activity encapsulated in the sub-process. The following section will present an instantiation of our methodology on a real-life example.

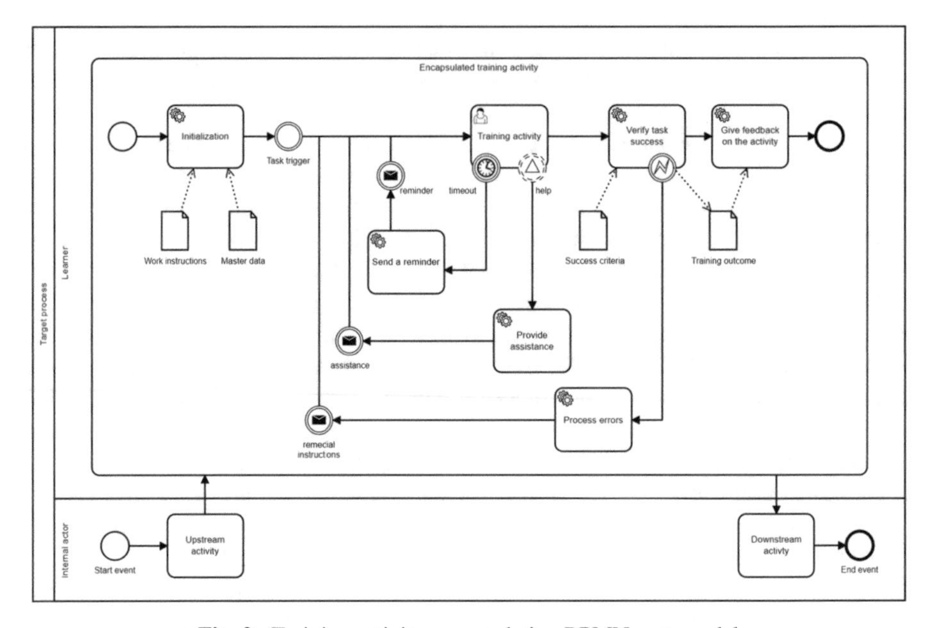

Fig. 2. Training activity encapsulation BPMN metamodel

5 Example Instantiation of the Methodology

This instantiation consists of a serious game for on-the-job training and onboarding in a service company. Employees can learn to execute the company's core processes through various game scenarios that have been created from the company's business processes. These simulated business cases (gamified realistic work scenarios) run directly in the company's Odoo ERP software. In this article, we focus specifically on one type of BPMN activity: user tasks. Indeed, these are the activities that are carried out by the employees with the assistance of a software application, in our case, the Odoo ERP software.

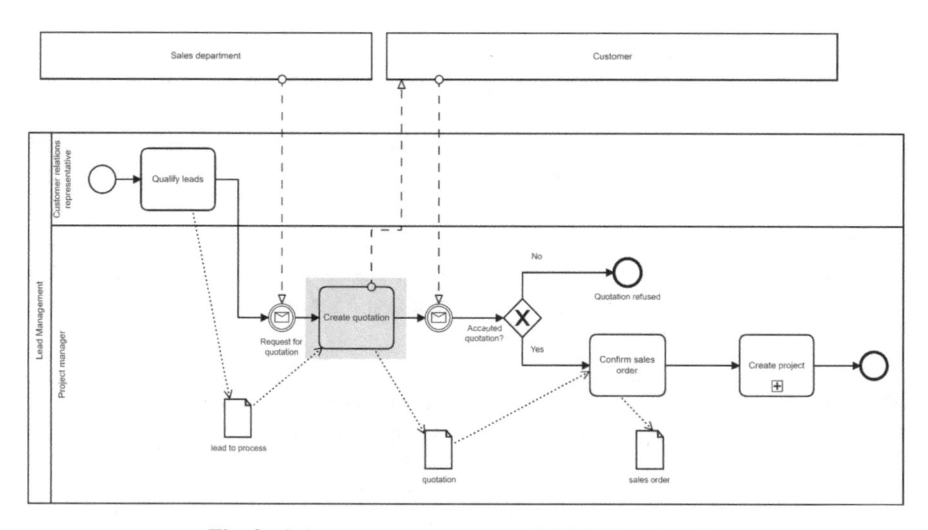

Fig. 3. Sales management process initial BPMN model

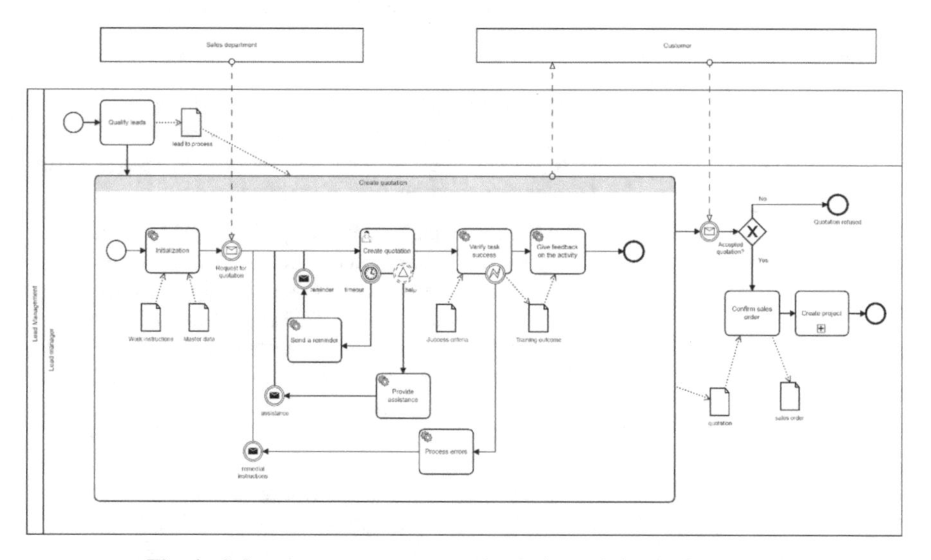

Fig. 4. Sales management process simulation training BPMN model

We conducted interviews with the company's managers to identify the main business processes that employees should master for their job. Based on the material provided by the company and the results of the interviews, we created a first business process model in order to validate our understanding of these processes with the stakeholders. After this first phase, we observed experienced employees performing the different process activities directly in the Odoo ERP. This allowed us to generate a data set containing all the objects necessary for the realization of the user tasks of the business processes and to have a video capture of these interactions to be able to return to them later if necessary. This approach led us to cover five business processes: lead management, sales management, project management, time and attendance management, expenses management.

In the following section we show how we applied our methodology to transform the **sales management – create a quotation** activity into an on-the-job training simulation serious game. This activity is relevant because project managers can be assigned to follow up on a lead based on their area of expertise. A customer relations representative can ask a project manager from a given area of expertise to prepare a quotation based on the customer's needs. This activity is a user task and requires the creation of an object in the sales management system.

Model the business process	**Sales management process**. This is the process initiated after the successful qualification of a lead by a customer relations representative. It starts with the creation of a quotation. Figure 3 shows the initial BPMN model for this example
Target the training activity	**Create a quotation**. For a given customer, create a quotation based on the customer's needs collected by the customer relations representative
Identify the participating actors	**Customer relations representative**. A representative in charge of qualifying a lead from a given customer **Customer**. A customer whose needs are known **Project manager**. An employee tasked with the creation of a quotation resulting from a qualified lead
Frame the task	The employee must be able to open the right lead and create a new quotation based on the customer's needs recorded in the lead logs by the customer relations representative *"Welcome aboard! I'm Steven from customer relations. I just had Bigeasy Corp, one of our regular customers, on the phone. They need a new mobile app for their support teams. Can you create a quotation based on the lead I created in the CRM? You will find the customer's needs in the logs of the lead."* The task is successful if a quotation is linked to the given lead and contains the right service lines and amounts
Provide incentives	*"Thank you, I'll send them the quotation. We'll see if they sign it! In any case, I will keep you posted"* If the learner has completed the scenario without fail, they receive the maximum number of points. If there were errors, the scenario is scheduled for a new instantiation later and the number of points is divided by the number of unsuccessful attempts

(*continued*)

(*continued*)

Manage exceptions	If a quotation is created but not linked to the given lead, we inform the learner to link the quotation or to try again: *"Hi, I saw that you were busy creating a new quotation. Unfortunately, it's not related to the lead of Bigeasy Corp. I don't know if you were working on another lead or if this is a mistake. Can you please take another look?"* If the quotation does not have the requested elements: *"I'm sorry, but I checked the quotation before sending it to Bigeasy Corp and noticed that their needs were not covered with the current positions. Can you please take another look at it?"* If the learner continues to submit an incorrect quotation or ask for help, we display a series of tooltips to guide them through the different ERP screens and show the information that needs to be recorded in the quotation *"I'm sorry, but it still doesn't look right. You can follow my lead, I will show you how to get to the correct quotation. Let's try together!"* If the learner does not respond within the allotted time, the simulator sends a message to remind the user to complete the task *"Hi, it's Steven again. I promised to send an offer today to Bigeasy for their mobile app project. Can you look into it as soon as possible?"*
Check data objects	**Lead**. The data object is a lead from the customer in the qualified state with all its attributes **Quotation**. A quotation based on the customer's needs detailed in the lead logs
Encapsulate the training activity	The activity is triggered by a message from the customer relations representative that is sent to the learner and ends with the saving of a correct quotation in the ERP. Multiple leads should be inserted in the ERP during the initialization step. A reference quotation is used as a success criterion for the validation task. Figure 4 shows the resulting BPMN model

This example shows the instantiation of the methodology in the context of creating serious games to support on-the-job training and employees onboarding on business processes. These scenarios are part of a larger game comprising nine individual task-related scenarios and one longer tutorial comprising a sequence of nine tasks.

6 Evaluation

Our scenarios were tested with 281 learners. They had no prior experience of the business environment and the Odoo ERP and they used the serious game during a single session. The activity was divided in two steps. First, they had to follow a nine-step scenario to configure the master data of the company in Odoo (suppliers, raw materials, bill of materials, finished products, customers) and then to create a cash-to-cash cycle (MPS, purchases, production, sales). This scenario is used to discover the software (windows, screens, manipulation of the interface,…) and to learn how to execute the main activities of the business processes they are involved in. At the end of this first phase of the game, the learners had to carry out six randomly selected training activities in a pool of nine scenarios, allowing them to test their ability to perform activities related to the company's various business processes. These scenarios are also used to check if the learners are able to perform the tasks related to the business processes satisfactorily.

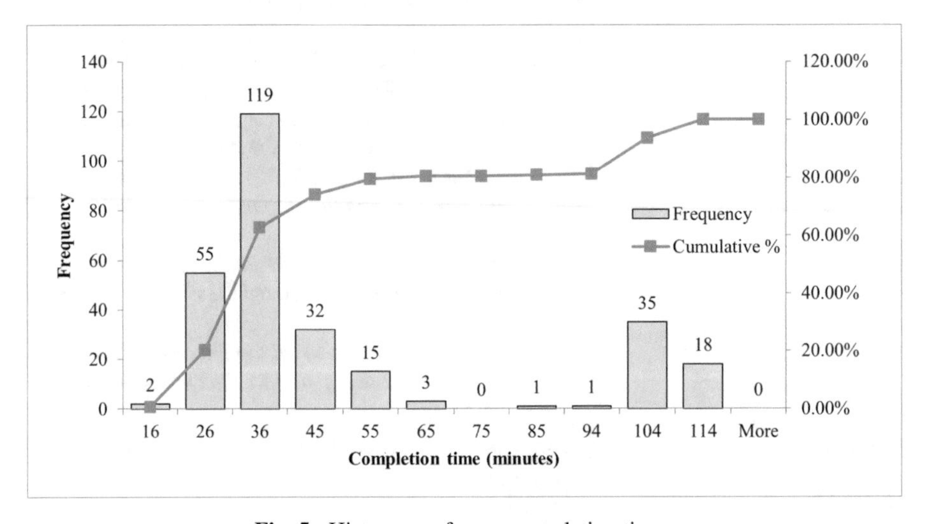

Fig. 5. Histogram of game completion times

The results in Table 1 indicate that of the 281 learners, 259 (92.17%) were able to successfully finish the first scenario. In the context of this scenario, this means that they managed to perform all activities correctly, with or without the help of the UAS. The distribution of completion times for successful games, see Fig. 5, shows that more than 75% of participants were able to complete the game successfully in less than 45 min.

Table 1. Game duration and number of learners based on the outcome of the scenario

Game outcome	Frequency	Percentage	Duration (h:min:s)			
			Minimum	Median	Average	Maximum
Success	259	92.17%	00:15:47	00:32:34	00:41:00	01:53:56
Fail	22	7.83%	01:35:26	01:43:25	01:43:34	01:48:27
Total	**281**	**100.00%**				

After the first phase of the game, from the 259 learners who finished the first scenario, 253 were able to successfully complete the second part of the game, consisting in executing six randomly selected tasks from the main business processes of the company. Table 2 shows the scores achieved by the learners on the different tasks. Each task was awarded a maximal score based on the type of the activity. Scenarios consisting in reading information from the ERP and providing this information back to the UAS were awarded one success point (inventory, profit, qty_sold, so_number). Scenarios consisting in updating information in the ERP were awarded two success points (change_bom, price_change, validate_po). Finally, scenarios consisting in creating a new record in the ERP were awarded three points (new_mo, new_so).

The first takeaway from this experiment is that from 281 learners using a serious game to perform on-the-job training scenarios, 253 were able not only to learn how to use the ERP and execute specific tasks, but also to properly mobilize these new skills in a verisimilar environment afterwards. We must recall that there was no specific training on the use of the system other than that provided by the UAS through the nine learning tasks of the learning scenario. Moreover, given that the awarded scores were reduced by one for each error (when the verification task raised an exception message to the learner), we see in Table 3 that the majority of users have performed the scenario correctly during their first attempt with the exception of the new_mo scenario where 77.65% of the learners had to take two attempts in order to validate the task. Secondly, it can be seen that the results of the tasks that have not been previously the subject of a training task are significantly worse than for the other tasks. In Table 3, we see that both the profit and quantity_sold scenarios perform more than 10% worse than the other scenarios. The tasks behind these scenarios (retrieve the net income from the balance sheet and find the total quantity sold of a specific product from the sales statistics) are not part of the nine learning tasks from the first step. This is a satisfactory result, supporting the impact of learning scenarios on learners' ability to complete their tasks. However, the current data does not yet allow us to draw conclusions from these numbers.

Table 2. Score of the learners on the different scenarios

Scenario	Number of players	Minimum score	Average score	Maximum score
scenario_change_bom	181	0	1.80	2
scenario_inventory	188	0	0.76	1
scenario_new_mo	170	2	2.22	3
scenario_new_so	151	0	2.48	3
scenario_price_change	155	0	1.73	2
scenario_profit	141	0	0.57	1
scenario_qty_sold	206	0	0.64	1
scenario_so_number	152	0	0.75	1
scenario_validate_po	174	2	2.00	2

Table 3. Distribution of scores for the various scenarios

Success rate by scenario	Score	0	1	2	3
scenario_change_bom		7.18%	6.08%	86.74%	N/A
scenario_inventory		23.94%	76.06%	N/A	N/A
scenario_new_mo				77.65%	22.35%
scenario_new_so		9.93%	9.93%	2.65%	77.48%
scenario_price_change		4.52%	18.06%	77.42%	N/A
scenario_profit		42.55%	57.45%	N/A	N/A
scenario_qty_sold		35.92%	64.08%	N/A	N/A
scenario_so_number		25.00%	75.00%	N/A	N/A
scenario_validate_po				100.00%	N/A

7 Discussion and Conclusion

To respond to the search for efficiency and adaptation to a competitive environment, companies are pushed to constantly improve their business processes and increase their integration with digital technologies. This requires the implementation of regular training for employees to enable them to integrate these changes and achieve the objectives set by the company. Of all the possibilities available for training employees in this context, this research is focusing on on-the-job training, which is considered a valid solution to this problem (Sree et al., 2019).

In order to develop on-the-job training scenarios, we began by identifying the key elements of controlled learning. Authors such as Gunter et al. (2008) propose models to interweave pedagogical elements in a game structure because learning through serious games is a recognized concept and is currently used in the educational field (de Freitas & Oliver, 2006; Gee, 2005) and leads to better performance than traditional training approaches (Baxter et al., 2017; Hamari et al., 2016). In their RETAIN model, Gunter et al. (2008) emphasize the need for a seamless integration of educational content into the game setting. In this way, the learner progresses through a sequential development that favors the adaptation of their knowledge and that invites them to transfer it to new situations. Another important element is that these games should offer an environment similar to the real world in order to facilitate the learner's immersion in the context (Beranič & Heričko, 2022). This can be achieved through the use of real-life artifacts. For the development of our training scenarios, we chose to use the company's ERP and to extend the functionalities of the integrated chatbot which are found to allow for greater interaction, feedback and personalization, and in this sense are pedagogical levers (Maedche et al., 2016).

To support scenario designers in developing scenarios with some confidence that learning objectives will be met, we have developed a new methodology for designing on-the-job training scenarios that can be implemented in generic business tools like

ERP softwares. The methodology not only allows the creation of an interactive learning scenario, but it will also, thanks to the consideration of the corresponding theories, guarantee that the learners will acquire the skills expected by the set learning objectives. The resulting eight-step methodology takes an activity from a business process as input and generates a sub-process of this activity integrating all components of the serious game scenario, ready to be instantiated.

The methodology was then used to create 18 game scenarios that were integrated in the Odoo ERP in order to train learners on the main business processes of a service company. The scenarios were created around key activities of the business processes, selected for their importance in the company's workflows. This first step of scenario creation allowed us to validate that the methodology is effective in supporting the creation of interactive scenarios to be integrated in an enterprise software through conversational agents. In order to test the effectiveness of the methodology in addressing the learning objectives and not just in entertaining learners, we devised an experiment where learners were first introduced to the business processes and the ERP through nine learning scenarios and later assessed through nine control scenarios. The results obtained are encouraging in that they show that learners who have never encountered either the business processes or the software are able to learn to perform their tasks autonomously with the help of learning scenarios integrated into their ERP. Furthermore, the results show that learners are able to successfully complete the control tasks after an average of 32 min of on-the-job training.

Building on these encouraging results, we will now need to evaluate the learning outcomes over the long term, to determine the error rate in task completion by employees who learned their tasks through the serious game scenarios. One limitation of our study is that it does not allow us to estimate the real impact of learning on the ability of employees to perform their tasks in the long run. However, the fact that a learning scenario can be repeated in the future should allow employees who are rarely confronted with a specific task to quickly get up to speed. Finally, since there was only one experiment to test the scenarios, we were not able to test different implementations of the same scenario. We will take advantage of these results to perform A/B testing between scenario variants to determine a set of best practices for implementing our methodology in the future.

8 Future Research

Following the work presented in this paper, this research opens new avenues for integrating on-the-job training simulation directly into enterprise software production environments. In terms of methodology and evaluation, more research can be performed to get a more detailed understanding of the learning mechanisms and outcomes explaining the success of our approach.

Acknowledgement. We are grateful to the Swiss Innovation Agency which provided partial funding for this work under grant number 41006.1 IP-ICT.

References

Abdullah, N., Hanafiah, M.H., Hashim, N.: Developing creative teaching module: business simulation in teaching strategic management. Int. Educ. Stud. **6**, 95–107 (2013). https://doi.org/10.5539/ies.v6n6p95

Acar, M., Tekinerdogan, B.: Analyzing the Impact of Automated User Assistance Systems : A Systematic Review (2020)

Arnab, S., et al.: Mapping learning and game mechanics for serious games analysis. Br. J. Edu. Technol. **46**(2), 391–411 (2015)

Ayyagari, R., Grover, V., Purvis, R.: Technostress: technological antecedents and implications. MIS Q. **35**, 831–858 (2011). https://doi.org/10.2307/41409963

Azadegan, A., Riedel, J.C.K.H., Hauge, J.B.: Serious games adoption in corporate training. In: Ma, M., Oliveira, M.F., Hauge, J.B., Duin, H., Thoben, K.-D. (eds.) SGDA 2012. LNCS, vol. 7528, pp. 74–85. Springer, Heidelberg (2012). https://doi.org/10.1007/978-3-642-33687-4_6

Baporikar, N.: Business process management. Int. J. Product. Manag. Assessm. Technol. **4**, 49–62 (2016). https://doi.org/10.4018/IJPMAT.2016070104

Baxter, R., Holderness, D., Wood, D.: The effects of gamification on corporate compliance training: a partial replication and field study of true office anti-corruption training programs. J. Forens. Account. Res. (2017). https://doi.org/10.2308/jfar-51725

Beranič, T., Heričko, M.: The impact of serious games in economic and business education : a case of ERP business simulation. Sustainability, **14**(2), 683 (2022). https://doi.org/10.3390/su14020683

Bilal, H., Ahmad, A., Bibi, P., Ali, S.: Evaluating on-the-job training effect on employees' satisfaction the SMEs sector of Malakand Division, Pakistan. Rev. Appl. Manag. Soc. Sci. **4**(1), 145 (2021). https://doi.org/10.47067/ramss.v4i1.107

Bloom, B.S.: Taxonomy of Educational Objectives: The Classification of Educational Goals. Longmans, Green (1956)

Botte, B., Matera, C., Sponsiello, M.: Serious games between simulation and game. A Propos. Taxonomy **5**, 11–21 (2009)

vom Brocke, J., Hevner, A., Maedche, A.: Introduction to Design Science Research, pp. 1–13. Springer, Cham (2020). https://doi.org/10.1007/978-3-030-46781-4_1

Campbell, J.P.: Modeling the performance prediction problem in industrial and organizational psychology. In: Handbook of Industrial and Organizational Psychology, vol. 1, 2nd edn, pp. 687–732. Consulting Psychologists Press (1990)

Cascio, W., Montealegre, R.: How technology is changing work and organizations. Annu. Rev. Organ. Psych. Organ. Behav. **3**, 349–375 (2016). https://doi.org/10.1146/annurev-orgpsych-041015-062352

Caseau, Y.: Processus et Entreprise 2.0 Innover par la collaboration et le lean management (Dunod) (2011)

Clarke, E.: Learning outcomes from business simulation exercises: challenges for the implementation of learning technologies. Educ. Train. **51**, 448–459 (2009). https://doi.org/10.1108/00400910910987246

Crookall, D., Oxford, R., Saunders, D.: Towards a reconceptualization of simulation: from representation to reality. Simulat. Games Learn. **17**, 147–171 (1987)

de Freitas, S., Oliver, M.: How can exploratory learning with games and simulations within the curriculum be most effectively evaluated? Comput. Educ. **46**, 249–264 (2006). https://doi.org/10.1016/j.compedu.2005.11.007

Diederich, S., Brendel, A., Kolbe, L.: On Conversational Agents in Information Systems Research : Analyzing the Past to Guide Future Work (2019, février 24)

Donovan, L.: The Use of Serious Games in the Corporate Sector A State of the Art Report (2012). https://doi.org/10.13140/RG.2.2.22448.28169

Eicker, S., Kochbeck, J., Schuler, P.: Employee competencies for business process. Management **7**, 251–262 (2008). https://doi.org/10.1007/978-3-540-79396-0_22

Fischer, M., Imgrund, F., Janiesch, C., Winkelmann, A.: Strategy archetypes for digital transformation: defining meta objectives using business process management. Inf. Manag. **57**(5), 103262 (2020). https://doi.org/10.1016/j.im.2019.103262

Gagne, R.M.: The Conditions of Learning and Theory of Instruction (Subsequent edition). Wadsworth Pub Co. (1985)

Garris, R., Ahlers, R., Driskell, J.: Games, motivation, and learning: a research and practice model. Simul. Gaming **33**, 441–467 (2002). https://doi.org/10.1177/1046878102238607

Gee, J.P.: Learning by design: good video games as learning machines. E-Learn. Digit. Media **2**(1), 5–16 (2005). https://doi.org/10.2304/elea.2005.2.1.5

Goel, K., Bandara, W., Gable, G.: A typology of business process standardization strategies. Bus. Inf. Syst. Eng. **63**(6), 621–635 (2021). https://doi.org/10.1007/s12599-021-00693-0

Greenblatt, M.: Making learning fun : A taxonomy of intrinsic motivations for learning (1987). https://www.academia.edu/3429898/Making_learning_fun_a_taxonomy_of_intrinsic_motivations_for_learning

Gregor, S.: The nature of theory in information systems. Manag. Inf. Syst. Q. **30**(3), 611–642 (2006)

Gregor, S., Chandra Kruse, L., Seidel, S.: Research perspectives: the anatomy of a design principle. J. Assoc. Inf. Syst. **21**(6), 2 (2020)

Gregor, S., Hevner, A.R.: Positioning and presenting design science research for maximum impact. MIS Q. **37**(2), 337–356 (2013)

Gunter, G.A., Kenny, R.F., Vick, E.H.: Taking educational games seriously: using the RETAIN model to design endogenous fantasy into standalone educational games. Educ. Tech. Res. Dev. **56**(5), 511–537 (2008). https://doi.org/10.1007/s11423-007-9073-2

Hamari, J., Shernoff, D.J., Rowe, E., Coller, B., Asbell-Clarke, J., Edwards, T.: Challenging games help students learn: an empirical study on engagement, flow and immersion in game-based learning. Comput. Hum. Behav. **54**, 170–179 (2016). https://doi.org/10.1016/j.chb.2015.07.045

Haračić, M., Tatic, K., Haračić, M.: The Improvement of Business Efficiency Through Business Process Management, vol. XVI, pp. 31–43 (2018)

Hevner, A., Chatterjee, S.: Design science research in information systems. In: Hevner, A., Chatterjee, S. (eds.) Design Research in Information Systems: Theory and Practice, pp. 9–22. Springer, Boston (2010). https://doi.org/10.1007/978-1-4419-5653-8_2

Hevner, A.R.: A three cycle view of design science research. Scand. J. Inf. Syst. **19**(2), 4 (2007)

Hrabal, M., Tuček, D., Molnár, V., Fedorko, G.: Human factor in business process management : Modeling competencies of BPM roles. Bus. Process. Manag. J. (2020)

ISO 9000. ISO 9000:2015, Systèmes de management de la qualité—Principes essentiels et vocabulaire (2015). https://www.iso.org/obp/ui/#iso:std:iso:9000:ed-4:v2:fr

Jacobs, R.: Structured on-the-Job Training, pp. 272–284 (2014)

Keller, J.M.: Development and use of the ARCS model of instructional design. J. Instr. Dev. **10**(3), 2 (1987). https://doi.org/10.1007/BF02905780

Krath, J., Schürmann, L., von Korflesch, H.F.O.: Revealing the theoretical basis of gamification: a systematic review and analysis of theory in research on gamification, serious games and game-based learning. Comput. Hum. Behav. **125**, 106963 (2021). https://doi.org/10.1016/j.chb.2021.106963

Maedche, A., et al.: AI-based digital assistants. Bus. Inf. Syst. Eng. **61**(4), 535–544 (2019). https://doi.org/10.1007/s12599-019-00600-8

Maedche, A., Morana, S., Schacht, S., Werth, D., Krumeich, J.: Advanced user assistance systems. Bus. Inf. Syst. Eng. **58**(5), 367–370 (2016). https://doi.org/10.1007/s12599-016-0444-2

Michael, D.R., Chen, S.: Serious Games: Games that Educate, Train and Inform. Thomson Course Technology (2006)

Morana, S., Pfeiffer, J., Adam, M.T.P.: User assistance for intelligent systems. Bus. Inf. Syst. Eng. **62**(3), 189–192 (2020). https://doi.org/10.1007/s12599-020-00640-5

Nguyen, T.Q., Nguyen, A.T., Tran, A.L., Le, H.T., Le, H.H.T., Vu, L.P.: Do workers benefit from on-the-job training? New evidence from matched employer-employee data. Financ. Res. Lett. **40**, 101664 (2021). https://doi.org/10.1016/j.frl.2020.101664

Niati, D., Siregar, Z., Prayoga, Y.: The effect of training on work performance and career development : the role of motivation as intervening variable. Budapest Int. Res. Critics Inst. (BIRCI-Journal): Human. Soc. Sci. **4**, 2385–2393 (2021). https://doi.org/10.33258/birci.v4i2.1940

Peffers, K., Rothenberger, M., Tuunanen, T., Vaezi, R.: Design science research evaluation. In: Peffers, K., Rothenberger, M., Kuechler, B. (eds.) DESRIST 2012. LNCS, vol. 7286, pp. 398–410. Springer, Heidelberg (2012). https://doi.org/10.1007/978-3-642-29863-9_29

Peffers, K., Tuunanen, T., Rothenberger, M.A., Chatterjee, S.: A design science research methodology for information systems research. J. Manag. Inf. Syst. **24**(3), 45–77 (2007)

Piaget, J.: Psychologie et Epistémologie. Pour une Théorie de la Connaissance (1970)

Roh, S., et al.: Goal-Based Manufacturing Gamification: Bolt Tightening Work Redesign in the Automotive Assembly Line, vol. 490, pp. 293–304 (2016). https://doi.org/10.1007/978-3-319-41697-7_26

Rosemann, M., vom Brocke, J.: The six core elements of business process management. In: vom Brocke, J., Rosemann, M. (eds.) Handbook on Business Process Management 1. IHIS, pp. 105–122. Springer, Heidelberg (2015). https://doi.org/10.1007/978-3-642-45100-3_5

Shu, Q., Tu, Q., Wang, K.: The impact of computer self-efficacy and technology dependence on computer-related technostress: a social cognitive theory perspective. Int. J. Hum.-Comput. Interact. **27**(10), 923–939 (2011)

Sree, V., Rabiyathul, S., Basariya, S.R.: On the job training implementation and its benefits. **6**, 210–215 (2019)

Stainton, A.J., Johnson, J.E., Borodzicz, E.P.: Educational validity of business gaming simulation: a research methodology framework. Simul. Gaming **41**(5), 705–723 (2010). https://doi.org/10.1177/1046878109353467

Tsagkani, C., Tsalgatidou, A.: Process model abstraction for rapid comprehension of complex business processes. Inf. Syst. **103**, 101818 (2022). https://doi.org/10.1016/j.is.2021.101818

Vaishnavi, V.K., Kuechler, W.: Design Science Research Methods and Patterns: Innovating Information and Communication Technology, 2nd edn. CRC Press, Inc. (2015)

vom Brocke, J., Hevner, A., Maedche, A.: Introduction to design science research. In: vom Brocke, J., Hevner, A., Maedche, A. (eds.) Design Science Research. Cases, pp. 1–13. Springer, Cham (2020). https://doi.org/10.1007/978-3-030-46781-4_1

Weske, M.: Business Process Management: Concepts, Languages, Architectures, 2nd édn. Springer, Cham (2012)

Complementation Between Usage Schemes and Mental Actions in an Approach to the Definite Integral Using GeoGebra

Mihály A. Martínez-Miraval[1](✉) and Martha L. García-Rodríguez[2]

[1] Pontificia Universidad Católica del Perú, Av. Universitaria 1801, San Miguel 15088, Perú
martinez.ma@pucp.edu.pe

[2] Instituto Politécnico Nacional CICATA-Legaria, Calz Legaria 694, Col. Irrigación, Miguel Hidalgo, 11500 Ciudad de México, CDMX, México
mlgarcia@ipn.mx

Abstract. In the process of approximating the area of a region by means of Riemann sums a relationship between different variables that change simultaneously takes place, and although digital technologies help to visualize the process, it is important to understand how an individual reasons with the software. This study aimed to analyze how the interaction between usage schemes and mental actions of an individual occurs when solving an assignment about the definite integral study using GeoGebra. Theoretical approaches of both covariational reasoning and instrumental approach were used in qualitative research using action research as a method. Students mobilized their usages schemes related to variation, summation, and limit notions, which were run using the GeoGebra commands slider and RectangleSum, these commands were used in order to represent and coordinate the changes in the values of the variables: number of rectangles and sum of their areas; this allowed identifying a complementation between these usage schemes and mental actions, strengthening the analysis of the information. Finally, it is concluded that the complementation of two theoretical approximations provides elements that are essential when analyzing the phenomenon studied, in order to understand how a student reasons covariationally when using GeoGebra during the learning process.

Keywords: Usage scheme · Mental action · GeoGebra · Definite integral

1 Introduction

Literature in mathematics education related to the concept of definite integral includes studies that deal with its study associated with approximation processes to make sense of this concept as a technique that allows determining the area of a region through the use of rectangles [1], to define it as the limit of a sum of terms of real value [2] or applied to the field of physics giving meaning to the product, sum, limit and function stages involved when studying the definite integral by means of Riemann sums [3]; or related with processes of accumulation of areas, whether it is applied in a mathematical

A. Rocha et al. (Eds.): WorldCIST 2023, LNNS 800, pp. 395–404, 2024.
https://doi.org/10.1007/978-3-031-45645-9_37

context to give meaning to the integral function [4], or in the field of engineering related to characterizing the filling of a swimming pool from changes in the speed of entry of water [5]. Throughout these processes, change phenomena are worked on, so it is important to understand the functional relationships in terms of a covariation between the values of two simultaneously varying quantities [6].

How an individual coordinate the changes between the values of two variables can be explained from the behaviors externalized by the individual, which are the representations of his mental actions associated with covariational reasoning [7]. The study of this kind of reasoning can be based on identify the number of variables that the student relates simultaneously when developing a task, and the complexity of coordinating changes between these [8]. Another possibility is to analyze the ability to reason covariationally of a student at a given level [7], or by identifying their mental actions related to their covariational reasoning when solving a task with GeoGebra [9].

Dynamic geometry systems function as a means of running the ideas and notions that are present in the thinking of an individual. They allow Dynamic changes between the values of two quantities, so that covariational reasoning can be analyzed even when working with digital technologies [6]. Moreover, these dynamic changes can be made by associating sliders to other tools, in order to generate a dynamic, interactive, and visual learning environment [10].

Based on the review of these studies, the importance of covariational reasoning and of the digital technologies in the study of the definite integral emerge. This led to the present research, which aims *to analyze how the interaction between usage schemas and mental actions of an individual takes place when solving a task related to the definite integral through GeoGebra.*

2 Theoretical Approaches

This research is supported by the set of mental actions identified by [7], from the theoretical construct developed by [9], in the work of students who solved a task on the concept of definite integral using Riemann sums with GeoGebra (Table 1).

Through the use of GeoGebra, it was possible to focus on the usage schemes mobilized by the students, related to ideas, concepts and mathematical processes, run with artefact tools, namely, they include a set of specific actions and activities geared toward the artefactual part [11]. By using these elements, which correspond to the theoretical approaches of covariational reasoning and the instrumental approach, will provide tools to describe how a student reasons when using GeoGebra to manage his/her learning.

Table 1. Mental actions related to each level and associated behaviors

Level	Behaviors
Smooth continuous covariation	MA6: Extend the simultaneous coordination idea between the base measure of each rectangle and the sum of the areas of the inscribed or circumscribed rectangles when the number of rectangles tends to infinity
Chunky continuous covariation	MA5: Extend the simultaneous coordination idea between the number of rectangles and the sum of the areas of the inscribed and circumscribed rectangles with the intention of approximating the area of the region R, by increasing the number of rectangles
Coordination of values	MA4: Construct a new element representing the coordination of the individual values of two quantities
Gross coordination of values	MA3: Express verbally how they coordinate changes between the number of rectangles and the sum of their areas in terms of increases or decreases
Pre-coordination of values	MA2: Establish asynchronous relationships between the values of various magnitudes, for example, at first, think in the number of rectangles, then, in the sum of areas of rectangles
No coordination	MA1: Perform arithmetic operations to give meaning to the values displayed in GeoGebra, such as the sum of areas of rectangles

Note: Adapted from [9]

3 Methods and Procedures

The methodology is qualitative and in order to contribute to both educational practice and theory, generating a more reflective educational process, the method chosen was action research [12]. An action research cycle presents four stages: planning, action, observation, and reflection. Two cycles whereby it was possible to lead to the current research were developed (Fig. 1).

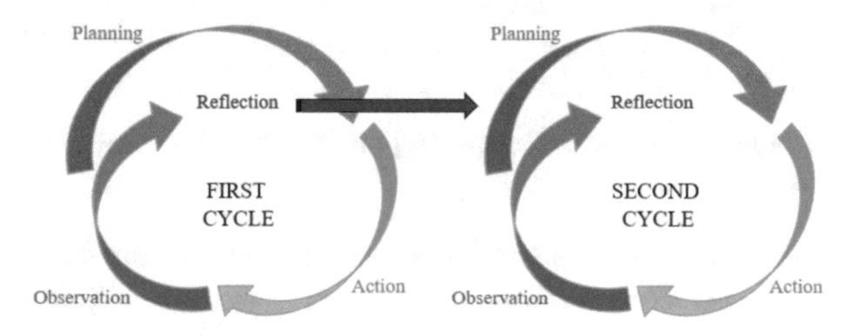

Fig. 1. Two cycles of action research

3.1 Development of the First Cycle of Action Research

The planning stage began with the identification of difficulties observed in the students when introducing the concept of definite integral by means of Riemann sums, which were manifested at the moment of coordinating the changes between variables: number of rectangles, measure of base and sum of areas of rectangles, led to the choice of the theoretical construct of covariational reasoning of [7], to describe the reasoning ability of students, and was complemented by the work of [13] who developed a theoretical approach oriented to the mathematical problem solving processes characterization through the use of GeoGebra, where group discussion prevails in order to see how knowledge emerges. Since it was planned to work with four university students, who had no knowledge regarding GeoGebra, it was decided to design applets with sliders, so that the students, in pairs, could manipulate them.

In the action stage, the activity was designed (Fig. 2) and the data collection instruments were defined: physical sheets, GeoGebra files, audios and videos, and a semi-structured interview that was videotaped.

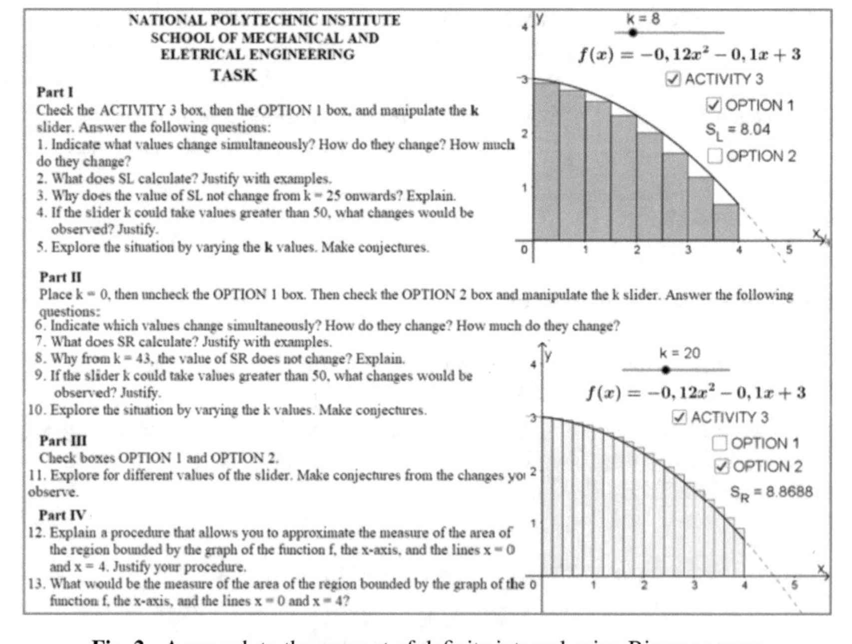

NATIONAL POLYTECHNIC INSTITUTE
SCHOOL OF MECHANICAL AND
ELETRICAL ENGINEERING
TASK

Part I
Check the ACTIVITY 3 box, then the OPTION 1 box, and manipulate the **k** slider. Answer the following questions:

1. Indicate what values change simultaneously? How do they change? How much do they change?
2. What does SL calculate? Justify with examples.
3. Why does the value of SL not change from k = 25 onwards? Explain.
4. If the slider k could take values greater than 50, what changes would be observed? Justify.
5. Explore the situation by varying the **k** values. Make conjectures.

Part II
Place k = 0, then uncheck the OPTION 1 box. Then check the OPTION 2 box and manipulate the k slider. Answer the following questions:

6. Indicate which values change simultaneously? How do they change? How much do they change?
7. What does SR calculate? Justify with examples.
8. Why from k = 43, the value of SR does not change? Explain.
9. If the slider k could take values greater than 50, what changes would be observed? Justify.
10. Explore the situation by varying the k values. Make conjectures.

Part III
Check boxes OPTION 1 and OPTION 2.

11. Explore for different values of the slider. Make conjectures from the changes you observe.

Part IV

12. Explain a procedure that allows you to approximate the measure of the area of the region bounded by the graph of the function f, the x-axis, and the lines x = 0 and x = 4. Justify your procedure.
13. What would be the measure of the area of the region bounded by the graph of the function f, the x-axis, and the lines x = 0 and x = 4?

Fig. 2. Approach to the concept of definite integral using Riemann sums

In the observation and analysis stage, the mental actions of the students were identified for each of the levels of covariational reasoning (shown in Table 1), as well as the type of questions that triggered these mental actions. The reflection stage helped to make important modifications in the approach to the concept of the definite integral with the students for the second cycle of action research: the number of questions was reduced because of redundancy, it was considered that students should make their own constructions with GeoGebra, instead of designing applets, since these showed signs of covariational reasoning, students were asked to work individually, due to pandemic restrictions, the theoretical approach was changed, instead of the work of [13], the instrumental approach of [11] was chosen because of the way in which the experimental part was carried out.

These reflections led to the beginning of the second cycle of action research.

3.2 Development of the Second Cycle of Action Research

As part of the planning for this cycle, the reflections of the previous cycle were considered: working with five students individually, reducing the number of items and providing a GeoGebra link for them to perform their explorations and develop the task, previously, there were several weeks of familiarization with the software in other topics of calculus such as functions, limits and derivatives. In addition, the theoretical approaches to covariational reasoning of [7], the mental actions identified by [9], and the notion of usage scheme proposed by [11] were reviewed together, in order to identify strong ideas to complement these theoretical aspects and deepen the analysis: on one side, it is identified the usage schemes that students generate or mobilize, related to different mathematical notions or concepts, which they execute with GeoGebra tools, and, on the other side, it is identified the variables that students mobilize and how they coordinate their values in the development of the task, in other words, how they use their covariational reasoning.

For the action stage, four tasks were designed, which addressed the study of the definite integral in processes of approximation to the area of a region and in processes of accumulation of areas. In the present research the task shown in Fig. 3 is presented and analyzed, the information was collected in written documents, images and GeoGebra files, videos, and a semi-structured videotaped interview.

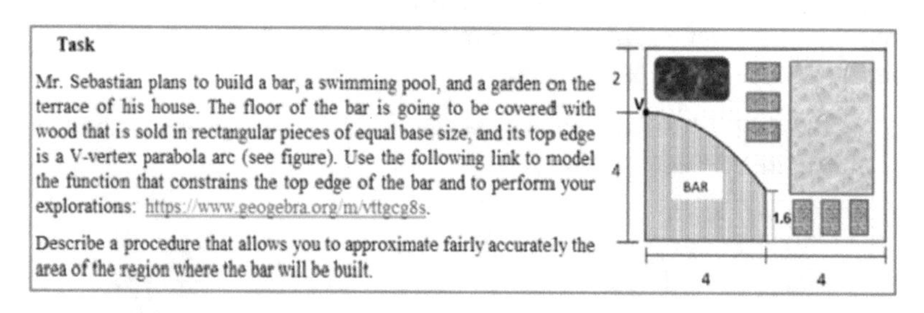

Fig. 3. Approach to the concept of the integral using GeoGebra

In the observation stage, the answers of the student, whom we will call Rodrigo, and the respective analysis are presented. The answer of the student was given in written form: "First we find the function, in this case it would be $(-0.15x^2 + 4)$, $0 \le x \le 4$. After having found the function, we put the points to join the polygon and get an approximation of the area of the bar, the result is 12.797", and as an image of what was done in GeoGebra (Fig. 4).

Rodrigo mobilized his usage scheme geometric figure, associated with the characteristics of a polygon such as having vertices, sides and area, and mobilized his usage

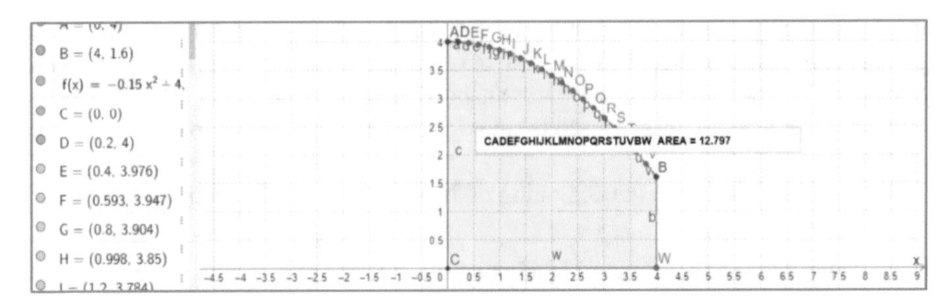

Fig. 4. Approximation of the area with the polygon tool of GeoGebra.

scheme area, to calculate the area of the geometric figure, actions performed by using GeoGebra's polygon and area tools.

When drawing a polygon, variables that can be manipulated appear such as number of points on the graph, distance between consecutive points, regions of the bar floor not covered by the polygon, and area of the polygon. The student did not draw another polygon to approximate the area of the region, so there was no coordination between the values of the variables, making a MA1 visible. This process is considered static, because, on the one hand, only another polygon was drawn to approximate the area, and, on the other hand, in the case that another approximation had been made, this would be done manually, and would hide the changes. The semi-structured interview made Rodrigo realize that he could have used another procedure to solve the task, involving rectangles inscribed or circumscribed to the region.

1. Researcher: Why did you use a polygon to approximate the area?
2. Rodrigo: It was the first thing that came to my mind. I also used rectangles to compare the results, but I did not report it in my answer.
3. Researcher: How did you place points so close together on the graph?
4. Rodrigo: I zoomed in with the mouse.

Rodrigo sent by WhatsApp the images showing how he approximated the area of the region with rectangles (Fig. 5).

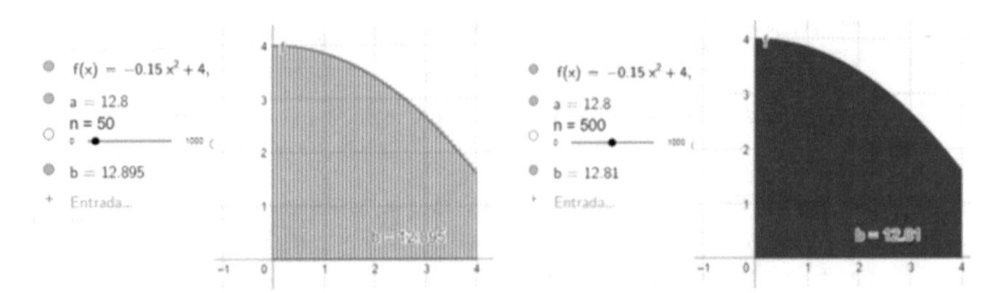

Fig. 5. Approximation of the area with the RectangleSum tool of GeoGebra.

The algebraic view shows the order Rodrigo followed to solve the task. He mobilized his defined integral usage scheme, related to area calculation, afterwards, he mobilized his slider usage scheme, related to variation processes, finally, to deter-mine the sum of the areas of the rectangles, he mobilized his usage scheme Sum, related to the concept of summation, all of these were made effective with the integral, slider and RectangleSum tools of GeoGebra respectively. The asynchronous coordination between the number of rectangles and the sum of their areas is a characteristic of a MA2.

Rodrigo generated his usage scheme Simultaneous Sum, with which he mobilized features related to the notion of Riemann sum, and which was run by associating the slider and RectangleSum tolos of GeoGebra. He expressed verbally that as the number of rectangles increased, the sum of their areas decreased, because the rectangles were circumscribed to the region. Behavior associated with an MA3.

This new element allows to determine the sum of the areas of the rectangles from an approximation process by varying the number of rectangles from their initial value to their final value, and simultaneously approximating the area of a region. These are behaviors of a MA4. With the purpose of inquiring about the relationship between the value of the definite integral and the value of the sum of the areas of the rectangles, the following questions were asked:

5. Researcher: Is it possible to determine the area of the region with rectangles?
6. Rodrigo: Yes, it is possible, with the rectangles you can find the area of the function in a certain way.
7. Researcher: How would you do it?
8. Rodrigo: Considering many rectangles, and the limit of the sum of all is the integral. The two images show that, with 50 rectangles, the sum is 12.895 and when there are 500, the area is 12.81, although the whole region is already painted, as if the rectangles were lines that paint the region from one end to the other. When the number of rectangles used goes up, we can get a better picture of what the integral is.

According to the student's answers, as the number of rectangles with the slider increased, there was a refinement of the usage scheme Simultaneous sum, maintaining the coordination between both variables (MA5), as a result, Rodrigo mobilized the notion of limit associated to the process of approximation to the area of a region, which allowed him to consider that the variables: base measure of each rectangle and sum of areas of the rectangles change simultaneously, and that all the values of these variables can be considered as if they were continuous variables, which do not depend on discrete changes in the number of rectangles, behaviors associated with a MA6.

Figure 6 shows a diagram of how the interaction between the usage schemas and mental actions occurred in the two responses presented by Rodrigo.

In Fig. 6, the one-way arrows indicate the sequence of the construction, the dotted rectangle represents the GeoGebra tool association, the green and red boxes represent the usage schemes that mobilized and the mental actions that became visible, and the two-way arrow means that there were several increments of the maximum value of the slider, which we interpret as a refinement in the approximation process.

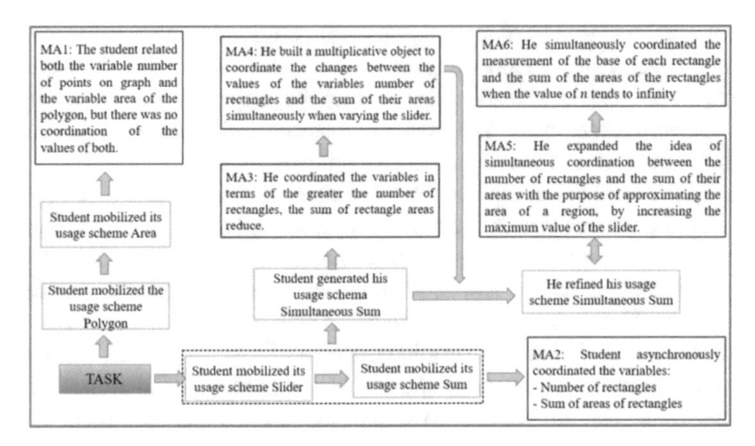

Fig. 6. Interaction of usage schemas and mental actions during Rodrigo's task completion.

4 Discussion

This study examined from the theoretical approach of the instrumental approach, how a student uses GeoGebra to apply the ideas and notions he has in his mind about the process of approximating the area of a region, and from the theoretical approach of covariational reasoning, different levels were identified in which the student coordinates the changes in the values of two variables, associated with his mental actions.

The student showed two different ways of approximating the area of a region: by using a single polygon with points very close to each other on the graph of a function, and by using rectangles that increased in number. In the first case, a single approximation was made to the area of the region with a polygon, which did not allow observing the dynamic changes between the values of the variables; in comparison with the second case, where a dynamic process was carried out, which led the student to approach the definition of the integral as the limit of a sum. In this regard, we agree with [14], when they state that developing notions of variation, covariation and accumulation in students are crucial to conceptualize the definite integral.

The student showed an evolution in his mental actions due to the construction process with GeoGebra. Representing the variables and their characteristics with GeoGebra tools, involves configuring them individually, and therefore there is an asynchronous relationship between them (MA2), the association of these tools means creating a multiplicative object that allows to coordinate simultaneously the individual values of the variables: number of rectangles and sum of their areas, when varying a slider (MA4), the search for a better approximation implies increasing the maximum value taken by the slider and maintaining the coordination between both variables (MA5), finally, focusing on the coordination of the values of two continuous variables, such as the base measure of each rectangle and the sum of the areas of the rectangles, as the number of rectangles grows unlimitedly, makes one think that there is a continuous covariation between these variables, where the limit represents the completion of the approximation process (MA6).

In this respect, [9] associate the evolution of mental actions related to students' covariational reasoning when working with GeoGebra, both to their interaction with the software and to the triggering questions that activate the mental actions, where questions oriented to explain how and how much one variable varies with respect to another, bring into play mental actions of lower levels of covariational reasoning, in contrast, hypothetical questions that induce students to raise conjectures and justify them, place mental actions at higher levels.

The usage schemes and mental actions complemented each other and helped to make the analysis more complete. Actions such as associating the slider and RectangleSum tools of GeoGebra, in order to mobilize features related to the notion of Riemann sum that are located in a student's mind, are achieved by mobilizing the student's usage scheme Simultaneous Sum; but the choice of these tools was made to represent two different variables: number of rectangles and sum of areas of these rectangles, and to build a multiplicative object of them, that is, there is an evolution in the way of coordinating the changes between the values of these two variables, which means an evolution in their mental actions. Similarly, when interval refinements are performed to justify the trend of the sum of the areas of the rectangles, it is observed that the usage scheme Simultaneous Sum activates a MA5, but this in turn, activates the same scheme to refine the interval again, this interaction activates a MA6.

There is agreement with [15], when they state that sometimes a single theoretical perspective is not enough to address a phenomenon related to the teaching and learning of mathematics, due to its complexity; they suggest approaching the phenomenon from different theoretical approaches based on some connecting strategy.

5 Conclusions

Dynamic geometry systems such as GeoGebra can be used in a number of ways to address Calculus topics. Designing applets or familiarizing students with the software tools before the activity, so that they can design their own resolution procedure, are strategies that work as long as they are complemented with tasks that generate a high level of cognitive demand, and that a close relationship between the task and the reasoning used by the student to solve it is generated, thus giving the student the opportunity to mobilize their previous knowledge as well as to generate new knowledge.

The method used allowed for a more reflective educational process, because it allows to draw conclusions that provide elements to improve the proposal or rethink it, based on a reflection on various aspects involved in the process, which are related to the proposal of the task, related to the way of using a dynamic geometry system, on the way in which the experimental part was performed, related to the theoretical approaches that support the research, among other aspects that lead to reflection.

The complementation of two theoretical approaches provides elements that are essential when analyzing the phenomenon under study, in order to understand how a student reasons covariationally using GeoGebra during the learning process. A student's covariational reasoning is different when working with a dynamic geometry system, due to the information obtained from the software and visible on its interface. Perceiving the dynamic, simultaneous and instantaneous changes that are generated between the values of two variables in GeoGebra influences their covariational reasoning.

This way of characterizing the covariational reasoning of a student who uses GeoGebra to solve a task can be extended to the use of functions that model rates of change, such as speeds, so that the study of the definite integral can cover other fields such as Physics or Engineering; Likewise, the concepts of functions, limits and derivatives are generally defined involving change phenomena, an opportunity that can be considered to carry out an analysis of the covariational reasoning of students on these topics when digital technologies are used.

Acknowledgment. The authors thank the Pontificia Universidad Católica del Perú and the Instituto Politécnico Nacional for the support provided in the development of this research.

References

1. Aranda, C., Callejo, M.: Construcción del concepto de integral definida usando geometría dinámica utilizando distintos sistemas de representación. Paradigma **41**(2), 305–327 (2020)
2. Martínez-Miraval, M., García-Cuéllar, D.: Estudio de las aprehensiones en el registro gráfico y génesis instrumental de la integral definida. Formac. Univ. **13**(5), 177–190 (2020)
3. Sealey, V.: A framework for characterizing student understanding of Riemann sums and definite integrals. J. Math. Behav. **33**(2014), 230–245 (2014)
4. Aranda, M., Callejo, M.: Construcción de la función integral y razonamiento covariacional: dos estudios de casos. Bolema **31**(58), 777–798 (2017)
5. Harini, N., Fuad, Y., Ekawati, R.: Students' covariational reasoning in solving integrals' problems. J. Phys. Conf. Ser. **947**(1), 1–7 (2018)
6. Carlson, M., Jacobs, S., Coe, E., Larsen, S., Hsu, E.: Applying covariational reasoning while modeling dynamic events: a framework and a study. J. Res. Math. Educ. **33**(5), 352–378 (2002)
7. Thompson, P., Carlson, M.: Variation, covariation, and functions: foundational ways of thinking mathematically. In: Cai, J. (ed.), Compendium for Research in Mathematics Education, pp. 421–456. National Council of Teachers of Mathematics, Reston (2017)
8. Kouropatov, A., Dreyfus, T.: Learning the integral concept by constructing knowledge about accumulation. ZDM Mathematics Education **46**(4), 533–548 (2014)
9. Martínez-Miraval, M., García-Rodríguez, M.: Razonamiento covariacional de estudiantes universitarios en un acercamiento al concepto de integral definida mediante Sumas de Riemann. Formac. Univ. **15**(4), 105–118 (2022)
10. Zengin, Y.: Incorporating the dynamic mathematics software GeoGebra into a history of mathematics course. Int. J. Math. Educ. Sci. Technol. **49**(7), 1083–1098 (2018)
11. Rabardel, P.: Los hombres y las tecnologías: visión cognitiva de los instrumentos contemporáneos. (Trad. por M. Acosta). 1ra edn. Universidad Industrial de Santander, Colombia (2011)
12. Elliot, J.: La Investigación-Acción en Educación, 1ra edn. Morata, Madrid (1990)
13. Santos-Trigo, M., Camacho-Machín, M.: Framing the use of technology in problem solving approaches. Math. Enthusiast **10**(1–2), 279–302 (2013)
14. Thompson, P., Dreyfus, T.: A coherent approach to the fundamental theorem of calculus using differentials. In: Göller, R., Biehler, R., Hochsmuth, R. (eds.) Proceedings of the Conference on Didactics of Mathematics in Higher Education as a Scientific Discipline, pp. 355–359. KHDM, Hannover (2016)
15. Prediger, S., Bikner-Ahsbahs, A., Arzarello, F.: Networking strategies and methods for connecting theoretical approaches: first steps towards a conceptual framework. ZDM Mathematics Education **40**, 165–178 (2008)

Reviewing Learning Software Engineering Through Serious Game Development

Manal Kharbouch[1](✉), José Alberto García-Berná[1], Juan Manuel Carrillo de Gea[1], Joaquín Nicolás Ros[1], José Ambrosio Toval Álvarez[1], Ali Idri[2], and Jose Luis Fernández-Alemán[1]

[1] Software Engineering Research Group, Department of Computing and Systems, Faculty of Informatics, University of Murcia, Murcia, Spain
manal.kharbouch@um.es

[2] Software Project Management Research Team, Department of Web and Mobile Engineering, ENSIAS, Mohamed V University in Rabat, Rabat, Morocco

Abstract. Serious games (SGs) are known as games designed for educational purposes rather than mere entertainment. Although their integration into software engineering (SE) education is a relatively new trend, it is spreading gradually, and thus gaining growing attention in the academic field. In this paper, we first investigated how SGs' design and development have been used as a pedagogical approach in SE education. Second, we analyzed the perceived complexity and potential effectiveness of this approach from the literature. Last, we addressed its resulting challenges and opportunities in the realm of SE education. The search for studies was carried out in leading search resources namely, Science Direct, IEEE Xplore, ACM, Scopus, as well as Wiley resulting in a set of six selected studies. Our study's results show that the design and development of SGs could be used to further motivate students, improve their SE learning experience and make it more enjoyable, and promote the use of these SGs. Moreover, this pedagogical approach was considered effective for a holistic understanding of SE courses, and better knowledge and skills acquisition.

Keywords: Serious Game · Software Engineering · Education

1 Introduction

The widespread use of SGs has had increasing popularity, not least in the world of education [1]. Thereby, SGs have many advocates for the role they can play in formal education. The latter consider that SGs are not only engaging and interactive, but also offering immersive activities, and having a real potential to be important teaching tools [2]. Good evidence of their potential in teaching are studies that have shown a 31% increase in the completion rate of Massive Open Online Courses (MOOCs) that incorporate SGs in their teaching process, compared to those that do not [3]. Even though SGs have such motivational power and high student retention, studies have shown that there are no known forms of education as effective as a professional human tutor [11].

A. Rocha et al. (Eds.): WorldCIST 2023, LNNS 800, pp. 405–414, 2024.
https://doi.org/10.1007/978-3-031-45645-9_38

A good strategy for achieving high SGs' effectiveness should focus on developing trust between target users, the technical development team, and educators as implementers of SGs [12]. A reasonable way in which these criteria can be met is when students can play both the role of target users and technical development staff.

Just like the rest of the computer science disciplines, SE is a difficult venture at its very best [4]. Yet, the difficulty of teaching SE to undergraduates is even more problematic [5]. Since SGs are associated with pleasure and fun [3], some approaches based on Game-Based Learning (GBL) in the field of SE education have been proposed [6–8]. Moreover, a systematic mapping study was carried out by Pedreira et al. [9], on the gamification gaps found in the SE field and the potential benefits that gamification can bring to it. Furthermore, Rodríguez et al. [10] conducted a multifocal literature review regarding SGs for teaching agile methods. This study identified novel methods to assess SGs, application domains in which assessment has been carried out, and the main characteristics considered to assess the pedagogical advantages of these SGs. In Addition, many studies were investigating how to learn/teach SE through playing SGs [13], but there are only a few that address learning/teaching SE through SGs development. Developing a SG is an important topic, especially when the process has a pedagogical objective in its background [14], There are studies in which students not only designed SGs but also examined potential issues and complexities involved in the developing phase and incorporated them within a teaching curriculum [15]. In a study by Robertson and Howells [16], they mentioned that there is research that argues that not only the use of games but also their design and construction as part of learning strategies contribute to learning, as learning becomes an active experience. Thus, games are attractive not only to players but also to game developers [17]. From this perspective, SGs' design and development could be used to motivate and enhance SE learning [18]. This paper reviews this novel perspective from which SGs are developed for teaching SE.

The rest of the paper is structured as follows, Sect. 2 describes the study design and the guidelines we followed to conduct this review, next Sect. 3 answers the research questions and further discusses the study results, and last Sect. 4 highlight the current trends and the gaps found in the literature.

2 Method

The aim of this study is to investigate SGs design and development as a novel teaching approach in SE, identify the areas of SE in which this approach is used, analyze its complexity and effectiveness, and identify potential pitfalls and opportunities for future research. This leads us to the main research question that drove this study: how are SGs' design and development adopted in SE teaching activities?

2.1 Search Process

The search for studies in which SGs' design and development were used as a teaching approach in SE followed the recommendations of Kitchenham and Charters [19] to ensure the accuracy and completeness of the search and retrieval process. The literature search was held between December 2020 and May 2021. The planning of this review consisted of the following steps: identifying the research questions, defining the data sources and research strategy, as well as the selection. Any disagreements and discrepancies were discussed until a consensus was reached.

2.2 Research Questions

To achieve the objective of this study a set of 4 research questions was elaborated: (RQ1) How SGs' design and development get adopted into SE teaching activities? (RQ2) What SWEBOK areas have been targets of SGs' design and development in SE teaching? (RQ3) To what extent is the usage of SGs' design and development in SE teaching considered complex? (RQ4) What are the cited pitfalls and potential opportunities of incorporating SGs' design and development in SE teaching?

2.3 Data Sources and Search Strategy

The leading search resources namely, Science Direct, IEEE Xplore, ACM, Scopus, and Wiley were considered to perform the search for candidate studies. The PICO criteria [20] was used to define the search string. The Population considered was that of SE studies. The Intervention focuses on SGs. Given that the primary focus of the present study was not comparative, the Comparison criterion was disregarded. With respect to the last criterion, all existing outcomes regarding SGs in the field of SE were of interest for this study, therefore, no term was selected for the outcome aspect in order not to restrict the result set. The identified keywords are SGs, SE, and the SWEBOK areas which were grouped into sets, and their synonyms were considered to formulate the search string in accordance with the recommendations of Petersen et al. [21]. Search terms were constructed using the steps described in [22], and the search was mainly focused on the Title, Abstract, and Keywords of the studies. This search string was reviewed and agreed on among all authors. Unfortunately, some databases did not allow using the full search string as defined. Therefore, slightly adapted and simplified search strings were used instead to suit the specific requirements of the different search interfaces.

2.4 Study Selection

After obtaining the potentially relevant primary studies, they were evaluated for their real relevance to the scope of our study. Three inclusion criteria were applied in the first phase and four exclusion criteria in the second phase. Specifically, in phase 1: IC1 (studies investigating SGs), IC2 (studies in the field of SE), and IC3 (studies written in English). In phase 2: EC1 (studies that are duplicates of other studies or duplicate articles of the same research in different databases), EC2 (papers not accessible in full text), EC3 (articles available only as abstracts or posters), and EC4 (studies that focus on the use of SE to build SGs with no pedagogical approach). The study only considered articles that met all the inclusion criteria and discarded those that met any of the exclusion criteria, finally identifying 125 studies. These papers considered relevant were reviewed by the first and last author by reading the full text and discussed among the rest of the authors, discarding any study whose main focus was not the design and development of SGs in SE teaching. The final number of articles selected was 6.

3 Results and Discussion

While SGs are not a recent topic, nor is their use in educational contexts [23], the frequency of publication as shown in Table 1 bellow has been very low up to the present, with one or no publication each year. Thereby this novel teaching approach has not yet aroused the interest of the scientific community, and the impact it can have on the teaching of the different areas of SE remains to be discovered. Table 1 also shows that the main target publication channel for SG in SE studies is international conferences with a rate of 66.66%, followed by publications in scientific journals and workshops with an equal share of 16.66% of the selected studies. The studies do not present solid empirical evidence showing the effectiveness of their proposals and their impact on SE learning. Moreover, systematic methodology for incorporating SGs design and development into SE teaching is lacking, making most of the proposals difficult to replicate successfully for other studies. This has shown that research on SGs design and development applied to SE teaching is immature. Table 1 also shows that 66.66% of selected studies are validation research, and the remaining 33.33% are opinion papers. The first publication was an opinion paper presenting preliminary results. Later, more validation papers were published. The relatively high percentage of validation studies developed is noteworthy leading to an interesting conclusion that this research line is promising. In this section the selected studies are presented and analyzed, responding to the research questions defined in Sect. 2.2.

Table 1. Selected Studies Overview

Title	Publishing Channel	Year	Type of research	SWEBOK Area
[24]	Conference	2011	Opinion article	Software engineering process
[17]	Conference	2013	Validation	Software construction
[26]	Workshop	2014	Validation	Software design
[18]	Conference	2015	Validation	Software construction
[25]	Conference	2017	Opinion article	Software design
[27]	Journal	2019	Validation	Software design

3.1 RQ1. How SGs' Design and Development Get Adopted into SE Teaching Activities?

Exploring how SE courses are using the design and development of SGs as a teaching tool. The study by Asuncion et al. [24] introduced the software process of designing and developing SGs to undergraduate students who are passionate about game development, have relevant programming skills, and yet not much experience in game development. In order to enable them to experience the entire production cycle firsthand, the development schedule was designed to fit in a 10-week academic quarter. Although each team developed a different SG, all participants used the same development process and adopted best practices from the video game industry. Blokhuis and Szirbik [25] investigated whether students' learning in systems engineering design courses can be influenced by the development of SGs. The teams had to design socio-technical chicken-and-egg systems. In this type of system, it is desired to develop a technology-intensive infrastructure for specific customers, but the funding for its development would only be available if the use of the infrastructure would allow users to pay for it. As a complementary task, the teams could develop a SG that mimics the growth of the system to be designed. In the SG, the different team members acted as designers, customers, investors and coordinators, with different interests, and had to interact to develop the architecture of the system. Dörner and Spierling [26] proposed the development of a SG to teach entertainment technology, multidisciplinary collaboration, and group work. The authors presented the results of four courses of different formats, with a duration of one semester (14 weeks), which occupied one-third of the students' workload during the semester. The courses were evaluated by means of a questionnaire that collected students' impressions of the knowledge acquired in the course, their level of motivation, and the classroom atmosphere. Garneli et al. [18] explored the effects of a Problem Based Learning approach in the area of programming. In particular, they analyzed the differences in the programming styles of students following three approaches: (1) the teacher introduces the theory within the framework of a project that consists of developing a SG that teaches a phenomenon, (2) the tutor introduces the theory within the framework of a project that consists of simulating a phenomenon, and (3) the teacher first teaches the theory and then the students put the theoretical concepts into practice by solving a problem, following a traditional learning strategy. The phenomenon exercised in the first two approaches is associated with the acquisition of certain learning objectives. Giannakos et al. [17] empirically evaluated an intensive course that they both developed and taught regarding game prototyping and development. The course progressed through 3 stages, it started with a hands-on session in which participants experimented with the tools that will be used throughout the course, and worked in small groups to better know each other in a creative motivating way. Later, one-half of the remaining sessions were informatory including presentations on the background of the topics, while the second half included hands-on creative sessions in which participants developed SGs. Santana-Mancilla [27] conducted a study on 40 undergraduates studying their third year as a SE major. The experiment was not a controlled trial. Still, it was between subjects in Human-Computer Interaction, in which students were taking the course for the first time. The learning in this course was mixed. The two approaches were adopted by both groups in different stages of the course. Each group was only exposed to either learning with utility software or learning with video game design at a time. To learn the competencies for the usability assessment, the students used a summative evaluation following the methodology called "IHCLab Usability Test" for SGs.

3.2 RQ2. What SWEBOK Areas Have Been Targets of SGs' Design and Development in SE Teaching?

This research question aims to track the distribution of studies addressing the usage of SGs design and development as a teaching tool in SE throughout the different SWEBOK knowledge areas and detect the main targeted areas by the identified teaching approach.

As illustrated in Table 1, Software Design was found to be the most reported SWEBOK knowledge area, which makes up half of the total selected studies, followed by Software Construction representing 33.33% of these studies. Finally, the Software Engineering Process ranks last with a minimum share of 16.66%. In contrast, a significant gap has risen regarding the usage of SGs' design and development as a teaching tool in SE courses that tackle the remaining SWEBOK knowledge areas. Subjects within areas such as Software Testing, Software Quality, and Configuration Maintenance have great potential for their teaching to be guided by SGs' development.

3.3 RQ3. To What Extent is the Usage of SGs' Design and Development in SE Teaching Considered Complex?

Selected studies have found that teaching SE concepts through SGs design and development can be quite challenging. Asuncion et al. [24] identified the inflexible academic term schedule, the workload of full-time students, and their technical skills as the main constraints in determining SGs implementation features and technologies. The first two constraints hindered the application of the Scrum methodology and required some adjustments to spring lengths and how progress evaluations and backlog refinements were conducted. Giannakos et al. [17] expressed difficulties in carrying out the evaluation due to the tight schedule of the course and a low teacher-student ratio. On the other hand, Dörner and Spierling [26] have found that the development of the game was time-consuming and that there was no time left within the course schedule for conducting user tests or playtests on developed SGs. In the study carried out by Blokhuis and Szirbik [25], students who followed the SG development approach found game development to be the most difficult aspect of the course, although it was the aspect they enjoyed the most. Students also pointed functional thinking, the high workload, and the development of a functional architecture as other difficulties of the course. Similarly, Garneli et al. [18] highlighted that teachers observed that the group of students who developed the SGs were more stressed at the beginning, and demanded more help, especially when they had coding errors. This factor influenced the students to not adequately handle some theoretical components such as variables and operators, due to the pressure to solve the problem. In addition, the complexity of the SG development process and the SG itself may have diverted the students' attention from the fundamental programming concepts, which were the backbone of the course. In addition, it was observed that 35% of the students in the SG development group could not successfully complete the required tasks versus 11% of the students in the control group.

3.4 RQ4. What Are the Cited Pitfalls and Potential Opportunities of Incorporating SGs' Design and Development in SE Teaching?

Given that 83.33% of selected studies have reported that adopting SGs' design and development in SE teaching was challenging if not complex, the scope of this research question is to weigh the pitfalls and opportunities that come with this teaching approach in SE courses. The study by Asuncion et al. [24] revealed through an informal retrospective with the SGs teams that students have had an overall positive experience in developing the games. The teams have surprisingly managed to avoid some of the challenges in typical game development projects, deliver quality products, and quickly learn new tools despite the course schedule and the lack of previous experience. According to the author, these results can be attributed to following the video game industry practice of devoting a significant percentage of planning to pre-production, with one-third of the time spent planning the project, and matching the game's features to the established schedule and the talents of the team members. In the study by Blokhuis and Szirbik [25], students who chose originally to develop SGs as a learning approach considered game development as the most-liked aspect of the course. In addition, students perceived that they had a greater ability to take a holistic view of a software development project. This characteristic was not mentioned in any case by students who did not develop a game. Dörner and Spierling [26] highlighted that these courses were well received by students and that they fostered a positive prospect of interdisciplinary work, also that the learning opportunities in these courses were significantly high. It was noted that students were particularly motivated to acquire additional knowledge when they were perceived as the experts on a particular topic and developing their own ideas. Thereby, they stressed the importance of students identifying themselves with the game idea and the game concept and forming a mental model of the game development process. In this light, the authors are against providing the students with a complete game specification at the beginning of the course since it hides crucial phases of the game development processes from the students restraining their learning opportunity and blocking their motivation. It should be noted that the subjects that included SGs development were in the top 10% of the university's top-rated subjects. The authors state that during SGs development, students learn to organize themselves, acquire knowledge about tools and technologies related to game development, learn SE methods, and exercise requirements specification and project management. According to the study by Garneli et al. [18], the findings suggest that a SG programming context provides an enjoyable experience that stimulates learning, and encourages students' creative expression. Students who developed SGs in the course were more productive and engaged in the programming activity after completing the course. The SG development group improved in all parts of the programming curriculum (sequential and iterative composition, synchronization and event handling, etc.) except conditional composition, variables, and operators. In addition, their code contained fewer errors, observing a positive influence on their programming habits. In the study by Giannakos et al. [17], students showed a high predisposition to participate in similar learning activities in the future. A survey based on motivational factors for technology indicated that the degree of satisfaction of the participants and the usefulness of the activity are the factors that most influence the intention to repeat a similar activity in the future. Through the results of this survey, they found that the ease with which students perform the course activities has a positive impact, while anxiety has a negative impact, on the students' perception of the course. As a solution to improve their progress and achieve a more enjoyable experience, it was proposed to train the participants on the software before the start of the course, to reduce their anxiety motivated by the pressure of an intensive one-day course. Moreover, the use of tangible materials and sensor boards that encourage the creative development of SGs prototypes increases students' willingness to participate again in similar courses. According to Santana-Mancilla [27], all participants reported that they enjoyed the course. Ninety-five percent of the students in the experimental group felt that they had learned the necessary skills through video game design and development. A total of 100% indicated that the knowledge acquired would have been less if the teaching process had not been guided by video game design. The motivating effect of the approach with these projects and the lessons learned in the course prompted nine undergraduate students to create a computer game as the final project of their engineering studies. In addition, students and faculty from the courses published more than 15 research papers in books, journals, and conferences in the field of video games and SGs.

4 Conclusion

Studies have shown that the video game industry has maintained constant growth and an important position in the market, being one of the most successful application areas in the history of interactive systems [27]. The design of video games for educational purposes is a relatively new discipline, which combines learning design with game mechanics and logic [28]. There exist arguments against learning with games. However, the latter is centered upon the lack of empirical evidence to support their effectiveness [28]. In our study, incorporating SGs design and development in SE teaching was not only considered an enjoyable and motivating learning experience, but also an effective approach toward a holistic understanding, and comprehensive knowledge and skills acquisition in SE courses. Moreover, students who participated in these courses showed a high intention to take other courses in the future with similar teaching approaches. Therefore, this type of courses not only supports students in developing their skills, but also fosters more widespread use of SGs, as they provide students with prospects to meet more learning goals than a regular game development course [26], or courses that do not use a game approach in the teaching process [27].

Acknowledgments. This research is part of the OASSIS-UMU (PID2021-122554OB-C32) project (supported by the Spanish Ministry of Science and Innovation), the BIZDEVOPSGLOBAL-UMU (RTI2018-098309-B-C33) project (supported by the Spanish Ministry of Science, Innovation, and Universities), and the Network of Excellence in Software Quality and Sustainability (TIN2017-90689-REDT). These projects are also founded by the European Regional Development Fund (ERDF).

References

1. Vargas, J.A., García-Mundo, L., Genero, M., Piattini, M.: A systematic mapping study on serious game quality. In: ACM International Conference Proceeding Series (2014). https://doi.org/10.1145/2601248.2601261
2. Wassila, D., Tahar, B.: Using serious game to simplify algorithm learning. In: 2012 International Conference on Education and e-Learning Innovations, ICEELI 2012 (2012). https://doi.org/10.1109/ICEELI.2012.6360569
3. Thirouard, M., et al.: Learning by doing: Integrating a serious game in a MOOC to promote new skills. In: Proceedings of the European Stakeholder Summit on Experiences and Best Practices in and Around MOOCs, pp. 92–96 (2015)

4. Mann, P.: Why is software engineering so difficult? Br. Telecom Technol. J. **10**, 18–27 (1992). https://doi.org/10.1007/978-1-4612-4720-3_10
5. Richardson, W.E.: Undergraduate software engineering education. In: Lecture Notes in Computer Science (including subseries Lecture Notes in Artificial Intelligence and Lecture Notes in Bioinformatics). pp. 121–144. Springer, New York (1988). https://doi.org/10.1007/BFb0043595
6. Rodrigues, P., Souza, M., Figueiredo, E.: Games and gamification in software engineering education: a survey with educators. In: IEEE Frontiers in Education Conference, FIE, pp. 1–9. Institute of Electrical and Electronics Engineers Inc. (2019). https://doi.org/10.1109/FIE.2018.8658524
7. Souza, M.R.D.A., Veado, L., Moreira, R.T., Figueiredo, E., Costa, H.: A systematic mapping study on game-related methods for software engineering education (2018). https://doi.org/10.1016/j.infsof.2017.09.014
8. Caulfield, C., Xia, J.C., Veal, D., Paul Maj, S.: A systematic survey of games used for software engineering education. Mod. Appl. Sci. **5**, 28–43 (2011). https://doi.org/10.5539/mas.v5n6p28
9. Pedreira, O., García, F., Brisaboa, N., Piattini, M.: Gamification in software engineering - A systematic mapping. In: Information and Software Technology, pp. 157–168 (2015). https://doi.org/10.1016/j.infsof.2014.08.007
10. Rodríguez, G., González-Caino, P.C., Resett, S.: Serious games for teaching agile methods: a review of multivocal literature. Comput. Appl. Eng. Educ. **29**, 1931–1949 (2021). https://doi.org/10.1002/CAE.22430
11. Ismailović, D., Haladjian, J., Köhler, B., Pagano, D., Brügge, B.: Adaptive serious game development. In: Proceedings of the 2012 2nd International Workshop on Games and Software Engineering: Realizing User Engagement with Game Engineering Techniques (GAS 2012), pp. 23–26 (2012). https://doi.org/10.1109/GAS.2012.6225922
12. Dimitriadou, A., Djafarova, N., Turetken, O., Verkuyl, M., Ferworn, A.: Challenges in serious game design and development: educators' experiences. Simul. Gaming **52**, 132–152 (2021). https://doi.org/10.1177/1046878120944197
13. Calderón, A., Ruiz, M., O'Connor, R.V.: A multivocal literature review on serious games for software process standards education. Comput. Stand. Interfaces **57**, 36–48 (2018). https://doi.org/10.1016/j.csi.2017.11.003
14. Darwesh, D.A.M.: Concepts of serious game in education. Int. J. Eng. Comput. Sci. (2016). https://doi.org/10.18535/IJECS/V4I12.25
15. Kapralos, B., Fisher, S., Clarkson, J., van Oostveen, R.: A course on serious game design and development using an online problem-based learning approach. Interact. Technol. Smart Educ. **12**, 116–136 (2015). https://doi.org/10.1108/ITSE-10-2014-0033
16. Robertson, J., Howells, C.: Computer game design: opportunities for successful learning. Comput. Educ. **50**, 559–578 (2008). https://doi.org/10.1016/j.compedu.2007.09.020
17. Giannakos, M.N., Jaccheri, L., Morasca, S.: An empirical examination of behavioral factors in creative development of game prototypes. In: Lecture Notes in Computer Science (including subseries Lecture Notes in Artificial Intelligence and Lecture Notes in Bioinformatics), pp. 3–8. Springer, Heidelberg (2013). https://doi.org/10.1007/978-3-642-41106-9_1
18. Garneli, V., Giannakos, M.N., Chorianopoulos, K., Jaccheri, L.: Serious game development as a creative learning experience: lessons learnt. In: Proceedings of the 4th International Workshop on Games and Software Engineering (GAS 2015), pp. 36–42. Institute of Electrical and Electronics Engineers Inc. (2015). https://doi.org/10.1109/GAS.2015.14
19. Kitchenham, B., Charters, S.: Guidelines for performing systematic literature reviews in software engineering. Tech. report, Ver. 2.3 EBSE Tech. Report. EBSE, vol. 5, pp. 1–65 (2007)

20. Stone, P.W.: Popping the (PICO) question in research and evidence-based practice (2002). https://www.sciencedirect.com/science/article/pii/S0897189702000101
21. Petersen, K., Vakkalanka, S., Kuzniarz, L.: Guidelines for conducting systematic mapping studies in software engineering: an update. Inf. Softw. Technol. **64**, 1–18 (2015). https://doi.org/10.1016/j.infsof.2015.03.007
22. Brereton, P., Kitchenham, B.A., Budgen, D., Turner, M., Khalil, M.: Lessons from applying the systematic literature review process within the software engineering domain. J. Syst. Softw. (2007). https://doi.org/10.1016/j.jss.2006.07.009
23. Alhammad, M.M., Moreno, A.M.: Gamification in software engineering education: a systematic mapping. J. Syst. Softw. **141**, 131–150 (2018). https://doi.org/10.1016/j.jss.2018.03.065
24. Asuncion, H., Socha, D., Sung, K., Berfield, S., Gregory, W.: Serious game development as an iterative user-centered agile software project. In: Proceedings of the International Conference on Software Engineering, pp. 44–47 (2011). https://doi.org/10.1145/1984674.1984690
25. Blokhuis, M., Szirbik, N.: Using a serious game development approach in the learning experience of system engineering design. In: IFIP Advances in Information and Communication Technology, pp. 279–286. Springer, New York (2017). https://doi.org/10.1007/978-3-319-66926-7_32
26. Dörner, R., Spierling, U.: Serious games development as a vehicle for teaching entertainment technology and interdisciplinary teamwork: Perspectives and pitfalls. In: Proceedings of the 2014 ACM International Workshop on Serious Games, Workshop of MM 2014 (Serious-Games 2014), pp. 3–8. ACM Press, New York (2014). https://doi.org/10.1145/2656719.2656724
27. Santana-Mancilla, P.C., Rodriguez-Ortiz, M.A., Garcia-Ruiz, M.A., Gaytan-Lugo, L.S., Fajardo-Flores, S.B., Contreras-Castillo, J.: Teaching HCI skills in higher education through game design: a study of students' perceptions. Informatics **6**, 22 (2019). https://doi.org/10.3390/informatics6020022
28. Lameras, P., Arnab, S., Dunwell, I., Stewart, C., Clarke, S., Petridis, P.: Essential features of serious games design in higher education: linking learning attributes to game mechanics. Br. J. Educ. Technol. **48**, 972–994 (2017). https://doi.org/10.1111/bjet.12467

Enriching Software Engineering Gamification Environments with Social Information

Oscar Pedreira[1], Félix García[2], Mario Piattini[2], José Luis Fernández-Alemán[3], and Manal Kharbouch[3](✉)

[1] Facultade de Informática, Universidad de Coruña, Elviña S/N, 15071 A Coruña, Spain
[2] Universidad de Castilla-La Mancha, Paseo de la Universidad, 4, 13071 Ciudad Real, Spain
[3] Fac. Informática, Universidad de Murcia, Campus Univ., 32, 30100 Murcia, Spain
Manal.kharbouh@um.es

Abstract. Motivation: There has been a recent focus on the potential benefits of using gamification techniques to increase motivation and improve performance in Software Engineering environments. Problem: While many gamification proposals in Software Engineering utilize techniques such as points, levels, and rankings, there is some skepticism about the effectiveness of these basic techniques and some authors explored other complementary alternatives. Approach: This paper presents ongoing research on using social information analysis techniques as a complementary alternative to enhance gamified Software Engineering environments. Result: the inclusion of social information into gamified environments was proposed. To accomplish this, the relevant information to be delivered to users and the data sources for this information were identified. A simple infrastructure for integrating and processing social information was also proposed. Impact: If successful, this approach could potentially have a significant impact on the motivation and performance of people in Software Engineering environments.

Keywords: Software Engineering · Gamification · Social Information

1 Introduction

Software Engineering (SE) gamification has attracted a great deal of interest to foster desired behaviors in software engineers by applying techniques from or inspired by games [1]. In our previous systematic mapping study regarding gamification in SE [1], which spans the papers published up to the year 2014, it was observed that although a good range of software lifecycle stages were covered, gamification application in SE was still quite preliminary, and solely based on basic extrinsic motivators known as BPL (Badges, Points, Leaderboards). These BPL mechanisms, named in some studies as "pontification" and which usually reward players directly on their individual actions, are configured in terms of the performance level shown by players in the development of their tasks throughout the gameplay. Directly rewarding specific activities with points can improve the players' motivation when performing those actions, levels can indicate a feeling of progress, and leaderboards can increase players' engagement by fostering

A. Rocha et al. (Eds.): WorldCIST 2023, LNNS 800, pp. 415–421, 2024.
https://doi.org/10.1007/978-3-031-45645-9_39

competitiveness. An update of this Systematic Mapping Study (SMS) with 103 primary studies (retrieved up to January 2020) obtained similar conclusions which evidenced no significant advances in the applied mechanisms and stressed the need to properly evaluate them [2]. However, the effectiveness of BPL has been questioned and additional challenges have arisen. In fact, in the last few years, some advances are being observed in the mentioned gaps. As a representative sample of this, for instance in [3], authors conducted an interdisciplinary work involving Social and Human Factors (SHF) and gamification. They identified the SHF involved in change resistance in Software Process Improvement initiatives such as motivation, commitment and team cohesion, and applied gamification elements to promote them, such as progression, narrative, rewards, leaderboard, and challenges (missions). Furthermore, feedback and achievements were identified as enablers of team cohesion. Teams' engagement in software development teams in a complex sector such as the automobile industry, was the focus in [4]. The relationship between developer engagement and job satisfaction (as a predictor of employees' intention to stay in their job) is addressed in [5], highlighting the need to build more robust theoretical models, as the relationships between gamification variables of interest may not be straightforward. In the SMS by [6], the evaluation of gamification was tackled focusing on: the strategy and its relationship with user experience and perceptions; the outcomes and effects on users and context, reinforcing also the need for both qualitative and quantitative data analysis approaches, given the subjective and objective mixed nature of the input. The analysis of behavioral effects on software developers by studying individuals contributing to GitHub conducted in [7] pointed out the highly varied behavior in response to gamification elements that users can have and how gamification can be used as a powerful tool for social influence. The study on other collaborative platforms such as StackOverflow confirmed the need to go beyond BPL [8], as even having gamified incentive mechanisms, developers were mainly motivated by intrinsic aspects. Furthermore, it is also remarked how different developer profiles conduct to lower or higher motivation with intrinsic or extrinsic motivations, although both motivators are needed to obtain high-quantity and high-quality contributions. In line with these last advancements in the field of SE gamification, we support the idea of going beyond individual interactions fostered with BPL, especially promoting teamwork, which is an increasingly important aspect in today's software engineering projects, following the strong influence of paradigms such as Agile, Management 3.0, DevOps, among others. Therefore, we present some ongoing work to illustrate how social relationships in software engineering environments can be a key success factor for gamification programs. The next section offers a general view of how gamified software engineering environments can benefit from the analysis and delivery of social information. Finally, Sect. 3 presents the conclusions and lines for future work.

2 Enriching Gamified Environments with Social Information

We propose to enrich gamified SE environments by incorporating social information that can be delivered to the players. This information can help them complete their tasks and develop social relationships at work, something that we believe can help improve their motivation and performance. In this section, we consider different types of players, key data sources from which we can analyse interactions between players, a tentative technological solution to retrieve this social information, and a suggestion for an adapted integration of this information into the gamification delivered to the players.

2.1 Players: Team Members but also Customers or Other Stakeholders

Our proposal considers a range of participants in the gamified environment. In some companies, only the members of the development teams share tools that imply some form of social interaction. However, in other organizations, other stakeholders such as customers also use these tools, as it usually happens with task/ticket-management tools such as JIRA or Redmine. Even if we focused only on internal participants, the organization's employees do not all have the same characteristics from the point of view of the social information that could be of interest to them, the reason why clustering these participants according to their shared social interaction is needed. In this line, current trends in teamwork in software engineering, following approaches such as Agile and Management 3.0 [9], suggest that the design of gamified SE environments should be oriented to teams rather than to individuals. A gamification solution based on social information should consider different types of players depending on which teams they belong to and distinguish the information that could be delivered to each of them:

- *Customers*: we consider members from the customers' team who actively participate in software development projects and work with the developers' team using the same social interaction tools for requirements gathering or task/issue management, for example.
- *Managers*: some internal members will have management responsibilities of different levels. This initial proposal considers them all at the same level. They could potentially receive relevant information on various aspects of the organization, such as communities where people have strong social ties and relevant social interactions. It can also be of interest for these players to receive sentiment analysis-based information to identify elements such as employee satisfaction or a desire to leave the company.
- *Developers*: this type of player includes the different members of the development teams, which can have different roles and responsibilities (requirements management, design, development, testing, etc.), but who all work at the same level in the development of the project.
- *Onboarding Members*: within the aforementioned category, we believe it is important to define a different type of player for newly employed developers, that is, people still in the onboarding process. Specific information can be valuable for these players, such as contact suggestions to help them discover their community.

2.2 Target Data Sources

We consider that a gamified SE environment can gather information regarding interactions between different players from all the tools in which those interactions are recorded. Without the intention of being exhaustive, some examples include:

- *Project and task management*: tools such as JIRA or Azure Boards involve all the participants in the project and allow us to gather information about the work that a player has assigned to other players and different comments and interactions between developers (sometimes including customers).
- *Issue management* tools may involve internal team members and often include customers, who register issues and may comment on their resolution process.
- *Version control software*: modern version control solutions go beyond just version control and provide the developers with functionalities that imply interactions between them. In some companies, tools such as GitLab can be the most important information source regarding players' interactions.
- Other software engineering tools, such as those of requirements engineering, support multiple users working in a collaborative mode [10].
- *Collaboration tools* such as Confluence can also be a relevant source of information interaction.
- *Chat or videoconferencing* platforms: in some cases, the information shared through these tools can be analysed to identify social interactions in the organization. This poses a challenge from the point of view of employee privacy. However, even an analysis of the intensity of the interaction through chat platforms can reveal the most interesting interactions in the workplace.

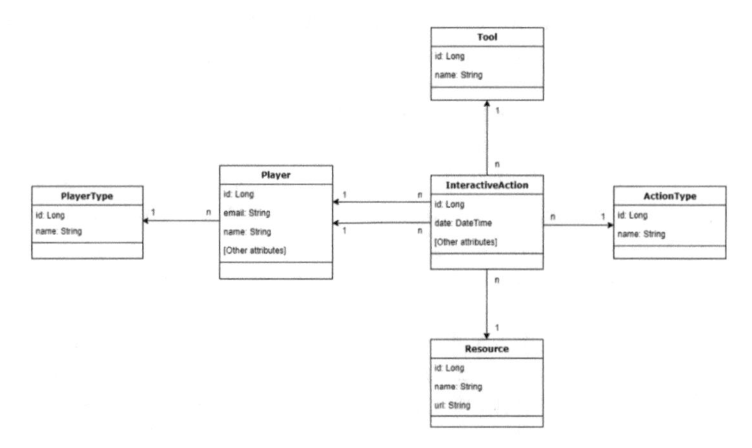

Fig. 1. Data model of the module for social information gamification.

In Fig. 1, we show a model that would allow us to integrate this information into a single data source. This model would allow us to keep the players' basic data and the actions they perform that imply interaction with other players. Also, we could conserve information regarding tools used in these interactions and their *links* to the resources involved. The key pieces of this data model are the players and the actions, which would be the nodes and edges of the organization's interaction graph respectively.

2.3 Technological Infrastructure

As we pointed out in previous sections, a key factor for this proposal to work is gathering and integrating interactions that happen in the different tools of the organization, since only this way can we get an accurate picture of the organization's social network.

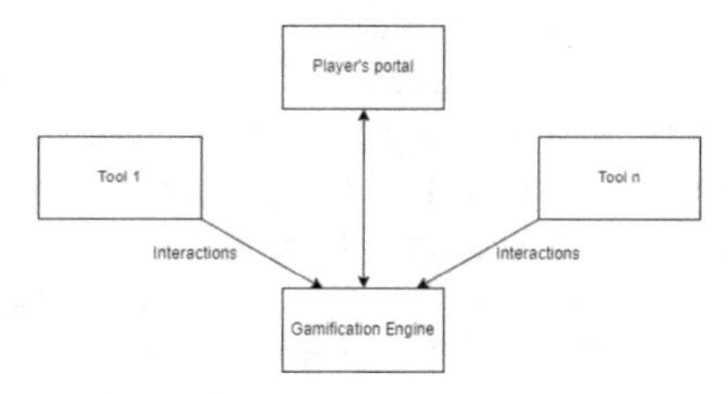

Fig. 2. Technological architecture to integrate and process the social information generated in the organization's tools (adapted from [11]).

Figure 2 shows a technological infrastructure that would meet this requirement, adapted from previous work [11]. The players work on different tools that generate records for actions that imply some form of interaction between them. This information should be integrated into a gamification engine which should be directly communicating with the other social interaction tools, or periodically gathering that information from their databases. The gamification engine would also analyse this information and generate the items that could be delivered to individual players in the gamified environment.

2.4 Information to be Delivered to the Participants

The analysis of the interaction information gathered in the gamification engine by applying well-known network analysis techniques (described in [12], for example) would let us come up with information that could be delivered to the players, such as:

- *Communities*: most software organizations organize their activity based on projects, which in some way create teams that can be considered communities on their own. Something similar happens with other organizational structures such as departments, areas, or business units. However, people often move from one of these defined communities to others, and personal and social communities emerge from the personal relationships between the players. Different algorithms can be applied to detect these communities from the data of the interaction network stored in the gamification engine.
- *Hub users*: in the case of small or medium companies, the number of employees is usually so small that the players know each other, even if communities may also exist, and they may feel more confident within their community. However, in larger organizations with hundreds or thousands of employees, it can be more challenging to know if a person that belongs to a certain community has a strong relationship with players from other communities. This information can be helpful for all the players, especially those with management responsibilities.

- *Contact suggestions*: can be helpful for all sorts of internal players, but we consider them especially important for onboarding players. Based on the information about communities, the player's front-end can suggest onboarding players with interesting contacts in the organization and vice versa, that is, it may also suggest established employees contact recent new employees.

Further study will consider other information items that could be obtained from the network (e.g. social-technical congruence [13]), but we believe these three are the most basic ones. From the preliminary experiments we have carried out, we have some preliminary evidence that sentiment analysis techniques can be successfully applied to the interactions between players to get the polarity (positive, negative, or neutral) of those interactions. In this way, we could also detect positive and negative relationships between players that can lead to conflicts in the workplace.

3 Conclusions and Future Work

In this short paper, we described ongoing work towards enriching gamified software engineering environments by gathering data about interactions between players and generating social information that can be valuable for players with different levels of responsibility. Incorporating this social information in the gamified environment goes beyond basic gamification techniques that imply a direct reward to players based on successful actions (points, levels, and rankings). It also allows the organization to improve the player's motivation and make their work easier by revealing useful information such as contact suggestions or identifying communities and hub users. Future work implies further analysing the social information items we could extract from the interaction network by applying techniques such as sentiment analysis. Also, a key line for future work implies deploying a gamified SE environment enriched with social information in a real company to obtain evidence of its potential benefits and drawbacks and both emphasize and compare the novelty of this proposed approach with non-gamified approaches. Another line for future work consists in exploring how the analysis of social information of the organization can help us to decide the best techniques and strategies to be applied in a gamified environment, that is, we believe the analysis of social information can be an input to the design of the gamified environment and not only a resource the environment can deliver to its players. Also, we are currently considering the data management policies that should be applied to ensure the employee's privacy in the analysis of social information at the workplace.

Acknowledgments. This work has been supported by OASSIS project (PID2021-122554OB-C31, PID2021-122554OB-C32, and PID2021-122554OB-C33 funded by MCIN/AEI/10.13039/501100011033/FEDER, EU).

References

1. Pedreira, O., García, F., Brisaboa, N., Piattini, M.: Gamification in software engineering – a systematic mapping. Inf. Softw. Technol. **57**, 157–168 (2015). https://doi.org/10.1016/J.INFSOF.2014.08.007

2. Porto, D. de P., Jesus, G.M. de, Ferrari, F.C., Fabbri, S.C.P.F.: Initiatives and challenges of using gamification in software engineering: a systematic mapping. J. Syst. Softw. **173**, 110870 (2021). https://doi.org/10.1016/j.jss.2020.110870
3. Gasca-Hurtado, G.P., Gómez-Álvarez, M.C., Machuca-Villegas, L., Muñoz, M.: Design of a gamification strategy to intervene in social and human factors associated with software process improvement change resistance. IET Softw. **15**, 428–442 (2021). https://doi.org/10.1049/SFW2.12045
4. Mounir, M., Badr, K., Sameh, S.: Gamification framework in automotive SW development environment to increase teams engagement. In: Communications in Computer and Information Science, pp. 278–288. Springer Science and Business Media Deutschland GmbH (2021). https://doi.org/10.1007/978-3-030-85521-5_18
5. Stol, K.J., Schaarschmidt, M., Goldblit, S.: Gamification in software engineering: the mediating role of developer engagement and job satisfaction. Empir. Softw. Eng. **27**, 1–34 (2022). https://doi.org/10.1007/S10664-021-10062-W/TABLES/9
6. Barbosa Monteiro, R.H., De Almeida Souza, M.R., Bezerra Oliveira, S.R., Dos Santos Portela, C., De Cristo Lobato, C.E.: The diversity of gamification evaluation in the software engineering education and industry: trends, comparisons and gaps. In: Proceedings of the International Conference on Software Engineering, pp. 154–164 (2021). https://doi.org/10.1109/ICSE-SEET52601.2021.00025
7. Moldon, L., Strohmaier, M., Wachs, J.: How gamification affects software developers: cautionary evidence from a natural experiment on Github. In: Proceedings of the International Conference on Software Engineering, pp. 549–561 (2021). https://doi.org/10.1109/ICSE43902.2021.00058
8. Bornfeld, B., Rafaeli, S.: Gamifying with badges: a big data natural experiment on Stack Exchange. First Monday **22** (2017). https://doi.org/10.5210/fm.v22i6.7299
9. Appelo, J.: Management 3.0: Leading Agile Developers, Developing Agile Leaders. Addison-Wesley (2010)
10. Carrillo De Gea, J.M., Ebert, C., Hosni, M., Vizcaino, A., Nicolas, J., Fernandez-Aleman, J.L.: Requirements engineering tools: an evaluation. IEEE Softw. **38**, 17–24 (2021). https://doi.org/10.1109/MS.2021.3058394
11. Pedreira, O., Garcia, F., Piattini, M., Cortinas, A., Cerdeira-Pena, A.: An architecture for software engineering gamification. Tsinghua Sci. Technol. **25**, 776–797 (2020). https://doi.org/10.26599/TST.2020.9010004
12. Barabási, A.-L.: Network Science. Cambridge University Press (2016)
13. Sierra, J.M., Vizcaíno, A., Genero, M., Piattini, M.: A systematic mapping study about socio-technical congruence. Inf. Softw. Technol. **94**, 111–129 (2018). https://doi.org/10.1016/J.INFSOF.2017.10.004

Something is Wrong with STEM Graduation Students: Teachers' Involvement in Their Citizenship Skills

Paula Amaro[1](✉) and Dulce Mourato[2]

[1] Polytechnic of Guarda (IPG), Guarda, Portugal
paula.amaro@ipg.pt
[2] Higher Advanced Technological Institute (ISTEC), Lisbon, Portugal
dulce.mourato@my.istec.pt

Abstract. Explain the apparent contradiction and technological education gap that exists and moves away in opposite directions: embedded technologies, synchronous media, an e-learning solutions as an emerging trend was the main goal, far away from past and present face-to-face teaching, similar to middle-aged monologue. It was possible to reflect also, on a years-long rewind observation, on the way teachers teach and students learn in STEM degree courses, after a year of confinement, in two Portuguese Higher Education Institutions, due to COVID-19 Pandemic restrictions. The first perception challenge to use was Charles Peirce's scientific approach as well as PBL and Learnability, always regarding science, technique and the evolution of knowledge, which gives consistency to the literature review. The qualitative methodology used in the case study analyses and discusses: why prospective research is unusual and not used very often in terms of scientific publications. Why not use students' perceptions to be the technologies for all starting points, to create a potential curriculum unit or course, both for teachers' training and students' learning?

The case study findings reflect students' concerns about old-fashioned *curricula* and smart technologies as well as less evolution that blocked the potentialities of Learnability growth and a semiotic way of life to promote personal and professional development.

The conclusion is still an open question as it is a risk to create a proposal for technological semiosis, in terms of prospective scientific research. Does it work? It is impossible to know yet, but it is urgent to do something!

Keywords: PBL · Learnability · STEM · Semiotic · Citizenship

1 Introduction

The red flag in pedagogical and interaction has been raised, due to the one year of academic interruptions in Science, Technology, Engineering and Mathematics (STEM) higher education in-person classes, perpetuated by the COVID-19 pandemic Portuguese restrictions. Teachers, trainers, researchers and students, despite being equipped with

A. Rocha et al. (Eds.): WorldCIST 2023, LNNS 800, pp. 422–431, 2024.
https://doi.org/10.1007/978-3-031-45645-9_40

all the technological tools used in face-to-face classes, such as collaborative and constructivist teaching-learning instruments and environments, the sudden transition to e-Learning has brought back the Middle Age classrooms, but at distance and online.

The throwback of Portuguese Universities and Institutes, which provided a distance learning platform where students met in virtual classrooms, only to listen and rarely interact, became usual and with no solution. Isolation, fear, social, and psychological competencies decadence, knowledge regression, and curricular units transformed into a massive melting pot of theoretical exercises and teachers' monologues, from 2020 March to 2021 December.

Assessments were minimized, how could we ask for more? Those students, from a general point of view, in that year on, mirrored the weaknesses of the 'Z generation': less reading, less writing, culture or art, less political involvement, less scientific or formal thinking, less solidarity and citizenship activities.

The main goal of this study, supported by a literature review and original case study methodology in two Portuguese Institutes, was to describe and characterize some of the measures to be adopted to foster critical thinking, Learnability, and a Semiotic attitude with Project Based Learning (PBL) directed issues. It was urgent and very important to question everyday life problems, as well as the way that technologies were worked and enjoyed, after graduation until the future profession time comes. The previous findings may conduce to content creation development and a PBL guide, to be aggregated to a technological and humanist nature curricular unit or a teaching training course materials development, to be held in the future to improve students' interaction in the cognition process.

2 Literature Review

It was difficult to see that the long path taken to improve teaching and learning, with the introduction of constructivist technologies and PBL inspired by Charles Peirce's scientific attitude, and the use of Siemens's [1] Connectivism theory, which popularized and stimulated collaborative work and synchronous technological means, used in the face-to-face classroom and in online teaching, interestingly, were not used in e-learning during the confinement period (in teaching STEM Portuguese graduate courses at the Polytechnic Institute of Guarda (IPG) and at the Higher Advanced Technologies Institute of Lisbon (ISTEC).

In this article, the use of the word Semiotic was a connotation as the science that studies the production of the meaning in different codes and languages, worked by Charles Sanders Peirce (1858–1914) - one of the most prominent scientists of his time, inspiring technological thinking since then and beyond, and this particular case study. In addition, the interpretation depends on each country's own language, each image or reality indicates different languages, reliant on the context and culture. A sign, according to Peirce [2] is instead of something, in the place of something else, that is, it signals. For example, the use of rainbow colors flags (the original flag featured eight colors, each having a different meaning. At the top was hot pink, which represented sex, red for life, orange for healing, yellow signifying sunlight, green for nature, turquoise to represent art, indigo for harmony, and finally violet at the bottom for the spirit [3] as linked to the

LGBT community and in World events, such as the Football World Cup held in Qatar in 2022, it served to draw attention to human rights violations, particularly freedom restrictions, against girls and women. Figures matter and become icons depending on context and culture, particularly in technology terms.

Peirce [4] founded Pragmaticism supported by representations that are real and verified independently of any particular mind (whatever the thinking subject) or a finite set of minds, so they are representations virtually efficient in their possibility of causing real effects. For example, COVID-19 virus vaccination during the Pandemic confirmed its effectiveness in stopping the epidemic, despite being contradicted by the negationists.

Peirce defended a scientific metaphysic, which privileged more general conclusions of our collective knowledge than its premises and based on initial and subjective impressions. Another of Peirce's starting points was the creation of syntheses (previous conclusions) that made it possible to simplify their dissemination and communication, as well as the pursuit of experience from a phenomenology based on a set of relationships between three basic and universal categories: quality (or possibility), reaction (or existence) and mediation (or representation). "The progress of science cannot go far except by collaboration; or, to speak more accurately, no mind can take one step without the aid of other minds" [5, p. C.P.2.220].

Charles Peirce thought the knowledge of everything we know is a sign, which is the principle of World and Life understanding, as well as scientific activity, is all intelligence action capable of learning from experience, a continuum semiosis inspired by the 'First Rule of Reason' [5, p. C.P.1.135]: "in order to learn you must desire to learn, and in so desiring not to be satisfied with what you already incline to think, Genuine doubt, like genuine belief, requires reasons". Several kinds of reasoning and classes of arguments are available: abduction or retroduction, deduction, and induction.

For Peirce, the real is not made up of discrete particles deterministically distributed in space-time but rather encompasses a continuous synthesis of meanings, which uses communion, connection, sharing of experiences and meanings in a community as a method of communication and clarification of ideas (co-mind) and is one of the principles present in the current network society and connectivity in social networks.

The use of PBL activities in higher education technologies curriculum units [6, p. 674], to encourage students in their own construction of knowledge, and to combine Learnability with common sense is also based on the scientific perspective inspired by Peirce. Science includes a continuous mode of latent curiosity, the relations of the mind with the Universe, and a proper articulation in the scientific process of thoughts. Peirce [5, p. C.P.1.43] explain the scientific attitude defining the type of Men: "If we endeavor to form our conceptions upon history and life, we remark three classes of men. The first consists of those for whom the chief thing is the qualities of feelings. These men create art. The second consists of the practical men, who carry on the business of the world. They respect nothing but power and respect power only so far as it [is] exercised. The third class consists of men to whom nothing seems great but reason. If force interests them, it is not in its exertion, but in that, it has a reason and a law. For men of the first class, nature is a picture; for men of the second class, it is an opportunity; for men of the third class, it is a cosmos, so admirable, that to penetrate to its ways seems to them the only thing that makes life worth living. These are the men whom we see as possessed by a passion to learn, just as other men have a passion for teach and disseminating their influence. If they do not give themselves over completely to their passion to learn, it is because they exercise self-control. Those are the natural scientific men, and they are the only men that have any real success in scientific research".

Although, for Osztián et al. [7] Computational Thinking (CT) and Diagrammatic Reasoning (DR) are important competencies from the perspective of both Computer Science and Engineering education. They invited students to participate in a CT and DR test and their results confirmed that CT and DR are closely related abilities. Osztián et al. [7] consider that "computational thinking and Diagrammatic Reasoning can be considered essential factors in every person's life in our digital era. With the help of these key abilities, students can learn to think differently and they can also gain a better experience in problem-solving and abstract reasoning" [7]. These authors [7] refer to the graph logic introduced by Peirce and according to him there is a duality in DR due to the fact that it can be considered as a tool to generate knowledge but also as a 'solution of problems of Logic'.

That kind of thinking could transform dreams into actions. Students should be able to get generic skills to build technologies for all, with user-friendly interfaces, and hypermedia documents, inspired by the notion of Solidarity Technologies [8, p. 2]. In this way, it is possible to stimulate students (future professionals in technology) to promote more well-being applications to solve community or citizenship problems, with a semiosis perspective (*continuum* knowledge in action) and a sign semiotic method for reality analysis, to promote Learnability [6] and professional generic skills. The term generic is used to emphasize that these skills are not linked to a specific context [9] and are useful for different situations and activities, also outside of the university (not only formal but informal and non-formal higher education).

The choice of a theoretical model mix, that substantiated the creation of the fundamental principles based on the development of subjects, to be operationalized through PBL, which focuses on stimulating Learnability, Critical Thinking, and Citizenship, brought even more doubts about the path to be taken and to go through, detailed in the next sections.

3 Methods and Materials

What determined the choice of this article subject was the student's concerns similarity from the degrees in Computer Science Engineering and Multimedia Engineering at ISTEC Lisbon and Students at Polytechnic of Guarda, a city in the interior of the country, with some dynamic in industrial and technological terms.

"Qualitative research is an approach for exploring and understanding the meaning individuals or groups ascribe to a social or human problem", according to Creswell [10, p. 32] and was the methodology followed by this research, because "involves emerging questions and procedures, data typically collected in the participant's setting, data analysis inductively building from particulars to general themes, and the researcher making interpretations of the meaning of the data. The final written report has a flexible structure. Those who engage in this form of inquiry support a way of looking at research that honors an inductive style, a focus on individual meaning, and the importance of rendering the complexity of a situation" [10, p. 32].

The case study approach was chosen to adapt students' perceptions to customized research as was proposed by George and Bennett [11], Gerring [12], Yin [13], Gomm and Hammersley [14], and others authors.

"Case studies are generally strong precisely where statistical methods and formal models are weak. We identify four strong advantages of case methods that make them valuable in testing hypotheses and particularly useful for theory development: their potential for achieving high conceptual validity; their strong procedures for fostering new hypotheses; their value as a useful means to closely examine the hypothesized role of causal mechanisms in the context of individual cases; and their capacity for addressing causal complexity" [11, p. 19].

The students' options regarding the professional future coincided with behaviours and perceptions, concerning the forthcoming certification, reflecting the absence of general, flexible, and transversal skills. These abilities which allow a greater knowledge about the World and also the humanist component, are essential for those who develop technologies. This is a great research challenge.

From March 2020 to March 2021 using observations, forum participation, collaborative works, and perceptions inquiry, was possible to compile students' answers and perceive how could be used to build a curriculum menu. From April 2021 to July 2021 it was possible to compile all the data, using the perceptions inquiry, regarding gender, citizenship, technological issues, and Learnability topics to optimize PBL dimensions. From September 2021 until December 2021, both teachers presented their previous conclusions about the several features that emerged from all the collected data. In the Results chapter, it was possible to show some of the raw material, on which the potential course or curricula unit framework development was based and inspired by Solidarity Technologies [8, 15]: a Virtual Learning Environment (VLE) creation, using Google Universe (Docs, Development tools) in the first phase.

As Mourato [8] explains, the goal of this VLE was to encourage, from a Peirce Semiotic viewpoint, the intervention of Learning Communities, collaboration, and communication practices, operationalising Solidarity Technologies, concerning their attributes of universality and accessibility, in practical activities and Open Source Digital Resources development. Mourato specifies the VLE: "Where devices and resources specifications must be tailored to shape students' progression, like alternative assignments (audio or tutoring reader), captioning & transcription for multimedia files, tagging images, simplifying digital documents and electronic texts that could be read by OCR software (optical character recognition), using PDFs with text-to-speech programs, making any text accessible, regardless of one's personal needs" [8, p. 5].

In ethical terms, the students answered the surveys and inquiries anonymously, knowing that all their answers, would be seen in a constructivist sense to encourage their individual performance and as a class improvement, that could be applied in future studies or projects. According to Dias de Figueiredo [16], an eminent Portuguese technologist, the synthesis of these uncertainty scenarios should focus on projective research "Ideal for the increasingly frequent problems of complex social contexts characterized by uncertainty and human disagreement Ideal for contexts of accidental discovery, innovation, and creative leaps".

In VLE, almost all the free applications based at Google Universe could be used as the first level of understanding, instead of each student going to use specific software development, they could deal with PBL tutorials on complex problems and solve all in collaborative groups, sharing several disciplines knowledge, incorporate social issues, professional practice, and citizen responsibility. They brainstormed on every path, that could be taken because complex projects or problems do not usually have only one definite solution. Deal with numerous hypotheses with uncertainty, looking forward creatively, defining what is important, making decisions, and developing generic and scientific skills were the central objectives.

In the face of these previous assumptions, provided by the use of a Virtual Learning Environment where topics or problems were worked, it was possible to build a Goals Framework (Table 1) that could be transformed into a potential curriculum unit inspiration for students or a training course for teachers.

Table 1. Goals Framework [17]

How the World/Country/Region/Community could be transformed by my action as a person or as a technologies student?	Covering only relevant issues, analysis of reality according to Peirce's vision; dismantling the meaning of the Sign, applied to technology and science-specific (STEM) matters
How to make ideas clear, according to logical, semiotic and flexible thinking, through PBL?	Working humanist skills of free will, and reflection explaining meanings, contexts and cultures. Pursuit simple writing methods, philosophic thinking, artistic point of view and mathematical analysis as well as new kinds of time and spatial dimensions exploration and their repercussion in one's life
How to prepare students to be able to answer climate changes, ecological transformations, technological accessibilities, consumer needs and desires and market volatility?	Use curiosity and self-knowledge to understand how to solve problems and build projects with value for the community, participating and giving voice to an active citizenship
How to stimulate an innovative, provocative, prospective and "techno-desiring" vision? - If it doesn't work that way, why not get another way to do it?	Bringing some of the most current solutions and leaving to the imagination, how it is possible to transform everyday objects into new forms of interaction Using the past experiences of Aristoteles, Leonardo da Vinci, and Peirce who inspired others such as Heisenberg, Ted Nelson, John Von Neumann and others technologies well-known until nowadays

4 Results and Discussion Results

The purpose of the study carried out for the development of the article on Learnability and PBL [6], in the academic year 2020/2021, data were collected from 12 female students and 126 male students classes, from last year's Graduation Computer Science and Multimedia Engineering Degree in ISTEC Lisbon, Portugal. However, this is not the scope of this research. Some considerations were made about the learning influence on students' professional life transition, the environment, and how technologies impact their life and their personal and social development.

To obtain the real dimension of their words, it was decided to create some concept reflections (Table 2) based on the results of their statements.

Table 2. Analysis of students' perceptions about teaching and learning and the relationship with technology dimension [6, p. 16]

Concepts/Reflections	Common Student Concerns
Students Profile The teaching-learning big gap	"Teachers teach us programming languages, building a Database, and creating 3D environments, however, they do not prepare us to understand the problems as a whole and deal with people. Sometimes there is a tangle of thoughts, which overlap my will and overflow with whys" "It is hard for me to understand what teachers want in exams, so even if I try, I get poor grades. I wish I knew how to interpret what I read, there never seems to be enough time to think about anything. I do not know how to relate things and get the big picture" "I like math, programming and video games playing. My life as a student only has two directions: home-ISTEC, ISTEC-Home. Thinking causes anguish and this is not good for our emotional stability. Maybe that is why the parent's house is the only safe place and it is only too late that we set out on our own, forced by the circumstances of the moment"

(*continued*)

Table 2. (*continued*)

Concepts/Reflections	Common Student Concerns
Knowledge obtained by the curricular units and the urgent need of long life learning education	"There are many errors in Computer and Multimedia Engineering teaching because they (teachers) perpetuate what happened in secondary school: courses with a scientific nature no longer have those subjects that required reading and learning how to think, such as Portuguese and Philosophy curricular units" "The Global Project delivering - monograph that certifies the skills acquired during my degree, I discovered that I did not read enough, I do not know how to argue, I do not know how to organize ideas or structure the speech or present my opinions orally, because I feel the terror of people criticizing me. After this pandemic, I am sure it will be worse"
The critical change in teaching training and learning strategies – technologies embedded	"So many subjects that could be transported to real life and continue to be transmitted by the monochord speech of teachers. Where is the evolution of technologies, if in a movie from the last century, you see the same classroom in which the Teacher remains in front of the students, as it happens now? Nothing changed" "Where are the technologies in face-to-face classes, the innovative methods that turn students into builders of their own knowledge? Although some professors are willing to change their strategy and use online means and new dynamics, there is still a lot of resistance from higher education institutions" "I have been to ERASMUS and I have seen very different approaches. In Nordic countries, the pleasure of discovering and solving real-life problems is privileged and technologies are clearly embedded in education"

Both the findings related to Table 1 (Goals Framework) and those obtained through Table 2 (Analysis of students' perceptions about teaching and learning and the relationship with the technology dimension) brought to the discussion: how to professionally integrate students and prepared them to absorb and update themselves constantly, taking into account the social and personal dimensions of their lives, could be stimulated with PBL contents. Maybe it could contribute to creating the desire for social, civic and political participation of students in their communities, caring about issues that concern everyone.

The implications above mentioned, must be followed by Universities or Institutes as an aspiration that could be held to improve and find out, what are the different goals of their students and to understand what will make a difference in their future and professional lives. A prior change is shifting the way how subjects' *curriculum* could be taught and seeking what kind of intervention in the business and social community will be relevant. The idea of setting up a course or curricular unit based on PBL inspired by Charles Peirce's semiotic perspective, perhaps a clue for other studies, however, it remains unclear whether its implementation will have real repercussions in terms of individual, professional, and citizen relations. This will be an observation to consider, hoping this article could have implications for other research topic generators.

5 Conclusion

Today technologies are intuitive and could be manipulated like a puzzle. No need to learn to code to create some very special applications or websites, even so, the essential STEM graduation degrees specialized skills make people forget some important issues: collaboration and cooperative work, and real-world experience to understand the utility of technologies.

Peirce's work suggests how to interpret, and read the World and the people's expectations, to be able to create, suggest, or stimulate technological innovations, to convert emotions and feelings into actions, and to get more diversity of lifelong backgrounds. This semiotic approach, based on historical aspects of science, technique, philosophical, psychological, and social principles, made sense at the moment that students revealed what worried them and what could be changed.

The main goal of this course, or curricular unit menu, was to place students in different and active positions: What is important to follow? How could people's knowledge be transformed? In the context of the World technological mutations, organizations changing and natural resources? How sustainable software, multimedia products, or environment-friendly devices could be built? How can their future business move forward, according to well-being and a good life for all?

This case study is based on two very fragile options: the first is based on prospective research, which is unusual and not used very often in terms of scientific publications however seems the most consistent addressing this case study; the second option is based on a course proposal, whose guiding principles for the content construction rise from the analysis of the student's perceptions and semiotic Peirce approach.

This article was a calculated risk considering the participant's small sample too, which may not be representative and may not correspond to legitimate expectations of future users of those contents, however, something has to change to bring graduate students closer to their employers, to improve their continuous professional conversion and adaptation, and provide their training as citizens, involved in an increasingly demanding World.

References

1. Siemens 2005 Connectivism a Learning Theory for the Digital Age (no date) Scribd. https://www.scribd.com/document/382203853/siemens-2005-connectivism-a-learning-theory-for-the-digital-age. Accessed 19 Nov 2022
2. Peirce, C.S.: The Collected Paper. 2ª Reimp- 3ª edição de 2000. https://16763326039401149306.googlegroups.com/attach/8f46c4fdd0450345/PEIRCE-charles-semiotica.pdf?part=0.1&view=1&vt=ANaJVrFWlvEOvBXSHPrbIowz7mmQcx3IxBoOLcXF9oYyiUpGSolHA2qkvEVfunPtC_mX5cZ7HUGzg8-1SY4m1wWCrlVnyou3MO309OBJR9zFlsoCVWtBygs
3. Morgan, T.: How Did the Rainbow Flag Become an LGBT Symbol? HISTORY, Jun. 02, 2017. https://www.history.com/news/how-did-the-rainbow-flag-become-an-lgbt-symbol
4. Hartshorne, C., Weiss, P. (eds.): The Collected Papers of Charles Sanders Peirce reproducing, vols. I–VI. Harvard University Press, Cambridge (1931–1935). Burks, A.W. (ed.), vols. VII–VIII (same publisher, 1958)
5. Peirce, C.S.: Collected Papers of Charles Sanders Peirce. Thoemmes Press (1899)
6. Osztián, P.R., Kátai, Z., Osztián, E.: (1AD) On the computational thinking and diagrammatic reasoning of first-year Computer Science and Engineering Students. Frontiers. https://doi.org/10.3389/feduc.2022.933316/full. Accessed 19 Nov 2022
7. Mourato, D.: Solidarity Technologies Article – KRIATIV-tech (n.d.). http://www.kriativ-tech.com/wp-content/uploads/2020/09/DulceMouratoSolidarityTechnologies.pdf. Accessed 19 Nov 2022
8. Mourato, D.: Aprendizagem por Problemas, Identidade de Género e Learnability em projetos finais nas Licenciaturas em Engenharia Multimédia e Informática (ICITS 2022 – RISTI), p. 674, (2022), ISSN: 1646–9895, No. E45. http://www.risti.xyz/issues/ristie45.pdf. Accessed 23 Dec 2022
9. Bennett, N., Dunne, E., Carré, C.: Patterns of core and generic skill provision in higher education. High. Educ. **37**, 71–93 (1999)
10. Creswell, J.W.: Research Design_ Qualitative, Quantitative, and Mixed Methods Approaches-SAGE Publications, Inc. (1) (2013)
11. George, A., Bennett, A.: Case Study and Theory Development in the Social Sciences. MIT Press, Cambridge (2005)
12. Gerring, J.: Case Study Research: Principles and Practices. Cambridge University Press, Cambridge (2007)
13. Yin, R.: Case Study Research: Design and Methods, 3rd edn. Sage, Thousand Oaks (2003)
14. Gomm, R., Hammersley, M., Foster, P. (eds.): Case Study Method: Key Issues, Key Texts. Sage, London (2000)
15. Mourato, D.: As Tecnologias Solidárias: Do Investimento no Conhecimento ao Desenvolvimento Pessoal, Repositório da Universidade de Lisboa: Página principal (2011). https://repositorio.ul.pt/handle/10451/4196. Accessed 19 Nov 2022
16. Figueiredo, D.: Projective research-a design-based research approach. https://www.researchgate.net/publication/358700125_Projective_Research_-_A_Design-Based_Research_Approach. Accessed 02 Jan 2023
17. Mourato, D., Amaro, P.: In: Notes, U.P. (ed.) Classroom and Tutorials Observations. (2022)

Qualitative Analysis on Psychological and Social Effects of Distance Education During SARS-CoV2 - COVID-19 on Students at Greek Public Institutes of Vocational Training

Anagnostou Panagiotis(✉) and Nikolova Nikolina

Sofia University "St. Kliment Ohridski", Sofia, Bulgaria
{anagnostu,nnikolova}@fmi.uni-sofia.bg

Abstract. During the pandemic, distance education was transformed from an adjunct to a main educational method, due to the social distancing required to limit the spread of the disease. This change brought to the educational community heaps of issues to be resolved and issues particularly important for the effectiveness of learning processes. The aim of the present research is to investigate the psychological and social impact of the SARS-CoV2 - COVID-19 coronavirus pandemic in the educational community of Greek Public Institutes of Vocational Training (PIVT), using qualitative analysis. The interview was carried out on a sample of seventeen (17) students of PIVT, with the aim of investigating their opinions and the attitude they demonstrate towards distance education, as well as a sample of 17 teachers with the aim of investigation the attitude of teachers towards the implementation of distance education, its benefits, or negative consequences for the educational community of the PIVT. The results of the qualitative analysis demonstrate that insufficient material and technical infrastructure, reduced skills of instructors, previous experience of students in such educational models and difficulties in laboratory courses, were key factors according to the students' responses, for their reduced performance. Although the participants recognize the positive elements of distance teaching, such as and the familiarity they have acquired with new technologies, they choose the standard in-person teaching. The lecturers argue that the students' performance showed a decline, since the students were removed from the physical learning environment and the degree of communication decreased.

Keywords: vocational training · distance learning · qualitative analysis · SARS-CoV2 - COVID-19

1 Introduction

The last two years have been marked by a specific change in the field of instruction because of the episode of Covid-19. The new educational scene and desires brought about by online instruction on such a large scale have presented a challenge to the global instructional community [1–3]. Even though it is widely acknowledged that the decision

A. Rocha et al. (Eds.): WorldCIST 2023, LNNS 800, pp. 432–441, 2024.
https://doi.org/10.1007/978-3-031-45645-9_41

to provide separate instruction in these circumstances was a one-way street, significant concerns have been raised about the quality of the instructional material provided as well as the achievement of the educational objective. Due to the nature of e-learning and the current pandemic's rapid and necessary uptake in the educational community, there are insufficient studies to investigate its effectiveness in vocational education, and an understanding of potential obstacles, difficulties, or benefits of online learning is critical (Kerres, 2020) [4].

Numerous studies have shown that online learning techniques are effective in vocational education as well as increasing student satisfaction rates through interaction [5, 6]. In 2014, CIGDEM et al. [7] donated the Hung et al. [8] Online Learning Readiness Scale (OLRS) to 725 vocational students to assess the effectiveness of online learning. OLRS consists of 18 elements organized into five factors. Computer/Internet self-efficacy (CIS), self-directed learning (SDL), student control (LC), motivation to learn (ML), and self-efficacy online communication are all concepts that have been studied (OCS). According to the study, students were generally prepared for online learning but needed to improve their skills in the CIS and OCS to succeed in online learning. Student characteristics (PC ownership, department, type of high school graduation) have a significant impact on students in some OLRS dimensions, particularly the CIS dimension. Following the spread of Covid-19, many studies have been conducted around the world to investigate the impact of the pandemic outbreak on the educational process of vocational education. Because of the effects of the Covid-19 pandemic, Syauqi et al. [9] studied the perceptions of Indonesian vocational students in the field of engineering in relation to online learning in 2020. According to the findings of this study, teachers in charge of online learning did not meet students' expectations. Han et al. [10] conducted a large-scale study in China with the participation of 270,732 vocational students and discovered that the online learning process was successful in that country and that the institutions met the challenges posed by the pandemic.

Even though a significant number of in-country studies on the level of satisfaction of vocational students with the process of online learning have been conducted, no corresponding study has been conducted in the case of Greece. The purpose of this paper is to investigate the degree of satisfaction of students and teachers at vocational schools in Greece with the process of online education, as well as the extent to which the set educational goals and the challenges posed by the pandemic outbreak were met.

2 Methods and Tools

2.1 Participants

A content analysis approach was used in this qualitative study. In case of student participants in this study were carefully chosen using the purposive sampling method [11]. Being a PIVT student and being willing to be interviewed were the inclusion criteria. There were 17 interviews in total, with 9 men and 8 women ranging in age from 18 to 26 years. Similarly, in case of teachers 17 participants (10 males and 7 females) selected using the purposive sampling method (Fig. 1).

The age range for the participant teachers was from 35 to 62 years. Regarding the participants' profession - specialty, there is diversity, since we find one participant each

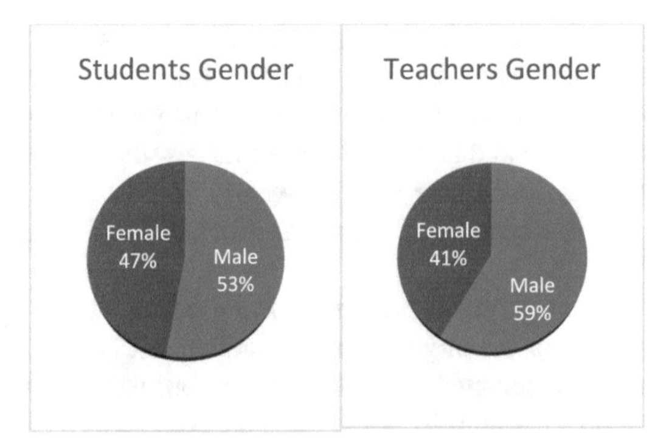

Fig. 1. Participants Gender

with the specialty of Economics, Lawyer, Chemist and Philologist. In addition, five (5) participants with the specialty of Cook, one (1) Medical Doctor, four (4) Trainers and one (1) Engineer participated in the interview (Fig. 2).

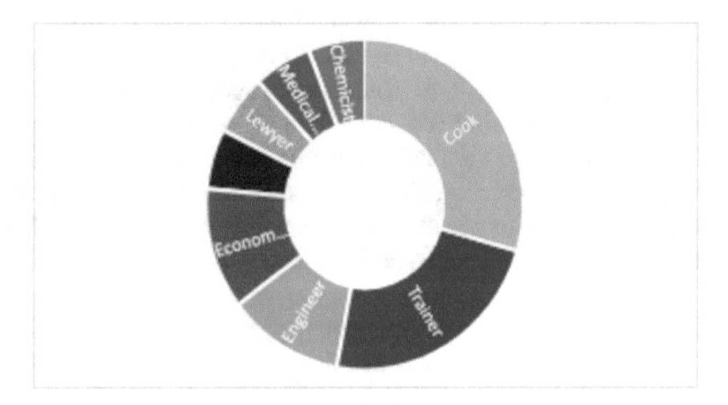

Fig. 2. Teachers profession - specialty

2.2 Data Collection and Analysis

Data was collected over a six-month period in 2021. Semi-structured interviews and field notes were initially used as data collection strategies. Purposeful sampling was continued until saturation, which meant that no additional data on the topics of interest were obtained. Each interview lasted 45 min on average. All interviews were recorded with a voice recorder and transcribed immediately following their completion. A list of general questions was included in the interview guide. The lead researcher listened to the interviews multiple times to get a sense of the data. Ambiguities were resolved by reviewing the transcripts with the participants immediately following or during the interview. After that, the units of meaning that were directly related to the research question were highlighted and chosen. Subsequent analysis was carried out in accordance with Graneheim and Lundman [12]. Data collection and analysis were carried out concurrently during the study period, as in other qualitative studies.

3 Results

3.1 General Axis

The first part of the interview contained questions about the changes brought about by the SARS-CoV2 - COVID-19 coronavirus pandemic and how they affected the Public Institutes of Vocational Training (PIVT) educational community. The respondents answered questions about the policies adopted by the PIVT management during the pandemic and whether they will be able to be implemented after it ends as well as questions related to the time required for the learners to spend to adapt to the exclusive application of distance education. During the collection and analysis of the data obtained from the interviews with the PIVT trainees, it was initially recorded how the SARS-CoV2 - COVID-19 coronavirus pandemic affected the educational community. According to the responses of the trainees, most of them, 13, were negatively affected by the SARS-CoV2 - COVID-19 coronavirus pandemic, while only 3 of them consider that the educational process was not significantly affected. Examining the answers provided by the participants, it becomes clear that there are no great differences in their answers.

A significant correlation is also shown in the responses of the trainees, regarding the negative influence of the pandemic on the main laboratory courses they attend at PIVT. The opinions of both S7 and S8 trainees converge in that the SARS-CoV2 - COVID-19 coronavirus pandemic negatively affected the conduct of laboratory courses, mainly due to the lack of logistical infrastructure and the insufficient technological training of teachers and students. Likewise, participant S8 refers to the lack of logistical equipment as well as the insufficient training of the teachers of the educational organization. It is also worth mentioning that the difficulties in teaching the laboratory courses mentioned by the participants during the SARS-CoV2 - COVID-19 pandemic period are also related, as they stated, to their increased absences from distance laboratory courses education. A large percentage of the trainees, S3, S4, S5 and S6, referred to the students' absences from the laboratory courses, mainly due to the insufficient logistical infrastructure of the educational organization and the inability of the teachers to cope with the increased demands of the laboratory courses. Specifically, the following were mentioned:

S3: The frequent absences of students from online laboratory courses were a result of the teachers' inability to cope with the increased demands of online learning
S4: The exclusion from the laboratories due to the coronavirus pandemic - COVID -19 and the difficulty in replacing the live laboratory courses with distance courses, caused student abstinence
S5: Many difficulties in the practical courses of the PIVT, mainly due to a lack of digital skills of the teachers
S6: The coronavirus pandemic COVID-19 affected the educational community quite a bit, mainly in the way of conducting the courses that required the laboratory application of the theory

Participating trainees also refer to the fear and insecurity caused by the SARS-COV2 - COVID-19 coronavirus and affecting the PIVT educational process. Participants S11 and S12 highlight the fear of both them and their fellow students for possible infection by the virus even after the end of confinement. It is also considered appropriate to mention the answer of S14 trainee, which reinforces the opinion that the SARS-CoV2 - COVID-19 coronavirus pandemic increased the trainees' anxiety, due to their fear of the possibility of getting sick.

It is worth mentioning the responses of students S15, S16 and S17, who consider that the SARS-CoV2 COVID-19 pandemic and the full implementation of distance teaching during the worst period of the pandemic did not affect the educational PIVT community, because most of the teachers and trainees possessed the necessary digital skills and technological knowledge, so that they could cope with the demands and the new data of distance education. The responses of the trainees are as follows:

S15: The excellent technological training of PIVT teachers helped to ensure that the educational community was not affected during the period of the coronavirus pandemic - COVID-19.
S16: It was certainly not a pleasant process, but it did not affect the educational process of the PIVT.
S17: Students' adequate knowledge of technology helped so that the educational community was not affected.

The next question deals with the conditions required for the optimal use of distance education. The answers of the participating PIVT trainees that were gathered, were categorized resulting in three main pillars. The participants referred to three basic conditions, which must be satisfied to utilize distance education to the optimal extent. Most of the participating trainees, S2, S3, S4, S7, S8, S10, S11, S12, S13 and S14, referred to the required technological equipment that the educational organization must have, so that distance education is efficient for all the students. Participant responses are consistent in content, making it clear that logistical infrastructure is a key condition for making the most of distance education.

3.2 Psychological/Spiritual/Social Axis

The second axis of the interview explores the psychological, spiritual and social ramifications of the pandemic caused by the COVID-19 virus in the PIVT educational community. The first question aimed to investigate whether the current practices implemented during the period of distance education due to the pandemic affected the development of interpersonal social relationships between learners and teachers. Most of the trainees, 16 in particular, commented that the development of interpersonal relationships between the trainees and the trainers were affected to a great extent as characteristically reported by S5, S6, SP7.

S5: I think so and to a large extent
S6: Definitely yes… by a lot and negatively
S7: They prevented them from being created and developed

Participants S1, S8, S9 and S10 confirm the previous point of view by adding that the pandemic caused by the COVID-19 virus clearly affected the development of interpersonal relationships, but also unfortunately weakened the existing ones, as stated by participant S1.

S1: With the implementation of distance education, social relations and therefore also the relationship with fellow students and teachers weakened

In addition, the participants characterize the interpersonal relationships of the educational community during the pandemic period as impersonal, since, as stated by participant S13, the feeling of the group in the educational community has been lost, a point of view also embraced by candidates S11, S12 and S15.

S13: To a large extent, the relationships between trainees and teachers became more impersonal during the pandemic period, the feeling of the group was lost...

Analyzing the responses of the participants, it is visible their negative view on developing interpersonal relationships between students and teachers during the period of the pandemic caused by the COVID-19 virus and given the current practices implemented by the PIVT. The next question of the psychological and social axis of the interview refers to the possibility of an increase in anxiety and depression experienced by students during distance education during the COVID-19 virus pandemic. Most of the participants thirteen (13) in particular, agreed that the level of anxiety and depression of the students increased during the distance education period. The participants, based on their personal experience from distance education S1, S4, S8, S9, S10, S11 and S12, answered positively with short answers and the use of affirmative adverbs. Participants S17 and S15 mainly referred to the stress of students during the pandemic period, as a consequence of social exclusion and the implementation of universal distance education, without mentioning the possibility of depression. Participants report:

S15: The removal of students from their educational environment is the main reason for the increase in students' anxiety, as is the insecurity for conducting laboratory courses
S17: Students' anxiety about conducting online courses and the course of the educational process has clearly increased, but I don't think depression...

In addition, it is considered appropriate to mention the opinions of S4, S14 students where their answers touch on the issue of the age of the students. The participants consider that the level of anxiety and depression increased in students of mainly older age groups, compared to the younger ones. They justify their response to the reduced digital skills of older students and the lack of previous experience in the use of modern technological tools.

3.3 Learning Axis

The next questions asked to the participants concern the learning axis and aim to explore the views of the students on the educational policies and practices implemented by the PIVT during the period of the pandemic caused by the COVID-19 virus. The first question of the learning axis invites the participating students to evaluate their educational performance, during the period of the pandemic and the universal application of distance education, based on the practices implemented by the Directorate of PIVT. Almost all participants, with their answers, stated that the performance of students worsened during the pandemic period, due to their bad psychology, the difficulties they faced during distance education and the inability of teachers to respond to the increased educational requirements of the courses, especially the laboratory ones. The responses that follow highlight the reasons, which the participants consider having determined the poor performance of the students, reflecting on the current practices implemented by the PIVT.

S8: The decline in our performance is a function of our bad psychology and not of the practices implemented by the PIVT
S9: Difficulties were too many during the distance courses and this was also the main reason for our poor performance.

The participating students, after evaluating their performance in the online courses during the distance education period, were then asked to express their opinion on whether they believe that the degree of their previous experience and familiarity with technological applications was a determining factor for their performance in classes during the COVID-19 pandemic and whether the distance education that took place widened the social and educational gap between citizens. Most of the participants (16), agreed with this point of view and characterized the students' previous experience and familiarity with technological applications as crucial and important, for the successful conduct of distance courses and their high performance.

3.4 Teachers Opinion

In response to the question posed to the teachers *Where do the online courses take place?* The answers given were uniform. After grouping them, we can conclude that most teachers (11) stated that they worked from their homes, while only 5, conducted the online courses from the professional space they had. The next question asked faculty to evaluate their previous experience of using distance education programs. Most of them (10) specifically stated that their experience in distance learning programs of synchronous and asynchronous education was average both as instructors and learners. While 5 instructors stated that they have good experience since they have worked as instructors in distance learning courses and have themselves participated in training programs using distance learning method. It is also appropriate to mention that only 2 participants have several years of experience. The participants' questions are characterized by brevity, as indicated below:

T8: The experience I have, it's not particularly great, I would say average...
T10: I can say good... I have attended some training courses and I have taught online, but not many hours
T16: Perennial...

The interview question that followed relates to the electronic devices, such as desktop or laptop computer, tablet, or mobile phone, used by the teachers (individually or in combination) for tele-education. In addition, they were asked to record what potential problems, such as connectivity or management, they encountered when using them.

Regarding the synchronous education platform used by the lecturers to conduct the distance learning courses, the majority considers that it contributed to a better understanding of the material by the students and helped to improve the quality of the educational process. It is worth mentioning the negative opinion of 2 participants, who argue that formal teaching is irreplaceable, and any other form of education could not replace it. In the question posed to the lecturers, regarding the response, participation of students in distance learning and whether there was improvement or deterioration in their performance, the answers of the lecturers were uniform, the totality of them mentioned the deterioration of the students' performance although the participation and response, as they stated, was satisfactory. Afterwards, participants asked to express their opinion on whether they would use a blended learning program in the future. Most of the lecturers stated that they would use a blended learning program, not overlooking the negative opinion of 3 participants. The participants justified their opinion by highlighting the positive aspects of this training method, such as saving time and money on travel, as well as in cases of student or teacher sickness. In addition, they consider that the implementation of a combined training program will work in combination and complementary to formal teaching and will improve the result. Concluding the interview questions, it was deemed appropriate to investigate whether modern tele-education met the need for enhancing life skills such as adapting to new situations, cooperation, improvement in solving problem situations, taking initiatives, acting and increasing creativity of the trainees during their attendance at the PIVT.

4 Discussion and Conclusion

The interview was carried out on a sample of seventeen (17) students of PIVT as well as seventeen (17) teachers of PIVT, with the aim of investigating their opinions and the attitude they demonstrate towards distance education. The results shows that the sudden imposition of distance teaching caused multiple difficulties for both students and teachers, especially in laboratory courses, an opinion that is justified during the interview. Most of the participants mentioned the difficulties they encountered during the teaching of the laboratory courses, due to the lack of appropriate and modern equipment by the teachers and students and the limited knowledge of the latter in New Technologies and their applications. The findings of the present research verify the corresponding findings from other research, during the period of implementation of distance teaching and that of Apostolou et al. [13], who considered, that the lack of equipment and the reduced digital skills of instructors are decisive factors in the implementation of distance learning. From the responses of the participants comes that the educational community was unprepared to respond to the new conditions in which training is more necessary than ever. The students of PIVT, although they mentioned the difficulties, they were asked to face during distance teaching, characterized the policies implemented by the educational unit of PIVT during the pandemic as satisfactory. In his answers, the students highlight the maximum effort made by the teachers to meet the increased demands of the online courses and cope with the daily difficulties they faced due to the technical difficulties. For this reason, they believe that the policies implemented by the PIVT could be applied even after the end of the pandemic, mainly in theoretical courses. This point of view is confirmed by earlier research by Karaiskou et al. [14], who concluded that the blended learning model can be constructive for the students of PIVT. Most of the students considers that the pandemic and the universal application of distance education affected the development of interpersonal social relationships between students and teachers. They report that the relationships became impersonal, weakened because of the educational process being a personal process and not a collaborative process, based on dialogue and interaction. The students' answers are consistent with the opinion that the sudden changes in the field of education, the inability of students and teachers to cope with the technological difficulties they faced every day, as well as the fear of infection by the virus, are the main causes of increasing anxiety which the research of Ksafakos et al. [15]. Contrary to the stress rates, which the students consider having increased during the period of distance learning, they do not consider that they were causes that will lead the students to depression or the manifestation of fears, about wasting time and money. Insufficient material and technical infrastructure, reduced skills of instructors, previous experience of students in such educational models and difficulties in laboratory courses, were key factors according to the students' responses, for their reduced performance [16]. It is worth mentioning that from the answers of the participants it becomes clear that they recognize the positive elements of distance teaching, such as the saving of money and time and the familiarity they have acquired with New Technologies, but they choose the standard in-person teaching, since most of them do not identify any negative points. The previous experience of the teachers in the implementation of distance education and the technological tools required, according to the teachers' responses, is characterized as Moderate, which undoubtedly affects the quality of training of distance education services. According to reports in the global literature, instructors should have the ability to promote and develop basic skills and competencies, alongside the teaching of up-to-date professional, prior experience is a key factor in this endeavor and in promoting learning.

References

1. Teymori, A.N., Fardin, M.A.: COVID-19 and educational challenges: a review of the benefits of online education. Ann. Mil. Health Sci. Res. **18**, e105778 (2020). https://doi.org/10.5812/amh.105778
2. Khlaif, Z., Salha, S.: The unanticipated educational challenges of developing countries in Covid-19 crisis: a brief report. Interdisc. J. Virtual Learn. Med. Sci. **11**(2), 130–134 (2020). https://doi.org/10.30476/ijvlms.2020.86119.1034
3. Toquero, C.M.: Challenges and opportunities for higher education amid the COVID-19 pandemic: the Philippine context. Pedagogical Res. **5**(4), 0063 (2020)
4. Kerres, M.: Against all odds: education in Germany coping with Covid-19. Postdigit. Sci. Educ. **2**(3), 690–694 (2020). https://doi.org/10.1007/s42438-020-00130-7
5. Belaya, V.: The use of e-learning in vocational education and training (VET): systematization of existing theoretical approaches. J. Educ. Learn. **7**(5), 92–101 (2018)
6. Bignoux, S., Sund, K.: Tutoring executives online: what drives perceived quality? Behav. Inf. Technol. **37**(7), 1–11 (2018)
7. Cigdem, H., Yildirim, O.G.: Effects of students' characteristics on online learning readiness: a vocational college example. Turk. Online J. Distance Educ. **15**(3), 80–93 (2014). https://doi.org/10.17718/tojde.69439

8. Hung, M., Chou, C., Chen, C., Own, Z.: Learner readiness for online learning: Scale development and student perceptions. Comput. Educ. **55**, 1080–1090 (2010)
9. Syauqi, K., Munadi, S., Triyono, M.B.: Students' perceptions toward vocational education on online learning during the COVID-19 pandemic. Int. J. Eval. Res. Educ. **9**, 881–886 (2020)
10. Han, X., Zhou, Q., Shi, W., Yang, S.: Online learning in vocational education of china during COVID-19: achievements, challenges, and future developments. J. Educ. Technol. Dev. Exch. (JETDE) **13**(2) (2021). https://doi.org/10.18785/jetde.1302.0
11. Etikan, I., Musa, S.A., Alkassim, R.S.: Comparison of convenience sampling and purposive sampling. Am. J. Theoret. Appl. Stat. **5**(1), 1–4 (2016)
12. Graneheim, U.H., Lundman, B.: Qualitative content analysis in nursing research: concepts, procedures and measures to achieve trustworthiness. Nurse Educ. Today **24**(2), 105–112 (2004)
13. Apostolou, N.: The implementation of distance education in Vocational Institutes. Int. J. Educ. **3**(1), 85–94 (2021)
14. Karaiskou, V., Georgiadi, E.: Implementation of blended learning model, by using the Moodle platform, in the Public Institute of Vocational Training of Korydallos. In: Proceedings of the 8th International Conference in Open & Distance Learning, Athens, Greece (2015)
15. Ksafakos, G., Tzilou, G., Pasiopoulos, G.: Perceived sense of stress of postgraduate students in distance learning. In: Proceedings of the 9th International Conference in Open & Distance Learning, Athens, Greece, November 2011
16. Tsekeris, C.: State, society and media in the era of the coronavirus. Inspection Soc. Res. **154**, 109–128 (2020)

How to Help Teaching in a Virtual Environment

Bertil P. Marques[1](✉), Marílio Cardoso[2], and Rosa M. Reis[1]

[1] GILT/ISEP/IPP, Porto, Portugal
{bpm,rmr}@isep.ipp.pt
[2] SIIS/ISEP/IPP, Porto, Portugal
joc@isep.ipp.pt

Abstract. COVID-19 required rapid change in teaching, using virtual environments, as well as at the entire educational level. At the time, teachers were unfamiliar with virtual teaching environments and needed to learn quickly and effectively how these environments worked and how they could be used in courses successfully, regardless of scope. The tips that are presented, provide an insight into the practice of teaching in virtual environments, from the design of courses, through the involvement of students, to evaluation practices, also considering the maximization of the potential that technology can provide both the teacher and students. In addition, these tips present virtual pedagogical practices for all levels of experience.

Keywords: Education · virtual learning · online reviews · faculty development · cognitive load theory · COVID-19

1 Introduction

The reality of the COVID-19 pandemic has called for a major and rapid change in teaching strategies and throughout educational programs globally. Specifically, the need for physical distancing severely limited traditional face-to-face classes and teaching strategies and encouraged a shift to online and virtual teaching strategies [1–3].

Virtual education environments present a unique set of opportunities and challenges that have been increasingly used over the last decade for distance education and courses using mixed learning models [4]. All schools had to provide resources and training for their teachers to develop skills and knowledge for providing courses in virtual education environments.

2 Improvement Suggestions

This knowledge base has been called for the rapid transition to teaching in virtual environments during the COVID-19 pandemic. In this context and based on experiences of teacher development and support during COVID-19 which affected the rapid transition of teaching in March and April 2020, 12 tips for teaching in virtual environments are

A. Rocha et al. (Eds.): WorldCIST 2023, LNNS 800, pp. 442–450, 2024.
https://doi.org/10.1007/978-3-031-45645-9_42

presented. These tips have been designed for teachers who are relatively new to virtual education, as well as to those who have previous experience, with the latter potentially benefiting from new and different perspectives on the application and effects of teaching through the technology presented below.

2.1 Review Learning Objectives to Align with Virtual Environments

The development of online courses and programs should include a careful analysis of learning objectives. Wiggins and McTighe [5] encourage educators to start the design process by identifying the evidence needed for students to demonstrate the level of proficiency they want before planning teaching and learning experiences, that is, through constructive alignment [6]. The learning objectives will therefore remain in place, but the context in which they are carried out will become virtual environments. Educators should think critically about what can be achieved realistically in a virtual environment, since, given the nature of the matter, clinical spaces are an imperative context that could be difficult to replace completely. Pausing and reflecting on learning objectives is an important step in deciding what is feasible and how best to align learning objectives with the reality of teaching in virtual environments.

2.2 Review and Adapt Resources for Teaching in a Virtual Environment

The interdependence between content, pedagogy and technology is a unique feature of online teaching. Learning in a digital context takes place through discussion, reflection and collaboration with students who are prepared to engage in active learning with a peer community. A wide range of tools and technologies is available and the choice of which to use depends on two key factors: their support for teaching objectives, and their unique offerings and potential learning benefits [7]. For example, learning management systems are designed to foster social and collaborative learning in both synchronous and asynchronous environments; social networks can increase social presence and build networks; and video conferencing platforms can allow live meetings and presentations, private conversations, and escape rooms for group work [8]. In addition, there is a growing list of tools to create digital learning content, such as interactive images and videos, as well as collaborative platforms that support messaging and content sharing across multiple platforms [9]. The effective integration of digital tools and technologies is facilitated by teachers who explore and learn technologies and new pedagogical practices, as well as the support of leaders, technology support staff and, most importantly, students.

2.3 Explore Strategies to Virtually Engage with Students

When designing the program, it is important to consider not only the content, but also how students are planned to engage with that content. Involvement refers to the time and energy spent on learning activities, and includes reading, practice, obtaining feedback, analyzing material, and solving problems [10]. Positive connections with instructors have proven to play a significant role in the satisfaction, persistence, and success of students [11]. For example, a simple welcome video, recorded on your phone and uploaded to the program page, will go a long way to make your course feel more personalized [12]. Starting and maintaining discussions through emails and forums, as well as virtual office hours (or a casual delivery or registration for specific slots) facilitates positive virtual engagement with students. If students feel welcome to connect with the teacher, they are more likely to look for answers to your questions. It is important to remember that this form of engagement can take some time to grow, however, the development of mutual trust will increase the effectiveness of online collaborative learning [13].

2.4 Design Educational Content for Virtual Environments

Consider the cognitive load theory (CLT) in the course design for remote instruction during COVID-19, since increased stress during the pandemic can have an additional negative impact on working memory [14, 15]. The CLT assumes that these three types of load information processing influence learning, impact the working memory capacity to effectively manage the learning task [16]. Specifically, the CLT is based on the premise that we have a limited working memory space, so the goal is to maximize the use of working memory. Teachers can do this by removing things that distract from the learning process (e.g., images that do not relate to the learning objective), making learning appropriate at the learning level, and maximizing tasks associated with learning processes. For remote instruction and teaching in virtual environments, we suggest focusing on specific strategies. Teachers can reduce the burden in several ways:

A. Most instructions should have written and verbal explanations in a single integrated source, especially for beginners [17]. When verbally integrating with writing, avoid dividing students' attention between multiple windows on the screen (e.g., chat box, video stream, slides, whiteboard, etc.) [18, 19].
B. Providing worked examples, either showing the student how a problem is solved or giving the student an idea of what the final intended of the product is like [20].
C. Teachers can plan activities that focus on the practice of recovery, that is, retrieving the information learned to apply it.
D. Allow students to 'self-explain' the material being presented, or summarize key points, either verbally, or by using the chat functionality available in many technological solutions. Escape rooms can also be used to allow students to self-explain the contents to small groups.
E. Using active learning activities they ensure adequate variability in their skills, so solving each case is a unique exercise in recovery practice.

3 Consider Using Various Teaching Strategies

The changes for new media provide opportunities to engage with our students in new ways and allow meaningful interaction with the content, the professor other students, as well as the ability to access information and lectures anywhere. However, it is important to be aware of what technologies students have access to. For example, do you have Internet connections to support synchronized technologies such as breakout rooms, video conferencing, chat areas, and collaboration software? Otherwise, consider asynchronous technologies that also facilitate engagement and communication, such as social networks (Twitter, YouTube), email, and discussion forums.

Using a variety of learning technologies will not only help you appeal to different student learning preferences but can also maximize their engagement and increase collaborative learning. Learning technologies give us the important opportunity to move our students from a Web 1.0 environment where they are consumers of online content to a Web 2.0 environment where they are responsible for creating online content (e.g., blogs, videos, webinars, wikis) [13]. This advantage can be considered an additional or extra benefit of the virtual course design.

3.1 Maximize What Technology Can do for a Teacher

While remote or online learning may seem like a deterrent to mixing assessment with learning activities, it can also offer an opportunity to consider practices that can be difficult to implement in a larger conference room with fixed furniture. For example, punctuating an online lecture with escape rooms where students collaboratively complete quizzes (these quizzes may or may not be classified), which are then discussed as a large group, can offer valuable and timely formative evaluation feedback and feedback for the student and for the teacher. In addition, escape rooms during an online lecture can also facilitate the development and maintenance of learning relationships. This example can also be extended to an asynchronous environment, where the lecture can be provided as a video, quizzes are completed individually or by student groups offline, or through discussion forums, and then interrogated as a scheduled lecture, or follow-up video based on learning by quizzes. Providing these types of assessments also facilitates opportunities for students to self-assess their knowledge and adjust their preparations for the scheduled lecture so that they can maximize their learning. Teachers can link follow-up resources to specific test questions to help with this preparation, which can be provided through the management and learning system.

3.2 Determine the Purpose of the Assessment and Ensure Alignment

When moving to online learning environments, the goal of assessments should be at the forefront. Assessments should reflect the intent, level of mastery and depth of understanding needed to achieve learning objectives.

In addition, it is necessary to consider whether the assessments are providing formative feedback or a summary note, and whether they are realistically achievable. In this regard, the online definition will only change the format, but not the substance of an evaluation. Authentic assessments, which mirror workplace contexts where students apply their knowledge [21], are engaging options for virtual environments, and facilitate formative feedback [22]. Using authentic reviews and other training assessments in online contexts also facilitates greater interaction and engagement with feedback delivery through online discussions, peer reviews using shared documents, written feedback, and video in synchronous or asynchronous mode.

3.3 Refine Assessments to Reflect Virtual Environments

If changes to assessments are needed with the move to online delivery, it is important to remember that there is a consistent need for students receiving quality feedback in time to improve their learning experiences [12]. Therefore, when considering that changes need to occur in an impacted environment covid-19, it is valuable to consider the level of mastery that each evaluation addresses, in addition to the evaluation format [23]. One way to assess whether assessments need to be refined is to develop an assessment map, where learning objectives are listed and a link to each assessment is articulated. Each assessment can then be analysed to see if it can continue as it is in a virtual environment, or if refinement is needed. Refinements for virtual environments can include combining or splitting assessments to ensure that they cover the level of mastery required for each learning goal. This assessment map provides a visual check to ensure that each learning objective could appear within an assessment so that the student could build and demonstrate their level of mastery as the course progresses in the virtual environment.

3.4 Familiarize the Teacher and Students with Virtual Environment Previously

It is important to guide the teacher and students with the technology and software that encompass the virtual teaching environment before the start of the course. McLeod *et al.*[24] focuses on the form of guidance that should be prepared to facilitate a simple knowledge of the operation of the chosen software in a way even at the last minute. These materials should be easily accessible, such as a one-page document, a short video, or a short podcast. The professors are suggested to test the technology in an environment that is conducive to making mistakes and exploring the features of the software before the start of the course [25]. This exploration will facilitate the technological competence necessary to lead the course and prepare the professor for unexpected challenges. We also suggest that the guidance materials be distributed in advance to the students, with incentive for them to play with the technology beforehand. In addition, it is recommend using the first 10–15 min of the initial lecture to review the orientation materials with the students, to ensure that everyone is on the same page and address any issues and concerns.

3.5 Keep the Virtual Teaching Environment Safe and Respectful

McLeod *et al.* [24] highlight several unexpected issues with teaching in virtual environments, classifying them as visual, curricular, and auditory exhibitions. These issues revolve around a different, and unique set of social circumstances implied through videoconferences and virtual environments, specifically people who feel exposed and uncomfortable with their persistent on-screen image, facilitators and support staff being exposed to sensitive teaching materials, and unintended auditory commentary for the environment. To compensate for these potential exhibitions, we suggest the inclusion of a section in the above-discussed guidance materials that explicitly discusses common issues associated with visual, curricular, and auditory exposures and sensitizes both the professors and the students. This discussion should take place at the beginning of the course, in addition to being outlined in the guidance materials, and should facilitate a respectful teaching environment [26]. Common etiquette suggestions include muting the microphone unless you're talking, using the hand-raising feature in video conferencing software to indicate that you'd like to talk and clarify minor problems with chat functionality.

3.6 Make Backup Plans

Making a back-up plan is a good practice, regardless of whether you teach in person or in a virtual environment. However, teaching in virtual environments can increase the potential for unknown and unpredictable factors that impact the session, since students heavily on technology to facilitate communication. Being familiar with the technology and software used to provide the course can facilitate greater flexibility and adaptability if something goes wrong. Consultation with IT experts in advance can also obviate technological issues as they can advise on common issues and successful solutions.

It is suggest also developing strategies and plans if there are problems with technology (software, Internet connection, hardware, etc.). For example, if technology issues prevent a virtual lecture, consider switching to an inverted classroom approach [27, 28], where materials can be provided to students outside the lecture and discussion can be resumed once the matter is resolved. Also consider flexibility in evaluation practices and, potentially, the inclusion of independent attributions since they may be more resistant to technological issues than online real-time assessments [29].

3.7 Maintaining Student's Understanding and Compassion with the Transition to Remote Learning

Stress is already an important component of students' lives throughout their education [9, 30], and increased with the COVID-19 pandemic [31, 32]. In this regard, it is important to maintain compassion and understanding for students who may be feeling an increase in anxiety, depression and stress related to exposure to COVID-19 (themselves, friends, family), self-isolation, lack of employment and/or funding, for international students who may need to leave the country, or some combination of the above. Research suggests that social support and self-compassion, which is being able to recognize that failures and disappointments, are important to increase resilience to stress and anxiety [33, 34]. Social support in remote learning contexts can be facilitated through regular group study tasks that encourage small group actions and cooperation and can lead to off-course socialization. Self-compassion is more challenging to encourage in virtual environments but can be developed through more frequent and lower betting assessments and/or self-assessment. Above all, maintaining compassion and understanding encompasses making time and space to listen to students' concerns and being flexible in response to their unique situations and contexts.

4 Conclusions and Final Considerations

The onset of the COVID-19 pandemic has forced a rapid shift to virtual education environments. The tips presented are intended to facilitate this rapid change in the educational community. In this regard, these tips reflect on the practice of education, commenting on the preparation and development of the course, engaging with students, designing relevant assessments, and working with, not against, technology. It is verified that tips are a starting point for the change from traditional face-to-face teaching to virtual environments, reinforcing considerations of virtual teaching environments as viable and attractive educational tools, and encouraging the discussion of techniques and strategies that contribute to the success of virtual classrooms. These virtual classrooms can be used in addition to face-to-face teaching in a non-pandemic situation.

As final considerations we can point out:

- Consider how available technologies can make it easier to deliver courses in virtual environments while maintaining course objectives.
- Use teaching strategies based on learning theory for virtual teaching environments.
- Refine assessments to reflect best practices in virtual environments and maintain alignment with learning objectives.
- Take care of students and maintain compassion, since they are also in transition to virtual education environments.

References

1. Carlson, E.R.: COVID-19 and educational involvement. J. Oral Maxillofac. Surg. (2020). https://doi.org/10.1016/j.joms.2020.04.033
2. DeFilippis, E.M., Stefanescu Schmidt, A.C., Reza, N.: Adapting the educational environment for cardiovascular fellows in training during the COVID-19 pandemic. J. Am. Coll. Cardiol. **75**(20), 2630–2634 (2020). https://doi.org/10.1016/j.jacc.2020.04.013
3. Prem, K., Liu, Y., Russell, T.W., Kucharski, A.J., et al.: The effect of control strategies to reduce social mixing on the results of the COVID-19 epidemic in Wuhan, China: a modeling study. Lancet Public Health **5**, e261–e270 (2020). https://doi.org/10.1016/S2468-2667(20)30073-6
4. Bonk, C.J., Graham, C.R.: The mixed learning manual: global perspectives, local drawings. San Francisco, CA: John Wiley & Sons, Hoboken (2012). https://tinyurl.com/bonk-graham-2012. Accessed 14 Oct 2022
5. Wiggins, G., McTighe, J.: Understanding by Design, 2nd edn. Association for Supervision and Curriculum Development, Alexandria (2005)
6. Biggs, J.: Constructive alignment: a guide for busy academics. LTSN Generic Center (2002). http://www.heacademy.ac.uk/resources.asp. Accessed 15 Oct 2022
7. Ng, W.: Affordances of new digital technologies in education. In: New Digital Technology in Education, pp. 95–123. Springer, Heidelberg (2015). https://doi.org/10.1007/978-3-319-05822-1_5. Accessed 12 Oct 2022
8. Dunlap, J.C., Lowenthal, P.: Tweeting the night away: using twitter to enhance social presence. J. Inf. Educ. Syst. **20**, 129–135 (2009)
9. Ang, R.P., Huan, V.S.: Academic expectations underline the inventory: development, factor analysis, reliability and validity. Educ. Psychol. Meas. **66**(3), 522–539 (2006). https://doi.org/10.1177/0013164405282461
10. Robinson, D.C., Hullinger, H.: New references in higher education: involvement of students in online learning. J. Educ. Bus. **84**(2), 101–109 (2008). https://doi.org/10.3200/JOEB.84.2.101-109
11. O'Shea, S., Stone, C., Delahunty, J.: I feel like I'm in college, even though I'm online. Explore how students narrate their involvement with higher education institutions in an online learning environment. Dist. Educ. **36**(1), 41–58 (2015). https://doi.org/10.1080/01587919.2015.1019970
12. Reyna, J.: Twelve tips for COVID-19 friendly learning design in medical education. MedEdPublish **9** (2020). https://doi.org/10.15694/mep.2020.000103.1

13. Sanders, M.J.: Classroom design and student involvement. In: Annual Proceedings of the Human Factors and Ergonomics Societ, pp. 496–500y. SAGE Publications Sage CA, Los Angeles (2013). https://doi.org/10.1177/1541931213571107
14. Klein, K., Boals, A.: The relationship of life event stress and working memory capacity. Appl. Cogn. Psychol. **15**(5), 565–579 (2001). https://doi.org/10.1002/acp.727
15. Hubbard, K.K., Blyler, D.: Improving academic performance and working memory in undergraduate health sciences students using progressive muscle relaxation training. Am. J. Occup. Therapy **70**(6), 7006230010 (2016). https://doi.org/10.5014/ajot.2016.020644
16. Young, J.Q., Van Merrienboer, J., Durning, S., Ten Cate, O.: Theory of cognitive burden: implications for medical education: AMEE Guide No. 86. Prof. Med. **36**(5), 371–384 (2014). https://doi.org/10.3109/0142159X.2014.889290
17. Tindall-Ford, S., Chandler, P., Sweller, J.: When two sensory modes are better than one. Am. Psychol. Assoc. **3**(4), 257–287 (1997). https://doi.org/10.1037/1076-898X.3.4.257
18. Ayres, P., Sweller, J.: The principle of divided attention in multimedia learning. Cambridge Multimedia Learn. Manual **2**, 135–146 (2005)
19. Chen, C.-M., Wu, C.-H.: Effects of different types of video lectures on sustained attention, emotion, cognitive load and learning performance. Comput. Educ. **80**, 108–121 (2015). https://doi.org/10.1016/j.compedu.2014.08.015
20. Van Merriënboer, J.J., Sweller, J.: Theory of cognitive burden in professional health education: principles and design strategies. Med. Educ. **44**(1), 85–93 (2010). https://doi.org/10.1111/j.1365-2923.2009.03498.x
21. Beck, L., Hatch, T.: Authentic assessment. In: Clauss-Ehlers, C. (ed.) Encyclopedia of Psychology of the Transcultural School, pp. 135–137. Springer, Boston (2020). https://tinyurl.com/y34xfm9r. Accessed 10 Oct 2022
22. Villarroel, V., Bloxham, S., Bruna, D., Bruna, C., et al.: Authentic evaluation: creation of a project for the design of courses. Eval. Eval. High. Educ. **43**(5), 840–854 (2018). https://doi.org/10.1080/02602938.2017.1412396
23. Amin, H.A., Shehata, M.H., Ahmed, S.A.: Step-by-step guide to creating skills. In: Assignments as An Alternative to Traditional Summary Assessment. MedEdPublish **9** (2020). https://doi.org/10.15694/mep.2020.000120.1
24. MacLeod, A., Cameron, P., Kits, O., Tummons, J.: Exposure technologies: distributed videoconferencing medical education distributed as sociomaterial practice. Acad. Med. LWW **94**(3), 412–418 (2019). https://doi.org/10.1097/ACM.0000000000002536
25. Kanhadilok, P., Watts, M.: Adult play-learning: observing informal family education in a science museum. Stud. Adult Educ. **46**(1), 23–41 (2014). https://doi.org/10.1080/02660830.2014.11661655
26. Terry, R., Taylor, J., Davies, M.: Successful teaching in virtual classrooms. In: Daniels, K., Elliot, C., Finley, S., Chapman, C. (eds.) Learning and Teaching in Higher Education: Perspetives from a Business School, pp. 211–220. Edward Elgar Publishing, Cheltenham (2019). https://books.google.co.za/books?hl=en&lr=&id=9oC4DwAAQBAJ&oi=fnd&pg=PA211&ots=9O6wLkj. Accessed 23 Oct 2022
27. Tolks, D., Schäfer, C., Raupach, T., Kruse, L., et al.: An introduction to the inverted/upset classroom model in advanced education and training in medicine and health professions. GmS J. Med. Educ. **33**(3) (2016). https://doi.org/10.3205/zma001045
28. Hew, K.F., Lo, C.K.: The inverted classroom improves students' learning in the education of health professions: a meta-analysis. BMC Med. Educ. **18**(1), 38 (2018). https://doi.org/10.1186/s12909-018-1144-z
29. Rachul, C., Collins, B., Ahmed, M., Cai, G.: Twelve tips for designing missions that foster independence in learning. Prof. Med. 1–5 (2020). https://doi.org/10.1080/0142159X.2020.1752914

30. Bedewy, D., Gabriel, A.: Examining perceptions of academic stress and its sources among university students: the perception of the Academic Stress Scale. Open Health Psychol. **2**(2), 1–9 (2015). https://doi.org/10.1177/2055102915596714
31. Cao, W., Fang, Z., Hou, G., Han, M., et al.: The psychological impact of the covid-19 epidemic on university students in China. Psychiatry Res. **287**, 112934 (2020). https://doi.org/10.1016/j.psychres.2020.112934
32. Sahu, P.: Closure of universities due to coronavirus disease 2019 (COVID-19): impact on education and mental health of students and academic staff. Cureus **12**(4), e7541 (2020). https://doi.org/10.7759/cureus.7541
33. Neff, K.D., Hsieh, Y.-P., Dejitterat, K.: Self-compassion, achievement goals and dealing with academic failure. Self Identity **4**(3), 263–287 (2005). https://doi.org/10.1080/13576500444000317
34. Poots, A., Cassidy, T.: Academic expectation, self-compassion, psychological capital, social support and student well-being. Int. J. Educ. Res. **99**, 101506 (2020). https://doi.org/10.1016/j.ijer.2019.101506

The Transformation of the Digital Communication of Educational Organizations in the Pandemic Period: New Practices and Challenges

Inês Miguel[1](✉) and Márcia Silva[2]

[1] ISCTE-IUL, IADE, Lisbon, Portugal
inesvazaomiguel@gmail.com
[2] Communication and Society Research Centre, Braga, Portugal

Abstract. The digital transformation caused by COVID-19 broke with classic communication models and brought about profound changes in institutional and marketing communication in educational organizations. In order to respond to this transformation, educational organizations had to quickly change and adapt different communication strategies in view of the current pandemic context. This article specifically focuses on the description of the transformation of digital communication that took place in two private education educational organizations in Portugal, before the pandemic (January–March 2020) and during the pandemic/confinement (March–June 2020). Following a quantitative and qualitative methodology, in the form of a multiple case study, we focused on listening through semi-structured interviews with directors, IT and communication technicians and questionnaire surveys with parents (n = 172) of both educational institutions. The results indicate that the crisis context raised by the COVID-19 pandemic has reinforced the role of communication in educational organizations and has profoundly changed the way of communicating. We conclude that regularity, proximity, clarity, and sincerity of communication increase trust in relation to the service provided. It is recommended that educational organisations adopt strategic co-communication plans and that they favour the use of social media and other online media, with a view to streamlining and flexibilities communication.

Keywords: Digital Communication · Educational Organizations · Pandemic · COVID-19

1 Introduction

The COVID-19 pandemic has had a significant impact on businesses and organisations around the world, and educational organisations are no exception. In the face of the COVID-19 pandemic, triggered in March 2020, the Directorate-General of Health implemented a set of public health rules (e.g., social distancing, school closures, mandatory confinement). This implied an unexpected, forced, and sudden transition to the virtual environment, very different from the classic models of teaching and communication.

A. Rocha et al. (Eds.): WorldCIST 2023, LNNS 800, pp. 451–460, 2024.
https://doi.org/10.1007/978-3-031-45645-9_43

With the outbreak of the COVID-19 pandemic, educational organisations, like other organisations, faced the need to intensify and/or diversify the use of digital communication tools and adapt their content to the current crisis context. Basically, COVID-19 brought about an extensive and sudden digital transformation in education (Fidan 2021) with strong implications for the communication tools used and a concern for the types of content. Social media, such as Facebook, YouTube, Instagram, and WhatsApp, are tools that have been increasingly used in the context of educational organisations, both in learning and in the communication strategies of the organisations themselves (Margaritoiuiu 2020; Mesquita et al. 2020).

As Portugal is one of the eight OECD countries with less effective integration of Information and Communication Technologies (OECD 2019), difficulties are added regarding the adequacy of responses to the challenges posed by digital practices. In this way, this article focuses specifically on the description of the transformation that occurred with the change in communication practices in two educational organizations of private education in Portugal, before the pandemic (January–March 2020) and during the pandemic/lockdown (March–June 2020).

2 Literature Review

The COVID-19 pandemic has led to an acceleration in the digitalization of education (Maier et al. 2020) and a change in communication models and practices (Moreira et al. 2020). As communication plays a role in the functioning of educational organizations (Cano 2003; Costa et al. 2000), the literature has focused on the numerous challenges associated with changes in digital communication practices, particularly in a period of crisis, as was the period of confinement caused by COVID-19 (Dias et al. 2020; Serhan 2020). Several studies highlight the importance of seeking to develop new organizational structures and systems that promote effective forms of digital communication (Nobre et al. 2021). As such, they suggest that they should be clear, timely, complete (Brammer and Clark 2020; Nobre et al. 2021; Ruão 2020), informal (Charoensukmongkol and Phungso-onthorn 2020), empathetic, transparent, truthful, with a focus on preventing and reducing alarmism (Knight 2020; Wong et al. 2021) to help individuals, interpret information (Ruão 2021).

Given that it is up to each organization to analyse digital communication strategies and instruments that best fit its reality (Mesquita et al. 2020), it is important to keep in mind that a crisis, such as the COVID-19 pandemic, is "a sudden and unexpected event that threatens to affect, or even interrupt, the life of organizations or other social groups" (Ruão 2021, p. 95). Therefore, it is inevitable that there will be a set of psychological reactions such as panic or fear since they are emotionally charged phenomena (Hall and Wolf 2019; Ruão 2021). Thus, this choice must be guided with a view to contributing to a greater management of uncertainties (Charoensukmongkol and Phungsoonthorn 2020) and to minimizing the impact of different stakeholders. It is important to be aware that the messages conveyed can impact organizational attitudes, such as trust (Guzzo et al. 2021). However, Ruão (2021) points out that "the literature on crisis management seems more attentive to the need to improve the effectiveness of systems, processes and models, than to deepen the human, social and community aspects associated with a critical event" (Ruão 2021, p. 95).

Literature has been focusing on the need to implement reason and emotion in the communication of messages (Ruão 2021). It is about considering "fundamental factors in the understanding of reality, in the construction of a collective meaning for events and in the co-creation of positive experiences in crisis situations" (Ruão 2021, p. 100). Several studies show that the expression of genuine rather than rational emotions have a greater impact on the public in times of crisis and on the reputation of organizations (Claeys et al. 2013; De Waele et al. 2020). Therefore, the use of social networks that allow people to be seen, even from a distance (e.g., Facebook. Youtube), the use of images, icons and the placement of the voice end up having a stronger effect on the public in times of crisis than just verbal statements (Ruão 2021).

This period of crisis, in addition to having given rise to new communication models and practices, contributed to highlighting the importance of communication as a strategy. If organizations, including educational organizations, compete strongly for the attention, admiration, and loyalty of their customers, it is recommended the use of strategic communication that aims at solutions of affirmation and risk management by strengthening persuasive, integrated and instrumental communication (Ruão 2020).

3 Methodology

To answer the starting question "How have educational organizations adapted their communication in times of a pandemic?", we chose to follow a quantitative and qualitative methodology, in the form of a multiple case study (Yin 2005). The object of study covers two private elementary schools of in Portugal, with different types of teaching and which we call school 1 and school 2. The school 1 before the pandemic only used email and face to face and the school 2 already used email, face to face and social media.

Using a quantitative methodological approach, a survey questionnaire was conducted, inspired by the study conducted by Ramello (2020) on the challenges of organisational communication developed in Brazil and the study by Domingues (2017) in Portugal. Composed of 33 closed and 4 open questions it was divided into before the pandemic and during the confinement. Within the scope of this text, only the analysis of open questions was considered such as the reasons of communication between the different actors, which positive aspects and challenges of internal communication were faced, and what could be improved. Before data collection, we carried out a pre-test in a public elementary school. The questionnaire was sent to guardians during the month of May 2021, by email through a link on the Qualtrics platform. Subsequently, the open-ended responses were analysed using a thematic analysis of the responses.

The semi-structured interviews were directed to the administration and the department or the person responsible for communication and IT at both schools. In school 1, it included 1 member of the board (dean), 1 person responsible for information technology and 1 person responsible for communication. In school 2, interviews were carried out with a member of the board, 1 person responsible for information technology and 1 person responsible for communication. The interviews were conducted in person and online, due to the pandemic situation we were going through during the months of January 2021 and May 2021 and lasted between 40 and 60 min.

The interview guide was divided into three parts - before the pandemic, during confinement and in the back-to-school phase. Different versions were prepared and adapted to each interviewee. In the first part, the following questions were posed - main interlocutors, communication channels used by the different actors, their frequency, and reasons for communication and, finally, the evaluation of the effectiveness of communication during this period. In the section during confinement, the previous questions were repeated, adding others such as to understand if there were awareness actions and which ones existed in both schools, which new channels were used, as well as their positive and negative aspects, to know if there were monitoring, training and support, which are the positive factors of internal communication, as well as the challenges during this phase.

After its non-integral transcription, the NVivo tool was used for the organization, coding, and analysis of the information. Thus, a thematic analysis was carried out, understood as a method to analyse qualitative data, in a flexible, useful way, contributing to the representation of data in a rich and detailed way (Gonçalves et al. 2021).

4 Results

Before the COVID-19 pandemic, communication, particularly in school 1, was not a worry or concern, or, in a way, a priority. It is a school that does not have a communication department, unlike school 2. The administration states that "communication is the most precious thing we have, and we do not want to diminish it" (Administration of the school 1), but that the main concern of the school focused on education and not on communication.

> In fact, what needs to be developed is how can we educate children, how can we do better. Our concern is not to develop communication, but education. The first is an instrument of the second. Knowing how to communicate well is almost a consequence. (Administration of School 1)

In the understanding of the person responsible for communication, due to the existing informality, communication ended up not being effective, since "[the] Dean ends up sending some information [and] the person responsible for communication ends up meeting the timetable". In this case, the person responsible for communication "was more concerned with external communication, with no availability to plan and implement internal communication actions", understanding communication on social networks as "a way to reach parents, show the school and enhance engagement" (responsible for communication at school 1).

On the contrary, school 2 had a communication department that carried out annual communication campaigns. These campaigns were aimed at attracting leads, contacts to visit the school and register. This communication was made to the students' guardians themselves, but also to those interested in their enrolment. These were campaigns designed in terms of design, publicized on different social networks and websites, in paid campaigns on google, to appear in searches. In practice, it is a school with a consolidated communication strategy, adapted to different audiences.

> The use of Facebook is more focused on parents and educators, to have access to activities carried out at the school [It aims to communicate] to a more adult audience (…). Instagram is for a younger target, communication is made in a

> lighter and more visual way. Internal newsletters are also sent to parents about the activities carried out (Communication Department of School 2).

Although the confinement prevented face-to-face communication, which was one of the most used means in the pre-pandemic period, the truth is that educational organizations quickly realized that it was "possible to continue teaching even without a physical school" (responsible for school communication 1). If "before the pandemic it was the basics. There were no online platforms" (Information Department at School 2), both schools under study quickly adapted to the new reality.

> The truth is that in the period of confinement other forms emerged. Instagram, social networks, and communication through images are also very good ways of communication. We were not used to communicating through these means, but it was interesting to communicate through videos and photographs, showing what was being done (Administration of school 1).

The board of school 1 recognizes that before the pandemic they did not give much importance and that the bet on "more things online, a more sophisticated website, was not valued. On this occasion they became precious".

> Basically, what happened pre-COVID was that everything was seen in the 'normal' channels of communication. (…) Post-COVID, everything happened, classes, teacher initiatives, communications to parents, communications to students, information about tuition fees, challenges, games, readings, information about the calendar, everything lived in the email, on YouTube, on Zoom, on Instagram, on the blog (…). It started telling stories on Instagram every day. It started to have a blog where the creative and cultural life of students and teachers could exist. It started to have an even more assiduous presence on social networks, with portraits of classes, with initiatives, challenges, games, and proposals; started having parent meetings on Zoom; started to adopt new didactics that adapt to Zoom; started to have even more creative and open teachers; teachers who are even more "technological" (Communication Department at School 1).

In the same way, school 2 mobilized and bet on the intensification of digital communication, on the website, on social networks, Facebook, and Instagram and on sending internal newsletters and videos to parents.

> Intensification of the digital communication process, through the creation of channels, use of platforms and social networks (share-point already existed, sharing platform, activities carried out, content to publish, timeline of all cycles until ninth grade). One of the functions is to communicate with the entire school. (Communication Department of School 2).

This period was of enormous complexity and forced schools to mobilize to bring families and students closer together. From the point of view of the internal functioning of educational organizations, the challenges were constant and similar between both schools, namely in terms of the volume of requests and the need for an effective and clear answer to the questions of parents. Furthermore, the difficulties of "managing to maintain the sense of online community" (Administration of school 2) and of "anticipating unknown scenarios" (Communication Department of School 2) are mentioned.

As the head of communication at school 2 said, "we had many fronts", so it became "difficult to anticipate doubts that could arise, managing the emotional part, not being able to meet expectations". He also adds that he would like to "have had more time for the transmission of internal knowledge, standardization of the message" (Communication Department of School 2).

This situation was similarly felt by school 1. The head of communication at school 1 pointed out difficulties in dealing with the "diversity of questions asked". As an example, she pointed out questions about "Will there be online classes? Will there be face-to-face meetings? Will we pay full tuition? Etc. These are the biggest concerns of parents." However, she said that "everything was managed in a very thoughtful and attentive way, with a timely response to each father and mother". She also added the growing awareness of "potentially controversial or sensitive issues, when to be silent and when to publish" (Communication Department at School 1).

Both schools were concerned about the quality of the message transmitted. For school 2's communication department, one of the main challenges in terms of communication was to convey a positive message, clearly and easily understood.

> To think, a whole morning to write a post, that is, to think very well about what you want to say, the perception that the message will have, what impact on the people who are at home, susceptibilities cannot arise, it is a very extended universe, parents, some of them, would be going through very difficult periods of their lives (…) In the adaptation and in the time it is necessary to think about what we are going to say, what is our objective and we also always try to give an optimistic tone to our communication, the black scenario we faced had already arrived. (Communication Department of School 2).

In practice, there was "greater care in the selection of what we are going to say, the maintenance of the normality of the school and that everything continues to work, albeit in different ways" (…) (Communication Department of School 2). Therefore, "communication had to be much more thought out", in such a way that it implied "translating what is said by the entities and summarizing what is necessary" (Administration Department of School 2). This was operationalized as follows:

> The anticipation scenario was very well thought and worked out with all kinds of questions and doubts, encompassing all points of communication with the school (telephone, website, message that we will tell parents). All doubts were mapped, FAQ, very interactive document, guide for all parents, to clarify all doubts. There was a very important objective behind it, to aggregate and channel in a single physical environment, so that there was no dispersion of information. It was very important to create this document and release it directly to parents. Creation of an SOS anticipation document, which would answer all questions and doubts about the pandemic (Communication Department of School 1)

In relation to school 1 "there were many Zooms, many meetings, even with the parents. In the first phase, almost daily emails were sent to parents, and many phone calls were made. Parents had access to classes. They were much more present in the activities" (Director of School 1). However, he considers that in the confinement phase, the communication strategy went through:

> (...) try not to give too many opinions. The question is, given what is happening, and with the regime we have, how can we continue to educate our students. Often, we try to direct our communication towards what we consider essential, not wasting time on communicating with things that are not relevant. (Administration of School 2)

There were also changes in terms of leadership in terms of communication practices in both schools. School 2 stated that the "direction had to assume an aggregating role, a unique communication for the parents". In this way, it bet on a "communication that aimed to be clear, institutional, not advertising, but affectionate".

In relation to School 1, in addition to a greater appreciation of online communication, the management recognizes greater proximity, that is, "we are all at the same distance. Communication tends not to be hierarchical anymore (…) [and] it came to reveal that it is possible to work in dialogue, work in common". Furthermore, it is mentioned that "although it was a time that affected external and internal community life, it was also a time of great exposure and openness. In a way, there were things that brought it up (Administration of School 1).

> (…) it was good to realize how essential our relationships are, how it is worth inventing ways of seeing ourselves in them. It was a challenging, enriching time of great learning. (Administration of School 1)

Such transformation is also recognized by the person responsible for communication.

> The way we communicate is also changing communication between colleagues and managers – it is less formal, we use technology more, a meeting no longer has to be in person, a workday no longer has to be in the office; brand communication is also closer, more spontaneous and informal (lives…); people are increasingly looking for authenticity and genuine connections. (Communication Department of School 1)

Despite all the challenges pointed out during this period, the parents of both schools recognized, in the overwhelming majority, the clarity of communication and its regularity. In school 1, when asked about the main positive aspects in terms of internal communication, of the 35 responses obtained in an open response, 18 pointed out the clarity of communication as positive aspects and 18 its regularity, while 4 were considered invalid. In school 2, of the 41 valid answers, 31 pointed to regularity, 7 to clarity and 3 others. This information is in line with the communication practices endorsed in the pre-pandemic period, in which we can observe that they were more regular in school 2, compared to school 1.

It is recognized by the guardians of school 1, "the speed at which the school adapted to the context", "the constant communication between the school and the families", the ability of the school to "maintain communication with the parents from the first moment when the students were sent home, advising the parents of the measures so that the classes would continue to take place even if the distance". "It had a fundamental guiding role", the "daily availability of communication" and the "continuous involvement of families during the course of the online classes", are some of the aspects highlighted by the guardians. A guardian adds that the school simply "remained identical to what was already happening, they even became more present and available, closely following the families, despite the distance."

In school 2, communication was able to "keep parents more informed, preventing alarmism. The quality and clarity of communication was visible in terms of the development of the pandemic "the school was concerned with keeping parents updated on the evolution of procedures related to the pandemic", said one respondent (Guardian of the School 2), but also in terms of the development of classes, as seen in: "keep parents informed about the development of the student during distance learning" (Guardian of School 2). He also added the availability of teachers, that is, "teachers available at all times, even at night when my daughter had questions."

This regularity and proximity gave rise to confidence in the quality of the adaptation of the process. This aspect was also highlighted in school 1, as they assumed that communication was based on "honesty" and "clarity" (Guardian of School 1). One respondent assumed that "regular communication, humility in saying that they did not know the answer to some question when they effectively did not know and the affirmation without shame or fear of what moves the school" are considered important aspects that contributed to the "transmission of trust" (Guardian of School 1). In both schools, the new communication practices adopted were understood as an "opportunity to broaden the horizons of communication (…) [to] gain new tools (blog, greater planning of the communication task, etc.), new skills, new friends" (Communication Manager at School 1). In addition, to allow "Gain respect for technologies and for social networks and digital platforms that allowed us all to be connected, without being isolated and keeping our students company" (Communication Department of School 2). It is about "opening horizons [and] producing and using more and better resources" (Communication Department of School 2).

5 Conclusion

This article aimed to describe the transformation of digital communication that took place in two private educational organizations in Portugal, before the pandemic (January–March 2020) and during the pandemic/confinement (March–June 2020). As we observed in the survey, both schools recognize that the pandemic has reinforced and revolutionized the role of digital communication in educational organizations. We noticed that although school 1 in the pre-pandemic period did not value the role of communication in the educational organization, during the confinement they recognized the potential of digital communication to disseminate, communicate and involve all stakeholders.

Like the conclusions of the study by Mesquita et al. (2020), we found that it is not enough to be present on the internet, but audiences are increasingly looking for organizations to listen, engage and respond with a view to create relationships and interactions. This interaction with audiences in a digital environment can influence the attitude towards a service or product and its image and reputation. In addition, the results show the importance of the contents and the quality of the message transmitted. And if, before the pandemic, communication in both schools was essentially focused on a marketing aspect, with a view to attracting enrolments, in the period of confinement, communication focused on a need for a more affectionate, close communication, which ended up being recognized and valued by the guardians of both schools. In this sense, clarity, regularity, and sincerity ended up being valued and highlighted aspects.

Aware that communication is an important tool in a phase of change or crisis, we consider it essential to define strategic plans for crisis communication in educational organisations, based on guidelines for prevention, preparation and crisis management. We recommend the use of social media and other online means with a view to speeding up communication between the different stakeholders and making it more flexible.

References

Brammer, S., Clark, T.: COVID-19 and management education: reflections on challenges, opportunities, and potential futures. Br. J. Manag. **31**, 453–456 (2020)

Cano, J.: La comunicación en las organizaciones escolares. In: González, M. (ed.) Organización y gestión de centros escolares. Dimensiones y processos, pp. 107–127. Pearson Educación SA (2003)

Charoensukmongkol, P., Phungsoonthorn, T.: The interaction effect of crisis communication and social support on the emotional exhaustion of university employees during the COVID-19 crisis. Int. J. Bus. Commun. **59**, 269–286 (2020)

Claeys, A.-S., Cauberghe, V., Leysen, J.: Implications of stealing thunder for the impact of expressing emotions in organizational crisis communication. J. Appl. Commun. Res. **41**(3), 293–308 (2013)

Costa, J.A., Mendes, A.N., Ventura, A.: Liderança e estratégia nas organizações escolares. Universidade de Aveiro (2000)

De Waele, A., Schoofs, L., Claeys, A.S.: The power of empathy: the dual impacts of an emotional voice in organizational crisis communication. J. Appl. Commun. Res. **48**(3), 350–371 (2020)

Dias-Trindade, S., Correia, J.D., Henriques, S.: O Ensino Remoto Emergencial Na Educação Básica Brasileira E Portuguesa: A Perspectiva Dos Docentes. Revista. Tempos Espaços em Educação **13**(32), e-14426 (2020)

Domingues, M.: Desafios da Comunicação Interna numa Creche. Proposta para a definição de um plano de Comunicação Interna numa Creche no Concelho de Lisboa. Master´s thesis, Instituto Universitário de Lisboa (2017)

Fidan, M.: COVID-19 and primary school 1st grade in Turkey: starting primary starting primary school in the pandemic based on teachers' views. J. Prim. Educ. **3**(1), 15–24 (2021)

Gonçalves, S.P., Gonçalves, J., Marques, C.G.: Manual de Investigação Qualitativa. Pactor, Lisboa (2021)

Guzzo, R., Wang, X., Madera, J.M., Abbott, J.: Organizational trust in times of COVID-19: hospitality employees affective responses to managers communication. Int. J. Hosp. Manag. **93**, 102778 (2021)

Hall, K., Wolf, M.: Whose crisis? pandemic flu, 'communication disasters' and the struggle for hegemony. Health **25**(3), 1–17 (2019)

Knight, M.: Pandemic communication: a new challenge for higher education. Bus. Prof. Commun. Q. **83**(2), 131–132 (2020)

Maier, V., Alexa, L., Craciunescu, R.: Online education during the COVID19 pandemic: perceptions and expectations of Romanian students. In: European Conference on e-Learning, pp. 317–324 (2020)

Margaritoiu, A.: Student perceptions of online educational communication in the pandemic. Jus et civitas **1**, 93–100 (2020)

Mesquita, K., Ruão, T., Andrade, J.G.: Transformações da Comunicação Organizacional no contexto digital: novas práticas e desafios nas mídias sociais. In: Dinâmicas Comunicativas e Transformações Sociais. Atas Das VII Jornadas Doutorais Em Comunicação & Estudos Culturais, 281–303 (2020)

Moreira, J.A., Henriques, S., Barros, D.: Transitando de um ensino remoto emergencial para uma educação digital em rede, em tempos de pandemia. Dialogia **34**, 351–364 (2020)

Nobre, A., Mouraz, A., Goulão, M., Henriques, S., Barros, D., Moreira, J.A.: Processos De Comunicação Digital No Sistema Educativo Português Em Tempos De Pandemia. Revista Práxis Educacional **17**(45), 1–19 (2021)

OECD: OECD Skills Outlook 2019 (OECD Skills Outlook). OECD (2019). https://doi.org/10.1787/DF80BC12-EN

Ramello, C.A.: Desafios da Covid-19 para a Comunicação Organizacional. Aberje (2020)

Ruão, T.: A Comunicação enquanto Estratégia. In: Félix, J. (ed.) Comunicação Estratégica e Integrada, pp. 29–39. Rede Integrada Editora (2020)

Ruão, T.: Introduction. How far can we take strategic communication? the sky is the limit. In S. Balonas, T. Ruão, Carrillo, M.-V. (eds.) Strategic Communication in Context. Theoretical Debates and Applied Research, pp. 9–19. CCES-UMinho Editora (2021). https://doi.org/10.21814/uminho.ed.46

Wong, I.A., Ou, J., Wilson, A.: Evolution of hoteliers' organizational crisis communication in the time of mega disruption. Tour. Manag. **84**(6), 104257 (2021)

Yin, R.: Estudo de caso: planejamento e métodos. Artmed, Porto Alegre (2005)

Influencing Factors in Perceived Learning Are Mediated by Satisfaction in Post-Pandemic University Students

Olger Gutierrez-Aguilar[1](✉), Ygnacio Tomaylla-Quispe[2], Lily Montesinos-Valencia[1], and Sandra Chicana-Huanca[2]

[1] Universidad Católica de Santa María, Arequipa, Peru
{ogutierrez,lmontesi}@ucsm.edu.pe
[2] Universidad Nacional de San Agustín, Arequipa, Peru
{itomaylla,schicanah}@unsa.edu.pe

Abstract. This research aims to determine the degree of influence exerted by factors such as planning, the relationship with the students, the teaching methodology, and the evaluation and academic monitoring of the perceived learning mediated by the satisfaction of the students in times of post-pandemic. The methodology used for the study corresponds to a non-experimental investigation; a questionnaire was applied to a convenience sample of 555 university students ($n = 18$; $\alpha = 0.930$ $\omega = 0.933$), using Factor Analysis as a test of validity and reliability. Exploratory and confirmatory, through the Modeling of Structural Equations of Partial Least Squares PLS-SEM. The results have shown a statistically significant effect between the evaluation and academic follow-up, the teacher's methodology, and the perceived learning mediated by the student's satisfaction; Similarly, student satisfaction positively influences perceived learning.

Keywords: Perceived learning · student satisfaction · teacher methodology · evaluation · educational planning · relationship with students

1 Introduction

Post-pandemic educational conditions bring with them a redefinition of their purposes and roles. The idea of meaningful learning as a tangible result of education today revolves around new conceptualizations, such as perceived learning and student satisfaction, in such a way that the educational processes that evaluation implies are a fundamental aspect and include different actions and measures of quality. Some factors can directly or indirectly influence the perception of learning, such as the teaching methodology, the teacher's planning, and the relationship with the students, recognizing different rhythms and learning styles in each student.

Academic evaluation and follow-up are crucial success factors in educational tasks today, considering the learning journey in perceived learning. Additionally, the educational context and social skills learning require interactive and collaborative teaching

A. Rocha et al. (Eds.): WorldCIST 2023, LNNS 800, pp. 461–470, 2024.
https://doi.org/10.1007/978-3-031-45645-9_44

methods in educational environments [1]. Academic follow-up reasonably implied the presence of the teacher and perceived learning, especially with the Internet, in such a way that online surveillance presupposes the willingness of individuals to orient their cognition towards online content and communication and exploit this content constantly [2]. Consequently, the relationship between teacher presence and perceived learning is most likely moderated by other factors [3], such as different levels of student satisfaction in perceived achievement in online discussion forums [4]. On the other hand, social support and career relationships are significantly and positively related to perceived learning [5]. Studies based on virtual learning environments have used the concept of perceived learning [6].

H1 There is a degree of influence between evaluation and academic follow-up on perceived learning.

H2 There is a statistically significant effect between academic evaluation and follow-up and student satisfaction.

Incorporating ICT into educational processes implies the implementation of different methodologies for different results in the learning process [6]. Teaching methods that promote problem-solving are more attractive than traditional methods, so students' preference for the teaching method and understanding must be considered [7]. The generation of expectations and the satisfaction of the students mediated by the interaction with the teaching staff are highly valued aspects in the educational context [8]. Therefore, it is necessary to study the influential factors in knowledge management with methodologies such as M-learning [9], In the same way, the control and enjoyment of positive emotions and their influence on school satisfaction with social networks, such as methodological strategy [10] and affective support for students, are mediated in constructivist environments for learning [11], In addition, the educational task is strengthened in co-teaching with parents moderated with tools such as WhatsApp, especially in the initial and primary academic levels and for students with disabilities, used in the pandemic due to COVID-19 [12].

H3 There is a degree of influence between the teaching methodology and perceived learning.

H4 The teaching methodology positively influences student satisfaction.

Teacher planning in an educational program involves understanding individual differences in information processing, thinking, learning, and studying in students [13], recognizing that all students have the potential to learn effectively if their preferred learning styles are considered and addressed [14] and, based on this diagnostic idea, planning to learn tasks in such a way that the student achieves satisfaction in the learning experiences and there is a perception of learning.

H5 The teacher's planning positively influences perceived learning.

H6 Teacher planning positively influences student satisfaction.

The relationship with students through university tutoring programs is necessary and very important so that the teacher also assumes the role of facilitator and mediator in situations of learning with students, in such a way that promoting collaborative learning strategies among students implies, in the same way as affective support [15], the mediation between affectivity and cognition will result in significant learning and the perception of learning through situations of metacognitive. The effect of ICT and its relationship with academic achievement in the educational context is characterized by positive and negative linear relationships [16], in the same way as the relationship between the acceptance of technology and self-directed learning and the preponderant role of mediation of positive emotions and technological self-efficacy [17].

H7 There is a positive relationship between the relationship with students and perceived learning.

H8 There is a positive relationship between student relationships and student satisfaction.

A topic of interest in educational research and its relationship with technology is the relationship between satisfaction with academic service, perceived learning, and retention [18]; Likewise, the strength of the relationship between social presence and satisfaction is a bivariate relationship that is moderated by course length, discipline area, and the scale used to measure social reality, similarly to online courses [19]. An essential factor in student satisfaction is related to the learning environment and the level of tutors [20]. The role of the teacher or instructor, content management, evaluation, and schedule are predictive variables of student satisfaction in Massive Online Open Courses (MOOC) [21]. Similarly, perceived learning, motivation, resilience, and collaborative learning are critical factors for a learning journey.

H9 There is a statistically positive effect on student satisfaction and perceived learning.

2 Methodology

The studied sample was made up of 555 university students from the professional careers of Industrial Engineering, Veterinary Medicine, and Zootechnics and Advertising, with 39.9% men and 60.5% women, whose ages were between 16 and 25 years old; the total average was 18.46 (SD = 1.779). The participants were selected for convenience. The application of the instrument was carried out in the month of May 2022. The tool applied is an adaptation of the teacher performance evaluation model in COVID-19 [22] and a validation model to establish the relationship between teacher performance and student satisfaction [23], Additionally, statistical tests were carried out, such as exploratory and confirmatory factor analysis, respectively, to guarantee the robustness of the instrument.

3 Results

The instrument's reliability analysis result was an Alpha Cronbach ($\alpha = 0.930$) and a McDonald's Coefficient ($\omega = 0.933$), which means excellent measurement. In addition, the commonalities, which are the sum of the squared factorial weights for each row, were also evaluated. The result is between 0.682 and 0.924, equivalent to saying that the items would explain the model by 68.2%, and in the best cases, they would obtain a 92.4.%. As for the total variance explained, the results obtained from the five components would explain the model by 78.98%. For the exploratory analysis, the following criteria were used: the primary component extraction method, with a fixed number of 5 factors; and for the verification of assumptions, a Kaiser-Meyer-Olkin test (KMO = 0.962), which indicates a reasonable adjustment of the analyzed items with their elements. Bartlett's Sphericity Test had the following results: $\chi^2 = 7665.562$; gl = 136, and p = < 0.000, whose assessment is reasonably significant.

The confirmatory analysis was performed with the SPSS Amos software (v:28) using the Structural Equation Modeling Covariance-Based (CB-SEM) method with the exogenous variables of the model, being: 1 Planning (PLA); 2 Relationship with students (REL); 3 Methodology (MET), 4 Student satisfaction (SAT) and 5 Evaluation and academic follow-up (EVA). The following results were obtained from the model's fit: Consequently, the results show robustness and reliability. CFI = 0.981; Root Mean Square Error of Approximation (RMSEA) = 0.049; the goodness of fit index (GFI) = 0.949; the Tucker-Lewis index (TLI) = 0.976; Chi-square/df = 2.345; cmin = 255,577: p = 0.000; df = 109. See Fig. 1.

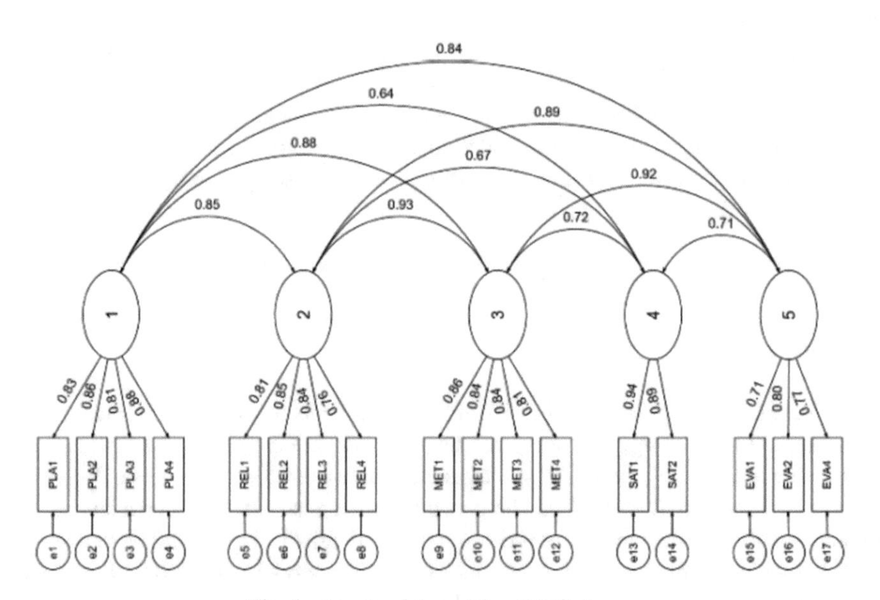

Fig. 1. Ajustes del modelo - SPSS-Amos

In Fig. 2, the measurement model results are obtained by performing the calculation with the PLS algorithm in SmartPLS; in addition, the trajectory coefficients are obtained using 300 iterations in the exploratory analysis, with a criterion of the stop from 10-7. The results of the hypothesis tests indicated that EVA has a positive effect on SAT (β EVA → SAT = 0.195, t = 3.497, p < 0.000), which supports H2; MET has a positive effect on SAT (β MET → SAT = 0.327, t = 3.86, p < 0.000), values that support H4;

PLA has a significant effect on AP (β PLA → AP = 0.117, t = 2.094, p < 0.018), which supports H5; SAT has a significant effect on AP (β SAT → AP = 0.679, t = 13.038, p < 0.000), results that support H9. On the other hand, EVA has a positive effect on AP (β EVA → AP = 0.013, t = 0.309, p < 0.379), results that do not support H1; MET has a positive effect on AP (β MET → AP = 0.064, t = 0.889, p < 0.187), results that do not support H3; PLA has a positive effect on SAT (β PLA → SAT = 0.099, t = 1.53, p < 0.063), results that do not support H6; REL has a positive effect on AP (β REL → AP = 0.031, t = 0.584 p < 0.280), results that do not support H7, and REL has a positive effect on SAT (β REL → SAT = 0.111, t = 1.588 p < 0.056), results that do not support H8.

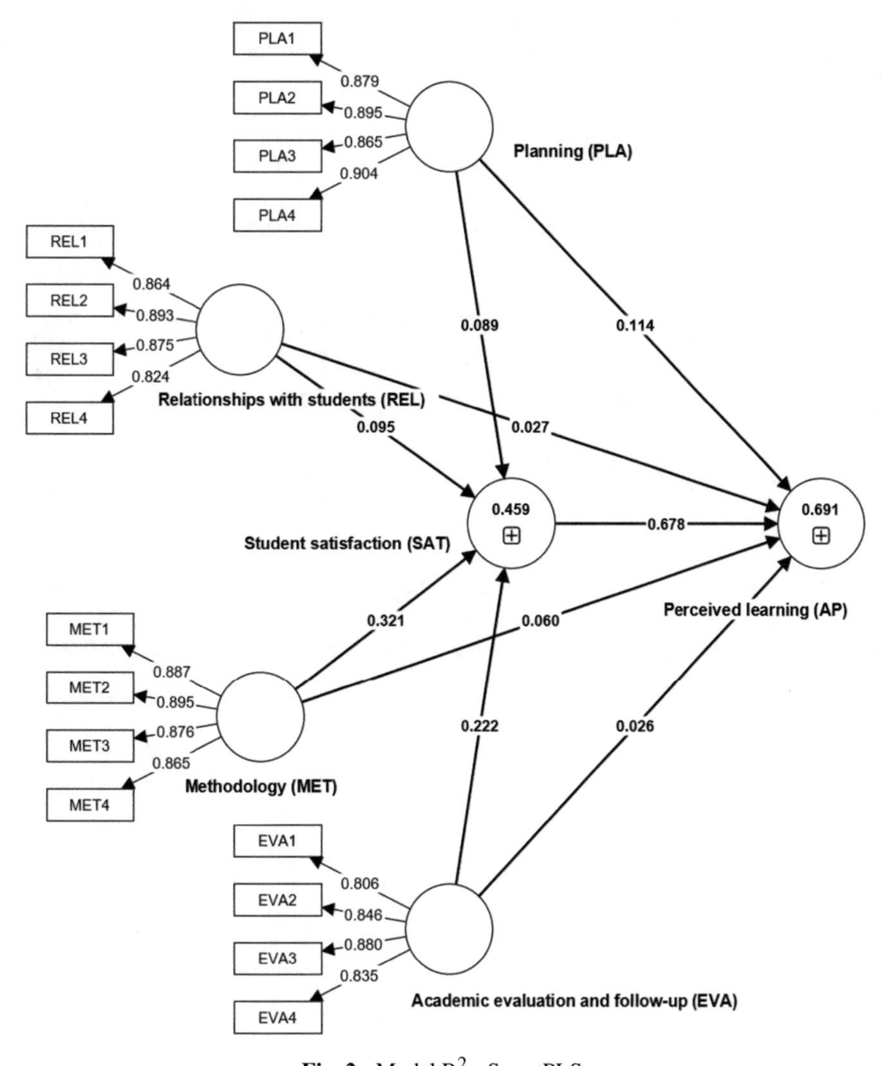

Fig. 2. Model R^2 - SmartPLS.

Table 1 shows the results of the reliability and construct validity tests; for this, we take into account the correlation coefficients, the reliability, and construct validity that is expressed in Cronbach's alpha; the values obtained vary between 0.805 and 0.915, being quite acceptable. The values of the average variance extracted from Average Variance Extracted (AVE) are between 0.719 and 0.921; these results exceed the recommended value of 0.500 [24], In this sense, it is concluded that the convergent validity of the model is acceptable in its model components. The application of this is essential for the analysis of composite reliability [25, 26], Values that should exceed 0.6 are recommended as acceptance criteria, thus demonstrating reasonable levels of internal consistency reliability for each of the variables. The values obtained vary between 0.885 and 0.959. The coefficient (rho_A) is used to verify the reliability of the values obtained in the construction and design of the PLS [27]. The results in the (rho_A) are recommended as acceptance criteria, with values greater than 0.7. The results vary between 0.810 and 0.916.

Table 1. Reliability and construct validity

	Cronbach's Alpha	rho_A	Composite reliability	(AVE)
Academic evaluation and follow-up (EVA)	0.805	0.810	0.885	0.719
Methodology (MET)	0.904	0.904	0.933	0.776
Planning (PLA)	0.908	0.908	0.936	0.784
Relationships with students (REL)	0.887	0.893	0.922	0.747
Student satisfaction (SAT)	0.915	0.916	0.959	0.921

Table 2 shows the results of the bootstrapping test through a run of 10,000 resamples with substitution of the original sample [28]. Considering the level of significance for the P Value ($p < 0.05$), the hypotheses are admitted: H2; H4; H5, and H9. The following hypotheses are rejected: H1; H3; H6; H7; H8.

Table 2. Hypothesis test – bootstrapping

Hypothesis	β	Sample mean (M)	Standard deviation (STDEV)	t-Statistics (|O/STDEV|)	P Value
H1 Academic evaluation and follow-up (EVA)-> Perceived learning (AP)	0.013	0.016	0.043	0.309	0.379
H2 Academic evaluation and follow-up (EVA)-> Student satisfaction (SAT)	0.195	0.196	0.056	3.497	0.000
H3 Methodology (MET)-> Perceived learning (AP)	0.064	0.061	0.072	0.889	0.187
H4 Methodology (MET)-> Student satisfaction (SAT)	0.327	0.322	0.085	3.86	0.000
H5 Planning (PLA)-> Perceived learning (AP)	0.117	0.116	0.056	2.094	0.018
H6 Planning (PLA)-> Student satisfaction (SAT)	0.099	0.101	0.065	1.53	0.063
H7 Relationships with students (REL)-> Perceived learning (AP)	0.031	0.03	0.053	0.584	0.280
H8 Relationships with students (REL)-> Student satisfaction (SAT)	0.111	0.114	0.07	1.588	0.056
H9 Student satisfaction (SAT)-> Perceived learning (AP)	0.679	0.68	0.052	13.038	0.000

4 Conclusions

After a pandemic, the roles and goals of education change. This means that planning, the relationship between the teacher and the students, the teacher's method, evaluation, and academic follow-up affect how the student learns. In this case, satisfaction with educational services would be a mediating variable. The research has shown a statistically significant link between evaluation and academic follow-up, the teacher's method, and the student's perception of learning, which is mediated by the student's satisfaction. Learning experiences like the use of discussion forums and the interaction with teachers through tutoring in virtual learning environments, as well as the mediation of affective situations that make people feel good, play a more critical role than evaluation and academic follow-up. On the other hand, evaluation and academic follow-up are essential parts of the educational task.

In the same way, the teacher's method and relationship with the students would not affect how much they learned. In the same way, planning and getting along with students would not be good predictors of student satisfaction. As future work, it is important to rethink educational conditions after the pandemic, especially in the teaching tasks related to academic evaluation and monitoring, so that the results can be used to make timely decisions, reorient teaching methods, and set up a planning process that focuses on learning achievements and a timely and effective relationship with students.

References

1. Virtanen, A., Tynjälä, P.: Pedagogical practices predicting perceived learning of social skills among university students. Int. J. Educ. Res. **111**, 101895 (2022). https://doi.org/10.1016/j.ijer.2021.101895
2. Reinecke, L., et al.: Permanently online and permanently connected: development and validation of the Online Vigilance Scale. PLoS ONE **13**(10), e0205384 (2018). https://doi.org/10.1371/journal.pone.0205384
3. Caskurlu, S., Maeda, Y., Richardson, J.C., Lv, J.: A meta-analysis addressing the relationship between teaching presence and students' satisfaction and learning. Comput. Educ. **157**, 103966 (2020). https://doi.org/10.1016/j.compedu.2020.103966
4. Choi, H.M., Tsang, E.Y.: Students' satisfaction and perceived attainment in the use of an online discussion forum: a follow-up study in the OUHK. In: Studies and Practices for Advancement in Open and Distance Education: Proceedings of the 28 th Asian Association of Open Universities Conference, pp. 265–279 (2015)
5. Thomas, L.J., Parsons, M., Whitcombe, D.: Assessment in Smart Learning Environments: Psychological factors affecting perceived learning. Comput. Hum. Behav. **95**, 197–207 (2019). https://doi.org/10.1016/j.chb.2018.11.037
6. Martínez-Borreguero, G., Naranjo-Correa, F.L., Cañada Cañada, F., González Gómez, D., Sánchez Martín, J.: The influence of teaching methodologies in the assimilation of density concept in primary teacher trainees. Heliyon **4**(11), e00963 (2018). https://doi.org/10.1016/j.heliyon.2018.e00963
7. Singh, R., Gupta, N., Singh, G.: Learning style and teaching methodology preferences of dental students. J. Anat. Soc. India **65**(2), 152–155 (2016). https://doi.org/10.1016/j.jasi.2017.02.009
8. Carrero-Planells, A., Pol-Castañeda, S., Alamillos-Guardiola, M.C., Prieto-Alomar, A., Tomás-Sánchez, M., Moreno-Mulet, C.: Students and teachers' satisfaction and perspectives on high-fidelity simulation for learning fundamental nursing procedures: a mixed-method study. Nurse Educ. Today **104**, 104981 (2021). https://doi.org/10.1016/j.nedt.2021.104981
9. Chicana-Huanca, S., Gutierrez-Aguilar, O., Ticona-Apaza, F., Calliñaupa-Quispe, G., Chicana-Huanca, B.: Influential factors in knowledge management in the acceptance of M-learning in university students. In: Iberian Conference on Information Systems and Technologies, CISTI 2022

10. Gutierrez-Aguilar, O., Escobedo-Maita, P., Calliñaupa-Quispe, G., Vargas-Gonzales, J.C., Torres-Huillca, A.: The use of social networks, usefulness and ease of use, enjoyment through positive emotions and their influence on school satisfaction mediated by school achievement. In: Iberian Conference on Information Systems and Technologies, CISTI 2022
11. Gutiérrez-Aguilar, O., Duche-Pérez, A., Turpo-Gebera, O.: Affective support mediated by an on-line constructivist environment in times of Covid-19. In: Smart Innovation, Systems and Technologies 2022, pp. 458–468 (2022)
12. Gutierrez-Aguilar, O., Rodriguez-Rios, M., Patino-Abrego, E., Cateriano-Chavez, T.: Co-teaching and the use of WhatsApp as a mediation tool between parents with children with disabilities and their teachers. In: Proceedings - 2021 4th International Conference on Inclusive Technology and Education, CONTIE 2021, pp. 47–52 (2021)
13. Deng, R., Benckendorff, P., Gao, Y.: Limited usefulness of learning style instruments in advancing teaching and learning. Int. J. Manage. Educ. **20**(3), 100686 (2022). https://doi.org/10.1016/j.ijme.2022.100686
14. Garber, L.L., Hyatt, E.M., Boya, Ü.Ö.: Gender differences in learning preferences among participants of serious business games. Int. J. Manage. Educ. **15**(2), 11–29 (2017). https://doi.org/10.1016/j.ijme.2017.02.001
15. Aguilar, O.G., Martinez Delgado, M., Quispe, F.P.: Collaborative learning and its relationship with the classmates affective support with the COLLES questionnaire application. In: Proceedings of the 15th Latin American Conference on Learning Technologies, LACLO 2020 (2020)
16. Zhu, J., Li, S.C.: The non-linear relationships between ICT use and academic achievement of secondary students in Hong Kong. Comput. Educ. **187** (2022). https://doi.org/10.1016/j.compedu.2022.104546
17. An, F., Xi, L., Yu, J., Zhang, M.: Relationship between technology acceptance and self-directed learning: mediation role of positive emotions and technological self-efficacy. Sustainability (Switzerland) **14**(16) (2022). https://doi.org/10.3390/su141610390
18. Walker, S., Rossi, D., Anastasi, J., Gray-Ganter, G., Tennent, R.: Indicators of undergraduate nursing students' satisfaction with their learning journey: an integrative review. Nurse Educ. Today **43**, 40–48 (2016). https://doi.org/10.1016/j.nedt.2016.04.011
19. Richardson, J.C., Maeda, Y., Lv, J., Caskurlu, S.: Social presence in relation to students' satisfaction and learning in the online environment: a meta-analysis. Comput. Hum. Behav. **71**, 402–417 (2017). https://doi.org/10.1016/j.chb.2017.02.001
20. Cervera-Gasch, A., González-Chordá, V.M., Ortiz-Mallasen, V., Andreu-Pejo, L., Mena-Tudela, D., Valero-Chilleron, M.J.: Student satisfaction level, clinical learning environment, and tutor participation in primary care clinical placements: an observational study. Nurse Educ. Today **108**, 105156 (2022). https://doi.org/10.1016/j.nedt.2021.105156
21. Hew, K.F., Hu, X., Qiao, C., Tang, Y.: What predicts student satisfaction with MOOCs: a gradient boosting trees supervised machine learning and sentiment analysis approach. Comput. Educ. **145**, 103724 (2020). https://doi.org/10.1016/j.compedu.2019.103724
22. Aguilar, O.G., Duche Perez, A.B., Aguilar, A.G.: Teacher performance evaluation model in Covid-19 times. In: Proceedings of the 15th Latin American Conference on Learning Technologies, LACLO 2020 (2020)
23. Aguilar, O.G., Gutierrez Aguilar, A.: A model validation to establish the relationship between teacher performance and student satisfaction. In: Proceedings - 2020 3rd International Conference of Inclusive Technology and Education, CONTIE 2020, pp. 202–207 (2020)
24. Hair Jr, J.F., Hult, G.T.M., Ringle, C., Sarstedt, M.: A Primer on Partial Least Squares Structural Equation Modeling (PLS-SEM). Sage Publications, Thousand Oaks (2016)
25. Bagozzi, R.P., Yi, Y.: On the evaluation of structural equation models. J. Acad. Mark. Sci. **16**(1), 74–94 (1988)

26. Hair, J.F., Sarstedt, M., Pieper, T.M., Ringle, C.M.: The use of partial least squares structural equation modeling in strategic management research: a review of past practices and recommendations for future applications. Long Range Plan. **45**(5–6), 320–340 (2012)
27. Dijkstra, T.K., Henseler, J.: Consistent partial least squares path modeling. MIS Q. **39**(2), 297–316 (2015)
28. Henseler, J.: Partial least squares path modeling. In: Leeflang, P., Wieringa, J., Bijmolt, T., Pauwels, K. (eds.) Advanced Methods for Modeling Markets, pp. 361–381. Springer, Cham (2017). https://doi.org/10.1007/978-3-319-53469-5_12

Uncovering Dark Patterns - Learning Through Serious Gameplay About the Dangers of Sharing Data

Ingvar Tjostheim[1](✉), Chris Wales[1], and John A. Waterworth[2]

[1] Hauge School of Management, NLA, Oslo, Norway
ingvar.tjostheim@nla.no
[2] Umeå University, Umeå, Sweden

Abstract. Dark patterns refer to tricks used in websites and apps to make you do things that you do not intend to do. This paper presents the board-game *Dark Pattern*, in which players install apps, draw dark patterns cards, and make choices about the sharing of personal data. To win the game, a player must share as little data as possible and play cards that punish other players. Two groups, the first with 56 students and the second with 45 students, played the game and then answered a survey with questions controlling their knowledge about the dark patterns types featured in the game. In addition, a further 50 students answered the same survey without playing the game. In this paper we present key findings about the dark patterns knowledge generated by playing the game. Then we present an exploratory analysis using Partial Least Square – Structural Equation modelling (PLS-SEM). We analysed whether dark patterns knowledge and risk perception, the likelihood of negative incidents due to data sharing, could predict the players behavioural intention to take proactive privacy steps. The two PLS-SEM models have a variance explained (R^2) of 0.34 and 0.35 indicating that approximately 35% of the variance could be accounted for by the two variables included in the model. Taken together, the analyses indicated that playing the Dark Pattern game had a positive effect on behavioural intention to proactive privacy steps as a result of by playing the game.

Keywords: Serious games · learning · user-test · exploratory study · dark patterns · sharing of personal data · partial least square modelling

1 Introduction

In the digital economy you pay with your data, for instance by filling in information about yourself and letting service providers or third parties use that data. There are laws and regulations that should secure citizens greater control of their data. Although most citizens have smartphones full of apps and subscriptions to a number of digital services, privacy and data-protection are not important topics [1, 2].

There are several ways to learn about the digital economy, particularly why and how data are used by service providers and their partners. In this paper we present a board-game that builds upon data-sharing techniques and tricks companies use to gain consent from users [3]. The purpose of the game is to learn about why data are collected, which deceptive techniques are used to gain consent from users and inform

A. Rocha et al. (Eds.): WorldCIST 2023, LNNS 800, pp. 471–480, 2024.
https://doi.org/10.1007/978-3-031-45645-9_45

about potential negative consequences for users in sharing of data. The remainder of the paper is organized as follows: Sect. 2 outlines a description of the game, followed by Sects. 3 where we present the key findings from the data-collection, the research questions and data analysis. In the final section, we conclude with a discussion of the results and their implications for future research on serious games.

2 Description of the Game

Hartzog [4] defines design as the "processes that create consumer technologies and the results of their creative processes instantiated in hardware and software." A dark pattern is a term used to explain how designers use their knowledge of human behaviour and psychology, along with the desires of end users, to implement deceptive functionality that is not in the user's best interest [3]. Often, it is the recurrent configuration of elements in digital interfaces that leads the user to make choices against their best interests and towards those of the designer. The design induces false beliefs either through affirmative misstatements, misleading statements, or omissions. Mathur et al. [5] evaluated 1983 websites with dark patterns. They found that many dark patterns exploited cognitive biases, such as the default and framing effects. The website www.deceptive design (darkpatterns.org) lists 12 common types of dark patterns. Some are very conventional and easy to identify while others are less familiar and more subtly deceptive. For further information about dark patterns, see Luguri & Strahilevitz [6] who present a table of existing dark patterns taxonomies.

The Dark Pattern game was developed in collaboration by the company Serious Games Interactive and the Norwegian Computing Center, a non-profit private foundation. [18, 19].

It was inspired by the master thesis written by K. M. Nyvoll, Serious Interactive Board Games: Increasing Awareness of Dark Patterns in Teenagers [7]. To the best of our knowledge, we are not aware of other dark patterns games aimed specifically at this age group. More details about the game are presented in Tjostheim et al. 2022 [18] In the next section we describe how the game is played.

Dark Pattern is a board-game for 3–5 players. The game is targeted at students aged 16–18, but any adult with a smartphone able to download apps can play the and discover how to make good choices while playing the game. The game has a duration of 20–30 min per round. The players, 3–5 individuals, have to choose one of the following roles: the Shopper, the Gamer, the Influencer, the Lover and the Healthy. The format is a board-game with cards, but the focus of play is upon apps and installing apps. The first information players receive is: *"You just got a brand-new phone. You all want your type of apps on the phone (dating, SoMe, games, health, shopping) without giving away too much data about yourself*." The players choose apps according to their role; the Shopper chooses and installs shopper apps on the game-board, the Gamer installs gaming apps on the game-board, and so on. For further information about the game, we refer to Tjostheim et al. 2022 [18, 19].

Figure 1 shows the role the Influencer, two Dark Patterns card, the instruction on how to calculate points and the 4 types of data-cubes. Figure 2 shows two apps and the text on two Dark Patterns card. The names can be associated with genuine apps, but no real app names are used. To win the game a player must install several apps while attempting to have as few data-cubes as possible, either by discarding or avoiding them.

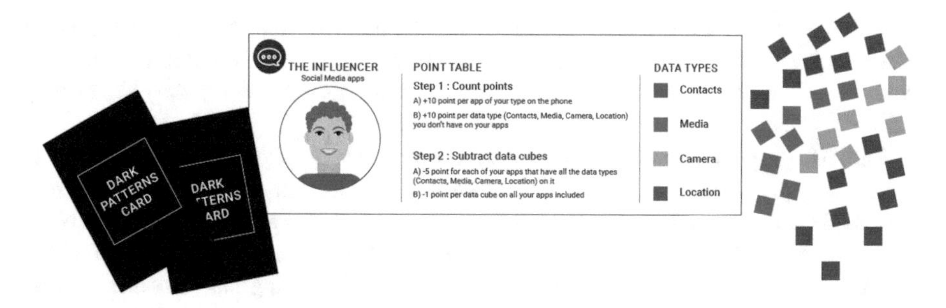

Fig. 1. A role in the game and the 4 data-types

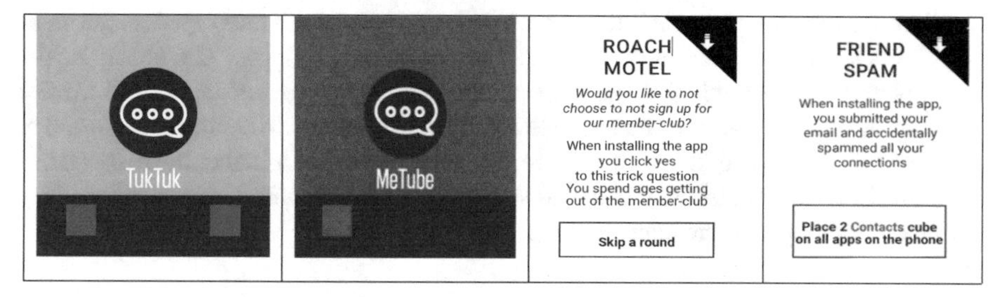

Fig.2. The apps *TukTuk* and *MeTube,* and two Dark Pattern cards

By reading the Dark Pattern cards and making choices about apps, the players learn about what can happen in the digital economy, along with the consequences of data-sharing. In Norway, learning about the digital economy and developing digital judgment skills is part of the national curriculum for students at upper secondary schools.

3 Data Collected, Research Questions and Data-Analysis

This research project was designed to understand how older teenagers respond to issues of personal data security, through playing the Dark Pattern game, which involved them interacting with a smaller group of their peers. A questionnaire was designed to understand their experiences of playing and any knowledge gained. We formulated the following research questions targeted at students playing the game.

RQ1 To what extent did the game convey the meaning of the types of dark patterns presented in the game to the players?

RQ2 To what extent is length of playing time associated with intention to protect personal data?

To test the game and collect data that we could analyse, we contacted five upper secondary schools: one in Copenhagen, Denmark and four in Oslo, Norway. Students at this educational level are in their 11th - 13th years of schooling. Class teachers divided the students into groups of 3, 4 or 5, some groups with mixed genders and some single gender. After a short introduction to the game and having watched a 90-s video published on Vimeo, the students took part in a rehearsal-session that lasted approximately 15 min.

After the rehearsal, the students started to play the game. After the game had finished the students calculated the score for each player and, subsequently, based on the game score, the winner of the game. When the students had finished playing, they were asked to fill in a questionnaire on their own mobile phone.

The 101 players comprise two groups. The first group with 56 students took part in a 60-min play-session. The second group with 45 students took part in a 75-min session. Based on observing the way players asked questions about the game in round 1 we decided to increase the time from 60 min to 75 min for round 2. Subsequently, the players in the second round had time for a rehearsal, playing the game for the first time and time for a second game. In round 2 not all finished their second game before answering the survey, but they had more time to learn the rules and read the game-cards before answering the survey compared to the group in the first round.

The questionnaire presents 7 actual dark patterns types plus 3 fake types. In the game the players get information of the 7 types, but not the 3 fake types. The questionnaire is designed to find out whether the respondents have a basic understanding of these 7 types, or can explain what they mean, followed by questions about intention to reinstall or replace apps. In the second half of the questionnaire, we asked whether the respondents felt there to be any detriment if negative consequences occurred, and whether they expected these negative incidents to happen.

Table 1. Questions about dark patterns – Non-players (N = 50)

Dark patterns are activities to trick internet users into doing things they might not otherwise do. How much do you know about the following types of dark patterns?	Don't know what it means vs I have a basic understanding, or I can explain the expression
Uninstall Shaming (mean: 1.40)	86%–14%
Malware Message (mean: 1.66)	78%–22%
Roach Motel (mean: 1.14)	98%–2%
Disguised Ads (mean: 2.46)	52%–**48%**
Trick Questions (mean: 2.86)	30%–**70%**
Friend Spam (mean: 2.30)	56%–**44%**
Privacy Zuckering (mean: 1.62)	78%–22%
Baking Story* (mean: 1.30)	92%–8%
Intuitive Jobs* (mean: 1.44)	90%–10%
Data Scaling* (mean: 1.60), * fake types	96%–4%

Table 1 presents three types of dark patterns in bold: disguised ads, trick questions and friend spam. These are the only three where a relatively high percentage of respondents (44%, 48% and 70%) report that they understand the expression. We included three fake types. We hypothesized that knowledge of the 7 real types would be higher than the fake types. In a survey some respondents might guess that they have seen an expression, a certain dark pattern type. It is therefore an expected result that a low percentage answered that they know or know about a fake type.

Table 2. Questions about dark patterns – Players round 1 (N = 56)

Dark patterns are activities to trick internet users into doing things they might not otherwise do. How much do you know about the following types of dark patterns?	Don't know what it means vs I have a basic understanding, or I can explain the expression
Uninstall Shaming (mean: 2.09)	61%–39%
Malware Message (mean: 2.11)	63%–37%
Roach Motel (mean: 1.43)	86%–14%
Disguised Ads (mean: 2.57)	43%–**57%**
Trick Questions (mean: 2.95)	29%–**71%**
Friend Spam (mean: 2.43)	54%–46%
Privacy Zuckering (mean: 1.82)	77%–33%
Baking Story* (mean: 1.46)	88%–12%
Intuitive Jobs* (mean: 1.70)	82%–18%
Data Scaling* (mean: 2.16), (* fake types)	64%–36%

Table 3. Questions about dark patterns – Players round 2 (N = 46)

Dark patterns are activities to trick internet users into doing things they might not otherwise do. How much do you know about the following types of dark patterns?	Don't know what it means vs I have a basic understanding, or I can explain the expression
Uninstall Shaming (mean: 2.31)	53%–47%
Malware Message (mean: 2.31)	57%–43%
Roach Motel (mean: 1.38)	93%–7%
Disguised Ads (mean: 3.22)	11%–**89%**
Trick Questions (mean: 2.98)	29%–**71%**
Friend Spam (mean: 2.56)	42%–**58%**
Privacy Zuckering (mean: 2.49)	42%–**58%**
Baking Story* (mean: 1.56)	84%–16%
Intuitive Jobs* (mean: 1.62)	84%–16%
Data Scaling* (mean: 1.71), (* fake type)	80%–20%

Table 2 and 3 present some of the types of dark patterns in bold. For the first group these are disguised ads and trick questions. In the second group these are disguised ads, trick questions, friend spam and privacy zuckering. These are the types which a high percentage of respondents' report that they understand the expression, at least 50%. We included three fake types. We hypothesized that knowledge of the 7 real types would be higher than the fake types, but in a survey some respondents might guess that they have seen an expression, a certain dark patterns type. It is additionally important to note that not all cards are used when playing the game. It is therefore unlikely that 90% or more will report that they know an expression. However, it is also possible that a type named in the game will receive a low percentage. For example, Roach Motel has a very low percentage, where only 14% and 7% answered that they understood the term. It is unlikely that the card has not been used in the game, but it seems in any case that the meaning of the term was not communicated to the players.

For the first research question we conclude that the game conveyed the meaning of the dark patterns types with one exception, roach motel. For roach motel, the knowledge is on the level with the fake types. For the second research question we developed a set of questions about harm due to data-sharing and to what extent expect respondents felt that an incident related to how data were used or mis-used would ensue. What happens with our data, how they are used in marketing, in algorithms, and in building profiles of users are often not communicated. There are many reasons for this. Some companies see this information as a trade secret; information that competitors should not have access to.

We were interested in whether the players plan to take proactive steps to protect their data and how likely it is that the expected that misuse will happen. One might problematise how it is possible to answer the question, how likely is it that something will happen? This is a question with many possible answers, but it in the context of dark patterns, it might indicate concerns or awareness about data-sharing, rather than actual knowledge of what companies do. We used Likert scales for the answers, from never (1) to always (5), from no harm (1) to extreme harm (6), and from never (1) to always (7). Table 4 shows the plan to take proactive steps, the behavioural intentions variable for the two groups. The table shows that majority did not plan to take proactive steps.

We have analysed the data by using Partial Least Square – Structural Equation modelling (PLS-SEM) We refer to Tjostheim et al. 2022 [18] for further information about (PLS-SEM) [10–13] and [15, 16].

In the PLS-SEM models we used a variable named risk perception in addition to dark patterns knowledge. Risk perception was measured in two dimensions: probability, and harm. We asked for the dark patterns types, to what extent do you feel harm if you experience the dark patterns type, and a new set of questions: "How likely do you think it is that you will experience the following", with a list of the dark patterns types. Figures 4 and 5 present the results, where behavioural intention to protect personal data is the dependent variable.

the second group.

Figures 4 and 5 show that the variance explained (R^2) is 0.35 and 0.34. Chin [12] describes a R^2 of 0.34 or higher as medium explanatory power. Of the two constructs on the dependent side, dark patterns knowledge is the strongest predictor in both models. In the first model, risk perception does not contribute to the variance explained, whereas

Table 4. Proactive steps - behavioural intentions

How often do you intend to perform the following actions in the next 30 days?		
	First group (N = 56)	Second group (N = 45)
	Never or rarely vs sometimes or always	Never or rarely vs sometimes or always
Deleting and reinstalling an app to get rid of accumulated data the app had access to	75%–25%	89%–11%
Getting a VPN connection which encrypts your data traffic and establishes a more secure internet connection	75%–25%	87%–13%
Replacing apps for similar ones that require access to fewer data to function	86%–14%	93%–7%
Revising your app's permission settings so that the app is unable to access some data	66%–34%	73%–27%

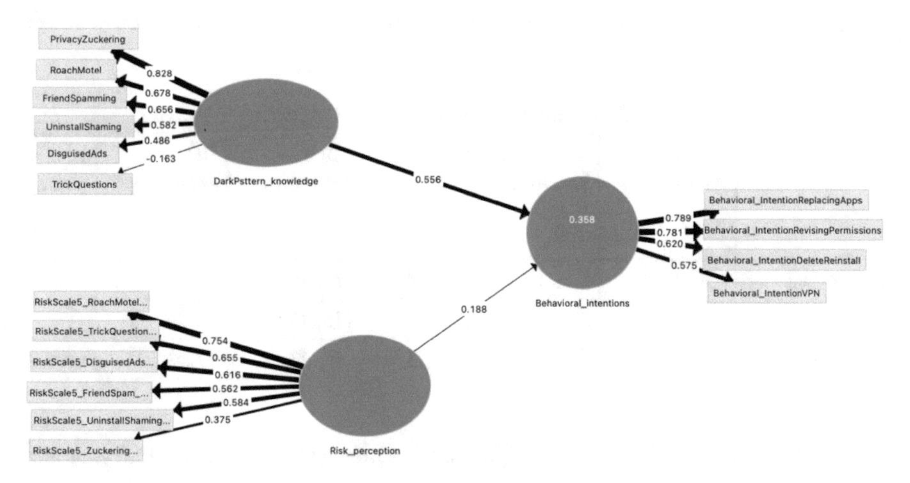

Fig. 4. A PLS-SEM model with behavioural intentions as the dependent variable - the first group

in the second model, it does. The path coefficient for risk perception in the first model is only 0.188, but in the second model rises to 0.358.

According to Tables 5 and 6, not all quality criteria for a sound PLS-models are met [15]. In Fig. 5 and 6, several of the lines are thin. This means that not all items are loading on the construct that they are connected to. In this exploratory research we have not delated any items. This is the main reason for not meeting all quality criterions for a sound PLS-model. At this stage of the research, we did not want delete items, and have included all seven types.

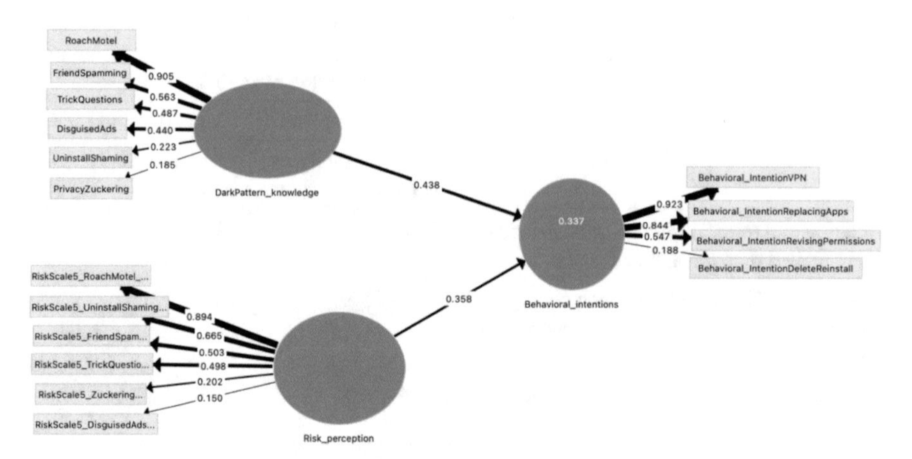

Fig. 5. A PLS-SEM mode with behavioural intentions as the dependent variable - the second group

Table 5. The construct reliability and validity, and discriminant validity of the PLS path model – the first group

	Construct Reliability and Validity			Discriminant Validity		
	Cronbach's Alpha	Composite Reliability	Average Var. Extracted (AVE)	Beh intentions	DP knowledge	Risk percep
Beh. Intent	0.638	0.788	0.487	**0.698**		
DPknowledge	0.657	0.711	0.363	0.438	**0.668**	
Risk percep	0.655	0.767	0.362	0.223	0.063	**0.602**

Table 6. The construct reliability and validity, and discriminant validity of the PLS path model – the second group

	Construct Reliability and Validity			Discriminant Validity		
	Cronbach's Alpha	Composite Reliability	Average Var. Extracted (AVE)	Beh intentions	DP knowledge	Risk percep
Beh. Intent	0.586	0.749	0.475	**0.698**		
DPknowledge	0.720	0.644	0.275	0.457	**0.525**	
Risk percep	0.703	0.669	0.301	0.381	0.052	**0.549**

4 Discussion and Concluding Remarks

Bellotti et al. [18] write, "*Serious games are designed to have an impact on the target audience, which is beyond the pure entertainment aspect.*" One of the application areas for serious games is education. According to Zhonggen [17] serious games are reported to be effective in education, but some studies arrive at negative conclusions. We used survey questions to measure the impact, if any, on the players of the Dark Pattern game. We conclude from the responses that the game had an impact on the players' understanding and knowledge about dark patterns, especially for the group that took part in the 75-min session. The first research question (RQ1) was, "*To what extent did the game convey the meaning of the types of dark patterns presented in the game to the players?*" where approximately 50% the players or more reported that they understand or can explain what the term means – the exception being the roach motel type - see Table 2. We conclude that RQ1 was supported.

The conclusion for the second research question is based on the result from PLS-SEM analysis. The analysis is exploratory in nature; we did not use constructs developed and validated by other researchers, but we developed questions specifically targeted to the 7 types of dark patterns in the game. The second research question (RQ2) was, *To what extent are the length of the playing associated with intention to protect personal data?* The PLS-SEM analysis shows a medium association between the two independent variables and intentions to protect personal data. Because risk perception plays a role as a significant predictor next to dark patterns knowledge the second group, we conclude that RQ2 was supported.

The findings presented in the Table 2 and partly in 3 indicate that only playing the game in one session is not enough if the purpose of the game is to learn the meaning of all the dark patterns types and evoke a reflection on risk associated with these tricks. It is likely that a player intending to win the game would need to practise more than 2 or 3 times. Then the player would be able to make choices and think about which cards to play, and when. For the future, we think that a better approach to understand the question of learning would be to use the score (the result from the game) together with the answers to the survey-questions, after a player has played the game at least three times.

References

1. Acquisti, A., Brandimarte, L., Loewenstein, G.: Privacy and human behavior in the age of information. Science **347**(6221), 509–514 (2015)
2. Acquisti, A., et al.: Nudges for privacy and security: understanding and assisting users' choices online. ACM Comput. Surv. **50**(3), 44 (2017)
3. Brignull, H., Miquel, M., Rosenberg, J., Offer, J.: Dark Patterns - User Interfaces Designed to Trick People (2015). http://darkpatterns.org/
4. Hartzog, W.: Privacy's Blueprint: The Battle to Control the Design of New Technologies, p. 2018. Harvard University Press, Cambridge, MA (2018)
5. Mathur, A., et al.: Dark patterns at scale: findings from a Crawl of 11K shopping websites. Proc. ACM Hum.-Comput. Interact. **3**(CSCW), 1–32 (2019)
6. Luguri, J., Strahilevitz, L.: Shining a Light on Dark Patterns. U of Chicago, Public Law Working Paper 719 (2019)

7. Nyvoll, K.M.: Serious Interactive Board Games: Increasing Awareness of Dark Patterns in Teenagers, Master's thesis, Department of Computer Science at the Norwegian University of Science and Technology (NTNU) (2020)
8. Sarstedt, M., Ringle, C.M., Cheah, J.-H., Ting, H., Moisescu, O.I. Radomir, L.: Structural model robustness checks in PLS-SEM. Tour. Econo. **26**(4), 1–24 (2019). https://doi.org/10.1177/135481661882392
9. Henseler, J., Ringle, C.M., Sinkovics, R.R.: The use of partial least squares path modeling in international marketing. Adv. Int. Mark. **20**, 277–320 (2009)
10. Barclay, D.C., Higgins, C., Thompson, R.: The partial least squares approach to causal modeling: personal computer adoption and use as an illustration. Technol. Stud. **2**(2), 285–308 (1995)
11. Chin, W.W., Marcolin, B.L., Newsted, P.R.: A partial least squares latent variables modeling approach for measuring interaction effects: results from a Monte Carlo Simulation study and an electronic-mail emotion/adoption study. Inf. Syst. Res. **14**(2), 189–217 (2003)
12. Bagozzi, R.P., Yi, Y.: On the evaluation of structural equation models. J. Acad. Mark. Sci. **16**(1), 74–94 (1988)
13. Gefen, D., Straub, D.: A practical guide to factorial validity using PLS-Graph: tutorial and annotated example. Commun. AIS **16**(5), 91–109 (2005)
14. Fornell, C., Larcker, D.F.: Evaluating structural equation models with unobservable variables and measurement error. J. Mark. Res. **18**(1), 39–50 (1981)
15. Chin, W.W.: The partial least squares approach to structural equation modeling. In: Marcoulides, G.A. (ed.) Modern methods for business research, pp. 295–358. Lawrence Erlbaum Associates, Mahwah NJ (1998)
16. Bellotti, F., Kapralos, B., Lee, K., Moreno-Ger, P., Berta, R.: Assessment in and of serious games: an overview. Adv. Hum.-Comput. Interact. **2013**(1), 136864 (2013)
17. Zhonggen, Y.: A Meta-analysis of use of serious games in education over a decade. Int. J. Comput. Games Technol. **5**(1), 1–8 (2019)
18. Tjostheim, I., Ayres Pereira, V., Wales, C., Egenfeldt-Nielsen, S., Manna, A.: Dark pattern: a serious game for learning about the dangers of sharing data. In: Proceedings of the European Conference on Games Based Learning (ECGBL 2022), vol. 16, pp. 774–783 (2022). https://doi.org/10.34190/ecgbl.16.1.872
19. Tjostheim, I.: A serious game about apps, data-sharing and deceptive design. In: Guarda, T., Portela, F., Diaz-Nafria, J.M. (eds.) ARTIIS 2023. CCIS, vol. 1936, pp. 332–343. Springer, Cham (2024). https://doi.org/10.1007/978-3-031-48855-9_25

Training Computational Thinking to Leverage Citizens of Next Generation

Cristiana Araújo[1(✉)], Pedro Rangel Henriques[1], and João José Cerqueira[2]

[1] ALGORITMI Research Centre/LASI, University of Minho, Braga, Portugal
decristianaaraujo@hotmail.com, prh@di.uminho.pt

[2] Life and Health Sciences Research Institute (ICVS), University of Minho, Braga, Portugal
jcerqueira@med.uminho.pt
https://epl.di.uminho.pt/ cristiana.araujo/ ,
https://www.di.uminho.pt/ prh/ ,
http://www.icvs.uminho.pt/about-icvs/people/jcerqueira

Abstract. The world is constantly changing and the problems and challenges we encounter today concerning the work eco-system or concerning in personal life may not be the same in the future. Today's students are the next generation of people who will be exposed to these new problems. So, it is urgent that these students acquire knowledge, skills, and the ability to solve problems, to successfully overcome the evolution challenges. We argue that Computational Thinking (CT) can help students develop skills deemed necessary for citizens of next generation. For this, we built a web platform called Computational Thinking 4 All, which will be the workbench of any teacher who wants to prepare his students to acquire Computational Thinking skills. Computational Thinking 4 All contains a repository of Learning Resources – LaRaCiTa. These Learning Resources will allow students to develop CT skills. This article presents these two artefacts, the platform and the repository, as a solution to help teachers develop CT skills in their students, and consequently help them acquire the necessary knowledge and skills for citizens of the next generation.

Keywords: Computational Thinking · Learning Resources · Repository of Learning Resources · Next Generation

1 Introduction

Today's students are the ones who will: work in jobs that haven't been created yet, tackle social challenges we can't yet imagine, and use technologies that haven't been invented yet. It is important to prepare students to thrive in a world that is constantly changing.

For this it is necessary that young people acquire a set of strong skills. *OECD*[1] *Future of Education and Skills 2030*[2] project aims at building a common understanding of the knowledge, skills, attitudes and values that today's students need

[1] Organisation for Economic Co-operation and Development.

[2] Acessible at: https://www.oecd.org/education/2030-project/about/.

A. Rocha et al. (Eds.): WorldCIST 2023, LNNS 800, pp. 481–490, 2024.
https://doi.org/10.1007/978-3-031-45645-9_46

to acquire to be the next generation's citizens [11]. In this project, it is the *Learning Compass 2030* that defines the **knowledge, skills, attitudes and values** that students must have to reach their potential. Knowledge and skills are the *Learning Compass 2030 competencies* that are most relevant for the purpose of this paper, and so are those that we will explore in detail.

Knowledge is a key component of the OECD *Learning Compass*. Knowledge usually includes theoretical concepts and ideas, as well as practical understanding based on the experience of having performed certain tasks [10]. The *OECD Future of Education and Skills 2030 project* distinguishes four different types of knowledge: disciplinary, interdisciplinary, epistemic and procedural. Disciplinary knowledge involves subject-specific concepts and detailed content, such as what the student learns in the study of mathematics or geography. Interdisciplinary knowledge aims to relate concepts and contents of different subjects. Epistemic knowledge is the understanding of how expert practitioners of discipline think and work. This knowledge allows students to find the purpose of learning, understand the application of learning, and extend their disciplinary knowledge. Procedural knowledge is the understanding of how something is done, the set of steps or actions taken to achieve a goal. Some procedural knowledge is domain-specific, others are transferable between domains. Transferable knowledge is knowledge that students can use in different contexts and situations to identify solutions to problems [10].

Skills are the ability and capacity to carry out processes and be able to use knowledge to achieve a goal [12]. The *Learning Compass 2030* distinguishes three different types of skills: cognitive and meta-cognitive skills; social and emotional skills; and physical and practical skills. Cognitive and meta-cognitive skills comprise learning to learn (learning about learning itself and learning processes [9]), self-regulation, critical thinking, and creative thinking. Social and emotional skills involve self-efficacy (self-esteem, effectiveness and competence to face problems [3]), empathy, collaboration and responsibility. Physical and practical skills are intended to include the use of new information and communication technology devices [12].

Training in Computational Thinking (CT) can help to develop in today's students some of the competences referred to in the components of knowledge and skills. To help teachers train their students thinking computationally, acquiring CT skills, we developed the Web platform – *Computational Thinking 4 All*. Computational Thinking 4 All is composed of three modules, one of which is the Learning Resources. Learning Resources play a crucial role as they will allow students to acquire CT skills. These resources are stored on the *LaRaCiTa* Web platform, thus allowing the teacher to search for a specific Learning Resource to train one or more CT skills for a given grade.

This paper is organized into five sections. Section 2 presents Computational Thinking and the mission of Learning Resources in CT training. Section 3 presents the Web platform, *Computational Thinking 4 All*, and its organization by modules. Section 4 describes one of the modules of the *Computational Thinking 4 All* platform – *LaRaCiTa*, a repository of Learning Resources to train

Computational Thinking. Finally, in Sect. 5 discusses the conclusions and future work.

2 Computational Thinking and Learning Resources

The concept of Computational Thinking (CT) was first pronounced in 2006 by Jeannette Wing. Wing states that Computational Thinking is a fundamental skill for everyone, not just computer scientists. To writing, reading and arithmetic must be added CT to increase every child's analytical ability [15]. According to Wing, Computational Thinking *involves solving problems, designing systems and understanding human behavior, based on fundamental concepts of computer science* [15].

However, over the years, several definitions have emerged for the concept of Computational Thinking. These definitions presented by the various researchers fall into two categories: related to programming and computing concepts [5,7,14]; and concerned with the competencies needed in both domain-specific knowledge and general problem-solving skills [4,8,13,16].

Definitions related to programming and computing concepts describe CT as being: computational concepts (programming terms like sequences, loops, events...), computational practices (iteration, debugging, and abstraction), computational perspectives (expressing, connecting and questioning) [5]; data practices (collecting data, creating, manipulating, analysing and visualizing), modeling & simulation practices (computation models), computational problem-solving practices, and systems thinking practices [14]; and programming, documenting and understanding software, and designing for usability [7].

The definitions that describe CT as necessary competencies and general problem-solving skills, detail these competencies and skills in: abstraction [4,8,13,16]; decomposition [4,8,13,16]; algorithmic thinking [4,8,13,16]; evaluation [4,13]; logical thinking [4,8,16]; patterns [4]; and generalization [8,13]. These skills are supported and enhanced by dispositions or attitudes that are essential dimensions of Computational Thinking. These dispositions or attitudes include: persevering [4,8], collaborating [4,8], debugging [4,16], ability to deal with open ended problems [8], confidence in dealing with complexity [8], tolerance for ambiguity [8], tinkering [4], and creating [4].

For us, Computational Thinking is defined as a set of competencies that require students to develop domain-specific knowledge and problem-solving skills. We essentially adopt the definitions followed by ISTE[3] & CSTA[4] [8] and by Barefoot – Computing at School [4].

Relating the competences presented in the *OECD Future of Education and Skills 2030 project* with Computational Thinking, we can see that when students acquire the competences of Computational Thinking, they are at the same time acquiring those proposed by the *Learning Compass 2030* (specifically Knowledge and Skills). For example when students acquire skills in abstraction, algorithmic

[3] International Society for Technology in Education.

[4] Computer Science Teachers Association.

thinking, pattern recognition and generalization they are developing procedural knowledge. And when they acquire logic skills they are developing cognitive and meta-cognitive skills. On the other hand, they are also developing cognitive and meta-cognitive skills as well as social and emotional skills when CT skills are acquired through attitudes: tinkering, creating, persevering, and collaborating.

But to acquire these skills and capabilities it is necessary for students to train them. Only after that, they will be prepared to be the citizens of the next generation. Students can perform this training with Learning Resources suitable for developing Computational Thinking skills. The Learning Resource aims to: put in practice or train previous knowledge; develop new knowledge; encourage the process of understanding; contribute to the organization and synthesis of educational content; contribute to the logical reasoning, communication and interaction; contribute to the development of different skills and acquisition of student values; and contribute to the retention of desirable knowledge and attitudes [6].

In Sect. 4 we will present in detail a Learning Resources Repository that allows the teacher to select a specific resource to train the skills he wants, in a given grade. In the next section we will present the *CT4ALL* platform whose mission is to disseminate Computational Thinking training to prepare the next generation of citizens.

3 Disseminating – CT4ALL, the Web Platform

As mentioned earlier, Computational Thinking can help to develop knowledge and skills, which the *OECD Future of Education and Skills 2030* project considers mandatory for next-generation citizens.

To disseminate Computational Thinking training as a solution to prepare students for the next generation, we built the Web platform – *Computational Thinking 4 All (CT4ALL)*. The Web platform is composed of 3 modules to support the adoption of Computational Thinking programs at schools. In other words, the *CT4ALL* platform is the workbench of any teacher who wants to prepare his students to acquire Computational Thinking skills. The three modules that compose *CT4ALL* platform are described below.

Module 1: CT Learning Space

- **Computational Thinking Issues** – in this component the different concepts, approaches, techniques and instruments that describe and characterize Computational Thinking and related matters are presented. Some forums, and events on Computational Thinking, as well as initiatives to train it, will also be announced and pointed out.

This module can be seen as a kind of encyclopedia on Computational Thinking topics, providing definitions and news on the subject and related links where to look for information to learn more about the subject.

Module 2: OntoCnE

- **Ontology for Computing at School** – in this component, the OntoCnE is presented, its purpose and its layered structure. OntoCnE aims at providing a detailed and rigorous description of two knowledge domains and their interception: Computational Thinking and Computer Programming [1]. So, OntoCnE is intended to support the design of education programs syllabus on Computational Thinking as the basis for Computer Programming. This justifies the importance of making OntoCnE available to the teaching community using a friendly interface. This ontology is structured in three layers: what concepts are needed to train Computational Thinking (layer 1); what to train at each level of education and how deep to go (layer 2); and what material to use to train Computational Thinking at the various levels of education (layer 3).

  This component will also include the OntoCnE knowledge repository and conceptual navigation.

  Conceptual navigation allows the visitor to consult the ontology concepts following logical or semantic paths, according to the relationships that connect them. Since an ontology is a network of related concepts, is possible to move from concept to concept and to understand the hierarchy or relationship that exists among them. In this way it is possible to build easily (during the navigation) knowledge about the domain of Computational Thinking.
- **Computational Thinking Training Program** – this component contains Computational Thinking training programs. The curriculum for the training of Computational Thinking, containing programs for different grades and with different levels of complexity, will the accessible through this platform component.

Module 3: Resources

- **Learning Resource Repository** – this component contains created or reused Learning Resources, which aim to train Computational Thinking.
- **Repository of Tests & Surveys** – this component contains a set of created or reused Efficacy Tests that aim to measure the effectiveness of Computational Thinking training. This component also contains surveys or questionnaires to collect information that allows for measuring whether the training with the proposed Learning Resources has a positive impact on the way students acquire the ability to think computationally.
- **Guide to Assessing the Quality of Learning Resources** – this component contains a guide with a set of requirements to assess the quality of Learning Resources.

Figure 1[5] shows all the modules and their components that make up the *Computational Thinking 4 All* platform. In Fig. 1 it is possible to see that *OntoCnE*

[5] All the icons in Fig. 1 were taken from the website: "The Noun Project" (https://thenounproject.com/). Accessed: 2021-12-30.

is linked to two components. Actually, *OntoCnE* is also connected to *Learning Resource Repository* as it allows for classifying Learning Resources stored in the repository. With this classification it is possible to know which Computational Thinking concepts are trained by a given resource.

On the other hand, *OntoCnE* also links to *Computational Thinking Training Program*, because as mentioned before, layer 2 of the ontology aims to define/state which concepts to train at grade and the degree of depth. Layer 2 will play a very important role in defining the teaching objectives to train Computational Thinking.

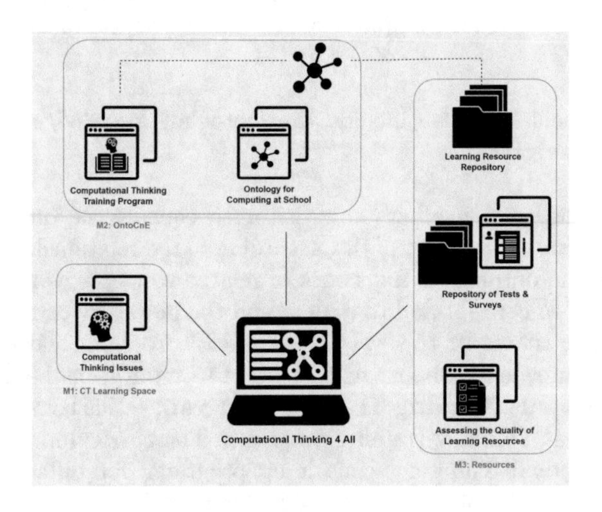

Fig. 1. *CT4ALL* – Architecture

CT4ALL platform is accessible at: https://computationalthinking4all.epl.di.uminho.pt/. It is important to note that *CT4ALL* is a work in progress and therefore the modules are not yet fully finished.

CT4ALL in addition to disseminating information on CT and supporting teachers to develop in students the knowledge and skills necessary for the next generation, also includes a repository of resources to carry out this training. In the next section we will present this repository of Learning Resources to train Computational Thinking.

4 Training – LaRaCiTa, a Web LR Repository

In the philosophy of disseminating Computational Thinking and how to train it, we decided to make categorized resources available by grade and skills they train. These categorized resources are stored and made available in the Learning Resources Repository (LR) of *CT4ALL* (Module 3 in Sect. 3), which we call *LaRaCiTa*. The LR categorization is performed based on the OntoCnE Ontology (Module 2 in Sect. 3). In order to categorize resources based on our ontology, *LaRaCiTa* was built according to OntoCnE and its layers.

LaRaCiTa is a Client-Server system. The Back-office contains a relational database (SQL), designed following the layered structure of OntoCnE. The interconnection between the client and the database is carried out through an API (Application Programming Interface) developed specifically for this platform (implemented in PHP).

The Front-office lists all stored resources and allows for: search resources (by trained concepts, grade and category); and add new resources. Through the Front-office it is also possible to login as Admin and so to manage the database and the specific ontology tables. It is important to point out that this user profile is only accessible to the platform administrator. Front-office was developed using HTML, CSS and JavaScript and communicates with the API in PHP with SQL and JSON.

To add a new resource to the *LaRaCiTa* repository, a form organized into four sections as follows, was constructed: resource identification, additional data, categorization, and attachments. This form was developed according to a LaTeX template we designed some time ago to describe and categorize LR [2].

In order to explain in more detail the attributes needed to describe the resource and how we relate it to the ontology, we demonstrate below the insertion of a LR in the *LaRaCiTa* platform. The resource we are going to add is called *Help Henrique to find the Plant of Life*. The LR was the adapted from the short story *The rewarded boy* extracted from the 4th grade Portuguese book. The student has to read the story and overcome the 3 challenges, regarding binary-code exercises, that are proposed along the story sections. The first challenge is to identify a selected object, write its name and convert each letter of the name into binary code, according to a given conversion table. The second challenge aims to decipher a sentence that is written in binary code using that conversion table. The third challenge is to identify a word based on clues, write that word and convert each letter into binary code. All conversions requested in the challenge are performed through a table that contains the binary code corresponding to each letter of the alphabet. It is important to note that all fields on the form will be completed in Portuguese because this platform is intended to provide resources for Portuguese schools.

Figure 2 displays the attributes that identify the resource: **Title**; **Short Description**; **Learning Objective**; **Description**; **Official Link**; and **Type of Resource** (unplugged, plugged or both). In Fig. 3 are presented the attributes that describe additional data of the resource: **Instructions**; **Materials** (e.g.: computer, paper and pencil, etc.); **Requirements**[6]; **Method Evaluation**; **Language**; **Duration**[7]; and **Author**. Figure 4 displays the attributes that categorize the resource, namely: **Category** (e.g.: game, worksheet, etc.); **Courses**[8] (e.g.: Math, History, Portuguese, etc.); **Grade (K1-K12)**; **Computational Thinking Skills**.

---

[6] Topics or skills necessary to execute that activity.

[7] Time necessary to complete the activity.

[8] It is important that a resource requires knowledge from different areas involving more than one course, promoting interdisciplinary.

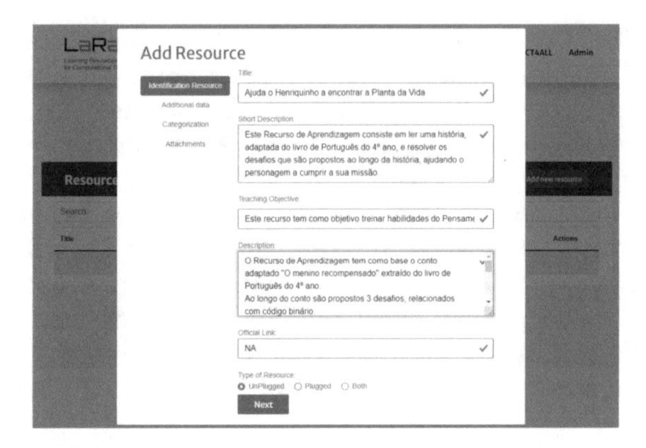

Fig. 2. Add Learning Resource – Resource Identification

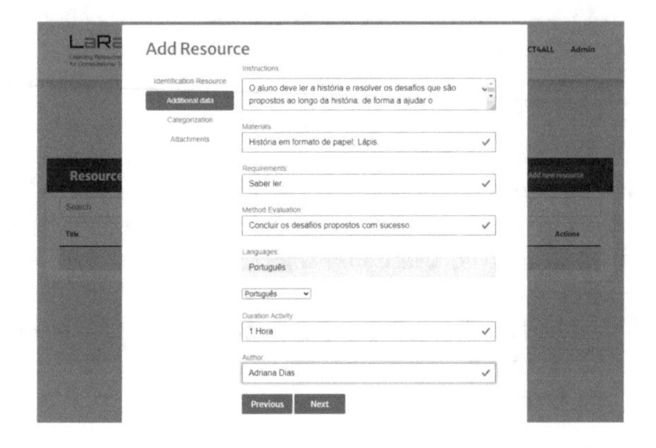

Fig. 3. Add Learning Resource – Additional Data

The *Grade* and *Computational Thinking Skills* attributes are what allow relating the resource to layers 1 and 2 of OntoCnE. When we select *Grade*, for example 4th grade, we are selecting layer 2 (what to train each year). As a result of this selection, the next attribute – *Computational Thinking Skills* – will only show layer 1 concepts that were defined in the ontology to train concepts associated with this grade. The purpose of only showing the concepts from the grades that we previously selected is to avoid the choice of concepts that shall not be trained in the selected grades. The concepts selected for the *Computational Thinking Skills* attribute relate precisely that resource to the concepts of layer 1 of OntoCnE. This description of the new LR, allows to populate layer 3 (the material to use for training) of OntoCnE linking the concepts of Computational Thinking and Computer Programming to LR. Through this classification, we managed to help the teacher to find a resource that trains, in a certain grade, the skills he needs.

Fig. 4. Add Learning Resource – Categorization

The last step is to load the attachments that are part of the resource: **Figures**; and **Files**. In this specific case, the file added was the story that the student has to read and solve the challenges. It should be noted that when we consult a specific resource on the platform, the presentation format follows the same four sections template. The platform is accessible at: https://computationalthinking4all.epl.di.uminho.pt/laracita/.

In the next section, we present the conclusions and future work.

5 Conclusion

It is important that next generation citizens acquire the skills presented by the *OECD Future of Education and Skills 2030 project.* One solution to develop some of these skills is to train Computational Thinking abilities. To put this solution into practice, we implemented *CT4ALL*, which allows disseminating Computational Thinking and provides tools to train it. These tools are called LR and are available in the *LaRaCiTa* repository. This repository was built following OntoCnE and its layered structure to store those LR categorized. This platform construction methodology allows us to ensure that the teacher selects an appropriate resource for a given level of teaching and to train the skills he/she wants.

As future work, we consider that it is necessary to finish layer 2 of OntoCnE. *LaRaCiTa* needs to be populated with more resources for the different levels of education, thus populating layer 3 of OntoCnE. After populating with more resources it is important to test these same resources with students in schools. To conduct these experiments it is necessary to build the repository of tests & surveys.

Acknowledgment. This team work has been supported by FCT - Fundação para a Ciência e Tecnologia within the R&D Units Project Scope: UIDB/00319/2020.

The Ph.D. work of Cristiana Araújo is supported by FCT - Fundação para a Ciência e Tecnologia, Research Grant, with reference 2020.09845.BD.

We are in debt to all our students of the Master's in Education and Informatics (eds. 21–22 and 22–23) for having developed the Learning Resources that made possible this paper.

References

1. Araújo, C., Lima, L., Henriques, P.R.: An Ontology based approach to teach Computational Thinking. In: 21st International Symposium on Computers in Education (SIIE), November 2019. C. G. Marques, I. Pereira, and D. Pérez, Eds., IEEE Xplore, pp. 1–6
2. Araújo, C., Rangel Henriques, P., José Cerqueira, J.: Creating Learning Resources based on Programming concepts. In: 15th International Conference on Informatics in Schools – A step beyond digital education (ISSEP 2022) (2022). A. Bollin and G. Futschek, Eds., CEUR-WS
3. Bandura, A.: Self-efficacy?: The Exercise of Control. W.H. Freeman, New York (1997)
4. Barefoot. Computational Thinking Concepts and Approaches (2021). Accessed: 2021-11-07
5. Brennan, K., Resnick, M.: New frameworks for studying and assessing the development of computational thinking. In: Proceedings of the 2012 Annual Meeting of the American Educational Research Association, Vancouver, Canada (2012), vol. 1, p. 25
6. Bušljeta, R.: Effective use of teaching and learning resources. Czech-Polish Historical Pedagogical J. **5**(2), 55–70 (2013)
7. Denner, J., Werner, L., Ortiz, E.: Computer games created by middle school girls: can they be used to measure understanding of computer science concepts? Comput. Educ. **58**(1), 240–249 (2012)
8. ISTE, and CSTA. Operational Definition of Computational Thinking for K-12 Education (2011). Accessed: 2021-11-07
9. OCDE. Embedding Values and Attitudes in Curriculum (2021)
10. OECD. Knowledge for 2030 concept note. Technical report, OECD – Organisation for Economic Co-operation and Development (2019)
11. OECD. Oecd future of education and skills 2030 concept note. Technical report, OECD – Organisation for Economic Co-operation and Development (2019)
12. OECD. Skills for 2030 concept note. Technical report, OECD – Organisation for Economic Co-operation and Development (2019)
13. Selby, C., Woollard, J.: Computational thinking: the developing definition
14. Weintrop, D., et al.: Defining computational thinking for mathematics and science classrooms. J. Sci. Educ. Technol. **25**(2), 127–147 (2016)
15. Wing, J.M.: Computational thinking. Commun. ACM **49**(3), 33–35 (2006)
16. Yadav, A., Mayfield, C., Zhou, N., Hambrusch, S., Korb, J.T.: Computational thinking in elementary and secondary teacher education. ACM Trans. Comput. Educ. **14**(1) (2014)

ePortfolios Based on Self-Regulated Learning Components in Flexible Learning for Engineers as a Vehicle for Academic and Career Development

Foteini Paraskeva(✉), Eleni Neofotistou, and Angeliki Alafouzou

Department of Digital Systems, University of Piraeus, Karaoli Dimitriou Street 80, 185 34 Piraeus, Greece
fparaske@unipi.gr

Abstract. Electronic Portfolios (ePortfolios) have become a popular pedagogical approach on the tertiary educational landscape worldwide. The ePortfolio development is based on the fact that the reflective practice of its creation allows students to document and track their learning, develop a coherent picture of their learning experience, and improve their self-understanding. This paper outlines the process of integrating an ePortfolio as a tool of professional development for students in tertiary education. As it is assumed that ePortfolios contribute to Self-Regulated Learning (SRL), the construction of the ePortfolio can be based on the Self-Regulated Learning strategies as well, as a parallel process that can lead learners to construct their artifacts and reflect on their prior knowledge. The evaluation of the research is based on qualitative and quantitative analysis. The results prove the correlation between the enhancement of students' self-regulation and the construction of an academic e-portfolio for the promotion of academic and career development.

Keywords: ePortfolio · self-regulated learning · Web 2.0 · academic development · career development

1 Introduction

Today's rapidly changing world and digital transformation demands higher education to prepare students with various competences and skills in order to face 21st century challenges. The interaction between the ICT and the educational system considers to be a prerequisite while, in parallel, has led to fundamental transformations to the learning process. In response to the needs of this changing environment, institutions of higher education are increasingly focusing on high- impact practices and they are trying to implement learning tools like ePortfolios to equip students with these necessary skills and competences.

The aim of this study was to develop the Self-Regulated Learning processes within the electronic ePortfolio content for engineer students' academic and career development.

A. Rocha et al. (Eds.): WorldCIST 2023, LNNS 800, pp. 491–500, 2024.
https://doi.org/10.1007/978-3-031-45645-9_47

As a result, this paper presents the design of the "ePSRL_system4students" for academic and career development. Moreover, it analyses the results of the implementation of SRL on an academic and system-oriented ePortfolio. Therefore, this paper is organized as follows. In the first section, it is described what an ePortfolio is while the affordances of an academic ePortfolio constructed by the pillars of SRL are also shown. The next section presents SRL as a foundational theory that espouse ePortfolios. The following section describes the research framework and presents the main research question. Finally, the authors conclude this paper by examining the results of this study.

2 ePortfolios for Academic and Career Development

EPortfolios support learners' self-monitoring, self-evaluation and sharing of learning. A generic definition of an e-portfolio is that it is "a purposeful aggregation of digital items - ideas, evidence, reflections, feedback which 'presents' a selected audience with evidence of a person's learning and/or ability" [1]. It is a digital collection of work that documents and showcases knowledge, skills and abilities and their growth over time and also a process of reflection on these artifacts. According to their purpose, there is a clarification which includes four main portfolio types in different learning contexts [2]:

- *Assessment Portfolio* which represents an alternative way of evaluation in which learners should provide evidence of their competence in particular subject areas.
- *Showcase/Career Portfolio* which highlights learners' work in specific areas, and it could be shown to potential employers.
- *Development Portfolio* supports learners' personal development planning and depicts the advancement of their skills over a period of time.
- *Reflective Portfolio* shows learner's accomplishments and how they are related to the learning goals.

According to Rhodes (2011), the use of ePortfolios in education can lead students to work digitally and in an organized, searchable, and transportable way. E-portfolios are also characterized as "personalized web-based collections of work" [3]. This personalization of learning is an increasing trend and as far as concerns the construction and the customization of the ePortfolios, it gives a sense of ownership to learners [4].

Regarding the benefits of the ePortfolios, the underling pedagogy draws on theories of constructivism (socio-cognitive dimensions), student-centered learning and authentic educational activities [2]. The artifacts are related to students' goals, achievements, ideas and experiences and they should reflect on their learning process, experiences, and skills. The e-portfolios can be also used for authentic assessment of learners due to their requirements [5]. They help students to organize and self-evaluate their work and as a result they are motivated to learn [6].

Consequently, ePortfolios can be successfully used in both learning and assessment in a number of disciplines including the Arts, Humanities and Social Sciences. Moreover, ePortfolios are mentioned as one of the most high-impact practices [7] that helps educational institutions to ensure access, equity, and quality in education [8]. As a result, computer science and engineering education faculties utilize ePortfolios in their curricula the recent years. Finally, except from the educational purposes, students have also mentioned the advantages of ePortfolio to showcase their work for employment purposes as well [9, 10].

2.1 ePortfolios' Affordances

ePortfolios are related to learner's ability to self-regulate his own learning and to enhance skills and abilities [11]. Moreover, self-regulation is considered to be one of the ePortfolio's affordances regarding the promotion of productive learning. According to literature, an effective e-portfolio should be designed in a way that it will incorporate the four affordances of ePortfolio assessment [12] in order to offer the development of the core characteristics of productive learning.

The first affordance of ePortfolio is *task authenticity* which represents the degree that artifacts capture the progress of the participants and showcase their achievements [13]. Moreover, they can be an alternative way of assessment as the representation of the learning outcomes occurs in a different way than writing and can depict the application of skills and academic knowledge in everyday situations [9]. Additionally, the new knowledge can be used through authentic tasks and also reflect on the prior knowledge as well [14]. Additionally, the sense of ownership of learning during the construction of e-portfolio can cause the increasement of learners' intrinsic motivation [15]. Learners that can take ownership of learning can be a driving force of their own learning. Eportfolios can empower and facilitate learners via the personalization of learning in the construction of ePortfolios [16]. For instance, the use of multimedia as artifacts in the online interface of ePortfolio can be a way of flexible and customizable representation of students' learning outcomes and experiences [17]. Moreover, the personalization of ePortfolios can also promote equitable assessment as it enables educators to deal with a range of different learning styles and needs [18]. Finally, the representation of the learning outcomes through the explicit guidelines on the construction of artifacts can be a path for learners to develop authentic tasks [19].

The second affordance of ePortfolio is *self-regulated and reflective learning* which means that the process of ePortfolio's construction allows students to follow their own learning path, set their goals and organize their action plans [20]. Moreover, learners should be provided with a facilitative learning environment which will enable them to be goal-oriented, self-managed, self-evaluated and self-improved by enhancing their independence during the learning process [21, 22]. When students are self- regulated, they have a clear understanding of the learning goals and as a result they are motivated by intrinsic interest in learning [23, 24]. Clear goals and intrinsic motivation guide learners to achieve their goals by formulating strategies and plans, seeking help and feedback, self-reflecting on current progress, and adjusting strategies for meeting goals, which are the self-regulation processes [25]. Finally, learners reflect in their learning experiences and also evaluated their progress concerning their goals, during the construction of artifacts [26].

The third affordance of ePortfolio is the existence of *constructive feedback*. Existing research has proved the effectiveness of timely and interactive feedback in learning, so as learners to be able through frequent interactions to overcome difficulties and misconceptions to improve continuously their strategies of tackling tasks [26]. The online ePortfolio system should facilitate learners to share their ePortfolios and obtain teacher and peer feedback [9]. Moreover, productive learning can be promoted via feedback interactions, instructor's guidance and students' contribution which supports peer-learning that enhances learners' cognitive process through the critical evaluation of peers' ePortfolios according to specific assessment standards [27]. By sharing artifacts and reflections of their work, learners seek and receive help, suggestions, and comments for both teachers and students in order to improve their performance [28]. However, this process can be extended beyond the classroom with the use of blogs and social networking for the construction of an ePortfolio [29]. Additionally, as a form of peer-learning, students can be asked to construct collaboratively ePortfolio's artifacts. This interaction through online collaboration will force learners to self-explain, communicate, elaborate, and integrate their ideas to a group [18]. As a result, group members should synthesize the collective information for achieving a maximum performance. However, learners will need clear guidelines for choosing what to include and what to reject in the construction of the e-portfolios and also in the way they should interact with peers in order their psychological well-being to be protected [30].

The final affordance of ePortfolios is *student autonomy*. This affordance represents the sense of independence during the process of construction of new knowledge [15]. Moreover, students' autonomy is enhanced in collaborative knowledge building [18]. During ePortfolio's assessment, students' artifacts can be used as resources of problem-solving which is an essential process of knowledge generation [14]. Learning plans and reflective artifacts lead learners to critically reflect on how problem- solving situations can contribute in the construction of new knowledge [18]. Consequently, learners' independence via feedback and social support can lead students to become autonomous learners and reduce their dependence on teacher support [31].

3 Self-Regulated Learning Theory

Self-regulation is defined as self-generated thoughts, feelings, and behaviors that are planned and cyclically adapted based on performance feedback to attain self-set goals [32]. Students' development as independent, inquisitive, and reflective learners, capable of lifelong intellectual growth, is increasingly included among the desired outcomes of the higher education experience [33].

Moreover, the self-regulation skills are required also in computer mediated environments in order learners to be motivated [34]. Zimmerman developed a cyclical model of self-regulation from social-cognitive theory. This model has been successfully applied to education [22, 32]. Zimmerman's (2000) cyclical model of self-regulation includes three phases:

1. *Forethought* phase which consists of processes that precede any effort, and they also involve the learners' beliefs and attitudes.
2. *Performance Control* phase which represents the processes that occur during the learning efforts.
3. *Self-Reflection* phase which occurs after learning or performance and it involves reflecting on the self-monitored information to evaluate one's performance and to adjust during future learning attempts.

In each of the phases mentioned above, cognitive, motivational, and social aspects could be depicted. Self-evaluation, attention-focus, task strategies, self-instruction, goal setting, as well as strategic planning require detailed preparation and specific knowledge while the definition of self-efficacy, expectations, task interest/value, goal orientation, self-satisfaction is related to motivation. Finally, peer-evaluation and help-seeking are some of the techniques that support the connection to the surrounding environment and support the social elements explored.

4 The Research Framework

The instructional design of the "ePSRL_system4students" was implemented in an undergraduate computer science program in tertiary education, in a course titled "IT-centric Professional Development" (ITcPD) which is delivered during the 6th academic semester. Its implementation aimed to the exploration of the following research question: *What is engineer students' perspective on the value of ePortfolios for their academic and career development?*

To deliver the course and explore this question, an educational site was built, which was designed by the pillars of SRL theory, enabling students to follow the theory's phases according to their needs and not on a strict linear sequence. As it happens during the actual ePortfolio creation process, forethought, performance, and self-reflection phases are accessed multiple times through the site's main menu. The relevant theory, strategies and techniques are also included, so as students to have a complete learning guide when it comes to their own performance.

The construction process was articulated by the SRL phases and guided by the ePortfolio key-elements that were introduced to students as characteristics that their ePortfolio should have in order to be successful [35]. These key-elements are the following:

- *Critical thinking through reflection* – Students should make sure that their ePortfolio includes all the artifacts they think could better describe their progress in the specific domain they have chosen, accompanied by reflective journals explaining their selection towards them.
- *Technical competency* – Technical details should be under careful consideration. The tools and multimedia features selected by the ePortfolio creator should be compatible to the platform and fully functional, while the navigation should be easy and far from perplexing.
- *Visual literacy* – When it comes to ePortfolio design, visual elements and the whole aesthetic should be carefully chosen in order to support creator's objectives and be pleasant and understandable to the viewer. Apart from the interface, student should present original and authentic content and respect any property rights.
- *Effective communication* – EPortfolio as a whole, illustrates the message that the creator needs to express. So, it is important to be coherent and support a specific identity, via the design and its original content.

The above characteristics were included as Key Performance Indicators in each e-course section. To be more specific, students had the opportunity via self-assessment multiple-choice forms to reflect on their progress while creating their ePortfolio in order to be able to make the necessary changes and improve their work.

These key-elements were operationally matched to the SRL phases, and both were supported by the e-learning system that was designed in order to host and deliver the e-course (Fig. 1).

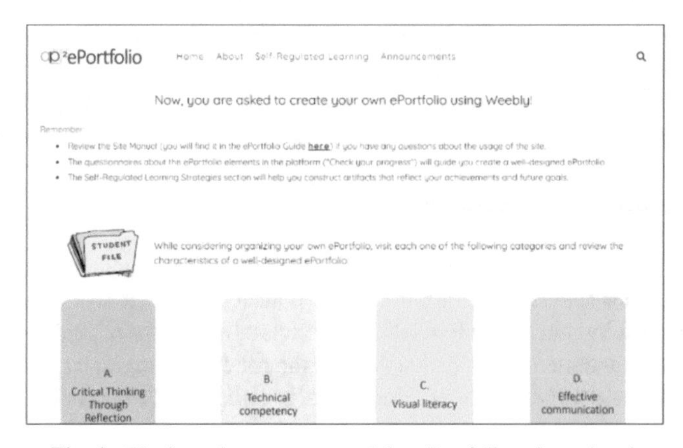

Fig. 1. The introductory page and the ePortfolio oriented path

The system designed has two main paths, in order to offer students, the opportunity to focus on the background theory, or on the practice, depending on their needs. The navigation is enabled through the main menu on top of the web- page and via hyperlinks and buttons in the webpages' main content, so it is clear and flexible. These paths are:

- The *SRL oriented path*, which includes necessary information about the SRL theory and the strategies that should be used in each phase.
- The *ePortfolio oriented path*, which includes the main guidelines in the creation process, the KPI assessment forms and useful paradigms.

So, the SRL phases and the ePortfolio key-elements were the main pillars in the e-learning system design procedure. This whole instructional framework attempted as well to support the ePortfolio affordances that were mentioned before (Table 1).

Table 1. E-course affordances in supporting ePortfolio's key-elements and affordances

ePortfolio key-elements	E-course affordances	ePortfolio affordances
Critical thinking through reflection	• Students are encouraged to collect original artifacts • Questionnaires that guide them to create well-designed ePortfolios • SRL strategies section that encourage reflection	Task authenticity Reflective and self-regulative learning
Technical competence Visual literacy	• Successful paradigms are included • Students are encouraged to find their identity • Students work on their own organizing the artifacts • Site's ease of navigation • Authentic graphics	Student autonomy
Effective communication	• Clear directions and guidelines • KPI self-assessment forms • Peer-review during the course	Constructive feedback

When it comes to ePortfolio creation, research has underlined that working on an e-learning basis has added value on the students learning as they have the chance to learn from each other, by observing their colleagues' ePortfolios', review the goals the have set and as a result exercise on reflecting on their own work as well [29].

5 Results

In order to collect data about students' perspective towards the usefulness of ePortfolios in their academic career and development, students were asked to fulfill specific questionnaires. The results were collected, elaborated and presented below (Table 2).

Table 2. The descriptive statistics of ePortfolio's affordances (N = 61)

	Mean	Std. Deviation
Task authenticity	4,3197	,46353
Reflective and SRL	4,1443	,47347
Constructive feedback	4,0820	,46210
Student autonomy	4,3484	,46610

The questionnaires were answered on a 5-point Likert scale [Never (1) – Rarely (2) – Sometimes (3) – Often (4) – Very often (5)], where higher scores indicate the most often affordances of the ePortfolio students are able to identify in their own progress. As it is clear from the above table, students agreed that by engaging in the ePortfolio creation process, were able to identify and use all of its affordances, but mostly student autonomy and task authenticity (Table 3).

Table 3. The correlation between the ePortfolio affordances (N = 61)

	Task authenticity	Reflective and self-regulative learning	Constructive feedback	Student autonomy
Task authenticity	1			
Reflective and self-regulative learning	,557**	1		
	,000			
Constructive feedback	,531**	,610**	1	
	,000	,000		
Student autonomy	,493**	,611**	,452**	1
	,000	,000	,000	

**. Correlation is significant at the 0.01 level (2-tailed).

There are statistically significant positive correlations at a high level of 0.01. This means that the total of the two correlations are positive and statistically significant and when one increases this results in the other one to increase as well, meaning that the ePortfolios' affordances are developed equally throughout the e-course.

Table 4. The correlation between the ePortfolio affordances and the SRL phases (N = 61)

	Forethought	Performance	Reflection	Task authenticity	Reflective and SRL learning	Constructive feedback	Student autonomy
Forethought	1						
Performance	,317*	1					
	,013						
Reflection	,548**	,577**	1				
	,000	,000					
Task authenticity	−,116	,047	,115	1			
	,374	,722	,378				
Reflective and SRL learning	-,146	,107	,062	,557**	1		
	,261	,412	,637	,000			
Constructive feedback	−,201	,051	,009	,531**	,610**	1	
	,121	,694	,944	,000	,000		
Student autonomy	−,029	,139	,118	,493**	,611**	,452**	1
	,826	,286	,367	,000	,000	,000	

*. Correlation is significant at the 0.05 level (2-tailed).
**. Correlation is significant at the 0.01 level (2-tailed).

Table 4 shows how the ePortfolio's affordances were scaffolded and developed in each phase of the SRL process, which was the main model on which the e-course's framework was designed. It is clear that student autonomy is highly related to constructive feedback, task authenticity and reflective learning, as these procedures encourage student's responsibility in learning throughout authentic environments, which is one of the main advantages of developing ePortfolios for academic and future professional fields.

6 Conclusion

This e-learning framework was developed in order to explore the affordances of ePortfolios for engineering students, for their academic and future professional needs. The theoretical basis was the Self-Regulated Learning theory, as it was found to be the most suitable for the ePortfolio creation process. From the research that was conducted it was clearly shown that these affordances covered the engineering students' needs and expectations and were activated in every phase of the SRL cycle. As long as the e-course was found successful towards the ePortfolio's value for engineering students, our future work should focus on exploring how important are specific SRL strategies in ePortfolio creation and on how do students perceive their competence, motivation, goal setting, learning strategies related to the ePortfolio development process.

Acknowledgment. This work has been supported by the Research Center of the University of Piraeus.

References

1. Barrett, B.: creating webpages as an electronic portfolio foundation for academic and life-long learning applications. In: E-Proceedings of the International Conference on Distance Learning, p. 197 (2019)
2. Stefani, L., Mason, R., Pegler, C.: The Educational Potential of e-Portfolios, Supporting Personal Development and Reflective Learning. Routledge, London (2007)
3. Namaziandost, E., Alekasir, S., Sawalmeh, M.H.M., Miftah, M.Z.: Investigating the Iranian EFL learners' attitudes towards the implementation of e-portfolios in English learning and assessment. Cogent Educ. **7**(1), 1856764 (2020)
4. JISC InfoNet: e-Portfolios, University Northumbria (2008). Accessed 15 June 2009. http://www.jiscinfonet.ac.uk/infokits/ePortfolios/future
5. White, A.: EPortfolios: integrating learning, creating connections and authentic assessments. In: Allan, C.N., Campbell, C., Crough, J. (eds.) Blended Learning Designs in STEM Higher Education, pp. 167–188. Springer, Singapore (2019). https://doi.org/10.1007/978-981-13-6982-7_10
6. Akçıl, U., Arap, I.: The opinions of education faculty students on learning process involving e-portfolios. Procedia Social Behav. Sci. **1**, 395–400 (2009). https://doi.org/10.1016/j.sbspro.2009.01.071
7. Kuh, G., O'Donnell, K., Schneider, C.G.: Hips at ten. Change Maga. Higher Learn. **49**(5), 8–16 (2017). https://doi.org/10.1080/00091383.2017.1366805
8. Martin, R.: Taking students success to scale. Change Maga. High. Learn. **49**, 38–47 (2017). https://doi.org/10.1080/00091383.2017.1265391
9. Heinrich, E., Bhattacharya, M., Rayudu, R.: Preparation for lifelong learning using ePortfolios. Eur. J. Eng. Educ. **32**(6), 653–663 (2007). https://doi.org/10.1080/03043790701520602
10. van Dinther, M., Dochy, F., Segers, M., Braeken, J.: The construct validity and predictive validity of a self-efficacy measure for student teachers in competence-based education. Stud. Educ. Eval. **39**(3), 169–179 (2013)
11. Mudau, P.K., Van Wyk, M.M.: E-portfolio alternative assessment strategy enhancing higher-order thinking skills in an open distance learning environment. Multidisc. J. Dist. Educ. Stud. **1**(1) (2022)
12. Barbera, E.: Mutual feedback in e-portfolio assessment: an approach to the netfolio system. Br. J. Edu. Technol. **40**(2), 342–357 (2009). https://doi.org/10.1111/j.1467-8535.2007.00803.x
13. Buzzetto-More, N.: Assessing the efficacy and effectiveness of an e-portfolio used for summative assessment. Interdisc. J. E-Learn. Learn. Obj. **6**(1), 61–85 (2010)
14. Davids-Latief, N., Kimani, L., Nelwamondo, M.: Exploring the pedagogy of eportfolio tool with reflection on transferable skills for business analysis students. In: Proceedings of ICERI2021 Conference, vol. 8, p. 9th (2021)
15. Barnett, R.: Assessment in higher education: an impossible mission? In: Boud, D., Falchikov, N. (eds.) Rethinking Assessment in Higher Education, pp. 29–40. Routledge, London (2007)

16. Romero, L., Saucedo, C., Caliusco, M.A., Gutiérrez, M.: Supporting self-regulated learning and personalization using ePortfolios: a semantic approach based on learning paths. Int. J. Educ. Technol. High. Educ. **16**(1), 1–16 (2019)
17. Dunbar-Hall, P., Rowley, J., Brooks, W., Cotton, H., Lill, A.: E-portfolios in music and other performing arts education: history through a critique of literature. J. Hist. Res. Music. Educ. **36**(2), 139–154 (2015)
18. Hoven, D., Walsh, P., Al-Tawil, R., Prokopetz, R.Z.: Exploring professional development needs and strategies for instructors/faculty facilitating eportfolios online. Irish J. Technol. Enhan. Learn. **6**(1), 154–176 (2021)
19. Handley, K., Williams, L.: From copying to learning: using exemplars to engage students with assessment criteria and feedback. Assess. Eval. High. Educ. **36**(1), 95–108 (2011)
20. Crisp, G.T.: Integrative assessment: reframing assessment practice for current and future learning. Assess. Eval. High. Educ. **37**(1), 33–43 (2012)
21. Boekaerts, M., Corno, L.: Self-regulation in the classroom: a perspective on assessment and intervention. Appl. Psychol. Int. Rev. **54**(2), 199–231 (2005)
22. Zimmerman, B.J.: Investigating self-regulation and motivation: Historical background, methodological developments, and future prospects. Am. Educ. Res. J. **45**(1), 166–183 (2008). https://doi.org/10.3102/0002831207312909
23. Cassidy, S.: Self-regulated learning in higher education: Identifying key component processes. Stud. High. Educ. **36**(8), 989–1000 (2011). https://doi.org/10.1080/03075079.2010.503269
24. Tucker, T., et al.: Transforming an engineering design course into an engaging learning experience using ePortfolios. In: 2020 ASEE Virtual Annual Conference Content Access (2020)
25. Alhitty, A., Shatnawi, S.: Using E-portfolios for writing to promote students' self-regulation. SSRN 3874465 (2021)
26. González-Mujico, F.: The impact of ePortfolio implementation on motivation, self-regulation and academic language development: the learners' and the teachers' perspectives. Asian ESP J. **16**(3), 209–242 (2020)
27. Nicol, D., Thomson, A., Breslin, C.: Rethinking feedback practices in higher education: a peer review perspective. Assess. Eval. High. Educ. **39**(1), 102–122 (2014)
28. Yang, M., Tai, M., Lim, C.P.: The role of e-portfolios in supporting productive learning. Br. J. Edu. Technol. **47**, 1276 (2015). https://doi.org/10.1111/bjet.12316
29. Chuang, H.-H.: Weblog-based electronic portfolios for student teachers in Taiwan. Educ. Tech. Res. Dev. **58**(2), 211–227 (2010)
30. Mok, J.: As a student, I do think that the learning effectiveness of electronic portfolios depends, to quite a large extent, on the attitude of students! Electron. J. e-Learn. **10**(4), 407–416 (2012)
31. Sadler, D.R.: Beyond feedback: developing student capability in complex appraisal. Assess. Eval. High. Educ. **35**(5), 535–550 (2010)
32. Zimmerman, B.J.: Development of self-regulated learning: which are the key subprocesses? Contemp. Educ. Psychol. **16**, 307–313 (1986)
33. King, P.M., Brown, M.K., Lindsay, N.K., Vanhecke, J.R.: Liberal arts student learning outcomes: an integrated approach. About Campus **12**, 2–9 (2007)
34. Hodges, B.: Self-regulation in Web-based courses: a review and the need for research. Q. Rev. Dist. Educ. **6**, 375–383 (2005)
35. Liu, J., Burt, R.: Introducing ePortfolios to construction management undergraduate students. In: 51st ASC Annual International Conference, Texas, USA (2015)

Smart Learning Affordances of Web Technologies from Future ICT Teachers' Perspectives

Vasiliki Karampa(✉) and Foteini Paraskeva(✉)

Department of Digital Systems, University of Piraeus, Karaoli Dimitriou Street 80, 185 34 Piraeus, Greece
{bkarampa,fparaske}@unipi.gr

Abstract. Information and Communication Technologies (ICT) have led to the evolution of the Web (from 1.0 to 4.0 with trends towards 5.0), offering more and better services to users worldwide. New opportunities to optimize learning now exist thanks to the application of these technologies in the field of education. The concept of smart learning has become increasingly popular since advanced web technologies are aligned with modern, innovative pedagogical approaches to make learning environments smarter. From the perspective of ICT students, and future teachers in a Technology-Enhanced Learning Environment (TELE), this study explores the educational potential of web technologies. Toward this end, twenty-nine ICT students participated in an e-course lab during an academic course. The e-course lab was designed based on constructivist approaches (Project based learning and collaborative strategies) and then implemented by integrating, among others, fifteen prominent web technologies in the form of web tools in an e-learning platform. As part of their e-course lab experience, learners recorded the advantages and disadvantages of each tool, along with their pedagogical value. Even though the tools serve different educational needs, a comparative analysis of the potential smart learning affordances was conducted based on this recording. The results are encouraging, providing empirical data for future research in the field of educational technology, as well as good practices for the design of smart learning environments (SLEs).

Keywords: Web Technologies · Smart Learning Affordances · Smart Learning

1 Introduction

The web evolves over the years as new technologies acquire increasingly more intelligent features to keep up with the needs of our time. Initially, web 1.0 was "read-only," followed by "social/interactive" web 2.0 and "semantic" web 3.0. With a current shift towards "symbiotic" 4.0 generation of "connected intelligences," users interact with machines [2], while trends towards an "emotive" 5.0 generation, involving emotional exchanges emerge between them [11]. Nevertheless, web 2.0 technologies are still leverage humans' everyday life, due to their core features of sociality and interactivity. In

A. Rocha et al. (Eds.): WorldCIST 2023, LNNS 800, pp. 501–511, 2024.
https://doi.org/10.1007/978-3-031-45645-9_48

educational contexts, the role of web 2.0 technologies is pivotal because they facilitate effective collaboration and communication between educators and learners, regardless of time and space constraints as well as technical expertise [7]. Compared with the technologies available in previous years, web 2.0 technologies are currently improved versions because they encompass new features due to 3.0 generation of web, which in turn has introduced smarter technologies for personalization enhancement, self-organization, and machine-readable content [9]. As a result, in this paper, the terms "web technologies" and "web tools" encompass both generations 2.0 and 3.0 respectively for web technologies and tools used in teaching and learning.

The evolution of the web and the rapid acceleration of technological achievements and infrastructure have transformed Technology-Enhanced Learning Environments (TELE), to e-learning, mobile, ubiquitous, and smart [1]. Smart Learning Environments (SLEs) are currently learning environments which provide access to widely varying resources and content, are context-aware, offer immediate, adaptive, and personalized learning, by addressing learners' needs [6]. From a technology perspective (without ignoring the role of pedagogy), an SLE performs three core functions: 1. It recognizes data from the context in which it is introduced, namely environmental data, users' profile, or additional information for their learning process. 2. Using logical reasoning and inference mechanisms, it analyzes the collected data. 3. It reacts with feedback, anticipating, intervening, advising, and supporting teaching and learning to meet the specific needs of its users [14].

Most e-courses designed nowadays are either developed on e-learning platforms, which can potentially provide smart learning with appropriate customization based on learning design or require further integration of external web tools to function as optimized learning environments. This study attempts to collect evidence in this direction. As such, it focuses on the educational capabilities of the web tools, which however, take advantage of advanced integrated technological features, for the enhancement of smart learning, through two perspectives. From the administrator's perspective, they could be considered as individual integrated learning environments. This case involves educators and learners using a web tool to experience the entire environment's functionality. As an alternative, web tools could be used as learning activities in an e-learning platform according to its learning design, namely from the user's perspective. In this case, the web tool's functionality is perceived usually by learners and is limited, due to it is based on the use case of the tool as learning activity, while it serves different educational needs. The rest of the paper is structured as follows: Sect. 2 presents the theoretical background for smart learning, explained in terms of its affordances in a SLE. This is followed by Sect. 3 on the research method, including research questions and their answers, followed by Sect. 4 on conclusions.

2 Theoretical Background

2.1 Affordances in Terms of Smart Learning

In the international literature, the term "affordance" is conceptualized and encountered from a variety of angles. Gibson [4] first introduced this term as the totality of all perceived action possibilities latent in the environment that are objectively measured, although independent of an individual's experience, knowledge, culture, or ability to recognize them [16]. Using Gibson's conceptualization, Salomon [13] described affordances as those properties of a thing that influence its use, primarily the functional properties that determine just how to make use of the thing. Norman [12] in turn defined them as actionable properties between actors and their world. In learning contexts, the term "educational affordances" is significantly discussed through the lens of technology and pedagogy. Certainly, there is a discrimination between technological, pedagogical, and educational affordances. Beginning from technological affordances, they could be defined as relational actions that occur among users and technologies [3], namely actions that educators and learners could carry out in the environment by interacting with a particular technological tool [8]. However, how a tool is used is quite different from how it is used in a meaningful way for teaching or learning. This exactly means that the pedagogical affordances of technology are more important than just their technological affordances, while they describe how emerging technologies can transform learning environments into dynamic, smart spaces for learning and development.

In this case, emphasis is placed on the so-called smart learning affordances (SLAs). In a SLE, they represent all the properties provided by the combination of innovative pedagogical approaches and intelligent technologies, which allow learners to learn, by detecting dynamically and acting in accordance with the circumstances, context, learning needs and styles, and the current state of their learning [10]. According to Uskov et al. [15], smart learning is comprised of features classified as six "Levels of Smartness" (LoS), namely adaptation, sensing, inferring, self-learning, anticipation, and self-organization. Learning environments can be instantly customized to meet the needs of learners or educators through adaptation, which is part of the educational process. Inference (logical reasoning) describes all the conditions for processing and producing information, evidence, and rules, whereas sensing (awareness) describes all the features that assist the user in recognizing the elements surrounding him or her and establishing the learning environment. Anticipation is all about predicting and managing different learning situations in the right manner. Self-learning involves using existing knowledge, experience, and behavior modifications for the purpose of improving functionality, processes, and learning. To maintain its integrity, a learning environment should have all the features that enable it to self-organize. In their systematic literature review, Tabuenca et al. [14] examined the affordances that make a learning environment smart. Accordingly, Karampa and Paraskeva [5] identified several SLAs arising from both technology and pedagogy and classified them into the LoS (Table 1).

Table 1. Codes for Smart Learning Affordances (SLAs)

Code	SLAs	Code	SLAs	Code	SLAs	Code	SLAs
AD_ACC	Accessibility	AD_ENG	Engagement	SE_ACT	Action	SL_TWC	Two-way communication
AD_EFF	Efficiency	SE_AWA	Context/ Location/ Situation awareness	INF_LR	Logical reasoning	SL_CL	Collaborative learning
AD_US	Usability	SE_MOB	Mobility	INF_MNT	Mentoring	ANT_SC	Scaffolding
AD_AFF	Affordability	SE_POR	Portability	SL_LC	Learner-centered	ANT_INT	Intervention
AD_ADP	Adaptability	SE_PERV	Pervasiveness	SL_SR	Self-regulation	ANT_AU	Autonomy
AD_PERS	Personalization	SE_OP	Openness	SL_CEN	Content enrichment	ANT_REF	Reflectiveness
AD_IND	Individualization	SE_AUTH	Authenticity	SL_INT	Interactivity	SO_INT	Interoperability
AD_MOT	Motivation	SE_INQ	Inquiry	SL_SI	Social Interaction	SO_FLX	Flexibility

Level of Smartness (LoS): Adaptation = {AD_ACC, AD_EFF, AD_US, AD_AFF, AD_ADP, AD_PERS, AD_IND, AD_MOT, AD_ENG}, **Sensing** = {SE_AWA, SE_MOB, SE_POR, SE_PERV, SE_OP, SE_AUTH, SE_INQ, SE_ACT}, **Inference** = {INF_LR, INF_MNT}, Self-learning = {SL_LC, SL_SR, SL_CEN, SL_INT, SL_CL, SL_TWC, SL_SI}, **Anticipation** = {ANT_SC, ANT_INT, ANT_AU, ANT_REF}, **Self-organization** = {SO_INT, SO_FLX}

3 Research Method

3.1 Research Questions

The aim of this study is to examine the potential of the SLAs of the web technologies in the form of web tools from ICT students' perspective. Particularly, the following research questions have been articulated:

RQ1. What advantages (+) and disadvantages (-) of the web tools utilized in an e-course do ICT students perceive?

RQ2. Based on ICT students' perspectives, which SLAs stand out among the utilized web tools?

3.2 ICT Students' Evaluation

To give answers to the research questions of this study, fifteen prominent web tools (Padlet, Lino, Twiddla, Google Docs, Google Sheets, Google Forms, Learningapps, Wordwall, Genially, Pixton, Canva, Quizizz, Wordart, Renderforest, PollEverywhere) were integrated into an e-course, which was delivered as an e-course lab over an academic semester to twenty-nine ICT students. Learning goals and educational needs guided the design of the e-course lab, focused primarily on problem-solving and collaboration processes. Learners had the opportunity to learn and apply pedagogical concepts (teaching and learning), through the design, development, implementation, and evaluation of computer-supported collaborative learning (CSCL) scripts. Further, they could relate educational methodologies and constructivist concepts (i.e., Project Based Learning, gamification, authentic assessment etc.) to digital tools and environments, and be able to orchestrate conceptual frameworks as integrated e-learning solutions. In other words, learners had the opportunity to get experienced of the proposed web tools, both from pedagogical and technological points of view. In addition, they could either use the web tools as administrators, to experience the functionality of their entire environment as future teachers, or a limited functionality on the part of learners who encounter a learning activity embedded in the e-course.

3.3 Advantages and Disadvantages of the Proposed Web Tools

After successfully completing all the procedures of the e-course lab and utilizing the fifteen proposed web tools both experientially in their learning and teaching practice as future teachers, learners were asked to fill in a form, recording the advantages and the disadvantages for each one of them. It was specifically requested that they consider both the technological capabilities and pedagogical approach. Data were collected and analyzed; thus, findings answered the RQ1 and are presented in Table 2. An advantage is indicated by a plus sign (+), while a disadvantage is indicated by a minus sign (−).

To present the results in a more compact and straightforward manner, qualitative factors were utilized. The question mark (?) highlights a non-mentioned by learners' qualitative factor.

Table 2. Advantages and disadvantages of web tools according to qualitative factors (QF)

1. Padlet, **2.** Lino, **3.** Twiddla, **4.** Google Docs, **5.** Google Sheets, **6.** Google Forms, **7.** Learningapps, **8.** Wordwall, **9.** Genially, **10.** Pixton, **11.** Canva, **12.** Quizizz, **13.** Wordart, **14.** Renderforest, **15.** PollEverywhere

QF	1	2	3	4	5	6	7	8	9	10	11	12	13	14	15
QF1	+	−	−	+	+	+	−	−	−	−	+	−	−	−	−
QF2	−	+	+	−	−	−	−	−	−	−	−	−	+	−	−
QF3	+	+	−	+	+	+	−	+	+	+	+	+	+	+	+
QF4	+	−	−	+	+	+	+	+	−	−	+	−	−	−	−
QF5	+	+	+	+	+	+	+	+	+	+	+	+	+	+	+
QF6	+	+	+	+	+	+	−	−	−	+	+	+	+	−	+
QF7	+	?	?	+	+	+	+	+	+	+	+	+	+	+	+
QF8	+	+	+	+	+	+	+	+	+	+	+	+	+	+	+
QF9	+	+	+	+	+	+	+	+	−	+	−	+	+	+	+
QF10	?	−	−	+	+	+	?	?	−	−	+	?	?	+	?
QF11	−	−	−	?	?	?	?	?	?	?	?	?	?	?	?
QF12	+	+	?	+	+	+	+	+	+	+	+	+	+	+	+
QF13	+	+	+	+	+	+	+	+	+	−	+	+	−	+	+
QF14	+	?	?	+	+	+	+	+	+	+	+	+	+	+	+
QF15	+	+	+	+	+	+	+	+	+	+	+	+	+	+	+
QF16	+	+	−	+	+	+	+	+	+	+	+	+	+	+	+
QF17	+	+	+	+	+	+	+	+	+	+	+	+	+	+	+
QF18	?	−	−	+	+	+	−	+	+	+	+	+	−	−	+
QF19	−	−	−	−	−	+	+	+	+	+	−	+	−	−	+
QF20	+	+	+	+	+	+	+	+	+	+	+	+	+	+	+
QF21	+	?	?	+	+	?	?	?	?	+	?	?	−	?	+
QF22	?	+	+	+	?	?	−	?	+	+	?	?	−	?	?
QF23	+	+	+	+	+	+	+	+	+	+	+	+	+	+	+
QF24	+	+	+	+	+	+	+	+	+	+	+	+	+	+	+
QF25	+	+	−	+	+	+	?	+	?	?	+	?	−	?	+
QF26	+	−	−	+	+	+	−	−	+	−	+	+	−	+	+
QF27	+	+	−	+	+	+	?	+	?	?	+	?	−	?	+

(*continued*)

Table 2. (*continued*)

1. Padlet, **2.** Lino, **3.** Twiddla, **4.** Google Docs, **5.** Google Sheets, **6.** Google Forms, **7.** Learningapps, **8.** Wordwall, **9.** Genially, **10.** Pixton, **11.** Canva, **12.** Quizizz, **13.** Wordart, **14.** Renderforest, **15.** PollEverywhere

QF	1	2	3	4	5	6	7	8	9	10	11	12	13	14	15
QF28	–	–	–	+	+	+	–	–	–	–	–	–	–	-	–
QF29	?	?	?	+	+	+	+	+	–	–	–	+	–	–	+
QF30	?	+	?	?	?	?	+	+	+	–	–	+	–	?	+
QF31	+	+	+	+	+	+	?	+	+	–	+	+	–	+	+
QF32	+	+	+	+	+	+	+	+	+	+	+	+	+	+	+
QF33	?	+	?	+	+	+	+	+	+	+	+	+	+	+	+
QF34	+	+	+	+	+	+	+	+	+	+	+	+	+	+	+
QF35	+	+	?	+	+	+	+	+	+	+	+	+	+	+	+
QF36	–	–	–	+	+	+	+	–	–	–	–	–	–	–	+

QF1 = Accessibility options, **QF2** = Immediate access without registration, **QF3** = Login with social media, **QF4** = Number of languages (> 15), **QF5** = User friendly, **QF6** = Easy to use/learn, **QF7** = Easy navigation, **QF8** = Content organization, **QF9** = Non-complex tools, **QF10** = Autosave/Revision options, **QF11** = Unlimited storage/capacity, **QF12** = Reusable content/switching templates, **QF13** = Variety of media/information, **QF14** = Content export/printable, **QF15** = Free of charge (limited), **QF16** = Customizable profile settings, **QF17** = Customizable content, **QF18** = Aesthetics, **QF19** = Gamification elements, **QF20** = Active involvement, **QF21** = Context awareness (Web camera, GPS, etc.), **QF22** = Realistic interface, **QF23** = Inquiry provoking, **QF24** = Interactive, **QF25** = Off-line access, **QF26** = Mobile application, **QF27** = Seamless work, **QF28** = Free/open/open source, **QF29** = Learning analytics, **QF30** = Task/Time management options, **QF31** = Collaborative work (workgroups/contributors), **QF32** = Communication options (Chat/Social media links), **QF33** = On-line community, **QF34** = Help page/tutorials & notifications, **QF35** = Reflective feedback, **QF36** = Compatible with platforms/tools

3.4 Comparative Analysis and Smart Learning Affordances

Taking into account the above-mentioned recorded advantages and disadvantages of each web tool, an attempt was made to answer RQ2 by conducting a comparative analysis regarding the SLAs (Table 1). The results are presented in Table 3.

Table 3. Comparative analysis of web tools regarding SLAs

1. Padlet, **2.** Lino, **3.** Twiddla, **4.** Google Docs, **5.** Google Sheets, **6.** Google Forms, **7.** Learningapps, **8.** Wordwall, **9.** Genially, **10.** Pixton, **11.** Canva, **12.** Quizizz, **13.** Wordart, **14.** Renderforest, **15.** PollEverywhere

Code	1	2	3	4	5	6	7	8	9	10	11	12	13	14	15
AD_ACC	1	0	0	1	1	1	0	0	0	0	1	0	0	0	0
AD_EFF	1	1	0	1	1	1	0	1	0	0	1	0	1	0	0
AD_US	1	1	?	1	1	1	1	1	1	1	1	1	1	1	1
AD_AFF	1	1	1	1	1	1	1	1	1	1	1	1	1	1	1
AD_ADP	0	0	0	0	0	0	0	0	0	0	0	0	0	0	0
AD_PERS	1	1	?	1	1	1	1	1	1	1	1	1	1	1	1
AD_IND	0	0	0	0	0	0	0	0	0	0	0	0	0	0	0

(*continued*)

Table 3. (*continued*)

1. Padlet, **2.** Lino, **3.** Twiddla, **4.** Google Docs, **5.** Google Sheets, **6.** Google Forms, **7.** Learningapps, **8.** Wordwall, **9.** Genially, **10.** Pixton, **11.** Canva, **12.** Quizizz, **13.** Wordart, **14.** Renderforest, **15.** PollEverywhere

Code	1	2	3	4	5	6	7	8	9	10	11	12	13	14	15
AD_MOT	?	-	-	?	?	1	?	1	1	1	?	1	0	0	1
AD_ENG	1	1	1	1	1	1	1	1	1	1	1	1	1	1	1
SE_AWA	1	?	?	1	1	?	?	?	?	1	?	?	0	?	1
SE_MOB	1	0	0	1	1	1	0	0	1	0	1	1	0	1	1
SE_POR	1	1	0	1	1	1	?	1	?	?	1	?	0	?	1
SE_PERV	1	1	0	1	1	1	?	1	?	?	1	?	0	?	1
SE_OP	0	0	0	1	1	1	0	0	0	0	0	0	0	0	0
SE_AUTH	?	1	1	1	?	?	0	?	1	1	?	?	0	?	?
SE_INQ	1	1	1	1	1	1	1	1	1	1	1	1	1	1	1
SE_ACT	1	1	1	1	1	1	1	1	1	1	1	1	1	1	1
INF_LR	?	?	?	1	1	1	1	1	0	0	0	1	0	0	1
INF_MNT	0	0	0	0	0	0	0	0	0	0	0	0	0	0	0
SL_LC	1	1	1	1	1	1	1	1	1	1	1	1	1	1	1
SL_SR	1	1	?	1	1	1	1	1	1	1	1	1	1	1	1
SL_CEN	1	1	1	1	1	1	1	1	1	1	1	1	1	1	1
SL_INT	1	1	1	1	1	1	1	1	1	1	1	1	1	1	1
SL_CL	1	1	1	1	1	1	?	1	1	0	1	1	0	1	1
SL_TWC	1	1	1	1	1	1	1	1	1	1	1	1	1	1	1
SL_SI	1	1	1	1	1	1	1	1	1	1	1	1	1	1	1
ANT_SC	1	1	?	1	1	1	1	1	1	1	1	1	1	1	1
ANT_INT	0	0	0	0	0	0	0	0	0	0	0	0	0	0	0
ANT_AU	0	0	0	0	0	0	0	0	0	0	0	0	0	0	0
ANT_REF	0	0	0	0	0	0	0	0	0	0	0	0	0	0	0
SO_INT	0	0	0	1	1	1	1	0	0	0	0	0	0	0	1
SO_FLX	0	0	0	0	0	0	0	0	0	0	0	0	0	0	0
TOTAL	**20**	**18**	**11**	**24**	**23**	**23**	**15**	**19**	**17**	**16**	**19**	**17**	**14**	**15**	**21**

Level of Smartness (LoS): Adaptation = {AD_ACC, AD_EFF, AD_US, AD_AFF, AD_ADP, AD_PERS, AD_IND, AD_MOT, AD_ENG}, **Sensing** = {SE_AWA, SE_MOB, SE_POR, SE_PERV, SE_OP, SE_AUTH, SE_INQ, SE_ACT}, **Inference** = {INF_LR, INF_MNT}, Self-learning = {SL_LC, SL_SR, SL_CEN, SL_INT, SL_CL, SL_TWC, SL_SI}, **Anticipation** = {ANT_SC, ANT_INT, ANT_AU, ANT_REF}, **Self-organization** = {SO_INT, SO_FLX}

For those tools, whose functionality described by advantages is in accordance with the consideration of SLAs, the number one (1) indicates this confirmation. In addition, disadvantages are indicated by the number zero (0), while SLAs that are not specified are presumed absent. For some web tools, however, a question mark (?) indicates a doubt about SLAs being provided, because there is not a clear picture of ICT students' perceptions, especially when the number of the advantages is equal to the number of the recorded disadvantages. Obviously, there is a sum of the SLAs for all the utilized web tools, namely an index of how powerful and effective these tools could be considered for smart learning. However, a detailed discussion follows to the next paragraph.

3.5 Concluding Remarks

Comparative analysis has revealed several significant issues to discuss. First, it is useful to list the comparative advantages or disadvantages of the web tools regarding SLAs. Starting with **accessibility**, Padlet, Google Docs, Google Sheets and Canva provide accessibility options, especially for visually impaired users. Padlet provides screen light contrast regulation, while Google tools offer options for working with screen readers, braille devices and screen magnification. In terms of **efficiency**, the choice of languages provided by the web tool is an important factor in reducing user's effort. Google is pioneer because it automatically updates the languages a user frequently uses to Google services. It should be noted, however, that Canva, Padlet, and Wordwall also offer over than 40 languages, while Quizizz, Wordart, and Polleverywhere do not provide any language option. Additionally, users can gain immediate access without logging in Lino and Twiddla for example, or directly signing in via social networks and other platforms' authentication methods, since this minimizes their effort. Several indicators are used to determine **usability**, such as easy navigation and direct access, easy-to-use tools in toolboxes, lightboxes or other visualizations, customizable and reusable content, such as templates or predefined activities for meaningful use, content organization, and variety of media. Therefore, all the proposed web tools provide various types of navigation such as horizontal/vertical, or side menus, breadcrumbs, sliders, tabs, and lists of buttons. Individual tools are placed suitable into toolboxes and reusable content is available through galleries and collection of templates, directly accessed by using tags (i.e., Pixton). Nevertheless, some web tools stand out for their functionality, content organization, and media variability during the learning process. Google tools excel due to the autosave function, the history of revisions and file exports. Wordwall and Pixton provide a lot of printables, Quizizz can import files, while Lino offers drag-and-drop functionality and direct connections to a calendar for further classification by dates. On the other hand, Genially and Canva were considered complex environments due to multiple functions and great volume of information, especially for young students. Content in most proposed web tools is organized in default categories, libraries, and collections often related to learning subjects, age/class levels and objectives (e.g., LearningApps, Quizizz, Pixton), as well as bookmarks and trends (e.g., LearningApps, Wordart) or folders created by users (e.g., Padlet, PollEverywhere). Moreover, content can be customized through (mini) editors by using drawing and formatting tools. Genially, Renderforest and Canva provide a rich collection of different types of information (including text representations such as symbols, formulas in Twiddla, 3D graphics in Renderforest, attachments in Lino, web camera shots and locations in Padlet and Google tools providing thus automatically **context** and **location awareness**). Hypermedia can be integrated both from external sources as well as spontaneously created by users during their learning, providing **content enrichment** and opportunities for creative expression. Unlike the proposed web tools, which are available for free in a limited version (**affordability**), the Google tools are free and fully functional, offering open access without restrictions (**openness**). Compared to other web tools, Canva and Genially' s profile settings offer a greater degree of customization as it has options for display (i.e., style, themes, brand, statistics), thereby options for **personalization**. Content is also personalized when the web tool gives users the right to select different layouts from a variety of templates, different types of information as well as a variety of options for formatting (e.g., background, fonts, icons, drawings etc.) according to their learning needs. All the proposed web tools offer this feature, but some of them, such as Canva, Genially, Renderforest, PollEverywhere, and Google tools are prominent, allowing users to create content from scratch. In addition, Wordwall excels due to its ability to automatically switch templates without work effort, suitable for adaptive teaching. **Motivation** and **engagement** were indicated by factors such as aesthetically pleasing interface with modern graphics and gamification elements. There is no doubt that Wordwall, Canva, and Genially graphics are modern and stylish. Padlet posts are also colorful and pleasant, in contrast to Lino, which was considered old-fashioned and quite unattractive. The elements of gamification appear strongly in Genially, Pixton, Quizizz, Wordwall. Pixton provides avatars and storytelling, while Genially embeds a whole category of gamification activities with games and quizzes tailored as escape rooms. At the same time Quizizz and Wordwall integrate points and leaderboards, which visualize learners' progress and give immediate feedback (a kind of learning analytics, therefore **logical reasoning**). Moreover, Quizizz offers memes, timers, power-ups, extra lives, hints, randomness, and music. All the proposed web tools demonstrate **authenticity** in the following ways: a) Logging into a web tool automatically enters users into a global learning community, where **social interaction** and **two-way communication** stems from various utilities of instant messages (i.e., Learningapps), social media direct linking (i.e., Genially), profile sharing (i.e., Quizizz), commenting and ratings (i.e., Padlet, Learningapps), forums (i.e., Wordart), reflective feedback (i.e., Wordwall) as well as simultaneous content editing and collaborative process (i.e., Lino, Twiddla, Google tools). b) Especially with Lino canvases and Genially board games, educators can simulate physical objects while students are able to directly complete tasks in virtual classes of Genially, Pixton, and Quizizz. c) Creative and problem-solving activities and tasks could be directly linked to real world problems, providing at the same time, **learner-centered, inquiry,** and **action** affordances; for example, Pixton could connect learning context to learners' lives through storytelling on real-world scenarios. Considering the proposed web tools as learning tasks, that promote simultaneously and real-time participation such as Padlet, Lino, Twiddla, Google tools, Wordwall, Genially, Quizizz and PollEverywhere demonstrate high degree of **interactivity** and **awareness** of learners' activity. **Collaborative learning** is achieved both when the web tool provides capabilities for group formation (e.g., Lino, Genially, Canva, Polleverywhere) and real-time collaborative activities are provided during the learning process (e.g., Padlet, Twiddla, Google tools, Wordwall). **Self-regulation** is indicated by reviewing feedback and ratings, seeking information and assistance through social interaction, organizing content, time-management, monitoring and planning through calendars, self-evaluation through assessment activities and so on. In addition, **scaffolding** is provided by all the proposed web tools since they dispose step-by-step guidance, support by video tutorials and notifications, reporting and help centers as well. In terms of **mobility**, all web tools can be accessed via web browsers from several devices, but only a few (i.e., Padlet, Polleverywhere) provide fully mobile applications. However, **portability** appears when users can work offline, such as when they use Google Tools, Wordwall, and Canva. As a result, **pervasiveness** can be achieved when web tools provide both mobility and portability for learners in every learning setting. Since all proposed web tools can be embedded in webpages, **interoperability** seems to be provided only by those who can be integrated in platforms as widgets or plugins (i.e., Learningapps can be integrated as SCORM, while Polleverywhere is compatible with other platforms). Finally, none of the proposed web tools provide **adaptability, individualization, mentoring**, **intervention**, **autonomy**, **reflectiveness,** and **flexibility**.

Among the thirty-two SLAs, one web tool, Google docs, is regarded as the most powerful and promising, offering twenty-four SLAs in total. In contrast, Twiddla provides the least number of eleven SLAs. Observing each level of smartness, individually, adaptation is succeeding by Padlet, Google tools and Canva, although without the significant affordances of adaptability and individualization. Sensing is also fully covered by Google's tools, while inference lags with few exceptions the tools that offer evaluation capabilities. Self-learning is achieved encouragingly by most tools. However, the results from future teachers' perceptions for anticipation and self-organization are substandard.

4 Conclusions

This paper attempted to investigate the potential of web technologies, utilized as web tools in an e-learning environment in terms of SLAs. Fifteen prominent web tools were integrated in an e-course lab designed for twenty-nine ICT students and future teachers' training in various pedagogical and technological issues. From a qualitative analysis, a set of concluding remarks emerged for future research in the field of educational technology, that could contribute to the design of SLEs. According to the perceptions of the learners, it was found that each of the proposed web tools provides possible SLAs, resulting mainly from the design of the user interface/experience, therefore the technological specifications and infrastructure of the tool as well as the pedagogical approach which concerns orientation to specific educational objectives and needs. By isolating the unique features of each tool that have given it some possible SLAs by classifying it into some Level of Smartness (LoS), either technologically or pedagogically, we can include them, if not in a framework for designing intelligent learning environments as guidelines, at least as good practices. Certainly, a SLE as a system should sense, analyze, and react [14], however the results revealed substantial shortcomings in web tools' affordances of adaptability, personalization, guidance, intervention, reflectiveness, and flexibility. These shortcomings are more about the "intelligence" of the technology that should be incorporated into web tools so that they can be considered stand-alone intelligent systems, rather than the pedagogical utilization. These shortcomings pave the wave to web tools evolution according to Web 4.0 and 5.0. Specifications. In addition, learners' perceptions are subjective therefore validity and reliability issues emerge. Nevertheless, they provide empirical evidence and indicators for scholars in the same research field as well as material for future study.

Acknowledgment. This work has been partly supported by the Research Center of the University of Piraeus.

References

1. Adu, E.K., Poo, D.C.: Smart learning: a new paradigm of learning in the smart age. In: Proceedings of the International Conference on Teaching and Learning in Higher Education (TLHE). National University of Singapore, Singapore. (2014)
2. Demartini, C., Benussi, L.: Do web 4.0 and industry 4.0 imply education X. 0? It Prof. **19**(3), 4–7 (2017)
3. Faraj, S., Azad, B.: The materiality of technology: an affordance perspective. Material. Organ. Social Interact. Technol. World **237**, 258 (2012)
4. Gibson, J. J.: The theory of affordances. R. Shaw and J. Bransford (eds.), Perceiving, Acting and Knowing. (1977)
5. Karampa, V., Paraskeva, F.: Design digital educational escape rooms with smart learning affordances. In: 2022 7th Panhellenic Scientific Conference Integration and Use of ICT in the Educational Process (HIUCICTE), pp. 461–474 (2022)
6. Koper, R.: Conditions for effective smart learning environments. Smart Learn. Environ. **1**(1), 1–17 (2014). https://doi.org/10.1186/s40561-014-0005-4
7. Krouska, A., Troussas, C., Sgouropoulou, C.: Usability and educational affordance of web 2.0 tools from teachers' perspectives. In: 24th Pan-Hellenic Conference on Informatics, pp. 107–110 (2020)
8. Leonardi, P.M.: Theoretical foundations for the study of sociomateriality. Inf. Organ. **23**(2), 59–76 (2013)
9. Miranda, P., Isaias, P., Costa, C.J., Pifano, S.: E-learning 3.0 framework adoption: experts' views. In: Zaphiris, P., Ioannou, A. (eds.) LCT 2016. LNCS, vol. 9753, pp. 356–367. Springer, Cham (2016). https://doi.org/10.1007/978-3-319-39483-1_33
10. Molina-Carmona, R., et al.: Research topics on smart learning. In: 8th International Conference on Technological Ecosystems for Enhancing Multiculturality, pp. 231–237 (2020)
11. Nedeva, V., Dineva, S.: Intelligent E-learning with new web technologies. IUP J. Comput. Sci. **16**, 1 (2022)
12. Norman, D.A.: Affordance, conventions, and design. Interactions **6**(3), 38–43 (1999)
13. Salomon, G. (Ed.): Distributed Cognitions: Psychological and Educational Considerations. Cambridge University Press, Cambridge (1997)
14. Tabuenca, B., et al.: Affordances and core functions of smart learning environments: a systematic literature review. IEEE Trans. Learn. Technol. **14**(2), 129–145 (2021)
15. Uskov, V.L., Bakken, J.P., Aluri, L.: Crowdsourcing-based learning: the effective smart pedagogy for STEM education. In: 2019 IEEE Global Engineering Education Conference (EDUCON), pp. 1552–1558. IEEE (2019)
16. Valanides, N.: Technological tools: from technical affordances to educational affordances. In: Problems of Education in the 21st Century, vol. 76, no. 2, p. 116 (2018)

Application of Flipped Learning for the Development of Autonomous Learning for Higher Education Students

Ronald Huacca-Incacutipa[1](✉), Luis Alberto Jesus Arenas-Rojas[1], and Ernesto Alessandro Leo-Rossi[2]

[1] Carver Research, 604 Courtland Street, Suite 131, Orlando, FL, USA
rhuacca@neumann.edu.pe
[2] Escuela de Posgrado Newman, 987 Bolognesi Avenue, Tacna, Peru

Abstract. The purpose of the study was to determine if flipped learning influences the autonomous learning of students of the professional career in accounting of the John Von Neuman Institute of Higher Technological Education - Tacna 2022. The research was of a quantitative approach, of a basic type, explanatory level causal and non-experimental design. The sample consisted of 70 students in the third cycle of the professional accounting career of the aforementioned institute, two questionnaires validated by experts were applied to measure the variables flipped learning and autonomous learning. The results were analyzed using the SPSS 25 statistical program. Non-parametric statistics were used, applying Spearman's Rho correlation coefficient with a significance level of 0.05.

Keywords: Flipped Learning · Autonomous Learning · Virtual Learning Environments · Accounting

1 Introduction

Higher education students in today's globalized environment must be equipped with the skills that our society will demand of them now and in the future. Much effort has been put into educational research to promote new teaching methods, including those that were previously discarded in favor of the teacher-centered method. The traditional role of the student in a higher education classroom is limited to taking notes and completing tasks (either individually or in small groups) given by the professor, who is responsible for documenting everything related to the subject or academic course in the blackboard (Sierra and Mosquera 2020).

In many cases, this situation leads students to perceive that education is still conventional, monotonous and even boring. On the other hand, we all know that the current educational system is in the midst of a technological revolution, in which information and communication technologies can favorably change teaching methods (Mercado 2020).

Given that Latin American societies are changing and evolving, particularly in regard to how students learn, it is important to seek teaching strategies and methodologies

A. Rocha et al. (Eds.): WorldCIST 2023, LNNS 800, pp. 512–518, 2024.
https://doi.org/10.1007/978-3-031-45645-9_49

that incorporate technology, understanding the latter as an essential component and a significant resource for training of students (Aguilera et al. 2017)

In this sense, teachers in many higher education institutions are trying to change the traditional teaching focused on meeting the progress from a curriculum to a teaching based on the needs of the students. The methodology that has aroused interest in students is the flipped learning classroom, as a methodology focused on transferring direct instruction outside the classroom in order to make use of and one-to-one interaction between classroom teacher and student. (Lemon et al. 2017).

Flipped teaching, or "flipped classroom" methodology, is based on the idea that individual instruction is more likely to be effective and efficient than group instruction. To do this, it is important to rethink the conventional classroom environment and implement new pedagogical approaches that make the most of face-to-face teaching to foster autonomous and collaborative learning, and that take advantage of the benefits of information and communication technologies to individualize instruction and ultimately improve student outcomes.

Due to the aforementioned, we should ask ourselves the following question: What is the relationship between flipped learning and the autonomous learning of the students of a technological higher education institute in the city of Tacna? For this reason, it has been proposed as an objective, to determine the relationship between flipped learning and the autonomous learning of the students of a technological higher education institute in the city of Tacna. In the same way, the following hypothesis was established, there is a significant relationship between flipped learning and the autonomous learning of students in a technological higher education institute in the city of Tacna.

2 Flip Learning

Based on constructivist theory of education, flipped classrooms place the student at the center of the experience and encourage independent, self-directed study. According to the active learning theory, students take charge of their own education and decide what they need to study and how they will study it (Dámaso 2022).

Part of this premise is that learning is an active process that encourages critical and synthetic thinking and gives the student the opportunity to direct their own education towards the search for concrete solutions in the real world. Autonomous learning is fostered through flipped learning, as are problem-solving skills and teamwork, both of which are crucial for academic success.

Flipped learning allows students to learn at their own pace by watching videos and using other resources provided at their own discretion, while encouraging interaction with their peers and teachers through ongoing online discussions. This is great for challenging better students and helping those who need to review the material multiple times to fully understand it (Aguilera-Ruiz et al. 2017).

3 Autonomous Learning

For Sovero et al. (2021), the promotion of learning environments is an important part of autonomous education. In addition, the promotion of autonomous learning will allow the training of individuals with clear objectives, capable of accessing, using and processing new information and developing new skills according to their style, rhythm and learning needs.

This will be a reflection of the behavior of a student that is mainly focused on responding to the knowledge requirements established by the instructor, where the student chooses for himself only the contextual conditions (of time, place, tools, etc.) that he considers necessary to formulate an appropriate response. That is, independence is granted with respect to the circumstances, but not with respect to the result or product of learning (León et al. 2020).

4 Virtual Learning Environments and Flipped Learning

The pedagogical approach consists of allocating class time to the most important and individualized learning activities, while the development of the course content takes place outside the classroom and through virtual learning environments, since students can access easily access all the technological resources and tools they have at home, the implementation of the "Flipped Learning" methodology in a virtual context is expected to be highly effective.

Students with different learning styles and backgrounds can benefit from Flipped Learning's emphasis on the use of online tools and resources. In addition, students are involved in the creation of learning resources (Ventosilla et al. 2021).

5 Methodology

The research was of a quantitative approach, of a basic type, causal explanatory level and non-experimental design. Likewise, it is cross-sectional because the data collection was carried out in a single moment, in a single time. The sample consisted of 70 students in the third cycle of the professional accounting career of the aforementioned institute, two questionnaires validated by experts were applied to measure the variables flipped learning and autonomous learning. Quantitative data were collected, processed and analyzed according to previously established variables.

6 Results

Table 1 shows that 70.0% perceive the application of flipped learning as a good methodology, 22.9% show a regular level of satisfaction, and 7.1% perceive the application of flipped as bad. Learning. Denoting acceptance of the applied methodology.

Table 1. Global analysis of the perception of Flipped Learning

Flipped Learning					
	Frequency		Percentage	valid percentage	Accumulated percentage
Valid	misperception	5	7.1	7.1	7.1
	Regular Perception	16	22.9	22.9	30.0
	good perception	49	70.0	70.0	100.0
	Total	70	100.0	100.0	

Table 2 shows that 72.8% show a high level of satisfaction regarding autonomous learning, 18.6% show a regular level of satisfaction and 8.6% a low level of satisfaction on the part of the students. Students, in relation to the autonomous learning developed. It can be mentioned that the level of satisfaction is high.

Table 2. Global analysis of the satisfaction of Autonomous Learning

Student Autonomous Learning					
	Frequency		Percentage	valid percentage	Accumulated percentage
Valid	Low level of satisfaction	6	8.6	8.6	8.6
	Fair level of satisfaction	13	18.6	18.6	18.6
	High level of satisfaction	51	72.8	72.8	100.0
	Total	70	100.0	100.0	

1) Contrast of the Hypothesis

H0: There is no significant relationship between flipped learning and the autonomous learning of students in a technological higher education institute in the city of Tacna.

H1: There is a significant relationship between flipped learning and the autonomous learning of students in a technological higher education institute in the city of Tacna.

2) Significance level

H0 is rejected for every probability value equal to or less than 0.05 and therefore H1 is accepted.

3) Rejection zone

H0 is accepted and H1 is rejected for any probability value greater than 0.05.

4) Test statistic

The non-parametric test of "Spearman's Rho" was chosen, since the data does not present a normal distribution and the data is obtained through ordinal categorical scales.

In Table 3, it is observed that the sig. (bilateral) is 0.000, which is why it is less than 0.05. Consequently, the decision is made to reject the null hypothesis (H0) and the alternate hypothesis (H1) is accepted, since there is a significant correlation between Flipped Learning and the Autonomous Learning of the students. Likewise, a very high positive statistically significant correlation coefficient was found (P=0.946), so the working hypothesis is verified.

Table 3. Spearman's rho of the hypothesis

correlations				
			Flipped Learning	Autonomous Learning
Spearman's Rho	Flipped Learning	Correlation coefficient	1,000	.946**
		Next (bilateral)		,000
		No	70	70
	Autonomous Learning	Correlation coefficient	.946**	1,000
		Next (bilateral)	,000	
		No	70	70
**. The correlation is significant at the 0.01 level (2 tails)				

7 Discussion

The purpose of this study was to determine the relationship that exists between flipped learning and the autonomous learning of students of the professional career in accounting of the Institute of Higher Technological Education John Von Neuman - Tacna 2022.

According to the results regarding the proposed hypothesis, it is evident at $r = 0.946$ that there is a positive and very strong relationship between the study variables. In this way, the alternative hypothesis is accepted and it is deduced that the better Flipped Learning in the students will also have a better autonomous learning of the students.

This result is corroborated by the research carried out by Touron et al (2014) who concludes that this pedagogical model is a great tool for higher education, because whether in a face-to-face or virtual setting, it helps students achieve their learning due to which is motivating, arouses interest in the topic to be developed and also because it strengthens the capacities for autonomous and collaborative learning.

Likewise, Villalobos et al. (2019) point out as a conclusion that in the analysis of the three learning experiences carried out, the use of this model generated that the student has a more active role, promoting their greater participation and interest in the subject, aspects that contribute to the students learning significantly. What is suggested by the teacher. Therefore, its use in the development of the subjects should be considered, considering that its execution must be planned and in accordance with the class purposes. to overcome deficiencies in academic results.

8 Conclusions

- Flipped learning significantly influences active learning in students of the professional accounting career of the John Von Neumann Institute of Higher Education in Tacna 2022, according to the results obtained, there was a positive and very strong relationship between the study variables, presenting a Spearman's Rho correlation coefficient of $r = 0.946$ and p-value (sig. = 0.000).
- Flipped learning as a methodology allows the use of different resources such as ICT, collaborative learning, but above all the emotional and affective part of the human being, in the same way part of the activities to be developed through Flipped Learning, before and During the class, they understand the use of technology, so a contribution is generated in the domain of virtual tools, under the guidance of the teacher and the strengthening of digital skills in students.
- It was verified that Flipped learning is an active methodology that guides the student, through the teacher, in the search and selection of information to achieve meaningful and effective learning, through socialization and continuous practice. It is evident that achieving the learning goals of the course is more effective through Flipped Learning, since before class time, the student begins his self-learning process, which will be strengthened with the support of the teacher through practical activities carried out in Classroom.

References

Aguilera-Ruíz, C., Manzano-León, A., Martínez-Moreno, I., Lozano-Segura, M.D., Casiano, Y.C.: The flipped classroom model. Int. J. Dev. Educ. Psychol. **4**(1), 261–266 (2017). https://www.redalyc.org/articulo.oa?id=349853537027

Damaso Rodriguez, R.M.: Use of flipped learning and meaningful learning in physical education students at the Universidad Nacional Mayor de San Marcos. Lima IGOBERNANZA **5**(18), 295–327 (2021). https://doi.org/10.47865/igob.vol5.n18.2022.197

Gaviria-Rodríguez, D., Arango-Arango, J., Valencia-Arias, A., Bran-Piedrahita, L.: Perception of the invested strategy in university settings. Mexican J. Educ. Res. **24**(81), 593–614 (2019). http://www.scielo.org.mx/pdf/rmie/v24n81/1405-6666-rmie-24-81-593.pdf

León-Pérez, F., Bas, M.C., Escudero-Nahón, A.: Self-perception of emerging digital skills in higher education students. Communicate **28**(62), 91–101 (2020). https://doi.org/10.3916/C62-2020-08

Limón, M., Cantera, E., Salinas, L.: Flipped learning: a teaching- learning proposal in a differential calculus class. Crit. Pedagogy J. **15** (2017). https://www.ecorfan.org/republicofperu/research_journals/Revista_de_Pedagogia_Critica/vol1num1/ECORFAN_Revista_de_Pedagog%C3%ADa_Cr%C3%ADtica_V1_N1_2.pdf

Market, E.P.: Limitations in the use of the flipped classroom in higher education. Transdigital **1**(1), 1–28 (2020). http://bit.ly/td11epml

Sierra, M., Mosquera, F.: The inverted classroom as a pedagogical strategy to improve learning in face-to-face education students. Thesis National Open and Distance University -UNAD. School of Education Sciences (ECEDU) (2020). https://bit.ly/3tOqo39

Sovero Vargas, G.I., Romero Diaz, A.D., Jimenez, O.C.S.: The Flipped Classroom and its influence on active learning in students of the business administration professional career of the Continental Higher Technological Institute - Huancayo 2019. IGOBERNANZA **4**(15), 466–495 (2021). https://doi.org/10.47865/igob.vol4.2021.146

Sosa, D.N.V., Relaiza, H.R.S.M., De La Cruz, F.O., Tito, A.M.F.: Flipped classroom as a tool for the achievement of autonomous learning in university students. Purp. Represent. **9**(1), e1043 (2021). https://doi.org/10.20511/pyr2021.v9n1.1043

Villalobos, G.M., Arciniegas, A.M., González, C.A.L.: Teacher training in ICT with the center for educational innovation CIER-SUR. Sci. Technol. Soc. Trilogy **8**(14), 65–80 (2016)

Touron, J., Santiago, R., Díez, A.: The foundations of the model why is a change in the school necessary? In: Touron, J., Santiago, R., Díez, A.: The Flipped Classroom. How to Turn the School into a Learning Space, pp. 5–16. Ocean Group (2014)

The Application of Artificial Intelligence in Recommendation Systems Reinforced Through Assurance of Learning in Personalized Environments of e-Learning

Fernando Fresneda-Bottaro[1(✉)], Arnaldo Santos[2], Paulo Martins[1,3], and Leonilde Reis[4]

[1] University of Tras-os-Montes and Alto Douro, Vila Real, Portugal
fc.fresneda.bottaro@gmail.com, pmartins@utad.pt
[2] Universidade Aberta, Lisbon, Portugal
arnaldo.santos@uab.pt
[3] Institute for Systems and Computer Engineering, Technology and Science, Porto, Portugal
[4] Polytechnic Institute of Setubal, Setubal, Portugal
leonilde.reis@esce.ips.pt

Abstract. Learning environments unquestionably enable learners to develop their pedagogical and scientific processes efficiently and effectively. Thus, considering the impossibility of not having conditions of autonomy over the routine underlying the studies and, consequently, not having guarantees of the learning carried out makes the learners experience gaps in the domain of materials adequate to their actual needs. The paper's objective is to present the relevance of the applicability of Artificial Intelligence in Recommendation Systems, reinforced through the Assurance of Learning, oriented towards adaptive-personalized practice in corporate e-learning contexts. The research methodology underlying the work fell on Design Science Research, as it is considered adequate to support the research, given the need to carry out the design phases, development, construction, evaluation, validation of the artefact and, finally, communication of the results. The main results instigate the development of an Adaptive-Personalized Learning framework for corporate e-learning, provided with models of Artificial Intelligence and guided using the Assurance of Learning process. It becomes central that learners can enjoy adequate academic development. In this sense, the framework has an implicit structure that promotes the definition of personalized attributes, which involves recommendations and customizations of content per profile, including training content that will be suggested and learning activity content that will be continuously monitored, given the specific needs of learners.

Keywords: Artificial Intelligence · Recommendation Systems · Assurance of Learning · Adaptive-Personalized Learning · Learning Analytics

A. Rocha et al. (Eds.): WorldCIST 2023, LNNS 800, pp. 519–529, 2024.
https://doi.org/10.1007/978-3-031-45645-9_50

1 Introduction

The development and launch of new training and qualification technologies are supporting organizations in optimizing the quality of the teaching and learning process, benefiting employees and collaborators, either in reducing costs and training time or in compensating for the lack of opportunities for initial and continuing improvements. Furthermore, it is a fact that professional qualification enables the development of skills to highlight learning and to privilege opportunities for innovation, which responds to the restructuring demanded of corporations, as a rule, especially in a knowledge-based society [1].

Furthermore, through Adaptive Learning (AL), it is possible to offer Personalized Learning (PL) and an experience stimulated by Digital Technologies [2]. Equally boosted by Information & Communication Technology (ICT) tools, e-learning is successfully used in multiple forms of corporate interaction. The benefits range from establishing ways of communicating and interacting to taking exams and evaluating progress. However, most of these environments are designed on a one-size-fits-all approach. Although effective, such environments are sometimes quickly abandoned by learners [3, 4].

In the evolution of this perspective, it is observed that the learner is utterly devoid of autonomy and control over the study routine. The experience starts to be imposed by predetermined contents and times, segmented tasks/jobs and codes of conduct. There are differences in the instruction process from learner to learner, as the interests and paces of learning differ. After his school cycle, the reality of the classical learner lies in the difficulty of mobilizing his acquired knowledge. Such setbacks will extend from personal to professional experience.

Planning to simplify intelligent learning, mechanisms and Recommendation Systems / Recommender Systems (RS) are designed to perform unique tasks where conveniences are proven relevant to learning outcomes. The challenges of e-learning are primarily because of the progressions in Content-Based (CB), Collaborative (CF), Hybrid (HF) filtering. Also, difficulties with Cold-Start, Sparsity, First Rater, Popularity Bias, Accuracy, Scalability. Some proposed solutions are considered to solve the problems, such as Cross-Domain recommendations, Context-Aware recommendations and Deep Learning (DL) techniques, among others [5–7].

Regarding the relationships between RS and e-learning tools, the references on performance indicators are reduced, including the lack of reports associated with student and faculty feedback [8]. Thus, to resolve the problems, the following actions are proposed: a) Identify Adaptive-Personalized Learning contexts; b) Plan metrics for the different contexts (indicators: performance, development, monitoring, satisfaction, training); c) Identify Artificial Intelligence (AI) methods and algorithms that enable content recommendations; d) Design a framework that explores the skills of employees.

2 Contribution

This document registers the intention to implement an Adaptive-Personalized Learning framework for corporate e-learning, provided with AI models (methods and algorithms) and guided using the learning management process [9–12]. Therefore, multiple attributions can be established, such as a) Investigations into the advantages and disadvantages of AI for RS and/or Adaptive-Personalized Systems, reinforced through *Assurance of Learning* (AoL); b) Checks of the leading AI algorithms and metrics, integrated with RS and/or Adaptive-Personalized Systems, currently used in e-learning platforms – Learning Management Systems (LMS)/Learning Content Management Systems (LCMS); c) Use of a framework and/or reference model for the application of AI in RS and/or Adaptive-Personalized Systems, reinforced through AoL in e-learning contexts; d) Findings of differentials in the practice of AoL in the application of AI in RS and/or Adaptive-Personalized Systems in e-learning environments.

According to the strategy adopted for the execution of the phases of this work, it appears that the research methodology to be applied is Design Science Research (DSR), which comprises a rigorous process to design artefacts in solving problems, making contributions, project evaluation and communication of results [13]. The methodology will support the research work due to the possibility of carrying out the design phases (distributed in different stages of the project), development, construction, evaluation and validation of artefacts, among others. Consequently, will be used the method proposed by Alturki and co-authors [14], in which the 3 (three) Hevner cycles [15] – Rigor, Relevance and Design - are considered, in addition to the recommendation of 14 (fourteen) activities foreseen in the process.

2.1 Motivation

E-learning has transformed educational didactics as an alternative to traditional teaching-learning processes, providing innovative trends and methodologies daily, which extensively influences and favours the education sector [16]. In terms of training, online learning becomes significantly profitable for employees to acquire knowledge, allowing them to learn at any time, at any pace – based on the student profile – and from anywhere. However, how to configure the appropriate content for learning applications (?) becomes an exceptional question, especially when each trainee has a distinct learning profile and, surprisingly, when the training material must also suit this work [17].

The following will be considered about the identified problem: a) Adaptive-Personalized Teaching: the teaching-learning process is static because "a single type of teaching" is offered. There is no adaptation and/or personalization according to the individual needs of the learners; b) Recommendation Systems: there are still specific weaknesses in traditional RS through filtering methods. Resources and content classifications are used, in which the learner's context is not contemplated; c) Learning Analytics: the systematic use of data and training indicators, which have been partially implemented and present summarized metrics (feedback, assessment); and d) Artificial Intelligence: the existing algorithms are, until now, insufficiently developed and minimally used to enable the recommendations of training content for learners.

In the understanding that systems involving AI almost always end up as protagonists in the context of adaptive and personalized learning and in the relevance of offering the employee/learner intelligent learning, this work aims to answer the following questions: (**RQ1**) What are the advantages and disadvantages of Artificial Intelligence for Recommendation Systems and/or Adaptive-Personalized Systems, reinforced through the Assurance of Learning? (**RQ2**) What are the leading Artificial Intelligence algorithms and metrics, integrated with the Recommendation Systems and/or Adaptive-Personalized Systems, currently used in e-learning platforms (LMS/LCMS)? (**RQ3**) What characteristics should a framework and reference model have for applying Artificial Intelligence for Recommendation Systems and/or Adaptive-Personalized, reinforced through the Assurance of Learning in corporate e-learning contexts? (**RQ4**) What are the differentials in using Assurance of Learning in applying Artificial Intelligence for Recommendation Systems and/or Adaptive-Personalized Systems in corporate e-learning contexts?

3 Related Works

Concerning related works, Adaptive-Personalized Learning – in different environments – and Assurance of Learning and Solutions/Frameworks are presented according to the study's pertinence.

3.1 Adaptive-Personalized Learning

The learning contents are analyzed in multiple, adaptive and personalized environments through recommendation methods, evaluation metrics, usability tests and attributes and cognitive aspects of learners [18]. An adaptive learning system, composed of an interactive and dynamic pedagogy, consists of several key features, with enough autonomy to keep learners engaged and motivated towards the objectives [19]. Furthermore, e-learning platforms include tools that adapt learning materials according to the learner's profile. The purpose is to offer unique learning materials in which it is possible to find solutions that support tutors in creating pedagogical content and learning objects adapted to the student's abilities and preferences [13].

The pedagogical model of adaptive learning can enable personalized and individualized learning. Continuous data collection on the user's general activities and actions release feedback that adapts to his pace and needs. By carrying a data-driven approach, it provides individualized learning paths. In this way, the information analysis mechanisms customize – in real time – the offer of e-learning materials according to the learners' performance level [20]. Furthermore, Peng and co-authors point out that "the development of current technologies has made Personalized Learning increasingly adaptive and Adaptive Learning increasingly personalized" [21].

3.2 Assurance of Learning

The appropriation of the Assurance of Learning (AoL) [22] process – in the precaution in quality education – is perceived by the systematic process of collecting and reviewing data on the results of training and qualifications, as well as in the adequacy of methodologies of learning (active, agile, immersive and analytical). It is continually used in developing and improving training and educational programs [12, 22]. Furthermore, using Learning Outcomes (LO) – through Bloom's Taxonomy or Taxonomy of Educational Objectives [23] – ensures the execution of the proposed learning. Similar to the

5 (five) steps of the Association to Advance Collegiate Schools of Business (AACSB) aimed at the learning management process [12, 22].

In some cases, Learning Objectives and Learning Outcomes are used interchangeably. However, in practice, the purposes are different. The 'objectives' indicate the purpose of the learning activity and the desired results. In the case of 'results', they show what the learner can accomplish when completing the proposed activities. Similarly, it is understood by statements of what is achieved and evaluated at the end of a cycle of studies [24]. According to Libba and coauthor-res [25], the expositions of learning – mission, objectives, outlines – in formations and programs are commonly disparate, even within the same study area. Such dissimilarity contributes to a difference in teaching materials and methodologies [24].

3.3 Solutions/Frameworks

According to a literature review, an expressive part of e-learning solutions and frameworks was identified, and used in this work, based on Recommendation Systems. However, most of the different solutions that this work seeks are to provide AI models. Specific frameworks are focused on Adaptive Learning, others on Personalized Learning, but limited frameworks are targeted explicitly at Adaptive-Personalized Learning. Likewise, about assurance of learning and corporate education. The details of each structure, including name, year and reference, are summarized and listed in Table 1.

Table 1. Summary of solutions and frameworks based on e-learning Recommendation Systems. The structure, year and reference details. The author.

Framework	Year	Reference
Recommender Systems in E-learning	2022	[26]
A Survey of Recommendation Systems: Recommendation Models, Techniques, and Application Fields	2022	[27]
Review and classification of content recommenders in an E-learning environment	2021	[28]
Adaptive E-Learning System	2021	[29]
A hybrid recommendation model in social media based on deep emotion analysis and multi-source view fusion	2020	[30]
Toward a Hybrid Recommender System for E-learning Personalization Based on Data Mining Techniques	2018	[31]
Personalized recommender system for e-Learning environment based on student preferences	2018	[32]
Good and Similar Learners' Recommendations in Adaptive Learning Systems	2016	[33]

4 Model Design

Below are the details of the DSR, as well as the draft of the model under development.

4.1 Design Science Research

The application of the DSR will allow the development of an artefact to support corporations in optimizing the quality of the teaching and learning process – reducing costs and training time for employees and collaborators. Table 2 illustrates the objective of this research work on the relevance of the applicability of Artificial Intelligence in Recommendation Systems, reinforced through Assurance of Learning, oriented towards adaptive-personalized practice in business e-learning contexts. A mechanism to promote the definition of personalized attributes involves recommendations and customizations of content by profile, including training content that can be recommended and content of learning activities that will be monitored, using the model developed by Alturki and co-authors [14].

Table 2. Alturki and co-authors' framework for the application of DSR. [14].

Activity	Description
1. Document the spark of an idea/problem	After the identified problems, it is recalled that the investigation's idea is to apply Artificial Intelligence in Recommendation Systems for personalized e-Learning environments
2. Investigate and evaluate the importance of the problem/idea	In addition to the listed problems, a Systematic Literature Review (SLR) will be made to verify and compare the work carried out with the proposed work
3. Evaluate the new solution feasibility	A paper (position paper) will be prepared to evaluate and verify the feasibility of the theme
4. Define the research scope	The constitution of an Adaptive-Personalized Learning framework with features for corporate e-Learning. Identified Adaptive-Personalized Learning contexts; To be projected metrics for the different contexts (indicators: performance, development, follow-up, satisfaction and training); To identify Artificial Intelligence methods and algorithms that enable content recommendations, and conceived, a framework that explores the abilities of the collaborators
5. Resolve whether within the DS paradigm	The work adheres to the DS perspective

(*continued*)

Table 2. (*continued*)

Activity	Description
6. Establish type (IS DS versus IS DSR)	The Assurance of Learning process is accredited by AACSB [22]. Furthermore, he remembers the use of DSR for the research methodology, which comprises a rigorous process for designing artefacts in problem-solving, making contributions, evaluating projects and communicating results [34] and adopting the 3 (three) Hevner cycles [15]. For the SRL, the guideline proposed by Kitchenham [35] will be used, which brings a careful analysis of the quality of the literature to be selected
7. Resolve the theme (construction, evaluation or both)	The investigation will be about the construction and evaluation of an artefact
8. Define requirements	Tools, including the Word or Writer text editor, should be used to elicit the requirements. The investigation technique will be the interview for data collection (qualitative)
9. Define alternative solutions	Not applicable at this point in the project
10. Explore knowledge 11. Base support of alternatives	Not applicable at this point in the project
12. Prepare for design and/or evaluation	The development and/or evaluation plan must be prepared concurrently with the evolution of the project
13. Develop (construction)	It will be carried out concurrently with the evolution of the project
14. Evaluate	
"Artificial" evaluation	Initial tests should be conducted in a laboratory context, as the intention is to prepare the framework for the most varied contexts
"Naturalistic" evaluation	Other tests in different contexts are necessary to prove the structure's robustness and acquire confidence in what is proposed. A company or several actual companies will be needed for the tests of the developed proposal (data collection and validation of what the research is coming to contribute, with a particular distinction to existing works
15. Communicate findings	It will be informed to the scientific community through the publication of articles

Table 2 emphasizes the activities and descriptions underlying the authors' views framed in the DSR.

4.2 Proposed Model

Figure 1 presents the draft of the model under development, demonstrating the flow of actions to be undertaken implicit in the learning process.

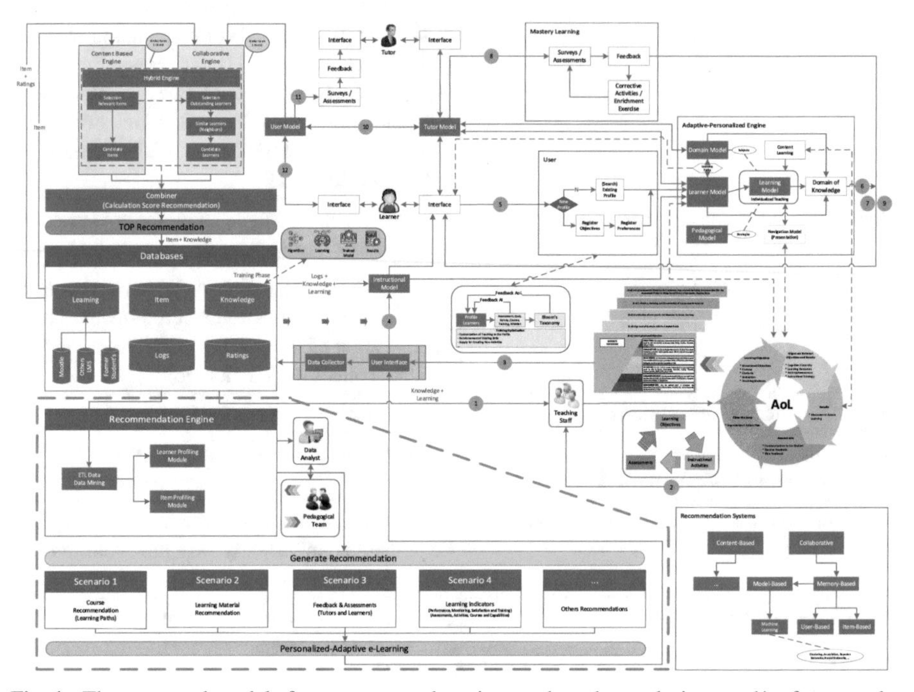

Fig. 1. The proposed models for corporate e-learning are based on solutions and/or frameworks by Bourkoukou & Bachari [31], Fazazi et al. [32], Jiang et al. [30], Joy & Pillai [28], Ko et al. [27], Nurjanah [33], Zhang et al. [26].

In summary, the learning process, shown in Fig. 1 contains: a) Processes and mechanisms: user, adaptive-personalized learning, learning mastery (learners must reach a level to advance), including compliance with the AoL process cycle; b) Learner interface: presents the result of course and material recommendations, according to the profile. Handles logic and events during training (surveys, quizzes, feedback, assessments, etc.); c) Tutor interface: presents contents and tips to guide and clarify doubts about the studies. Partial performances and events during training (surveys, feedback, etc.); d) Databases: storage of learning information (behaviors used in building/conducting the learning profile. Sources: Moodle, other LMS, former students), items (information on materials, objects and learning activities), logs and mainly knowledge extracted from the list of materials, learning path and performance. The 'combiner' (relevant recommendations) retrieves data from the recommendation's engine (of types) (content-based, collaborative, hybrid) and later ETL and DM; d) Recommendation engine: the central part, where the device trains the recommendation methods for knowledge generation. Contains the learner and item profile modules. Implements recommendation methods (involving the pedagogical team and the data analyst). The results (classification, measurement and combination: learners or items) are generated as a sorted list of items for decision-making. Recommendations: a) Scenario 1 - Course Recommendation (Learning Paths): a learner who is looking for a course and the skills he would like to improve; b) Scenario 2 - recommendation of learning material: learners who are enrolled in one or more courses, however, looking for learning materials that help them advance in knowledge; c) Scenario 3: feedback and evaluations (tutors and students); d) Scenario 4: Learning Indicators (performance, monitoring, satisfaction and training: assessments, activities, courses and training); e) Other Recommendations: Different data about students/items are used by the RS, depending on the recommendation scenario. It is stressed that ethical and privacy issues should be taken into consideration when considering such recommendation frameworks, in particular involving AI. Mechanisms must ensure the security of user information.

5 Future Work

In this context, however, it is possible to highlight the strategic planning of organizations, which can guarantee the practical success of e-learning – in some instances, transformed into corporate academies/universities – listed as one of the means to achieve the business goals. The domains to be addressed, as are-as of challenges, in future works are i) Training (training), competence (skill) and awareness (experience) of tutors in the preparation and use of digital technologies; ii) Appropriation process for adaptability and customizable intelligent learning systems (digital pedagogy and educational or instructional design); and iii) Acceptance of intelligent learning, as well as active methodologies (innovative education) in general, by learners and tutors.

References

1. Bell, D.: The coming of the post-industrial society. Educ. Forum **40**, 574–579 (1976). https://doi.org/10.1080/00131727609336501
2. Anton, C., Shikov, A.: The method of personalized corporate e-learning based on personal traits of employees. Procedia Comput. Sci. **136**, 511–521 (2018). https://doi.org/10.1016/J.PROCS.2018.08.253
3. Alomair, Y., Hammami, S.: A review of methods for adaptive gamified learning environments. In: 2020 3rd International Conference on Computer Applications & Information Security (ICCAIS), pp. 1–6. IEEE (2020)
4. Samoylenko, N., Zharko, L., Glotova, A.: Designing online learning environment: ICT tools and teaching strategies. Athens J. Educ. **9**, 49–62 (2021). https://doi.org/10.30958/aje.9-1-4
5. Gahier, A.K., Gujral, S.K.: Cross domain recommendation systems using deep learning: a systematic literature review. SSRN Electron. J. (2021). https://doi.org/10.2139/ssrn.3884919
6. Gogo, K.O., Nderu, L., Mutua, S.M., et al.: Context aware recommender systems and techniques in offering smart learning: a survey and future work. In: ACSE (2020)

7. Srivastav, G., Kant, S.: Review on e-learning environment development and context aware recommendation systems using deep learning. In: 2019 3rd International Conference on Recent Developments in Control, Automation & Power Engineering (RDCAPE), pp. 615–621. IEEE (2019)
8. Jannach, D., Zanker, M.: Value and Impact of Recommender Systems. In: Recommender Systems Handbook, pp. 519–546. Springer, New York (2022). https://doi.org/10.1007/978-1-0716-2197-4_14
9. Jump, A., Goodness, E., Hare, J., et al.: Emerging Technologies and Trends Impact Radar: Artificial Intelligence, 2021. Gartner, Inc (2021). https://www.gartner.com/en/documents/4006010. Accessed 20 July 2022
10. Goasduff, L.: The 4 Trends That Prevail on the Gartner Hype Cycle for AI, 2021. In: Gartner, Inc (2021). https://www.gartner.com/en/articles/the-4-trends-that-prevail-on-the-gartner-hype-cycle-for-ai-2021. Accessed 20 July 2022
11. Afini Normadhi, N.B., Shuib, L., Md Nasir, H.N., et al.: Identification of personal traits in adaptive learning environment: Systematic literature review. Comput. Educ. **130**, 168–190 (2019). https://doi.org/10.1016/j.compedu.2018.11.005
12. Ching, H.Y., Gross, A., Vasconcellos, L.: Gestão da aprendizagem: casos práticos. Atlas, São Paulo (2020)
13. Talaghzi, J., Bennane, A., Himmi, M.M., et al.: Online adaptive learning: a review of literature. In: ACM International Conference Proceeding Series, pp 115–120. ACM, New York (2020)
14. Alturki, A., Gable, G.G., Bandara, W.: A design science research roadmap. In: Jain, H., Sinha, A.P., Vitharana, P. (eds.) Lecture Notes in Computer Science (including subseries Lecture Notes in Artificial Intelligence and Lecture Notes in Bioinformatics), pp. 107–123. Springer, Berlin, Heidelberg (2011). https://doi.org/10.1007/978-3-642-20633-7_8
15. Hevner, A.R., March, S.T., Park, J., Ram, S.: Design science in information systems research. MIS Q. Manag. Inf. Syst. **28**, 75–105 (2004). https://doi.org/10.2307/25148625
16. Cano, P.A.O., Alarcón, E.C.P.: Recommendation systems in education: a review of recommendation mechanisms in e-learning environments. Rev. Ing. Univ. Medellín. **20**, 147–158 (2020). https://doi.org/10.22395/rium.v20n38a9
17. Zhong, L., Wei, Y., Yao, H., et al.: Review of deep learning-based personalized learning recommendation. In: Proceedings of the 2020 11th International Conference on E-Education, E-Business, E-Management, and E-Learning, pp 145–149. ACM, New York (2020)
18. Raj, N.S., Renumol, V.G.: A systematic literature review on adaptive content recommenders in personalized learning environments from 2015 to 2020. J. Comput. Educ. **9**, 113–148 (2022). https://doi.org/10.1007/s40692-021-00199-4
19. Foley, P.: Transactional distance and adaptive learning, planning for the future of higher education. J. Interact. Media. Educ. **2019** (2019). https://doi.org/10.5334/jime.542
20. Paramythis, A., Loidl-Reisinger, S.: Adaptive learning environments and e-learning standards. Electron. J e-Learn. **2**, 181–194 (2004)
21. Peng H, Ma S, Spector JM (2019) Personalized Adaptive Learning: An Emerging Pedagogical Approach Enabled by a Smart Learning Environment. In: Lecture Notes in Educational Technology. pp 171–176
22. AACSB International (2007) AACSB Assurance of Learning Standards: An Interpretation (2013)
23. Bloom, B.S.: Taxonomy of Educational Objectives: The Classification of Educational Goals, Parts 1–2. David McKay Co Inc., New York (1972)
24. Shah, A.A., Syeda, Z.F., Shahzadi, U.: Assessment of higher education learning outcomes of university graduates. Glob. Educ. Stud. Rev. **1**, 72–83 (2020). https://doi.org/10.31703/gesr.2020(V-I).08

25. McMillan, L., Johnson, T., Parker, F.M., et al.: Improving student learning outcomes through a collaborative higher education partnership. Int. J. Teach. Learn. High. Educ. **32**, 117–124 (2020)
26. Zhang, Q., Lu, J., Zhang, G.: Recommender systems in E-learning. J. Smart Environ. Green Comput. (2022). https://doi.org/10.20517/jsegc.2020.06
27. Ko, H., Lee, S., Park, Y., Choi, A.: A survey of recommendation systems: recommendation models, techniques, and application fields. Electronics **11**, 141 (2022). https://doi.org/10.3390/electronics11010141
28. Joy, J., Pillai, R.V.G.: Review and classification of content recommenders in E-learning environment. J. King Saud Univ. – Comput. Inf. Sci. **34**, 7670–7685 (2022). https://doi.org/10.1016/j.jksuci.2021.06.009
29. Sweta, S.: Adaptive E-learning system. In: Modern Approach to Educational Data Mining and its Applications, pp. 13–24 (2021)
30. Jiang, L., Liu, L., Yao, J., Shi, L.: A hybrid recommendation model in social media based on deep emotion analysis and multi-source view fusion. J. Cloud Comput. **9**, 57 (2020). https://doi.org/10.1186/s13677-020-00199-2
31. Bourkoukou, O., El Bachari, E.: Toward a hybrid recommender system for e-learning personnalization based on data mining techniques. JOIV Int. J. Inf. Vis. **2**, 271 (2018). https://doi.org/10.30630/joiv.2.4.158
32. El Fazazi, H., Qbadou, M., Salhi, I., Mansouri, K.: Personalized recommender system for e-Learning environment based on student's preferences (2018)
33. Nurjanah, D.: Good and Similar learners' recommendation in adaptive learning systems. In: Proceedings of the 8th International Conference on Computer Supported Education, vol. 1, pp. 434–440. CSEDU, SciTePress (2016)
34. Peffers, K., Tuunanen, T., Rothenberger, M.A., Chatterjee, S.: A design science research methodology for information systems research. J. Manag. Inf. Syst. **24**, 45–77 (2007). https://doi.org/10.2753/MIS0742-1222240302
35. Kitchenham, B.: Procedures for Performing Systematic Reviews, Version 1.0. Department of Computer Science, Keele University, UK (2004)

Learning Performance and the Influence of Individual Motivation Through Social Networks

Olger Gutierrez-Aguilar[1] (✉), Ygnacio Tomaylla-Quispe[2], Lily Montesinos-Valencia[1], and Sandra Chicana-Huanca[2]

[1] Universidad Católica de Santa María, Arequipa, Perú
{ogutierrez,lmontesi}@ucsm.edu.pe

[2] Universidad Nacional de San Agustín, Arequipa, Perú
{itomaylla,schicanah}@unsa.edu.pe

Abstract. The research aims to find out how individual motivations, like altruism, affect how well people learn and how social networks help people form and share knowledge as one of their functions. The methodology used for the study corresponds to a non-experimental investigation. Reliability tests and a questionnaire were applied to a random sample of 130 university students ($n = 24$; $\alpha = 0.978$ $\omega = 0.978$). Through Partial Least Squares Structural Equation Modeling (PLS-SEM), the causal relationship between the exogenous and endogenous variables was used to design the model. This was done in addition to the exploratory and confirmatory factorial analyses. It has been shown that social networks are the best way for personal motivations like reputation and, significantly, altruism to affect the exchange of information and knowledge and the learning performance of college students in a way that supports the thesis. There is a statistically significant link between the altruistic sharing of knowledge, the formation of knowledge, and how well people learn.

Keywords: Individual motivation · social networks · altruism · learning performance · knowledge exchange · knowledge formation

1 Introduction

Social networks have played a predominant role in times of COVID-19; their usefulness and ease of use, enjoyment through positive emotions, and their relationship with school performance are significant in current education [1]. However, its reasoned use has made it possible to establish relationships that revolve around personality and creative thinking and are undoubtedly mediated by the reputation of those who participate in social networks [2]. Likewise, using social networks as a resource for teaching and learning allows students to establish reasonable levels of individual motivation, which could ultimately improve their school performance. Thus, social networks are tools that have changed how people interact and communicate, so social networks are becoming a valuable platform to facilitate the exchange of knowledge [3].

A. Rocha et al. (Eds.): WorldCIST 2023, LNNS 800, pp. 530–537, 2024.
https://doi.org/10.1007/978-3-031-45645-9_51

According to the theoretical foundations of social cognitive theory and connectivism, it is possible to frame the influence of social networks and individual motivation, primarily related to reputation and altruism, especially in knowledge sharing and learning performance [4]. The relationship between sharing knowledge to gain and improving reputation characterizes an altruistic individual [5], recently supported by Hoseini and others in the multidimensional model of knowledge-sharing behavior in mobile social networks [6]. The contributions of Choi, Ramírez, Gregg, and Lee are conclusive, supporting the thesis that altruism and reputation are individual motivational factors that significantly affect student knowledge exchange [7], Therefore, it would improve performance in their learning in a collaborative context due to the easy accessibility, affordability, and rapid interaction characteristic of social networks [8], in addition to immediacy, in a safe environment with identity and good digital citizenship [9].

H1 There is a degree of influence between altruism and learning performance.

The exchange of knowledge among students occurs through altruistic attitudes, which in educational situations improves interpersonal skills, reinforcing learning [10]. Shahzad et al. warn that the benefits students can obtain through social networks, and individual motivation will influence the exchange of knowledge and thus learning performance [11]. The high degree of interactivity of the networks between students and teachers strengthens the exchange of knowledge and learning performance [12]. The application of social learning theory makes it possible to inquire about the knowledge exchange behavior of the virtual community, as is the case of YouTube, which revealed that the attributes of individuals, such as self-efficacy and expectations of results, accelerate the intention to share knowledge [13].

Sharing or exchanging information and knowledge in social networks is a condition for participating in virtual communication and knowledge formation [14]. For the theory of connectivism, the exchange of information, virtual communication, and knowledge formation is positively associated with more significant learning and, obviously, with the exchange of knowledge [15]. Likewise, through effective communication, social networks such as Instagram, Facebook, Instagram, ResearchGate, Telegram, Twitter, WhatsApp, and YouTube create better opportunities for knowledge exchange [16].

H2 There is a statistically significant effect between knowledge sharing and altruism.

H3 There is a degree of influence between knowledge sharing and knowledge formation.

H4 There is a statistically significant relationship between knowledge sharing and learning performance.

Predictors such as reputation and altruism (individual motivating factors), resource sharing for learning, virtual communication, and knowledge formation (social network functions) undoubtedly influence knowledge sharing and learning performance [4]. On the other hand, the formation of knowledge is carried out through the creation of content; Nonaka and Takeuchi at the time believed that creation occurs at the time of writing or recording personal experiences, assignment problems, presentation slides, training, and development of an idea into a video, and uploading the content to any social networking site for the benefit of others [17]. For knowledge formation through social networks, individuals are more likely to improve knowledge sharing and overall performance in their online learning groups or networks [18]. Thus, knowledge formation has a positive association with knowledge sharing and student achievement.

H5 There is a degree of influence between knowledge formation and altruism.

H6 There is a meaningful relationship between knowledge formation and learning performance.

2 Methodology

The study units were made up of 130 university students in the professional career of industrial engineering; 52.3% were men, and 47.7% were women, aged between 18 and 25 years or older. The total mean was 20.88 (SD = 2.288), and the participants were randomly selected. The application of the instrument was carried out in June 2022. The applied instrument is an adaptation of the one proposed by Mosharrof Hosen et al. in a paper entitled Individual motivation and the influence of social networks on students' knowledge sharing and learning performance: Evidence from an emerging economy [4].

3 Results

The reliability tests were carried out with the Jamovi software (v:2.3.13); the result was an Alpha Cronbach ($\alpha = 0.924$) and a McDonald's Coefficient ($\omega = 0.933$), which means excellent measurement. Then the commonalities were evaluated, which are the sum of the squared factorial weights for each row; the result oscillates between 0.390 and 0.860, which is equivalent to saying that the items would explain 39% of the model and, at best, they would get 86%. As for the total variance explained, the results obtained from the four components would explain the model by 64.30%. For the exploratory analysis, the following criteria were used: the principal component extraction method, with a fixed number of 4 factors, and for the verification of assumptions, a Kaiser-Meyer-Olkin test (KMO = 0.831), which indicates a reasonable adjustment of the analyzed items with their factor. Bartlett's Sphericity Test had the following results: $\chi 2 = 1266.51$; gl = 171, and p = <0.001, which are significant.

The confirmatory analysis was performed with the SPSS Amos software (v:28) using the Covariance-Based Structural Equation Modeling (CB-SEM) method, with the independent variables of the model being: 1 Knowledge Formation (KF); 2 Altruism (AL); 3 Learning Performance (LP) and 4 Knowledge Sharing (KS). Using the fit criteria for the CB-SEM model, a comparative fit index CFI = 0.927 was obtained; root mean square error of approximation (RMSEA) = 0.068; the goodness of fit index (GFI) = 0.846; the Tucker-Lewis index (TLI) = 0.914; Chi-square/df = 1.589; cmin = 319.962: p = 0.000; df = 146. Therefore, the results show the robustness and reliability of the model. See Fig. 1.

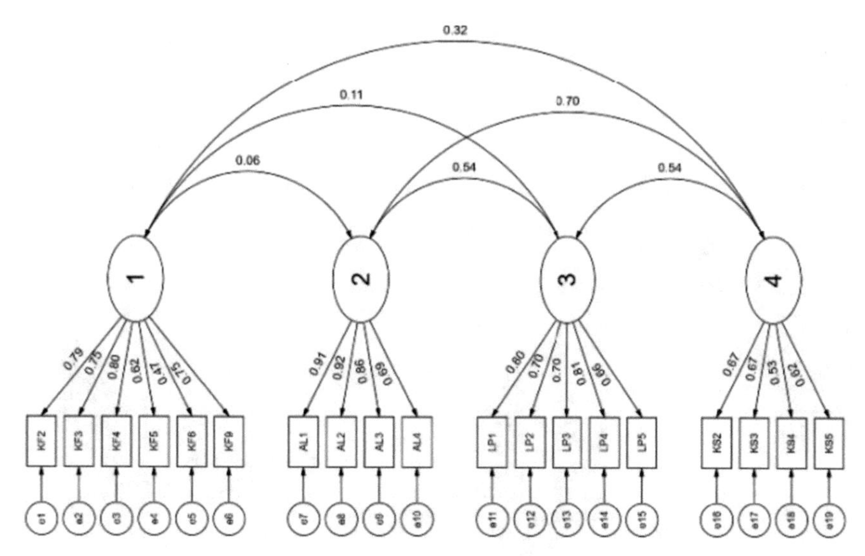

Fig. 1. CB-SEM Model Settings

According to the results obtained in the structural model analyzed by PLS-SEM, not all direct effects are significant (p 0.05), so the effects of knowledge exchange (KS) on altruism (AL) (= 0.66; p .000) are significant and support H2; knowledge sharing (KS) on knowledge formation (KF) (= 0.490; p .000) is significant and supports H3; and knowledge sharing (KS) on learning performance (LP) (= 0.504; p .000) is significant and supports hypothesis H4. In contrast, the following hypothetical relationships would not exert a positive effect: Altruism (AL) on Learning Performance (LP) (=0.141; p .099) would not support H1; Knowledge Formation (KF) on Altruism (AL) (=0.006; p 0.475) would not support H5, and Knowledge Formation (KF) on Learning Performance (LP) (=0.092; p .0174) would not support H6. These results suggest that not all the hypotheses raised were supported. See Fig. 2.

Table 1 shows the data obtained in the reliability and validity tests; in this sense, the correlation coefficients, reliability, and validity expressed by Cronbach's alpha were taken into account, and the resulting values are for the variable (AL) 0.906 and for (LP) 0.852, which is very acceptable. The values of the average variance extracted are: Average Variance Extracted (AVE) for (AL) is 0.783, and for (LP) is 0.628. Both cases exceed the recommended value of 0.500 [19], Therefore, it is concluded that the convergent validity is acceptable. Its application is essential for the analysis of composite reliability [20, 21], It is recommended as an acceptance criterion that its results exceed 0.6, thus demonstrating reasonable levels of internal consistency and reliability for each of the variables. The values obtained for (AL) are 0.935 and for (LP) 0.894. The coefficient (rho_A), whose use allows verifying the reliability of the values obtained in the construction and design of the PLS [22], The results in (rho_A) are for (AL) 0.909 and for (LP) 0.855, which are the minimum recommended.

The compound mediation model (Mode B) presents two variables in training mode; this knowledge formation (KF) and knowledge exchange (KS); as validity criteria, the

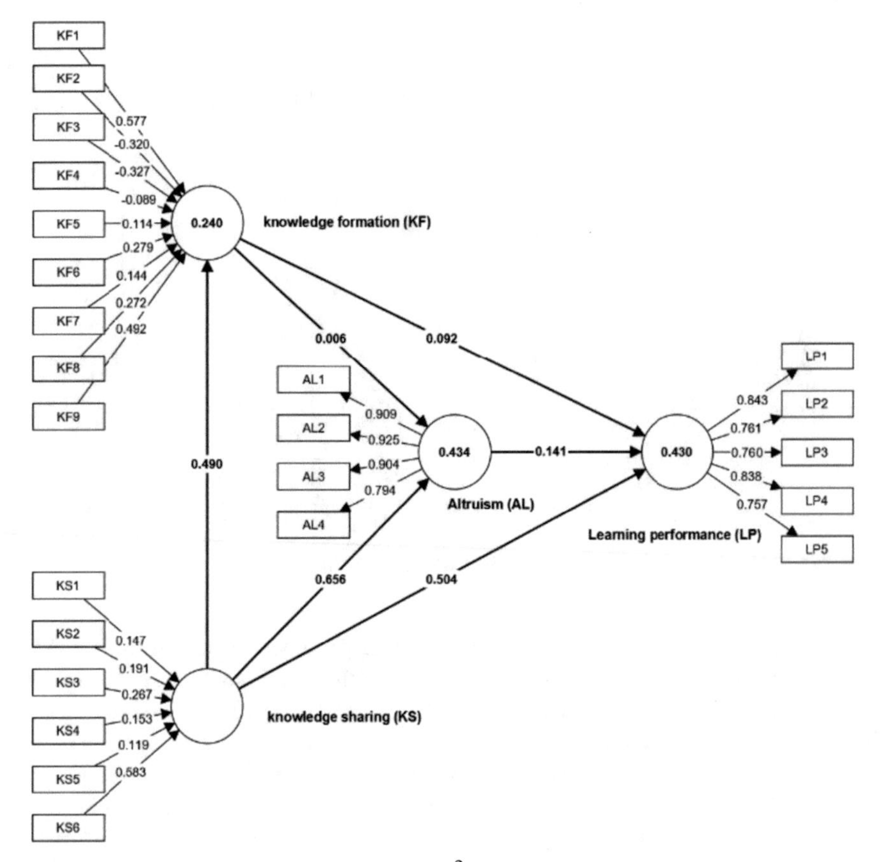

Fig. 2. Model R^2 - SmartPLS.

collinearity statistics (VIF) were used-model media external; for (KF), the values are between KF8 = 1.456 and KF2 = 2.386; for (KS) the values are between KS1 = 1.239 and KS5 = 1.525, according to the variance inflation factor (VIF) ≤ 3.3 there would be no collinearity [23].

Table 1. Reliability and construct validity

	Cronbach's Alpha	rho_A	Composite reliability	(AVE)
Altruism (AL)	0.906	0.909	0.935	0.783
Learning performance (LP)	0.852	0.855	0.894	0.628

Table 2 shows the hypothesis tests using the significance level for the P-Value ($p < 0.05$) as the acceptance criterion. The Bootstrapping test was used for this purpose, with 10,000 resampling with substitution of the original sample [24] recommended for this case. The results predict that the hypotheses are admitted: H2, H3, and H4. All three hypotheses were rejected: H1, H5, and H6.

Table 2. Hypothesis test – bootstrapping

Hypothesis	β	Sample mean (M)	Standard deviation (STDEV)	t-Statistics (|O/STDEV|)	P Value
H1 Altruism (AL) - > Learning performance (LP)	0.141	0.132	0.109	1.289	0.099
H2 knowledge sharing (KS)- > Altruism (AL)	0.656	0.668	0.081	8.107	0.000
H3 knowledge sharing (KS)- > knowledge formation (KF)	0.490	0.534	0.064	7.629	0.000
H4 knowledge sharing (KS)- > Learning performance (LP)	0.504	0.522	0.121	4.16	0.000
H5 knowledge formation (KF)- > Altruism (AL)	0.006	−0.002	0.097	0.063	0.475
H6 knowledge formation (KF)- > Learning performance (LP)	0.092	0.089	0.098	0.938	0.174

4 Conclusions

Individual motivation is related to reputation and altruism, and they maintain strong links with social networks, which is good because both concepts are essential for people today, especially students. So, people can share knowledge and improve their learning by coming up with collaborative and cooperative learning methods. According to the model suggested by this research, it has been shown that training and knowledge sharing with altruism have a statistically significant effect on knowledge sharing with altruism, knowledge training, and learning performance in university students. Students think of knowledge formation as a planned way of learning something important. This is done by sharing resources and talking to each other online using social media.

On the other hand, the research shows no positive relationship between altruism and learning performance, and knowledge formation, altruism, and learning performance do not affect each other. In order to help people, get to know each other better in virtual learning environments, the research results show that teaching and learning strategies based on collaboration and social networks need to be emphasized. Only then will significant learning be achieved in virtual environments, especially among university students.

References

1. Gutierrez-Aguilar, O., Escobedo-Maita, P., Calliñaupa-Quispe, G., Vargas-Gonzales, J.C., Torres-Huillca, A.: The use of social networks, usefulness and ease of use, enjoyment through positive emotions and their influence on school satisfaction mediated by school achievement. In: Iberian Conference on Information Systems and Technologies, CISTI 2022
2. Gutierrez-Aguilar, O., Torres-Palomino, D., Gomez-Zanabria, A., Garcia-Begazo, C., Argüelles-Florez, A., Silvestre-Almerón, F., Zeta-Cruz, V.: Personality and creative thinking in artists and the idea of their reputation on Social media. In: Iberian Conference on Information Systems and Technologies, CISTI 2022
3. Ahmed, Y.A., Ahmad, M.N., Ahmad, N., Zakaria, N.H.: Social media for knowledge-sharing: a systematic literature review. Telematics Inf. **37**, 72–112 (2019). https://doi.org/10.1016/j.tele.2018.01.015
4. Hosen, M., Ogbeibu, S., Giridharan, B., Cham, T.-H., Lim, W.M., Paul, J.: Individual motivation and social media influence on student knowledge sharing and learning performance: evidence from an emerging economy. Comput. Educ. **172**, 104262 (2021). https://doi.org/10.1016/j.compedu.2021.104262
5. Hsu, C.L., Lin, J.C.C.: Acceptance of blog usage: the roles of technology acceptance, social influence and knowledge sharing motivation. Inf. Manag. **45**(1), 65–74 (2008). https://doi.org/10.1016/j.im.2007.11.001
6. Hoseini, M., Saghafi, F., Aghayi, E.: A multidimensional model of knowledge sharing behavior in mobile social networks. Kybernetes **48**(5), 906–929 (2019). https://doi.org/10.1108/K-07-2017-0249
7. Choi, J.H., Ramirez, R., Gregg, D.G., Scott, J.E., Lee, K.H.: Influencing knowledge sharing on social media: a gender perspective. Asia Pac. J. Inf. Syst. **30**(3), 513–531 (2020). https://doi.org/10.14329/apjis.2020.30.3.513
8. Marc, L.W., Ling, L.A., Chen, P.C.S.: Toward a conceptual framework for social media adoption by non-urban communities for non-profit activities: Insights from an integration of grand theories of technology acceptance. Aust. J. Inf. Syst. **23** (2019). https://doi.org/10.3127/ajis.v23i0.1835
9. Salas-Valdivia, L., Gutierrez-Aguilar, O.: Implications of digital citizenship in social media to build a safe environment in the covid-19 situation. In: Proceedings - 2021 16th Latin American Conference on Learning Technologies, LACLO 2021, pp. 364–367 (2021)
10. Alamri, M.M.: Undergraduate students' perceptions toward social media usage and academic performance: a study from Saudi Arabia. Int. J. Emerg. Technol. Learn. (iJET) **14**(03), 61–79 (2019). https://doi.org/10.3991/ijet.v14i03.9340
11. Shahzad, F., Xiu, G., Khan, I., Shahbaz, M., Riaz, M.U., Abbas, A.: The moderating role of intrinsic motivation in cloud computing adoption in online education in a developing country: a structural equation model. Asia Pac. Educ. Rev. **21**(1), 121–141 (2020). https://doi.org/10.1007/s12564-019-09611-2
12. Blasco-Arcas, L., Buil, I., Hernández-Ortega, B., Sese, F.J.: Using clickers in class: the role of interactivity, active collaborative learning and engagement in learning performance. Comput. Educ. **62**, 102–110 (2013). https://doi.org/10.1016/j.compedu.2012.10.019

13. Zhou, Q., et al.: Understanding the use of YouTube as a learning resource: a social cognitive perspective. Aslib J. Inf. Manag. **72**(3), 339–359 (2020). https://doi.org/10.1108/AJIM-10-2019-0290
14. Glassner, A., Back, S.: Connectivism: networks, knowledge, and learning. In: Glassner, A., Back, S. (eds.) Exploring Heutagogy in Higher Education: Academia Meets the Zeitgeist, pp. 39–47. Springer, Singapore (2020)
15. Mpungose, C.B., Khoza, S.B.: Postgraduate students' experiences on the use of moodle and canvas learning management system. Technol. Knowl. Learn. **27**(1), 1–16 (2022). https://doi.org/10.1007/s10758-020-09475-1
16. Eid, M.I.M., Al-Jabri, I.M.: Social networking, knowledge sharing, and student learning: the case of university students. Comput. Educ. **99**, 14–27 (2016). https://doi.org/10.1016/j.compedu.2016.04.007
17. Nonaka, I.: The knowledge-creating company. In: The Economic Impact of Knowledge, pp. 175–187. Routledge (2009)
18. Barker, R.: Management of knowledge creation and sharing to create virtual knowledge-sharing communities: a tracking study. J. Knowl. Manag. **19**(2), 334–350 (2015). https://doi.org/10.1108/JKM-06-2014-0229
19. Hair, J.F., Jr., Hult, G.T.M., Ringle, C., Sarstedt, M.: A Primer on Partial Least Squares Structural Equation Modeling (PLS-SEM). Sage publications, Thousands Oaks (2016)
20. Bagozzi, R.P., Yi, Y.: On the evaluation of structural equation models. J. Acad. Mark. Sci. **16**(1), 74–94 (1988)
21. Hair, J.F., Sarstedt, M., Pieper, T.M., Ringle, C.M.: The use of partial least squares structural equation modeling in strategic management research: a review of past practices and recommendations for future applications. Long Range Plan. **45**(5–6), 320–340 (2012)
22. Dijkstra, T.K., Henseler, J.: Consistent partial least squares path modeling. MIS Q. **39**(2), 297–316 (2015)
23. Hair, J.F., Risher, J.J., Sarstedt, M., Ringle, C.M.: When to use and how to report the results of PLS-SEM. Eur. Bus. Rev. **31**(1), 2–24 (2019). https://doi.org/10.1108/EBR-11-2018-0203
24. Latan, H., Noonan, R. (eds.): Partial least squares path modeling. Springer, Cham (2017). https://doi.org/10.1007/978-3-319-64069-3

HIL in a Vehicle Washing Process Implementing MQTT Communication Network based in TCP/IP

Torres Bryan[1,3], José Varela-Aldás[2(✉)], Diaz Israel[3], S. Ortiz Jessica[3], Guamanquispe Andrés[3], and Velasco Edwin[3]

[1] SISAu Research Group, Facultad de Ingeniería, Industria y Producción, Universidad Indoamérica, Ambato, Ecuador

[2] Centro de Investigaciones de Ciencias Humanas y de la Educación - CICHE, Universidad Indoamérica, Ambato, Ecuador

josevarela@uti.edu.ec

[3] Universidad de las Fuerzas Armadas – ESPE, 171103 Sangolquí, Ecuador

{bstorres2,pidiaz2,jsortiz4,alguamanevelasco3, evelasco3}@espe.edu.ec

Abstract. This work describes the documentary synthesis of the design and adaptation of an industrial environment of an automatic car washing machine with tunnel type structure. The same one that has four stations properly distributed, with respect to the time and operation of the mechanisms of the washer with the objective of improving the operation of the system and saving water. This automated industrial process is virtualized in the graphic development platform Unity for its adaptation to a Hardware in the Loop (HIL) control loop. The whole process is carried out by means of a programming logic unit (PLC); considering an automatic control of the stations carried out in LOGOSOFT software. The communication of the network is made by means of an Ethernet architecture based on TCP-IP, in addition the use of the MQTT communication protocol is presented based on its fluidity, simplicity and lightness that makes it suitable for its use in programmable controllers, the communication network is designed in Node-Network.

Keywords: Virtualization · Hardware in the Loop · PLC · TCP-IP

1 Introduction

Nowadays, most industrial processes are represented by dynamic sequence systems, which is to say, they are discrete systems; in addition, there are mixed systems which combine the characteristics of discrete systems and continuous systems. Their variables are controlled, monitored and supervised, with the purpose of automating industrial processes, thus facilitating the degree of use of a product in service or the creation of novel ideas for the application of algorithms to industrial processes.[1]. With the passage of time the control techniques used in industrial processes, have gone through various

A. Rocha et al. (Eds.): WorldCIST 2023, LNNS 800, pp. 538–549, 2024.
https://doi.org/10.1007/978-3-031-45645-9_52

advances; the same that allowed to reach an almost complete automation of the system. Several authors mention that the industrial using mostly programmable controllers. This is due to the facilities provided by the same in the various industrial processes, in the process of the automatic washing machine, which aims to improve productivity in washing and especially reduce water consumption, as mentioned [2], which combine the technology of proximity sensors, motors, brushes and dryer to encompass it all in an automatic car wash system whose main objective is to focus on saving water, Gaikwad mentions an efficiency of 95% giving as a fully acceptable system.[3].The automation of the car wash process is advancing with great magnitude, according to [4], it is understood that car washing and maintenance accounts for 60% of the vehicle market, as well as customer service, water saving and new technologies such as IoT, which is increasingly entering the market.

In this context, it can be mentioned that there has been an increase in automobiles worldwide, therefore, various industries have focused their interest on improving their productivity through the automation of systems. Industrial processes have focused on automating their plants, in order to improve the productivity, quality and efficiency of the products produced [5]. Therefore, it is proposed to generate a service for vehicle users, this approach will design an automatic washing process. Designing a diagram that defines the operation between the stations involved in the system[6]. According to various works, industrial processes are controlled by programmable logic controllers, PLCs, which are connected to human-machine graphic interfaces. Whose purpose is to allow the user to control the process from an intouch screen. The development of car washes has gone through several processes starting from prototypes in which all washing stages are included in a didactic way [7, 8].

As described, the implementation of prototypes of industrial processes in the educational area represents a high cost. Reason why higher education institutions have sought strategies and tools that allow students to complement their education in the area of automation, control, instrumentation among other fields of study. Developing virtualized environments focused on industrial processes [9]. Unity 3D, is a program suitable to work with any type of communication architecture, among the most used is the parallel computing architecture, Unity 3D is capable of communicating in real time the programmed virtual environment with a programmable physical controller [10], It can also be used as a teaching engine for programmable controllers, to impose real-time control strategies on objects in the virtual environment. Another strategy that can be used is the hardware in the Loop technique, HIL, which allows the physical part that is the controller to communicate with the simulated part that would be the virtual work environment [11]. It allows the training of users in process control focused on learning with the development of a virtual environment for the creation of instruments such as equipment, animations and high-quality sounds.

This paper presents a research paper that develops a car wash plant, which has four main sequential stations: soaping, brushing, rinsing and drying, each with a waiting time of 4 seconds per station based on the American car wash standards [14, 15]. This virtualization of the process requires the design of the stations elaborated in the CAD software, in order to import the instruments to the Unity 3D graphic engine. The virtual environment of the automatic washing process will be connected to the PLC through

HIL, which allows connecting the real input/output signals of a controller; allowing the system to simulate the reality of the process, that is, the controller acts as if it were assembled with the real system[12, 13]. The proposed work not only aims to animate the plant, but also to study the key characteristics that allow the application of modern control techniques in the automation of these processes; A HIL prototype that can be used in the learning of manipulation of automation and control plants belonging to engineering careers to teach some concepts is suitable to implement a control in this plant.

The present work consists of five sections, in which the first one includes the introduction, works related to the topic and the general description of the project, in the second section is the modeling of the plant, as well as the conditions of sequences and the whole development process, the third section refers to the communication used in the system, then continues the section of results in which describes the final result of the work done and ends with the last section of conclusions.

2 Development of the Industrial Environment

This section describes the steps considered for the development of the automatic car wash process environment.

2.1 Plant Modeling

For the modeling of the plant begins with the design of the infrastructure of the washing machine, the same that is done in a CAD software, in this case the SolidWorks software is used, after the development of the 3D drawing of the plant must transform the modeling software file to an extension that is easy to recognize in the virtual environment engine Unity, Sketchup software is used for this purpose, this program can read corresponding binary files from SolidWorks and transform them into ideal files for the virtual environment (fbx), it should be noted that if dynamic elements are desired in the process they must be modeled separately (Fig. 1).

Fig. 1. Plant modeling process

2.2 Process Sequence Conditions

For the automatic washing of vehicles starts with the soaping station, before reaching the station the conveyor belt is activated until it makes contact with the presence sensor, once this happens the conveyor belt stops, the pump containing the soaping liquid is activated and proceeds to perform the operation which has a duration of 4 seconds, once the execution time is over, the conveyor belt is activated to take the vehicle to the next station, at the brushing station it is stopped in the same way by a presence sensor and then the motors that help brushing the cars are activated, once the time of 4 seconds is over, the conveyor belt takes the vehicle to the third station, rinsing, which is activated by the presence sensor and starts the water pump, once the waiting time is over the belt goes with the vehicle to the last drying station where the fans are activated and thus end the process, it should be noted that each station has its own independent presence sensor (Fig. 2).

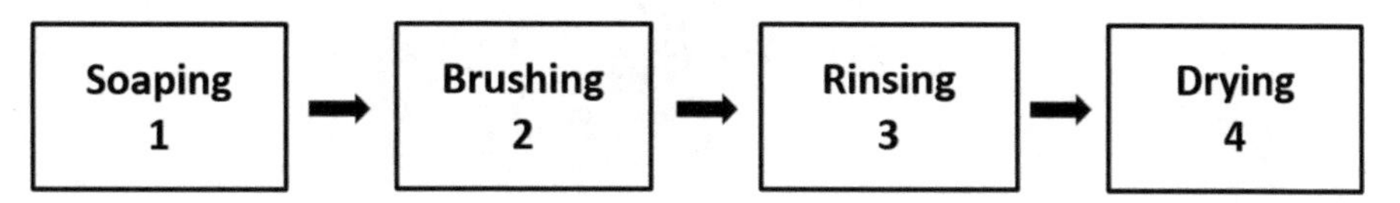

Fig. 2. Process that fulfills the washing system

3 HIL - Networks and Virtualization

We want to automate a vehicle washing machine which has four stations, soaping, brushing, rinsing and drying. A conveyor belt will move the car to the different stations. It is worth mentioning that the activation of each station will be carried out during a certain time. In the soaping and rinsing stage, the belt must stop for a certain period of time and then move forward.

For the activation of each station there are distance sensors, which will send the signal of the presence of the vehicles for each station, the belt will stop at the end of the process.

3.1 Virtual Environment

For the virtualization of the plant, we use the environment provided by Unity, in addition to different software that help us to model the realistic environment of a car wash located in a warehouse as shown below, it is worth mentioning that some of the designs were obtained from the digital library Unity asset store.

Process

As a first point, the plant should be modeled as real as possible in solidworks, real according to size.

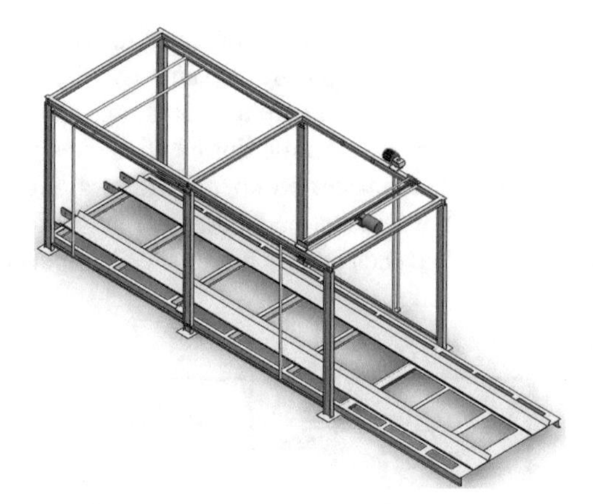

Fig. 3. Structure designed in solidworks

The file must be converted to stl, the binary stl file is opened in the SketchUp program, and converted to an fbx file, which can be read in Unity 3D.

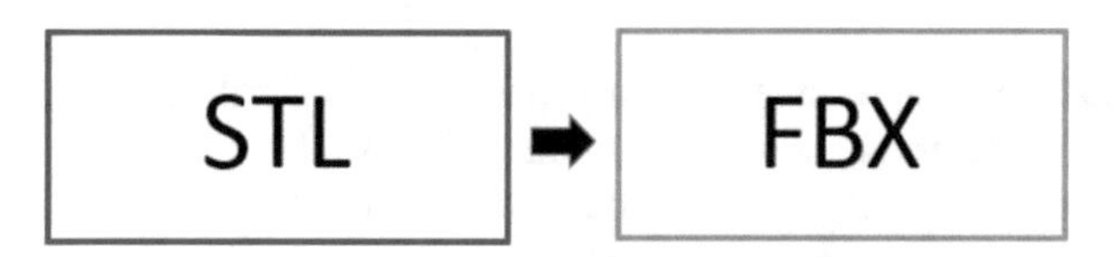

Fig. 4. Process of converting files for Unity 3D

The last step is to complement the other spaces or environments with elements available in the unity store, the virtualized environment looks like this (Fig. 5):

Fig. 5. Virtual environment of the car wash machine

The following shows the results on the different workstations (see Fig. 6). The development of the automatic car wash process is presented, for which in part a) the soaping station is shown, in which car soap is emitted through the pipes, in section b) it is found the brushing part, in which the actuators perform the cleaning of the vehicle body, then in part c) is the rinsing section and ends with the vehicle drying station, it should be noted that each station has a time determined that can be changed in the programmable controller.

Fig. 6. Automatic washing process, 4 stations. a) soaping station, b) brushing station, c) rinsing station, d) drying station.

3.2 HIL Communication

To start with work station one, the 3D modeling is done in a design program such as Solidworks, by means of sketchup the binary file of the CAD program is transformed to the extension recognized by the virtual environment Unity. To complement with the other elements of the virtualized plant, you can go to the Asset store, where you can find vehicles, pipes, pumps, keys, buildings, etc. And we proceed to the creation of an avatar, which is the element that will control the process. The next step is the virtual environment in the upper part of the virtual environment you can see the scene, so that this can work requires commands among them we have inputs and outputs, among the inputs are the presence sensors, and outputs the actuators, the whole process that is detailed in the picture of the system starts with a star button, and begin to run the processes. In the hardware in the Loop, we find the physical part of the programmable controller, in which the ladder diagram is detailed and finally the communication section, made in node red architecture TCP-IP which is the protocol used in the communication of the entire system (Fig. 7).

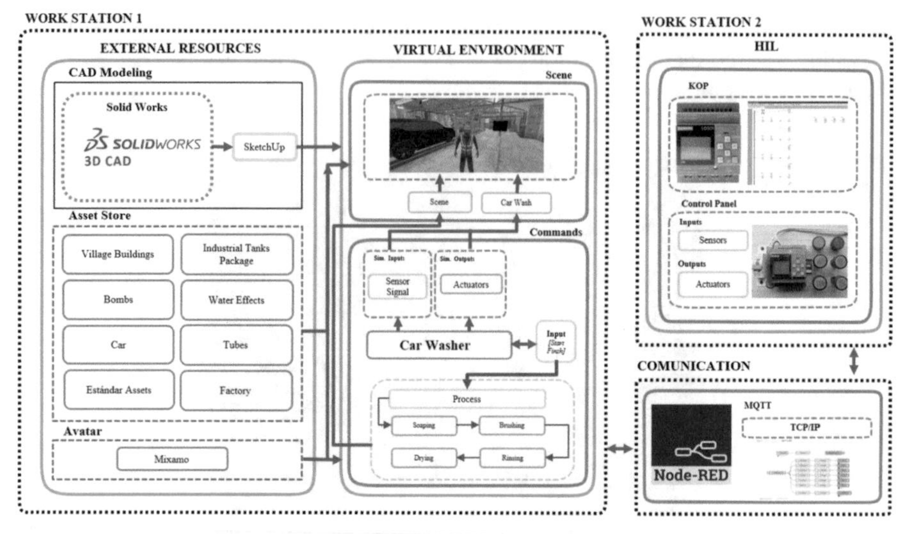

Fig. 7. HIL communication diagram for a vehicle washing machine.

3.3 Communication Protocol

For communication between the controller and the simulation (virtualization) it is necessary to establish a protocol that allows us to make use of the inputs and outputs in both in order to ensure proper operation.

However, some of the controllers cannot be directly connected to the simulation (virtualization), which represents a problem in the case of not having a suitable controller, LOGO Siemens is one of these, so it is necessary to establish another communication protocol.

The MQTT is a Machine-to-Machine communication protocol of message queue type. MQTT is based on the TPC/IP stack as the basis for communication Figure 3, the operation is described as a publisher/subscriber messaging service. That is, a client can publish a message to a certain topic. Other clients can subscribe to this topic, and the broker will forward the subscribed messages to them.

For the correct communication it is necessary to make use of the Node Red, where the different inputs and outputs established in the Ladder diagram of the Logo must be placed, the HIL diagram implemented for the car wash is shown in Figure 4 (Fig. 8).

Process

To start with the connection, run the "Node.js" application (Fig. 9).

Then type the command "node-red" (Fig. 10).

When typing the command an IP address is loaded, the same one used for MQTT communication, the IP address is copied and pasted into the Node-network dialog box (Fig. 11).

Then run the Unity 3D program. Verify that you can send and receive signal in the MQTT object inspector. And the communication is done (Fig. 12).

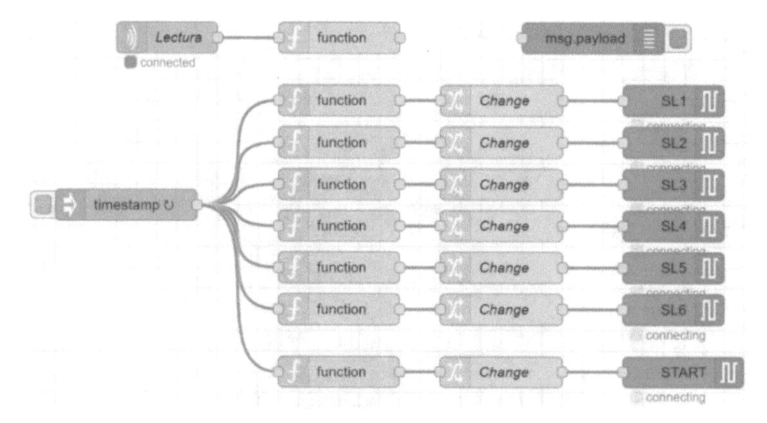

Fig. 8. TCP-IP network design in node red

Fig. 9. Node application

```
Seleccionar node-red
Your environment has been set up for using Node.js 16.17.0 (x64) and npm.

C:\Users\Legion>node-red
1 Sep 06:20:59 - [info]

Welcome to Node-RED
===================

1 Sep 06:20:59 - [info] Node-RED version: v3.0.2
1 Sep 06:20:59 - [info] Node.js  version: v16.17.0
1 Sep 06:20:59 - [info] Windows_NT 10.0.22000 x64 LE
1 Sep 06:21:01 - [info] Loading palette nodes
1 Sep 06:21:03 - [info] Settings file  : C:\Users\Legion\.node-red\settings.js
1 Sep 06:21:03 - [info] Context store  : 'default' [module=memory]
1 Sep 06:21:03 - [info] User directory : \Users\Legion\.node-red
1 Sep 06:21:03 - [warn] Projects disabled : editorTheme.projects.enabled=false
1 Sep 06:21:03 - [info] Flows file     : \Users\Legion\.node-red\flows.json
1 Sep 06:21:03 - [info] Server now running at http://127.0.0.1:1880/
1 Sep 06:21:03 - [warn]
```

Fig. 10. IP Address

4 Results

The results were as expected, communication through TCP-IP architecture generates a fast response time, because it is accompanied by an MQTT protocol, which makes your connection stable and does not generate noise. It should be noted that the communication process is simple, it is explained in detail in the communication protocol segment (Fig. 13).

In the virtualization section, it can be seen that the results are in accordance with what an industrial plant needs, in the present case of a car washer it can be seen that

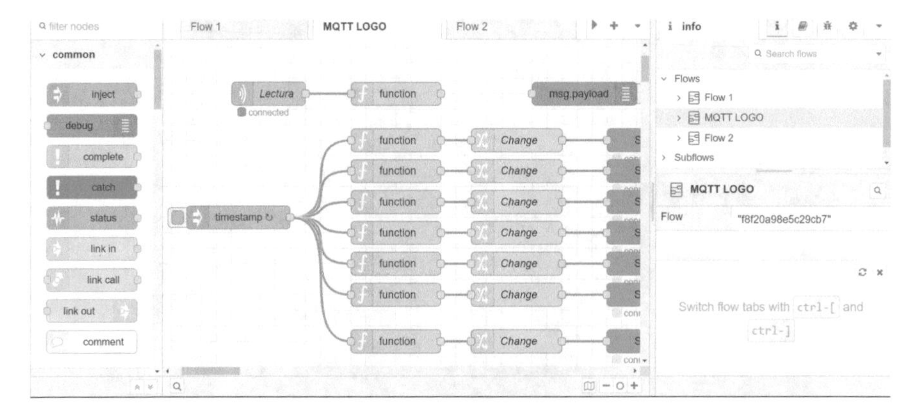

Fig. 11. Communication setup with MQTT and Node network

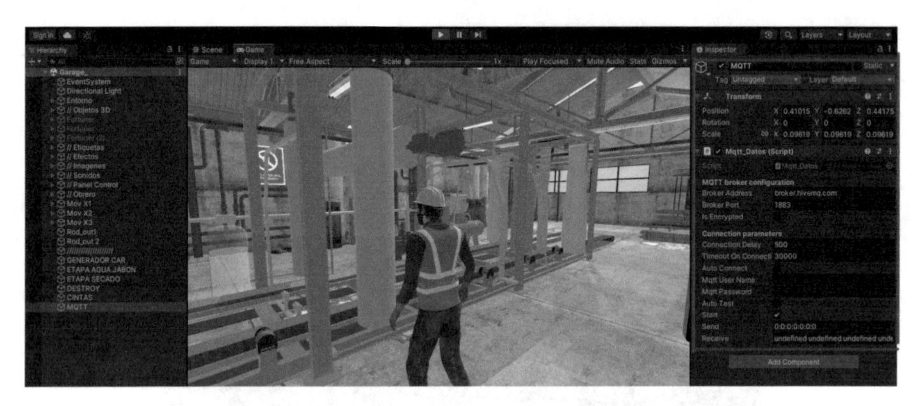

Fig. 12. Unity runs with the successful connection

Fig. 13. Simulated environment with special effects

the car enters each of the stations, it is generated in the same way special effects that make virtualization more real. The CAD design is correctly done, due to this there are no errors in the virtual simulation. Next, the result in action of the system and the indicators in the HMI that is located in the virtual environment and in the PLC are presented.

HMI Display

In the section of the HMI that is shown in the virtualized environment, there are the indicators of each process, which is being developed in real time (Fig. 14).

Fig. 14. Avatar with the HMI interface

The physical prototype is shown below in which it consists of the PLC, the indicators of each section of the washing process, the start and emergency stop of the process and a switch that turns on the entire process (Fig. 15).

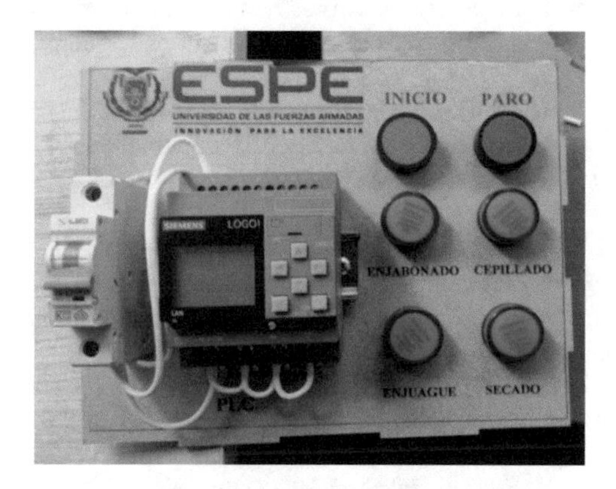

Fig. 15. LOGO connection

The communication time depends on the internet connection because the MQTT protocol is used, the average communication speed is 100ms to 200ms.

5 Conclusion

The virtualization was a success, the design made in the CAD program did not generate interferences within the virtual environment, which leads to a correct development of the program, the HIL control loop is adequate, using a TCP-IP architecture with MQTT was quite satisfactory because the connection time is relatively good generating a speed of 100 to 200 ms, which leads to a successful communication in real time, the PLC as well as the indicators of each station have no failures so that encompasses a correct operation of the virtual automatized plant.

Acknowledgments. The authors would like to thank the ARSI Research Group and SISAu Research Group for their support in the development of this work.

References

1. Castilla, M., Rodríguez, F., Álvarez, J.D., Donaire, J.G., Ramos-Teodoro, J.: A hardware-in-the-loop prototype to design benchmarks for automation and control education$_{**}$this work has been funded by teaching innovation project named 'desarrollo de un prototipo hardware-in-the-loop de bajo coste para la ensenanza de automatizacion en estudios de ingenierı́ia' belonging to the call for the creation of teaching innovation groups and good teaching practices at the University of Almerı́ia (Biennial 2019–20). IFAC-Pap. **53**(2), 17314–17319 (2020). https://doi.org/10.1016/j.ifacol.2020.12.1815
2. Gaikwad, R., Kharat, S.M.M., Thakur, J.: PLC based automatic car washing system using proximity sensors. In: 2017 IEEE International Conference on Power, Control, Signals and Instrumentation Engineering (ICPCSI), pp. 1875–1878 (2017). https://doi.org/10.1109/ICPCSI.2017.8392041.
3. Gurung, T.R., Stewart, R.A., Beal, C.D., Sharma, A.K.: Smart meter enabled water end-use demand data: platform for the enhanced infrastructure planning of contemporary urban water supply networks. J. Clean. Prod. **87**, 642–654 (2015). https://doi.org/10.1016/j.jclepro.2014.09.054
4. Zhong, S., Zhang, L., Chen, H.-C., Zhao, H., Guo, L.: Study of the patterns of automatic car washing in the era of internet of things. In: 2017 31st International Conference on Advanced Information Networking and Applications Workshops (WAINA), pp. 82–86 (2017). https://doi.org/10.1109/WAINA.2017.132.
5. M. Zhilevski and V. Hristov, "Design of an Automated Car Washing System with Verilog HDL," in *2021 3rd International Congress on Human-Computer Interaction, Optimization and Robotic Applications (HORA)*, Jun. 2021, pp. 1–5. doi: https://doi.org/10.1109/HORA52670.2021.9461184.
6. Mumtaz, F., Saeed, A., Nabeel, M., Ahmed, S.: Autonomous car washing station based on PLC and HMI control. In: 4th Smart Cities Symposium (SCS 2021), vol. 2021, pp. 103–107 (2021). https://doi.org/10.1049/icp.2022.0322.
7. Sărăcin, C.G., Ioniţă, A.: Educational platform for simulating the operation of an automatic car wash machine. In: 2020 International Symposium on Fundamentals of Electrical Engineering (ISFEE), pp. 1–4 (2020). https://doi.org/10.1109/ISFEE51261.2020.9756172.

8. Abid, A., Hasan, T., Baig, T., Jadoon, A.: Design and development of automatic carwash system. In: 2016 IEEE Conference on Systems, Process and Control (ICSPC), pp. 58–63 (2016). https://doi.org/10.1109/SPC.2016.7920704.
9. M. Abrams, E. H. Page, and R. E. Nance, "Simulation program development by stepwise refinement in UNITY," in *1991 Winter Simulation Conference Proceedings.*, Dec. 1991, pp. 233–242. doi: https://doi.org/10.1109/WSC.1991.185620.
10. Wang, H., Wang, Z.: Research on PLC simulation teaching platform based on unity. In: 2020 International Conference on Intelligent Design (ICID), pp. 15–18 (2020). https://doi.org/10.1109/ICID52250.2020.00011.
11. Pruna, E., Balladares, G., Teneda, H.: 3D virtual system of a distillation tower, and process control using the hardware in the loop technique. In: De Paolis, L.T., Arpaia, P., Bourdot, P. (eds.) AVR 2021. LNCS, vol. 12980, pp. 621–638. Springer, Cham (2021). https://doi.org/10.1007/978-3-030-87595-4_45
12. Tan, X., Tang, L.: Application of enhanced coagulation aided by UF membrane for car wash wastewater treatment. I: 2008 2nd International Conference on Bioinformatics and Biomedical Engineering, pp. 3653–3656 (2008). https://doi.org/10.1109/ICBBE.2008.415.
13. Lamb, F.: Industrial Automation: Hands-On. McGraw-Hill Education (2013). Accessed 01 Sept 2022. https://www.accessengineeringlibrary.com/content/book/9780071816458
14. Galali, A.M.J. et al.: 6th International Visible Conference on Educational Studies and Applied Linguistics Book of Proceedings (2022). Accessed 01 Sept 2022. https://www.academia.edu/22125004/6_TH_INTERNATIONAL_VISIBLE_CONFERENCE_ON_EDUCATIONAL_STUDIES_AND_APPLIED_LINGUISTICS_BOOK_of_PROCEEDINGS
15. Burdon, M.: The smart world is the collected world (2020). https://doi.org/10.1017/9781108283717.002.

Simulation of an Automated System with Hardware in the Loop Using a Human Machine Interface (HMI) for the Process of Bottling Liquids Employing Unity 3D

Rivas Fabian[3,1], José Varela-Aldás[2], Cofre Nadia[3], Jessica S. Ortiz[3(✉)], Guamán Dennis[3], and Guzmán Edwin[3]

[1] SISAu Research Group, Facultad de Ingeniería, Industria y Producción, Universidad Indoamérica, Ambato, Ecuador
[2] Centro de Investigaciones de Ciencias Humanas y de La Educación - CICHE, Universidad Indoamérica, Ambato, Ecuador
josevarela@uti.edu.ec
[3] Universidad de las Fuerzas Armadas – ESPE, 171103 Sangolquí, Ecuador
{farivas,nacofre,jsortiz4,dsguaman,eaguzman}@espe.edu.ec

Abstract. This document shows the development of the simulation of an automated bottle filling and sealing system in order to monitor and supervise the behavior of each stage of the process using the Unity 3D software. In addition, the implementation the Modbus RTU network, which will be communicated to the PLC, applying Ladder programming through the Tia Portal V17 software, in order to determine the states of the input and output variables, as well as the behavior that occurs in the event of failures. In addition, the use of hardware in the loop that through an HMI system which allows us to establish the desired communication with the Unity 3D software, knowing the process of bottling and sealing bottles that is very common in the industry.

Keywords: PLC · Modbus · unity 3D · network · sealing · filling

1 Introduction

Technological advances have evolved rapidly, today it is very important to continuously update knowledge about technology; especially in the field of automation [1]. Automation is a field of engineering that consists of the computerization and computerized control of industrial processes or machinery due to the need to improve, increase and optimize the production of a factory in a more profitable way [2, 3].

A few years ago, virtualization was a relatively unknown concept in the industrial sector. As of the year 2020, 75% of the industrial sector have adopted virtual server platforms to run SCADA, HMI, Historian, MES among others [4]. With virtualization comes the promise of cost savings, simplified management, greater efficiency, greater agility, and better system availability [5]. Process simulation is a useful tool for designing,

A. Rocha et al. (Eds.): WorldCIST 2023, LNNS 800, pp. 550–557, 2024.
https://doi.org/10.1007/978-3-031-45645-9_53

analyzing and evaluating each phase of said process, thus facilitating its optimization and monitoring, and allowing its behavior to be predicted in certain situations [6]. Through programming in the Tia Portal V17 software that facilitates graphical programming in a Ladder diagram [7]. For the implementation, the siemens PLC communication is used and through the connection of the *max232* module [8]. In order to develop the Unity interface for the virtualization of the HMI of the bottling system process, which corresponds to a process industry and The industries related to automation are basically the manufacturing industry and the process industry, which consists of 3 stages: filling, sealing, product output, which meets all the requirements for production and thus guarantees a final product [9, 10].

HIL testing is a technique where actual signals from a controller are fed into a test system that simulates reality, tricking the controller into thinking it is in the assembled product. Testing and design iteration is performed as if using the real-world system [11, 12]. Industrial communication networks are the backbone of any automation system architecture as it has provided a powerful means of data exchange, data controllability, and flexibility to connect multiple devices [13]. Through the development of the industry, the demand for better industrial communication networks increased. Many manufacturers began to develop smart solutions to offer the best way to connect various devices on the same platform [14]. Among the solutions, the following networks stand out: Profinet, Industrial Ethernet, Profibus, As-i, Modbus, etc. [15].

Currently, the control of automated processes helps ensure the durability and quality of the automation factor in industrial development [16]. Since the location of production lines is distributive because it integrates all the components into a single operating unit, for which the correct distribution manages to minimize production costs and optimize the quality of the process.

In this document, the virtualization will be carried out through Unity, the same one that will have a communication with the PLC using Tia portal, for programming, HIL for the verification of the industrial network used. On the other hand, the system consists of a Modbus RTU communication network, a virtualized sealing and bottling plant in Unity 3D, communication with Arduino for the inputs required by the plant, also known as HIL, which allows interaction between human and machine (HMI).

2 Developing

The development consists of three different stages for the operation of the Modbus RTU network with HIL in Unity 3D, these stages play an important role in obtaining optimal results. In the first place, the selection of the correct parameters for the simulation is established, followed by the physical connection of the HIL, both its inputs and outputs, to establish communication with the network.

2.1 Configuration of Parameters for the Simulation

The parameters for each stage of the bottling process are configured in the Unity 3D virtual environment. This can be seen in Fig. 1A. The local IP is configured within the NetToPLCsim program as indicated in Fig. 1B and 1C shows the complete virtualization of the process where each stage of the process is observed. And finally, the connection is verified with the Tia Portal program and the Unity software. Serial communication is important to establish the addresses with which you work and finally proceed to verify the connection and communication between Tia Portal, Unity 3D and HIL in such a way that there is adequate operation.

Fig. 1. Parameter Settings.

2.2 Connection and Industrial Network to be Used

Figure 2, highlights what refers to the connections made for the bottling processes. In Fig. 2A you can see the HIL connection where the real signals from the controller are connected to the PLC to perform tests that simulate reality, which tricks the controller into thinking that it is in the assembled product. Figure 2B shows the connection of the Modbus Serial network which will be used for the industrial process. An HMI was implemented in Unity, this is shown in Fig. 2C. The operation of the Modbus network and the connection with the logo can be seen in Fig. 2D which shows the operation of the bottling process.

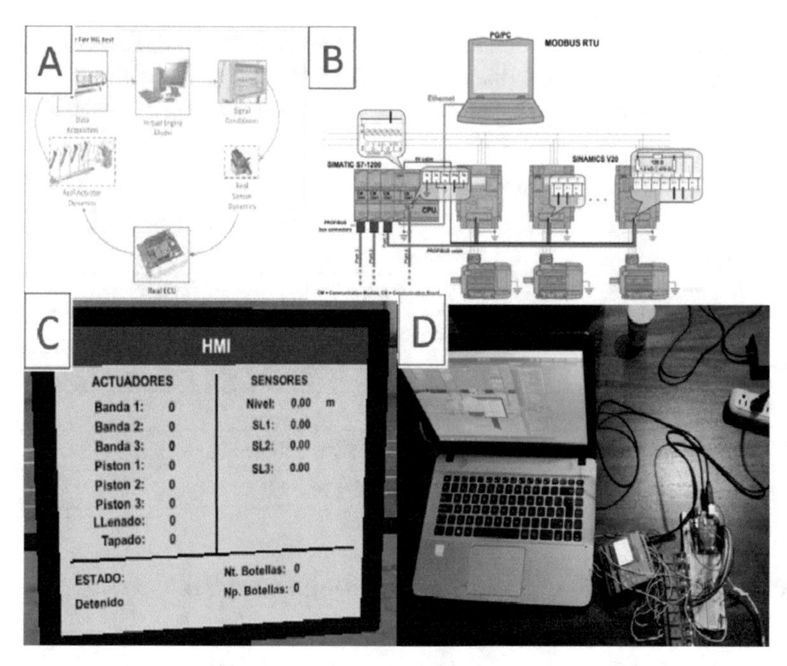

Fig. 2. Modbus connection and network

Figure 3, shows the Modbus RTU protocol, the exchange of data between the programmable logic controllers (PLC), the Arduino module is used both to send information from the computer (PC) to the PLC and to receive information from it. On side a) of Fig. 3. It is observed that the necessary libraries are inserted to be able to carry out serial communication and the address (100) that the data sending frame will have. On side b)

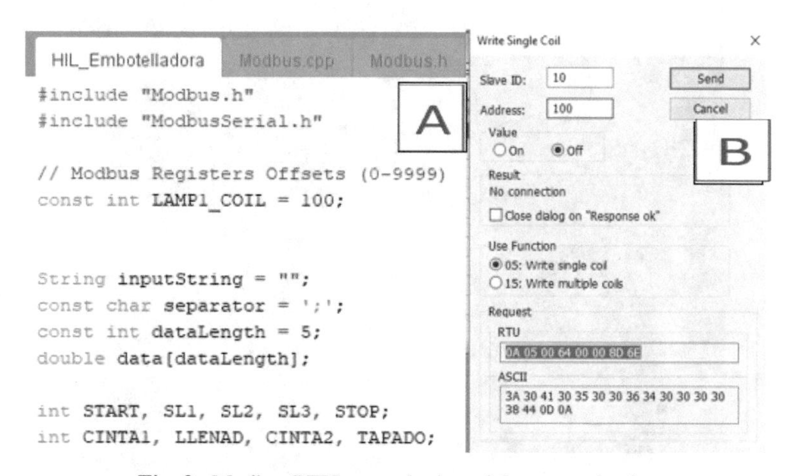

Fig. 3. Modbus RTU network via serial communication

of Fig. 3, with the help of the Modbus POLI program, the RTU data frame that is being used is displayed, which is described below: **0A 05 00 64 00 00** 8D 6E.

In the Table 1. We find the meaning of the data frame for a better understanding.

Table 1. Meaning of frame data

Heading level	Meaning
0A	Slave address
05	Modbus function code
00 64	It is the number (100)
00 00	Send data
8D	CRC error check (occupies 16 bits)
6E	Least significant bit

2.3 Process of Virtualization and Simulation with Unity 3D

The operation of the HMI and all the stages can be seen in Fig. 4. The operation of the HMI is presented in Fig. 4A where the bands, pistons, sensors and the states of each stage of the process are found. The process begins when the band is activated and the presence sensor detects the bottle and it goes to the filling stage. Once the bottle is filled,

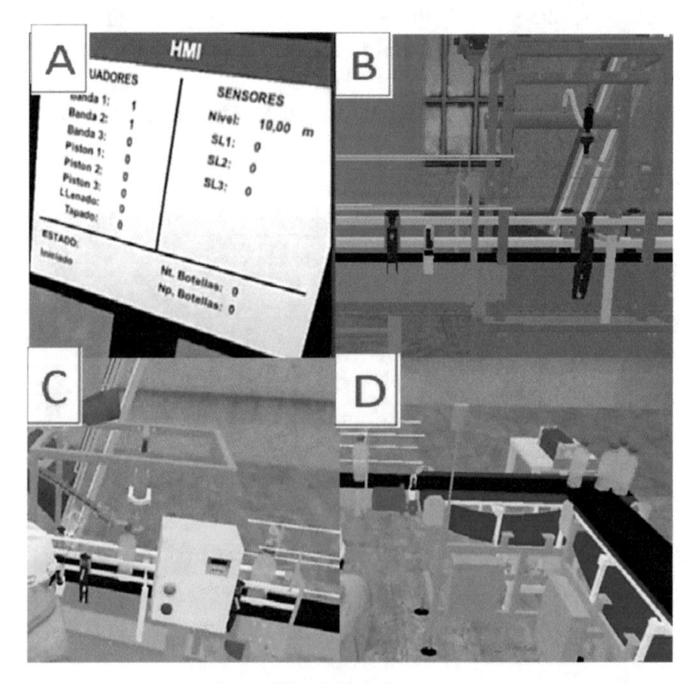

Fig. 4. Liquid bottling process

it advances and goes to the sealing stage. This can be seen in Fig. 4B and 4C. To finish the process, it reaches the end of the band where it will wait until there are 4 bottles and the piston will be activated to move the bottles. This can be seen in Fig. 4D.

3 Results

In the implementation, the connection time of the PLC to the Unity software is not immediate, since the IP address must be entered correctly to obtain communication from the Modbus network. Modbus has become a standard for the link between devices because both intelligent control elements and field elements from different manufacturers can be incorporated into the network without any problem, facilitating communication between control systems and the HMI.

The simulation carried out allows the threads of interest to be followed step by step, as can be seen in Fig. 4A, 4B and 4C, allowing the behavior of the system to be observed under normal conditions or in emergency situations where the indicators of each stage show whether there is a fault, this is indicated in Fig. 4D. With virtualization, the automation of the process can be seen taking into account that the time it takes to connect between Unity and Tia Portal will depend on the computer's processor, data transmission being faster or slower, which affects the simulation. However, with the help of the virtualization of the process environment plus the programming carried out in Tia Portal, they provide the liquid bottling process as a solution (Fig. 5).

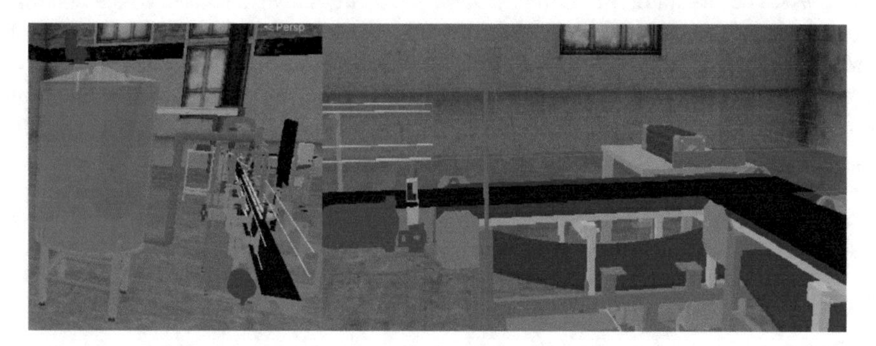

Fig. 5. Virtualized plant in Unity 3D

One of the advantages of carrying out PLC communication through Tia software Portal V17 is the programming of the process using the Ladder language, where it is possible where it is possible to monitor and configure the process control parameters, as well as the time of each stage time of each stage, being 0.72 s. the ideal to obtain an adequate process in the different stages (Fig. 6). The number of bottles obtained is related to the height of the liquid found in the 10 m tank, since for each full bottle it occupies 0.01m, the range of liquid level in the tank from 10 m to 0.02 m is determined for the correct operation, since if it is below the range of 0.02 m the process stops as there is not enough liquid for filling.

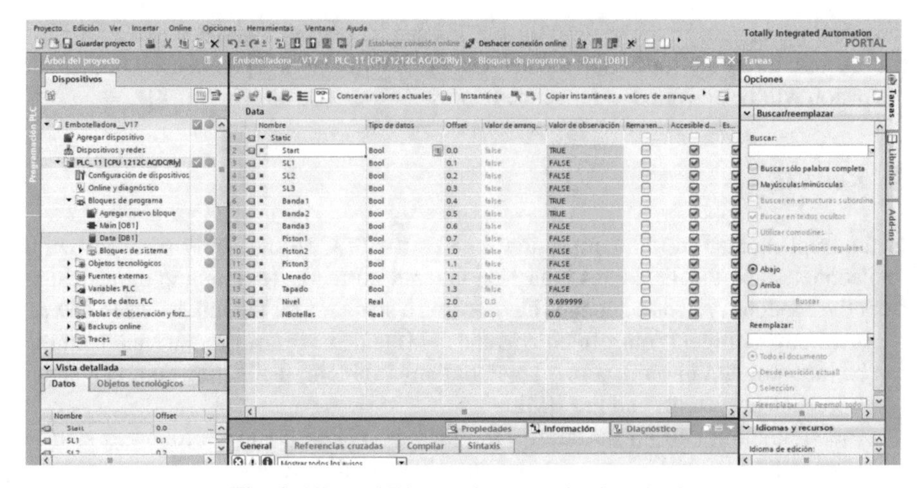

Fig. 6. HIL and Tía portal communication check.

4 Conclusions

Modbus has become a fairly common protocol, frequently used by many manufacturers in many industries. Thus, this communication system is generally used to transmit signals from instrumentation and control devices to a main controller or data collection system (SCADA). The simulator is a suitable and low-cost alternative that allows understanding, supervising and predicting the behavior that the real plant would have, in terms of the threads of interest. The time that the filling stage takes must be precise because when the bottle advances, a period of time must be waited until the bottle reaches the filling stage, if the time is not precise, it will cause a liquid spill. For this reason, the time established for the process is 0.72 s.

Acknowledgments. The authors would like to thank the ARSI Research Group and SISAu Research Group for their support in the development of this work.

References

1. Zhuhuijuan: Virtual roaming system based on unity 3D. Comput. Syst. Appl. **21**(10), 36–40 (2012)
2. Zhangdongjie, F.: Three-dimensional modeling and analysis of hydraulic support based on Pro/E. Sci. Technol. Innov. Herald **24**, 45–46 (2011)
3. Zhengxiaowen, Z.: Study on hydraulic support's working state based on virtual reality. Coal Mine Mach. **33**(10), 72–74 (2012)
4. Wuyafeng: Game Developing Details & Typical Cases by Unity 3D. Beijing:Posts & Telecom Press (2012)
5. Qiujiansong: Research on real-time virtual simulation system based on unity 3D. Pract. Electron. 2012(12) (2012)
6. Mario, D.: Comunicaciones en entornos industriales. http://fing.uncu.edu.ar/investigacion/institutos/IAEI/Cursos2.html

7. Sam, M.: Moving Ethernet to plant floors. http://www.isa.org/journals/ic/feature/1,1162,541,00.html
8. Schneider Automation Modbus/TCP protocol specification. http://www.modicon.com/openmbus
9. O'Murchu, L.: Last-minute paper: an indepth look into Stuxnet. In: Proceedings of the 20th Virus Bulletin International Conference. Vancouver, BC, Canada (2010)
10. Valdes, A., Cheung, S. Intrusion monitoring in process control systems. In: Proceedings of the 42nd Hawaii International Conference on System Sciences (2009)
11. Modbus Organization: Modbus application protocol specification (2006). http://www.modbus.org/docs/Modbus_Application_Protocol_V1_1b.pdf
12. Trexon Inc.: Modbus protocol (2000). http://irtfweb.ifa.hawaii.edu/~smokey/software/about/sixnet/modbus/modbus_protocol.pdf
13. Hanselmann, H.: Hardware-in-the-loop simulation testing and its integration into a CACSD toolset. In: IEEE International Symposium on Computer-Aided Control System Design (1996)
14. Hanselmann, H.: Advances in desktop hardware-inthe-loop simulation. SAE Paper 970932 (1997)
15. Nabi, S.: Development of an integrated test environment for ecu software validation. dSPACE User Conference (2002)
16. Eckel, B.; Design patterns in Python. In: 2002 Software Development West Conference, San Jose, CA, USA (2002)

And Now What (Part 1)? Language Teaching: How We Met the ERT Challenge?

Joaquim Guerra(✉)

CIAC/Algarve University, Campus de Gambelas, 8005-139 Faro, Portugal
jguerra@ualg.pt

Abstract. Foreign language teaching in general has been thought of from a dynamic perspective, in face-to-face context learning, where learners and teachers interact immediately and simultaneously in the classroom. COVID-19 and the successive lockdowns were a complex test for teaching and learning in general. The pandemic forced a discourse and methodological change with an urgent integration of ICT in the teaching and learning process. These allowed the incorporation of applications that ensure immediate contact and at the same time are within the reach of all teachers and learners. Based on a survey that led to the development of a questionnaire addressed to teachers, this study seeks to take stock of the teaching of foreign languages, in particular, post-lockdowns. The questionnaire had both closed and open-ended questions, and the data were analyzed statistically and through content analysis. This paper presents the social, demographic, and academic background of the participants and what they considered their difficulties. In short, they revealed that they faced similar pedagogical and technological difficulties as other university teachers, leading one to believe that the Emergency Remote Teaching highlighted the general lack of preparation of university teachers to deal with emergency situations, such as the one experienced during the COVID-19 lockdowns, regardless of the country and teaching context.

Keywords: Foreign Language Teaching · COVID-19 · Emergency Remote Teaching · Consequences · Technology in Education · Teaching Methodology · Digital Literacy · Higher Education

1 Introduction

The health emergency caused by the pandemic of the COVID-19 virus has led us to an unprecedented situation, causing the confinement of thousands of citizens around the world. As a result, educational institutions, including higher education, have started to teach classes remotely, leading to the adoption of digital tools that could replace face-to-face teaching with remote teaching, changing the dynamics of the teaching and learning process and leading to the learning of digital skills that are expected to last over time, even after the lockdowns are over.

The adaptation to remote teaching and the consequent digital transformation of the teaching and learning process brought with it several logistical and attitudinal challenges

A. Rocha et al. (Eds.): WorldCIST 2023, LNNS 800, pp. 558–566, 2024.
https://doi.org/10.1007/978-3-031-45645-9_54

for some authors [1] or technological, pedagogical, and social challenges for others [2]. To a greater or lesser extend, experiments have been overcome by teachers and students, hoping that the digital learning acquired will not be relegated or even abandoned after the lockdowns. These researchers argue that "[...] new challenges and opportunities at social and technological level may emerge. It is an experiment that enables us to reflect on the different approaches and lessons learned in different countries and additionally provides an opportunity to find new solutions" [2].

This article intends to contribute to the understanding of the challenges and opportunities encountered by language teachers, using an online survey that was sent to the teachers of my institution. The aim is to delimit to highlight the digital solutions found by the teachers, as well as both the difficulties encountered and the opportunities to include these solutions in face-to-face classes after the end of the lockdowns. For now, the study presents the participants and what difficulties they faced during the ERT.

2 Research Literature Background

The emergency remote teaching (ERT) led students and teachers to the intensive use of digital applications that, for the most part, they were not used to. In fact, "[t]he global acceptance of social distancing policy, as announced by WHO [World Health Organization] as a measure to curb the spread of COVID-19, has forced schools to close their doors, and this has caused unexpected disruption of traditional teaching and learning methods" ([3], p. 3). The main goal for the authorities was to enable their teachers and students to continue to teach and learn while keeping them safe [4][1]. Face-to-face teaching moved to a screen-mediated environment dependent on numerous external (Internet connection, connection speed, etc.) and internal (digital competence, learning styles, easy access to necessary resources) factors. The university staff have started several trainings in digital competence for teachers, trying to support them in this abrupt change to a virtual reality for which most of them were not prepared. However, systematic coaching was practically impossible. Trainings were organized according to the difficulties of the moment, considering that he implementation of new teaching and learning technologies was, in most cases, decided on the spot to meet immediate needs [5, 6]. In this context, a great diversity of digital applications and software has erupted in the teaching and learning process to try to respond in the best way to ERT. Many of these were recent and cover uses and needs other than the most common ones (e.g., [7, 8]), while others were unknown or not used in educational contexts until the adoption of ERT, thus increasing the complexity of the planning and organizing of the instructional situation.[2] Consequently, all students and teachers developed new skills in the context of COVID-19, in academic or nonacademic situations, such as problem solving, uncertain web surfing in search of solutions to unfamiliar problems, learning

[1] This reflection will focus on emergency remote teaching and distance learning [4], caused by the pandemic health crisis, which does not correspond to the consolidated approaches, in design and methodology, of teaching and learning related to e-Learning environments (EaD).

[2] A list was progressively constructed by the European Center for Modern Languages of the Council of Europe and can be consulted at https://www.ecml.at/ECML-Programme/Programme2012-2015/ICT-REVandmoreDOTS/ICT/tabid/1906/Default.aspx.

and using new technologies, relying on the resilience and strength of their family and the community around them. Some research pointed to this variety of digital resources as an advantage for online education (e.g., [9]) that should be preserved for other educational contexts.

Thus, the time has come to focus on a post-COVID scenario and to try to understand how we can use the lessons learned from digital language teaching to inform our future practice. As stated, "[h]opefully the COVID-19 threat will soon be a memory. When it is, we should not simply return to our teaching and learning practices prior to the virus, forgetting about ERT." ([4] p. 11). Not only because, as Hodges says, it must "become part of a faculty member's skill set" ([4], p. 11), but also to extend the use of digital knowledge in-presence, making them more dynamic, interactive and, perhaps, more interesting for our net-generation/Z-students, or at least more correlated with their digital life [10, 11]. One year later, the same authors pointed out that some institutions and teachers remain reluctant to use the knowledge acquired in lockdowns to plan lessons in multimodal learning ecosystems, while others assume that the things are different now, supporting their assumptions with testimonials from teachers from colleges and universities, who are moving on to "learning modalities and technologies forming a rich learning ecosystem to afford flexibility, access, continuity, and resilience" [12]. As stated, the growth of the pedagogy of active learning, in particular in higher education contexts, "has intensified the movement toward technology-enhanced education, which has the benefits to increase student participation, improved learning outcomes [...]" [9] but it still had to be adapted to the ERT and now to the situation of face-to-face teaching.

3 Methodology

3.1 Sample

The data were collected from language university teachers at my institution who taught during the lockdown periods. Sixteen of the 30 teachers responded to the survey. It should be noted that some of the 30 were on extended sick leave, which started before March 2020, and others were on sabbatical leave, so they did not respond to the survey. For that reason, this study considers the population sample to comprise all the respondents (n = 16), excluding those who did not answer the questionnaire.

Table 1. Social-demographic characteristics of the participants

Characteristics	
Gender	
Male	2 (14.5%)
Female	14 (87,.5%)
Age range	
30–34	1 (6.3%)
40–44	2 (12.5%)
45–49	3 (18,.8%)
50–54	4 (25%)
55–59	5 (31.3%)
60–65	1 (6.3%)

Based on the answers obtained, Table 1 shows the demographic characteristics of the population and Table 2 shows the academic background of the participants.

Table 2. Academic background of the participants

Academic background	
Degree	
Master	6 (37.5%)
Ph.D	10 (62.5%)
Post-graduate: pre-service teacher training	
yes	10 (62.5%)
no	6 (37.5%)

3.2 Research Instrument, Administration, and Analysis

The questionnaire was constructed based on others referenced in the literature (e.g., [13, 14]) and is divided into two parts. The first part collected sociodemographic and academic data, such as education, years of experience, and what language(s) the respondent teaches. The second part contains questions about their teaching experience during their lockdowns (difficulties, challenges, and opportunities) and the relevance of this experience to their current practice in the classroom.

The questionnaire was sent via email to language teachers and faculties and high school boards so that it could be disseminated to all teachers who teach languages in my institution. Responses were obtained online through a Google Form.[3]

The e-survey data were analyzed using SPSS for the statistical data and content analysis for the open-ended questions.

4 Results and Discussion

The results obtained show us that most of the teachers are female (Table 1) and mainly over 45 years old (Table 2), which may lead us to believe that they have had more difficulty in adapting to remote teaching due to generational issues. An older population may not be as digitally prepared because digital tools have not been part of their regular teaching practice, nor has their pedagogical training included digital literacy issues related to teaching and learning processes. Nevertheless, the teaching experience (*cf.* Table 3) and the progressive digitalization of some academic tasks (the use of Moodle, for example, for communication and distribution of documents and assignments) may have facilitated the shift to digital environments.

Table 3. Years of experience teaching languages

Years of experience	
0 to 10	2 (12.5%)
11 to 20	4 (25%)
21 to 30	9 (56.3%)
40 or +	1 (6.3%)

The languages taught by the respondents are presented in Table 4. English is the language that stands out the most regarding the main language taught, followed by Portuguese as a foreign language (PLE). French and Spanish followed, with two teachers each, and finally Galician, and one teacher who replied that he or she had not taught a foreign language during this period, or at least not all the time. At my institution, it is common for a teacher of a foreign language or mother tongue to have to teach a second language to fill in gaps or complete teaching hours. This explains why many of the teachers who answered one main language also teach another, such as PLE.

Most of the participants felt it challenging or extremely challenging to move to ERT in terms of digital skills (Table 5). This may be interlinked with what is mentioned above with regard to their age and the period when they did their pedagogical training, in which resources still relied heavily on paper documentation. It can also be deduced that it is likely that the planned language lessons were not based on active pre-teaching methodologies, such as task or project methodology.

[3] A print version of the e-survey can be accessed here: pos-remote class study questionnaire - Google Forms - cópia.pdf (please note that the questionnaire is in Portuguese).

Table 4. Languages taught by the participants

Languages	
Principal language taught	
Spanish	2 (12.5%)
French	2 (12.5%)
Galician	1 (6.3%)
English	8 (50%)
Portuguese as a Foreign Language (PLE)	3 (18.8%)
None	1 (6.3%)
Second language	
German	1 (6.3%)
Galician	1 (6.3%)
English	2 (12.5%)
Portuguese as a foreign language (PLE)	7 (43.75%)
Portuguese (mother tongue)	1 (6.3%)
None	4 (25%)

Table 5. How they rate the move to ERT in terms of digital skills

Neither challenging nor discouraging	3 (18.8%)
Challenging	9 (56.3%)
Extremely challenging	4 (25%)
total	16 (100%)

The main difficulties[4] expressed by the participants are, on the one hand, associated with their work in planning and organizing teaching activities and, on the other, getting the students to participate (namely, orally) in the activities, preventing them from becoming distracted and losing motivation. Adapting the resources they had, to have the time to do so and, at the same time, to discover new digital instruments and to know how they work in planning the remote classes, as well as enthusing the students to participate in them were the challenges expressed. Assessment, and especially making sure there is no cheating, was also one of the concerns expressed. This apprehension is also found in other studies ([2, 3, 9], for example). The most curious thing is that the main concern was not how to evaluate in remote mode (in what ways and with what digital resources), but how to ensure that students did not cheat, as if the difficulties with digital resources were not relevant to this issue. Finally, for one of them the principal concern was about the urgency to transform the domestic space into a working and teaching space. Many

[4] The answers to open-ended questions can be found at this link: Open_questions_1.docx (Please note that they are written in Portuguese and Galician and not translated into English).

of us experience this difficulty, especially when the context also kids and other adults having to share space to work and attend classes remotely.

When we confront the participants with some suggestions with regard to which aspects they had more difficulties in, the answers are not divergent of what they wrote before. In fact, Table 6 shows us that.

Table 6. Which aspects were more difficult

Aspects	
The lack of control in the moments of evaluation	10 (62.5%)
The lack of control over the work of the class and the students	4 (25%)
The lack of control over the real presence of students in class	6 (37.5%)
Keeping students focused	10 (62.5%)
Getting students to turn on the camera	9 (56.3%)
Stimulating the oral participation of the students during the lessons	5 (31.3%)
The absence of digital materials	3 (18.8%)
The poor knowledge regarding the necessary digital technologies (teacher and/or students)	1 (6.3%)

The items on the list have been taken randomly from the specialty literature regarding the challenges. From what can be observed, the participants focused on the aspects related mostly to the students and not to themselves. So, they found challenging the lack of control in the assessment periods, keeping the students focused on the lecture and learning activities, and making them turn on the camera in Zoom sessions. These choices are those found in other researches and they do not constitute a surprise. However, it is peculiar hat they do not highlight the lack of digital skills of either the respondents or their students as being a challenge alongside assessment or student concentration. The lack of digital literacy in the context of the teaching and learning process is probably the biggest challenge present. Indeed, knowing the possible digital tools for the different teaching and learning contexts would solve many of the problems expressed.

5 Conclusion and Future Works

ERT has allowed universities not to close and to continue to provide learning opportunities to their students by converting face-to-face classes into online classes. However, this situation has not been innocuous, bringing into the pedagogical realm numerous difficulties and challenges that students and teachers have had to face and seek to resolve. In the case of my Institution, the teachers, probably due to their age range, manifested several difficulties in the transition from face-to-face to online classes. The technological and pedagogical challenges were not much different from those experienced by others in universities, countries, or continents: the use of unknown or uncommon digital resources in classrooms, to make an online class more dynamic, to carry out evaluation moments, or simply to have the students present and with the camera on.

In future research, how these difficulties were overcome will be analyzed along with what strengths and weaknesses are considered to have existed. Finally, we will try to find out what remains of the learning experiences of the teachers during this period and whether they continue to use digital tools to prepare and plan the teaching and learning process in-presence.

References

1. Ribeiro, R.: How university faculty embraced the remote learning shift. EDTech Magazine (2020). https://edtechmagazine.com/higher/article/2020/04/how-university-faculty-embraced-remote-learning-shift. Accessed 22 Nov 2022
2. Ferri, F., Grifoni, P., Guzzo, T.: Online learning and emergency remote teaching: opportunities and challenges in emergency situations. Societies **10**(4), 86 (2020). https://doi.org/10.3390/soc10040086,lastaccessed2022/11/15
3. Adedoyin, O.B., Soykan, E.: Covid-19 pandemic and online teaching learning: the challenges and opportunities. Interact. Learn. Environ. (2020). https://doi.org/10.1080/10494820.2020.1813180,lastaccessed2022/11/10
4. Hodges, C., Moore, S., Lockee, B., Trust, T., Bond, A.: The difference between emergency remote teaching and online learning. EDUCAUSE Review (2020). https://er.educause.edu/articles/2020/3/the-difference-between-emergency-remote-teaching-and-online-learning. Accessed 10 Oct 2022
5. Guppy, N., Verpoorten, D., Boud, D., Lin, L., Tai, J., Bartolic, S.: The post-COVID future of digital learning in higher education: views from educators, students, and other professionals in six countries. Br. J. Educ. Technol. **53**(6), 1750–1765 (2022). https://doi.org/10.1111/bjet.13212. Accessed 10 Oct 2022
6. Oliveira, G., Teixeira, J., Torres, A., Morais, C.: Na exploratory study on the emergency remote education experience of higher education students and teachers during the COVID-19 pandemic. Br. J. Edu. Technol. **52**(4), 1357–1376 (2021). https://doi.org/10.1111/bjet.13112,lastaccessed2022/10/12
7. Guerra, J.: Web-based language class activities: contexts of uses and background methodologies. In: Uzunboyiu, H., Ozdami, F. (eds). Procedia – Social and Behavioral Sciences. 2nd World Conference on Educational Technology Research – 2012, vol. 83, pp. 117–124. Elsevier, London (2013)
8. Guerra, J., Olkhovych-Novosadyuk, M.: Multimedia as an efficient web-based tool for the development of communicative competence of students studying English for specific purposes. World J. Educ. Technol. **6**(3), 273–277 (2014)
9. Fuchs, K.: Perceived satisfaction of emergency remote teaching: more evidence from Thailand. Int. J. Learn. Educ. Res. **20**(6), 1–15 (2021). https://doi.org/10.26803/ijlter.20.6.1. Accessed 07 Nov 2022
10. Berguerand, J.-B.: An introduction to teaching Gen Z students. Trends Education (2022). https://hospitalityinsights.ehl.edu/teaching-gen-z-students. Accessed 22 Nov 2022
11. Schukei, A.: What you need to understand about generation Z students. The Art of Education – University (2020). https://theartofeducation.edu/2020/12/14/what-you-need-to-understand-about-generation-z-students/. Accessed 22 Nov 2022
12. Moore, S., Trust, T., Lockee, B., Bond, A., Hodges, C.: One year later… and counting: reflections on emergency remote teaching and online learning. EDUCAUSE Review (2021). https://er.educause.edu/articles/2021/11/one-year-later-and-counting-reflections-on-emergency-remote-teaching-and-online-learning. Accessed 24 Oct 2022

13. Nikade, E.C., Chukwudi, E.M.: Language teaching and learning in post-Covid-19 teaching and learning. Presentation at 2nd Faculty of Humanities International Conference. Ignatus Ajuru University of Education, Rumuolumeni Port Harcourt (2022)
14. ECML: The Future of Language Education in the Light of Covid. Lessons Learned and Ways Forward. https://www.ecml.at/ECML-Programme/Programme2020-2023/Thefutureoflanguageeducation/tabid/5491/language/en-GB/Default.aspx. Accessed 14 Oct 2022

Virtual Laboratory as a Strategy to Promote Reading in 5-Year-Old Children

Paola Carina Villarroel Dávila, Lucy Deyanira Andrade-Vargas, María Isabel Loaiza Aguirre, Paola Salomé Andrade Abarca, and Diana Elizabeth Rivera-Rogel(✉)

Universidad Técnica Particular de Loja, San Cayetano Alto, Calle Marcelino Champagnat, Casilla, 11-01-608 Loja, Ecuador

{pcvillarroel,ldandrade,miloaiza,psandrade,derivera}@utpl.edu.ec

Abstract. Learning to read during the first years of life is a subject of extensive research and intervention at the level of public, social and institutional policies. The objective of this research is to design a prototype of a virtual learning laboratory, integrating didactic, technological and trans-media inclusive resources that contribute to the development of reading competence in 5-year- old children. A mixed quantitative-qualitative, exploratory-descriptive design was applied, assessing the interactivity and usability of the virtual laboratory "LecturLab", using as instruments the LORI-AD questionnaire [1] with the following criteria: quality of content, correspondence with the objective, feedback and adaptation, motivation, design and presentation, interaction and usability, accessibility, reusability; and an ad-hoc observation sheet to know the children's skills in relation to the designed resource. It is concluded that "LecturLab" allows children to handle technological resources independently and intuitively, awakening motivation and active participation, as well as the development of phonetic, syntactic and semantic awareness, which are integrated when using visual and auditory resources. In the results of the teachers, 57.1% indicate that the resource is good, the design of this type of laboratories expands the access and use of digital tools with a variety of activities to initiate the reading process in face-to-face or virtual spaces. The creation and use of digital resources makes visible new forms of teaching- learning in early childhood and the need for further research, since the reading process is not limited to the exclusive use of concrete material as it was a few decades ago.

Keywords: Educación Preescolar · technological resources · reading

1 Introduction

In the field of education, the reading process during the first years of life is a subject of extensive research that includes several aspects such as: (a) autonomous motivation for early childhood literacy [2], (b) phono- logical awareness, (c) letter and pseudoword reading [3], (d) rhythm, pre-literacy and auditory processing in early childhood [4], e) the direct and indirect effects of the literacy environment, stories and arithmetic [5], f) knowledge about the role of interactive read aloud as a pedagogical tool [6] and g) the design of resources through stories [7].

A. Rocha et al. (Eds.): WorldCIST 2023, LNNS 800, pp. 567–576, 2024.
https://doi.org/10.1007/978-3-031-45645-9_55

At the level of public policies or cultural promotion, reading in early childhood has several initiatives, among them the National Reading Plan and the Chile crece contigo program [8]. The National Reading and Writing Plan 'Reading is my story' of the Colombian Institute of Family Welfare [9], Bookstart, an English program that seeks to introduce children to the world of books at a very early age, with the aim of stimulating their cultural, educational, social and emotional life [10], Uruguay's reading and language education program "Reading is embracing us with words" (Ministry of Education and Culture, 2021), Marathon of the Story in Ecuador, proposed by the Ministry of Education and Culture [11], and the National Reading and Writing Program "Reading is embracing us with words" of Uruguay (Ministry of Education and Culture, 2021), the National Reading and Writing Plan "Reading is my story" of the Colombian Institute of Family Welfare [12] by Girándula, the Asociación Ecuatoriana del Libro Infantil y Juvenil, is a space where children enjoy stories read by the authors themselves [11].

Although early childhood reading leads to various reading experiences, it is necessary to specify the approach from which they are generated. Approaches have evolved over time according to theoretical contributions and practical evidence that have transcended from the narrow domain of the code to consider a broad and diverse set of mental processes, linguistic and cultural knowledge, communicative practices and ways of thinking, derived from the use of letters in a community [12], hence three approaches to teaching reading can be distinguished: linguistic, psycholinguistic and sociocultural [13].

In the linguistic approach to reading, there is a superficial understanding of texts. This approach draws attention to the idea that the meaning of the text is the sum of the meaning of all words and sentences [14]. For its part, the psycholinguistic approach considers the importance of phonological knowledge, in terms of syllabic and phonemic knowledge [15], alphabetic knowledge, speed of nomination [16], as well as the development of thinking and the construction of knowledge of the environment [8], through the understanding of the meaning of words and phrases presented in different contexts.

The socio-cultural approach focuses its attention on the construction of meanings within a cultural context and the need to understand the vision of the other, it is a process that allows to search and offer meaning to voices, events, objects, linguistic and iconic signs and dynamics that underlie reality, it allows to signify texts and similarly to existence [17].

The research is aligned with the psycholinguistic approach, it is related to the methodologies for teaching reading that are grouped into two major currents, phonic methods that begin by presenting the child with the correspondence letter or phoneme, the teaching of reading is based on the alphabetic principle on which our writing system is based, and global methods that start from words, assuming that the simple exposure to written material, in a context in which this material makes sense and is understood by the child, is sufficient for the child to discover the alphabetic code [18]. Nowadays, the aforementioned methodologies must be reconsidered due to the use of technology, since the child has access to images, texts, audios, which enriches his psycholinguistic skills and his communicative capacity. The sophisticated architecture of most of the digital resources with a friendly interface, which includes several itineraries according to the users, with various levels of difficulty and which offers various types of help to the learner, allows us to solve our doubts at the moment and to learn in an easy way much more autonomous [19].

The initiatives that integrate technology to early reading development are multiple, some of them are: digital storytelling combines the ancient art of storytelling with a range of contemporary tools to weave stories together with the author's narrative voice, including digital images, graphics and music [20]. The use of the mobile app as an innovative strategy that accompanies and mobilizes the parent towards maintaining reading routines with their children [21]. The design of early childhood creation spaces in the framework of positive technological development [22].

2 Methodology

The research followed a mixed quantitative-qualitative, exploratory-descriptive design, which made it possible to identify how a phenomenon occurs within a real text [23]. In order to validate the virtual laboratory "LecturLab", we worked with two different samples (a) teachers and (b) students. The participants were randomly selected taking into account the appreciation of the resource from two points of view, teaching by teachers and learning by children.

The group (a) was made up of 28 teachers, 35.7% in early education and 64.3% in basic education, who provide reading and writing training to children aged 5 years and the corresponding training is 75% at the undergraduate level and 25% at the graduate level; the experience of the teachers is divided between 53.6% with 11 to 15 years and more and 46.4% with 1 to 10 years. The population of selected professors are members of the network of educational institutions where the Universidad Técnica Particular de Loja, Ecuador, carries out community internships and linkage, 78.6% of the institutions where these professors work are public and 21.4% are private and are located in the urban sector of the city of Loja-Ecuador.

Group (b) consisted of 18 children 5 years of age, 33.3% male and 66.7% female, who at the time of observation were in the first grade of basic education in one of the educational institutions of the network of practice centers. Public institution, located in the urban sector of the city of Loja-Ecuador.

In this section it is necessary to consider that the number of teachers and students sampled does not allow us to generalize the results. However, it presents a first evaluation of the LecturLab virtual laboratory.

2.1 Instruments

For the group of teachers, the instrument used was the LORI-AD questionnaire [1], which validated the following categories: quality of content, correspondence with the objective or competence, feedback and adaptation, motivation, design and presentation, interaction and usability, accessibility, reusability, according to the proposal of the author of the instrument, which was applied preserving its original structure and content, without making any modifications; the evaluation criterion of compliance with standards related to whether the resource is defined with metadata according to the specifications of international standards was not considered, because the virtual laboratory was not yet in application at the time of validation. For the quantification of results, according to the author of the instrument, a rating scale was applied that assigns a discrete score of one point for the descriptors in each category. The total score is calculated by adding the 8 criteria: cct + co + ra + m + dp + iu + a + r.

Since 8 criteria were used, the rating scale proposed by [1] was adapted as follows: 30–49 poor, 50–69 acceptable, 70–79 good and 80 very good.

Taking into account the adjustments, a reliability analysis of the internal consistency of the scale was carried out, the result of the Cronbach's Alpha value for the reliability level is Good (0.636) considering the classification of reliability levels according to Cronbach's Alpha detailed in the study by [24].

For the group of students, an ad-hoc observation form was used, designed and elaborated by the researchers to know the abilities and emotions of the children when interacting with the digital resources. The card consists of 10 criteria that were the reference to observe the path and learning that the children developed in the laboratory, these were: ease of navigation, beginning and end of the task/game, attention, motivation, movements, active listening and repetition, path, gestural or verbal responses, successes and errors.

2.2 Procedure

The following phases were established for the development of the virtual laboratory "LecturLab": (1) define the reading skills required for 5-year-old children and contemporary methodologies for teaching reading (2) perform a benchmarking, identifying the best existing platforms, applications and resources for the development of reading competence in childhood (3) extract and curate videos and educational resources according to the analyzed reading skills (4) develop new audiovisual courses based on the identified reading profile and context (5) design the virtual laboratory platform based on existing technologies (6) develop, test and improve the laboratory.

The design of a laboratory has a pedagogical and a technological basis. The pedagogical design, as shown in Fig. 1, is aligned with the psycholinguistic approach that integrates two currents of the methodology of initiation to analytical and synthetic reading to develop linguistic awareness considering phonological, semantic, lexical and syntactic elements through various literary resources that awaken the child's interest thanks to the relevant use of images, sounds, colors, movement. The process takes into account diversity, inclusion, pertinence and aesthetics as transversal axes in order to recognize and value the context and individual learning differences, as well as to generate child readers who discover the pleasure of learning and having fun through reading throughout life.

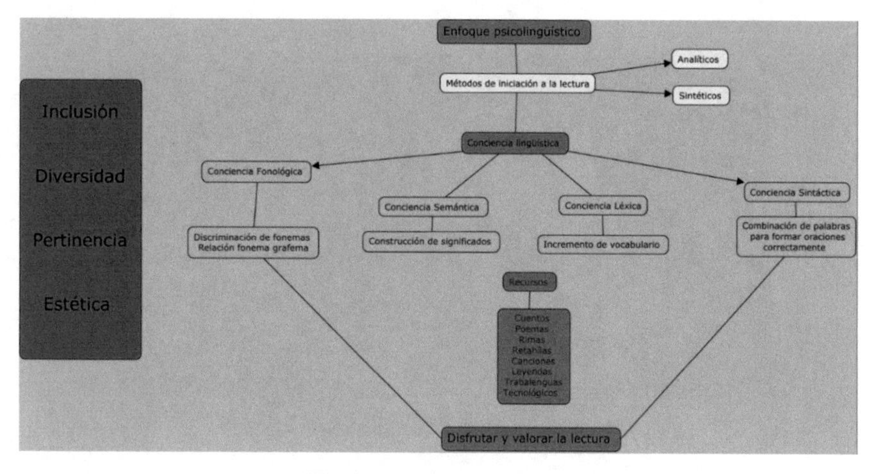

Fig. 1. Pedagogical Design.

Likewise, the technological design of the resource repository website is developed based on a layered architecture using the Angular Web library. For the storage of the site, Digitalocean digital cloud services are used to deliver the web content as shown in Fig. 2.

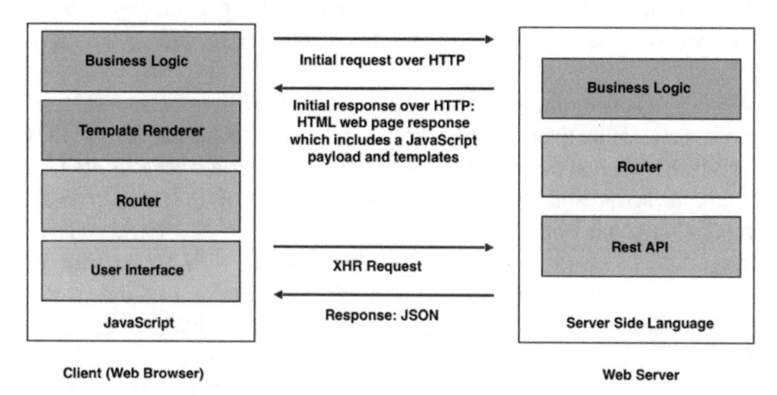

Fig. 2. Technological Design. Adapted from Isomorphic Go: Learn how to build modern isomorphic web applications using the Go programming language, GopherJS, and the Isomorphic Go toolkit, por K. Balasubramanian, Packt Publishing (2017).

3 Results

As a result of the first five stages described in the procedure, we obtained a laboratory platform integrated by open educational resources (OER) developed by the authors, which allow to promote reading and several OER extracted from the web, in addition, it was determined that the virtual laboratory has the pedagogical and technological criteria for the development of reading competence (Fig. 3).

Fig. 3. LecturLab Prototype home page.

To determine whether the laboratory met these quality criteria, a usability test was applied to real users in order to determine whether the platform can be used by school children and teachers and whether its resources have educational value to promote reading.

After the application of the LORI-AD validation instrument, 57.1% of the participating teachers rated LecturLab as a Good resource, followed by 57.1% who rated LecturLab as a Good resource, followed by 57.1% who rated LecturLab as a Good resource 28.6% rated it as Very good and 14.3% as Acceptable in the eight criteria considered.

The average results of the teachers' evaluations in the different categories of the validation process show that the criterion Correspondence with the Objective (CO) reaches the highest score in the validation (10 Very good). The criteria Accessibility (A) Motivation (M) and Interaction and Usability (IU), with averages of (9.86%, 9.76% and 9.64% respectively) also score Very Good. However, the criteria Feedback and Adaptation (RA), Design and Presentation (DP) and Reusability (R) with scores between 9.46% and 9.29% reach the Good level. Finally, the criterion Quality of Content reaches the lowest score on the rating scale (8.50%), placing it at the acceptable level. In detail, the results show that:

The correspondence with the objective is very good, that is to say, alignment is observed in the instructional design of the resource, the activities and contents proposed in the laboratory allow reaching the stated goals and showing the user their level of achievement. The accessibility of the resource is very good and, therefore, it can be accessed by any user who wishes to take it, in addition, the design of the controls and the presentation of the information are adapted for people with special sensory and motor abilities or needs in an accessible way using special devices allowing its use, finally, the resource can be accessed from mobile devices that facilitate its consultation.

The motivation generated by the laboratory is high, since it presents content relevant to the interests and personal goals of the users, generates interest in the topic presented and offers a representation of its contents based on reality through multimedia, interactivity, humor, and/or challenges, games, unpublished stories linked to the country's context that stimulate the interest of the children. The interaction and usability with the resource is very good, the interface has an implicit design that informs users how to interact with it, it is a predictive (intuitive) interface for the user and the instructions for use are clear.

The feedback and adaptation in the resource allows a good user interaction, the options to move forward, backward and end the process facilitate the interaction and the decision buttons offer feedback according to the answers. The design and presentation of the laboratory is good, the information shown in graphic, textual or audiovisual form favors processing, the organizational structure facilitates the effective identification of the elements present in the resource, the style and design of the resource allow the user to learn in a good way.

The reusability of the resource is good, it can be reused by different courses and/or contexts, it has the licensing of use, it can be downloaded from its source site and related through its link address. The quality of the content of the laboratory is acceptable, the content of the resource is free of errors and is presented without bias, it is easy to access and certainly the use of digital tools is one of the difficulties when interacting with the resource; however, the laboratory has a variety of activities that allow starting the reading process.

An individual observation was made with the students to see their reactions to the platform. It was observed that 50% of the children used the platform at a level of excellent in each of the following criteria: they easily understand the page navigation and the games, they start and finish each game, video or song, and they repeat orally what they hear in the games. Also, 44.4% of the children maintain attention during the play period and complete the activities correctly.

With respect to whether they are motivated during the play route, a high percentage of children (72.2%) express that their motivation is excellent. On the scale above 60% and at an excellent level, children control manual movements for the use of resources (66.1%) and give gestural or verbal responses to successes or errors in the games (66.7%). Finally, 44.4% of the children identify very well how to return to each game or route and 72.2% identify very well the route of the resource without any difficulty, which is related to the children's comments when they said "I got a little lost on the way".

In general terms, the results show that the interaction of children with the practices developed in LecturLab allows them to handle technological resources independently and intuitively, an indispensable skill in the current learning context.

4 Discussion and Conclusions

Teachers value Lecturlab as a useful resource to generate learning with respect to reading in 5-year-old children, since they consider aspects that they mention as follows [25–27] for whom the ideal in learning to read is to integrate oral and written language, to propose activities such as auditory and visual comprehension, auditory and visual association, verbal and motor expression, etc., to develop the vocabulary in the initial stage considering the sounds that form the words.

The criteria of content quality, feedback and adaptation, design and presentation reach the good level. Special attention is given to the result on the quality of content, which has a lower evaluation, although it was considered what [28] indicates that a child between 4 and 7 years of age needs to develop the following linguistic skills: phonological competence, identification of known written words, analysis of pseudo-words, verbal short-term memory, rapid automatic nomination of objects, vocabulary and knowledge of the name and sound of the letter, which involve a series of changes where a wide range of cognitive functions intervene [29]. It could be inferred that the teaching logic responds to didactics rather than linguistics and, on the other hand, there is technology, in which teachers are not experts [30].

Likewise, the results show a high interest on the part of the children to develop the activities proposed in the laboratory, focused on improving reading, since they are structured with an essential factor such as the game, gamification awakens the motivation of children, in such a way that they enjoy the initiation of the reading process, another element is the stories and tales, which corroborates what was stated by [7] who mentioned that the design of resources through stories is important for the reading process during the early years.

The category most valued by teachers at the time of validating the virtual laboratory is the correspondence with the proposed learning objective, proving that the activities, contents and resources designed in the laboratory under a psycholinguistic approach [13] will allow students to achieve the proposed learning objectives, which goes hand in hand with the research of [8, 31, 32] who state that the reading process is related to several cognitive, linguistic, metalinguistic functions, integrating oral, written language and audiovisual media for the construction of knowledge.

In this same sense, in the validation of the students, gestural or verbal responses to the successes or errors were observed when the children were interacting on the platform, which proves that the game designed is considered a challenge for the one that offers a different learning experience to its users, developing children's skills, awakening interest and making each child participate in their learning, which allows taking advantage of the resources in an unlimited way [33].

For their part, the children easily understood the navigation of the site and the games, started and finished the games successfully and were motivated. This is mainly due to the fact that children are enthusiastic about multimedia platforms and a friendly architecture. Along the same lines, authors such as [19] state that the sophisticated architecture of most of the digital resources with a friendly interface, which includes various itineraries according to the users, with various levels of difficulty and which offers various types of help to the Internet user, makes it possible to resolve doubts and learn in a much more autonomous way.

It has been evidenced that the design of the "LecturLab" laboratory has considered the different methods of teaching reading either from a synthetic approach such as the alphabetic, phonetic, syllabic method that focus their attention on learning letters, phonemes and syllables, and the analytical approach that considers words, phrases or texts, as a starting point to promote the habit of reading. In this context, it is important to promote the use of virtual laboratories in educational centers, since they allow the child to participate autonomously, achieving a meaningful learning experience.

Considering the research carried out, it is concluded that laboratory practices as a didactic strategy in the teaching of reading are fundamental for children, this is validated by the results obtained from the survey of information with students and teachers. However, it was observed that, despite the motivation and the great effort of the researchers, there are elements that still need to be improved, one of them is the content, for which it would be necessary to expand the knowledge on linguistic aspects as pointed out by the studies on teacher action of [34] that suggest the need to integrate linguistic and literary training in early childhood programs, as well as formative assessment to promote reflective learning.

Finally, from the point of view of researchers in the field of early childhood education, it is important to develop and analyze the benefits of virtual laboratories focused on children's learning and the development of new skills, linking in their lines of research the technological axis as a complement to an interdisciplinary training and highly demanded in the current context.

References

1. Adame, S.: Instrumento para evaluar Recursos Educativos Digitales, LORI – AD. Revista CERTUS **12**, 56–67 (2015)
2. Erickson, J., Wharton-McDonald, R.: Fostering autonomous motivation and early literacy skills. Reading Teacher **72**(4), 475–483 (2019)
3. Romero, S., Jiménez, E., Macarena, P.: Word reading: the influence of phonological processing and the reading-writing method. J. Fuentes **21**(1), 11–24 (2019)
4. Bonacina, S., Huang, S., White-Schwoch, T., Krizman, J., Nicol, T., Kraus, N.: Rhythm, reading, and sound processing in the brain in preschool children. Sci. Learn. **6**(1), 1–11 (2021)
5. Salminen, J., Khanolainen, D., Koponen, T., Torppa M., Lerkkanen, M.: Development of numeracy and literacy skills in early childhood—A longitudinal study on the roles of home environment and familial risk for reading and math difficulties. Front. Educ. **6** (2021). https://doi.org/10.3389/feduc.2021.725337
6. Frejd, J.: Children's encounters with natural selection during an interactive read aloud. Res. Sci. Educ. **51**(1), 499–512 (2019). https://doi.org/10.1007/s11165-019-09895-9
7. UNICEF, Series "Tales that care". https://www.unicef.org/argentina/informes/serie-cuentos-que-cuidan. Accessed 30 Nov 2022
8. Herrera, M.: Reading promotion in kindergarten education. Analysis and intervention program in kindergartens in Valparaíso, Chile. Doctoral thesis, University of Granada (2017)
9. ICBF.: Reading Festival. https://www.icbf.gov.co/programas-y-estrategias/primera-infancia/fiesta-de-la-lectura. Accessed 28 Nov 2022
10. Guarín, S.: Books for Early Childhood, Return on an Investment in the Country. Bogotá, Fundalectura (2015)
11. CERLALC: Current: Girándula, Ecuadorian Association of Children's and Young Adults' Books. https://cerlalc.org/directories/girandula-asociacion-ecuatoriana-del-libro-infantil-y-juvenil/. Accessed 28 Nov 2022
12. Cassany, D.: Provisional autobiography. Current axes for research on Didactics of Reading and Writing. In: Ballestar, J., Ibarra, N. (coord.) Between Reading, Writing and Education, pp. 57–74. Narcea (2020)
13. Cassany, D.: Behind the lines. On Contemporary Reading. Anagrama, Barcelona (2006)
14. Núñez, J.: Notions for Reading Comprehension Development. Edelvives, Madrid (2017)
15. González, M.: Learning to read and phonological awareness. J. Study Educ. Dev. **19**(76), 97–108 (1996)

16. Sellés, P.: Current status of the assessment of predictors and skills related to early reading development. Open Classroom **88**, 53–72 (2006)
17. Ramírez, C., Castro, D.: Reading in Early Childhood. Bogotá, Regional Center for Book Promotion in Latin America and the Caribbean (2005)
18. Alegría, J.: For a psycholinguistic approach to learning to read and its difficulties 20 years later. Child. Learning **29**(1), 93–111 (2006)
19. Hernández, D., Cassany D., López, R., Reading and Writing Practices in the Digital Age. Brujas, México (2018)
20. Rahiem, M.D.H.: Storytelling in early childhood education: time to go digital. Int. J. Child Care Educ. Policy **15**(1), 1–20 (2021). https://doi.org/10.1186/s40723-021-00081-x
21. Vargas, V., Sánchez, J., Delgado, A., Aguirre, L., Agudelo, F.: Dialogic reading in the promotion of cognitive, emotional and behavioral profiles in early childhood. Ocnos **19**(1), 7–21 (2020)
22. Bers, M., Strawhacker, A., Vizner, M.: The design of early childhood makerspaces to support positive technological development: two case studies. Libr. Hi Tech **36**(1), 75–96 (2018)
23. Creswell J., Poth, C.: Qualitative Inquiry and Research Design: Choosing Among Five Approaches. Saga Publications, Canadá (2018)
24. Tuapanta, J., Duque, M., Mena, A.: Cronbach's alpha to validate a questionnaire of ICT use in university teachers. J. mktDescubre **10**, 37–48 (2017)
25. Arnáiz, P., Castejón, L., Ruiz, M.: Influence of a psycholinguistic skills development program on access to reading and writing. J. Educ. Res. **20**(10), 189–208 (2022)
26. Villalón, M.: Initial Literacy. Editions UC, Chile (2008)
27. Mediavilla, A., Fresneda, R.: Reading and Reading Difficulties in the 21st Century. Octaedro, Spain (2020)
28. Porta, M., Ison, M.: Towards a comprehensive approach to initial language learning as a cognitive process. Iberoamerican J. Educ. **55**, 243–260 (2010)
29. Puche, R., Orozco, M., Orozco C., Correa, M.: Child development and Competencies in Early Childhood. Ministry of National Education, Bogotá (2009)
30. Zorro, A.: Teaching strategies mediated by new information and communication technologies to strengthen the teaching of foreign languages. Research Project, Universidad de la Salle (2012)
31. Velarde, E., Canales, R., Meléndez, M., Lingán, S.: Cognitive and psycholinguistic approach to reading Design and Validation of a pre-reading skills test (THP) in children from the constitutional province of Callao. Peru. J. IIPS **13**(1), 53–68 (2010)
32. Jolibert, J.: Children building the power of reading and writing. Conference Universidad Tecnológica de Pereira (2010)
33. Alexiou, C.B.: Using VR technology to Support e - Learning: The 3D Virtual Radiopharmacy Laboratory. In: 24th International Conference on Distributed Computing Systems Workshops, pp. 268–273. IEEE, Tokyo (2004)
34. Trigo, E., Rivera, P., Sánchez, S.: Reading aloud in the initial training of early childhood education teachers at the University of Cadiz. Ikala **25**(3), 605–624 (2020)

Evaluating the Graduation Profiles of a University Primary Education Program: A Virtual Evaluation from the Perspective of Employers and Graduates

Carla Vasquez(✉), Elvis Gonzales Choquehuanca, Roberto Cotrina-Portal, Estela Aguilar, and Ivan Iraola-Real

Universidad de Ciencias y Humanidades, Lima 15314, Peru
{cvasquez,egonzales,rcotrina,eaguilar,iiraola}@uch.edu.pe

Abstract. The evaluation of graduate profiles is essential for the innovation of curricular proposals. Thus, the present study aims to evaluate the graduate profiles of a primary education program at a private university in Lima, Peru, from the point of view of employers and graduates. Using an evaluative qualitative methodology, semi-structured interviews were conducted with three directors and a focus group with six graduates. Then, through open coding and triangulation processes, emerging categories were generated. Finally, the evaluation showed that graduates manage pedagogical proposals by applying teaching and evaluation strategies, but they need to reinforce curriculum approaches. They participate in curriculum management but not in administrative management. They develop tutoring for students and parents, but need to reinforce these strategies. And they do not show evidence of participating in research projects.

Keywords: Education Evaluation · Teacher Evaluation · Teacher Training · Professional Profile · Primary Education

1 Introduction

It is undeniable that in the 21st century teachers must be able to work collaboratively, organized and planned in learning environments; focusing their attention on the student (Espinoza 2020). In this sense, this research reflects the graduate profile of a university program in Primary Education, consisting of competencies that have to do with managing a pedagogical proposal, leading educational management processes, promoting tutorial action and participating in research projects. It should be noted that, in Peru, competencies, abilities, attitudes and values are evaluated by the National System of Evaluation, Accreditation and Certification of Educational Quality (SINEACE). On the other hand, in Spain, various results of the level of achievement of competencies and the shared design of competency assessment are presented. In addition, in Europe and the United States, minimum standards of performance evaluations, progressive professional

A. Rocha et al. (Eds.): WorldCIST 2023, LNNS 800, pp. 577–591, 2024.
https://doi.org/10.1007/978-3-031-45645-9_56

practices, curricular variations, political reforms and transformations focused on professionalizing the teaching profession have been progressively installed (Gómez 2013). In addition, it should be mentioned that the aspiring teacher must effectively develop pedagogical and didactic performance (Konstantinou and Konstantinou 2021). It should be added that, the curricula of students in the education program should be comprehensible through multiple activities (Kennedy 2016). On the other hand, the graduate must know the problems of the school, especially those that strengthen the institutional mission (Valles-Ornelas et al. 2015). In addition, in teacher training for tutorial action, the future graduate must have soft skills such as empathy, listening skills and ethical behavior (Ministry of Education [MINEDU] 2021). As for, the evaluation of research qualities of the graduate of the primary education program, it should be oriented to research to improve the quality of teaching (Dijkema et al. 2019). After what has been presented, some of these competencies of the graduate profile of a university primary education program will be addressed below, based on a rigorous theoretical support.

1.1 Assessing the Profile of Exit in Primary Education Professionals

The evaluation of the graduate profile in teacher training programs at the primary level is important, since there must be coherence between the degree granted and the competencies of teacher training in Primary Education. Follow-up and evaluation processes must be in place to ensure the educational quality of future education professionals (Medina-Pérez and González-Campos 2021). On the other hand, the graduate profiles should seek the integration of the institutional dimensions that graduates should achieve according to the demands of the current context, so that the graduate profile every time it is implemented or modified should be socialized to evaluate and provide feedback on the competencies of future primary education teachers (Aravena et al. 2020). In Latin America, competencies and the set of skills, attitudes and values are evaluated for the accreditation of educational quality by different governmental entities, in the case of Peru, by SINEACE. Thus, each university assumes a curricular model that defines the curriculum or syllabus of each study program, considering a graduate profile based on certain competencies and, therefore, develops capabilities that allow achieving the educational objectives, this leads the university to achieve certain quality standards, specifically in the education programs.

In this sense, it is necessary to evaluate the graduate profile and evidence the achievement of the proposed competencies and educational objectives; for this, the following trends have been identified in these evaluation mechanisms, first of all, it has to do with the evaluation at the end of university studies, developing and applying instruments (Barrera 2009; Ortiz et al. 2015). Next, collecting the opinion of external stakeholders, which are employers and interested organizations and where graduates work (Fraga 2013; Olivos et al. 2015). Likewise, a comprehensive proposal for the evaluation of competency achievement by subject at the program and university level (García 2010; Ibarra et al. 2009). Finally, the evaluation of competencies in a progressive manner during the time of training (Cardona et al. 2018).

1.2 Evaluation of the Profile of Primary School Teachers' Graduation in Peru

In the Peruvian context, the Peruvian Ministry of Education, with regard to the teacher training graduate profile, considers *"Knowing how to be, Knowing how to live together, Knowing how to think, Knowing how to do"*, distinguishing between the Facilitator, the Researcher and the Promoter, each having certain characteristics according to these four skills (MINEDU 2022). In addition, there are three trends in the evaluation of the graduate profile, such as the coherence between the curriculum and its evaluation instruments, then, the progressive evaluation and that of the end of the career, and the evaluation by graduates with employers (Huamán et al. 2020). Related to this, the evaluation trends of the graduate profile in Peru are: 1) the application of instruments at the end of university studies; 2) evaluation of the graduate profile in the opinion of external stakeholders; 3) evaluation of the profile competencies progressively during training, and 4) proposal of a comprehensive system of evaluation of the competencies established for the subjects or modules, as well as the competency profile at the program and university level.

Regarding the Review and Contributions to the Initial Teacher Education Graduate Profile, a graduate profile is proposed based on generic and specific competencies, which are oriented to the realization as people in the development of soft skills and towards their professional training in teaching considering the performances found in the Framework of Good Teaching Performance (Instituto de Educación Superior Pedagógico Público de Educación Física [IESPP] 2018). Thus, it is required that the curricula implement a process of evaluation of achievement of the profile and a follow-up of students' performance is carried out. For this, there are accreditation standards for study programs, such as Dimension 1 of Strategic Management, Factor 2 is found, which refers to the management of the graduate profile, which indicates that standard 5 and 6, relevance and review of the graduate profile, respectively, must be met. In the same line, Dimension 4 of Results is Factor 12, which is Verification of the graduate profile, which complies with standards 33 and 34, achievement of competencies and follow-up of graduates and educational objectives, respectively. This shows that, in Peru, in order to be accredited a university or pedagogical training institute must meet a series of standards to be accredited, as a study program (SINEACE 2016).

1.3 Research Objectives

In accordance with the above, the objective of this research is to evaluate the graduate profile of a primary education program at a private university in Lima, Peru, from the point of view of employers and graduates. Specifically, to evaluate the graduate profile of future teachers in terms of the management of the pedagogical proposal, the leadership of educational management processes, the promotion of tutorial action and the participation in research projects. Thus, the novelty of the study consists in providing a qualitative methodology for the evaluation of professional profiles of primary education teachers. And with this, it provides solutions to problems such as decontextualization in the training of students in the elementary education program. Therefore, in coherence with the objective, the following study methodology is exposed:

2 Methodology

The present research assumes a qualitative approach, due to the fact that it is oriented to the analysis of the perceptions of the participants on a specific topic (Leavy 2017). These perceptions were obtained from interviews with employers and graduates of the primary education school. And consistent with this approach, the type of research is evaluative (Kushner 2017) and formative (Ngan 2020).

2.1 Participants

In accordance with the objectives of the research, it was decided to select the participants through a non-probabilistic purposive sampling procedure (Campbell et al. 2020). Thus, a sample was selected consisting of three employer principals (two women and one man) and six graduates of the professional school of elementary education who are currently working in their careers (four women and two men).

2.2 Instruments

Semi-structured interview guide: This is a technique for collecting information in a relatively orderly manner without depriving the informant of his or her freedom (Mahat-Shamir et al. 2021). Accordingly, a semi-structured interview guide with four open-ended questions was developed. Each question is related to each of the following categories of analysis: management of the pedagogical proposal, leadership of educational management processes, promotion of tutorial action and participation in research projects. For example, in the first category: *"What aspects do we need to improve in graduates to achieve the desirable competency profile? Why?"*

Focus group guide: This technique is useful when it is desired to guide a group discussion according to a particular topic (Stewart 2018); in this way, with the support of a moderator, the discussion was conducted with the six graduates of the professional career according to the research categories. Similar to the semi-structured interview, the same questions were used.

2.3 Procedures

First, the bibliography was reviewed and the instruments were constructed and validated by expert judgment. Then, the execution of the interviews with employers and graduates was coordinated. The virtual application of the interviews and the focus group was carried out using Zoom (2021). Due to research ethics criteria, the anonymity of the participants was preserved (Hoft 2021). Afterwards, the analysis of the information was initiated through a process of open or textual coding by extracting fragments of the responses and then axial coding generating emerging categories (Williams and Moser 2019), to then triangulate the information (Fusch et al. 2018). In this way, the responses were integrated with the emergent categories in the analysis and discussion of results.

3 Analysis and Discussion of Results

Having explained the methodology, the categories of analysis are analyzed below. These categories are: manages a pedagogical proposal, leads educational management processes, promotes tutorial action, and participates in research projects.

3.1 Manages a Pedagogical Proposal

This category is based on the first specific competence called "executes the teaching-learning processes" of the graduate profile. It is oriented to develop competencies and capacities of the child through innovative methodological proposals, pedagogical and didactic processes of the curricular areas, and the performances according to the learning standards. That, according to the Good Teaching Performance Framework, every teacher is expected to design curricular programming and learning sessions by applying motivational, teaching and evaluative strategies (MINEDU 2014). Thus, the analysis revealed the emergent category in which the informants expressed that it is necessary that:

A. **Graduates should anticipate by knowing the reality of the classroom**

This classroom knowledge is essential because it enables teachers to create, foster, and retain an effective learning environment (Wolff et al. 2021); thus, participants indicated that:

> Graduates must be able to anticipate, if I anticipate, I know the reality of my classroom, I know what my children are like, so I can foresee these types of activities [...]. Anticipation and flexibility are required for my students (Employer Director 1).
>
> When we enter a classroom we are very observant, knowing the emotions of the students, we are empathetic and we show that way with the students. We consider the students' realities because many times there is no love at home (Focus Group - Graduates).

Thus, according to these testimonies, the participants expressed that the graduates should know the students because this way:

B. **Curricular programs should be flexibly contextualized**

Contextualize the curriculum where the teacher engages students in learning (Jilin 2018), which requires graduates to be flexible and creative according to educational needs as they state below:

> It is important for graduates to be a little more flexible with respect to virtuality, flexibility to take with students [...] In addition; situations may arise in which students are not demonstrating the objectives that I am going to reach, so, for the time being, change the teaching strategies that meet the needs of these students [...]. Therefore, I will be flexible in terms of my scheduling to address this learning need (Employer Principal 1).

> I have observed that the graduates apply creative and innovative strategies that they adapt for learning in different areas. However, it is necessary to further develop discovery and creativity based on contextualization and the development of critical thinking (Employer Director 3).
>
> On a daily basis, we have been able to contextualize the teaching of the areas to the students' reality. Through the areas, we have been able to apply strategies to direct learning (Focus Group - Graduates).

After the graduates gained knowledge of the classroom and were able to contextualize the school curriculum, they have been able to develop the didactic processes that are evidenced in the following emergent category.

C. **Execute sessions, materials and evaluations according to educational needs**

This emerging category shows that graduates have developed the competencies set out in the Framework for Good Teaching Performance. In which it is detailed that all teachers must develop learning sessions, apply various strategies and materials, and finally evaluate learning (MINEDU 2014).

> During the learning sessions, we were able to identify the educational needs of each student according to their developmental stages, learning rhythms and affective needs. This helped us to prepare the following learning sessions with motivating educational materials (Focus Group - Graduates).
>
> We have been able to elaborate concrete motivational materials with the resources at hand. We also make use of unstructured materials, according to the needs of the students in different curricular areas (Focus Group - Graduates).
>
> At the beginning of the cycles, I had difficulties to understand the theories and gradually I did it, I understood the pedagogical approaches to put them into practice at the time of evaluation. Then, we managed to apply them in relevant evaluations according to the students' needs. Moreover, I managed to learn how to elaborate instruments such as rubrics, checklists, etc. (Focus Group - Graduates).
>
> The graduates have developed their digital competencies to carry out the learning sessions applying participatory methodologies with creative didactic strategies such as songs or puppets (Employer Director 3).

However, the participants expressed the need to reinforce in a theoretical and practical way the competencies and approaches of the National Curriculum, as shown in the following emerging category.

D. **The competencies and approaches of the National Curriculum should be reinforced**

The Peruvian Primary Education Curriculum Program proposes competencies, skills and approaches that provide theoretical and methodological aspects to guide the teaching processes for each grade (MINEDU 2016).

> At the professional level, we need to reinforce the cross-cutting themes of the National Curriculum by managing methods and techniques to develop competencies in students. The university prioritizes more theory and little practice; therefore, there are difficulties in understanding and applying the approaches in areas such as mathematics, science, social personnel, communication, etc. There is a need to ground the theory of the approaches in practice (Focus Group - Graduates).
>
> During the pandemic, the need to reinforce technological skills to conduct the learning sessions was evident. The difficulty was observed with the fact of having to use Zoom or Meet. This is what the university needs to reinforce in our training (Focus Group - Graduates).
>
> However, it is suggested that they master the approaches to didactic processes, technological tools and deepen their knowledge of formative assessment (Employer Director 3).

According to these responses, the need to emphasize approaches and competencies by areas of didactic processes can be seen. In addition, cross-cutting competencies to develop in virtual environments and manage learning autonomously (MINEDU 2016). However, in spite of these limitations, an employer principal stated that:

E. **In general, they have good teaching methodologies**

In this emerging category, there is evidence of a positive evaluation of the graduates, because they are perceived as teachers who develop methodologies to guarantee learning (Wolff et al. 2021), as shown in the following testimony.

> I see that the graduates come with good teaching methodology, work methodology and adequate strategies. They are good professionals (Employer Director 2).

In the evaluation of this category and its emerging categories, it is observed that teacher training manages to integrate future educators into the educational system by articulating the disciplinary domain and the use of didactic strategies (Konstantinou and Konstantinou 2021). Although there are limitations in terms of the competencies and approaches of the National Curriculum (MINEDU 2016), it can be seen that graduates apply teaching and evaluation strategies according to the level of the students (MINEDU 2014).

3.2 Leads Educational Management Processes

This category is linked to the second specific competency that proposes that the graduate of the professional career "evaluates curricular and administrative management processes". This category expresses the professional capacity to propose solutions to institutional problems in the public and private sector, considering contemporary management models and the needs of educational actors, in an active and innovative manner. This is consistent with the Framework for Good Teaching Performance, which proposes that all educators should participate in school management in cooperation with families and the community (MINEDU 2014).

A. **Actively participate in curriculum planning and implementation**

Planning coherently what one wants to achieve with students implies making decisions before the practice itself, considering a context, the proposed activities and the roles of the students (Gaviria et al. 2020). In this sense, participants indicated that:

> Regarding this competency, we have observed that the graduates manage in an innovative way, proposing activities according to the needs of the school [...]. However, monitoring should be strengthened. For example, when I see a need, that is why monitoring is important because it allows me to see how students enter and after the project how they leave, to see the input, the process and the output. The graduates have not strengthened practical management (Employer Director 1).
>
> They support management through cooperative work and propose improvement actions; some have a leading role in commissions that the school convenes (Employer Director 3).

This evaluation shows that the graduates plan and execute the curriculum, and the participants also stated that:

B. **They usually do not participate much in administrative management**

The processes and procedures of administrative management focus on the concrete actions of teachers within the classroom with the support of managers, thus allowing for better learning results in the school (Muñoz and González 2019). In relation to this, the participants mentioned:

> Usually, graduates do not have much access to administrative management. Only in one particular case was it possible to observe that one of them demonstrated leadership in the face of the school needs (Employer Director 2).
>
> It is necessary that they continue to understand the importance of flat meetings as spaces for organization and training, as well as the role they play in the construction and improvement of the educational proposal (Employer Director 3).

According to these testimonies, the participation of graduates in the different administrative processes and procedures is more necessary. In addition, that:

C. **It is necessary to strengthen the practice of educational management**

In educational management, specifically in the administrative area, it is necessary to have a series of competencies such as skills in the handling of specific tasks in the delivery of documentation, management of a vocabulary in pedagogical, administrative, financial management, among others; that allow to fulfill the tasks efficiently and effectively (Riffo 2019). In this regard, the participants said:

> We have taken courses in which we learned to prepare management documents, but in practice, the practical management action is executed more by the director (Focus Group - Graduates).

> During the courses, we learned how to prepare management documents, but due to time constraints, we would have liked to put them more into practice. The teacher taught us how to elaborate them and explained them to us with her experience, but we would have liked to execute them in the process. Although I took courses during the pandemic, in which we learned how to develop educational projects with the community to prevent Covid-19. They gave us the incentive to be able to develop ourselves to support the communities; this was achieved from the pre-professional practices (Focus Group - Graduates).
>
> Now in the schools as teachers we are required to participate in the management or administration of an educational institution. However, from the university we learned more theories, but the practical part was lacking. The university has helped us in our training to learn how to prepare management documents for the diagnosis of the educational institution, but the practice of management should be reinforced (Focus Group - Graduates).

According to the testimonies and the emerging categories, it is concluded that for the category leads educational management processes, it can be seen that the graduates participate very well in curricular management but very little in administrative management. Taking into account that, according to MINEDU (2014), a teacher is required to participate in school management in cooperation with families. However, when analyzing this category, the need to strengthen administrative management within pre-professional practices is observed. But it is also necessary to:

3.3 Promoting Tutorial Action

This third category is linked to the third specific competence that refers to the fact that the graduate, when teaching, "organizes the tutorial action". This category explains that the tutorial action is executed with the purpose of creating a climate of healthy and democratic coexistence; with an intercultural approach, considering their needs, strategies, methods, techniques, psychological and pedagogical theories. And according to this, the Tutoring Guide for Primary Education teachers states that the teacher must be competent and show a set of positive actions in the healthy coexistence with their peers and students in the school (MINEDU 2021). Thus, promoting an inclusive and intercultural environment in which linguistic, cultural and physical diversities are respected (MINEDU 2014). Accordingly, the participants responded that:

A. **The tutorial profile is god and should be further strengthened**

The teaching profile requires competencies and strategic abilities to solve cultural and ethnocultural problems (Neustroev et al. 2020), which is important that the tutor must know the cultural context of the educational community for the management of a healthy coexistence; this is evidenced below:

> I think the teachers are doing a good job in terms of tutoring. I see that they have understood that there must be a connection with the students, because if there is a proper connection there is learning. And mainly to manage a preventive tutoring promoting a healthy and democratic coexistence. We have even implemented many

> work projects in which we have been able to reinforce good treatment, good and healthy coexistence with students and parents. Therefore, with respect to this, we must continue to strengthen the tutorial action, this good work that is being done in a preventive way (Director Employer 1).
>
> They have excelled the most in this aspect. Practically the core part. Even knowing that the students have several difficulties, not only economic and emotional. This is where the graduate teachers have supported them; they have managed to prevent them from dropping out (Employer Director 2).
>
> At the tutorial level, they show concern for student learning. This is observed when they generate spaces of trust, affection, good treatment and respect involving the students (Employer Director 3).

As can be seen in this emerging category, the graduates present a set of skills to solve the problems of the educational community. In addition, they expressed the following opinion:

B. **We orient students for a good educational climate**

The impact of teacher-student relationships positively favors the school environment and thus learning (Smith 2020). Thus, the participants indicated that:

> In terms of interpersonal relationships as teachers, I have been able to generate spaces for adequate learning, fostering an appropriate educational climate. I have been able to participate with large groups of up to 40 students, the first thing I did was to observe and identify the students, and then little by little I achieved listening strategies from them. I taught them to listen and so I applied strategies for a good climate. Finally, I felt that out of 40 there were 30 who attended the class (Focus Group - Graduates).
>
> During our university training, we were able to learn from the experiences of teachers to generate a good educational climate. It was very useful to us because we applied it in our work. It has been gratifying to remember what each teacher told us and to be able to solve each situation (Focus Group - Graduates).

In this sense, this emerging category emphasizes the importance of the teacher-student relationship to promote learning. In addition, the principals added that tutorial action should be worked with parents.

C. **Guided parents and contributed to problem solving**

Orientations to families are relevant because they generate communication links to improve learning conditions (MINEDU 2016). In this regard, they mentioned that:

> They also provided support and guidance to the parents. Many parents have economic problems, sometimes they abandon their children, and although during the quarantine they have been with their children, after the quarantine they abandoned them because their social environment is very difficult, with poverty and extreme poverty. This is where the intervention and improvement was observed (Employer Director 2).

> They provide spaces for attention to the family, which makes their work very well perceived by parents. However, there is still a need to develop more strategies to guide parents (Employer Director 3).

According to these responses, there is a need to reinforce parent orientation strategies (MINEDU 2016). Also:

D. **The integral, democratic and participatory training approach must be reinforced**

Comprehensive training provides the development of attitudes that enable tolerance and equity. Therefore, it is required that future teachers should be able to solve problems of the educational community taking into account culturology and ethnoculturality (Neustroev et al. 2020). This is evidenced as follows:

> I want to contribute something... That the university should not lose its sociocultural focus, that is what identifies it. Because we are human and that in our training we should not lose the focus of integral, sociocultural, emotional training because it is what identifies us, but sometimes we have limitations in it (Focus Group - Graduates).
>
> It is necessary to improve the orientation of classroom assemblies, as well as to promote democratic and participatory coexistence in the classroom (Director Employer 3).

Regarding the evaluation of this category, it can be seen that the graduates develop very well the tutorial action, guide the students, promote an optimal educational climate, and have oriented the parents. However, there is a need to reinforce the strategies to guide parents and also to reinforce the integral approach. Thus, it complies with what was proposed by MINEDU (2021) regarding the needs of group and individual tutoring to work with families in coordination with teachers (MINEDU 2021). However, there are educational needs that demand answers in the professional training of future teachers.

3.4 Participates in Research Projects

This is the last category related to the specific competence that proposes that the graduate "carries out educational research". The category is because students participate in educational research to optimize educational processes in different social contexts, considering lines of research, problems, needs in the classroom and in the educational institution, applying different research approaches. Therefore, in the Good Teaching Performance Framework, the teacher must manage research domains to improve the pedagogical conditions of the collective and the academic performance of students, managing to end with the systematization of school experiences (MINEDU 2014). According to this category, principals expressed that:

E. **It is the least evidenced competency**

While it is true that in the educational field, achievements in research competencies are important because they must be based on critical, holistic, creative and reproductive thinking (Iskakova et al. 2021). Being the reason why it should be part of the profile of an education graduate. Thus, in relation to this, the principals mentioned that:

> I have not perceived from them that they say I am going to do a project, that they communicate us to the directors or to the teachers. If there had been, it would have been isolated, but it has not been perceived. We have to think that research in our country is an Achilles heel. It is their lowest point, but I cannot point it out as a deficiency because it is like that (Director Employer 2).
>
> We have to promote in the educational institution the development of research lines in a more systematic way (Employer Director 3).
>
> With respect to this competence, I think we have been working on it. We have different approaches to research. For example, in the research that we work with projects, it is the teachers who get involved with this first and thus better guide the students (Employer Director 1).

The evaluation of this category shows the need to reinforce research skills. Thus, to achieve what MINEDU (2014) established that every teacher should develop research skills to optimize teaching and learning processes. At the same time, to develop a critical, innovative and holistic vision of reality (Iskakova et al. 2021). Of course, these competencies are relevant to improve teaching and learning processes, but there is a lack of them, which demands solutions in the training process of future primary education teachers.

4 Conclusions and Future Works

According to the analyses, it is concluded that in relation to the category manages a pedagogical proposal, it can be seen that the graduates apply didactic and evaluative strategies according to the level of the students, but show limitations in the achievement of the approaches of the National Curriculum. Then, with respect to the category leads educational management processes, it is observed that the graduates participate very well in curricular management but very little in administrative management. Then, in the category promotes tutorial action, it was observed that the graduates develop tutorial action effectively, thus, they have oriented the educational community; however, they should reinforce the strategies to orient parents and also the integral approach. Finally, in the category of participation in research projects, it is perceived that the graduates do not demonstrate the achievement of research competencies. In view of this evaluation, it is expected to carry out the curricular redesign by means of participatory action research to implement strategies and pedagogical contents to improve the training of future teachers.

References

Aravena, M., Berrios, A., Figueroa, U.: Metodología de la evaluación de logro de las competencias de los estudiantes orientadas hacia el perfil de egreso. Revista on line de Política e Gestão Educacional, Araraquara, **24**(esp. 2), 1093–1103 (2020). https://doi.org/10.22633/rpge.v24iesp2.14334

Barrera, S.: Evaluación del perfil de egreso en programas de pedagogía, una experiencia piloto en la Universidad Católica (UCSH). En Revista Foro Educacional **16**, 85–120 (2009)

Campbell, S., et al.: Purposive sampling: complex or simple? Research case examples. J. Res. Nurs. **25**(8), 652–661 (2020). https://doi.org/10.1177/1744987120927206

Cardona, A., Velez, B., Jaramillo, S.: Metodología para la evaluación de competencias en un entorno de aprendizaje virtual. Revista Espacios **39**(23), 3 (2018). http://revistaespacios.com/a18v39n23/18392303.html. Accessed 07 July 2022

Carrera, C., Lara, Y., Madrigal, J.: Análisis curricular del perfil de egreso desde la experiencia de los usuarios. REDIPE **7**(10), 139–146 (2018). https://revista.redipe.org/index.php/1/article/view/603/572. Accessed 12 May 2022

Dijkema, S., Doolaard, S., Ritzema, E., Bosker, R.: Ready for take-off? The relation between teaching behavior and teaching experience of Dutch beginning primary school teachers with different educational backgrounds. Teach. Teach. Educ. **86**, 1–13 (2019). https://doi.org/10.1016/j.tate.2019.102914

Espinoza, E.: Características de los docentes en la educación básica de la ciudad de Machala. Transformación **16**(2), 292–310 (2020). http://scielo.sld.cu/pdf/trf/v16n2/2077-2955-trf-16-02-292.pdf. Accessed 14 July 2022

Fraga, J.: Los perfiles del egresado de la carrera de ingeniería en Comunicaciones y electrónica y su relación con los perfiles requeridos por el mercado de trabajo. Caso ESIME Zacatenco. Tesis de maestro. Lima: UNE (2013)

Fusch, P., Fusch, G., Ness, L.: Denzin´s paradigm shift: revisiting triangulation in qualitative research. J. Soc. Change **10**(1), 19–32 (2018). https://doi.org/10.5590/JOSC.2018.10.1.02

García, J.: Diseño y validación de un modelo de evaluación por competencias en la universidad. Universidad Autónoma de Barcelona, Tesis doctoral (2010)

Gaviria, K., Perez, M., Carriazo, C.: Planificación educativa como herramienta fundamental para una educación con calidad. Utopía y Praxis Latinoamericana **25**(3), 87–95 (2020). https://www.redalyc.org/articulo.oa?id=27963600007. Accessed 14 July 2022

Gómez, H.: Análisis crítico del discurso del currículum y de género de formación inicial docente de Pedagogía en Educación Básica en Chile. (Tesis de Magíster inédita). Pontificia Universidad Católica de Chile (2013)

Hoft, J.: Anonymity and Confidentiality. The Encyclopedia of Research Methods in Criminology and Criminal Justice, pp. 223–227 (2021). https://doi.org/10.1002/9781119111931.ch41

Huamán, L., Pucuhuaranga, T., Hilario, N.: Evaluación del logro del perfil de egreso en grados universitarios: tendencias y desafíos. Revista Iberoamericana para la Investigación y el Desarrollo Educativo **11**(21), 131–148 (2020). https://doi.org/10.23913/ride.v11i21.691

Ibarra, S., Álvarez, S.: Proyecto SISTEVAL. Recursos para el establecimiento de un sistema de evaluación del aprendizaje universitario basado en criterios, normas y procedimientos públicos y coherentes. Madrid: Programa de Estudios y Análisis. Secretaría de Estado de Universidades e Investigación. Dirección General de Universidades (2009)

Konstantinou, Ch., Konstantinou, I.: Determinants of teacher effectiveness: pedagogical and didactic training. Open J. Educ. Res. **5**(1), 11–24 (2021). https://doi.org/10.32591/coas.ojer.0501.02011k

Instituto de Educación Superior Pedagógico Público de Educación Física Lampa.: Revisión y aportes al perfil del egresado de formación inicial docente. Lampa (2018)

Iskakova, L., Amirova, A., Ospanbekova, M., Zhumabekova, F., Ageyeva, L., Zhailauova, M.: Developing the future primary school teachers intellectual skills in Kazakhstan. Int. J. Instruct. **14**(3), 755–770 (2021). https://doi.org/10.29333/iji.2021.14344a

Jilin, L.: Four areas of constructing the contextualized curriculum. In: Jilin, L. (ed.) Curriculum and Practice for Children's Contextualized Learning, pp. 41–86. Springer, Heidelberg (2018). https://doi.org/10.1007/978-3-662-55769-3_2

Kennedy, M.: ¿Cómo mejora el desarrollo profesional la enseñanza? Revista de Investigación Educativa **86**, 945–980 (2016). https://doi.org/10.3102/0034654315626800

Kushner, S.: Evaluative Research Methods: Managing the Complexities of Judgment in the Field. Information Age Publishing, New York (2017)

Leavy, P.: Research Design: Quantitative, Qualitative, Mixed Methods, ArtsBased, and Community-Based Participatory Research Approaches. The Guilford Press, London (2017)

Mahat-Shamir, M., Neimeyer, R., Pitcho-Prelorentzos, S.: Designing in-depth semi-structured interviews for revealing meaning reconstruction after loss. Death Stud. **45**(2), 83–90 (2021). https://doi.org/10.1080/07481187.2019.1617388

Medina-Pérez, J., González-Campos, J.: Construcción del perfil de egreso: propuesta para la formación inicial docente en Chile. Investigación Valdizana **15**(3), 195–202 (2021). https://doi.org/10.33554/riv.15.3.811

Ministerio de Educación del Perú: Guía de tutoría para docentes de educación primaria (2021). https://hdl.handle.net/20.500.12799/7605. Accessed 11 Nov 2022

Ministerio de Educación del Perú. Perfil de egresado de formación docente (2022). http://www.minedu.gob.pe/normatividad/reglamentos/xtras/perfil_cuadro.pdf. Accessed 07 July 2022

Ministerio de Educación del Perú: Marco del Buen Desempeño Docente (2014). http://www.minedu.gob.pe/pdf/ed/marco-de-buen-desempeno-docente.pdflast. Accessed 07 July 2022

Ministerio de Educación del Perú.: Programa Curricular de Educación Primaria. Lima: MINEDU (2016)

Muñoz, N.delC., González, E.: Reflexión de la gestión administrativa para mejorar los resultados académicos de la Comuna de Ovalle. Revista Scientific **4**(Ed. Esp.), 136–152 (2019). https://doi.org/10.29394/Scientific.issn.2542-2987.2019.4.E.8.136-152

Neustroev, D., Neustroeva, N., Shergina, A., Kozhurova, A.: Tutor support in the ungraded school of the North. Propósitos y Representaciones **8**(SPE3), e708 (2020). https://doi.org/10.20511/pyr2020.v8nSPE3.708

Ngan, J.: Implications of summative and formative assessment in Japan: a review of the current literature. Int. J. Educ. Literacy Stud. **8**(2), 28–35 (2020). https://doi.org/10.7575/aiac.ijels.v.8n.2p.28

Olivos, E., Voisin, S., Fernández, A.: Evaluación del perfil de egreso de profesores de francés de parte de los empleadores: propuestas de mejora y desarrollo. En Revista Actualidades investigativas en educación **15**(1), 1–16 (2015)

Ortiz, A., Venegas, M., Espinoza, M.: Diseño de un sistema para la verificación del desarrollo de una competencia del perfil del egresado. En: FEM **18**(1), 71–77 (2015)

Pintor, P., Jiménez, F., Navarro-Adelantado, V., Hernández, J.: Experiencias de evaluación desde un enfoque competencial y formativo en la educación superior: la percepción de los estudiantes ante la adquisición de las competencias (2014). https://cutt.ly/hmiyc4p. Accessed 12 June 2022

Rifo, R.: Administrative and quality management in schools in the Chorrillos. Revista Scientific **4**, 153–172 (2019). https://doi.org/10.29394/Scientific.issn.2542-2987.2019.4.E.9.153-172

Sistema Nacional de Evaluación, Acreditación y Certificación de la Calidad Educativa (SINEACE), Modelo de Acreditación para Programas de Estudios de Educación Superior Universitaria. Lima. Perú (2016)

Smith, K.: Perceptions of school climate: views of teachers, students, and parents. J. Invit. Theory Pract., 5–20 (2020). https://files.eric.ed.gov/fulltext/EJ1282606.pdf. Accessed 10 Feb 2021

Stewart, W.: Focus groups. In: Frey, B. (ed.) The SAGE Encyclopedia of Educational Research, Measurement, and Evaluation, vol. 2, pp. 687–692. Sage Publications, Thousand Oaks (2018)
Tena, M., Tricas, J.: Un sistema de evaluación de competencias centrado en el estudiante. La implicación del profesor y el rol del estudiante no como participante sino como responsable de su aprendizaje. Universitat Ramon Llull, Girona (2009)
Valles-Ornelas, M., Viramontes-Anaya, E., Campos-Arroyo, A.: Retos de la formación permanente de maestros. Ra Ximhai **11**(4), 201–212 (2015)
Williams, M., Moser, T.: The art of coding and thematic exploration in qualitative research. Int. Manage. Rev. **15**(1), 45–55 (2019)
Wolff, E., Jarodzka, H., Boshuizen, A.: Classroom management scripts: a theoretical model contrasting expert and novice teachers' knowledge and awareness of classroom events. Educ. Psychol. Rev. **33**, 131–148 (2021). https://doi.org/10.1007/s10648-020-09542-0
Zoom: Comprehensive Guide to Educating Through Zoom (2021). https://www.zoom.us/docs/en-us/childrens-privacy.html. Accessed 10 Feb 2021

Dropout in Computer Science, Systems Engineering and Software Engineering Programs

Sussy Bayona-Oré(✉)

Vicerrectorado de Investigación, Universidad Autónoma del Perú, Panamericana Sur Km. 16.3, Villa EL Salvador, Lima, Peru
sbayonao@hotmail.com

Abstract. Student dropout as a decision-making process is complex and conditioned by various factors. Despite the existence of regulatory frameworks and the efforts made by universities to implement programs to retain students, the objectives do not always achieve. The Computer Science, Systems Engineering and Software Engineering careers are no strangers to this problem. This article presents the results of a bibliometric analysis to determine the aspects related to student dropout in these careers. The analysis of keywords co-occurrence shows that, out of a total of 282 keywords corresponding to 48 articles, 41 met the condition. Five clusters were formed relating to the prediction of student desertion, teaching systems, retention models, careers and desertion in higher education. The factors conditioning student desertion highlighted in the studies include communication among teachers, usefulness of the degree, cognitive gain, university entrance exam score, gender, place of residence, number of siblings, family income, English qualification, mathematics qualification and administration, among others. Understanding the factors involved in student dropout will allow these to be considered in the strategies for student dropout prevention and student retention.

Keywords: Dropout · Higher education · Software engineering · Computer science · ICT

1 Introduction

Student dropout is a phenomenon that has gained the interest of the academic community, especially in the period from 2019–2022, which coincides with the impact of COVID-19 a period in which universities worldwide closed their doors to make room for virtual classes in a disruptive way. During the pandemic, student dropout rates increased [1] for various reasons, from not having the necessary infrastructure to continue classes to stress and anguish [2].

Leaving a study program before attaining a diploma or certificate has been called student dropout [3] and is considered a problem that impacts the country's educational system and its economy [4, 5]. The decision to drop out of school is long, complex and conditioned by several factors [6]. The impact of student dropout not only negatively

A. Rocha et al. (Eds.): WorldCIST 2023, LNNS 800, pp. 592–599, 2024.
https://doi.org/10.1007/978-3-031-45645-9_57

affects the student, but also affects the university system worldwide [7, 8], which is why it is considered an important education research topic [9].

Entering college is an experience that generates many expectations for students and their families [9]. Even more so when, in the university admission process, thousands of students are left without university admission [10] and only a percentage manage to reach a vacancy. Of the total number of entrants, a percentage follow careers related to Computer Science, Software Engineering and Systems Engineering, among others.

Moving from high school to university is characterised by a change in the student's level of independence and responsibility. Attrition in Computer Science is a well-known phenomenon [11] and several studies argue that the first years of college are determinant [12, 13], where the academic history of the student prior to college has been considered by some a predictor of the continuity of studies [14].

The various causes of student dropout have been extensively studied and described in the scientific literature. This has resulted in it being considered a multifactorial phenomenon [15]. During the first years, a series of factors can condition a student to drop out, such as the level of preparation [16], vocation [17, 18] academic integration [19], motivation for the program [12] and academic performance—this latter being one of the main causes [16, 20]. In the first semesters, lack of study organization can translate into low academic performance [6]. The factors mentioned are organizational or institutional [7], individual [7], economic [7], academic, motivational and psychological [8]. Other factors mentioned include aspects of health and performance in English [21].

One of the earliest and most recognised models [22] on retention and dropout was proposed by Tinto [19]. Tinto argues that, if the benefit to a student of staying at an institution is perceived as greater than the personal cost, he or she will remain in the institution [3]. Given this situation, several studies have been conducted to predict student dropout using data mining [23].

Although the phenomenon of dropout affects all careers, the aim of this study is to investigate the factors of dropout in careers related to Computer Science, Systems Engineering and Software Engineering. The study of dropout in these fields has received greater attention due to the need for professionals in these careers. In this regard, [24] argue that, although enrolments are increasing in Computer Science courses, retention remains a challenge. A bibliometric review was conducted using the Web of Science database in order to understand the trends and factors that condition student dropout.

The article has been structured in five sections, including this introduction. Section 2 describes the related works. Section 3 presents the methodology used. Section 4 presents the results and discussion. Finally, Sect. 5 presents the conclusions.

2 Factors in Students Dropout

Several factors have been mentioned in studies on student dropout. The authors classify the factors into (1) institutional factors relating to the university [9, 17], (2) individual factors [9, 21], (3) academic factors [18, 21, 25] related to the student's academic information, (4) economic factors [9, 21], (5) vocational factors [17, 18], (6) motivational factors [18] and (7) psychological factors [22].

Another study that studies the factors that influence student dropout, using the Scopus database [26], includes age, gender, sense of belonging, academic performance, performance in previous studies, motivational, organizational factors, among others. Another aspect that should be taken into account is the difficulty that the student may face in a given career since a low performance in mathematics or programming constitute a risk of dropping out of studies [26].

3 Methodology

A bibliometric analysis was conducted to determine the factors that condition student desertion. The Web of Science database was used to conduct the study.

The terms used to search for scientific articles were 'dropout', 'drop out', 'retention', 'drop-out', 'dropouts', 'computer science', 'systems engineering', 'software engineering', 'higher education' and 'university students'. Articles should be written in English, Spanish or Portuguese.

A total of 79 studies were analysed and identified with the search string. Applying the inclusion criterion of being articles that include factors in student dropout, a total of 48 articles were selected. The relevant information extracted from the scientific articles was collected in an Excel sheet.

The general data of the article, the factors identified, and the impact of the factors were recorded. The software VOSviewer was used to create the keyword co-occurrence map.

4 Results and Discussion

The results show the interest of researchers in conducting studies relating to student dropout in Computer Science, Systems Engineering or Software Engineering careers. The search results show an increasing trend of publications in the period 2019–2021, a period that coincides with the pandemic due to COVID-19.

Globally, COVID-19 affected all sectors, including the education sector [2], and has increased student dropout rates [13]. The teaching-learning process underwent a disruptive change by moving from face-to-face to virtual classes. Spain, the United States, South Korea and Australia are the countries with the largest number of publications.

4.1 Co-occurrence Map

The relationship between the keywords was analysed using the co-occurrence map. All keywords were considered as the unit of analysis; and the full counting method was used, with a minimum of 2 occurrences of the keyword. Out of a total of 282 keywords corresponding to 48 articles, only 41 met the condition. The keywords with the highest occurrence were *higher education, retention, motivation, higher-education, dropout* and *computer science*. Figure 1 shows the co-occurrence map. The 41 keywords formed 7 clusters: Cluster 1 with 12 items (red), Cluster 2 with 10 items (green), Cluster 3 with 10 items (blue), Cluster 4 with 7 items (yellow) and Cluster 5 with 2 items (purple).

Among the keywords of Cluster 1, we can mention: *artificial neural network*, *educational data mining*, *learning analytics*, *machine learning*, *performance* and *prediction*, in addition to the keywords *academic performance*, *computer science*, *dropout* and *attrition*. The importance of the phenomenon of student desertion has motivated the interest in conducting studies related to the prediction of student desertion using different techniques.

Cluster 2 groups together the following keywords: *blended learning, science education, e-learning, education, engagement, persistence, STEM* and *university*. Cluster 3 groups together the following keywords: d*istance education, drop-out, higher-education, model motivation, predictor, retention* and *students*. In Cluster 4, the keywords are grouped as follows: *academic performance, computer-science, distance-learning, engineering, gende*r, *ICT* and *science*. In Cluster 5 the following keywords appear: *dropouts* and higher *education*.

4.2 Most Cited Studies

Alonso et al. [27] proposes changes in the teaching/learning system of a course related to Software Engineering (Faculty of Informatics), due to high dropout rates (25%–35%) and low student performance. They propose the use of learning technology, supported by a moderate constructivist instructional model and a blended learning approach. The results of the study showed that the means obtained with the new model is significantly higher than means obtained with the traditional teaching. The study has 46 citations.

The importance of student dropout motivates the use of prediction techniques. Lacave et al. [8] use Bayesian networks with 383 records representing the academic and social data of students enrolled in the Computer Science career. The findings show that the number of subjects passed by the student and on the highest course in which they are enrolled are relevant factors for dropping. The risk of dropout depends on the maximum number of times that a student has been examined in any subject. The gender, the fact of being working and the family's residence province are not relevant factors.

The first year of studies has been considered by various authors key in student dropout [6, 12, 13, 16]. In order to identify the factors that condition student dropout, Kori et al. [28] conducted a quantitative study to determine the factors that influence student dropout. Three of the factors found in both evaluations had similar effects: 1) personal contact with IT (Information Technology) based on the student perception about the knowledge of IT, job opportunities or salary, 2) the reputation of the IT field, 3) development of IT, 4) motivation when choosing the career y 5) motivation during studies.

Kori et al. [29] argue that the unmet demand for IT specialists is due to low retention rates in higher education IT studies and that therefore universities need to increase their retention rates. The Tinto integration model to analyse the profiles of 509 Estonian first-year students was used. They distinguish four profiles considering aspects of social integration, academic integration, confidence to graduate and professional integration. They consider important the collaboration between universities and IT companies for integrating students professionally. They argue that self-efficacy is associated with academic performance, however, self-efficacy may change during the three years of higher education.

Agrusti et al. [23], in their study, conduct a literature review on data mining techniques for predicting student dropout. Among the main classification techniques used are Decision Tree, Bayesian Classification, Neural Networks, Logistic regression and Support vector machines. The most used data mining tools are WEKA, SPSS, R and Rapid Miner. They conclude that the prediction of dropout is relevant and of interest to the academic community.

Lagesen et al. [30] presents inclusion strategies designed to reduce the higher education gender gap in engineering and ICT (Information Communication Technology) engineering. Inclusion efforts are critical for obtaining greater participation of women in ICT. Work is needed to achieve a higher proportion of female faculty to obtain more sustainable recruitment and retention of women in ICT.

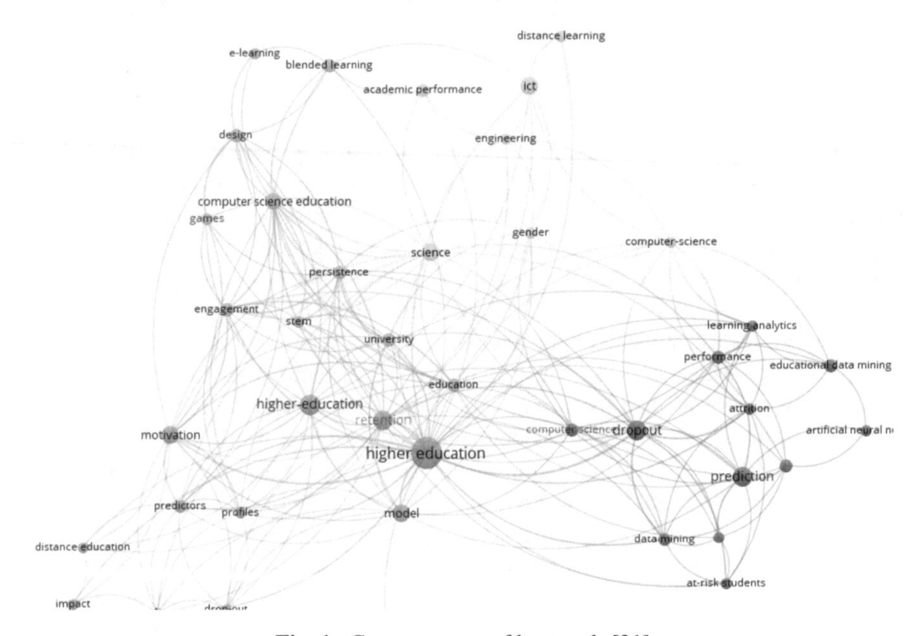

Fig. 1. Co-occurrence of keywords [31]

4.3 Student Dropout

Kori et al. [29] argued that the lack of ICT workers is a global problem and dropout rates are high. In Estonia, many ICT students do not complete their studies. Communication between teachers and students in a friendly and helpful way, as well as communication between the university and students, are factors for student retention. Another relevant factor is student satisfaction.

Lacave et al. [8] argued that the grade students obtain in the university entrance exam is a relevant factor in predicting dropout. In turn, gender, the student's province of family residence or whether the student works while studying influence dropout. They concluded that the profile of the student who drops out is an 18-year-old male whose first choice was Computer Science and obtained grade 5.

Kinnunen et al. [32] assessed the expectations of first-year computer science students in three countries and found that students pursue this career because they are attracted to programming, software development and its role in society. However, there is a mismatch between what students want to study and what they think they need to study. In addition, there is a need to understand that computer science develops soft and hard skills, as students aspire to work for themselves or to form small businesses; thus, teachers are required to go beyond developing technical skills and develop others, to ensure career success.

Febro [5] used a university database to study the factors associated with student dropout in first year students. The study found that apart from the first semester gap, student retention in college was positively correlated with the following predictors, college entrance exam score (mathematics, language use, aptitude and science category), number of siblings, family income, English score and mathematics grade.

Shmeleva and Froumin [33] conducted a study in Russian universities in Engineering and Computer Science programs and confirmed that (1) the role of student academic integration which includes student-teacher interaction and attendance to classes, prevents dropout, (2) students drop out in the first year of studies where mathematics and programming courses are emphasized, and (3) there is a dependency between the successful student in engineering and having been a good student in mathematics at school.

Student dropout is a topic that continues to be of interest to researchers at a general level and by area; as in this case when we analyse student dropout in Computer Science, Systems Engineering or Software Engineering careers. A growing trend of publications is observed both in the Web of Science database and in ScopusS, especially in the period of the pandemic due to COVID 19.

When analysing the information from the scientific articles of the Web of Science database, using the Vosviewer Software (co-occurrence map), it is observed that the keywords of Cluster 1 are related to *machine learning techniques, data mining* and *prediction.* This result is consistent with a previous study [26] using the Scopus database. Student desertion has been studied both in classes using a face-to-face modality and using a virtual modality.

The importance of retaining students is highlighted, and the fact that dropout is a problem is mentioned in some cases, as the latter influences the lack of IT specialists. The importance of promoting the participation of women, especially in STEAM careers, is also highlighted. Of the factors mentioned, communication between teachers and students, usefulness of the degree, reputation of the ICT field, entrance exam and failures in the mathematics course are highlighted. In previous research [34] on the factors that influence student dropout, the low expectation of the career in the future was considered a determinant for student dropout.

5 Conclusions

A phenomenon that negatively affects students, the family and the university is student dropout. Based on a bibliometric analysis, the situation and trends in this topic are presented. The results show an increase in publications on dropout in areas related to

ICT, especially during the period of the COVID-19 pandemic. The results show that there are a set of factors that condition student desertion in Computer Science, Systems Engineering and Software Engineering careers, such as the usefulness of the degree, motivation of student, student satisfaction, system of teaching and learning, reputation of ICT field, academic performance, university entrance exam score, family income, communication between teachers and students, qualifications, among others. The factors found here are similar to those identified in other dropout studies. Knowing the specific factors of careers related to ICT will make it possible to establish more appropriate policies and strategies. For future work, a review of the factors in the Scopus and Web of Science databases which establishes a model for determining how these factors influence student dropout is proposed.

References

1. Flores, V., Heras, S., Julian, V.: Comparison of predictive models with balanced classes using the SMOTE method for the forecast of student dropout in higher education. Electronics **11**(3), 1–16 (2022)
2. Jacobo-Galicia, G., Máynez, A., Cavazos-Arroyo, J.: Fear of COVID, exhaustion and cynicism: its effect on the intention to drop out of college. Eur. J. Educ. Psychol. **14**(1), 1–18 (2021)
3. Himmel, E.: Modelo de análisis de la deserción estudiantil en la educación superior. Revista Calidad de la Educación **17**, 91–108 (2002)
4. Sandoval-Palis, I., Naranjo, D., Vidal, Gilar-Corbi, R.: Early dropout prediction model: a case study of university leveling course students. Sustainability **12**(22), 1–17 (2020)
5. Febro, J.: Utilizing feature selection in identifying predicting factors of student retention. Int. J. Adv. Comput. Sci. Appl. **10**(9), 269–274 (2019)
6. Behr, A., Giese, M., Teguim, H., Theune, K.: Motives for dropping out from higher education—an analysis of bachelor's degree students in Germany. Eur. J. Educ. **56**(2), 325–343 (2021)
7. Behr, A., Giese, M., Teguim, H., Theune, K.: Dropping out of university: a literature review. Rev. Educ. **8**(2), 614–652 (2020)
8. Lacave, C., Molina, A., Cruz-Lemus, J.: Learning Analytics to identify dropout factors of Computer Science studies through Bayesian networks. Behav. Inf. Technol. **37**, 993–1007 (2018)
9. Aina, C., Baici, E., Casalone, G., Pastore, F.: The determinants of university dropout: a review of the socio-economic literature. Socioecon. Plann. Sci. **79**, 1–16 (2022)
10. Virkki, O.: Computer science student selection? A scoping review and a national entrance examination reform. In: 52nd ACM Technical Symposium on Computer Science Education. Proceedings, pp. 654–659 (2021)
11. Takács, R., Kárász, J.T., Takács, S., Horváth, Z., Oláh, A.: Successful steps in higher education to stop computer science students from attrition. Interchange **53**, 1–16 (2022)
12. Kori, K., et al.: First-year dropout in ICT studies. In: IEEE Global Engineering Education Conference EDUCON, Estonia, pp. 437–445. IEEE (2015)
13. Chong, Y., Soo, H.: Evaluation of first-year university students' engagement to enhance student development. Asian J. Univ. Educ. **17**(2), 13–121 (2021)
14. Vásquez, J., Miranda, J.: Student desertion: what is and how can it be detected on time? In: García Márquez, F.P., Lev, B. (eds.) Data science and digital business, pp. 263–283. Springer, Cham (2019). https://doi.org/10.1007/978-3-319-95651-0_13

15. Morelli, M., Chirumbolo, A., Baiocco, R., Cattelino, E.: Academic failure: individual, organizational, and social factors. Psicología Educativa **27**(2), 167–175 (2021)
16. Sacală, M., Pătărlăgeanu, S., Popescu, M., Constantin, M.: Econometric research of the mix of factors influencing first-year students' dropout decision at the faculty of agri-food and environmental economics. Econ. Comput. Econ. Cybern. Stud. Res. **55**(3), 203–220 (2021)
17. Ambiel, R., Cortez, P., Salvador, A.: Predição da Potencial Evasão Acadêmica entre Estudantes Trabalhadores e Não Trabalhadores. Psicologia Teoria y Pesquisa **37**, 1–10, (2021)
18. Rodríguez-Pineda, M., Zamora-Araya, J.: College student dropout: cohort study about possible causes. Uniciencia **35**(1), 19–37 (2021)
19. Tinto, V.: Dropout from higher education: a theoretical synthesis of recent research. Rev. Educ. Res. **45**(1), 89–125 (1975)
20. Améstica-Rivas, L., King-Domínguez, Gutiérrez, D., González, V.: Efectos económicos de la deserción en la gestión universitaria: el caso de una universidad pública chilena. Hallazgos **18**(35), 209–231 (2021)
21. Aldahmashi, T., Algholaiqa, T., Alrajhi, Z., Althunayan, T., Anjum, I., Almuqbil, B.: A case-control study on personal and academic determinants of dropout among health profession students. High. Educ. Stud. **11**(2), 120–126 (2021)
22. Giannakos, M.N., et al.: Identifying dropout factors in information technology education: A case study. In: 2017 IEEE Global Engineering Education Conference (EDUCON), pp. 1187–1194. IEEE (2017)
23. Agrusti, F., Bonavolontà, G., Mezzini, M.: University dropout prediction through educational data mining techniques: a systematic review. J. e-learning Knowl. Soc. **15**(3), 161–182 (2019)
24. Ebert C., Duarte, C.: Requirements engineering for the digital transformation: industry panel. In: 2016 IEEE 24th International Requirements Engineering Conference Proceedings, China, pp. 4–5. IEEE (2016)
25. Miranda, M., Guzmán, J.: Análisis de la deserción de estudiantes universitarios usando técnicas de minería de datos. Formación Universitaria **10**(3), 61–68 (2017)
26. Bayona-Oré, S.: Student dropout in information and comunications technology careers. In: 2022 17th Iberian Conference on Information Systems and Technologies (CISTI), pp. 1–6. IEEE (2022)
27. Alonso, F., Manrique, D., Martínez, L., Viñes, J.: How blended learning reduces underachievement in higher education: an experience in teaching computer sciences. IEEE Trans. Educ. **54**(3), 471–478 (2010)
28. Kori, K., Pedaste, M., Altin, H., Tõnisson, E., Palts, T.: Factors that influence students' motivation to start and to continue studying information technology in Estonia. IEEE Trans. Educ. **59**(4), 255–262 (2016)
29. Kori, K., Pedaste, M., Must, O.: The academic, social, and professional integration profiles. ACM Trans. Comput. Educ. **18**(4), 1–19 (2018)
30. Lagesen, V., Pettersen, I., Berg, L.: Inclusion of women to ICT engineering–lessons learned. Eur. J. Eng. Educ. **47**(3), 467–482 (2021)
31. Van, N., Waltman, L.: Software survey: VOSviewer, a computer program for bibliometric mapping. Scientometrics **84**, 523–538 (2010)
32. Kinnunen, et al.: Understanding initial undergraduate expectations and identity in computing studies. Eur. J. Eng. Educ. **43**(2), 201–218 (2018)
33. Shmeleva, E., Froumin, I.: Factors of attrition among computer science and engineering undergraduates in Russia. Educ. Stud. Moscow **3**, 110–136 (2020)
34. Bayona-Oré, S.: Student dropout in higher education as perceived by university students. World Trans. Eng. Technol. Educ. **20**(2), 82–88 (2022)

Multimedia Systems and Applications

Ridge Gap Waveguide Based Array Antenna for 5G/WiFi Applications

A. M. M. A. Allam[1], Hesham kamal[2(✉)], Hussein Hamed Mahmoud Ghouz[2], and Mohamed Fathy Abo Sree[2]

[1] Faculty of Information Engineering and Technology, German University in Cairo (GUC), Cairo, Egypt

[2] Arab Academy for Science, Technology, and Maritime Transport, Cairo, Egypt
hesham.kamal2151994@gmail.com, Mohamed.fathy@aast.edu

Abstract. This paper presents an efficient antenna configurations design with low dispersion, low loss antenna with high gain. It is based on SIGW technology. The feeding network is composed of two layers, the top layer comprises the top ground implemented on dielectric material Rogers RO4003C and the bottom layer includes the ridge and the surrounding periodic structure of vias implemented in dielectric material Rogers RO4350B. The ridge is branched to form power divider to feed the antenna array. The two slot antennas are implemented on the top ground of the top layer. It is analyzed using the Finite Deference Time Domain (FDTM) analysis (CST microwave studio) and fabricated using PCB technology. A triple band is achieved for WiFi application. It operates at 5.9, 6 and 6.1414 GHz with reasonable gain. The structure is fabricated and measured which finding reasonable agreement between both results. This antenna can also be used for energy harvesting at these bands, once connected with a rectifying circuits.

Keywords: SIGW · EBG · Antenna Array · WiFi Applications

1 Introduction

Most of all commercial wireless communication devices today like smartphones, tablets, laptops and wearable devices are important to uphold multiple communication standards like WiFi, Bluetooth and long-term evolution (LTE). The increased trend for robust wireless connection to achieve the development of the increasing and evolving cloud-based applications and mobile devices and to spread the extent and business in WiFi technology connected devices, which operate at 2.4/3.6/4.9/5/5.9/6/60 GHz.

On the other hand, Gap Wave Guide (GPW) technology has received increasing attention within the past decade because it has excellent properties in comparison with microstrip transmission lines, bulky metallic waveguide, coplanar waveguide (CPW) and substrate integrated waveguide (SIW) [1–5, 6–10]. I gives low radiation loss, planar profile without metal contacts, and low dispersion due to propagation of quasi TEM inside the gap over the ridge. GW can be classified into Ridge Gap Waveguide (RGW), Groove Gap Waveguide (GGW) and Microstrip Gap Waveguide (MGW) [–]. Substrate

A. Rocha et al. (Eds.): WorldCIST 2023, LNNS 800, pp. 603–614, 2024.
https://doi.org/10.1007/978-3-031-45645-9_58

Integrated Gap Waveguide (SIGW) is a significant type of MGW, which reduces industrial complications and gives structure that is more firm with extra design facilities. It is a hopeful alternative to replace waveguides and microstrip lines particularly for microwave and millimeter applications [11–15].

In this paper, SIGW based array antenna of two elements is designed and fabricated for WiFi applications at 5.9 GHz, 6 GHz and 6.1414 GHz**.** A power divider is firstly implemented with its matching unit using set of quarter wave transformers based on the same technology. This network could play a key role in the realization of the future 5G wireless communication systems development for future smart antenna systems and beam forming networks, [16, 17].

The paper is organized as follows: Sect. 2 presents the design of SIGW feeding network (unit cell, super cell and ridge lines) and the antenna array. Section 3 presents simulated and measured results. Finally, Sect. 4 concludes that work.

2 Design of Antenna and SIGW Feeding Network

The process of designing the proposed antenna starts with the design of electromagnetic band gap (EBG) structure unit cell, supper cell, and the SIGW based feeding line operates from 5 GHz to 20 GHz.

2.1 Unit Cell

The structure unit cell and Eigen mode analysis are shown in Fig. 1, where all the required dimensions are depicted in Table 1. It is composed of two layers; the printed SIGW mushroom which represents the bottom layer. It has a squared patch grounded by a via and surrounded with a dielectric substrate. The second layer is the top ground loading dielectric gap. Rogers RO4350B substrate with the dielectric constant of 3.66 and thickness 1.524 mm is used for the bottom layer while Rogers RO4003C with the dielectric constant of 3.55 and thickness 0.203 mm is used for the top layer. The factors that affect the band gap frequency range are the die-electric layers thicknesses, the periodicity, and mushroom dimensions which have been optimized to provide the required band gap. Using Eigen Mode Solver in CST microwave Studio, the dispersion diagram of the SIGW unit cell is calculated and shown in Fig. 1. One can see that an EBG band is generated at frequency band from 5 GHz to 20 GHz.

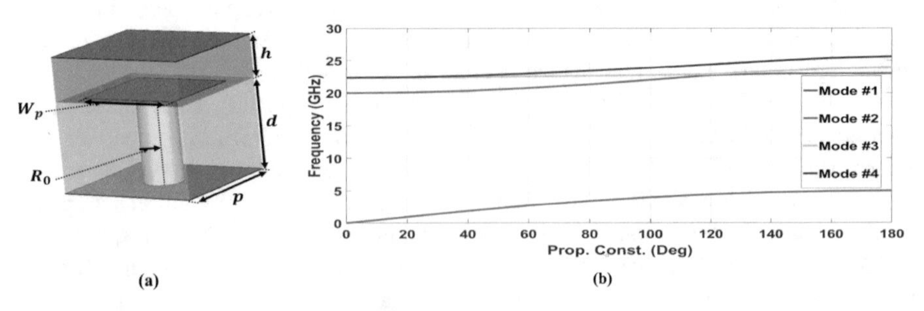

Fig. 1. Unit cell (a) Structure. (b) Dispersion diagram.

The supper cell consists of central unit cell with ridge and four unit cells on both sides. The wave passes over the ridge in the dielectric gap between ridge and top ground and showed an onset of a quasi-TEM propagating mode along the ridge. It is clear that the wave is evanescent and stops on the right and left unit cells. Figure 2 represents structure and dispersion diagram of supper cell.

Table 1. Dimensions of unit cell

Parameters	Value in (mm)
Substrate thickness, d	1.524
Patch width, W_p	3
Radius of via, R_o	0.5
Air gap, h	0.203
Unit cell width W	3.5
Copper thickness t	0.015

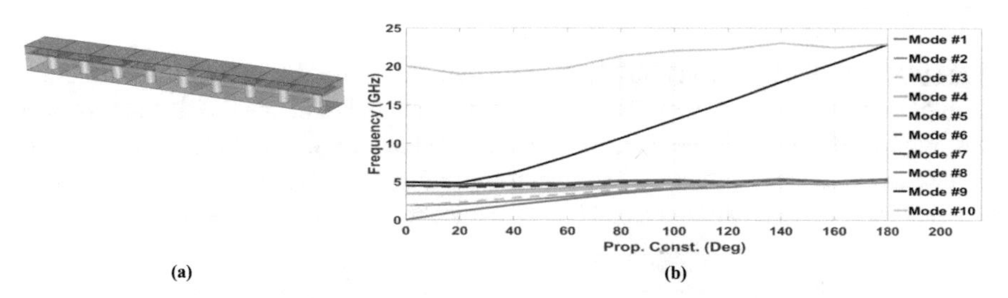

Fig. 2. Super cell. (a) Structure. (b) Dispersion diagram.

Figure 3a presents the structure of SIGW based transmission line where the transmission and reflection coefficients are presented in Fig. 3b. It is clear that the wave is transmitted along the ridge for the operating band without reflection. Also one can see that the current is confined along the ridge at a frequency inside the band, while the current can pass along the ridge and the periodic cells in a frequency out of the band as depicted in Fig. 3c,d.

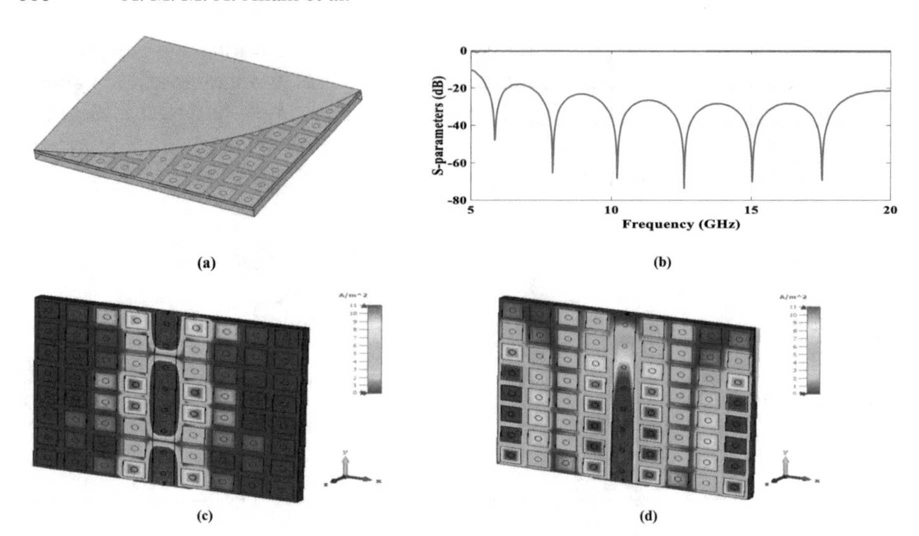

Fig. 3. SIGW transmission line. SIGW line structure (b) Scattering parameters (c) The current distribution inside the band (d) The current distribution outside the band

2.2 SIGW Feeding Network

Figure 4a represents the structure of power divider. It is a 3dB divider of SIGW structure. A set of three quarter wave transformers are implemented to match the ridge to the feed port. Figure 4b represents the matching network. Table 2 shows the dimensions of power divider and its matching units. S-parameter of divider is shown in Fig. 4c. One can see that the power is equally divided and it is well matched over the required operating band.

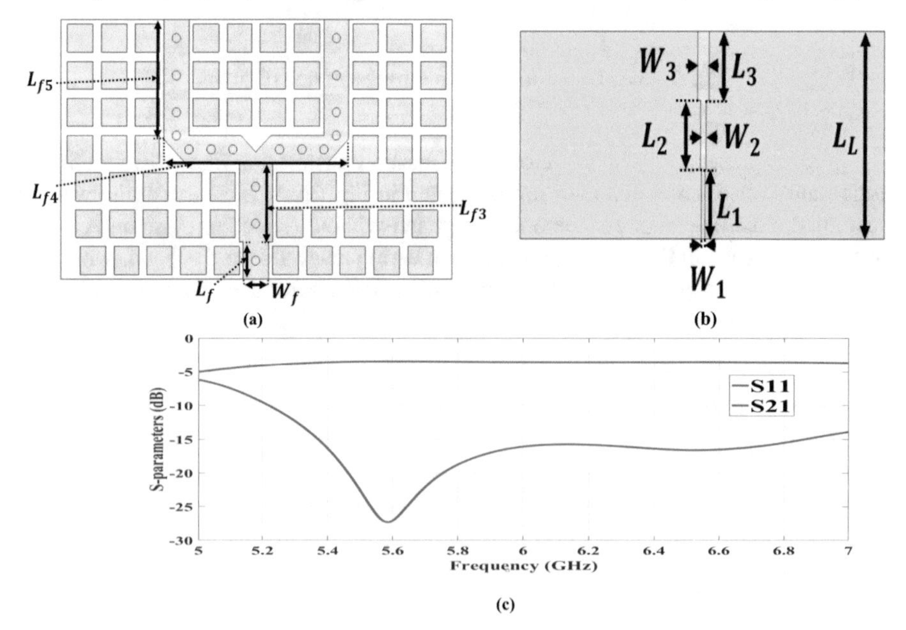

Fig. 4. The power divider and matching unit (a) Structure of power divider. (b) Matching transformers (c) S-parameter of power divider

The matching sections are implemented based on the following equations [18] (1), (2). The impedance of the quarter wave transformer Zo can be determined as (Fig. 5):

$$\frac{Z_{in}}{Z_0} = \frac{Z_0}{Z_L} \tag{1}$$

Fig. 5. Quarter wave impedance of transformer

Table 2. Dimensions of power divider

Dimension	Value in (mm)
L_f	4
W_f	3
L_{f3}	8.5
W_{f3}	4
L_{f4}	16
W_{f4}	4
L_{f5}	12.5
W_{f5}	4
L_3	6.5
W_3	1.5
L_2	6.5
W_2	0.75
L_1	6.5
W_1	0.44

To obtain the equivalent impedance of the ridge of SIGW structure, one can use the strip line formula [18] in (2)

$$Z_{0,RGW} = \frac{1}{2}\left[\frac{W}{2h} + 0.441\right]^{-1} \tag{2}$$

An optimization process is performed to obtain a better matching level for the required operating band. Also smoothing the edges and a trapezoidal shape is cut at the connection between branches of the power divider. This power divider is used to feed the array antenna designed in subsection C.

2.3 Antenna Array

An array of two hexagonal shape slot antennas is depicted in Fig. 6. It is implemented using FTDM analysis method (CST MICROWAVE STUDIO. They are etched on the top ground. It is fed via the designed SIGW feeding network. The length of each slot is chosen to satisfy the resonance of a complement of a hexagonal patch antenna of approximately half wavelength taking into consideration the effect of the relative dielectric constant of the dialectic gap of SIGW structure. Table 3 shows the dimension of the proposed slot antenna array. To validate the proposed design of array antenna, it is fabricated using PCB technology as shown in Fig. 7. The measurement set up for reflection coefficient is illustrated in Fig. 8 using N52271A phase network analyzer. For the measurement of gain and radiation pattern, Fig. 9 illustrates the measurement set up of an anechoic chamber.

3 Results and Discussion

The simulated and measured reflection coefficient of the antenna is shown in Fig. 9. One notices that the antenna achieves three operating band at 5.9, 6 and 6.1414GHz. The matching bandwidth level is below −10 dB in the required frequency bands for each frequency. It covers the bandwidths of 75 MHz at 5.9 GHz, 20 MHz at 6 GHz and 20 MHz 6.1414 GHz that meets the standard WiFi applications. One can notice a good agreement between the simulation and measurements (Fig. 10).

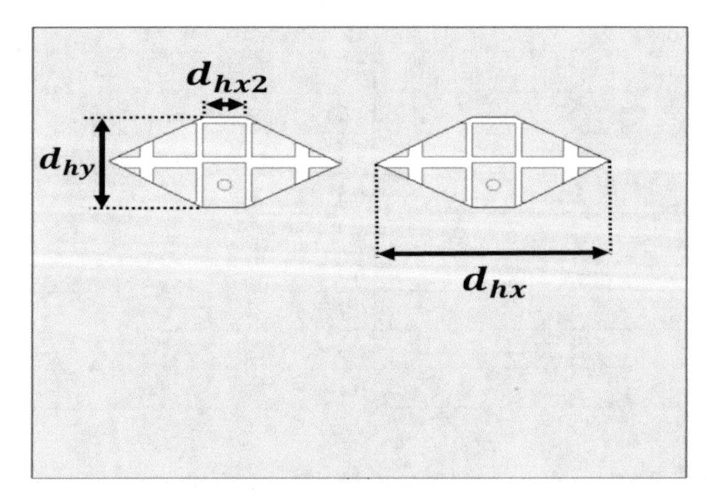

Fig. 6. Antenna array of slots.

Table 3. Dimensions of the slot

Dimension	Value in (mm)
d_{hx}	16.557478
d_{hy}	7.998782
d_{hx2}	3

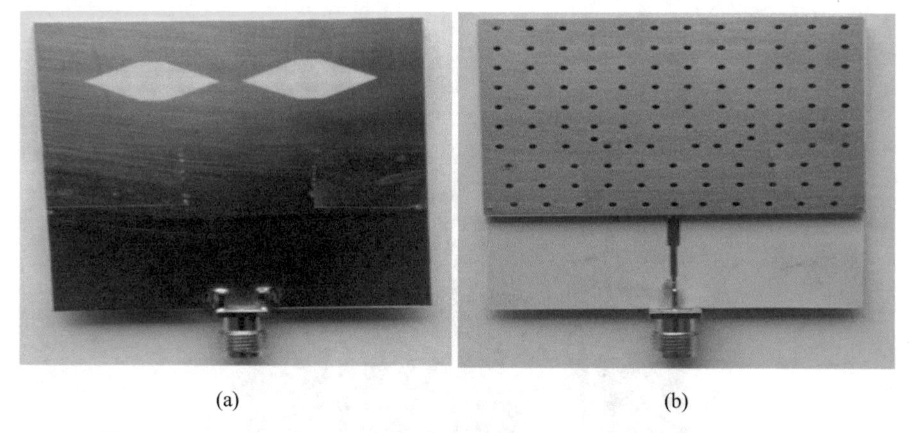

(a) (b)

Fig. 7. Fabrication of proposed antenna array (a) Top Layer (b) Bottom layer

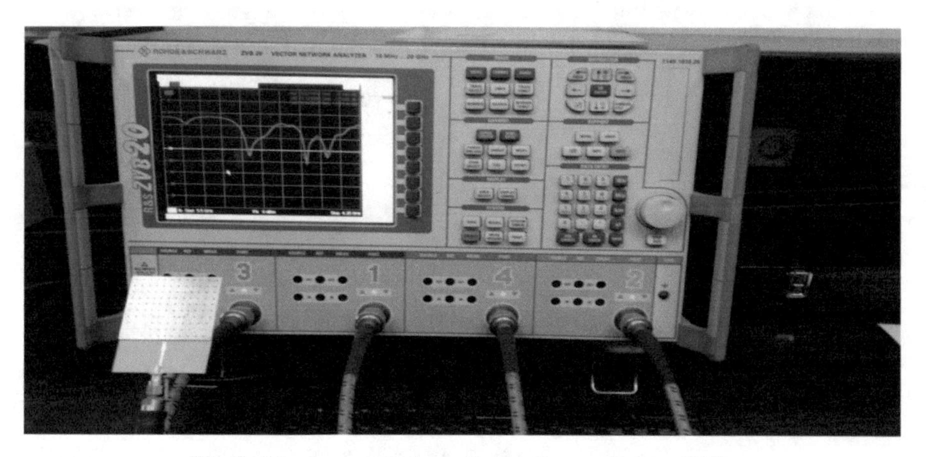

Fig. 8. Measurement set up of reflection coefficient (S11).

The simulated and measured gain at 5.9 GHz, 6 GHz and 6.1414 GHz is illustrated in Fig. 11. The two-dimensional radiation pattern at required three WiFi center frequencies 5.9 GHz, 6 GHz and 6.1414 GHz in the two principle planes (H-Plane and E-Plane) are depicted in Fig. 12. One notices the good agreement between the simulated and measured results. Figure 13 illustrates the current over the antenna array and their feeding lines at the three main frequencies of the proposed antenna in Fig. 13. It is clear that the current is outlined over the edge and the resonant structure of the antennas.

Fig. 9. Measurement of the radiation pattern and gain in anechoic chamber

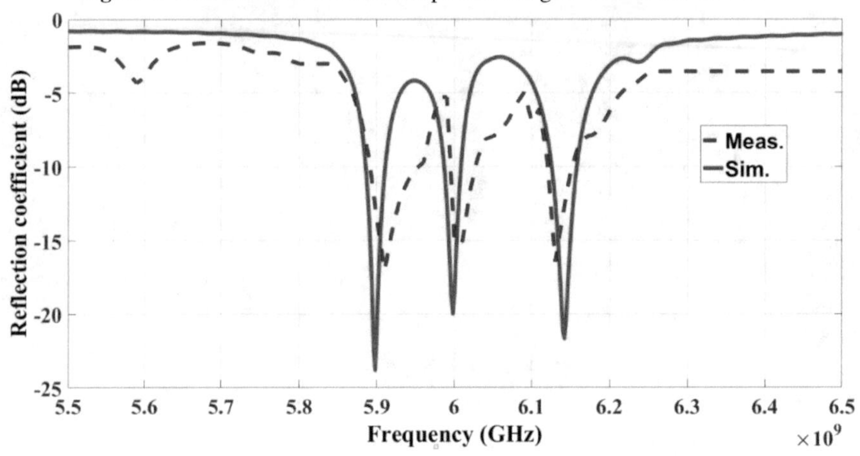

Fig. 10. Simulated and measured reflection coefficient

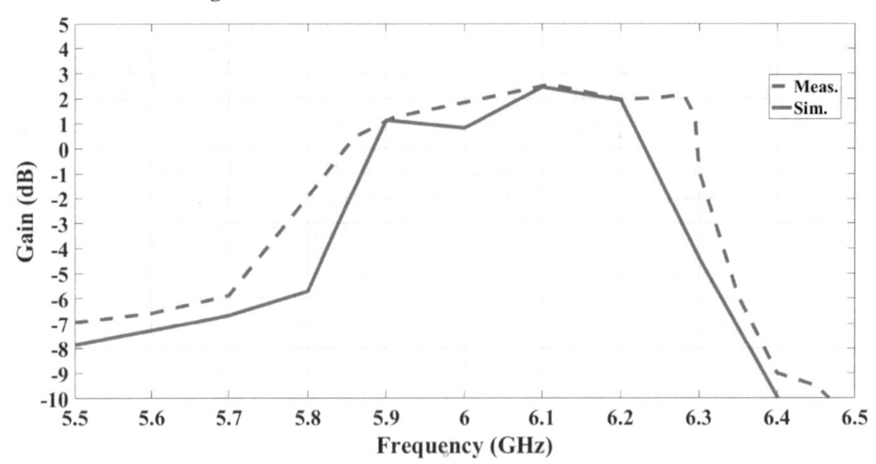

Fig. 11. Simulated and measured gain.

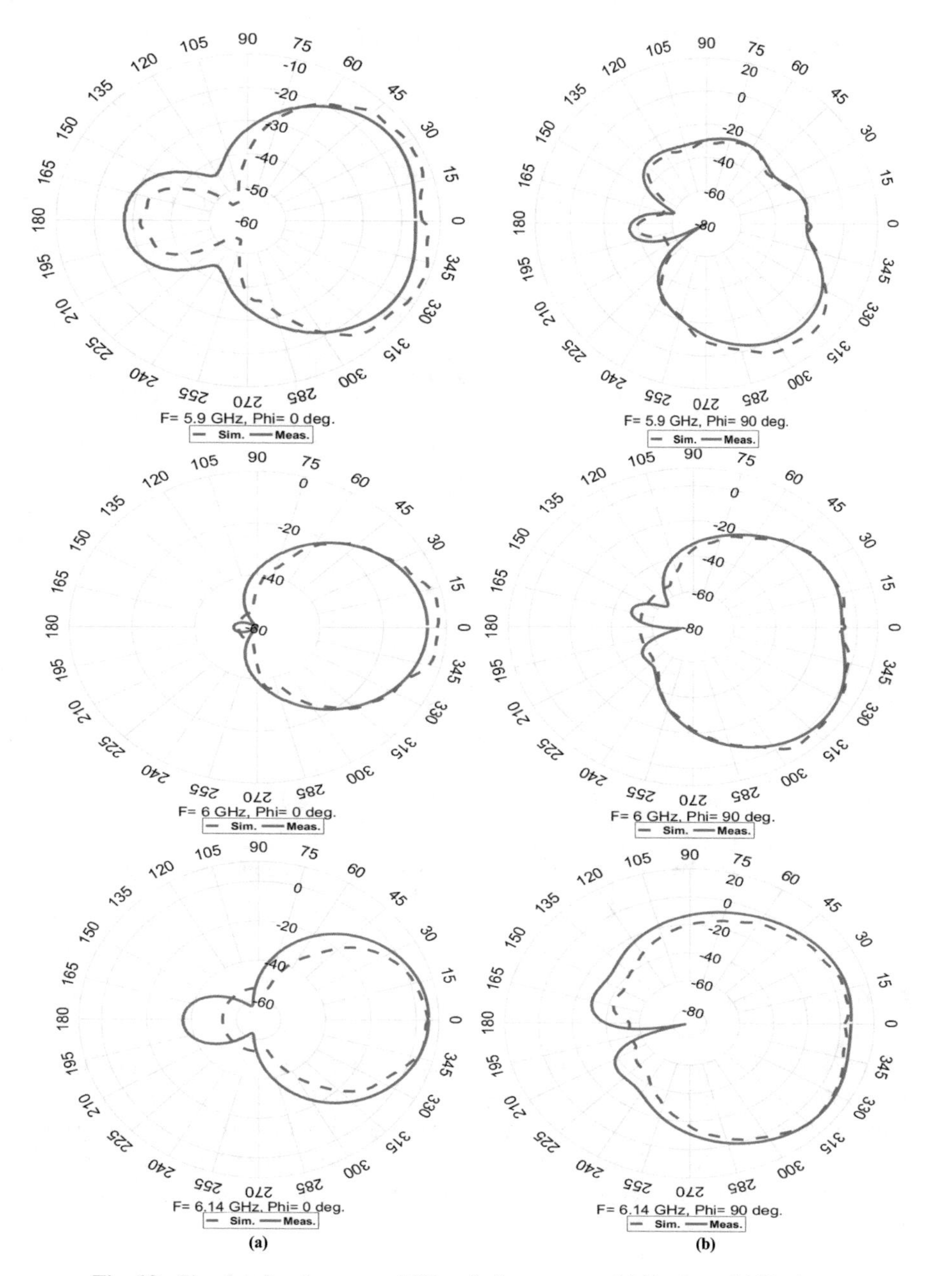

Fig. 12. Simulated and measured 2D radiation patterns (a) E- plane (b) H-plane.

4 Comparison with Other Related Works

A comparative study with related works is done and summarized in Table 4. The study took in consideration the type of GW, feeding method and antenna, as well as the overall performance of the antenna in terms of bandwidth, gain and radiation efficiency. The result of this study shows that the proposed design offers good gain and efficiency in comparison to other related works.

Table 4. Comparison with other related works

Ref	GW type	Antenna	BW (%)	Gain (dB)	Efficiency
[19]	RGW	8 × 8	19.5–21.5, 29–31	26–29.5	80%
[20]	GGW	16 × 16	14% (57 to 66 GHz)	32.3	70%
[4]	RGW	16 × 16	16% (57 to 66 GHz)	32.5	70%
[21]	MGW	16 × 16	17% (54–64 GHz)	30.5	55%
[22]	GGW	16 × 16	18% (57 to 67)	32.5	80%
[23]	RGW	4 × 4	13.6% (27.5 to 31 GHz)	19.24	80%
[24]	SIGW	4 × 4	35.3% (3.5 to 5.5 GHz)	8.8	80%
Proposed work	SIGW	1 × 2	50% (5.9 to 6.14GHz)	4 (fabricated)	75%

5 Conclusion

The presented paper proposed design and implementation of SIGW based array antenna of two slots for WiFi applications. It is implemented using CST Tool. It operates at three bands of the standard WiFi operating bands at 5.9, 6 and 6.1414 GHz and with frequency band width of 75 MHz, 20 MHz and 20 MHz respectively. The design comprises the SIGW based feeding network of low loss and dispersion. It is found that the measured and simulated results exhibit fair matching.

References

1. Kildal, P.-S.: Definition of artificially soft and hard surfaces for electromagnetic waves. Electron. Lett. **24**(3), 168 (1988). https://doi.org/10.1049/el:19880112
2. Bayat-Makou, N., Kishk, A.A.: Realistic air-filled TEM printed parallel-plate waveguide based on ridge gap waveguide. IEEE Trans. Microw. Theory Tech. **66**(5), 2128–2140 (2018)

3. Rajo-Iglesias, E., Kildal, P.-S.: Numerical studies of bandwidth of parallel-plate cut-off realised by a bed of nails, corrugations and mushroom-type electromagnetic bandgap for use in gap waveguides. IET Microw. Antennas Propag. **5**(3), 282–289 (2011)
4. Zarifi, D., Farahbakhsh, A., Uz Zaman, A., Kildal, P.-S.: Design and fabrication of a high-gain 60-GHz corrugated slot antenna array with ridge gap waveguide distribution layer. IEEE Trans. Antennas Propag. **64**(7), 2905–2913 (2016)
5. Sorkherizi, M.S., Kishk, A.A.: Fully printed gap waveguide with facilitated design properties. IEEE Microw. Wirel. Compon. Lett. **26**(9), 657–659 (2016)
6. Salah El-Din, M.S.H., Shams, S.I., Allam, A.M.M.A., Gaafar, A., Elhennawy, H.M., Sree, M.F.A.: Design of an E-sectoral horn based on PRGW technology for 5G applications. Int. J. Microw. Wirel. Technol. **15**(6), 1082–1090 (2022). https://doi.org/10.1017/S1759078722001076
7. Salah El-Din, M.S.H., Shams, S.I., Allam, A.M.M.A., Gaafar, A., Elhennawy, H.M., Sree, M.F.A.: SIGW based MIMO antenna for satellite down-link applications. IEEE Access **10**, 35965–35976 (2022). https://doi.org/10.1109/ACCESS.2022.3160473
8. El-Din, M.S.H., Salah, S.I., Shams, A.M.M.A., Allam, M.F.A.S., Gaafar, A., El-Hennawy, H.: Bow-tie slot antenna loaded with superstrate layers for 5G/6G applications. In: 2021 IEEE International Symposium on Antennas and Propagation and USNC-URSI Radio Science Meeting (APS/URSI), pp. 1561–1562. IEEE (2021)
9. Taha, A., Allam, A.M.M.A., Wahba, W., Sree, M.F.A.: Printed ridge gap waveguide based Radome antenna for k-ka applications. In: 2021 15th European Conference on Antennas and Propagation (EuCAP), pp. 1–5. IEEE (2021)
10. Sree, M.F.A., Allam, A.M.M.A., Mohamed, H.A.: Design and implementation of multi-band metamaterial antennas. In: 2020 International Applied Computational Electromagnetics Society Symposium (ACES), pp. 1–2. IEEE (2020)
11. Wossugieniri, O., Faezi, H., Fallah, M.: Compact 2-way H-plane power dividers for a rectangular waveguide in Ku band. In: 2018 22nd International Microwave and Radar Conference (MIKON), Poznan, pp. 298-301 (2018)
12. Xu, S., Tian, X., Xu, L.: A 4-way broadband power divider based on the suspended microstrip line. In: 2018 International Conference on Microwave and Millimeter Wave Technology (ICMMT), Chengdu, pp. 1–3 (2018)
13. Yang, T.Y., Hong, W., Zhang, Y.: Wideband millimeter-wave substrate integrated waveguide cavity-backed rectangular patch antenna. IEEE Antennas Wirel. Propag. Lett. **13**, 205–208 (2014). https://doi.org/10.1109/LAWP.2014.2300194
14. Shams, S.I., Kishk, A.A.: Wide band power divider based on Ridge gap waveguide. In: 2016 17th International Symposium on Antenna Technology and Applied Electromagnetics (ANTEM), Montreal, QC, pp. 1–2 (2016)
15. Ahmadi, B., Banai, A.: A power divider/combiner realized by ridge gap waveguide technology for millimeter wave applications. In: 2016 Fourth International Conference on Millimeter-Wave and Terahertz Technologies (MMWaTT), Tehran, pp. 5–8 (2016)
16. Liu, D., et al.: User association in 5G networks: a survey and an outlook. IEEE Commun. Surv. Tutor. **18**(2), 1018–1044 (2016)
17. Araniti, G., Condoluci, M., Scopelliti, P., Molinaro, A., Iera, A.: Multicasting over emerging 5G networks: challenges and perspectives. IEEE Netw. **31**(2), 80–89 (2017)
18. Pozar, D.M.: Microwave Engineering, 4th Edn., pp. 72–75. John Wiley & Sons, New York (2011)
19. Ferrando-Rocher, M., Herranz-Herruzo, J.I., Valero-Nogueira, A., Bernardo-Clemente, B.: Full-metal K-Ka dual-band shared-aperture array antenna fed by combined ridge-groove gap waveguide. IEEE Antennas Wirel. Propag. Lett. **18**(7), 1463–1467 (2019)

20. Ferrando-Rocher, M., Valero-Nogueira, A., Herranz-Herruzo, J.I., Teniente, J.: 60 GHz single-layer slot-array antenna fed by groove gap waveguide. IEEE Antennas Wirel. Propag. Lett. **18**(5), 846–850 (2019)
21. Liu, J., Vosoogh, A., Zaman, A.U., Yang, J.: Design and fabrication of a high-gain 60-ghz cavity-backed slot antenna array fed by inverted microstrip gap waveguide. IEEE Trans. Antennas Propag. **65**(4), 2117–2122 (2017)
22. Farahbakhsh, A., Zarifi, D., Zaman, A.U.: 60-GHz groove gap waveguide based wideband h-plane power dividers and transitions: for use in high-gain slot array antenna. IEEE Trans. Microw. Theory Tech. **65**(11), 4111–4121 (2017)
23. Ferrando-Rocher, M., Herranz-Herruzo, J.I., Valero-Nogueira, A., Bernardo-Clemente, B.: Single-layer sequential rotation network in gap waveguide for a wideband low-profile circularly polarized array antenna. IEEE Access **10**, 62157–62163 (2022). https://doi.org/10.1109/ACCESS.2022.3182336
24. Li, T., Chen, Z.N.: Wideband sidelobe-level reduced ka -band metasurface antenna array fed by substrate-integrated gap waveguide using characteristic mode analysis. IEEE Trans. Antennas Propag. **68**(3), 1356–1365 (2020). https://doi.org/10.1109/TAP.2019.2943330

Citizen Engagement in Urban Planning – An EPS@ISEP 2022 Project

Carla G. Cardani[1], Carmen Couzyn[1], Eliott Degouilles[1], Jan M. Benner[1], Julia A. Engst[1], Abel J. Duarte[1,2], Benedita Malheiro[1,3(✉)], Cristina Ribeiro[1,4], Jorge Justo[1], Manuel F. Silva[1,3], Paulo Ferreira[1], and Pedro Guedes[1,3]

[1] ISEP, Polytechnic of Porto, Rua Dr. António Bernardino de Almeida, 4249-015 Porto, Portugal

[2] REQUIMTE/LAQV, ISEP, Polytechnic of Porto, Rua Dr. António Bernardino de Almeida, 4249-015 Porto, Portugal

[3] INESC TEC, Campus da Faculdade de Engenharia da Universidade do Porto, Rua Dr. Roberto Frias, 4200-465 Porto, Portugal
epsatisep@gmail.com

[4] INEB, Rua Alfredo Allen, 208, 4200-135 Porto, Portugal
https://www.eps2022-wiki2.dee.isep.ipp.pt/

Abstract. Involving people in urban planning offers many benefits, but current methods are failing to get a large number of citizens to participate. People have a high participation barrier when it comes to public participation in urban planning – as it requires a lot of time and initiative, only a small non-diverse group of citizens take part in governmental initiatives. In this paper, a product is developed to make it as easy as possible for citizens to get involved in construction projects in their community at an early stage. As a solution, a public screen is proposed, which offers citizens the opportunity to receive information, view 3D models, vote and comment at the site of the construction project via smartphone – the solution was named Parcitypate. To explain the functions of the product, a prototype was created and tested. In addition, concepts for branding, marketing, ethics, and sustainability are presented.

Keywords: Engineering Education · European Project Semester · Citizen Engagement · Participation · Urban Planning · Public Screen

1 Introduction

The European Project Semester[1] (EPS) offers students with engineering or other backgrounds the opportunity to carry out a project in a multidisciplinary team at a partner university, under scientific supervision, within one semester. This project was performed at the Instituto Superior de Engenharia do Porto[2] by Carla, Carmen, Eliott, Julia, and Jan.

[1] http://europeanprojectsemester.eu/.
[2] http://isep.ipp.pt/.

A. Rocha et al. (Eds.): WorldCIST 2023, LNNS 800, pp. 615–624, 2024.
https://doi.org/10.1007/978-3-031-45645-9_59

People's participation in urban planning has several advantages, but present strategies are unable to engage a sizeable number of individuals. When it comes to public involvement in urban planning, there is a significant participation barrier. Only a small, non-diverse group of residents participate in local consultation projects since it takes a lot of time and initiative. This fact prevents citizen participation and leads to dissatisfaction among the population. Even smart cities have not yet been able to come up with a comprehensive solution.

The project's purpose is to help solve the above described problem. Citizen engagement will be enabled via a combination of a public screen with a visual experience and active interactivity via a smartphone website. A public screen is installed at the site of a prospective project development in order to pique the interest and participation of passers-by. To give visualisation, official data, and public feedback (comments, ideas, and likes), the screen will display 3D models and project information. The population's interaction with the screen should take place via smartphone and a web application. Once the app is installed, it is possible to access the website just by reading a QR code from the screen. While several citizens can connect with the website at the same time to learn more about the project and provide their feedback in the form of comments or likes, active involvement with the public screen is limited to one person at a time. In this situation, the citizen gets to explore and animate the many 3D models of the project on the screen for a short time window, preventing long occupation. Experience from past research should be considered to ensure the project's success. The overall goal of this approach is to lower the barriers to participation and inspire residents to participate in urban planning in a simple and enjoyable manner.

Following this brief introduction to the problem, in the sequel of this document, first, the state of the art is briefly explained and, after, the solution approach and the development of a prototype is presented.

2 Preliminary Studies

This section presents the "Smart City", "Public Display" and "Citizen Participation" concepts and analyses solutions for citizen participation in urban planning.

Smart City is an increasingly popular term in scientific and political discussions [3]. Initially, it was limited to the application of information and communication technologies (ICT) to the efficient management of urban infrastructure and services, but today it refers to almost any form of technological innovation to improve the urban management efficiency [9].

A Public Display is, by definition, an electronic device that broadcasts visual content to a large number of people or an area. It is also a sub-segment of public electronic signage for this project, which is "centrally managed and individually addressable for display of text, animated or video messages for advertising, information, entertainment, and retailing to targeted audiences" [11].

"Involvement of citizens in governmental decision-making processes" is what public participation entails. It varies from "being given notice of public hearings" to "active participation in decisions that affect communities" [8].

More and more governments at various levels of power are implementing citizen participation projects in an effort to involve citizens in decision-making. To begin with, increased citizen participation in urban planning enables local governments to satisfy actual demands and boost acceptance of public efforts. Typically, city planners do not listen to, or give equal weight, to the opinions of all stakeholders. Furthermore, tiny public spaces, such as playgrounds, are sometimes disregarded during the larger-scale construction of cities. Thus, citizen participation can aid in identifying both general and specific needs, ranging from the building of community centers and parks to the placement of new accessibility ramps and litter bins. Cities will be able to adapt closer to community needs and become more inclusive in this way. Furthermore, citizen participation strives to reconnect residents with local governments, hence increasing confidence [2].

The following subsections describe the shortcomings, warnings, and restrictions associated with traditional methods of citizen engagement.

2.1 Related Work

As demonstrated, involving residents in planning can boost local democracy and increase the quality of projects. However, smart cities continue to have difficulties in involving citizens [2]. This research takes into account both conventional and contemporary approaches to offer a public engagement solution for smart cities.

Traditional methods have been utilized for a long time, but they have accessibility, time requirements, and dissemination issues that can be solved by using ICT [2,10,12]. ICT, which is being extensively investigated by current means, enables quickly and easily accessing a big number of citizens. Additionally, anyone can join at any moment anonymously online. Using e-planning, citizens can participate in the planning process at an early level and create a new richness of data. Nonetheless, modern methods of citizen participation still have their open challenges, which are sometimes similar to traditional methods [6]. Table 1 summarises the most representative citizen participation approaches found in the literature.

This state-of-the-art review identified the following criteria for a successful solution: (*i*) reach as many citizens as possible; (*ii*) provide a two-way information flow; (*iii*) be free and simple for users; (*iv*) ensure barrier-free participation; (*v*) be transparent, inclusive, and fair; (*vi*) promote interaction; (*vii*) support participation at the location; (*viii*) allow anonymous participation; (*ix*) foster serious, responsible participation; and (*x*) support the simultaneous participation of several users.

2.2 Ethics

In a society, ethical and deontological concerns are crucial. They inspire regulations that go above and beyond the law. Companies, especially those that strongly rely on brand image, stand to lose from ethical problems, namely, numerous businesses have lost customers, markets, or even collapsed. With the

Table 1. Comparison of different approaches for citizen participation.

Authors	User Feedback	Devices	Strengths	Weaknesses
Claes *et al.* (2018) [1]	Smileys	Screen Sensors Buttons	Environmental awareness[1]	Limited interaction
Steinberg *et al.* (2014) [13]	Voting	Screen Buttons	Buttons	Limited interaction
Hosio *et al.* (2012) [5]	Comments Pictures	Screen Camera Touchscreen	Feedback	Limited Information
Muelhaus *et al.* (2022) [7]	Quest 3D Models Comments	Smartphone Tablet	Augmented Reality Feedback Information	Hurdle (application) Difficult to adapt to different projects
Wilson *et al.* (2019) [14]	Comments Voting	Smartwatch Smartphone	Feedback Notifications	Hurdle (application) Limited information
Du *et al.* (2020) [4]	Comments Voting	Screen Smartphone	Feedback Information	Hurdle (application) Distance to location

growth of social media in our society in recent years, this influence might be even more significant.

The team prioritise the ethical issues regarding team members – must follow personal ethical standards, engage in open dialogues and avoid spreading false information – and the solution – must take into account the impact of the technical, marketing, and environmental decisions.

2.3 Marketing

A marketing plan outlines objectives and methods to help a firm reach its core purpose. It plays a role in that company's success, going beyond simply promoting a product. The objective of the product and its brand should be identified, anticipated, and integrated into this plan.

Parcitypate relies on a smart interactive display to encourage citizens to contribute their opinions and thoughts on city choices. Moreover, it allows private participation through a website. The solution is intended for city governments that want citizen participation in urban planning. This concept differs from those of Claes [1], "vote with your feet" [13], or ubinion [5]. The marketing strategy is built from the 3 pillars identity, object and response.

Identity: Parcitypate was created to represent the idea of citizens interaction in the city planning and the logo creates a connection to the city of Porto. The strong and modern typography and the representative geometry show the power and importance of the city.

Object: Parcitypate consists of two products: the screen environment and the web/app design. To accomplish both successfully, it is needed to be robust and connected with the visual identity of the brand. The product itself was designed to draw attention of citizens to participate in the city decision-making.

Response: A consolidated human resource is needed to be in contact with distributors and suppliers and make sure the product is functional and not harmful to its consumers.

2.4 Sustainability

The United Nations first outlined sustainability as "a development that meets the needs of the present without compromising the ability of future generations to meet their own needs" in 1987. It can be accomplished by taking into account the social, economic, and environmental impact of Parcitypate. The social aspect comes first. In fact, the major goal of Parcitypate is to create a system that enhances social sustainability in urban design. The second consideration is financial. In this regard, the team made sure to create a sustainable business plan that does not increase its market price substantially. The environmental factor comes in third. By detecting and minimising the impact of this product on the environment at every stage of its life cycle, starting with the resources, moving through production, and ending with its end of life. Each of these processes benefited from changes made to the design and materials.

3 Proposed Solution

3.1 Design

The goal of the product's design is to draw attention. It creates a setting in which users can participate in the suggested experience and advance the city. The Porto-based origin is reflected in the Parcitypate logo (Fig. 1a). From Porto's renowned structures, one can see the combination of geometric shapes. Parcitypate, which combines the words "city" and "participate", symbolises the notion of community participation in city planning. Adopting a powerful, contemporary typography and integrating a symbolic geometric shape, it conveys the importance and might of the city. There are the vertical and horizontal logo versions. The logo monogram can also be used as a minor feature to harmonise the design of applications.

The product design includes the screen, a roof to protect users and a sitting bench (as depicted on Fig. 1b). The roof is made of a thin sheet of metal, cut according to the Parcitypate logo, with a transparent plate on top, both 2.50 m wide). This structure shields the user and the screen from precipitation and sunshine and hides the cabling. Under the roof, a support holds the screen, protecting the screen and cabling. Four steel pillars support all of these components. They are hollow to be light and contain the cables and fastened to the ground through brackets.

(a) Parcitypate logo variants (b) 3D-Model of the public screen

Fig. 1. Parcitypate design

3.2 Control

Users can control the public display through their smartphones, combining public and private engagement. Some interaction features and outcomes, such as comments or likes, are quickly and visibly reflected on the screen. Others, like choosing the preferred model, are done in private. This style of semi-public interaction aims to bring together the low participation barrier of personal gadgets with the enhanced attention of public displays.

The user connects the smartphone to the screen through Wi-Fi or QR code scanning. Once the connection is successful, both interfaces provide visual feedback and the user can start to navigate the app. The main navigation feature is to use a simple 3D cube on the app to navigate the 3D model on the screen.

4 Prototype Development

The development of a proof-of-concept prototype allows the team to identify and correct problems and verify the correct operation of the main features. The key Parcitypate features include the communication between the screen and app and screen energy-saving measures.

4.1 Assembly

For implementing the prototype, the following physical components were used: (*i*) a Raspberry Pi 4 mini-computer (2 GB of RAM); (*ii*) a General Purpose Input/Output (GPIO) extension board; (*iii*) resistors; (*iv*) distance (HC-SR04) and light (BH1750FVI) sensors; (*v*) High-Definition Multimedia Interface (HDMI) monitor; and (*vii*) a smartphone.

4.2 Control

The mini-computer uses open source solutions to access and control the assembled components. Specifically, it runs Ubuntu 22.04 and uses GNOME as desktop environment. The control code was developed in Python.

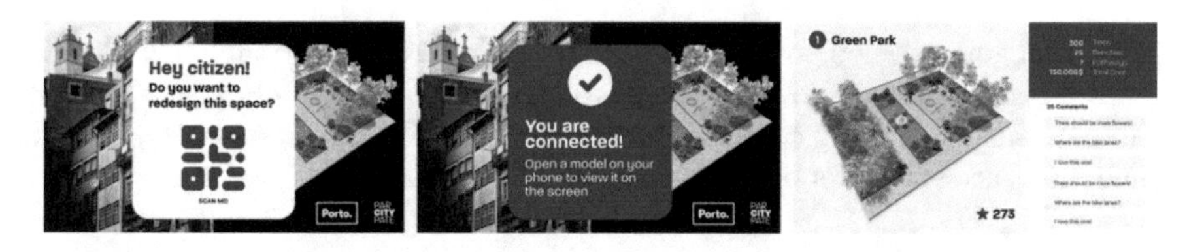

Fig. 2. Public display user interfaces

Fig. 3. Mobile user interfaces

Since the screen is connected via HDMI, it is impossible to adjust the screen brightness. Depending on the ambient brightness, the prototype switches between a "dark" and "light" mode of the operating system for demonstration. The screen is switched on an off based on the readings of the distance sensor, *i.e.*, whether a person has been detected in front of the screen in the last minutes.

4.3 Web Application

Software. The software consists of four components: (*i*) the front-end web-application for the public screen (Fig. 2); (*ii*) the mobile-optimised web-application for the user smartphone (Fig. 3); (*iii*) the websocket-based communication between the screen and the smartphone applications, which is handled by a shared server; and (*iv*) the database holding the information on the 3D models, including likes, comments and votes.

To reduce costs, all software components rely on open-source technologies. Both front-end applications utilise the JavaScript framework Vue.js[3] and the database is implemented with the NoSQL database platform MongoDB[4]. The websocket server uses the JavaScript server Node.js[5]. A key challenge for the websocket server was to distinguish between queries arriving from the smartphone app and those coming from the screen. Additionally, whenever there is already one user operating the screen, the server must warn and refuse requests from other clients. When a client successfully connects to the screen, the screen

[3] https://vuejs.org/.
[4] https://www.mongodb.com.
[5] https://nodejs.org/.

application gets notified to start providing visual feedback. Finally, clients have an unique identifier to allow the server to push information only to the relevant recipient, thus optimising communication efficiency. On the server-side, all active connections are stored as objects in an array.

Tests. Two tests were carried out to assess the proper operation of the user interface, as well as the technical performance of the applications.

First, the interaction flow was tested with the quality assurance platform Rainforest[6], which allows for automated interface testing based on pixel and text recognition algorithms. To test the web application in a mobile setting, a virtual machine simulating an iPhone 11 Pro Max and Safari version 14 was used. The automated simulation followed a predefined interaction path, which included all views of the mobile application and all possible interaction functionalities. The connection to the screen is automatically established on page load and disconnected after an idle time of 61 s. The test results show that all views and functionalities are visible and accessible on a mobile device, and lead to the correct interaction outcome.

Additionally, the performance of the software was assessed using the Google Developers service Page Speed Insights[7], which sends a request to the page and measures response and load times. The performance scores of the mobile and screen applications were both high, with 90% and 92%, respectively. The small reduction in performance scores is due to the large data size of the 3D plugin. Table 2 displays the test results.

Table 2. Page Speed Insights results for the front-end applications

Performance Indicator	Phone App	Screen App
Overall Score (%)	90	92
First Contentful Paint (ms)	700	700
Time to Interactive (ms)	1500	400
Speed Index (ms)	1100	1000
Total Blocking Time (ms)	20	10
Largest Contentful Paint (ms)	2000	2000
Cumulative Layout Shift (ms)	0	2

[6] https://www.rainforestqa.com.

[7] https://developers.google.com/speed/docs/insights/v5/about.

5 Conclusion

5.1 Achievements

After analysing traditional concepts of citizen participation and considering modern approaches, it was possible to identify existing problems and propose a new solution – Parcitypate. The listed product objectives were met.

For the team, working in a multinational and multidisciplinary team proved to be a challenge. Due to different cultural and academic backgrounds, the ways of working and the way of thinking about the project were very different. Ultimately, this has led to the need for human and professional development in order to adopt different perspectives and learn from other team members. These experiences can be seen as extremely positive and may prove useful in the future.

5.2 Ideas for Future Developments

As the complete physical construction of the project would have exceeded the scope of the EPS, only a prototype of Parcitypate was developed to demonstrate the idea and general functionality. Therefore, the next step in potentially further development of the project would be to construct a life-sized prototype. With this prototype, more realistic user tests could be conducted and the feasibility of the construction and packaging solutions could be examined in more detail.

As far as the software is concerned, a potential further addition would be a front-end application for city governments, where 3D models and the corresponding information could be uploaded. Moreover, it could include a functionality to collect and export user data and interaction statistics to provide further information on participation demographics.

Acknowledgements. This work was partially financed by National Funds through the Portuguese funding agency, FCT – Fundação para a Ciência e a Tecnologia, within project UIDB/50014/2020.

References

1. Claes, S., Coenen, J., Moere, A.V.: Conveying a civic issue through data via spatially distributed public visualization and polling displays. In: Bratteteig, T., Sandnes, F.E. (eds.) Proceedings of the 10th Nordic Conference on Human-Computer Interaction, Oslo, Norway, 29 September–3 October 2018, pp. 597–608. ACM (2018). https://doi.org/10.1145/3240167.3240206
2. Clarinval, A.: Citizen participation in smart cities: facilitating access through awareness, open government data, and public displays. Ph.D. thesis, University of Namur (2021)
3. Council on Foreign Relations: A smarter planet: The next leadership agenda (2008). https://www.cfr.org/event/smarter-planet-next-leadership-agenda. Accessed 22 Mar 2022
4. Du, G., Kray, C., Degbelo, A.: Interactive immersive public displays as facilitators for deeper participation in urban planning. Int. J. Hum.-Comput. Interact. **36**(1), 67–81 (2020). https://doi.org/10.1080/10447318.2019.1606476

5. Hosio, S., Kostakos, V., Kukka, H., Jurmu, M., Riekki, J., Ojala, T.: From school food to skate parks in a few clicks: using public displays to bootstrap civic engagement of the young. In: Kay, J., Lukowicz, P., Tokuda, H., Olivier, P., Krüger, A. (eds.) Pervasive 2012. LNCS, vol. 7319, pp. 425–442. Springer, Heidelberg (2012). https://doi.org/10.1007/978-3-642-31205-2_26
6. Münster, S., et al.: How to involve inhabitants in urban design planning by using digital tools? An overview on a state of the art, key challenges and promising approaches. Procedia Comput. Sci. **112**, 2391–2405 (2017). https://doi.org/10.1016/j.procs.2017.08.102
7. Muehlhaus, S.L., Eghtebas, C., Seifert, N., Schubert, G., Petzold, F., Klinker, G.: Game. UP: gamified urban planning participation enhancing exploration, motivation, and interactions. Int. J. Hum.-Comput. Interact. **39**, 1–17 (2022). https://doi.org/10.1080/10447318.2021.2012379
8. Porta, M., Last, J.: A Dictionary of Public Health. Oxford University Press, Oxford (2022). Accessed 06 Apr 2022
9. Praharaj, S., Han, H.: Cutting through the clutter of smart city definitions: a reading into the smart city perceptions in India. City Cult. Soc. **18**, 100289 (2019). https://doi.org/10.1016/j.ccs.2019.05.005
10. Sameh, H.M.M.: Public participation in urban development process through information and communication technologies. Master's thesis, Ain Shams University (2011)
11. Schaeffler, J.: Digital Signage: Software, Networks, Advertising, and Displays: A Primer for Understanding the Business. NAB Executive Technology Briefings. Taylor & Francis (2012)
12. Smørdal, O., Wensaas, K., Lopez-Aparicio, S., Pettersen, I., Hoelscher, K.: Key issues for enhancing citizen participation in co-constructing city futures. In: Proceedings of Fourth International Workshop on Cultures of Participation in the Digital Age - CoPDA 2016. CEUR (2016)
13. Steinberger, F., Foth, M., Alt, F.: Vote with your feet: local community polling on urban screens. In: Proceedings of The International Symposium on Pervasive Displays, PerDis 2014, pp. 44–49. Association for Computing Machinery, New York (2014). https://doi.org/10.1145/2611009.2611015
14. Wilson, A., Tewdwr-Jones, M., Comber, R.: Urban planning, public participation and digital technology: app development as a method of generating citizen involvement in local planning processes. Environ. Plan. B Urban Anal. City Sci. **46**(2), 286–302 (2019). https://doi.org/10.1177/2399808317712515

On the Convenience of Using 32 Facial Expressions to Recognize the 6 Universal Emotions

Miquel Mascaró-Oliver(✉), Ramon Mas-Sansó, Esperança Amengual-Alcover, and Maria Francesca Roig-Maimó

Universitat de les Illes Balears, Ctra. de Valldemossa, Km. 7.5, 07122 Palma, Spain
{miquel.mascaro,ramon.mas,eamengual,xisca.roig}@uib.es

Abstract. Emotion and facial expression recognition are a common topic in artificial intelligence. In particular, main efforts focus on constructing models to classify within the six universal emotions. In this paper, we present the first attempt to classify within 33 different facial expressions. We define and train a simple convolutional neural network with a low number of intermediate layers, to recognize the 32 facial expressions (plus the neutral one) contained in an extension of UIBVFED, a virtual facial expression dataset. We obtained a global accuracy of 0.8, which is comparable to the 0.79 accuracy we got when training the neural network with only the six universal emotions. Taking advantage of this trained model, we explore the approach of classifying the images within the six universal emotions, translating the facial expression predicted by the model into its associated emotion. With this novel approach, we reach an accuracy level of 0.95, a value comparable to the best results present in the literature with the plus of using a very simple neural network.

Keywords: Facial Expression Recognition · Emotion Recognition · Affective Computing Dataset · Convolutional Neural Network

1 Introduction

For people, facial expression and emotion recognition is a key element of non-verbal communication. For artificial intelligence, mastering the ability to identify facial expressions, or emotions, remains a challenge.

Recently, there have been many technological advances regarding facial recognition, mainly focused on identification and security matters [1]. However, the possibility of using this resource to detect and measure the emotions expressed through a face is somewhat more difficult to address. Many researchers working in the field of artificial intelligence have dedicated efforts to address this task, by using various modeling and classification techniques, among which convolutional neural networks (CNN) are very popular [2, 3], due to the fact that they can be fed directly by images.

Data is an integral part of the existing approaches in facial expression recognition. For this purpose, a relevant number of datasets are available [4]. These datasets have been

A. Rocha et al. (Eds.): WorldCIST 2023, LNNS 800, pp. 625–634, 2024.
https://doi.org/10.1007/978-3-031-45645-9_60

used in many studies in different areas such as psychology, medicine, art and computer science. In Mollahosseini et al. [5] the state of the art of the main facial expression datasets is presented.

We are mainly interested in the synthesis and recognition of facial expressions and in their possible applications. In previous work we presented UIBVFED, a virtual facial expression database [6]. To the best of our knowledge, available datasets contain between 6 and 8 labeled expressions, except EmotioNet [7] which has 23 identified expressions. The UIBVFED database presents 32 facial expressions described in terms of Action Units (AUs) [8].

In contrast to the most usual efforts present in the literature that focus on constructing models to classify the universal emotions [8–11], we present the first attempt, to the best of our knowledge, to obtain a CNN model to recognize 32 different facial expressions. Besides presenting the results of our 32 facial expressions classification model, we analyze how this 32-class classification could be used to obtain a six universal emotion classification. This approach of translating facial expressions into their associated emotion can lead to a better recognition of human emotions when compared to the traditional approach of training the CNN using a reduced set of 7 or 8 classes.

This paper is organized as follows: Sect. 2 presents the procedure to train and test our CNN model, together with the description of the used dataset. In Sect. 3 we discuss the results we get. And, finally, Sect. 4 shows our conclusions and future work.

2 Materials and Methods

This section contains description of the data, the data pre-processing, and the procedure followed to train and test the CNN model.

2.1 Data Description: Extension of the UIBVFED Dataset

In this paper, we use an extension of the UIBVFED dataset. The original UIBVFED dataset is composed of twenty gender-balanced avatars from different ethnicities and ages, performing 32 facial expressions. In this extended dataset, the number of avatars has been expanded from 20 to 100 (50 men and 50 women, aged between 20 and 80, from different ethnicities) using the online interactive tool Autodesk Character Generator.

The UIBVFED dataset is organized according to the six universal emotions following Faigin's nomenclature [12]: *Anger*, *Disgust*, *Fear*, *Joy*, *Sadness*, and *Surprise*. The list of original universal emotions described by Ekman [13] were the same six plus the *Contempt* emotion. This added emotion is accepted by a lot of but not all the experts. UIBVFED reflects two facial expressions associated with the emotion of *Contempt*: Disdain (associated with *Disgust* emotion) and Debauched Smile (associated with *Joy* emotion).

The extended dataset is composed of 3300 facial images from 100 avatars each reproducing 32 facial expressions plus the neutral facial expression (100 avatars $\times$ 33 facial expressions $=$ 3300 images). Facial expressions are classified based on the six universal emotions, plus the neutral one (see Table 1). Figure 1 shows the 32 facial expressions plus the neutral facial expression of one of the 100 avatars in the database.

Table 1. Emotions associated to each facial expression, and the number of images per facial expression in the dataset

Emotion	Facial expression	Total
Anger	Enraged Compressed Lips, Enraged Shouting, Mad, Sternness Anger	400
Disgust	Disdain, Disgust, Physical Repulsion	300
Fear	Afraid, Terror,Very Frightened, Worried	400
Joy	False Laughter 1, False Smile, Smiling closed mouth, Smiling open mouthed, Stifled Smile, Laughter, Uproarious Laughter, False Laughter 2, Abashed Smile, Eager Smile, Ingratiating Smile, Sly Smile, Melancholy Smile, Debauched Smile	1400
Neutral	Neutral	100
Sadness	Crying closed mouth, Crying open mouthed, Miserable, Nearly crying, Sad, Suppressed sadness	600
Surprise	Surprise	100

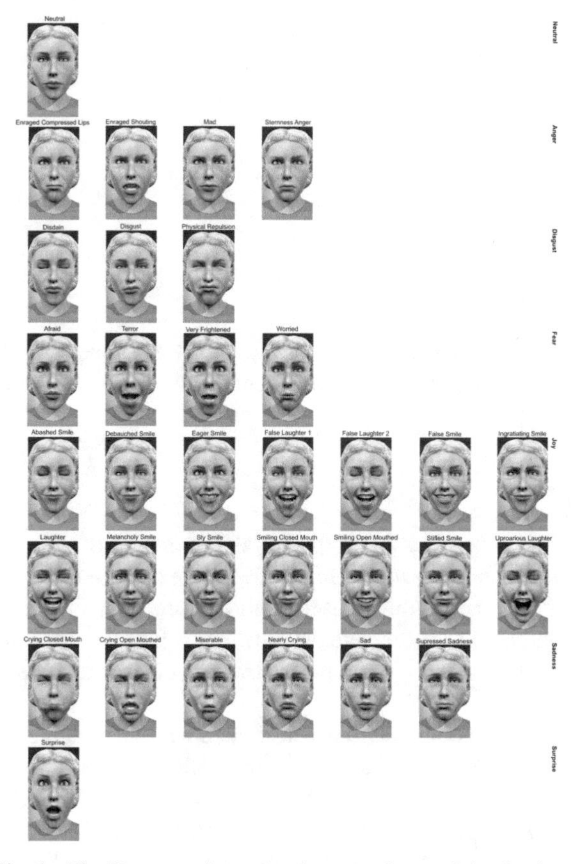

Fig. 1. The 32 expressions plus the neutral one of the database.

2.2 The Convolutional Neural Network

Our convolutional neural network takes a 128 × 128 input grayscale image and successively applies 32, 64 and 128 filters using convolutions to detect the main characteristics of the image. All the filters are 3 × 3 sized and are followed by a Rectifier Linear Unit (ReLu) activation function to select only the meaningful nodes and by a max_pooling to reduce the number of parameters. Finally, the last layer is flattened in a dense layer. The last dense layer provides the 33 outputs corresponding to each class of the facial expressions. The network architecture can be seen in Fig. 2.

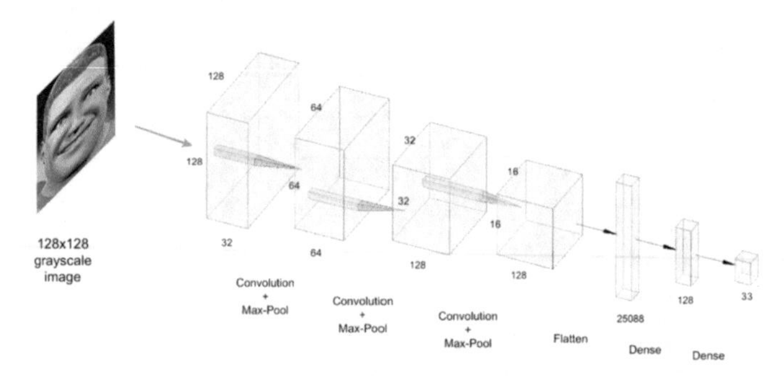

Fig. 2. The CNN architecture.

2.3 Data Pre-Processing

The pre-processing steps included the conversion of the image to grayscale and its resizing to fit the 128x128 pixels size of the input data in the CNN (see facial expression image in Fig. 2).

2.4 Procedure

After completing the pre-processing step over all the images of the dataset, we created the training and testing dataset: we collected 80% of the data for the training dataset, and the remaining 20% for the testing dataset. Both datasets contained a class distribution that was representative of the complete dataset: for each of the 33 facial expressions, a random set of 80 images were selected for the training dataset, and the remaining 20 for the testing dataset.

We trained the CNN model previously described with the training dataset. Then, the model was tested with the testing dataset and the evaluation metrics, in terms of global accuracy and confusion matrix, were computed.

At a later stage, we applied LIME (Local Interpretable Model-agnostic Explanations) to a subset of our input images to get an explanation of which areas of the images are taken into account by the CNN to predict the result. LIME is one of the most used eXplainable Artificial Intelligence (XAI) techniques to explain the predictions of a classifier when dealing with input images [14]. LIME gives us an insight of which parts of the image most influence the predicted result.

3 Results and Discussion

The CNN model had a global accuracy of 0.8. This value of accuracy is very similar to the one we get when training the CNN using only seven classes: six universal emotions plus the neutral one (0.79) [15]. Figure 3 shows the confusion matrix detailing the accuracy obtained per facial expression. A confusion matrix is a table layout that depicts the performance of a classifier: each row of the matrix represents the instances in a predicted class while each column represents the instances in an actual class. Therefore, facial expressions labeled correctly by the CNN are displayed in the diagonal of the matrix. The darker the color, the higher the performance (accuracy) of the classification.

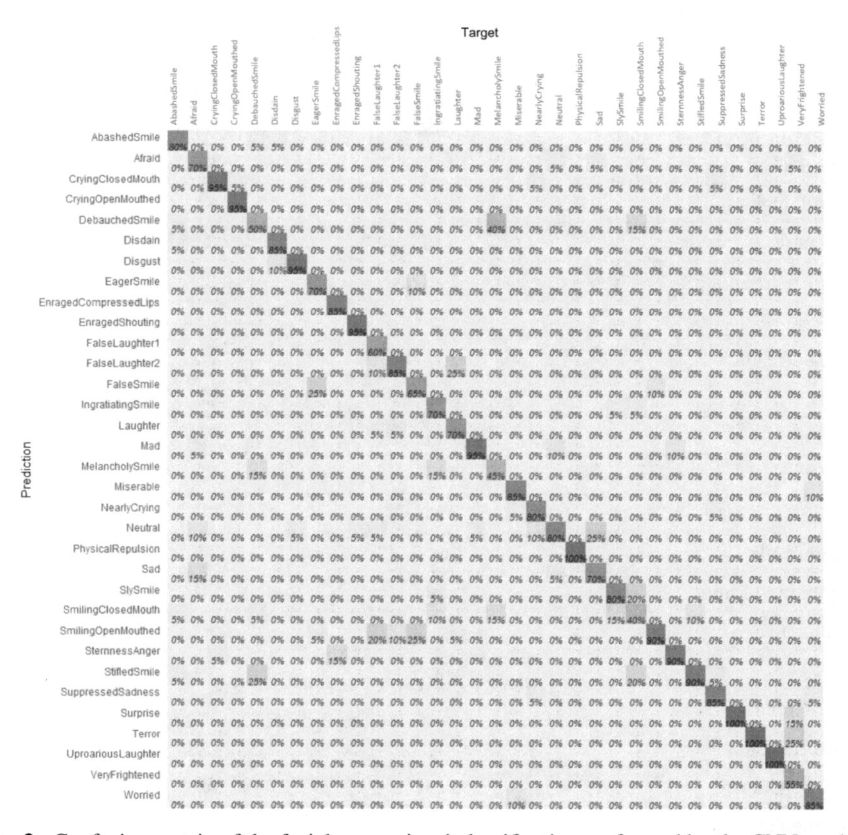

Fig. 3. Confusion matrix of the facial expressions' classification performed by the CNN model.

Table 2 shows the accuracy obtained per class ordered descending by accuracy. From Table 2, we can deduce that 22 of the facial expressions (66.67% of the total) get an

Table 2. Accuracy obtained per facial expression (ordered by accuracy)

Facial expression	Emotion	Accuracy
Physical Repulsion	Disgust	100
Surprise	Surprise	100
Terror	Fear	100
Uproarious Laughter	Joy	100
Crying Closed Mouth	Sadness	95
Crying Open Mouthed	Sadness	95
Disgust	Disgust	95
Enraged Shouting	Anger	95
Mad	Anger	95
Smiling Open Mouthed	Joy	90
Sternness Anger	Anger	90
Stifled Smile	Joy	90
Disdain	Disgust	85
Enraged Compressed Lips	Anger	85
False Laughter 2	Joy	85
Miserable	Sadness	85
Suppressed Sadness	Sadness	85
Worried	Fear	85
Abashed Smile	Joy	80
Nearly Crying	Sadness	80
Neutral	Neutral	80
Sly Smile	Joy	80
Afraid	Fear	70
Eager Smile	Joy	70
Ingratiating Smile	Joy	70
Laughter	Joy	70
Sad	Sadness	70
False Smile	Joy	65
False Laughter 1	Joy	60
Very Frightened	Fear	55
Debauched Smile	Joy	50
Melancholy Smile	Joy	45
Smiling Closed Mouth	Joy	40

average accuracy of a 0.9 and that only 6 of the 33 facial expressions (18.2% of the total classes) get an accuracy lower than 0.7.

The facial expressions *Physical Repulsion*, *Surprise*, *Terror* and *Uproarious Laughter* obtained a perfect accuracy (100%). These facial expressions have exaggerated facial features (like very opened or very closed eyes, or very opened or distorted mouth) (see Fig. 4), which make them easily distinguishable from the other facial expressions.

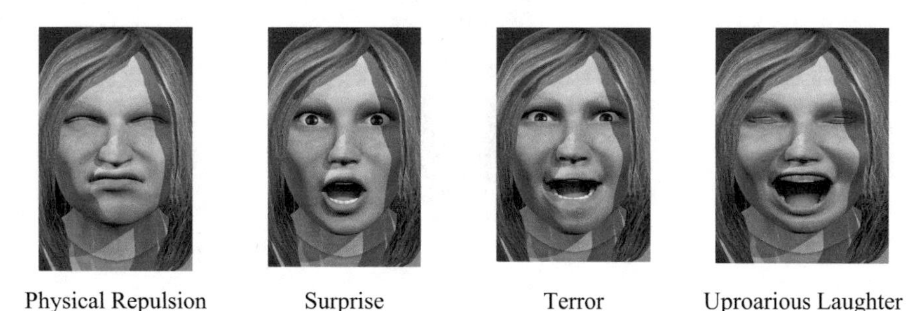

Fig. 4. An avatar performing the facial expressions *Physical Repulsion*, *Surprise*, *Terror* and *Uproarious Laughter*.

Looking at the results reported by LIME for a representative image of these four facial expressions (see Fig. 5), we can see in blue the areas of the images that most influence the predictions. Specifically, the mouth is the facial feature most important for the CNN model to classify these images: for example, for the *Physical Repulsion* facial expression, it is important the upper lip raised in a sneer; for the *Surprise* facial expression, it is important the mouth opened in an oval shape; for the *Terror* facial expression, it is important the mouth opened and widened; and for the *Uproarious Laughter* facial expression, it is also important the mouth opened and widened but with most of upper teeth revealed.

Fig. 5. Results of applying the XAI method LIME to the images of Fig. 4 correctly labeled by the CNN model.

In general, after applying LIME to a set of predictions, the explanations provided by LIME confirm that the interest regions for the CNN model in most of the images were located in the surroundings of the moth, the eyes and the eyebrows (see Fig. 6), which correspond to the facial regions that contain the facial features that determine a facial expression according to Faigin [12].

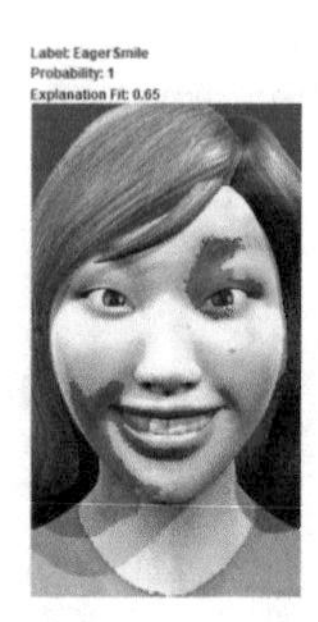

Fig. 6. Areas of interest of a facial expression image as suggested by LIME.

If we carefully analyze the mislead facial expressions in the confusion matrix (see columns in Fig. 3), we can see that the facial expression *Afraid* is mostly confused (15%) with the *Sad* facial expression (sadness emotion), which is a plausible expected result according to Faigin [12], and with the *Neutral* facial expression (10%).

Interestingly, all the facial expressions related to the emotion of Joy (*Eager smile, Ingratiating smile, Laughter, False smile, False laughter1, Debauched smile, Melancholy smile* and *Smiling closed mouth*) are only confused with other facial expressions of the emotion of Joy. For instance, the *Debauched smile* is confused with the *Stifled smile* (25%), the *Melancholy smile* (15%) and the *Abashed smile* (5%) facial expressions.

The facial expression *Sad* (sadness emotion) is confused with *Neutral* (25%) and with *Afraid* (5%), both of which are also stated as probable confusions in Faigin studies. Finally, the facial expression of *Very frightened* (fear emotion) is confused with *Terror*, which is also a facial expression of fear emotion, and with *Surprise* (15%), another Faigin expected confusion.

Taking advantage of these very good classification results, we constructed a confusion matrix translating the facial expression classes into their associated emotion (see two first rows in Table 1). For example, the facial expressions classes *Enraged Compressed Lips, Enraged Shouting, Mad* and *Sternness Anger* were all translated into the emotion class *Anger*. After applying this 33-to-7 class translation to the predictions of the testing dataset, we obtained a global accuracy of 0.95.

Figure 7 depicts the confusion matrix detailing the accuracy obtained per emotion (six universal emotions plus the neutral emotion). With this approach, the worst classified emotion was *Neutral,* with an accuracy of 80%, followed by *Fear* emotion (85%); the remaining emotions achieved an accuracy above 90%, with a 100% of accuracy for the *Surprise* emotion.

This novel approach of emotion classification from facial expression classification, reaches an accuracy level of 0.95, which is as good as the reported results in the literature (usual machine learning models' recognition rates for emotions range from 70% to 95% [16]), but with a very simple neural network.

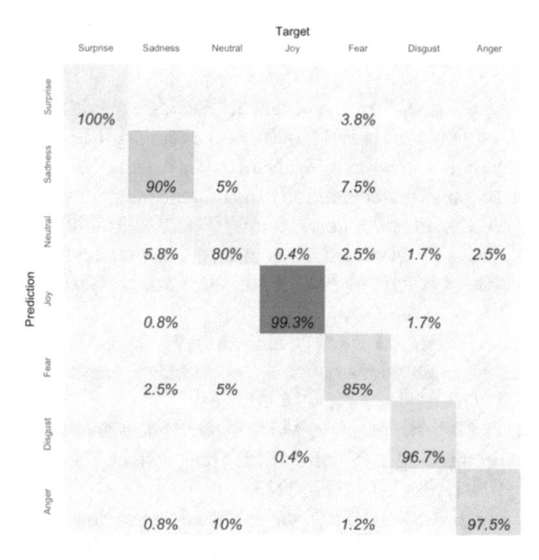

Fig. 7. Confusion matrix showing emotions' classification derived from the facial expression classification obtained with the CNN model.

4 Conclusions and Future Work

Facial expression and emotion recognition is a popular topic in artificial intelligence as it provides a huge amount of information about non-verbal communication.

Traditionally, researchers invested their efforts in creating models to classify within 7 classes corresponding to the six universal emotions plus the neutral one. The reason behind stands onto two pillars: (1) most of available datasets contain between 6 and 8 labeled expressions, and (2) it is often easier to classify into a smaller number of classes.

We trained and tested a simple CNN model onto an extension of the UIBVFED dataset that contains avatars performing 32 facial expressions. Therefore, our CNN model classifies within 33 classes. We obtained a global accuracy of 0.8, which is not only a very good result for a 33-class classifier, but, to the best of our knowledge, the classifier of facial expressions with the biggest number of classes.

Besides, due to the good results obtained, we used this model to classify within the six universal emotions, translating the facial expression predicted by the model into its associated emotion. This approach of classifying an emotion from a 32 facial expression classification, reached an accuracy level of 0.95, which is as good as the reported results in the literature with the added advantage of using a very simple neural network.

As future work, we plan to validate the proposed approach using datasets with real images instead of synthetic avatars. Therefore, it is necessary to allocate a dataset with real images that categorizes up to 32 facial expressions.

Acknowledgments. The authors acknowledge the Project EXPLainable Artificial INtelligence systems for health and well-beING (EXPLAINING) funded by PID2019-104829RA-I00 / MCIN/ AEI / https://doi.org/10.13039/501100011033. We also thank the University of the Balearic Islands, and the Department of Mathematics and Computer Science for their support.

References

1. Kaur, P., Krishan, K., Sharma, S.K., Kanchan, T.: Facial-recognition algorithms: a literature review. Med. Sci. Law **60**(2), 131–139 (2020). https://doi.org/10.1177/0025802419893168
2. Bisogni, C., Castiglione, A., Hossain, S., Narducci, F., Umer, S.: Impact of deep learning approaches on facial expression recognition in healthcare industries. IEEE Trans. Ind. Inform. **18**(8), 5619–5627 (2022). https://doi.org/10.1109/TII.2022.3141400
3. Sun, X., Zheng, S., Fu, H.: ROI-attention vectorized CNN model for static facial expression recognition. IEEE Access **8**, 7183–7194 (2020). https://doi.org/10.1109/ACCESS.2020.2964298
4. Khan, G., Samyan, S., Khan, M.U.G., Shahid, M., Wahla, S.Q.: A survey on analysis of human faces and facial expressions datasets. Int. J. Mach. Learn. Cybern. **11**(3), 553–571 (2020). https://doi.org/10.1007/s13042-019-00995-6
5. Mollahosseini, A., Hasani, B., Mahoor, M.H.: AffectNet: a database for facial expression, valence, and arousal computing in the wild. IEEE Trans. Affect. Comput. **10**(1), 18–31 (2019). https://doi.org/10.1109/TAFFC.2017.2740923
6. Oliver, M.M., Alcover, E.A.: UIBVFED: virtual facial expression dataset. PLoS ONE **15**(4), e0231266 (2020). https://doi.org/10.1371/journal.pone.0231266
7. Benitez-Quiroz, C.F., Srinivasan, R., Martinez, A.M.: EmotioNet: an accurate, real-time algorithm for the automatic annotation of a million facial expressions in the wild. In: 2016 IEEE Conference on Computer Vision and Pattern Recognition (CVPR), Jun. 2016, pp. 5562–5570 (2016). https://doi.org/10.1109/CVPR.2016.600
8. Ekman, P., Friesen, W.V.: Facial Action Coding System: A Technique for the Measurement of Facial Movement. Consulting Psychologists Press (1978)
9. Jain, N., Kumar, S., Kumar, A., Shamsolmoali, P., Zareapoor, M.: Hybrid deep neural networks for face emotion recognition. Pattern Recogn. Lett. **115**, 101–106 (2018). https://doi.org/10.1016/j.patrec.2018.04.010
10. Jain, D.K., Shamsolmoali, P., Sehdev, P.: Extended deep neural network for facial emotion recognition. Pattern Recogn. Lett. **120**, 69–74 (2019). https://doi.org/10.1016/j.patrec.2019.01.008
11. Liu, S., Tang, X., Wang, D.: Facial expression recognition based on sobel operator and improved CNN-SVM. In: 2020 IEEE 3rd International Conference on Information Communication and Signal Processing (ICICSP), September 2020, pp. 236–240 (2020). https://doi.org/10.1109/ICICSP50920.2020.9232063
12. Faigin, G.: The Artist's Complete Guide to Facial Expression. Watson-Guptill (2012)
13. 'Contempt: Paul Ekman Group. https://www.paulekman.com/universal-emotions/what-is-contempt/. Accessed 08 November 2022
14. Ribeiro, M.T., Singh, S., Guestrin, C.: Why should i trust you?": Explaining the predictions of any classifier. In: Proceedings of the 22nd ACM SIGKDD International Conference on Knowledge Discovery and Data Mining, New York, NY, USA, Aug. 2016, pp. 1135–1144 (2016). https://doi.org/10.1145/2939672.2939778
15. Carreto Picón, G., Roig-Maimó, M.F., Mascaró Oliver, M., Amengual Alcover, E., Mas-Sansó, R.: Do machines better understand synthetic facial expressions than people? In: Proceedings of the XXII International Conference on Human Computer Interaction, New York, NY, USA, pp. 1–5, September 2022. https://doi.org/10.1145/3549865.3549908
16. Ramis, S., Buades, J.M., Perales, F.J., Manresa-Yee, C.: A novel approach to cross dataset studies in facial expression recognition. Multimed. Tools Appl. **81**(27), 39507–39544 (2022). https://doi.org/10.1007/s11042-022-13117-2

Gender and Use of Digital Banking Products in Cameroon: Is There a Gap?

Jean Robert Kala Kamdjoug[1(✉)], Arielle Ornela Ndassi Teutio[2], and Diane Ngate Tchuente[2]

[1] ESSCA School of Management, 1 Rue Joseph Lakanal, 49003 Angers, France
jean-robert.kala-kamdjoug@essca.fr

[2] Faculty of Social Sciences and Management, GRIAGES, Catholic University of Central Africa, 11628 Yaoundé, Cameroon

Abstract. The development of new technologies is putting more pressure on banks to offer digital solutions to their customers. This issue is even more significant in developing countries where digital investments are growing. In this study, we examine whether the gender digital divide persists in emerging banking environments such as Cameroon, where the digitalization of services is becoming a focal point of marketing strategy. We analyze the gender gap in the influence of perceived ease of use, social influence, perceived security, and perceived trust on the use of digital banking products, as well as the influence of this use on bank notoriety. The partial least square structural equation modeling (PLS-SEM) method is used to test a sample of 347 bank customers, including 221 men and 126 women. The results show that the positive influence of perceived ease of use and social influence on the use of digital banking products is more vital for women than men. Further, the influence of perceived trust is more significant for men than women. There is no gender gap in the influence of perceived security, as well as the impact of the use of digital banking products on bank notoriety. The results also show that women are more open to adopting digital banking products ($R^2 =$ 49.5%) than men ($R^2 = 43\%$).

Keywords: Gender gap · digital transformation · banking products · developing countries

1 Introduction

The digital revolution in the banking sector is increasingly becoming the cornerstone of today's and tomorrow's financial world [1]. Affecting human behavior and the competitiveness criteria of banks implies increased digital reforms and numerous technological investments [2, 3]. In this context, banks in sub-Saharan Africa, primarily underdeveloped infrastructures, are not left out. Enormous upheavals linked to the success of fintech companies in the region and the Covid-19 crisis have affected many banks, which still rely on the physical bank branch for all their activities [4]. To respond to the challenges they face, African banks are accelerating their process of digitalization of services. They

A. Rocha et al. (Eds.): WorldCIST 2023, LNNS 800, pp. 635–644, 2024.
https://doi.org/10.1007/978-3-031-45645-9_61

are increasingly calling on companies specialized in digital design, artificial intelligence, and cyber security [5]. With this transformation, many banking customers are adopting digital solutions to perform their operations. Although the use of digital banking technologies is made possible by the increasing adoption of mobile phones and much more good internet access, the issue of the gender digital divide remains in this part of the world. Studies have shown that the popularity of digital banking products is much more significant among men [6, 7]. Men in developing countries have easier access to financial services than women, and are more likely to embrace new technologies [8].

On the other hand, women are less open to technology because they are more skeptical, have difficulty using it, have less access to essential banking services, and have low levels of financial and digital literacy [9]. However, when women adopt and use digital banking products on par with men, does this gender gap remain? This study aims to examine the gender difference in the use of digital banking products. Regarding gender, we are considering people who are biologically born (sex) male or female. The case of Cameroonian bank customers is relevant insofar as women's banking inclusion is still a problematic issue in an environment where banking structures are underdeveloped but open to digitalization. This study contributes to the state of the art by showing that banked women are more willing to adopt digital banking solutions, contrary to what the literature claims [10]. The paper is divided as follows: Sect. 2 presents the conceptual framework and the development of hypotheses. Section 3 describes the methodology used. Section 4 presents the results obtained, followed by discussions and implications in Sect. 5. Finally, limitations and future research orientations are presented in Sect. 6.

2 Conceptual Framework and Hypotheses Development

This study highlights four revealing concepts from the digital gender divide literature on the Use of digital products: Perceived Ease of Use [11] from the Technology Acceptance Model (TAM), Social Influence [12] from the Unified Theory of Acceptance, and Use of Technology (UTAUT), Perceived Security as an essential feature of digital technologies [13], and Perceived Trust as a cognitive factor of the user. We also examine the impact of the Use of digital banking products on Bank Notoriety. Based on the research model presented in Fig. 1, we hypothesize that there is a difference in perceptions related to the use of digital banking products between men and women regarding the causal relationships highlighted.

Perceived ease of use is the degree to which an individual believes using technology will require no effort [11]. Studies that investigate the effect of gender on technology acceptance show that men have a higher degree of perceived ease of use of technologies than women [14]. This is also true in developing countries where women are later adopters of technology. Hence, we hypothesize:

H1: There is a gender gap in the positive influence of perceived ease of use on the use of digital banking products.

Social influence represents the degree to which a person values others' opinions regarding using a technology [12]. This is essential in discerning social pressure's role in adoption behavior. Depending on the forms of communication, beliefs, and cultures, some studies show that women are more open to others' opinions than men, who act more independently and autonomously [15]. We hypothesize the following:

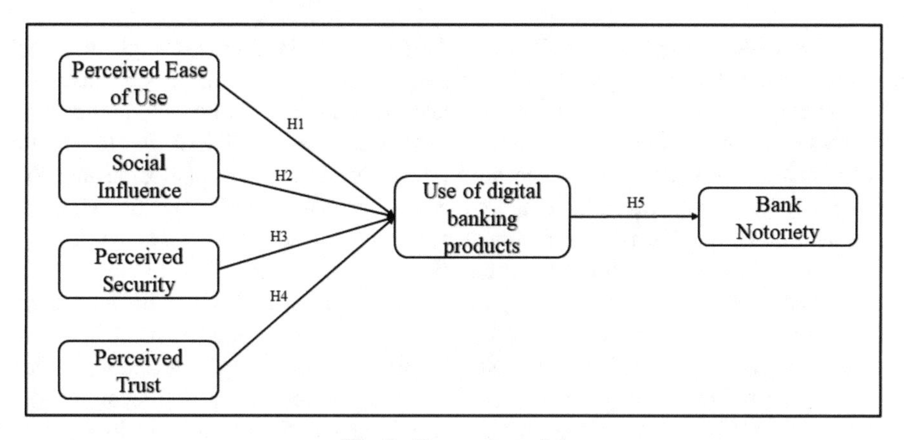

Fig. 1. Research model

H2: There is a gender gap in the positive influence of social influence on the use of digital banking products.

Perceived security here refers to the degree to which users perceive it is safe to disclose personal and financial information while interacting and transacting with the bank's digital channels [16]. Security is widely perceived as one of the main barriers to adopting technological innovations. Women seem more alert to protecting their personal information than men and are therefore more sensitive to security failures in a technology [17]. We hypothesize the following:

H3: There is a gender gap in the positive influence of perceived security on the use of digital banking products.

Perceived trust here refers to the expectation of efficiency and reliability of the technical devices and systems developed by the bank [18]. Given the gender gap, the cognitive processes between men and women evolve differently. Men build more trust after a positive evaluation of digital technology than women, who are more alert to the risks [19]. We, therefore, hypothesize the following:

H4: There is a gender gap in the positive influence of perceived trust on the use of digital banking products.

Notoriety represents in this study the level at which male and female users recognize or continuously remember their bank's brand, symbol, logo, and digital offerings [20]. Indeed, good notoriety guides customers in their choices and positions the company against the competition significantly when it increases the overall quality of the product in a banking environment. We, therefore, put forward the following hypothesis:

H5: There is a gender gap in the positive impact of the use of digital banking products on bank notoriety.

3 Methodology

The present study adopts a quantitative methodology based on partial least squares structural equation modeling (PLS-SEM). With a consistent literature review, we selected twenty-four (24) items to measure each of the six (06) constructs of the research model. A questionnaire was designed, and a 7-point Likert scale ranging from "strongly disagree" to "strongly agree" was adapted to all measurement items. A pre-test was conducted to ensure the questionnaire's comprehension, accuracy, and clarity. For this purpose, six (06) master's students in banking and finance were selected. Their task was to read the questionnaire thoroughly and to detect any ambiguities. During this phase, the questionnaire proved to be understandable and coherent. Subsequently, the questionnaire was illustrated on Google Forms and administered online via social networks (WhatsApp, LinkedIn, and Facebook) to easily reach our targets and allow them to answer it on their phone or computer. Our main targets have been Cameroonian banking customers. We informed them about the anonymity of the questionnaire and the use of the data for research purposes only. With the first fifty (50) responses, we conducted a pilot test to ensure the reliability of the measurement items. The results met the recommended threshold of 0.7 [21], from which we concluded the stability of the model constructs and continued the data collection. At the end of the data collection phase, which took place from October 2020 to January 2021, we obtained a total of three hundred and forty-seven (347) responses after a thorough data check. The data were analyzed using the SmartPLS 4 software [22]. The assessment of the PLS results was carried out in two phases: (1) the assessment of the measurement model and (2) the assessment of the structural model. Then, the multigroup analysis was performed.

4 Results

4.1 Demographic Profile

The group distribution of the demographic characteristics of our study sample is presented in Table 1. We have 221 men (Group 1) and 126 women (Group 2). For all groups, most respondents are between 20 and 30 years old and have experience with digital banking products ranging from 6 months to 1 year and 2–3 years.

Table 1. Groups repartition

Groups description		Man	Woman	Total
Gender	Number	221	126	347
Age	Less than 20	13	2	15
	20–30	111	72	183
	31–40	57	31	88
	41–50	26	18	44
	More than 50	14	3	17
	Total	**221**	**126**	**347**
Experience with digital banking products	Less than 6 months	37	14	51
	6 months-1 year	70	45	115
	2–3 years	63	40	103
	4–5 years	23	13	36
	More than 5 years	28	14	42
	Total	**221**	**126**	**347**

4.2 Research Model Assessment

First, we analyzed the measurement model, ensuring the reliability and validity of the constructs. The criteria to be examined are the outer loadings, which indicate the reliability of the reflective measurement items. Next, Cronbach's alpha, rho_A, and composite reliability (CR) assess the internal consistency reliability of the model. These criteria must have a value greater than or equal to 0.7 [21]. Then, the average variance extracted (AVE), which allows us to verify the convergent validity of each of the constructs, must also be examined and presented with a value greater than or equal to 0.50 [21]. Finally, the discriminant validity is reviewed through the heterotrait-monotrait (HTMT) correlation ratio, whose values must respect the threshold of 0.90 [21]. After running the PLS algorithm, the results showed that all the above criteria values respect the recommended threshold values. Hence, we concluded that the model is stable. Table 2 presents the constructs' reliability and validity measures. Table 3 shows the discriminant validity according to the HTMT ratio.

Table 2. Constructs' reliability and validity/R^2

Constructs	Cronbach's alpha	rho_A	CR	AVE	R^2 Overall sample	R^2 Man	R^2 Woman
BN	0.885	0.935	0.927	0.808	0.064	0.054	0.115
CT	0.902	0.905	0.939	0.836	-	-	-
PEOU	0.922	0.930	0.940	0.722	-	-	-
PS	0.853	0.869	0.901	0.694	-	-	-
SI	0.762	0.858	0.834	0.566	-	-	-
USB	0.876	0.888	0.914	0.726	0.429	0.430	0.497

BN = Bank Notoriety, CT = Customer Trust, PEOU = Perceived ease of use, PS = Perceived Security, SI = Social Influence, USB = Use of digital banking products

The second step in assessing the PLS results is the structural model assessment. We examined the constructs' inner VIFs to detect collinearity problems [21]. These VIFs all presented values less than 3, indicating that collinearity is not a problem in the data. Next, we examined the explanatory power of the model (R^2). The results show (Table 2) that USB has an R^2 value of 0.429 (42.9%), and BN has a weak R^2 value of 0.064 (6.4%) for the overall sample. Subsequently, the multigroup analysis was performed [23, 24]. First, we ensured that the sample size for each group (Group 1 = Man and Group 2 = Woman) was not less than the minimum sample size (n = 85 responses; obtained via the G*power 3.1.9.2 software [25]) required to test the model. We then generated the data groups and performed the measurement invariance (MICOM) test. By running the permutation test (1000 permutations and a significant level of 0.05), the results show that the compositional invariance (Step 2) and the composite equality (Step 3a) are correctly established [24]. But, in step 3b, only the permutation p-value of USB is not higher than 0.05 (p = 0.048). Because the original USB difference fell within the 95% confidence

Table 3. Discriminant validity- heterotrait-monotrait ratio (HTMT)

	BN	CT	PEOU	PS	SI	USB
BN						
CT	0.286					
PEOU	0.329	0.613				
PS	0.099	0.126	0.075			
SI	0.225	0.207	0.218	0.171		
USB	0.264	0.645	0.576	0.171	0.264	

BN = Bank Notoriety, CT = Customer Trust, PEOU = Perceived ease of use, PS = Perceived Security, SI = Social Influence, USB = Use of digital banking products

interval of the lower (2.5%) and upper (97.5%) boundaries, and the permutation p-value is close to 0.05, we then concluded that full measurement invariance was supported. Table 4 shows that the original correlation is greater than the 5% quantile. Table 5 shows that both mean and variance differences fall between the 2.5% and 97.5% boundaries.

Table 4. MICOM-Step 2

Constructs	Original correlation	Correlation permutation mean	5.0%	Permutation p value (>0.05)
BN	0.993	0.994	0.984	0.178
CT	1.000	1.000	0.999	0.955
PEOU	1.000	0.999	0.998	0.690
PS	0.989	0.945	0.767	0.689
SI	0.995	0.956	0.856	0.875
USB	0.999	0.999	0.998	0.495

BN = Bank Notoriety, CT = Customer Trust, PEOU = Perceived ease of use, PS = Perceived Security, SI = Social Influence, USB = Use of digital banking products

Finally, we ran the bootstrap multigroup analysis (MGA) to test the multigroup comparison [24]. The results in Table 6 show that differences between men and women exist in the H1 ($P < 0.1$), H2 ($P < 0.05$), and H4 ($P < 0.05$) relationships. On the other hand, they have no difference for the H3 ($p = 0.149$) and H5 ($p = 0.257$) relationships.

Table 5. MICOM-Step 3

Constructs	Original difference	Permutation mean difference	2.5%	97.5%	Permutation p value (>0.05)
Step 3a-Mean					
BN	0.036	0.003	−0.206	0.227	0.716
CT	−0.081	0.000	−0.240	0.220	0.467
PEOU	0.093	0.003	−0.220	0.227	0.414
PS	−0.083	−0.003	−0.219	0.234	0.454
SI	−0.024	−0.003	−0.219	0.214	0.818
USB	−0.096	0.003	−0.215	0.230	0.403
Step 3b-Variance					
BN	0.166	0.003	−0.211	0.209	0.113
CT	0.417	0.008	−0.422	0.453	0.060
PEOU	−0.030	0.009	−0.416	0.472	0.892
PS	0.077	0.003	−0.238	0.248	0.537
SI	−0.047	−0.000	−0.257	0.269	0.716
USB	0.378	0.005	−0.343	0.406	**0.048**

BN = Bank Notoriety, CT = Customer Trust, PEOU = Perceived ease of use, PS = Perceived Security, SI = Social Influence, USB = Use of digital banking products

Table 6. MGA analysis

Hypotheses		P values (Overall sample)	p value (Man)	Significance influence	p value (Woman)	Significance influence
H1	PEOU -> USB	0.000****	0.001****	Yes	0.000****	Yes
H2	SI -> USB	0.000****	0.206 n.s	No	0.000****	Yes
H3	PS -> USB	0.025**	0.401 n.s	No	0.005***	Yes
H4	CT -> USB	0.001****	0.000****	Yes	0.101 n.s	No
H5	USB -> BN	0.000****	0.001****	Yes	0.000 ****	Yes

Group difference (Man vs Woman)

Hypotheses		Difference (Man - Woman)	1-tailed (Man vs Woman) p-value	2-tailed (Man vs Woman) p-value	Sign. Level	Conclusions
H1	PEOU -> USB	−0.175	0.955	0.091	*	Accepted
H2	SI -> USB	−0.160	0.981	0.037	**	Accepted
H3	PS -> USB	0.149	0.075	0.149	n.s	Rejected
H4	CT -> USB	0.293	0.010	0.020	**	Accepted
H5	USB -> BN	−0.107	0.872	0.257	n.s	Rejected

**** $P < 0.001$; *** $P < 0.01$; ** $P < 0.05$; * $P < 0.1$; n.s. not significant; BN = Bank Notoriety, CT = Customer Trust, PEOU = Perceived ease of use, PS = Perceived Security, SI = Social Influence, USB = Use of digital banking products

5 Discussions and Implications

The results of this study reveal some interesting points about the gender digital divide in developing countries like Cameroon. The positive influence of PEOU on USB is more robust for women than men. This indicates that women, contrary to what is argued in the literature [10], are becoming more open to digital technologies and find them easy to use. However, as the significance of this hypothesis is low (*), we can also note that the perception of the ease of use of digital banking products is a factor that is almost equally approved between genders.

The results also show that women are more influenced by others' opinions than men. This corroborates with the literature, concluding that women are more attentive to social cues than men [26]. Also, it is worth noting the role played by cultural factors. In sub-Saharan African countries such as Cameroon, women generally evolve or gather in groups or associations and hold, for the most part, the roles of guides and advisors within the groups and communities [27]. As a result, through their leadership, women are more likely to listen to other people's opinions and make this an essential criterion in their decision-making.

On the other hand, the results show no perceived difference between men and women regarding the positive influence of security on the use of digital banking products. However, concerning the positive influence of perceived trust, the results show that men are more trusting than women. Thus, women are sensitive to the security provided by digital banking products. Still, they do not cognitively dismiss the possibility of any risk or loss. As a result, they do not readily trust.

Finally, the results show that gender difference is not a significant criterion for the bank's notoriety regarding the use of digital banking products. Moreover, the explanatory power of USB on BN is very low ($R^2 = 6.4\%$). Nevertheless, it is essential to underline that women are more receptive to using digital banking products than men. Among women, USB is explained at 49.7%, while among men at 43%. Similarly, BN is described at 11.5% for women and 5.4% for men.

In terms of implications, this study contributes to the literature on gender in the use of digital technologies. Examining the case of Cameroon contributes to a better understanding of the behavioral evolution of banked individuals in Sub-Saharan Africa towards digital banking technologies. This study also shows that banked women are becoming more open to digital technologies. This aspect can be used by banks to include more unbanked women and use their social openness to encourage the adoption of digital channels. It is also useful for banks to improve the personalization of their digital products and increase communication about risk prevention measures and the reliability of the devices that support the technology to reassure female customers.

6 Limitations and Future Research Orientations

However, our study reveals some limitations. We didn't measure the respondents' digital literacy, even though they have a good experience with digital banking products. We encourage future research to analyze the digital literacy as a moderator or mediator factor to highlight the gender gap. We only considered bank customers located in urban areas. Future research can study the behaviors of rural banked individuals to conduct comparative analyses based on gender and geographic location. We also suggest future research on the users' technological readiness to understand the gender digital divide. In developing countries where digital transformation is a key growth driver, understanding the extent to which the propensity to want to use digital banking technologies continues to evolve across genders and generations in sub-Saharan Africa would be useful.

References

1. Gabor, D., Brooks, S.: The digital revolution in financial inclusion: international development in the fintech era. New Polit. Econ. **22**(4), 423–436 (2017). https://doi.org/10.1080/13563467.2017.1259298
2. Malar, D.A., Arvidsson, V., Holmstrom, J.: Digital transformation in banking: exploring value co-creation in online banking services in India. J. Glob. Inf. Technol. Manag. **22**(1), 7–24 (2019). https://doi.org/10.1080/1097198X.2019.1567216
3. Ndassi Teutio, A.O., Kala Kamdjoug, J.R., Gueyie, J.-P.: Mobile money, bank deposit and perceived financial inclusion in Cameroon. J. Small Bus. Entrepreneursh. **35**, 1–19 (2021). https://doi.org/10.1080/08276331.2021.1953908
4. Litt, W.H.: COVID-19, the accelerated adoption of digital technologies, and the changing landscape of branch banking. Choices **36**(3), 1–4 (2021)
5. Myovella, G., Karacuka, M., Haucap, J.: Digitalization and economic growth: a comparative analysis of Sub-Saharan Africa and OECD economies. Telecommun. Policy **44**(2), 101856 (2020). https://doi.org/10.1016/j.telpol.2019.101856
6. Dzogbenuku, R.K., Amoako, G.K., Kumi, D.K., Bonsu, G.A.: Digital payments and financial wellbeing of the rural poor: the moderating role of age and gender. J. Int. Consum. Mark. **34**(2), 113–136 (2022). https://doi.org/10.1080/08961530.2021.1917468
7. Kulkarni, L., Ghosh, A.: Gender disparity in the digitalization of financial services: challenges and promises for women's financial inclusion in India. Gend. Technol. Dev. **25**(2), 233–250 (2021). https://doi.org/10.1080/09718524.2021.1911022
8. Antonio, A., Tuffley, D.: The gender digital divide in developing countries. Future Internet **6**(4), 673–687 (2014). https://doi.org/10.3390/fi6040673
9. Kusimba, S.: "It is easy for women to ask!": gender and digital finance in Kenya. Econ. Anthropol. **5**(2), 247–260 (2018). https://doi.org/10.1002/sea2.12121
10. Alozie, N.O., Akpan-Obong, P.: The digital gender divide: confronting obstacles to women's development in Africa. Develop. Policy Rev. **35**(2), 137–160 (2017). https://doi.org/10.1111/dpr.12204
11. Davis, F.D.: Perceived usefulness, perceived ease of use, and user acceptance of information technology. MIS Q. **13**(3), 319–340 (1989). https://doi.org/10.2307/249008
12. Venkatesh, M., Davis, D.: User acceptance of information technology: toward a unified view. MIS Q. **27**(3), 425 (2003). https://doi.org/10.2307/30036540
13. Lim, S.H., Kim, D.J., Hur, Y., Park, K.: An empirical study of the impacts of perceived security and knowledge on continuous intention to use mobile fintech payment services. Int. J. Human Comput. Interact. **35**(10), 886–898 (2019). https://doi.org/10.1080/10447318.2018.1507132
14. JoséLiébana-Cabanillas, F., Sánchez-Fernández, J., Muñoz-Leiva, F.: Role of gender on acceptance of mobile payment. Ind. Manag. Data Syst. **114**(2), 220–240 (2014). https://doi.org/10.1108/IMDS-03-2013-0137
15. Abima, B., Engotoit, B., Kituyi, G.M., Kyeyune, R., Koyola, M.: Relevant local content, social influence, digital literacy, and attitude toward the use of digital technologies by women in Uganda. Gend. Technol. Dev. **25**(1), 87–111 (2021). https://doi.org/10.1080/09718524.2020.1830337

16. Balapour, A., Nikkhah, H.R., Sabherwal, R.: Mobile application security: role of perceived privacy as the predictor of security perceptions. Int. J. Inf. Manage. **52**, 102063 (2020). https://doi.org/10.1016/j.ijinfomgt.2019.102063
17. Michota, A.: Digital security concerns and threats facing women entrepreneurs. J. Innov. Entrepreneursh. **2**(1), 7 (2013). https://doi.org/10.1186/2192-5372-2-7
18. Oliveira, T., Faria, M., Thomas, M.A., Popovič, A.: Extending the understanding of mobile banking adoption: When UTAUT meets TTF and ITM. Int. J. Inf. Manage. **34**(5), 689–703 (2014)
19. Shao, Z., Zhang, L., Li, X., Guo, Y.: Antecedents of trust and continuance intention in mobile payment platforms: the moderating effect of gender. Electron. Commer. Res. Appl. **33**, 100823 (2019). https://doi.org/10.1016/j.elerap.2018.100823
20. Dell'Atti, S., Trotta, A., Iannuzzi, A.P., Demaria, F.: Corporate social responsibility engagement as a determinant of bank reputation: an empirical analysis. Corp. Soc. Responsib. Environ. Manage. **24**(6), 589–605 (2017). https://doi.org/10.1002/csr.1430
21. Hair, J.F., Risher, J.J., Sarstedt, M., Ringle, C.M.: When to use and how to report the results of PLS-SEM. Eur. Bus. Rev. **31**(1), 2–24 (2019). https://doi.org/10.1108/EBR-11-2018-0203
22. Ringle, C., Wende, M.S., Becker, J.-M.: SmartPLS 4. SmartPLS (2022)
23. Matthews, L.: Applying multigroup analysis in PLS-SEM: a step-by-step process. In: Latan, H., Noonan, R. (eds.) Partial Least Squares Path Modeling, pp. 219–243. Springer International Publishing, Cham (2017). https://doi.org/10.1007/978-3-319-64069-3_10
24. Cheah, J.-H., Thurasamy, R., Memon, M.A., Chuah, F., Ting, H.: Multigroup analysis using smartpls: step-by-step guidelines for business research. Asian J. Bus. Res. **10**(3), 1–19 (2020)
25. Schoemann, A.M., Boulton, A.J., Short, S.D.: Determining power and sample size for simple and complex mediation models. Soc. Psychol. Personal. Sci. **8**(4), 379–386 (2017). https://doi.org/10.1177/1948550617715068
26. Venkatesh, V., Morris, M.G.: Why don't men ever stop to ask for directions? Gender, social influence, and their role in technology acceptance and usage behavior. MIS Q. **24**(1), 115–139 (2000). https://doi.org/10.2307/3250981
27. Titi Amayah, A., Haque, M.D.: Experiences and challenges of women leaders in Sub-Saharan Africa. Africa J. Manage. **3**(1), 99–127 (2017). https://doi.org/10.1080/23322373.2017.1278663

Parallel Corpus of Somatic Phrasemes

Katerina Zdravkova(✉) and Jana Serafimovska

Faculty of Computer Science and Engineering, Skopje, North Macedonia
katerina.zdravkova@finki.ukim.mk,
jana.serafimovska@students.finki.ukim.mk

Abstract. Multilingual dictionary of Slavic somatic phrasemes was frequently used by the linguists from the Faculty of Philology in Skopje. Outdated versions of the development tools, frequent migrations and the lack of support were the main reason to withdraw it. Due to its importance for the researchers, who regularly surveyed and enlarged the existing repository of these phrasemes, it was crucial to either reengineer or to develop it from scratch. New technologies, application programming interfaces and architectural styles that enable smooth interaction were the main motivation to reinforce the second alternative. This paper introduces the new parallel corpus of somatic phrasemes, which includes all the functionalities of the previous site enabling further extension to new languages, additional multiword expressions and their meanings. After presenting several similar projects, the paper gives an overview of the major system components, the platform, the programming language and technologies, and finally the distributed server for data search. The portal is illustrated with the most important components of its interface. The paper concludes with the impact of the portal for the linguists worldwide and its further extensions.

Keywords: Multiword expressions · Parallel corpora · Somatic phrasemes

1 Introduction

According to Goddard, phrasemes are compounds or derived words, represented by a finite set of identifiable expressions [1]. They have a distinctive syntax and semantics [2]. These recognizable multiword expressions (MWEs) are language specific and sometimes well known to native speakers only. Therefore, their good translation depends on the translator's phraseological competence of the source language [3].

Most of the traditional myths of machine translation, which point toward phrases, quotes and sentences such as: "invisible idiot", and "the whisky is strong, but the meat is rotten", as a back translation of "out of sight, out of mind" and "the spirit is willing, but the flesh is weak" [4] have already been more or less successfully avoided, mainly by implementing the phrase-based statistical machine translation [5]. However, there are still far too many multiword expressions that need a special attention to avoid the "unexpected lexical, syntactic, semantic, pragmatic and/or statistical properties" during translation [6]. The paper by Krstev and Savary [6] presents 70 such examples collected as part of the games within the COST action PARSEME [7].

A. Rocha et al. (Eds.): WorldCIST 2023, LNNS 800, pp. 645–654, 2024.
https://doi.org/10.1007/978-3-031-45645-9_62

The phrasemes that include some body parts of internal organs are called somatic phrasemes, or body idioms [8]. For 20 years, the researchers from the Faculty of Philology in Skopje have systematically studied their occurrence, meaning, examples and behavior during translation [2]. Starting from the initial corpus of 560 Macedonian somatic phrasemes created by Veljanovska [9], the corpus was extended to Czech, French, Polish, and Russian. Each phraseme is associated with a brief explanation. Whenever a translation in another language exists, it is coupled with the original one.

In 2016, the first multilingual dictionary of Slavic somatic phrasemes was launched at the Faculty of Computer Science and Engineering [2]. It was developed using the content management system Drupal [10]. The search engine was supported by Apache Solr Server [11]. The application was hosted on a virtual machine configured by Acquia Dev Desktop [12]. Frequent server migrations and the lack of support for several components were the main reasons to withdraw the site. This action prevented the researchers who put a lot of effort into its creation from continuing their research and extending the corpus. In order to support their activities, there were two alternatives: to reengineer it or to develop it from scratch using the existing databases. New development technologies, application programming interfaces and architectural styles that enable smooth interaction were the main motivation to force the second alternative.

The paper continues with the second section, which introduces several similar projects, comparing them with the old dictionary and the newly developed portal. Third section gives a detailed overview of the major system components, the platform, the programming language and technologies, and finally the distributed server for data search. Fourth section presents the main components of user interface. The paper concludes with further extensions and impact of the portal for the linguists worldwide.

2 Parallel Corpora of Multiword Expressions

Among computational linguistics, phrasemes are usually called multiword expressions (MWEs) or idiomatic phrases [1]. Therefore, search of existing parallel corpora with similar functionalities was mainly concerned with the multilingual corpora of MWEs and idioms.

The first ambitious initiative of collecting and researching MWEs was COST action IC1207 PARSEME (PARSing and Multi-word Expressions) [13]. As a result of the action, a corpus of 62 thousand annotated verbal MWEs in 18 languages has been created [14]. Another successful project exploring Chinese and German translations to English resulted in a collection of around 150000 MWE pairs [15]. To investigate MWEs in argumentative writing, Nam and Park explored the collected essays in three repositories: the South Korean UNIARG, and two American: LOCNESS-ARG, and MICUSP-ARG [16]. In total, they represent around 5000 MWE Korean and English tokens, categorized in 638 different types. While the first collection predominantly contains verbal MWEs, nominal and prepositional MWEs are predominant in the last two essays written by nonnative speakers. For a collection of 90 different languages Nivre et al. [17] developed a cross-linguistically consistent treebank annotation within Universal Dependencies v2 that include certain types a MWEs.

None of the abovementioned projects presents the whole collection of MWEs. On a contrary, many collections of idiomatic phrases are open and searchable, such as *The large idioms dictionary* [18], which consists of 154 pages of English idioms divided into around 180 different topics. The collection of more than 700 German idioms translates them literally (with the same words) or properly (with the exact meaning) into English [19]. For example, literal translation of German idiom "Alles hat ein ende, nur die wurst hat zvei" is "Everything has an end. Only a sausage has two." while the proper translation is: "All good things must come to an end". A similar collection of French idioms also exists [20]. Only a few are translated into English [21] in the same manner like in the German portal, showing the literal and the real meaning. For example, literal translation of the idiom: "Être blanc comme neige" is "To be as white as snow", while the real meaning is "To be completely innocent".

There are several multilingual collections of idiomatic phrases and MWEs. They were the main inspiration for the creation of the parallel corpus of somatic phrasemes. The following subsection introduces and compares their functionalities.

2.1 Comparison of Parallel Idioms and MWEs Corpora

The multilingual library signage (MLS) has a collection of the common library phrases that are translated into 55 different languages [22]. Some of the phrases consist of a single word only, such as: address, book, closed, exit, fiction, and all the languages the phrases can be translated to. They are alphabetically ordered to bypass search. No explanation or meaning for any of them is provided.

A multilingual linked idioms dataset (LIdioms) consists of around 800 idioms in five different languages: English, German, Italian, Portuguese, and Russian [23]. It enables searching through the set of idioms with the same meaning and translations across the stored languages. The online repository seems to be abandoned [24].

One of the largest collections of English idioms is available from Cambridge Dictionary (CD) [25]. For each MWE or idiom, the dictionary provides its meaning, associated with a thesaurus with synonyms, antonyms, and examples. For example, the idiom "to be on cloud nine" is explained as "to be extremely happy and excited", illustrated with an example: "She was on cloud nine!". The thesaurus unites the words and phrases with a common meaning "feeling or showing pleasure". For each word or phrase, CD presents the corresponding translation into Arabic, Catalan, Chinese, Czech, Danish, Dutch, Filipino, French, German, Hindi, Indonesian, Italian, Japanese, Korean, Malay, Norwegian, Portuguese, Russian, Spanish, Thai, Turkish, Ukrainian and Vietnamese. Some of the translations are literal ("soyez sur Cloud Nine", the French translation of "be on cloud nine"), while others are equivalent idioms ("cases sa pipe", the French translation of "kick the bucket").

Collins English Dictionary (CED) has similar functionalities as CD [26]. Although the list of target languages is limited to Chinese, French, German, Italian, Hindi, Japanese, Korean, Spanish and Portuguese, this dictionary enables multilingual search. For example, the French translation of the idiom "to be on cloud nine" is "être aux anges". Its opposite translation is "to be over the moon". Moreover, CED automatically selects many examples from the Collins Corpus.

The withdrawn dictionary of Slavic somatic phrasemes (DSSP) had many functionalities existing in the abovementioned repositories. Table 1 compares them all. Although the richest with various languages, multilingual library signage (MLS) offers no searching, no explanation, and examples. Moreover, it has no embedded corpus. All the other dictionaries are searchable in the pivot language only. MLS and LIdioms offer no explanations of the idiomatic phrases and no examples. They do not have an embedded corpus too. It seems that the explanations in CD are in English only. CED and DSSP are multilingual, similarly as the embedded corpora. The only inflected language in this sample is Macedonian. Although computationally demanding, DSSP enabled searching of the inflectional forms of all the words existing in the corpus. Inflectional search is also enabled by CD and CED, later supporting multilingual search of different word forms. This search is less challenging, because the languages are slightly inflected.

Table 1. Comparison of the five repositories of phrasemes.

Repositories and features	MLS	LIdioms	CD	CED	DSSP
Pivot language	English	English	English	Multilingual	Macedonian
Languages	55	5	24	10	5
Searchable	No	Yes	Yes	Yes	Yes
Inflectional search	No	No	English	Multilingual	Macedonian
Explanation	No	No	English	Multilingual	Multilingual
Examples	No	No	English	Multilingual	Multilingual
Embedded corpus	No	No	Yes	Yes	Yes

2.2 Building of the Multilingual Corpus of Somatic Phrasemes

Automatic building of multilingual MWEs corpora is a rather complex task. It has been thoroughly examined by Han and Smeaton [14]. Their approach depends on machine learning techniques for disambiguation of MWEs meaning. They first try to efficiently recognize continuous and discontinuous expressions, i.e. those MWEs where the group of words have no gaps versus those where some words are inserted. Filtering of those candidate MWEs with no translation was manually examined [14].

The multilingual corpus of somatic phrasemes that was the bases of the portal was manually created by the researchers from the Faculty of Philology in Skopje, predominantly from the Chamber of Slavic languages [2, 9]. It is intended to be additionally extended to all the languages that are thought the Faculty.

The collection process started with thousands of somatic phrasemes in Macedonian. Each phraseme has a title and an associated meaning. All phrasemes are aligned with translation equivalent phrasemes in at least one target language. Moreover, they are accompanied with examples from Macedonian literary works. Those phrasemes that have similar meaning are clustered.

The creation and the illustration of the portal will be explained in more detail in the next two sections of the paper.

3 Functionalities of the Portal of Somatic Phrasemes

Each phraseme in the portal is represented with the following information (see Fig. 1):

1. Title of the phraseme in Macedonian ("Го зеде на заб", which can be translated with the English idiom "Take in stride")
2. Meaning in Macedonian ("Некој некого прогонува бидејќи му смета некоја негова постапка или размислување" = "To persecute someone because some actions or thoughts bother you")
3. Example from the Macedonian corpus ("Ќе земе на заб, та којзнае дали не ќе ми попречи во завршните испити, дали не ќе повлијае кај оној мојон…" = "It's going to take him in stride, so who knows if it won't interfere with my final exams, if it won't affect my guy…")
4. Source of the Macedonian example (Столе Попов, Доктор Орешковски, стр 123 = Stole Popov, Doctor Oreshkovski, page 123)

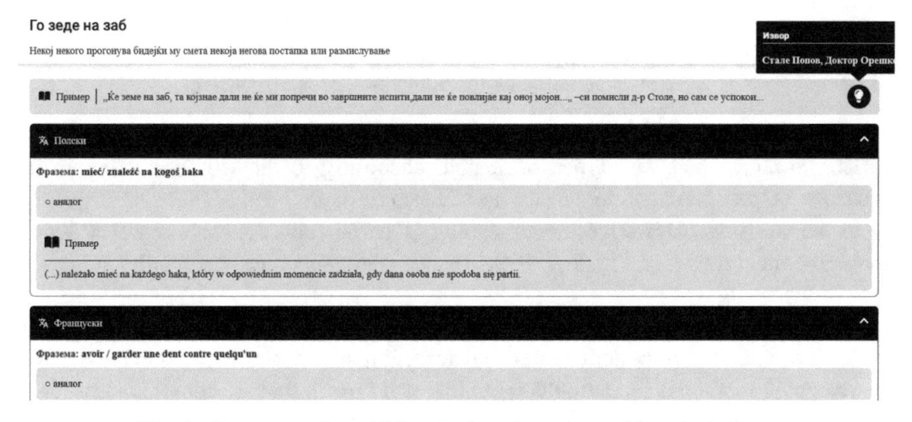

Fig. 1. Representation of Macedonian phraseme and its translations.

5. Translations with an example from the corpus, if an example exists:
   a. Fully equivalent target phraseme, which exists in Polish: "mieć/znaleźć na kogoś haka" = "have/find a hook on someone") with an example from the Polish corpus: "Należało mieć na każdego haka, który w odpowiednim momencie zadziała, gdy dana osoba nie spodoba się partii" = "You had to have every hook that would work at the right moment when a person didn't like the party")
   b. Partially equivalent target phraseme, which exists in Czech: "mít někoho na mušce; mít na někoho pifku" = "to target someone; have a crush on someone") with an example from the Czech corpus: "Ale ona na mě měla pifku za ten deník, za všechno, co v něm bylo napsaný, kouření cigaret,že mě stejně vylila, u komise řekla, že prej jsem opisovala, že sama bych to prej nikdy nevěděla" = "But she had a crush on me for that diary, for everything that was written in it, smoking cigarettes, that she dumped me anyway, she said at the commission that I had copied it before, that I would never have known it myself..".

   c. Analog translation, which occurs in French: "avoir/garder une dent contre quelqu'un" = "to have/keep a grudge against someone"), but does not have a corresponding example in the French corpus.
6. List of similar phrasemes (see Fig. 2)

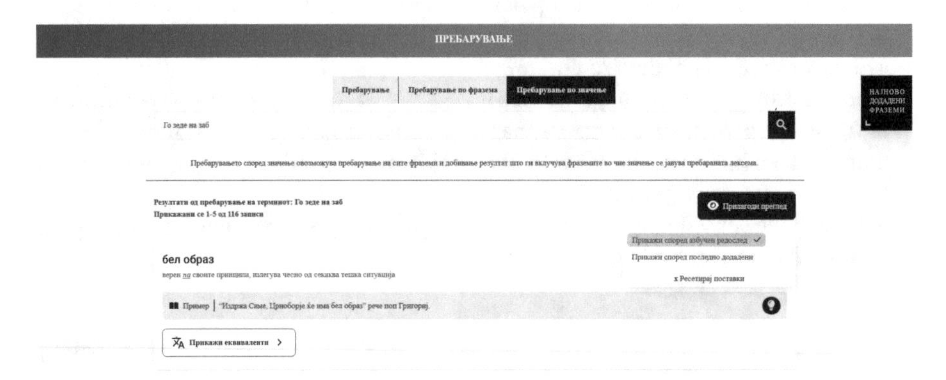

Fig. 2. Results of searching for similar Macedonian phrasemes.

The portal enables two types of search: according to the words existing in the phrasemes (Пребарување по фразема = Search through the phrasemes), and according to the words that describe their meaning (Пребарување по значење = Search though the meanings (see Fig. 2.)). If the search is not specified, the results unite both types of searching.

So far, searching is enabled in the pivot language only. It can be done using the whole phraseme, its component words, and according to any of the words that exist in it. The searching method is inflectional, including word forms that are completely different from the lemma, such as: луѓе (people) and човек (man) (see Fig. 3.).

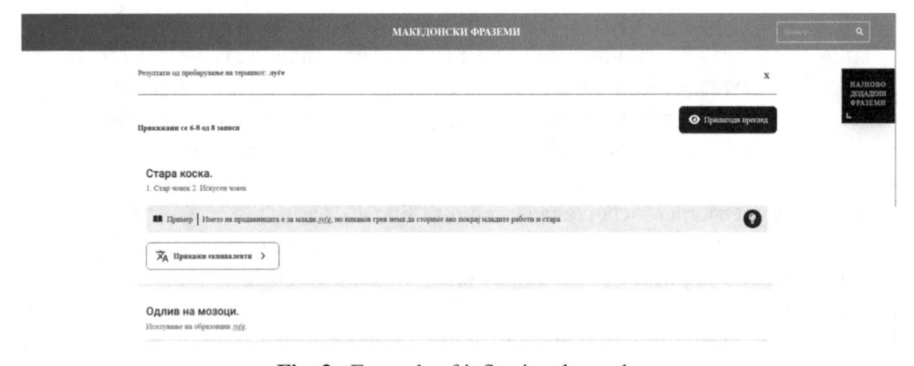

Fig. 3. Example of inflectional search.

4 Technical Details of the Portal

The technical implementation of the portal can be described as a well working machinery consisting of the three main components:

1. Web application through which the user can access the portal,
2. RESTFul API component serving as an endpoint for data retrieval,
3. Distributed server for data indexing and search

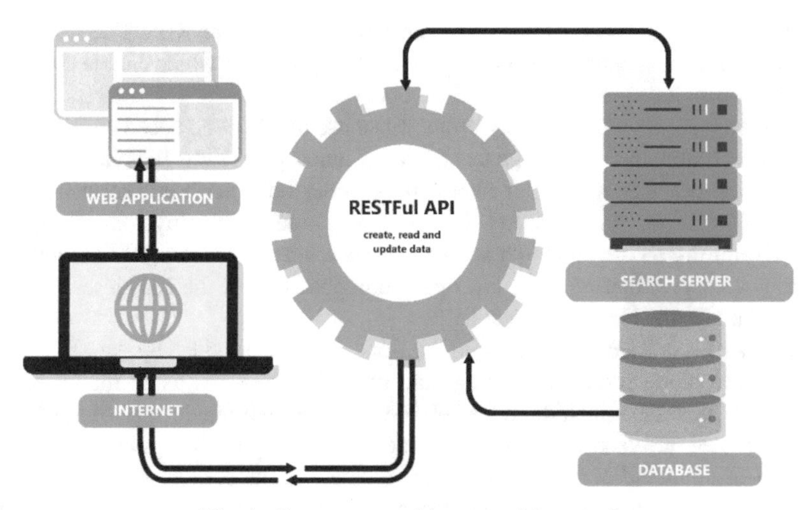

Fig. 4. Component architecture of the portal

These components and the existing database communicate as follows (see Fig. 4.):

- Step 1: The user attempts to access the portal and search for a given phraseme.
- Step 2: The web application acknowledges user's request and communicates with the RESTFul API component to announce user's request for search.
- Step 3: The RESTFul API component parses the request received from the web application and sends another request to the search server to get the actual data from the phrasemes.
- Step 4: The search server performs the relevant search operation on the already indexed phrasemes, and sends back a response to the RESTFul API component, which contains the requested phrasemes.
- Step 5: The RESTFul API component receives the search results from the search server and sends a direct response to the web application containing this result.
- Step 6: The web application displays the results to the user.

4.1 Technologies Used to Develop the Portal

Following the latest trends and aiming for smooth and fast user interaction, a few technologies and frameworks were used to successfully implement the technical solution for the portal. Giving the solution a top-down approach, the web application was fully implemented using the Angular framework, which is the third leading technology used for the development of web applications in 2022 [27]. Angular is a TypeScript based platform that incorporates concepts such as models, services, components and themes, and therefore enables development of dynamic single page web applications [28]. Applying the advantages that Angular provides, the simple and elegant user interface of this portal was completely achieved using the theming provided by the TailwindCSS styling framework and the DaisyUI theme library.

Moving one step down, the RESTFul API component was implemented using the Spring Boot open source, a microservice-based Java framework [29]. Benefiting from the large number of dependencies that this framework supports, the RESTFul API component was built implementing the layered architecture principle, which enables separation of responsibility between the defined component layers.

The search server relies entirely on the Elasticsearch Lucene based engine for data indexing and search [30]. Given the fact that as of January 2021, the Elasticsearch engine is no longer under the open-source license of Apache, the OpenSearch service provided by Amazon was used for the successful implementation and deployment of the search server. Exploiting the OpenSearch service we were able to successfully implement an Elasticsearch server that runs the last open-source version of the Elasticsearch engine that is still under the license of Apache.

This portal is currently hosted using the cloud services provided by Amazon, as well as the Heroku cloud service providing platforms [31]. Namely, using the "Free Tier" offered by Amazon that includes a freemium version for a limited usage of the Amazon web services for one year, we were able to successfully deploy the search server and the web application included in the software architecture of the portal.

The RESTFul API was deployed using the free services provided by Heroku and the underlying Heroku dyno containers that run the hosted application's processes. These services are currently free, but as of November 28th this year, Heroku will only offer a paid version of its services [32].

5 Conclusions and Further Work

The parallel multilingual portal that was introduced in this paper presents the repository of 560 Macedonian somatic phrasemes, their meanings, examples and three types of translations: full, partial and analog equivalent. Almost all Macedonian phrasemes and the majority of their translations are illustrated with examples from the embedded multilingual corpus. The portal enables two types of search, according to group of words in the phrasemes and according to their meaning. If the search criterion is not specified, the results unite both, the phrasemes that contain some of the keywords in the title and the phrasemes that contain the keywords in their meaning. The search is not limited to lemmas.

The portal is currently hosted on Amazon Simple Storage Service (Amazon S3) cloud [33]. During the following weeks, the linguists from the Faculty of Philology will thoroughly test all the functionalities in all the five languages: Czech, French, Macedonian, Polish and Russian. The additions will be made by the portal administrator within the existing database, to minimize the risk of compromising the stored data.

The extension of the portal will need a substantial work, predominantly by the linguists. They will enlarge the repository of Macedonian somatic phrasemes, and fill in the missing explanations and examples. Reorganization of the clusters of phrasemes according to their meaning will be polished, because currently, the clusters are too broad. The clusters of phrasemes with opposite meanings are also intended. All these activities are very ambitious and time consuming.

After completing the repository in the pivot language and the embedded corpus, the linguists from the departments of Slavic and Romance languages will have to complete their collections too. Step forward would be the inclusion of the researchers from the departments of Albanian, English, German, Italian and Turkish.

The technical extension of the portal will be directed towards the inflectional search of all the languages. There are several open corpora within CLARIN (Common Language Resources and Technology Infrastructure) that will significantly facilitate this task [34]. Once completed, the parallel corpus of somatic phrasemes will become a fruitful tool for all the linguists interested in researching phraseology, multiword expressions and idioms. By presenting the richness and peculiarities of the embedded languages, this parallel portal will enable various multilingual research activities, embracing quality machine translation, intelligent text processing, question answering, as well as creation of multilingual assistive technologies for people with communication disabilities. Creators of the portal cordially invite all the linguists worldwide to contribute to its massive multilingualism.

Acknowledgement. This work was supported in part by grants from the Faculty of Computer Science and Engineering, Ss. Cyril and Methodius University in Skopje.

References

1. Goddard, C.: Lexico-semantic universals: a critical overview. Linguist. Typol. **5**(1), 1–65 (2001)
2. Bekjkovikj, M., Markova E., Zdravkova, K.: Multilingual dictionary of Slavic somatic phrasemes. In: Proceedings of the 13th International Conference for Informatics and Information Technology, pp. 88–92 (2016)
3. Hallsteinsdóttir, E.: Phraseological competence and the translation of phrasemes. In: Multi-Lingual Phraseography: Second Language Learning and Translation Applications, 279, Schneider Verlag Hohengehren (2011)
4. Hutchins, J.: «The Whisky Was Invisible» or Persistent Myths of MT. MT News Int. **11**, 17–18 (1995)
5. Lample, G., et al.: Phrase-Based & Neural Unsupervised Machine Translation. EMNLP (2018)
6. Krstev, C., Savary, A.: Games on multiword expressions for community building, INFOtheca. J. Inf. Libr. Sci. **17**(2), 1–19 (2017)
7. PARSEME Homepage. https://typo.uni-konstanz.de/parseme/
8. Němcová, M.: Comparative analysis of English and French body idioms, LAP Lambert, Brno (2013)
9. Veljanovska, K.: Phraseological expressions in the Macedonian language: with reference to somatic phraseology, Makedonska riznica (2006). (in Macedonian)

10. Drupal Homepage. https://www.drupal.org/. Accessed 04 Jan 2023
11. Solr Homepage. https://solr.apache.org/. Accessed 04 Jan 2023
12. Dev Desktop Homepage. https://docs.acquia.com/resource/archive/dev-desktop/. Accessed 04 Jan 2023
13. PARSEME Homepage. https://typo.uni-konstanz.de/parseme/. Accessed 04 Jan 2023
14. Savary, A., et al.: PARSEME multilingual corpus of verbal multiword expressions. In: Multiword Expressions at Length and in Depth: Extended Papers from the MWE 2017 Workshop (2018)
15. Han, L., Jones, G., Smeaton, A.: MultiMWE: building a multi-lingual multi-word expression (MWE) parallel corpora. In: Proceedings of the 12th Language Resources and Evaluation Conference, pp. 2970–2979 (2020)
16. Nam, D., Park, K.: I will write about: Investigating multiword expressions in prospective students' argumentative writing. PLoS ONE **15**(12), e0242843 (2020)
17. Nivre, J., et al.: Universal Dependencies v2: An evergrowing multilingual treebank collection. (2020)
18. The idioms Homepage. https://www.theidioms.com/. Accessed 04 Jan 2023
19. Materhorn Languages Homepage. https://matterhornlanguages.com/german-idioms/. Accessed 04 Jan 2023
20. Expressio Homepage. https://www.expressio.fr/index.php. Accessed 04 Jan 2023
21. The French Experiment Homepage. https://www.thefrenchexperiment.com/learn-french/expressions. Accessed 04 Jan 2023
22. Multilingual library signage Homepage. https://multilingual-library-signage.sl.nsw.gov.au/. Accessed 04 Jan 2023
23. Moussallem, D., Sherif, M.A., Esteves, D., Zampieri, M., Ngomo, A.C.N.: LIdioms: a multilingual linked idioms data set. (2018)
24. LIdioms Homepage. https://github.com/dice-group/LIdioms. Accessed 04 Jan 2023
25. Cambridge Dictionary Homepage. https://dictionary.cambridge.org/dictionary/english/. Accessed 04 Jan 2023
26. Collins English Dictionary Homepage. https://www.collinsdictionary.com/dictionary/english. Accessed 04 Jan 2023
27. devjobsscaner Homepage. https://www.devjobsscanner.com/blog/the-most-demanded-frontend-frameworks-in-2022/
28. Angular Homepage. https://angular.io/. Accessed 04 Jan 2023
29. Spring Boot Homepage. https://spring.io/projects/spring-boot. Accessed 04 Jan 2023
30. Elastic Homepage. https://www.elastic.co/. Accessed 04 Jan 2023
31. HEROKU Homepage. https://www.heroku.com/. Accessed 04 Jan 2023
32. HEROKU Blog. https://blog.heroku.com/new-low-cost-plans. Accessed 04 Jan 2023
33. Amazon S3 Homepage. https://aws.amazon.com/s3/. Accessed 04 Jan 2023
34. CLARIN Homepage. https://www.clarin.eu/. Accessed 04 Jan 2023

Multimodal Data Fusion Architectures in Audiovisual Speech Recognition

Hadeer M. Sayed[1](✉), Hesham E. ElDeeb[2], and Shereen A. Taie[1]

[1] Faculty of Computers and Artificial Intelligence, Fayoum University, Fayoum, Egypt
{hms08,sat00}@fayoum.edu.eg
[2] Electronic Research Institute, Cairo, Egypt
heldeeb@mcit.gov.eg

Abstract. In the big data era, we are facing a diversity of datasets from different sources in different domains that describe a single life event. These datasets consist of multiple modalities, each of which has a different representation, distribution, scale, and density. Multimodal fusion is the concept of integrating information from multiple modalities in a joint representation with the goal of predicting an outcome through a classification or regression task. In this paper, the strategies of multimodal data fusion were reviewed. The challenges of multimodal data fusion were expressed. A full assessment of the recent studies in multimodal data fusion was presented. In addition, a comparative study of the recent research on this point was performed, and the advantages and disadvantages of each were illustrated. Furthermore, the audiovisual speech recognition task was expressed as a case study of multimodal data fusion techniques. Particularly, this paper can be considered a powerful guide for interested researchers in the fields of multimodal data fusion and audiovisual speech recognition.

Keywords: Multimodal data · Data fusion · Audiovisual Speech Recognition (AVSR) · Deep Learning

1 Introduction

The forms and sources of information that describe the daily events may vary. The human mind has the ability to analyze this information in its various forms and to draw interpretations and inferences from these events. Due to the ability of artificial intelligence science to simulate the way the human mind works in inference, interpretation, and prediction, it needs to be able to interpret the multimodal data together [1]. The combining of these data to form a single, unified representation, namely multimodal data fusion. Several works concentrated on displaying and perceiving events based on a single modality has been utilized [2]; however, they were ignoring different available modalities. This is deficient when the events to be identified are corrupted and not obvious.

A. Rocha et al. (Eds.): WorldCIST 2023, LNNS 800, pp. 655–667, 2024.
https://doi.org/10.1007/978-3-031-45645-9_63

Multimodal fusion has considerable benefits [3]. Firstly, exploiting various modalities to form a unified representation based on extracting the correlations among these modalities increases prediction accuracy. Secondly, when the machine learning algorithm relies on a single modality, the absence of this modality means the absence of the decision. In contrast, the presence of complementary modalities for the same event means the ability to take the decision even if one of these modalities is absent.

The contributions of this research are summarized in the following points:

- Illustrate the different strategies of multimodal data fusion with detailed illustrations of the strengths and limitations of each strategy.
- Demonstrate the challenges in multimodal data fusion.
- Perform a comparative study between the different machine learning architectures used for fusion through addressing their effect on performance in the audio-visual Speech recognition (AVSR) task.
- Extrapolate open issues through the limitations of the current studies and presenting these issues to guide new researchers.

The paper is organized as follows: Sect. 2 explains multimodal data fusion strategies with their capabilities and limitations. Section 3 discusses the challenges of multimodal data fusion. Section 4 provides a comparative study of deep neural network techniques in AVSR systems. Finally, Sect. 5 expresses the conclusion.

2 Multimodal Data Fusion Strategies

Depending on the fusion strategy, machine learning model architectures vary. There are four distinct strategies of data fusion according to the level of fusion: early/raw data fusion, intermediate/features fusion, late/decision fusion, and hybrid fusion. Early fusion is a concatenation of original or extracted features at the input level into one vector that is the input of a machine learning algorithm. In contrast, late fusion aims to perform fusion after the decision has been made by each modality, with the same model or different models for each modality. It uses one of the various decision fusion techniques, such as voting schemes [4], signal variance [5], averaging [6], and weighting based on channel noise [7]. A hybrid data fusion strategy is a combination of early fusion and late fusion strategies. It performs fusion on the decision that comes from different models that were trained on the concatenation of modalities. Finally, intermediate fusion aims to extract the correlations between different modalities as a shared representation of modalities. After the explanation of each strategy mechanism, we will compare them according to some attributes to illustrate the strengths and limitations of each strategy, as shown in Table 1.

Table 1. The strength and limitations of multimodal data fusion strategies

Attribute	Early Fusion	Late Fusion	Hybrid Fusion	Intermediate Fusion
Several models required	✗	✓	✓	✓
Multistage models	✗	✗	✗	✓
Prediction with missing modalities	✗	✓	✓	✓
Interaction between modalites	✓	✗	✓	✓
Exploitation of inter modalites correlations	✗	✗	✗	✓
Time Consuming	✗	✓	✓	✓
Prediction performance enhancement	✓	✓	✓	✓

3 Challenges in Multimodal Data Fusion

For a variety of reasons, data fusion is a challenging task. First, the data are produced by extremely complex systems that are controlled by several underlying processes and rely on a large number of unknown variables. Second, the quantity, kind, and scope of new research questions that might be presented are potentially quite significant as a result of the increased diversity. Third, It is not always easy to combine heterogeneous datasets in a way that maximises each dataset's advantages and minimises its disadvantages. In this paper, we will discuss the challenges in data fusion that relevant to data and model selection.

3.1 Data

- **Non-commensurability:** As a result, different instruments will report on various elements of the issue since they are sensitive to various physical events. Consequently, the data is represented by a variety of physical units. [8].
- **Different resolutions:** While having the same coordinates, two datasets may have extremely different resolutions.
- **Number of dimensions:** Datasets with various architectures may be produced by various acquisition techniques. Matrix versus higher-order tensors [9], for instance.
- **Noise:** All datasets contain unavoidable amounts of thermal noise, calibration imprecision, and other measurement quality degradation. Naturally, each measurement technique generates a variety of various amounts and types of errors in addition to the varied forms of sought data [10].
- **Missing data:** Several cases could be described by this word. First, defective detectors can result in some samples being unreliable, rejected, or simply missing. Second, a modality may occasionally report just on a portion of the system (w.r.t. the other modalities). Third, if samples from different modalities are not gathered at comparable sampling points [10] and we want to build a more comprehensive picture from the entire sample set, data may be viewed as structurally missing.
- **Conflicting:** Only when there are several datasets present can this issue arise.

3.2 Model Selection

Enabling modalities to completely communicate and inform one another is the main goal of data fusion. Therefore, choosing an analytical model that accurately captures the relationship between modalities and produces a meaningful combination of them without imposing phantom connections or suppressing existent ones is crucial. The fundamental links between various modalities are not well understood. Consequently, it is crucial to be as data-driven as possible. Models now in use can be loosely divided into two kinds. Modalities in one class of models share a specific component deterministically. Each dataset's unique set of variables is stored in a second class. The statistical correspondence between the latent variables is what causes the link in this instance. The diversity of currently available solutions and the fact that new solutions are consistently being developed for the data [11] suggest that the choice of model is a difficulty that has not yet been fully solved.

4 Case Study: Audio-Visual Speech Recognition (AVSR)

Automatic speech recognition (ASR) is a field that has seen extraordinary advances in the past few years. However, the capability to comprehend the discourse in noisy situations is one region where machines are still a slack distance behind their human rival. The conventional audio-only speech recognition models do not exploit the advantages of visual data to decode what has been said. This recognition has incited analysts to investigate the strategies of audio-visual speech recognition (AVSR), but this is still a generally modern field. AVSR could be a procedure that employments image processing capabilities in lip reading to help discourse recognition frameworks in recognizing nondeterministic phones or giving dominance among close likelihood choices. ASVR tasks can be categorized into two levels: word level recognition and sentence-level recognition.

4.1 Word Level Recognition

Ngiam et al. [12] proposed an ASR model based on the audio-only RBM model and evaluated this model on the CUAVE dataset. The model achieved an accuracy of 95.8% for clean audio and 79.6% for noisy audio. Nevertheless, for the lib reading task without audio, they proposed a Video Only Deep Auto-encoder, which achieved an accuracy of 69.7%. Regarding their first attempt to present an AVSR model, they proposed a concatenation between the audio-only RBM and the video-only Deep Autoencoder. This concatenation achieved an accuracy of 87% for clean audio and 75.5% for noisy audio, which means the model could not take advantage of the visual information to aid the speech recognition task. The reason behind that is the model's inability to find correlations between both the audio and the visual features. The second attempt was a bimodal deep autoencoder. It achieved an accuracy of 90% for clean audio and 77.6% for noisy audio. Although those results were better than the concatenation, it indicates to lip reading for speech recognition is inferior to the usage of audio-only. This fact inspired them to increase the role of the audio modality. Therefore, they merged the bimodal deep auto-encoder with the audio-only RBM. This merge was able to achieve the best accuracy for clean and noisy audio by 94.4% and 81.6% respectively, with a signal-to-noise ratio (SNR) equal to 0dB.

Rahmani et al. [13] worked on the same dataset as [12]. They are concerned with adjusting the contribution of each modality. They proposed an audiovisual feature fusion model based on a deep autoencoder and used a DNN-HMM hybrid as the speech classifier. So, their experiment included four schemas of a bimodal deep auto-encoder, each one a modified version of the previous one for performance improvement. The bimodal deep autoencoder (DNN-IV) outperformed the model proposed by Ngiam et al. [12] by 3.5% for clean audio and 1.1% for noisy audio.

Yang et al. [14] proposed the CorrRNN model for multimodal fusion of temporal inputs. The model includes GRUs to capture long-term reliance and time-related structure in the input. Their main contribution was a dynamic weighting mechanism. It permits the encoder to assign weights dynamically for each modality to adjust the role of each one. The primary cause is to focus on the modality that has a useful signal in a feature representation, which is computed especially when one of the modalities is corrupted with noise. The model was tested on CUAVE database.

Table 2 shows the performance of the three models [12,13] and [14] on CUAVE dataset. Yang [14] model accuracy outperformed with a big difference from the other models in case of noisy audio. As mentioned previously, the need for visual information occurs when the audio signal is corrupted. Thus, it is a great contribution from our point of view.

Table 2. Performance of different models on CUAVE dataset

	Clean Audio	Noisy Audio
Ngiam [12]	94.4%	82.2%
Rahmani [13]	97.9%	83.3%
Yang [14]	96.11%	**90.88%**

Petridis et al. [15] presented an end-to-end audiovisual speech recognition system using BLSTM. What distinguishes this system from others is its ability to perform a classification task with multiple lip views. The audio features were extracted directly using encoding layers instead of the Mel Frequency Cepstral Coefficient (MFCC). The system was evaluated on the OuluVS2 database that provides multiple lip views (five lip views between 0° and 90°). In order to investigate the profits of audiovisual fusion, they added a babble noise to the audio signal from the NOISEX database. The SNR varied from 0 dB to 20 dB.

Petridis et al. [16] proposed another end-to-end audiovisual speech recognition system based on the ResNets network and BGRU. They evaluated their architecture on the Lip Reading in the Wild (LRW) database, which is one of the largest lip reading databases. At this time, it was a very challenging dataset. The importance and strength of this database lie in three main points: 1) It contains more than 1,000 speakers and an expansive variety in head posture and light. 2) The number of words, 500, is the largest lip-reading database for word recognition. 3) It contains different words with similar pronounce (e.g. write/right). Moreover, audio features were extracted using the ResNet network and BGRU. The results showed that the end-to-end audiovisual model slightly outperformed a standard MFCC-based system (the end-to-end audio model) under clean conditions and low levels of noise. It also significantly outperformed the model that used MFCC features to train the two layers of BGRU in the presence of high levels of noise.

Yang et al. [17] provided a unified framework for joint audio-visual speech recognition and synthesis. Their methodology may transform language qualities that are based on visual or auditory cues into modality-neutral representations by developing cross-modal mutual learning procedures. With the use of such derivate linguistic representations, one is able to control the output of audio and visual data based on the desired subject identity and linguistic content information, in addition to performing ASR, VSR, and AVSR. They evaluated the framework on LRW and LRW-1000 datasets.

Sayed et al. [18] presented a bimodal variational autoencoder (VAE) for audiovisual feature fusion. They exploited VAE's abilities to generate new data and learn latent representations smoothly to construct a strongly unified representation of the audiovisual feature. They tested their proposed model on CUAVE dataset. They utilized different classifiers and different audio feature extraction techniques to prove the model's capability for generalization. Their model exceeded all the state-of-the-art models tested on the CUAVE dataset.

4.2 Sentence Level Recognition

Zhang et al. [19] presented a bimodal DFSMN model that considered the following AVSR task challenges. First, there is the ideal integration of audio and visual information. Second, a strong ability to deal with noisy environments. Third, strong modeling helps prevent the loss of visual information during testing and deployment. The architecture of DFSMN consists of three key components: an audio network, a visual network, and a joint network. The deep representation of acoustic features and visual features was generated by the audio net and the visual net, respectively. Then, these features were concatenated and fed into a joint net. During the model training, they used both audio and visual information concurrently. However, during model usage in practical applications, difficulties may occur when capturing the speaker's mouth. This disparity issue between training and test will cause great harm to the performance. Hence, they proposed a per-frame dropout regularization to improve the ability of the AVSR system to deal with this issue. The model was tested on the NTCD-TIMIT corpus. It contains the sound signals and the visual features of 56 Irish speakers and six noisy types (white, babble, car, living room, café, and street) with five variations of SNR (20dB, 15dB, 10dB, 5dB, 0dB). The evaluation was performed with two experimental setups: five hours of clean audio, and one hundred fifty hours of noisy audio. They achieved a phone error rate of 12.6% on clean conditions and an average phone error rate of 26.2% on all test sets (clean, various SNRs, various noise types) compared with DNN-HMM released by Kaldi toolkit [20].

Sadeghi et al. [21] developed a conditional variant auto-encoder (CVAE) for audio-visual speech enhancement, where the audio speech production process is conditioned on the visual features gained from lip movements. The idea here is to exploit visual information to act as a probabilistic Wiener filter. NTCD-TIMIT dataset was used for the training and testing phases, and some samples of the GRID dataset were used for the testing. In order to measure the enhancement scores, they used the signal-to-distortion ratio (SDR), the perceptual evaluation of speech quality (PESQ), and the short-time objective intelligibility (STOI) score. For each measure, they computed the difference between the enhanced speech signal and the noisy/unprocessed mixture signal. They used six levels of SNR −15 dB, −10 dB, −5 dB, 0 dB, 5 dB, and 15 dB. The proposed unsupervised audiovisual method, AV-CVAE, was compared with the audio-only A-VAE, the unsupervised nonnegative matrix factorization NMF method [22] and the supervised method proposed in [23]. AV-CVAE beat other methods clearly in terms of PSEQ and SDR except in the case of low-level noise. In another view, the supervised method in [23] needs to be trained with various noise types and noise levels to have a high capability of generalization. While AV-CVAE is trained on clean audio only with its corresponding visual information. However, the computational efficiency of testing in [23] is better than AV-CVAE. In addition, the supervised method [23] takes advantage of the dynamics of both the audio and visual data through the presence of convolutional layers, which is not the case of AV-CVAE which used fully connected layers.

Zhou et al. [24] considerate the importance of each modality. In which they proposed a multimodal attention-based method (MD_ATT) for AVSR that is able to automatically adjust the contribution of each modality in the AVSR task. In addition, the attention mechanism changes the weights over time based on which modality carries useful information in this observation. In order to investigate their contribution, they compared MD_ATT with four models: 1) Listen, Attend, and Spell (LAS) model [25] for audio-only speech recognition 2) Watch, Attend, and Spell (WAS) model [26] for visual-only speech recognition 3) Watch, Listen, Attend, and Spell (WLAS) model [26]. 4) AV_align model [27] for audiovisual speech recognition. The difference between the MD_ATT model and WLAS is that multimodal attention considers the selection of more trustworthy modal information over modalities' context vectors, while WLAS treats modalities equally by concatenation. In the case of AV_align, to acquire an enhanced representation used for an attention-based decoder, it uses audio features. The differences between AV-align and MD_ATT are: first, AV_align attention supplies the lip features associated with the current audio feature and occurs on the encoder side, while MD_ATT attention is integrated into the decoder side and gives out the preference of modalities by attention weights. Second, the MD_ATT model has the align-like process between two modals and it is performed by individual attention, which generates individual context vectors associated with the current output units. Finally, denoting T_a, T_v, and T_d as the length of the audio feature, visual feature, and decoding steps respectively. AV_align performs attention for every frame of audio representation, so, whose complexity is $O(T_aT_v)$. While MD_ATT operates on a context vector for every decoding step, so, whose complexity is $O(2T_d)$, which is much more efficient. LAS, WAS, WLAS, and Av_align were trained side by side with MD_ATT. Four copies of the MD_ATT model were trained and tested with clean and noisy audio at various SNRs. MD_ATT_MC denotes the MD_ATT model, which was trained and tested on multi-condition data. The experimental results showed that comparative improvements from 2% up to 36% over the audio modality alone are obtained depending on the different SNR.

Yu et al. [28] focused on the recognition of overlapped speech. They considered three issues for building AVSR. First, the design of the model. So, they compared different end-to-end systems for AVSR in order to arrive at a powerful basic architecture for AVSR. Second, the method of integrating acoustic and visual features, proposed two gated neural networks to perform dynamic fusion between the two modalities. Third, in contrast to a traditional pipelined architecture for overlapped speech recognition, which contains explicit speech separation and recognition components, they proposed a streamlined and integrated AVSR system architecture. It contains implicit speech enhancement and recognition components optimized consistently using the lattice-free maximum mutual information (LF-MMI) discriminative criterion. The experiments on overlapped speech simulated from the LRS2 dataset showed that the proposed AVSR system outperformed the audio-only baseline LF-MMI DNN system by up to 29.98% absolute in word error rate (WER) reduction, and produced recognition performance comparable to a more complex pipelined system.

Braga et al. [29] considered the scenario of multiple people existing on the screen at the same time. It is a big challenge in AVSR tasks. A traditional AVSR pipeline that attempts to handle this issue consists of a sequence of separate modules: face tracking, active speaker face selection, and the AVSR system. Rather than depending on two partitioned models for speaker face selection and AVSR on a single face track, they presented an attention layer to the ASR encoder that is able to soft-select the suitable face video track. In other words, instead of training the AVSR model with a single face track, they trained it with multiple tracks and learned how to select the correct track to aid speech recognition with an attention mechanism. From the advantages of an end-to-end model they focused on the following:

- Computational: Each of the A/V active speakers and the ASR models needs to rely on a separate visual frontend, and each by itself can easily dominate the total number of FLOPS of the entire model. However, low-level visual features are typically transferable between computer vision tasks, so it seems

redundant to have two separate sub-modules that are potentially playing similar roles.
- Simplicity: There is less coordination between subsystems with an end-to-end system.
- Robustness: Their system is able to soft-select the active face track, without making an early hard decision. The rest of the model is naturally allowed to adapt to that bad decision, even when a high probability is assigned to the wrong face, as the whole model is differentiable.

Their model was trained on 30k hours of YouTube videos and tested on YTDEV18 and LRS3-TED datasets.

Ma et al. [30] proposed an end-to-end audio-visual speech recognition system. They developed a hybrid CTC/attention model based on a ResNet-18 and a convolution-augmented transformer (Conformer). The audio and visual features were directly retrieved from the audio waveform and the raw pixel, respectively, and passed to conformers before being fused by a multi-layer perceptron (MLP). Their strategy prioritized end-to-end training rather than the commonly used pre-computed visual characteristics, a conformer rather than a recurrent network, and a transformer-based language model, all of which considerably improved the performance of the model. They evaluated the proposed architecture on LRS2 and LRS3 datasets.

Shi et al. [31] proposed a self-supervised AVSR framework based on Audio-Visual HuBERT (AV-HuBERT). In order to enable unsupervised learning of joint representations over audio and visual streams, they used the Audio-Visual HuBERT (AV-HuBERT) pre-training approach. This is the first attempt to create an AVSR model using a significant amount of audio-visual speech data that is not labeled. The model was evaluated on the LRS3 dataset.

Finally, we summarized all the work mentioned above in the following Table 3. Table 3 demonstrates a comparison between the recent research on multimodal data fusion in AVSR tasks according to the challenges considered in each study. According to the number of challenges it addresses, the model put forth by Braga et al. [29] may be regarded as the most robust. However, it lacks a dynamic weighting mechanism in order to adjust the contribution of each modality because of its effect on raising recognition accuracy. Furthermore, we provide some new research issues for multimodal data fusion to be guide for the next researchers. The new research issues can be listed as the following:

- Design new features for fusion frameworks with more powerful computing architectures.
- Combine the current compression strategies of free parameters for a single modality of feature learning to design new compression methods for multimodal deep learning.
- Design new deep learning models for multimodal data that take semantic relationships into consideration. The combination of deep learning and semantic fusion strategies may be a way to solve the challenges posed by the exploration of multimodal.

Table 3. Comparison between recent researches in multimodal data fusion in AVSR task according to the considered challenges in each research.

Authors	Year of Publication	Dataset	Types of Noise	Challenges					
				Head Poses		Overlap Speech	Multiple Person in Scene	Dynamic Weighting for Modalities	Recognition Level
				Frontal face only	Multiple Views				
Ngiam et al. [12]	2011	CUAVE	White Gaussian, SNR {0} dB	✓	✗	✗	✗	✗	Word
Petridis et al. [15]	2017	OuluVS2	Babble, SNR {20, 15, 10, 5, 0} dB	✓	✓	✗	✗	✗	Word
Yang et al. [14]	2017	CUAVE	White Gaussian, SNR {0} dB	✓	✗	✗	✗	✓	Word
Petridis et al. [16]	2018	LRW	Babble. SNR {20, 15, 10, 5, 0, -5} dB	✓	✓	✗	✗	✗	Word
Rahmani et al. [13]	2018	CUAVE	Babble, Factory and White, SNR {15, 10, 5, 0, -5} dB	✓	✗	✗	✗	✗	Word
Zhang et al. [19]	2019	NTCD-TIMIT	Babble, White, Car, Living room, Cafe and Street, SNR {20, 15, 10, 5, 0} dB	✓	✗	✗	✗	✗	Sentence
Zhou et al. [24]	2019	LRS	White Gaussian, SNR {10, 5, 0} dB	✓	✗	✗	✗	✓	Sentence
Yu et al. [28]	2020	LRS2	White Gaussian, SNR {10, 5, 0, -5} dB	✓	✗	✓	✗	✗	Sentence
Sadeghi et al. [21]	2020	NTCD-TIMIT	Babble, White, Car, Living room, Cafe and Street, SNR {15, 10, 5, 0, -5, -10, -15} dB	✓	✗	✗	✗	✗	Sentence
Braga et al. [29]	2020	YTDEV18 LRS3-TED	Babble, SNR {20, 15, 10, 5, 0} dB	✓	✗	✓	✓	✗	Sentence
Sayed et al. [18]	2021	CUAVE	White Gaussian, SNR {15, 10, 5, 0, -5, -10, -15} dB	✓	✗	✗	✗	✗	Word
Ma et al. [30]	2021	LRS2 LRS3	Babble, SNR {20, 15, 10, 5, 0, -5} dB	✓	✗	✗	✗	✗	Sentence
Yang et al. [17]	2022	LRW LRW-1000	–	✓	✗	✗	✗	✗	Word
Shi et al. [31]	2022	LRS3	Babble, SNR {10, 5, 0, -5, -10} dB	✓	✗	✗	✗	✓	Sentence

- Design online and incremental multimodal deep learning models for real-time data fusion.
- Currently, several deep learning models are focusing only on noisy single-modality data. With the explosion of low-quality multimodal data, deep learning models for low-quality multimodal data need to be addressed.

5 Conclusion

The diversity of data modalities from different sources that describe a single life event, motivates the researchers to exploit the relations between them in decision making. The need to find a unified representation of multiple modalities in order to capture the nonlinear correlations between them is a recent research direction. This paper fully assesses the most recent multimodal data fusion techniques in the past few years. Moreover, we demonstrated the difference in efficiency between the several fusion strategies for multimodal data. Furthermore, we discuss the challenges in mutimodal data fusion task in concept level apart from any applications. AVSR is one of the common domains that uses different modalities to enhance recognition accuracy, especially with corrupted audio. This paper exploited this problem as a case study to investigate the robustness of deep learning in multimodal data fusion based on the variations in accuracy and the number of challenges considered in each model. This full assessment will motivate and guide the researcher to develop in the open issues of this research direction.

References

1. Russell, S.J.: Artificial Intelligence a Modern Approach. Pearson Education Inc., London (2010)
2. Zaykovskiy, D.: Survey of the speech recognition techniques for mobile devices. In: Proceedings of DS Publications (2006)
3. Frey, B.J.: Graphical Models for Machine Learning and Digital Communication. MIT Press, Cambridge (1998)
4. Morvant, E., Habrard, A., Ayache, S.: Majority vote of diverse classifiers for late fusion. In: Fränti, P., Brown, G., Loog, M., Escolano, F., Pelillo, M. (eds.) S+SSPR 2014. LNCS, vol. 8621. Springer, Heidelberg (2014). https://doi.org/10.1007/978-3-662-44415-3_16
5. Evangelopoulos, G., et al.: Multimodal saliency and fusion for movie summarization based on aural, visual, and textual attention. IEEE Trans. Multimedia **15**(7), 1553–1568 (2013)
6. Shutova, E., Kiela, D., Maillard, J.: Black holes and white rabbits: metaphor identification with visual features. In: Proceedings of the 2016 Conference of the North American Chapter of the Association for Computational Linguistics: Human Language Technologies (2016)
7. Potamianos, G., Neti, C., Gravier, G., Garg, A., Senior, A.W.: Recent advances in the automatic recognition of audiovisual speech. Proc. IEEE **91**(9), 1306–1326 (2003)

8. Biessmann, F., Plis, S., Meinecke, F.C., Eichele, T., Muller, K.R.: Analysis of multimodal neuroimaging data. IEEE Rev. Biomed. Eng. **4**, 26–58 (2011)
9. Acar, E., Rasmussen, M.A., Savorani, F., Næs, T., Bro, R.: Understanding data fusion within the framework of coupled matrix and tensor factorizations. Chemometr. Intell. Lab. Syst. **129**, 53–63 (2013)
10. Van Mechelen, I., Smilde, A.K.: A generic linked-mode decomposition model for data fusion. Chemometr. Intell. Lab. Syst. **104**(1), 83–94 (2010)
11. Debes, C., et al.: Hyperspectral and LiDAR data fusion: outcome of the 2013 GRSS data fusion contest. IEEE J. Sel. Top. Appl. Earth Obs. Remote Sens. **7**(6), 2405–2418 (2014)
12. Ngiam, J., Khosla, A., Kim, M., Nam, J., Lee, H., Ng, A.Y.: Multimodal deep learning. In: ICML (2011)
13. Rahmani, M.H., Almasganj, F., Seyyedsalehi, S.A.: Audio-visual feature fusion via deep neural networks for automatic speech recognition. Digit. Signal Process. **82**, 54–63 (2018)
14. Yang, X., Ramesh, P., Chitta, R., Madhvanath, S., Bernal, E.A., Luo, J.: Deep multimodal representation learning from temporal data. In: Proceedings of the IEEE Conference on Computer Vision and Pattern Recognition, pp. 5447–5455 (2017)
15. Petridis, S., Li, Z., Pantic, M.: End-to-end visual speech recognition with LSTMs. In: 2017 IEEE International Conference on Acoustics, Speech and Signal Processing (ICASSP), pp. 2592–2596. IEEE (2017)
16. Petridis, S., Stafylakis, T., Ma, P., Cai, F., Tzimiropoulos, G., Pantic, M.: End-to-end audiovisual speech recognition. In: 2018 IEEE International Conference on Acoustics, Speech and Signal Processing (ICASSP), pp. 6548–6552. IEEE (2018)
17. Yang, C.-C., Fan, W.-C., Yang, C.-F., Wang, Y.-C.F.: Cross-modal mutual learning for audio-visual speech recognition and manipulation. In: Proceedings of the 36th AAAI Conference on Artificial Intelligence, Vancouver, BC, Canada, vol. 22 (2022)
18. Sayed, H.M., ElDeeb, H.E., Taie, S.A.: Bimodal variational autoencoder for audio-visual speech recognition. Mach. Learn. **112**, 1–26 (2021)
19. Zhang, S., Lei, M., Ma, B., Xie, L.: Robust audio-visual speech recognition using bimodal DFSMN with multi-condition training and dropout regularization. In: ICASSP 2019-2019 IEEE International Conference on Acoustics, Speech and Signal Processing (ICASSP), pp. 6570–6574. IEEE (2019)
20. Povey, D., et al.: The Kaldi speech recognition toolkit. In: IEEE 2011 Workshop on Automatic Speech Recognition and Understanding, no. CONF. IEEE Signal Processing Society (2011)
21. Sadeghi, M., Leglaive, S., Alameda-Pineda, X., Girin, L., Horaud, R.: Audio-visual speech enhancement using conditional variational auto-encoders. IEEE/ACM Trans. Audio Speech Lang. Process. **28**, 1788–1800 (2020)
22. Smaragdis, P., Raj, B., Shashanka, M.: Supervised and semi-supervised separation of sounds from single-channel mixtures. In: Davies, M.E., James, C.J., Abdallah, S.A., Plumbley, M.D. (eds.) ICA 2007. LNCS, vol. 4666, pp. 414–421. Springer, Heidelberg (2007). https://doi.org/10.1007/978-3-540-74494-8_52
23. Gabbay, A., Shamir, A., Peleg, S.: Visual speech enhancement. arXiv preprint arXiv:1711.08789 (2017)
24. Zhou, P., Yang, W., Chen, W., Wang, Y., Jia, J.: Modality attention for end-to-end audio-visual speech recognition. In: ICASSP 2019-2019 IEEE International Conference on Acoustics, Speech and Signal Processing (ICASSP), pp. 6565–6569. IEEE (2019)

25. Chiu, C.-C., et al.: State-of-the-art speech recognition with sequence-to-sequence models. In: 2018 IEEE International Conference on Acoustics, Speech and Signal Processing (ICASSP), pp. 4774–4778. IEEE (2018)
26. Son Chung, J., Senior, A., Vinyals, O., Zisserman, A.: Lip reading sentences in the wild. In: Proceedings of the IEEE Conference on Computer Vision and Pattern Recognition, pp. 6447–6456 (2017)
27. Sterpu, G., Saam, C., Harte, N.: Attention-based audio-visual fusion for robust automatic speech recognition. In: Proceedings of the 20th ACM International Conference on Multimodal Interaction, pp. 111–115 (2018)
28. Yu, J., et al.: Audio-visual recognition of overlapped speech for the LRS2 dataset. In: ICASSP 2020-2020 IEEE International Conference on Acoustics, Speech and Signal Processing (ICASSP), pp. 6984–6988. IEEE (2020)
29. Braga, O., Makino, T., Siohan, O., Liao, H.: End-to-end multi-person audio/visual automatic speech recognition. In: ICASSP 2020-2020 IEEE International Conference on Acoustics, Speech and Signal Processing (ICASSP), pp. 6994–6998. IEEE (2020)
30. Ma, P., Petridis, S., Pantic, M.: End-to-end audio-visual speech recognition with conformers. In: ICASSP 2021-2021 IEEE International Conference on Acoustics, Speech and Signal Processing (ICASSP). IEEE (2021)
31. Shi, B., Hsu, W.-N., Mohamed, A.: Robust Self-Supervised Audio-Visual Speech Recognition. arXiv preprint arXiv:2201.01763 (2022)

Digital Technologies in Marketing: The Case of University Tunas

Vitor Gonçalves(✉) and Bruno F. Gonçalves

CIEB, Instituto Politécnico de Bragança, Bragança, Portugal
vg@ipb.pt

Abstract. Digital technologies are currently present in all economic sectors of society and higher education as one of the basic sectors of contemporary society is no exception. Whether in the teaching-learning process, in other administrative and technical-pedagogical services, or in the student groups that make up the academic communities, technologies seem to play an important role in the development and operationalization of the various types of activities and processes that make up the different sectors of higher education institutions. It is in this context that the present research arises, which, in general, aims to identify the technologies adopted for the advertising and dissemination of the Tuna's brand and image, but also to understand how the team responsible for this area uses these technologies in this area of digital marketing. To carry out the study, the case study methodology is adopted, whose focus will be on the "RaussTuna - Tuna Mista de Bragança", Polytechnic Institute of Bragança (Portugal). The results suggest that this particular Tuna uses a set of digital technologies that help to operationalize its activity in the area of digital marketing. The results also show that each of the technologies has its own complexities, so the members who use them recognize having difficulties in using some of these tools. In this sense, it seems important to invest not only in the training of members in the area to acquire more digital skills, but also to invest in updating the various tools, platforms and software of the Tuna.

Keywords: Academia · Digital Technologies · Higher Education · Marketing · Tunas Universities

1 Introduction

The use of digital technologies in higher education is already part of the day-to-day academic life of institutions in its various dimensions: implementation of the teaching-learning process, posting of grades, signing of documents, research and development, dematerialization and document management, management of libraries and scientific repositories, administrative, financial and academic services, can-didactic applications to the various study cycles, registration, exam registrations, document requests, payment of tuition and other fees, booking of meals in canteens, promotion and dissemination of the brand and image of the institutions, among many others. Obviously, it wasn't only

A. Rocha et al. (Eds.): WorldCIST 2023, LNNS 800, pp. 668–677, 2024.
https://doi.org/10.1007/978-3-031-45645-9_64

higher education institutions and their services that needed to adopt technologies to be digitally modernized, but also the teachers, employees, and students themselves had to adapt to the digital operation of higher education. As technologies are like a snowball, so too have the bodies based in these institutions (research centers, business incubation centers, academic and student associations, student centers and university Tunas) felt the need to accompany this change and invest in their digital modernization to adapt to reality, becoming more current and better prepared to meet future challenges.

The present research focuses precisely on the study of how one of these associations, in this case, an associative group of students - the "RaussTuna - Tuna Mista de Bragança (TMB)" - uses the technologies for the promotion and dissemination of its image in the exercise of its activity.

2 Digital Technologies and Digital Marketing

Technologies, as mentioned in the previous section, are already part of the daily life of higher education institutions. They are, in fact, an indispensable tool for carrying out the other activities promoted by these institutions, from teaching, research, services, to the other associative groups that integrate these institutions. Technologies are even a useful and absolutely necessary pillar for the functioning of the other processes, activities and operation of the other organisms.

The technologies in digital marketing, especially with regard to editing and production of multimedia content (text, images, videos, animations, among others), but also to its advertising and dissemination on networks and other platforms are vital tools in a contemporary society. They are vital because organizations, companies, institutions, and associations need to promote their products and services so that the target audience acquires them, and through this, organizations can make a profit, attract even more customers, and gain value for themselves.

Kotler et al. (2012) state that the definition of marketing is currently related to identifying and understanding the needs of human beings as social beings. Marketing can then be considered "a social process, in which individuals and groups get what they need and what they want, through the creation, offering and exchange of products and services of value with others" (Kotler et al. 2012, p. 5). It is a way of doing business, taking into account the customer's knowledge, to develop a product or service that benefits all the elements that interact in the exchange process (Gomes and Kury 2013). It should be noted that the RaussTuna - Tuna Mista de Bragança (TMB) has no interest in doing business, only in promoting its music, activities, Tunas festivals in which it participates and events that it promotes annually of a social, cultural, musical and scientific nature. In general, the promotion of the Tuna has as main objective to make known the activities to the Tunas of the country, to the institutions of higher education, but also to the local inhabitants and the localities of the region of Bragança and North of Portu-gal. Obviously, the promotion of the Tuna has as main objective the attraction of new members for the association that can actively participate in the activities, play and sing in the other performances that are held throughout the country.

The human being, seen as a consumer agent is in constant evolution, building and deconstructing values, which makes the theories about consumer behavior a mandatory knowledge content of the marketing area (Glória 2009). It is, obviously, fundamental to understand the consumer behavior of a given product or service so that it is possible to respond, in the most adequate way, to the needs, desires, and tastes of the current and potential public. According to Maçães (2017), is a complex process, as it is influenced by a multitude of factors, these being: psychological factors; personal factors; cultural factors; social factors. These factors are also taken into consideration in the production of the various contents that are published on the web.

The conception of a marketing plan supported by the company's strategy is also a vital instrument for the operation of the organization, as it ensures greater organization in the execution of other tasks, from the capture of content, to its production, to its publication on the web. As McDonald and Wilson (2013) emphasize, marketing planning is the planned application of resources to achieve marketing objectives. Planning "consists in making a set of explicit decisions in advance, as opposed to improvising, which consists in making decisions at short notice, usually without explaining them formally" (Lendrevie et al. 2015, p. 502). The case study of our research is concerned with making a marketing plan at the beginning of each year in order to organize its activities, but also to distribute them among the members of the digital technologies team. Still regarding the marketing plan, it is important to mention that over the years several models of plans have been proposed by different authors who study themes related to marketing (Cohen 2005; Kotler and Armstrong 2010; Kotler and Keller 2006; Philip 2002; Wood 2007). It is, of course, up to each marketing team to elaborate its marketing plan taking into consideration its p's, thus defining its objectives, principles, target audience, among others that it eventually considers necessary.

3 Methodology

In this section we present the methodological framework and the study objectives, the data collection tools, as well as the data analysis.

3.1 Methodological Framework and Objectives of the Study

For the development of the study, the methodology of single case study was adopted, namely, in RaussTuna - Tuna Mista de Bragança (TMB), Polytechnic Institute of Bragança (IPB), Portugal. Through this methodology, it will be possible to be inside the case, to know the dynamics of the association's activity and to know the members, namely, the dynamics of the digital technologies team of the Tuna. In addition to these advantages, with this methodology we intend to identify the technologies used, but also understand how the team members interact and use this set of technologies in the area of digital marketing, particularly, in the promotion and dissemination of Tuna in the academic community, local and Tunae.

With the development of this research, it is intended to achieve three objectives:

- Identify the functions of the digital technologies team;
- To identify the technologies used by the team;
- To understand for what purposes the technologies are used;
- To identify constraints on the use of these technologies.

These investigative objectives were achieved with the support of a qualitative approach, which is explained in the following section.

3.2 Data Collection Instruments

Two data collection tools were used for this research: participant observation and interview.

The participant observation allowed us to observe and understand how the digital technologies team, namely its members, use the panoply of technologies to carry out their activities. In this sense, it was possible to identify the technologies and understand for what purpose they are used in various areas.

The interview was also applied to the members of the Tuna technologies team who use the various technologies, but also to the person responsible for the administration of all technologies. The semi-structured interview type was adopted since it does not require a rigid script and allows a greater openness in the conversations held with the interviewees and, consequently, a greater extraction of data that, eventually, may be important for the investigative reflections of this research.

3.3 Data Analysis

The analysis of the data from the two instruments was carried out in Microsoft Excel. Through participant observation, it was possible to identify the tools used by the team members and categorize and organize the various functions for which these tools were used. The semi-structured individual interview allowed not only to corroborate the conclusions drawn from the participant observation, but also to collect information about the skills of each team member, as well as to identify the constraints in the use of the various tools.

Thus, in general, we sought to perform a content analysis with the objective of identifying the various directions of response and, simultaneously, cross-referencing the data obtained from both data collection instruments, with the results presented in the following section.

4 Presentation of Results

This section presents the results based on the objectives defined in the previous section: functions of the digital technologies team, identification and use of technologies, and constraints on the use of technologies.

4.1 Functions of the Digital Technologies Team

The semi-structured interviews allowed us to understand the organization of the digital technologies team, but also to identify the responsibilities and roles assigned to the members of the team. These roles are identified below:
Team and technology manager:

- Leads and takes responsibility for the team;

- Manages, updates and maintains the official TMB website;
- Manages, updates and maintains the official website of the International Tunas Conference;
- Manages, updates and maintains the Information and Administrative Management System (SIGA);
- Manages the Massive Open Online Course (MOOC) - Tutorials Digital Space;
- Manages changes to the code and design of the Rauss Bar application;
- Manages all Tuna's digital platforms (Alexa, Tipeestream, Youtube, Spotify, DistroKid, Amen, Google apps, Udemy and Wordpress);
- Manages Tuna's social networks (Facebook and Instagram);
- Manages the social networks of the International Scientific Event of Tunas of the respective even-to (Facebook, Instagram and Youtube);
- Proposes investments in digital marketing for publicizing and disseminating the brand and image of the Tuna;
- Replaces the team members in their absence or impediment.

Content Producer:

- Video producer;
- Merchandising producer;
- Photo editor;
- Responsible for archiving and organizing digital content;
- Replaces team members in their absence or impediment;
- Performs any competence assigned by the Team Manager.

Social Media Manager:

- Posts on social media;
- Record videos;
- Coverage of events (photos and videos);
- Writes and corrects texts;
- Manages comments and messages;
- Replaces team members in their absence or impediment;
- Performs any competence assigned by the Team Manager.

Obviously, for each of these functions, there are various associated activities and tasks that are carried out by all the members of the digital technologies team. These tasks vary according to the type of activity that the Tunas promotes or participates in (concerts, festivals, release of singles or albums, tours, Tunas meetings, international meetings of Tunas, social causes, rehearsals, among other activities). Each of these activities has its own objectives and specificities that the digital technologies team studies carefully in order to produce the necessary content to transmit the message to the target audience.

4.2 Identification and Use of Technologies

Each of the tools identified above is used for very specific purposes that we will list below:

Content production tools

For each of the multimedia contents the Tuna recognizes that it uses several tools. Audio (Audacity); Image (Canva and Photoshop); Text (Microsoft Word and Google Docs); Video (Video Pad and Sony Vegas).

- Audacity: a tool for capturing and editing audio. For Tuna it is useful for recording voice over for promotional videos for social events and others, but also for recording lines of some musical instruments such as guitar, flutes, bass, accordion, among others. It should be noted that, after export, these sound clips are inserted into video tools to synchronize with other multimedia content, such as image and text;
- Canva: widely used for creating posters, invitations, banners and other content aimed at promoting a particular activity of the Tuna. This tool is widely used in the organization of the International Scientific Event of Tunas (scientific event held annually). It is also important to mention that the use of this tool has to do with the fact that it is very practical in the production of contents;
- Photoshop: it is a more complex tool, with a much higher level of detail, and is used by the team members to edit images that require a higher level of care and accuracy. It is a tool that members find difficult to use, both in terms of the wide range of options that the tool offers and the technical level. They also assume a lack of training in image editing and in the tool itself, and therefore recognize that they constantly resort to tutorials to acquire more skills in the area;
- Microsoft Word: is used to write texts for subsequent publication on websites and social networks. The use of Word has to do with the fact that writing, spelling and grammar corrections are more practical. Besides this aspect, it is possible to send the file to other colleagues in the team for verification and final revision of the text;
- Google Docs: this tool provides in general what Microsoft Word provides; however, it has the particularity of sharing directly with teammates, allowing simultaneous editing and collaboration. It also has the advantage that the work is stored in the Goo-gle's drive, which facilitates later access by team members;
- Video Pad: a tool used for editing and producing promotional and other types of videos. In addition to allowing the synchronization of various multimedia contents, it allows the addition of various types of effects, animations and transitions to all these contents. It is a very appealing tool for the user and technical skills are easily acquired in its use;
- Sony Vegas: like the previous one, Sony Vegas is a video editing program, however it is a bit more complex and requires greater digital dexterity in its use. It is a program that is used only by one member of the team in the production of videos that require greater care and a higher level of work and quality.

Content repository tools

In addition to the two websites that the Tuna has (Tuna's Website and Website of the International Scientific Event of Tunas) and the two social networks (Facebook and Instagram), the following content repository tools are used:

- Youtube: this tool is used since the Tuna was born and is very important for the dynamics of the dissemination of the brand and image of the Tuna around the world.

The Tuna has a channel that has 76 videos, almost all of them of good quality. Some correspond to performances recorded in auditoriums and theaters by professional cameramen, others are promotional videos of events that the Tuna organizes, and others come from projects recorded in the studio such as singles, albums, among others. It is a central platform of the Tuna, which allows us to get closer to other Tunas, but also to the community of the Polytechnic Institute of Bragança;
- Spotify: platform that includes the albums and singles released and has an average of 133 listeners per month. It is very useful in disseminating the Tuna's music, particularly the original songs it produces. Since it is a tool widely used by younger age groups, the Tuna understands that it is a good bet for the dissemination of their music and their image.

Social networks - platforms for promotion and dissemination.

Two social networks are used in the Tuna, namely Facebook and Insta-gram. Each of these networks serves two different purposes and, therefore, complement and articulate each other.

- Facebook (www.facebook.com/rausstunabraganca): is used because it aggregates an older public that goes back to the foundation of the Tuna and, therefore, it is very important to continue using it. This social network, among others, aggregates former students of the Polytechnic Institute of Bragança who joined the institution from the year 2007. Besides this aspect, it is through Facebook that the families of the members of Tuna follow the various activities from a distance, as well as former members or even entities, sponsors and some partners. We can say that it is a social network used to reach a very specific public and not so much the younger public and current students of the Polytechnic Institute of Bragança. The publications are not daily, but there is a regular update of what the Tuna is doing in their activities;
- Instagram (www.instagram.com/rausstuna.braganca): unlike Facebook, Instagram is adopted to reach students who have recently finished their studies or are currently attending the Polytechnic Institute of Bragança. It is a social network that allows us to complement Facebook and is oriented towards younger audiences and, therefore, we are concerned with keeping this social network updated on a daily basis, specifically through insta stories or even publications that are justified.

Website - promotion and dissemination platform

- WordPress (General Site of Tuna - www.rausstuna.pt): As a static space for information, the Tuna chose to build already in 2009 (foundation of the Tuna) a website in WordPress that allows to have updated information about the Tuna as, for example, social bodies, projects in development, bibliography, current contacts, discography released, among many others. It is a site that has a lot of information available to the Internet user, so it is very useful for students of the Polytechnic Institute of Bragança (target audience) to search for credible information about Tuna;
- WordPress (Site International Scientific Event of Tunas - www.jornadastmb.ipb.pt): this site was created to promote and disseminate a scientific event, called International Scientific Event of Tunas. This event aims at the construction of knowledge through the presentation of communications from the Tunas and external personalities who

are invited to contribute to the reflection and discussion of the themes of the event. It is intended to disseminate the best of what is done in the Tunas, aggregating the dimensions of associativism, traditions and values, music and higher education. Through this site you can find information about the event, but also download the abstract and poster templates, as well as consult dates and register for the event. It is important to mention that the page is in two languages: Portuguese and English. The target audience is clearly the Tunas, the academy in general of the Polytechnic Institute of Bragança, but also Tunas based in foreign higher education institutions.

Obviously, each of the tools has its own complexities and specificities, so it becomes a real challenge for the digital team members since most of the members have no training on these technologies. However, the members recognize that they are committed and willing to study the tools, to understand how they work to improve their skills and competencies in using them. Only in this way will it be possible to produce quality content, perceptible to the reader and that transmit to the target audience the correct, assertive and clear message that the student wishes to convey.

4.3 Constraints on the Use of Technologies

The use of technologies by the members of the Tuna has led to the emergence of a set of constraints that, in fact, were assumed throughout the development of this research and that we identify below:

- Difficulty in mastering different types of tools that have distinct purposes;
- Difficulty in keeping up, in a timely manner, with the latest technologies appropriate for publicizing and disseminating on the networks;
- Need to have training oriented to the acquisition of digital skills in the other tools;
- Need to have training oriented to the strategies to be adopted in digital marketing;
- Understand, in a timely manner, the aspirations, tastes and desires of the Tuna's target audience;
- Produce quality content (text, images, videos…) that meet the expectations of the public;
- Update, in real time, the social networks and the website of Tuna;
- Lack of equipment (hardware) to capture quality images;
- Lack of software that allows quality editing of images and production of sound and video;
- More time to produce quality content;
- Get feedback from the public about the work done;
- Costs associated with the use of some platforms.

All the constraints or obstacles identified above can naturally be overcome with investment in the acquisition and updating of software, in the acquisition of equipment for image, video and sound capture, in the provision of training in other technologies, but also in digital marketing, specifically in the adoption of more appropriate strategies for the advertising and dissemination of the Tuna's brand. It should be noted that the members of this team are undergraduate students at the Polytechnic Institute of Bragança (Portugal), and only one element attends a course related to arts and design. This means that 80% of the team has no qualifications, skills or training in the area. The learning of these elements occurs through tutorials and the experience coming from the use of other technologies (social networks, websites, editing tools and production of multimedia content, applications, among many others). Effectively there are times when team members recognize that it is not easy to overcome the difficulties that arise, however, with the support of research, in the conversations and reflections that they have in group, in the suggestions for improvement that other members of the Tuna present and with the support of all, it becomes easier to overcome the various challenges that arise.

5 Conclusions

The research allowed us to identify the technologies used by the members of the Tuna, understand for what purposes they were used, and identify the constraints detected in the use of these technologies.

Functions of the digital technologies team

Through the instruments of data collection it was possible to see that the members of the digital technologies team are responsible for a wide range of functions that they perform on a daily basis and whenever there are activities that the students organize, but also participate in. Each of these functions includes specific activities and tasks ranging from team leadership, management of websites and social networks, to the production of other types of digital content that serve to transmit certain messages and information to the target audience of the Tuna.

Identification and use of technologies

Several tools have been identified as vital to the development of the team's work in the area of editing, production, promotion and dissemination of the Tuna's content on the networks. The team members recognized learning through the tutorials available on the Internet, but especially, through experience as they use the other technologies. They assume that the quality of their work increases as they interact with the various technologies, and they assume the search for similar technologies that allow them to improve their practice and acquire more digital dexterity. They also recognize the importance of inspiration and creativity in the production of digital content, as well as the creation of a dynamic, active and collaborative environment among all team members. Also group reflections, small discussions and suggestions for improvement from outside the team are vital to improve the quality of the products.

Constraints on the use of technologies

As we approached the team members and analyzed the work produced, we could identify several constraints on the use of some technologies. Generally speaking these constraints have to do with the need for the Tuna to acquire new software that allows the production of contents with higher quality, but also the acquisition of equipment that allows the capture of sound, video and image. In addition, the members highlight the lack of training in the area of the tools they use, but also in the area of digital marketing, specifically in the most appropriate strategies to adopt in social networks, taking into account the constantly changing profile of the target audience, that is, the consumers. In this sense, it seems important to invest in the preparation of the Tuna members in general through training aimed at the acquisition and/or improvement of digital skills, as well as to invest in updating the various tools, platforms and software of the Tuna.

We cannot end without mentioning that although the results may not be clear, the truth is that the tuna's web presence has been favoring its promotion in the tunae context, as well as in the local and national or even international community.

References

Cohen, W.A.: The Marketing Plan. John Wiley & Sons, New York (2005)

Glória, M.P.: Mercado cor-de-rosa: um estudo voltado para o ideal publicitário refletido sobre a mulher brasileira (2009)

Gomes, M., Kury, G.: A Evolução do Marketing para o Marketing 3.0: o Marketing de Causa. Intercom–Sociedade Brasileira de Estudos Interdisciplinares Da Comunicação XV Congresso de Ciências Da Comunicação Na Região Nordeste, Mossoró (2013)

Kotler, P., Armstrong, G.: Principles of Marketing. Pearson Education, London (2010)

Kotler, P., Keller, K.L.: Marketing para o século XXI. Pearson Prentice Hall, London (2006)

Kotler, P., Keller, K.L., Brady, M., Goodman, M., Hansen, T.: Marketing Management. 14th Prentice Hall. New Jersey (2012)

Lendrevie, J., Lévy, J., Dionísio, P., Rodrigues, J.: Mercator da língua portuguesa. Teoria e prática do marketing: Casos de Angola, Cabo Verde, Moçambique, Portugal e exemplos de outros países de língua portuguesa. D. Quixote, Lisboa (2015)

Maçães, M.: Marketing e gestão da relação com o cliente. Edições Almedina, Lisboa (2017)

McDonald, M., Wilson, H.: Planos de marketing, vol. 7. Elsevier Brasil (2013)

Philip, K.: Marketing Management-Millennium Edition. Pearson Custom Publishing, London (2002)

Wood, M.B.: Essential Guide to Marketing Planning. Pearson Education, London (2007)

Correction to: Profiling the Motivations of Wellness Tourists Through Hotels Services and Travel Characteristics

Rashed Isam Ashqar and Célia M. Q. Ramos

Correction to:
Chapter 10 in: A. Rocha et al. (Eds.): *Information Systems and Technologies*, LNNS 800, https://doi.org/10.1007/978-3-031-45645-9_10

In the original version of the book, the following belated corrections were incorporated:

In chapter "Profiling the Motivations of Wellness Tourists Through Hotels Services and Travel Characteristics", the affiliation "Al Zaytona University of Science & Technology (ZUST), Salfit, Palestine" of author "Célia M. Q. Ramos" was changed to "ESGHT, University of Algarve, Faro, Portugal".

The book has been updated with the changes.

The updated version of this chapter can be found at
https://doi.org/10.1007/978-3-031-45645-9_10

A. Rocha et al. (Eds.): WorldCIST 2023, LNNS 800, p. C1, 2024.
https://doi.org/10.1007/978-3-031-45645-9_65

Author Index

A. Rocha et al. (Eds.): WorldCIST 2023, LNNS 800, pp. 679–681, 2024.
https://doi.org/10.1007/978-3-031-45645-9